Quasars and High-Energy Astronomy

QUASARS AND HIGH-ENERGY ASTRONOMY

Including the Proceedings of the Second Texas Symposium o Relativistic Astrophysics, December 15-19, 1964

Edited by

K. N. DOUGLAS   IVOR ROBINSON   ALFRED SCHILD
E. L. SCHUCKING   J. A. WHEELER   N. J. WOOLF

GORDON AND BREACH, SCIENCE PUBLISHERS

New York   London   Paris

150 Fifth Avenue, New York, N. Y. 10011

Library of Congress Catalog Card Number: 68-23993

Editorial office for Great Britain:

Gordon and Breach, Science Publishers Ltd.
12 Bloomsbury Way
London W. C. 1

Editorial office for France:

Gordon & Breach
7-9 rue Emile Dubois
Paris 14e

Distributed in Canada by:

The Ryerson Press
299 Queen Street West
Toronto 2B, Ontario

Printed in the United States of America

EDITORS' INTRODUCTION

Scientists see the universe with diverse eyes. They look with the two-hundred-inch pyrex mirror on top of Mount Palomar, with a hundred thousand gallons of cleaning fluid buried more than a mile underground, with scintillation counters flying in rockets and satellites, with a retina covering several square miles of the New Mexico desert at Volcano Ranch, with a steel bowl two hundred and ten feet across at Parkes in Australia, and with a gently swinging aluminum bar in Maryland, waiting patiently for the tremor of a gravitational wave.

Some of these eyes perceive only darkness, some have blurred and distorted vision, others detect a profusion of fine detail. Scientists have mapped the sky with meticulous care as they see it by radio waves, light, x rays, γ rays and cosmic radiation. There are some striking differences among these charts.

The finest picture is that available in the optical region. By comparison, the universe as seen in the radio spectrum is rather bare. A radio astronomer confined to strictly professional sources of information would not think of distinguishing between night and day. He discovered the sun by chance as late as 1943. He would have no reason to regard the quiet sun as an energy source of importance on the local scene; it is very faint besides Cygnus A which is 10^{18} times farther away. In a region where Palomar plates show millions of stars and thousands of galaxies, only one subject, insignificant on the photograph, may appear on the corresponding radio, x-ray, or γ-ray map.

It is only within recent years that the optical and radio maps of the skies have begun to merge. This "two-color" picture of radio and light reveals a new universe. It is not the smooth, quiet, placid world which the "black and white" map of conventional optical astronomy had shown. As staining revealed the mysteries of cell division and genes under the microscope and showed the dynamic nature of life, so the two-color maps of the sky, pinpointing hot spots of great energy events, are initiating a revolution in our view of the cosmos. Majestic galaxies, giving the impression of dull harmony, are now believed to evolve in a cataract of explosions that involve millions of solar masses.

Reprinted from PHYSICS TODAY, July 1965, 17-24.

Supernovae were the first indication that violent events play an important role in the life and death of a star. Now we see that such cataclysms occur on a galactic scale, beside which supernovae look like innocent firecrackers. High-energy particles, so difficult to obtain in laboratories on earth, are commonplace in the skies, and illuminate it with synchrotron light. Relativity theory, until recently believed to be unimportant for astronomy, provides the basic laws that govern these great events.

The new observations may give more than a new picture of galactic and cosmological evolution. The merging of the radio and light maps of the sky led to the discovery of the quasi-stellar and strong radio sources, whose tremendous energy output may well present a problem in basic physics.

At present, the other maps seem to be painfully unrelated to the optical picture. When they become sharper and more precise, when they begin to show common features, we can expect the new "many-color" view of our universe to lead to new revolutions in astronomy and in physics.

This will only be achieved by the concerted effort of scientists from many different fields. Optical astronomers will have to join forces with elementary-particle physicists, radio astronomers will have to talk to cosmic-ray experts. The Texas Symposia on Relativistic Astrophysics were conceived with the deliberate purpose of overcoming the traditional segregation of specialists in their narrow cubbyholes.

The first symposium was held in Dallas in December 1963.* It dealt mainly with optical and

* The proceedings of the first Texas symposium have been published under the titles *Quasi-Stellar Sources and Gravitational Collapse*, $10.00, and *Gravitation Theory and Gravitational Collapse*, $6.50, both by the University of Chicago Press, Chicago and London.

radio observations of quasi-stellar sources and with theories which offered promise of explaining the enormous energy releases.

The second symposium was held at the University of Texas in Austin, from December 15 to 19, 1964. It continued the discussion of quasi-stellar sources and reported on the progress made in the preceding year. Half of the symposium was devoted to the new maps of the sky, to x-ray, γ-ray, and cosmic-ray astronomy, to the search for cosmic neutrinos and to the possible large-scale implications of the breakdown of CP-invariance.

The opening session was chaired by P.A.M. Dirac of Yeshiva and Cambridge Universities. Participants were welcomed by C. H. Green of the Graduate Research Center of the Southwest, W. W. Heath of The University of Texas, John Connally, governor of the State of Texas, and E. L. Schucking, on behalf of the organizing committee.

Geoffrey Burbidge from the University of California, San Diego, was the first speaker. He reported that the strongest radio objects may be the result of events involving energies of 10^{64} erg. Such an energy, equivalent to the mass of five billion suns, is about a thousand times larger than the energy proposed a year earlier at the Dallas Symposium. Astrophysicists had then assumed that the fast-moving atomic particles in these sources were accelerated by a mechanism of almost 100 percent efficiency. Burbidge, assuming that the Lord was no better an electrical engineer than the terrestrial builders of the biggest nuclear accelerators at Brookhaven, CERN, and Berkeley, came up with an efficiency factor of only 0.03 percent for the cosmic machines.

Not everybody in the audience agreed. It was held against the argument that electrical engineers on earth are limited by the conductivity of copper, whereas the Lord, in the vast vacuum of space, might have other means at His command. Burbidge thought that the only way of reducing the energy estimates was to bring these radio sources implausibly near to us. In conclusion, he recalled the difficulties that had beset astronomy more than thirty years ago when nuclear energy, the energy source of the stars, had yet to be discovered. He suggested that our difficulty in understanding strong radio sources might point to a gap in our knowledge of basic physics.

John Bolton of CSIRO, Sydney, one of the founders of modern radio astronomy, an art that was developed largely at the antipodes, also discussed in detail the properties of the strong radio sources. With his beautiful new instrument, the 210-foot dish at Parkes near Sydney, he found evidence that not all of the volume of the huge clouds is evenly filled with high-energy particles. He said that his observations were more indicative of shell-like structures with bright spots. This would indicate that the energy, calculated by assuming a uniform distribution of particles over the emitting volume, had been overestimated. At this stage, the chairman of the meeting, Murray Gell-Mann of the California Institute of Technology, took the microphone and asked: "Would Dr. Burbidge please defend his volume?"

He did.

Among the beautiful new results presented by Bolton were measurements of the direction of polarization of the radiation from strong sources. Many of them look like dumbbells or hourglasses hundreds of thousands of light years long, leaky magnetic bottles containing high-energy cosmic rays. Bolton's observations show that in many cases the direction of the magnetic field is perpendicular to the long axis of the dumbbell with an accuracy of one percent.

Thomas Matthews, of Caltech's Owens Valley Radio Astronomy Observatory, had shown earlier in the day that these dumbbells exhibit bright edges at both ends. In these regions, it is believed, the gas smashes its way into the ambient medium.

Since astronomers have available for interpretation only static pictures of these cosmic explosions, it is often difficult to say how evolution is occurring, and at what stage it is being observed. Frequently they can only guess. Radio astronomers had believed that the biggest objects were the older generation in the radio community, while the quasi-stellar radio sources, smaller in size, were the new born. Bolton told the audience that he thought evolution might go the other way: the strongly energetic radio sources were not on the verge of radio death, but in fact were infants which later collapsed into the less energetic smaller objects, including quasistellars. "I am very happy that somebody is following this line of approach," commented Fred Hoyle amiably, "because I think it is wrong."

Although they are not the most spectacular radio transmitters, the quasi-stellar sources are objects of enormous and bewildering brilliance, brighter than a million million suns. Maarten Schmidt of the Mount Wilson and Palomar Observatories, who had identified the first of them in 1962, reviewed carefully the evidence that these

Table 1.* Properties of quasi-stellar sources with known redshift. The fluxes are computed under the assumption that the cosmos is a Friedman universe. The cosmological constant is assumed to be zero, the space curvature positive, the present value of the Sandage parameter $q_0 = -\ddot{R}R/\dot{R}^2 = 1$, and the Hubble constant 100 km/sec/Mpc.

Source	*Red shift* $z = \frac{\lambda' - \lambda}{\lambda}$	*log of intrinsic radio flux emitted at 10^9 Hz in $W(Hz)^{-1}$*	*log of intrinsic optical flux emitted at 10^{15} Hz in $W(Hz)^{-1}$*
3C 273	0.158	26.8	23.8
3C 48	0.367	27.5	23.0
3C 47	0.425	27.0	22.5
3C 147	0.545	27.9	23.1
3C 254	0.734	27.4	22.9
3C 245	1.029	27.5	23.5
CTA 102	1.037	27.6	23.5
3C 287	1.055	27.8	23.4
3C 9	2.012	28.1	23.6

* M. Schmidt, Astrophys. Journ. **141**, 1299 (1965).

were indeed the most distant objects in the universe, and were not being confused with nearby large celestial bodies. Disposing first of the possibility that the red shift is gravitational and that these objects are as heavy as the sun, smaller than Austin, and at a distance of fourteen kilometers, Dr. Schmidt led his audience step by step to the conclusion that they are at least several million light years distant, and probably much further away. His model of a quasi-stellar source is a shell-like structure, with an outer cloud, huge but very thin, filled with atomic particles of high energy which emit radio synchrotron radiation. It may be several thousand light years across, much smaller than an ordinary galaxy. Within this radio cloud lies a thin shell of glowing rarefied gas with a diameter of some ten light years. This region emits light predominantly at a few fixed frequencies. The extreme brilliance of the quasi-stellar comes from an inner, much smaller core, a hot superstar of perhaps a hundred million solar masses with a temperature in excess of 10 000°C. Its diameter may be as small as one light year.

Allan Sandage of Mt. Wilson and Palomar reported that, in the few weeks preceding the Austin Conference, he and his colleague, P. Veron, had found fifteen new quasi-stellar sources in a systematic search with the 48-inch Palomar Schmidt telescope. This brought the total count to 34, leaving some others whose identification is not yet certain. Sandage described how he found these objects by taking survey photographs of the sky, first with a violet, and then with an ultraviolet filter. After the first filter has been removed, the photographic plates are shifted so that the stellar image seen through the second filter is slightly displaced. This makes it possible to spot these objects which are more brilliant in ultraviolet light than in violet. There may be more than 80 000 of these "interlopers," as Sandage called them, which are possible candidates for identification with quasi-stellar radio sources. One of them has the same position as the second entry in the third Cambridge catalogue of radio sources, known for short as 3C2. This "star" of 20th magnitude is very probably located in the depths of space-time, much further out than anything seen before.* Its light, which reaches us now, may have been emitted before the solar system was formed (5×10^9 B.C.). It is running away from us at something very close to the speed of light, but the precise velocity has not yet been measured. Sandage reported a four-fold change in brightness for 3C2 over the last two years. This is many times more than the fluctuations previously observed in quasi-stellar radio sources.

Henry Palmer of the Jodrell Bank Radio Observatory near Manchester started the hunt for quasi-stellar sources nearly a decade ago. At the symposium in Austin he reported on the observations he had made with the world's largest scientific instrument, a radio interferometer a hundred and thirty-two kilometers long. Palmer revealed that some of the quasi-stellar sources have an apparent radio diameter of only a fraction of a second of arc. He predicted that the quasi-stellar source CTA 102 may have a radio diameter as small as one hundredth of a second of arc. This source was recently reported by Russian ra-

* Maarten Schmidt has just measured the red shift of 3C9. It corresponds to a special relativistic recession velocity of four-fifths that of light (May 1965). See Table 1.

dio astronomers to have variable radio brightness, which led some of them to suggest that it was a broadcasting station operated by extraterrestrial intelligence.

J. E. Baldwin of Cambridge University showed that the quasi-stellars are distinguished from other radio sources by their radio spectrum. W. W. Morgan, of the Yerkes Observatory, reviewed the properties of D-galaxies, supergiant systems which are predominant among the radio sources identified with optical objects.

The next sessions of the symposium were devoted to what may be called exotic astrophysics, the exploration of the skies through the observation of relativistic particles and high-energy radiation. The speakers were mostly physicists. This accounted for a quantum jump in terminology: the act of detecting the sun's visible radiation is an observation; that of observing its neutrino flux is an experiment. The first of these sessions was chaired by A. E. Chudakov of the Soviet Academy of Sciences. He presented some recent work by V. L. Ginzburg, who again was unable to attend the conference.

Bruno Rossi of the Massachusetts Institute of Technology reported on cosmic rays of the highest energy. At Volcano Ranch in New Mexico, the MIT group recorded about a dozen large air showers, each caused by a primary proton with energy larger than 10^{19} electron volts. Data from Mount Chacaltaya in Bolivia indicate that the cosmic-ray spectrum has a kink at about 10^{17} electron volts. Particles below this energy, Rossi suggested, come from our own galaxy, those above from outside. Fred Hoyle, on the other hand, believes that all the high-energy cosmic rays are emitted by strong radio sources.

Ken McCracken, of the Southwest Center for Advanced Studies, discussed the local cosmic-ray background. He said: "In the same way that an astronomer may be hindered in his work by peculiarities of his local environment, in that atmospheric dust, and man-made lights may limit the scope of his work, so is the cosmic-ray physicist limited in his ability to measure the properties of the cosmic radiation in the galaxy by 'bad seeing'—in this case, the reason for the bad seeing being the magnetic fields which pervade the solar system." He reviewed in detail the solar cosmic-ray component.

Peter Meyer of Chicago University, one of the discoverers of primary cosmic-ray electrons, discussed their origin. These electrons, together with protons, could be injected into the galaxy by supernova remnants. They may also arise, together with a larger number of positrons, from the collision of cosmic-ray protons with interstellar hydrogen. Meyer's observations on the electron-positron ratio in primary cosmic rays point to a supernoval origin.

George Clark of MIT reported on recent observations of cosmic γ rays, photons ranging in energy from 10^4 to more than 10^{15} electron volts. In a balloon experiment of July 1964, he observed a peak in the counting rates when the Crab Nebula was within the field of view of his scintillation detector, which was sensitive in the range 15 to 62 keV. He stated that the Crab Nebula, the only known source of x rays in this region of the sky, was also likely to be the source of the higher-energy radiation. This seems to be the first observation of a cosmic γ-ray source.

Philip Morrison of MIT gave a masterly review of the whole field of cosmic γ rays. He stressed the need for further observations in order to distinguish the relative importance among different production mechanisms. He pointed out that the isotropic component of the γ rays may originate from the collision of starlight with relativistic electrons. Through this inverse Compton effect, soft photons are converted into hard γ rays.

R. Giacconi, of American Science and Engineering, reported on x-ray observations with detectors flown in a rocket in October 1964. The x rays were in the range 0.5 to 15 Å. He and his coworkers resolved two new sources near the galactic equator. Each of these point sources has intensity less than 1/10 Scorpio (one Scorpio being the new intensity unit of the x-ray astronomer). It is the flux of the brightest x-ray source, 10^{-7} erg cm^{-2} sec^{-1} between 2 and 8 Å. This flux, in the visible range, would be that of a sixth magnitude star.

Herbert Friedman of the Naval Research Laboratory, with the energetic assistance of his chairman Hayakawa, wrote on the blackboard the locations of ten discrete x-ray sources in the sky, giving the latest results from observations made with Geiger counters aboard unstabilized Aerobee rockets. Astronomers in the audience took down the coordinates, and prepared to search for peculiar objects on their photographic plates. Friedman's observations showed conclusively that the x rays from the Crab Nebula in the constellation Taurus (Tau XR-1) did not have a point source. This disposed of the theory that they came from a neutron star. The x rays could be the high-energy tail of the synchrotron-radiation spectrum arising from the inner parts of the Crab cloud. The Scorpio source (Sco XR-1) has not been definitely identified with a known optical object. I. S. Shklovsky, who also was again unable to attend the conference, had suggested that it might be the remnant of a supernova which exploded about 50 000 years ago in our vicinity, some 150 light years away. Prehistoric astrologers must have greeted the sudden appearance of this object, as bright as the full moon, with something of the respect and bewilderment which the quasi-stellar objects have evoked in our own generation. Friedman concluded: "All of the x-ray sources observed lie rather close to the galactic plane and within plus and minus 90° of the galactic center. This distribution resembles that of galactic novae and suggests that all of the x-ray sources thus far observed may be associated with supernova remnants in our galaxy."

G. Wataghin of Turin University gave the results of new calculations on URCA processes and on equilibrium problems which include neutrinos at high temperatures and densities.

The neutrino session continued with a report by R. D. Davis of Brookhaven National Laboratory. He discussed the chances of catching neutrinos from the sun. The main part of his equipment will be a tank 20 ft in diameter and 48 ft long, filled with cleaning fluid (C_2Cl_4). This will be buried in a deep mine, under more than 4400 ft of rock. It is hoped that a few of the chlorine atoms in the fluid will undergo the reaction $\nu_e + {}^{37}Cl \rightarrow {}^{37}Ar + e^-$. Positive results would give the first direct proof that the sun is in fact a nuclear fusion device. If Davis' $600 000 neutrino eye does see solar neutrinos, it will actually be looking right into the sun's central region, which is inaccessible to all other scientific instruments.

In 1953, Frederick Reines from the Case Institute of Technology and his coworker C. L. Cowan were the first to detect neutrinos produced by a nuclear reactor. Reines described two neutrino telescopes. The first is to be buried this year 2000 feet deep in a salt mine near Cleveland. It is made of lithium and is designed to detect solar electron-neutrinos by the reaction $\nu_e + {}^7Li = {}^7Be + e^-$. The second telescope, located two miles underground in the East Rand Proprietary Gold Mines near Johannesburg, South Africa, is constructed to detect energetic muons produced by high-energy neutrinos (ν_μ) in the surrounding rock. If such neutrinos are seen, there won't be many of them. A friend told Dr. Reines: "You may possibly have a long-distance record for commuting to an experimental site, but you are one of the few under these conditions who can commute between counts." Dr. Reines remarked: "It is interesting that despite the size of this detector, it has too little sensitivity, by perhaps a factor of 10^3 or more, to detect an expected flux of true cosmic (that is, extraterrestrial) neutrinos of high energy. But we must take one step at a time and see whether, as we increase our sensitivity, nature is as we think it is or whether some surprises might not be in store for us."

John Bahcall from Caltech reviewed different possibilities for neutrino detection. He stressed that a neutrino-spectroscopic study of the solar interior should be included in the long-range program as a means of determining quantitatively conditions in the interior of the sun—in much the same way as astronomers have already studied its surface by photon spectroscopy. He discussed possible observations of neutrinos from strong radio sources and their connection with mass estimates of the hypothetical W^- boson. Bahcall also proposed that the military be persuaded to surrender their supplies of tritium to the neutrino enthusiasts to build detectors. Luis Alvarez from Berkeley, chairman of the session, expressed his doubts about the technical feasibility of this approach to nuclear disarmament. In the discussions, G. G. Zatsepin, delegate to the symposium from the Soviet Academy of Sciences, was in favor of building neutrino observatories on the moon. Hong-Yee Chiu of the Goddard Space Flight Center preferred a detector, consisting of 1000 tons of completely ionized ^{37}Cl, located at the outer limits of the solar system on the planet Pluto. He envisaged counting time of a few hundred years, and conceded that his Pluto experiment required a civilization more affluent than our own.

The next session was chaired with commendable firmness by Leopold Infeld of the Polish Academy of Sciences. R. K. Sachs of The University of Texas proposed tests which would enable astronomers to deduce the structure of the universe from their observations without begging the question by presupposing a particular cosmological model, such as a homogeneous isotropic Friedman universe. He pointed out that spherical galaxies or clusters at a great distance might all appear elliptical in shape because they are seen through ripples of gravitational waves which pervade the space-time ocean.

This gathering of physicists and astronomers

was a natural environment for the discussion of the possible new long-range force proposed by J. Bernstein, T. D. Lee, N. Cabibbo, J. Bell, and J. K. Perring. A hard apple had struck the heads of these physicists a few months earlier: the CP-invariance experiment of Christenson, Cronin, Fitch, and Turlay. J. H. Christenson of Columbia University described the experiment. Gerald Feinberg, also from Columbia, reported cautiously on the hypothetical new long-range force. This fifth force, much weaker than the gravitational interaction, was invented in order to preserve time-reversal symmetry in elementary-particle processes. It would affect matter and antimatter differently [*Physics Today*, April 1965, p. 88]. Limitations on the range and strength of the force are provided by measurements which verify Einstein's principle of equivalence. R. H. Dicke of Princeton University reviewed the Dicke-Eötvös experiments which show that the acceleration toward the sun of gold and aluminum are equal to within the impressive accuracy of one part in 10^{11}. He said: "Some idea of the required sensitivity can be obtained by noting that this requires the detecting of a relative acceleration as small as 6×10^{-12}cm/sec^2. Starting from rest a body would reach the enormous velocity of 1.2×10^{-4}cm/sec after being accelerated a whole year at this rate."

Freeman Dyson chaired the next session, devoted to general discussion and summaries. Thomas Gold of Cornell University discussed his model of a quasi-stellar source: an extremely dense cluster of stars where frequent collisions give rise to fluctuating emission of light. Harlan Smith of the University of Texas summed up the observational results on quasi-stellar sources. He proposed "stark", the astronomer's quark, as a new name for these intriguing objects. His motion was not seconded. "Quasar" received the most enthusiastic support. The vote was twenty ayes to some four hundred abstentions. Other summaries were given by S. Hayakawa of Nagoya University on x-ray and γ-ray astronomy, G. Cocconi of CERN on cosmic rays, W. A. Fowler of Caltech on neutrinos, R. Hanbury Brown of Sydney University on radio astronomy, G. Gamow of Colorado University on cosmology, J. Bjorken of Stanford University on CP-violation, and T. Page of Wesleyan University on optical astronomy.

The symposium concluded with a seminar on gravitational collapse, chaired by L. Gratton of the University of Rome and J. A. Wheeler of Princeton University. Short theoretical papers were presented by J. Bardeen and W. A. Fowler (Caltech), S. A. Colgate, M. May, and R. H. White (Livermore), C. W. Misner (Maryland), R. W. Lindquist (Texas), R. A. Schwartz (NASA and Columbia), D. H. Sharp, L. Shepley, and K. S. Thorne (Princeton), J. N. Snyder (Illinois), A. H. Taub (Berkeley), and L. Gratton (Rome). They dealt mainly with the collapse of large masses under their own gravitational weight and with the resulting release of energy which may be the origin of the strong radio sources. These papers revealed the impressive progress achieved during the preceding year in relativistic hydrodynamics applied to astronomical situations with strong gravitational fields. Fowler suggested that quasi-stellar sources consist of pulsating and rotating supermassive stars, energized by nuclear reactions, with radio and optical emissions from extended surrounding regions which the star excites with ultraviolet radiation and relativistic particles. Only after the exhaustion of nuclear energy would gravitational energy from collapse become available and the evolution of a quasi-stellar into an extended radio source become possible.

Astrophysics today draws on a wide range of talents from many countries. Scientists came to the symposium from Argentina, Australia, Brazil, Canada, England, Denmark, France, Germany, Holland, Hungary, India, Ireland, Israel, Italy, Japan, Mexico, Pakistan, Poland, Sweden, the USA, and the USSR. There was widespread regret at the absence of several distinguished Soviet scientists who are not permitted to attend meetings outside the Soviet Union, and of one well-known West-European physicist who was refused a US visa.

LIST OF PARTICIPANTS

ABELL, D. F., Alfred University
AITKEN, D. W., Stanford University
ALLEN, Lt. Col. R. G., AFSOR
ALVAREZ, L. W., University of California
ANGIONE, R. J., The University of Texas
ARP, H. C., Mt. Wilson & Palomar Observatories
ASHE, J. B., Texas Nuclear Corp.
ASHKIN, J., Carnegie Institute of Technology
ATKINSON, R. d'E., Indiana University
AXFORD, W. I., Cornell University
BABCOCK, H. W., Mt. Wilson & Palomar Observatories
BAGGERLY, L. L., Texas Christian University
BAHCALL, J. N., California Institute of Technology
BALDWIN, J. E., University of Cambridge
BARKER, E. S., The University of Texas
BARDEEN, J. M., California Institute of Technology
BARLOW, C. A., Jr., Texas Instruments, Inc.
BARTLEY, W. C., Graduate Research Center of the Southwest
BAŻAŃSKI, S. L., Syracuse University
BECK, C. A., Hdqrs. Strategic Air Command
BELINFANTE, F. J., Purdue University
BERGMANN, P. G., Syracuse University
BERGESON, H. E., University of Utah
BEVERAGE, D. G., University of Texas
BIRAM, B. M., The University of Chicago Press
BJORKEN, J. D., Stanford University
BLAKESLEE, A. H., Associated Press
BLANCHARD, P. A., Harvard College Observatory
BODANSKY, D., University of Washington
BOLDT, E. A., Goddard Space Flight Center (NASA)
BOLEY, F. I., Dartmouth College
BOLTON, J. G., California Institute of Technology
BOMKE, H. A., US Army Electronics Command
BOYER, R. H., University of Texas
BRACEWELL, R. N., Stanford University
BREHME, R. W., University of North Carolina
BRIGMEN, G. H., General Dynamics
BRILL, D. R., Yale University
BURBIDGE, G. R., University of California San Diego
CAMERON, A. G. W., Goddard Institute for Space Studies (NASA)
CANTRELL, W. G., Texas A & M University
CARMELI, M., Lehigh University
CARNAHAN, W. G. Texas Education Agency
CARTER, J. M., The University of Texas
CECCARELLI, M., Institut di Fisica "Augusto Righi"
CHASSON, R. L., University of Denver
CHIU, H., Institute for Space Studies (NASA)
CHRISTENSON, J. H., Columbia University
CHRISTY, R. F., California Institute of Technology
CHUDAKOV, A. E., P. N. LeBedev Institute of Academy of Sciences
CLARK, G. W., Massachusetts Institute of Technology
CLARKE, K. M., AFOSR
CLAYTON, D. D., Rice University
CLINE, T. L., Goddard Space Flight Center
COCCONI, G., CERN
COLGATE, S. A., New Mexico Institute of Mining and Technology
COLLINS, R. L., University of Texas
COX, J. P., Joint Inst. for Lab. Astrophysics
CRADDOCK, W. L., Rice University
DAVIS, R., Jr., Brookhaven National Laboratory
DEBNEY, G. C., Jr., The University of Texas
DEEMING, T. J., The University of Texas
DEHNEN, H. A., Universität Freiburg im Breisgau
DENT, W. A., University of Michigan
DESER, S., Brandeis University
DESSLER, A. J., Rice University
DE WITT, C. M., University of North Carolina
DICKE, R. H., Princeton University
DIECKVOSS, W. K. E., Hamburger Sternwarte
DIRAC, P. A. M., Yeshiva University
DOUGLAS, J. N., Yale University
DUNLAP, J. N., The University of Texas
DYBDAHL, A. W., University of Nebraska
DYSON, F. J., University of California
EDMONDS, F. N., Jr., The University of Texas
EDWARDS, G., Rice University
EHLERS, J., Southwest Center for Advanced Studies
ELIOSEFF, L. A., The University of Texas
ELLIS, D. V., Rice University
ELVIUS, A. M., Astronomical Observatory
ELWERT, G., i. Hse. Institut fur Theoretische Physik

ENGLE, P. R., Pan American College
EPSTEIN, E. E., Aerospace Corporation
ESTABROOK, F. B., California Institute of Technology
FARNSWORTH, D. L., The University of Texas
FAUL, H., Southwest Center for Advanced Studies
FAULKNER, J., California Institute of Technology
FAZIO, G. G., Smithsonian Astrophysics Observatory and Harvard University
FEHR, E. S., The University of Texas
FEINBERG, G., Columbia University
FEJER, J. A., Southwest Center for Advanced Studies
FICKLER, S. I., Wright Patterson Air Force Base
FIELD, G. B., Princeton University Observatory
FINKELSTEIN, D., Yeshiva University
FISHER, A., Scholastic Magazines, Inc.
FISHER, P. C., Lockheed Missiles & Space Company
FOWLER, W. A., California Institute of Technology
FRIEDLANDER, M. W., Washington University
FRIEDMAN, H., Naval Research Laboratory
FRISCH, O. R., University of Cambridge
FRYE, G. M., Jr., Case Institute of Technology
GAMOW, G. A., University of Colorado
GAPOSCHKIN, C. P., Harvard College Observatory
GAPOSCHKIN, P. J. A., University of California
GAPOSCHKIN, S., Harvard College Observatory
GAVENDA, J. D., The University of Texas
GELL-MANN, M., Institute for Defense Analyses
GETZE, G., The Times (Los Angeles)
GIACCONI, R., American Science and Engineering, Inc.
GILLESPIE, W., Jr., Manned Spacecraft Center
GLASER, H., Department of the Navy
GLASHOW, S. L., University of California
GOLD, T., Cornell University
GOLDBERG, J. N., Syracuse University
GOLDREICH, P. M., University of California at Los Angeles
GOLDSTEIN, S., Brandeis University
GOOD, W. B., New Mexico State University
GORRELL, Mjr. J. E., Air Force Office of Scientific Research
GOULD, R. J., University of Calif., San Diego
GRATTON, L., University of Rome
GREEN, C. H., Graduate Research Center of the Southwest
GREEN, L. C., Haverford College
GRENCHIK, R. T., Louisiana State University
GUNN, J. E., California Institute of Technology
HADDOCK, F. T., University of Michigan
HALLE, R. L., The University of Texas
HAMZEH, S. M., Texas Christian University
HANBURY-BROWN, R., University of Sydney
HANSON, H. P., The University of Texas
HARRINGTON, M. C., Air Force Office of Scientific Research
HARRINGTON, R. S., The University of Texas
HARRISON, B. K., Brigham Young University
HARRISON, M., Sam Houston State College
HAVAS, P., Lehigh University
HAYAKAWA, S., Nagoya University
HAYMES, R. C., Rice University
HAYNES, J. H., San Diego, California
HEESCHEN, D. S., National Radio Astronomy Observatory
HELFER, H. L., University of Rochester
HELMKEN, H., Harvard College
HERCZEG, T. J., Hamburger Sternwarte
HERZ, A. j., US Naval Research Laboratory
HILL, H. A., Wesleyan University
HLAVATY, V., Indiana University
HOBBS, R. W., U.S. Naval Research Laboratory
HOCHMAN, L. D., Queens College
HOFFMANN, B., Queens College
HOGARTH, J. E., Queen's University
HOLLINGER, J. P., U.S. Naval Research Laboratory
HOWARD, W. E., National Radio Astronomy Observatory
HOYLE, F., Cambridge University
HUBBARD, H. W., Newsweek Magazine
HUDSPETH, E. L., The University of Texas
HUGGETT, R. W., Louisiana State University
HUGHES, M. P., Radio Astronomy Station
HURLBURT, E. H., National Science Foundation
HURT, J. T., Texas A. & M. University
HYDER, S. B., The University of Texas
HYNEK, J. A., Northwestern University
IBEN, I., Jr., Massachusetts Institute of Technology
INFELD, L., The University of Warsaw
ISRAEL, W., University of Alberta
ITO, D., Louisiana State University
IVASH, E. V., The University of Texas
JACKSON, A. A., IV, North Texas State University
JACKSON, T. A. S., University of Liverpool
JANIS, A. I., University of Pittsburgh
JEFREYS, W. H., Yale University
JONES, F. C., Goddard Space Flight Center
JONES, Cpt. J., Jr., Air Force Office of Scientific Research
JORDAHL, P. R., The University of Texas
JUST, K. W., University of Arizona
KANTOWSKI, R., The University of Texas
KATO, S., Goddard Space Flight Center
KAUFMAN, M., Harvard University
KAYE, K. C., The University of Texas
KEATH, E. P., Graduate Research Center of the Southwest and North Texas State Univ.
KEIL, J. E., The University of Texas
KERR, R. P., The University of Texas

KINSEY, B. B., The University of Texas
KISSELL, K. E., Wright Patterson Air Force Base
KLARMANN, J., Washington University
KOBAYAKAWA, K., Louisiana State University
KOMAR, A. B., Yeshiva University
KRAUS, G. L., NASA, Manned Spacecraft Center
KRISHNAN, T., Stanford University
KRISTIAN, J., University of Wisconsin
KROGDAHL, W. S., University of Kentucky
KUNKEL, W. E., The University of Texas
LAMBERT, A. A., Grumman Aircraft Engineering Corp.
LANDOLT, A. U., Louisiana State University
LASTER, H. J., University of Maryland
LATTES, C. M. G., Pisa University
LENCHEK, A. M., Graduate Research Center of the Southwest
LENHART, J., The University of Texas
LICHNEROWICZ, A. L., College de France
LIMBER, D. N., Yerkes Observatory
LINDMAN, E. L., The University of Texas
LINDQUIST, R. W., The University of Texas
LISS, A. R., Academic Press
LITTLE, R. N., The University of Texas
LOCHBAUM, J., San Antonio Express and News
LOCKENVITZ, A. E., The University of Texas
LODHI, M. A. K., Texas Technological College
LÓPEZ-LÓPEZ, F. J., The University of Texas
LUNGERSHAUSEN, W. T., California Institute of Technology
LUYTEN, W. J., University of Minnesota
LYNDS, B. T., University of Arizona
LYNDS, C. R., Kitt Peak National Observatory
LYTTLETON, R. A., Cambridge University
MC CLAIN, E. F., Naval Research Lab.
MC CRACKEN, K. G., Graduate Research Center of the Southwest
MC CREA, W. H., University of London
MAC GREGOR, M. H., Lawrence Radiation Laboratory
MC LENAGHAN, R. G., Brandeis University
MC MAHON, A. J., Space Technology Labs., Inc.
MACE, R., Army Research Office
MALIK, G. M., The University of Texas
MARAN, S. P., Kitt Peak National Observatory
MARKOWITZ, W., U.S. Naval Observatory
MARSHAK, R. E., The University of Rochester
MARSHALL, L. C., Southwest Center for Advanced Studies
MARX, G., Stanford University
MAST, C. B., University of Notre Dame
MATTHEWS, T. A., California Institute of Technology
MAVRIDES, S., Institute Henri Poincare
MAY, M. M., University of California
MELVIN, M. A., Florida State University
MENGES, Lt. G. B., U. S. Air Force
MENON, T. K., National Radio Astronomy Observatory
MERCIER, A. P., University of Bern
MESSEL, H., University of Sydney
METZGER, A. E., Jet Propulsion Laboratory
MEYER, P., University of Chicago
MICHEL, F. C., Rice University
MICHELIS, C.-H. P., The University of Texas
MICHIE, R. W., Kitt Peak Observatory
MIDDLEHURST, B. M., University of Arizona
MIELNIK, B. S., Instituto Politecnico Nacional
MILLER, L. J., The University of Texas
MILLER, L. W., Los Alamos Scientific Laboratory
MILLETT, W. E., The University of Texas
MILLS, J. M., Jr., The University of Texas
MISNER, C. W., University of Maryland
MITCHELL, A. T., Graduate Research Center of the Southwest
MITLER, H. E., Smithsonian Institution
MOFFET, A. T., California Institute of Technology
MORGAN, T. A., University of Nebraska
MORGAN, W. W., University of Chicago
MORI, K., Louisiana State University
MORRISON, P., Massachusetts Institute of Technology
MORTON, D. C., Princeton University Observatory
MOTZ, L., Columbia University
MURTY, S. S. R. Purdue University
NAGEL, J. G., Universidad de Mexico
NARLIKAR, J. V., California Institute of Technology
NE'EMAN, Y., California Institute of Technology
NEY, E. P., University of Minnesota
NEYMAN, J., University of California
NOLLE, A. W., The University of Texas
NORMAN, R. D., Queen's University
NOVICK, R., Columbia University
NUTTALL, J., Texas A & M University
OAKES, M. E. L., The University of Texas
O'CEALLAIGH, C., Institute for Advanced Studies
ODA, M., Massachusetts Institute of Technology & Univ. of Tokyo
ORME-JOHNSON, N. R., The University of Texas
OSTERBROCK, D. E., University of Wisconsin
OZSVATH, I., Southwest Center for Advanced Studies
PACHOLCZYK, A. G., University of Colorado
PAGE, T. L., Wesleyan University
PALMER, H. P., Manchester University
PAPAPETROU, A., Princeton University
PARNELL, T. A., The University of North Carolina
PEBBLES, P. J. E., Princeton University
PETERS, P. C., University of Washington
PETERSON, R. W., University of California
PINKAU, K., Louisiana State University & Univ. of Kiel, Germany
PISHMISH, de R. P., University of Mexico
PLEBANSKI, J. F., El Centro de Estudios Avanzados del I.D.N.
PORTER, J. R., Graduate Research Center of the Southwest

POVEDA, A. R., Ciudad Universitaria
RAAPHORST, C., University of Texas
RAO, R. U., Graduate Research Center of the Southwest
RASH, J. L., The University of Texas
RASTALL, P., Graduate Research Center of the Southwest & Univ. of British Columbia
RAY, J. R., Auburn University
RAYCHUDURI, A. K., University of Maryland
REED, Lt. Col. C. K., Hq. OAR, AFOSR
REINES, F., Case Institute of Technology
REYNOLDS, J. M., Louisiana State University
RHEE, J. W., Rose Polytechnic Institute
RINDLER, W., Southwest Center for Advanced Studies
ROBERTSON, W. W., The University of Texas
ROBINSON, I., Southwest Center for Advanced Studies
ROGERS, H. H., The University of Texas
ROOSEN, R. G., The University of Texas
ROSEN, N., Technion-Israel Institute of Technology
ROSEN, S., Maritime College
ROSINO, L., University of Padova
ROSSI, B., Massachusetts Institute of Technology
ROSTOKER, N., University of California, LA Jolla
ROXBURGH, I. W., University of London
RUDERMAN, M. A., New York University
SACHS, R. K., The University of Texas
SALPETER, E. E., Cornell University
SAMARAS, D. G., Air Force Office of Scientific Research
SANDAGE, A. R., Mount Wilson and Palomar Observatories
SAVIN, N. J., The University of Texas
SCHATZMAN, E. L., Institut D'Astrophysique
SCHERR, C. W., The University of Texas
SCHILD, A., The University of Texas
SCHILD, D. H. M., Rice University
SCHLOSSER, J., University of Chicago
SCHLUTER, H., The University of Texas
SCHMIDT, M., California Institute of Technology
SCHÜCKING, E. L., The University of Texas
SCHWARTZ, M., New York University
SCHWARTZ, R. A., Institute of Space Studies & Columbia University
SCIAMA, D. W., Cambridge University
SCOTT, E. L., University of California
SEEGER, C. L., The University of Texas
SEGRE, E. G., University of California
SERSIC, J. L., Observatorio Astronomico Nacional
SETTI, G., Columbia University
SHAKESHAFT, J. R., Mullard Radio Astronomy Observatory
SHAPIRO, M. M., U.S. Naval Research Lab.
SHARP, D. H., Princeton University
SHAW, P. B., Rice University
SHEN, C. S., Purdue University
SHEPLEY, L. C., Princeton University
SHIVANANDAN, K., United States Naval Research Laboratory
SHOBBROOK, R. R., The University of Texas
SHRIVASTAVA, P. N., The University of Texas
SILBERBERG, R., U.S. Naval Research Laboratory
SILK, J. J., Harvard College Observatory
SIMON, R. L., University of Liege
SMITH, H. J., The University of Texas
SMITH, H. W., The University of Texas
SOFIA, S., Yale Observatory
SOLOMON, P. M., Princeton University Observatory
STEARNS, H. O., Gravity Research Foundation
STOVER, J. E., The University of Texas
STRAND, K. A., U.S. Naval Observatory
STRAWN, D. D., Texas Observer
STRECKER, J. L., General Dynamics
STRUGHOLD, H., Aerospace Medical Division
STURROCK, P. A., Stanford University
SUESSMANN, G., Frankfurt University
SULLIVAN, W. S., New York Times
SUTTON, R. B., Carnegie Institute of Technology
TALBERT, F. D., The University of Texas
TAMBURINO, L. A., Wright-Patterson Air Force Base
TANAKA, Y., University of Leiden
TAUB, A. H., University of California
TAUBER, G. E., Western Reserve University
TEMESVARY, S. C., Institute for Advanced Study (Princeton)
TEOH, H., The University of Texas
TERRELL, J., Los Alamos Scientific Laboratories
THOMAS, L. H., Columbia University
THOMAS, P. D., The University of Texas
THOMAS, T. Y., Indiana University
THOMPSON, J. C., The University of Texas
THOMPSON, W. B., University of Oxford
THORNE, K. S., Princeton University
TINSLEY, B. M., Southwest Center for Advanced Studies
TOOPER, R. F., IIT Research Institute
TOUPIN, R. A., IBM, Thomas J. Watson Research Center
TSURUTA, S., Smithsonian Astrophysical Observatory
TULL, R. G., The University of Texas
ÜBERALL, H. M., Catholic University
VAIDYA, P. C., Washington State University
VALLARTA, M. S., Comision Nacional de Energia Nuclear
VANDEKERKHOVE, E., Royal Observatory
VAN DER LAAN, H., The University of Western Ontario
VARDYA, M. S., Joint Institute for Laboratory Astrophysics
VOGL, J. L., TRW Space Technology Laboratories

VOLKOFF, G. M., University of British Columbia
VON HOERNER, S., National Radio Astronomy Observatory
VRANCEANU, G. C., University of Florida
WADDINGTON, C. J., University of Minnesota
WADE, C. M., National Radio Astronomy Observatory
WAGONER, R. V., Stanford University
WAHLQUIST, H. D., California Institute of Technology
WALKER, A. G., The University of Liverpool
WALKER, E. L., Texas A. & M. University
WALKER, M. F., University of California
WALT, M., Lockheed Missiles and Space Company
WATAGHIN, G., Instituto di Fisica dell'Universits di Torino
WEBER, J., University of Maryland
WEIDEMANN, V., Physik, Techn. Bundesanstact
WEIGOLD, E., Air Force Office of Scientific Research
WEINBERG, J. W., Western Reserve University
WEINSTEIN, D. H., University of Houston
WELLS, D. C., The University of Texas
WENNERSTEIN, D. L., Air Force Office of Scientific Research
WESTERHOUT, G., University of Maryland
WESTERVELT, P. J., Brown University
WEYMANN, R. J., University of Arizona
WHEELER, J. A., Princeton University
WHITE, R. H., Lawrence Radiation Laboratory
WHITFORD, A. E., University of California
WILDT, R., Yale Observatory
WILKINS, G. A., U.S. Naval Ordnance Test Station
WILLIAMS, J. O., The University of Texas
WINTERBERG, F. M., University of Nevada
WINICOUR, J. H., Wright-Patterson Air Force Base
WOLF, A. M., The University of Texas
WOOD, L. A., Air Force Office of Scientific Research
WRIGHT, J. P., Harvard-Smithsonian Astrophysical Observatory
WRIGHT, W. E., Office of Naval Research
WRUBEL, M. H., Indiana University
YOUNG, B., University of Chicago
YOUNG, D. M., Jr., The University of Texas
ZAPOLSKY, H. S., University of Maryland
ZATSEPIN, G. T., Lebedev Phys. Inst. of Academy of Sciences of U.S.S.R.
ZHARKOV, G. F., Academy of Sciences of the U.S.S.R.
ZIPOY, D. M., University of Maryland
ZISK, S. H., Stanford University
ZUND, J. D., University of North Carolina
ZUND, J. R., University of North Carolina

CONTENTS

Editors' Introduction v

List of Participants xi

Part 1.

RADIO OBSERVATIONS OF RADIO GALAXIES AND QUASARS

Radio Sources, J. G. BOLTON 5

Counts of Radiosources in the Revised 3C Catalogue, P. VERON 9

Identification of Radio Galaxies and Quasi-stellar Objects, J. G. BOLTON 13

The Angular Sizes of Discrete Radio Sources, H. P. PALMER 19

New Limits to the Angular Sizes of Some Quasars, R. L. ADGIE, H. GENT, O. B. SLEE, A. D. FROST, H. P. PALMER, and B. ROWSON 27

Radio Sources Having Spectra with a Low Frequency Cut-off, J. M. HORNBY and P. J. S. WILLIAMS 31

Quasi-stellar Sources: Variation in the Radio Emission of 3C 273, W. A. DENT 39

Quasi-stellar Source 3C 273 B: Variability in Radio Emission, GEORGE B. FIELD 43

3C 273: Variations in its 3.4-mm Flux, E. E. EPSTEIN, J. P. OLIVER, and R. A. SCHORN 45

Observations of 3C 273 and 3C 279 at 1 mm, F. J. LOW 47

A Model for Variable Extragalactic Radio Sources, H. Van Der LAAN 49

Part 2.

RADIO GALAXIES AND QUASARS, OPTICAL OBSERVATIONS AND MODELS

Models of Quasi-stellar Sources, MAARTEN SCHMIDT 55

The Supergiant Galaxies, W. W. MORGAN and JANET ROUNTREE LESH 61

Photoelectric Spectrophotometry of Quasi-stellar Sources, J. B. OKE 65

The Spectrum of 3C 273, F. L. LOW and H. L. JOHNSON 79

The Correlation of Colors with Redshifts for QSS Leading to a Smoothed Mean Energy Distribution and New Values for the K-Correction, ALLAN SANDAGE 81

Intensity Variations of Quasi-stellar Sources in Optical Wavelengths, ALLAN SANDAGE 93

The Change of Intensity, Color, Line Strength, and Line Position in the QSS 3C 466 during the 1966 Outburst, ALLAN SANDAGE, J. A. WESTPHAL, and P. A. STRITTMATTER 97

The Existence of a Major New Constituent of the Universe: The Quasi-stellar Galaxies, ALLAN SANDAGE 103

The Nature of the Fainter Haro-Luyten Objects, T. D. KINMAN 123

CONTENTS

Remarks on Evolution of Galaxies, THORNTON PAGE 133

Quasi-stellar Radio Sources as Spherical Galaxies in the Process of Formation, GEORGE B. FIELD 135

A Model of Quasi-stellar Radio Sources, P. A. STURROCK 147

A Quasar Model Based on Relaxation Oscillations in Supermassive Stars, WILLIAM A. FOWLER 151

Summary, R. HANBURY BROWN 165

Observed Characteristics of Quasars, HARLAN J. SMITH 167

Summary Remarks on Quasars, HARLAN J. SMITH 181

Part 3.

COSMIC RAYS

Cosmic-ray Particles of Highest Energy, BRUNO ROSSI 185

The Electron-Positron Component of the Primary Cosmic Radiation, PETER MEYER 193

Detection of Interplanetary 3- to 12-MeV Electrons, T. L. CLINE, G. H. LUDWIG, and F. B. MacDONALD 199

Origin of Cosmic Rays, G. COCCONI 203

Part 4.

X- AND γ-RAY ASTRONOMY

Recent Observations on Cosmic X-rays, R. GIACCONI, H. GURSKY, J. R. WATERS, B. B. ROSSI, G. W. CLARK, G. GARMIRE, M. ODA, and M. WADA 207

A Measurement of the Location of the X-ray Source SCO X-1, H. GURSKY, R. GIACCONI, P. GORENSTEIN, J. R. WATERS, M. ODA, H. BRANDT, G. CARMIRE, and B. V. SREE-KANTAN 215

On the Optical Identification of SCO X-1, A. R. SANDAGE, P. OSMER, R. GIACCONI, P. GORENSTEIN, H. GURSKY, J. WALTERS, H. BRADT, G. CARMIRE, B. V. SKREEKAN-TAN, M. ODA, K. ŌSAWA, and J. JUGAKU 223

Cosmic X-ray Sources, H. FRIEDMAN 233

Cosmic X-ray Sources, Galactic and Extragalactic, E. T. BYRAM, T. A. CHUBB, and H. FRIEDMAN 245

Night Sky X-ray Sources, PHILIP C. FISHER, WILLARD C. JORDAN, and ARTHUR J. MEYEROTT 253

Observational Work on Cosmic Gamma Rays, GEORGE W. CLARK 257

Extended Source of Energetic Cosmic Rays, ELIHU BOLDT, FRANK B. McDONALD, GUENTER RIEGLER, and PETER SERLEMITSOS 265

Evidence for a Source of Primary Gamma Rays, J. G. DUTHIE, R. COBB, and J. STEWART 269

Galactic X-ray Sources, G. R. BURBIDGE, R. J. GOULD, and W. H. TUCKER 275

X-rays from the Coma Cluster of Galaxies, J. E. FELTEN, R. J. GOULD, W. A. STEIN, and N. J. WOOLF 279

Summary of X-rays and γ-ray Astronomy, S. HAYAKAWA 283

CONTENTS

Part 5.

NEUTRINO ASTRONOMY

Solar Neutrinos, RAYMOND DAVIS, Jr., DON. S. HARMER, and FRANK H. NEELY 287

Cosmic Neutrinos, F. REINES 295

Limits on Solar Neutrino Flux and Elastic Scattering, F. REINES and W. R. KROPP 301

Evidence for High-Energy Cosmic-ray Neutrino Interactions, F. REINES, M. F. CROUCH, T. L. JENKINS, W. R. KROPP, H. S. GURR, G. R. SMITH, J. P. F. SELLSCHOP, and B. MEYER 305

The K. G. F. Neutrino Experiment, C. V. ACHAR, M. G. K. MENON, V. S. NARASIMHAM, P. V. RAMANA MURTHY, B. V. SREEKANTAN, K. HINOTANI, S. MIYAKE, D. R. CREED, J. L. OSBORNE, J. B. M. PATTISON, and A. W. WOLFENDALE 311

Observational Neutrino Astronomy: A ν-Review, JOHN N. BAHCALL 321

Comment, H. Y. CHIU 331

Research Facilities and Programs, Physics Today 333

Remarks on Statistics of Particles at High Temperatures, G. WATAGHIN, C. V. ACHAR, M. G. K. MENON, V. S. NARASIMHAM, P. V. RAMANA MURTHY, B. V. SREEKANTAN, K. HINOTANI, S. MIYAKE, D. R. CREED, J. L. OSBORNE, J. B. M. PATTISON, and A. W. WOLFENDALE 335

Observations in Cosmology, J. KRISTIAN and R. K. SACHS 345

Evidence for the 2π Decay of the K_2^0 Meson, J. H. CHRISTENSON, J. W. CRONIN, V. L. FITCH, and R. TURLAY 367

Two-pion Decay of the K_2^0 Meson, W. GALBRAITH, G. MANNING, A. E. TAYLOR, B. D. JONES, J. MALOS, A. ASTBURY, N. H. LIPMAN, and T. G. WALKER 371

The Equivalence of Inertial and Gravitational Mass, R. H. DICKE 375

CP Invariance and Possible New Long Range Interactions, G. FEINBERG 381

Part 6.

GRAVITATIONAL COLLAPSE

General Relativistic Hydrodynamics near Equilibrium, J. M. BARDEEN 387

The Equations of Relativistic Spherical Hydrodynamics, CHARLES W. MISNER and DAVID H. SHARP 393

Energy Transport by Radiative Diffusion in Relativistic Spherically Symmetric Hydrodynamics, CHARLES W. MISNER and DAVID H. SHARP 397

The Acceleration of Matter to Relativistic Energies in Supernova Explosions, STIRLING A. COLGATE and RICHARD H. WHITE 401

Gravitational Forces Accompanying Burst of Radiation, ROBERT A. SCHWARTZ, RICHARD W. LINDQUIST, and CHARLES W. MISNER 407

Relativistic Gas Spheres at Constant Entropy, J. N. SNYDER and A. H. TAUB 411

Preliminary Results of Hydrodynamic Calculations of General Relativistic Collapse, M. M. MAY and R. H. WHITE 425

The Resistance of Magnetic Flux to Gravitational Collapse, KIP S. THORNE 443

Can a Dust-filled Cosmology Bounce? L. C. SHEPLEY 451

Numerical Results on the Equilibrium of Highly Collapsed Bodies, L. GRATTON 469

PART 1

RADIO OBSERVATIONS OF RADIO GALAXIES AND QUASARS

RADIO SOURCES

J. G. Bolton
C. S. I. R. O., Radiophysics Laboratory, Sydney

We recognise four broad types of extragalactic objects as radio sources:

1. *Normal galaxies* whose radiation is on a scale similar to that of our own. These can only be detected out to relatively small distances; the spirals are the strongest emitters, the irregulars somewhat weaker, and the ellipticals below the limit of detection. Some of the spirals appear to possess only disk radiation, others may have coronae.

2. *Sc galaxies of high luminosity*. These are weak emitters compared to the radio galaxies, and the radiation is principally confined to a small region near their nuclei.

3. *The radio galaxies*, elliptical systems of very high luminosity. At present this class is dominant among the identified radio sources.

4. *The quasi-stellar objects*, dense compact systems of unprecedented luminosity.

Neglecting the normal galaxies, the number of identifications amongst the various classes are roughly: Sc's, 10; elliptical radio galaxies, ~70; and quasi-stellar objects, 10. These identifications have resulted from searches, principally on the 48-inch sky atlas plates of about 150 very precise radio positions and about 250 additional positions whose accuracy is better than 1'. All very faint identifications have naturally been made from the more precise positions. It is worth noting that particularly for the elliptical galaxies the radio positions used for final identification have generally had a higher degree of accuracy than really required, i.e., the position error has been comparable to the optical and much smaller than the radio size of the source. This reflects something of a reluctance in the past on the part of optical astronomers to use the large telescopes necessary for completing the optical part of an identification--the recognition of spectral peculiarities and determination of distance. The opposite situation may very well occur in the future where, particularly in the case of the quasi-stellar objects, there could be a tendency to "identify" on the basis of spectral peculiarity and only a rough radio position.

With one very marked exception, there are no clear interrelations between the optical and radio parameters of the elliptical radio galaxies. The galaxies may be classified on the basis of appearance into various subtypes—"D" galaxies, systems with bright nuclei and diffuse extended envelopes; E galaxies with less diffuse and obvious envelopes; "dumbbells" or double systems involving one or both types; and "N" galaxies with dominant nuclei and essentially no envelope. All subtypes, however, embrace the whole range of radio parameters with perhaps a suggestion that the "N" types are concentrated among the strongest intrinsic radio emitters (as indeed are the known quasi-stellars).

The radio galaxies are often characterized by the presence of strong emission lines, particularly those of [O II], and it is this feature which makes the measurement of their redshifts and distances an easier task than for a normal galaxy. In the case of Cygnus A almost half the total emission is in the line spectrum. However, there is not a one to one correlation between the line emission and the radio emission as is shown in Table I, based on data by Schmidt. In addition, it is possible that those with the greatest line emission show a preference for the median spectral index, though this may merely reflect the relation of Table I and a loose connection between radio magnitude and spectral index.

Table I. Relation between line strength and radio emission, Schmidt's class.

M_r (Radio Magnitude)	Line Strength			
	0-1	2-3	4-9	10-20
Fainter than - 26	3	1	....	....
-26 to -28.5	4	2	5	....
-28.5 to -31	2	4	2	6
Brighter than -31	....	2	1	4

The very definite connection between radio and optical characteristics is that, to be a radio source, the galaxy must apparently be of very high luminosity. The radio galaxies all have absolute magnitudes of ~-21 with little or no dispersion. Radio sources in clusters—and 50 per cent of them ~~are cluster members—are~~ exclusively the brightest member. One might suggest that the radio phenomenon is somehow connected with the limit of stability for a very massive system—a safety valve for excess energy.

This constancy of absolute luminosity for the ellipticals is now so well established that given an identification based on a radio position one can probably estimate distance to within 10 per cent on a measurement of apparent magnitude. Similarly, given a radio indication of distance such as angular diameter, one can predict the apparent magnitude of the optical counterpart--if it is of the above type. Such considerations enable one to form an estimate of what fraction of the total radio sources down to a given radio magnitude are of the elliptical type. For the identified E's we can determine the distribution of linear diameters and from interferometry we can determine the distribution of angular diameters. Knowing the absolute photographic luminosity we can then estimate the number of E identifications at each apparent magnitude. Comparing this with actual numbers from searches of the sky survey plates, I have concluded that probably not more than 50 per cent of sources are of this type. This fraction could be increased only if there exists a substantial fraction of E sources whose linear diameters are ≥ 1 Mpc (such that very distant sources would still give appreciable angular diameters) which remain unidentified. The fraction of Sc's is much easier to estimate because they are weak radio emitters; the optical counterparts are close and easily recognized. They represent perhaps only 1 or 2 per cent of the population.

What are the remainder? Quasi-stellar objects? The identified quasi-stellars for which distances are available have absolute photographic magnitudes of around -24; i.e., they are brighter than the E's. The known range of linear diameters is the same as that of the ellipticals (a feature which is itself quite remarkable), and thus one can apply the same arguments to the expected apparent magnitudes of quasi-stellars in searches of the sky survey plates. If the absolute magnitude of -24 persists, quasi-stellars should, on the average, be brighter than E's, and this is clearly not so. As before, we can circumvent this difficulty by having a large fraction with linear diameters in excess of 1 Mpc--or, what is more likely, having a wide distribution of absolute photographic magnitudes whose upper limit is -24 with the vast majority considerably fainter. Present mass estimates and rates of energy output for those around -24 suggest that this is the likely solution.

Characteristics of Sources

As the number of known quasi-stellar objects is relatively small, I shall restrict this discussion tion angles are within 1° of the position angle of the line of centers. These intrinsic vectors are in all cases deduced by extrapolating the position angle-wavelength relation back to zero wavelength to allow for Faraday rotation within our Galaxy. The source field, of course, is normal to the E vector and thus to the line of centers. In one other well-resolved object, NGC 5128, the magnetic field of one polarized source of a central pair with the visual region is along the line of centers and the field structure in the outer complex regions is highly variable. Some northern sources observed by interferometry also suggest a field alignment along the line of centers, while others have fields normal to the line of centers. Both sets of data show a marked lack of intermediate angle difference.

The spectral index of sources is another parameter which exhibits a fair degree of uniformity. Perhaps 80 per cent of all sources show spectra which can be represented by straight power laws over most of the observable radio spectrum. The median index at 75 cm is close to 0.75 with very little dispersion. Most of the odd forms of spectra can be associated with extreme values of other parameters. For example, strongly curved spectra with maxima in the radio range are generally objects of high surface brightness or temperature, and the curvature is thought to be due to synchrotron self-absorption setting in at the long-wavelength end of the spectrum. A small number of oppositely curved spectra are probably due to double sources, one of which has a very flat spectrum as in the case of the B component of the quasi-stellar 3 C 273. There is a suggested relationship between strong polarization and low brightness temperature and between strong polarization and index and/or lack of curvature. These relations together with the relation between curvature and surface brightness are mutually consistent. They suggest that sources of large linear extent have weak but well-aligned field structures; small sources of high emissivity have turbulent field structures.

Energy Considerations

One of the principal difficulties in the understanding of the radio sources is the energy supply. Calculation of the stored energy requirement is now too well known to bear repeating: in general a minimum estimate is made by assuming to the ellipticals. The association between radio sources and a very restricted luminosity class has already been noted. This singularity is so marked that it must surely contain information fundamental to our understanding of these objects. Luminosity presumably reflects mass, although whether one should regard radio emission as a direct consequence of the mass or whether both phenomena result from some further pre-existing condition is hard to determine. If indeed clues to the evolution of the sources exist, they should properly be found in the radio parameters of the sources. Many of these parameters cover wide ranges; alsolute luminosity, for example, has a range of $\sim 10^5$; linear size can vary by $\sim 10^3$, brightness temperature by perhaps 10^8, and so on. Interrelations between parameters such as these are thus ill defined—if they exist at all—and often merely reflect instrumental and other selection effects. Nevertheless there do exist features similar to the singularity in M_{pg} and the most marked of these is the structure of the source. Most sources have a distinct axis; this axis is defined by two components of the source; each component lies with its long axis along the line of centers. Not infrequently, there is a third weaker component more closely associated with the parent galaxy which is in general on the axis. Over 70 per cent of sources investigated by interferometry show this double character, and a fair proportion of the remainder, described loosely as disks or core and halo objects, could be accounted for by end-on or near end-on observation of a two- or three-component source. Thus once again we have a singular feature to aid in our understanding of the source formation.

The recent addition of the polarization parameter, available with increased observational facilities at the short-wavelength end of the radio spectrum, makes it clear that in many cases the source-distribution axis is also a magnetic axis. In four of the five cases of sources whose components can be resolved with the Parkes telescope (NGC 1316, IC 4296, Pictor A and 0453-30 + 0456-30), the position angle of the intrinsic electric vector of the radiation of each component lies close to the position angle of the line of centers. Three of these sources are well-resolved doubles with the radio components well outside the optical counterpart. IC 4296 is a well-resolved triple with one component coincident with the optical counterpart; however, the intrinsic E-vector posiequipartition between field and particle energy with some allowance for a heavy-particle content in addition to electrons. For sources such as Cygnus A this results in values of the order of 10^{61} erg--or the rest mass energy of 10^8 solar masses. Allowing nuclear energies as the source and assuming a 1 per cent efficiency of conversion into relativistic particles, we need the entire nuclear energy available from a very massive galaxy. Such considerations have led recently to a search for more efficient energy sources involving complete gravitational collapse of large masses. An alternate approach might be to question the whole basis of the energy supply. The idea of the parent galaxy as the source of energy is based on two things; the fact that the most obvious source of energy in a galaxy lies in the region of high mass and energy density near the nucleus and the static picture of radio sources in all sizes from a few kiloparsecs to a few megaparsecs. The latter is interpreted as a dynamic evolution from small to large objects formed from an explosion in the nucleus. The theory that such explosions can occur has been strengthened by the remarkable optical investigations of M82—even though this galaxy is not outstanding as a radio source. In fact, three alternatives are possible: (1) that the sources are evolving outward; (2) that they are static; and (3) that they are evolving inward. In case 1 we probably must seek the parent galaxy as the energy source, in the other two cases the energy may be drawn from vast regions of intergalactic space. More familiar examples of outward explosion or evolution are known to us in the solar flares or the supernova explosions; however, we know of at least one static case—the planet Jupiter.

Evidence in favor of an outward explosion might be found in both the polarization and spectrum variation with linear size. An expansion might be expected to introduce order into both the field system and the electron energy distribution. However, such features could equally apply to a static picture in which small sources within a galaxy would be strongly affected by local conditions. That the more energetic sources--both in terms of emission and storage--are the larger objects would favor a static or even inward-evolving model.

Whichever view is correct, it is very likely that the energy requirements can be reduced considerably by treating the sources as hollow-shell structures rather than the conventional filled volumes. Certainly this applies to many of the older supernova remnants in our own Galaxy. In these objects the flattening of spectral indices with age suggests a diffusion of high-energy electrons into high field regions and vice versa. A similar energy separation may well exist in the extragalactic sources. Evidence for this is becoming available from two types of measurement. The most direct is from interferometry at different wavelengths, which, although suggesting that the over-all dimensions of a source are the same at all wavelengths, suggests differences in the fine structures in the sense that the short-wavelength radiation is concentrated toward the leading edge of a source (e.g., Cygnus A and 3C 33 and perhaps even the jet of 3C 273). Some indirect evidence is available from the observed depolarization of sources. Depolarization varies considerably from source to source, yet it cannot be associated with either galactic coordinates or Faraday rotation. As sources at high galactic latitudes show no Faraday rotation, it is unlikely that depolarization is due to Faraday effect within a source and in any case estimates of Faraday depolarization within the source lead to very large masses of ionized matter. The depolarization observed must thus be intrinsic and presumably due to radiation as the longer wavelengths arising from regions of greater variation in field direction—as might arise in a shell structure or partial shell structure.

COUNTS OF RADIOSOURCES IN THE REVISED 3 C CATALOGUE

P. Véron
Observatoire de Meudon

It has been shown that the log N/log S curve for radiosources of high galactic latitude is a straight line with a slope of -1·85 ± 0·05—at least for fluxes greater than 9×10^{-26} W m^{-2} (c/s)$^{-1}$ at 178 Mc/s (ref. 1). (*N* is the number of sources brighter than the flux *S*.)

These radiosources are known to be extragalactic and to belong to two main classes of objects: the radiogalaxies and the quasi-stellar radiosources (quasars). The radiogalaxies are elliptical giant galaxies; the quasars are stellar objects with very large red-shifts[2-7]. The theoretical log *N*/log *S* curves for all the models of the universe, without any evolutionary effect, have a slope of -1·50 for the larger fluxes and this slope decreases numerically to zero with decreasing flux. Thus the observed slope of -1·85 for radiosources of high galactic latitude seems to show that there is really some evolutionary effect; if this is so, then the effect must be larger for the quasars than for the radiogalaxies, because on the average the quasars have red-shifts larger than those of the radiogalaxies. To verify this assumption, it is necessary to construct separate log *N*/log *S* curves for radiogalaxies and for quasars; to accomplish this a very large number of identifications is required.

A year ago, the number of identifications that had been published was very small, but since then many very accurate radio positions have become available[8-11]. Using these positions, Wyndham[12] and I have made a systematic search to identify any objects in the field of the radiosources in the *Revised* 3 *C Catalogue*[13] on the *Palomar Sky Survey* prints and plates. We were able to make many positive and tentative new identifications.

The *Revised* 3 *C Catalogue* provides an extensive survey of the sky north of $\delta = -5°$, for fluxes larger than 9×10^{-26} W m^{-2}(c/s)$^{-1}$, at $\nu = 178$ Mc/s, and it is thus suitable for statistical analysis.

The *Catalogue* deals with 328 sources: (a) 32 can be classified as galactic or possibly galactic sources on the basis of positive identifications (Crab Nebula = 3C 144, *IC* 434 (HII region) = 3 C 147·1, etc.), thermal spectrum, large angular size, radio structure and small galactic latitude; (b) 42 sources are probably extragalactic but in absorbed fields near the galactic plane, and identification is thus impossible or extremely difficult; (c) 254 sources occur in clear fields.

The sources in group *c* were sub-divided into six classes: 100 certain radiogalaxies; 44 possible radiogalaxies; 39 certain quasars; 21 possible quasars; 36 empty fields; 14 non-identified sources. Identification with a galaxy was considered to be 'certain' when there was a very good agreement between optical and radio positions, or when a spectroscopic confirmation had been obtained. It was considered 'possible' when the agreement was not so good. Identification with a quasar was considered 'possible' when a blue star (blue on the *Palomar Sky Survey* prints) was present at the radio position, and 'certain' when spectroscopic or photoelectric confirmation had been obtained[7,14,15]. An 'empty field' was considered to be one where nothing was visible at the limit of the *Palomar Sky Survey* plates ($\nu \sim 21$) at the radio position. A source was classified as 'not identified' when the radio position was not sufficiently accurate to classify it in any of the previous classes.
our classes

It was now possible to construct log *N*/log *S* curves for the radiogalaxies and for the quasars. Figure 1 shows these curves. (*a*) For all the

...ted from NATURE, 724, 211, 1966.

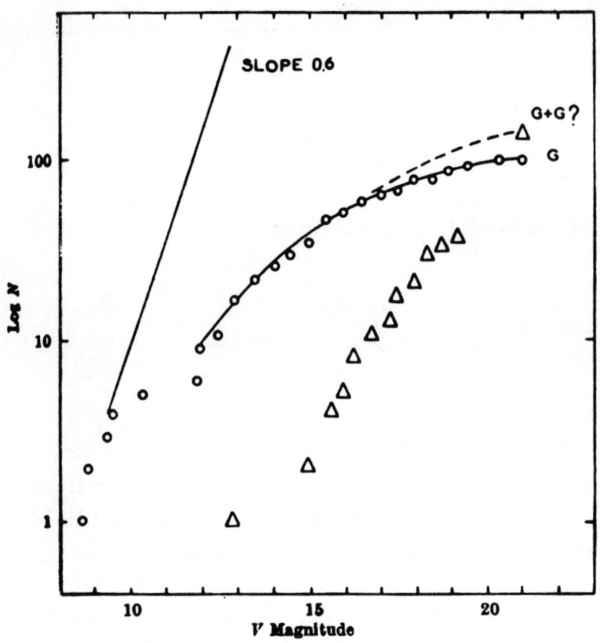

FIG. 1. Log N/log S curves for: (*a*) all the extragalactic radiosources in the *Revised* 3 *C Catalogue* (x); (*b*) radiogalaxies (O); (*c*) quasars (Δ); (*d*) the quasars, the empty fields and the 'non-identified' sources (+).

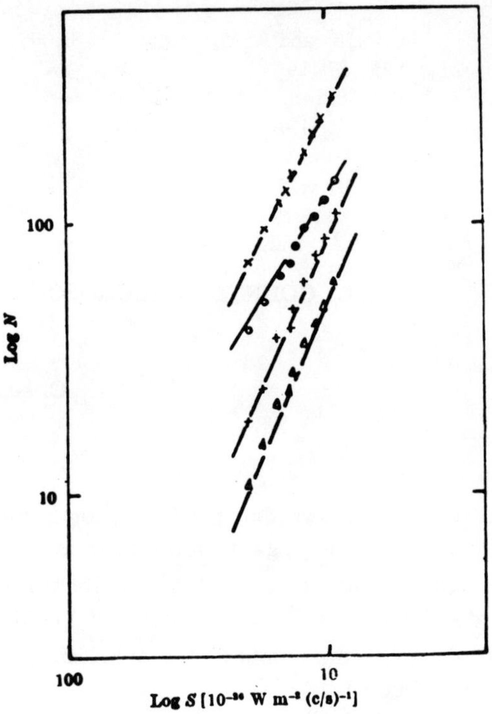

FIG. 2. Curves log N/V for the radiogalaxies (O) and the quasars (Δ) in the *Revised* 3 *C Catalogue*.

extragalactic sources in the *Revised* 3 *C Catalogue*, we found the already known slope of -1·85 ±0·05. (*b*) For 'certain' and 'possible' radiogalaxies, we found a slope of -1·55 ±0·05: this value is normal for objects as near as these galaxies. (*c*) For 'certain' and 'possible' quasars, we found the very large slope of -2·2 ±0·1. (*d*) For the quasars, the empty fields and the non-identified sources, we found the same slope, -2·2.

These findings would seem to indicate that the peculiar slope of -1·85 found for all the radiosources is due only to the quasars and that almost all the empty fields and 'non-identified' sources are quasars and not radiogalaxies.

Fig. 2 shows the curves for log N/V for 'certain' and 'possible' radiogalaxies and for the 'certain' quasers. (The visual magnitudes (v) of the quasars were measured by Sandage and those of the radiogalaxies either measured by Sandage or estimated on the *Palomar Sky Survey* prints[14,15].) The slope of the log N/V curve for the radiogalaxies at the limiting magnitude $V \sim 21$ is very small and this indicates that the number of galaxies associated with radiosources in the 3*C* revised catalogue, with magnitude greater than 21, is very small. The slope of the log N/V curve for the quasars is much larger and an extrapolation of this curve indicates that a number of quasars has still to be identified. This confirms my suggestion that almost all the 'empty field' and 'non-identified' objects are quasars rather than radiogalaxies.

I thank R. L. Adgie, B. G. Clark, I. I. K. Pauliny-Toth, C. M. Wade, D. S. Heeschen, A. R. Sandage and J. D. Wyndham for enabling me to see their results before publication.

This work was undertaken while I was a research assistant at the Mount Wilson Observatory.

[1] M. Ryle, and A. C. Neville, *Mon. Not. Roy. Astro. Soc.*, **125**, 39 (1963).

[2] J. G. Bolton, *Obs. CIT Radio Observatory, Owens Valley*, No 5 (1960).

[3] D. W. Dewhirst, *Proc. Paris Symp. Radioastronomy*, edit. by R. N. Bracewell, 507 (1959).

[4] T. A. Matthews, W. W. Morgan, and M. Schmidt, Astrophys. J., **140**, 35 (1964).

[5] B. Y. Mills, *Amer. J. Phys.*, **13**, 550 (1960).

[6] M. Ryle, and A. Sandage, *Astrophys. J.*, **139**, 419 (1964).

[7] M. Schmidt, *Astrophys. J.*, **141**, 1295 (1965).

[8] R. L. Adgie, and H. Gent, *Nature*, **209**, 549 (1966).

[9] B. G. Clark, and D. E. Hogg, (to be published).

[10] I. I. K. Pauliny-Toth, C. M. Wade, and D. S. Heeschen, (to be published).

[11] J. D. Wyndham, *Astrophys. J.*, **70**, 384 (1965).

[12] J. D Wyndham, (in the press).

[13] A. S. Bennett, *Mem. Roy. Astro. Soc.*, **68**, 163 (1962).

[14] A. Sandage, *Astrophys. J.*, **141**, 1560 (1965).

[15] A. Sandage, (personal communication).

IDENTIFICATION OF RADIO GALAXIES AND QUASI-STELLAR OBJECTS

J. G. Bolton
C. S. I. R. O., Radiophysics Laboratory, Sydney

About 2,000 radio sources between declinations +20° and -90° have been catalogued during the past four years with the 210 foot telescope at Parkes. The finding survey for the catalogue was made at a frequency of 408 Mc/s and additional observations were made of most sources at 1,410 and 2,650 Mc/s. The nominal lower limit of flux density for most of the survey was 2·5 flux units

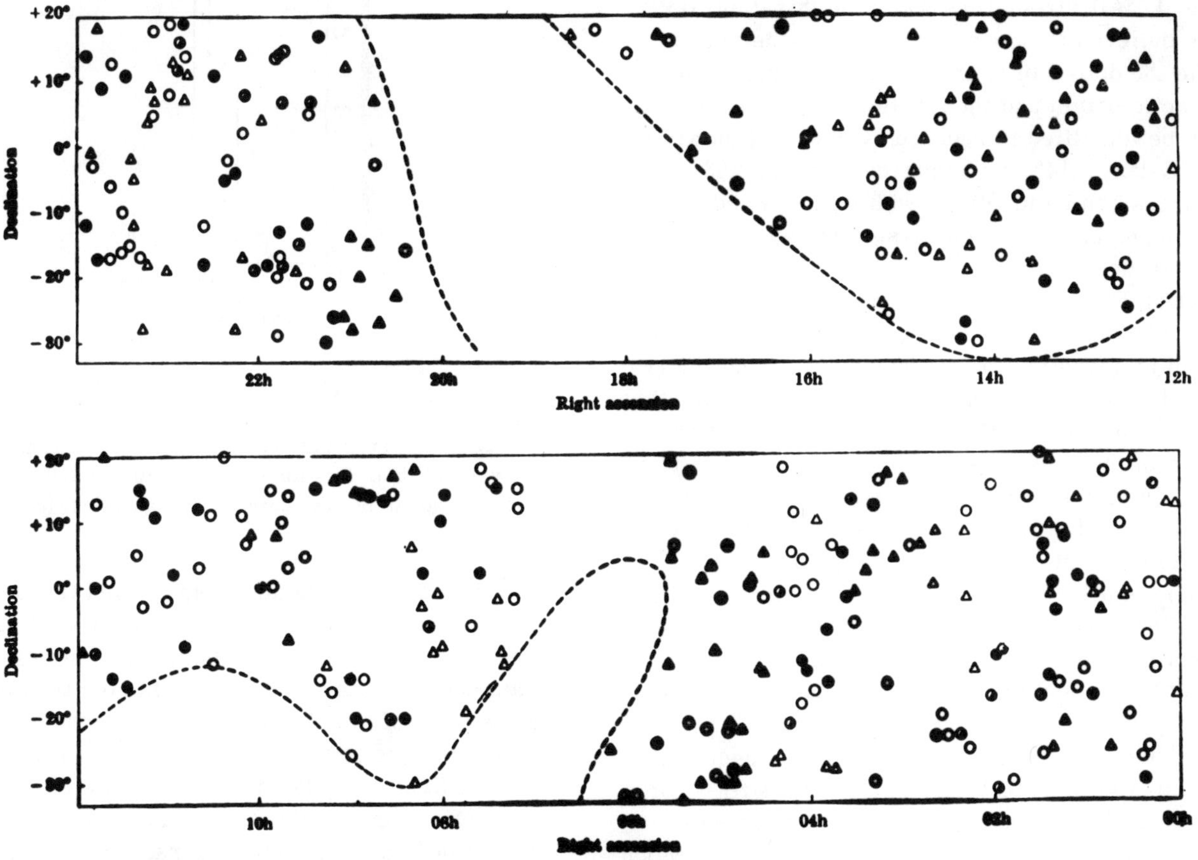

FIG. 1. Distribution of 383 radio sources between declinations 20° and -30° for which precise positions are available. O, Unidentified sources; Δ, galaxies; ●, confirmed quasi-stellar objects; ◑, possible quasi-stellar objects. Dashed line encloses area containing a fraction of galactic sources or regions of obscuration.

Reprinted from NATURE, 211, 917-920, 1966.

(10^{-26} W m^{-2} (c/s)$^{-1}$) and the accuracy of the source positions is estimated to be ± 1 minute of arc. More precise positions (± 12 seconds of arc) have been determined for 640 sources with flux densities greater than 1·6 flux units at 1,410 Mc/s. A search of the *Palomar Sky Survey* prints in the area between declinations 20° and -33° which is common to both the *Sky Survey* and the *Parkes Catalogue* has yielded about 400 identifications. The various zones of the *Catalogue*[1], the precise positions[2] and details of the identifications[3] are at present being published in the *Australian Journal of Physics*.

So far as identifications are concerned, the statistically most complete sample is that of 383 sources between declinations 20° and -30° for which accurate positions are available Their distribution in celestial co-ordinates and the identification status of each source is shown in Fig. 1. All the sources shown on this diagram are believed to be extragalactic. In the area inside the dotted line, the sources are partly of galactic origin and the fraction of sources which can be identified is reduced by obscuration within the Galaxy. 143, or 38 per cent, of the 383 sources can be identified with galaxies and 100, or 26 per cent, with quasi-stellar objects. About 60 per cent of the quasi-stellar objects have been confirmed by optical investigations. Somewhat similar results have been obtained by Wyndham[4] from an investigation of 300 sources from the revised 3*C* catalogue with flux densities greater than 9 flux units at 178 Mc/s. Of the 242 identifications in the present sample, sixty are either previously well-known identifications, or have been independently identified by Wyndham. Wyndham's ratio of quasi-stellar objects to galaxies is somewhat lower than that of the present results; this probably arises from the difference in the survey frequencies used in the 3*C* and *Parkes Catalogues* and the difference in radio spectra for quasi-stellar objects and galaxies.

This difference between the radio spectra of quasi-stellar objects and galaxies was first noted by A. J. Shimmins of this laboratory (private communication). It is illustrated in Fig. 2, which shows the distribution of the quasi-stellar objects, the galaxies and the unidentified sources in terms of the ratio of their flux densities at 1,410 and 2,650 Mc/s. Most of the galaxies lie well to the right of the line at a ratio of 1·6, whereas the peak of the distribution of quasi-stellar objects is to the left. Thus, the quasi-stellar objects can be distinguished from galaxies by their high frequency spectra and, on this basis, Fig. 2*c* strongly suggests that the majority of unidentified sources are galaxies which are presumably fainter than the plate limit of the *Sky Survey*. Two further arguments may be advanced to support this result.

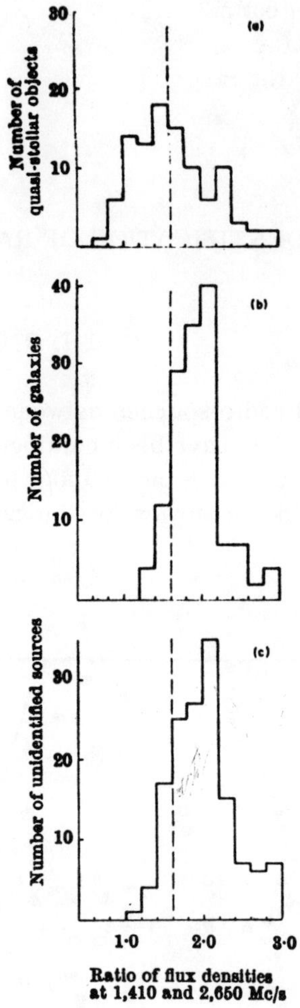

FIG. 2. Histograms of the number of sources with a given ration of flux density at 1, 410 and 2, 650 Mc/s for: (*a*) quasi-stellar objects; (*b*) galaxies; (*c*) unidentified sources.

The first depends on the distribution of radio and optical magnitudes for the identified sources shown in Fig. 3, in which the radio magnitude scale at 1, 410 Mc/s has been adopted such that 1 flux unit is equivalent to a radio magnitude of 10·5. Most of the optical magnitudes were estimated by comparison of their images on the *Sky Survey* plates with those of objects for which photometric magnitudes are available. For faint objeets these estimated magnitudes probably have quite large errors but not of sufficient extent radically to alter Fig. 3. Fig. 3*d* for the galaxies shows the well-known large range in optical

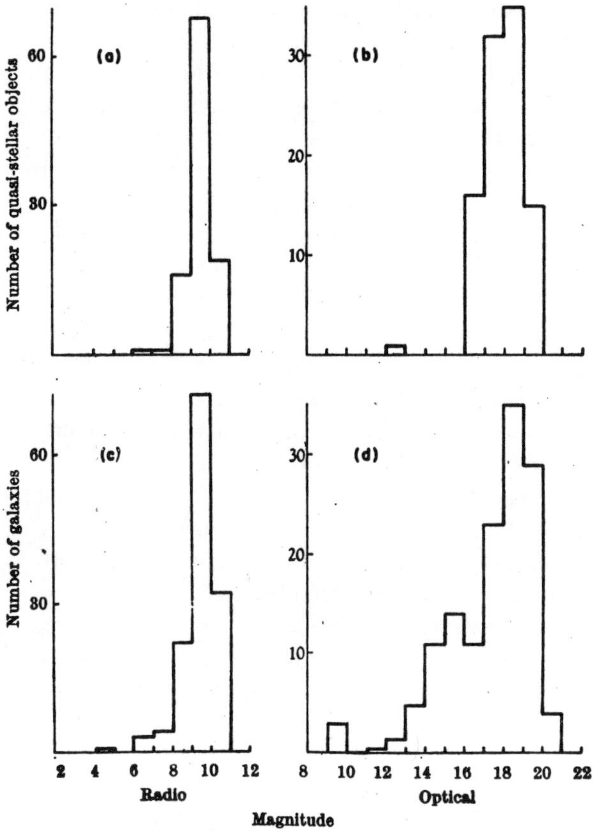

FIG. 3. Histograms of apparent radio and optical magnitudes for quasi-stellar objects and galaxies.

magnitudes corresponding to a small range in radio magnitudes. The distribution is markedly skew and strongly reflects the effect of the plate limit of the *Sky Survey*. The distributions for quasi-stellar objects show a much closer relationship between optical and radio magnitudes: there is no suggestion of a cut-off owing to plate limit, which would probably be at least one magnitude fainter for stellar objects than for galaxies. The sharp cut-off in the magnitude distribution for the galaxies suggests that the unidentified sources may well be galaxies beyond the *Sky Survey* plate limit. The magnitude distribution for quasi-stellar objects suggests that the identifications with quasi-stellar objects are essentially complete for the present sample, and that quasi-stellar objects between the nineteenth magnitude and the plate limit would be found among radio sources which are fainter than those of the present sample.

A second argument is provided by observations for interplanetary scintillations of about 200 of this sample of sources (due to R. D. Ekers of this laboratory). Such scintillation is an indication of the small angular size of a source (2 sec of arc or less). Ekers found that more than 50 per cent of the suggested quasi-stellar objects scintillate, whereas only 20 per cent of the galaxies fainter than the seventeenth magnitude do so. He found that only 25 per cent of the unidentified sources scintillate and, on the assumption that they are, in general, more distant than identified sources of the same class, his results strongly suggest that they are galaxies rather than quasi-stellar objects.

Source Counts. The anomaly of "too many faint sources" in the counts of radio sources with flux density is now well established. The slope of the graph of the logarithm of the number of sources (N) with a flux density greater than S against this value of log S is approximately -1·8, compared with an expected slope of -1·5 for a uniform distribution of sources in a Euclidean universe. This result has been interpreted either as indicating an evolving universe in which radio sources were more numerous or intrinsically brighter in its earlier history. or in terms of a local deficiency in a fraction of the population, which is presumed to be of galactic origin. Véron[5] and Longair[6] have independently pointed out that the slope of the log N-log S curve is significantly different for different classes of sources. They find the slope for galaxies is of the order of -1·5, for quasi-stellar objects -2·2, and even steeper for unidentified sources. The obvious inferences have been drawn--that the quasi-stellar objects and unidentified objects are responsible for the anomaly on the evolving universe model—or, for example, in a model by Sciama and Saslaw[7], that the unidentified sources and possibly some of the quasi-stellar objects for which red-shifts are not yet available represent a galactic component.

The present observations substantiate the results of Véron and Longair. The counts for the three classes at 1, 410 and 2, 650 Mc/s are shown in Fig. 4*a*, *b* and *c*. The differences between the slopes are not as extreme as those of Véron and Longair, possible because of the larger number of samples involved, but they are nevertheless quite pronounced. The log N-log S relationship for galaxies does not, in fact, have a constant slope but is markedly flatter towards the faint end.

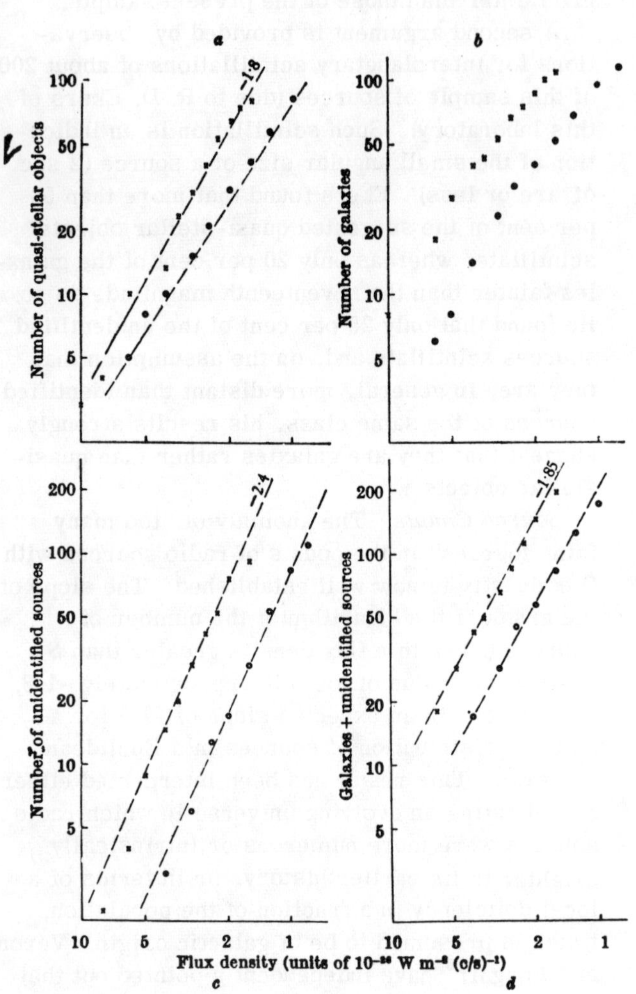

FIG. 4. Source counts at 1, 410 Mc/s (x) and 2, 650 Mc/s (O) for: (*a*) quasi-stellar objects; (*b*) galaxies; (*c*) unidentified sources; and (*d*) composite of galaxies and unidentified sources.

Such an effect would result from the cut-off imposed on identifications of galaxies by the plate limit of the *Sky Survey*, for the fraction of identifications possible will decrease with the flux density. If the previous deduction that the unidentified sources are predominantly galaxies is correct, then these sources should be included with the galaxies in the count. The log N-log S relationship for galaxies and unidentified sources is shown in Fig. 3d, where its slope can be seen to be indistinguishable from that of the quasi-stellar objects. It will be seen that the log N-log S slope for the quasi-stellar objects is slightly flatter at 2, 650 Mc s than at 1, 410 Mc s. No significance should be attached to this result; it arises from the high fraction of flat spectrum sources among the quasi-stellar objects, and the selection criteria for the source sample. The galaxy counts do not show this effect, as the fraction of flat-spectrum sources among them is negligible.

The result that the modified log N-log S slope for galaxies is the same as that for quasi-stellar objects is perhaps not surprising for, if the red-shifts of the quasi-stellar objects are cosmological, the quasi-stellar objects and galaxies in the present sample span much the same volume of space. Most of the sources under consideration have apparent radio magnitudes between 9 and 10. In this range seventy-eight sources can be identified with galaxies and eighty are unidentified. A few of the latter may be faint or unrecognized quasi-stellar objects and some are probably N-type or compact galaxies above plate limit--which are difficult to distinguish from stars on the *Sky Survey* prints. However, at least fifty are probably galaxies beyond plate limit. These will have absolute radio magnitudes equal to or brighter than $M_\gamma = -32$. The exact $M_\gamma = -32$ cannot be determined, but it would certainly extend to form of the luminosity distribution beyond $M_\gamma = -34$, because objects of this luminosity are known to exist (for example, Cygnus *A* and 3*C* 295). Galaxies in this range will have distance moduli between 41·5

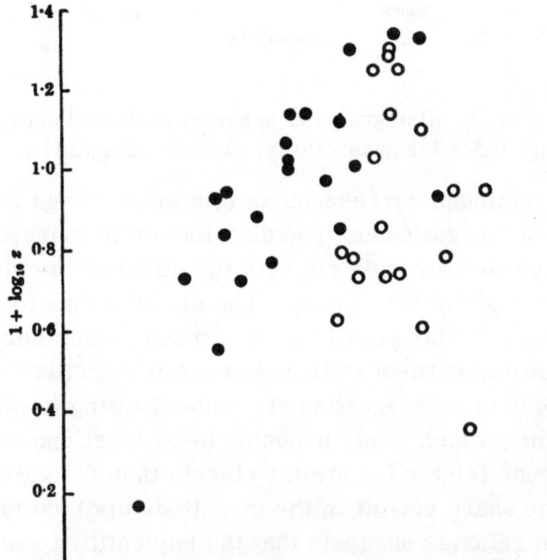

FIG. 5. Radio magnitude-red-shift diagram for quasi-stellar objects. ●, Objects with flat or relatively flat spectra; O, objects with steep radio spectra.

and 43·5, and the red-shifts of the most distant will be well in excess of unity. The similarity of the log N-log S for the galaxies and quasi-stellar objects within the same range of red-shifts may indicate a parallel evolutionary development.

Red-shift–Magnitude Relation for Quasi-stellar Objects. So far in this article it has been assumed that the red-shifts of the quasi-stellar objects are cosmological in origin. A number of alternative suggestions have been made to overcome difficulties associated with their energy supply and incompatibilities between their angular diameters and short period light variations. However, no really insurmountable objections to a cosmological origin have been raised, and the alternative hypotheses appear to provoke greater difficulties than they aim to overcome.

Hoyle and Burbidge[8] have recently questioned the cosmological red-shift on new grounds. They point out that there is no relationship between the red-shift and radio magnitude at low frequencies (178 Mc/s), and that even the relationship between red-shift and optical magnitude appears to be simply a scatter diagram. However, such a result does not necessarily vitiate a cosmological red-shift; it may merely be an indication of a large dispersion in the relevant absolute luminosities. To obtain a reasonable relationship demands an intrinsic property which is comparatively constant, as is the absolute optical magnitude for the radio galaxies. The large variations in the light output of some of the quasi-stellar objects would suggest that the optical magnitude is not such a property. About half the objects considered by Hoyle and Burbidge have fairly steep spectra, and quite a few of this class of quasi-stellar objects are known to have components of considerable angular and presumably linear size. Such components may be the results of singular events, and similar to the large physical components of radio galaxies--for which there is certainly a very large range in intrinsic luminosity and a correspondingly large scatter in their radio magnitude-red-shift diagram. It is possible that the radio emission from quasi-stellar objects with flat radio spectra, which might result from a continuous process in a small physical volume, could be a better indicator of intrinsic luminosity. In Fig. 5 the relationship between the radio magnitude at 1,410 Mc/s and the red-shift is shown, with different symbols for the objects with relatively flat and steep spectra. While there is no very distinct relationship, the scatter for the flat-spectrum sources is considerably lower than that in either of the diagrams given by Hoyle and Burbidge. There is a suggestion that radio spectra might be useful in assigning a luminosity classification, and that a search for objects with very large red-shifts could be most successful among faint radio sources with flat spectra.

I thank many of my colleagues at the Radiophysics Laboratory who provided the basic data used in this investigation, particularly Dr. Margaret Clarke and Mrs. Jennifer Ekers for their work on identifications. I also thank Dr. T. D. Kinman of the Lick Observatory for providing optical confirmation and accurate optical positions of many of the southern quasi-stellar objects, and Dr. Margaret Burbidge for providing red-shift data in advance of publication.

REFERENCES

[1] Parkes Catalogue of Radio Sources: Declinations -20° to -60°, J. G. Bolton, F. F. Gardner, and M. B. Mackey, Austral. J. Phys., 17, 340 (1964). Declinations 0° to 20°, G. A. Day, A. J. Shimmins, R. D. Ekers, D. J. Cole, Austral. J. Phys., 19, 30 (1966). Declinations 0° to -20°, A. J. Shimmins, G. A. Day, R. D. Ekers, and D. J. Cold, Austral. J. Phys. (to be published).

[2] A. J. Shimmins, M. E. Clarke, and R. D. Ekers, Austral. J. Phys. (in the press).

[3] Identification of Sources: Declinations -20° to -45°, J. G. Bolton, M. E. Clarke, and R. D. Ekers, Austral. J. Phys., 18, 627 (1965). Declinations -20° to -30°, J. G. Bolton, and J. Ekers, Austral. J. Phys., 19, 275 (1966). Declinations 0° to 20°, M. E. Clarke, J. G. Bolton, and A. J. Shimmins, Austral. J. Phys., June 1966, and J. B. Bolton, and J. Ekers, Austral. J. Phys., June 1966. Declinations 0° to -20°, J. G. Bolton, and J. Ekers, Austral. J. Phys., 19 (1966).

[4] J. D. Wyndham, Astrophys. J., 144, 459 (1966).

[5] P. Véron, Conference on Observational Aspects of Cosmology, Miami (1965).

[6] M. S. Longair, Conference on Observational Aspects of Cosmology, Miami (1965).

[7] D. W. Sciama, and W. C. Saslaw, Nature, 210, 348 (1966).

[8] F. Hoyle, and G. R. Burbidge, Nature, 210, 1346 (1966).

THE ANGULAR SIZES OF DISCRETE RADIO SOURCES

H. P. Palmer
Nuffield Radio Astronomy Laboratories, Jodrell Bank

I. Introduction

In the last few years evidence has accumulated that many of the discrete radio sources have complex angular structures. Some of the largest have features more than 1° across, with values of surface brightness temperature as low as 10° K, while a few sources are known which are smaller than 1'', and have brightness temperatures in excess of 10^{9}°K. The angular sizes of sources in this latter group may be investigated with interferometers consisting of two aerials separated by a suitably large number of wavelengths. In Sections II and III preliminary series of interferometer measurements are described which have enabled us to estimate the approximate angular size, or angular scale, of about 400 discrete sources. More detailed investigations of the angular structure of a few of the smallest sources are described in Section IV. The relationships between these measurements and other parameters of discrete sources are discussed in Section V.

II. Investigation of the angular scale of discrete sources

The angular sizes and structures of the more intense sources were investigated by several workers in 1949-1952, and shown to be larger than 1' (Jennison and Das Gupta 1951; Mills 1951; and Smith 1952). An extension of these measurements to sources of smaller angular size and lower flux density was reported by Morris, Palmer, and Thompson (1957). Observations were made at a wavelength $\lambda = 1.89$ m of seven sources with a series of E. W. base lines, one end being the 218 foot transit telescope at Jodrell Bank, the other a small transportable broadside array of area 36 m^2. For the base lines longer than $\lambda 500$ we developed phase swinging methods of controlling the output fringe speed of the interferometer so that it was not too fast for a standard pen recorder. We also developed a phase-coherent VHF radio link. With these facilities we were able to observe recognizable fringe patterns for three sources with an aerial separation of $\lambda 10600$ (20 km). The observations were interpreted to mean that these three sources, which were then unidentified, were unresolved, or only barely resolved, and therefore smaller than 12". We concluded that these sources had a surface brightness greater than 10^7 °K, which was comparable with that of the remote galaxy of Cygnus A. These sources are now known by their numbers in the *Third Cambridge Catalogue*, as 3C 147, 196 and 295. The first two have recently been identified as quasars. 3C 147 has a redshift of 0.545, which is the largest value which has been measured reliably so far, while no comparable measurements have yet been possible for 3C 196. 3C 295 was identified by Minkowski (1960) as an extremely remote galaxy having a redshift of 0.47. As this is the most remote (non-quasar) galaxy studied so far, these three sources provide considerable justification for the feeling that sources of small angular size are usually remote and often extremely interesting.

III. Transit observations of the angular scale of discrete sources

Shortly after the Mark I telescope was completed in 1957, an improved interferometer was used to make similar transit observations of approximately 300 sources at the same wavelength of 1.89 m. The Jodrell Bank Radio Telescope Mark I was used as one element of the interferometer, and cylindrical paraboloids as shown in Fig. 1 were erected, at distances $\lambda 2200$ (4.2 km) and $\lambda 9700$ (18.4 km) from Jodrell Bank. More than 300 sources were observed with each of these base lines. Many of them were partially or completely resolved by the longer base line, but about 10 per cent of the sources were found to be unresolved by either base line. During 1960 and 1961

FIG. 1. The interferometer aerials erected at site 4 in 1961.

the remote aerials were therefore moved to still more distant sites, λ32000 (60.3 km) and λ61100 (115.4 km) from the Radio Telescope Mark I. The observational and instrumental problems became increasingly difficult as the separation of the aerials was increased, so at the longest base lines we concentrated mainly on those sources which our previous work had shown to be most likely to give discernible fringe patterns.

In order to use these observations to derive information about the angular size of sources, one needs to compare the amplitude of the fringe pattern observed at a long base line with the amplitude which would have been observed with an interferometer which had a much smaller aerial separation, but was otherwise identical. The ratio of these amplitudes is known as the fringe visibility γ. It is not usually possible to make such a comparison directly, and differences between interferometers are therefore expressed by their calibration factor, K. If all the parameters of the systems are known sufficiently well, values of K may be calculated and the required fringe visibility can then be found by comparing the observed and theoretical signal-to-noise ratio for each source. However, with a complex system involving radio links, the parameters are not usually known well enough to allow this procedure. For comparatively short base lines it is sufficient to observe the signal-to-noise ratios for several sources which have previously been shown to be unresolved, either by observations at longer base lines or by statistical studies of a reasonably large sample of sources. For the observations at the longest base lines neither method could be used because very few sources gave large fringe amplitudes. As a first approximation we selected the highest apparent value of fringe visibility, and then calculated a calibration factor by assuming that the true value was $\gamma = 1.0$ for that source.

Transit observations with a few very different spacings give a useful indication of the approximate angular size, or angular scale of the sources, though they do not allow one to derive accurate values for the detailed angular structure of complex sources. An analysis of angular scale has been given by Allen, Hanbury Brown, and Palmer (1963) and Table I contains a summary of their results. Only five sources were found which appeared to have fringe visibilities $\gamma \geq 0.8$ at each base line, corresponding to an angular scale smaller than 1". The table shows that approximately 2 per cent of sources have major components smaller than 1". It must be remembered that, with an interferometer, errors of observation are not symmetrical in their effects, for, in general, they are much more likely to decrease the observing fringe amplitude than to increase it. This value of 2 per cent is therefore more likely to be too low than too high.

Table I. A summary of transit observations of radio sources with long base line interferometers at a wavelength λ = 1.89 m.

	Base Line d/λ	a (km)	No. of Sources Studied	$\gamma \geq 0.8$	Fringe Pattern Seen ($\gamma < 0.8$)	No Fringe Pattern Seen
(1)	2200	4.2	274	94	113	67
(2)	9700	18.4	305	40	124	141
(3)	32000	60.3	121	14	33	74
(4)	61100	115.4	187	5	17	165

IV. Observations with tracking interferometers

In a brief extension of the work at λ61100 a small fully steerable aerial was installed at the remote station. This permitted tracking observations of six sources for periods of 8-12 hours, (depending on their declinations) with consequent variations in effective resolving power from about λλ30000 to 61100 and associated changes in the position angle in which the resolution was obtained. The results for 3C 48 and 3C 295 are shown in Fig. 2, a and b. It will be seen that 3C 48 did not change significantly as the effective base line increased from λ30000 to λ61100. This provided additional support for our assumption that this source was unresolved, and could be used to calculate the calibration factor.

The observations of 3C 295 show one pronounced minimum near transit, and a second near H. A. 4^h30^m. The simplest source model which fits these observations consists of two parallel strips 1.7 × 1 inches separated by 4.2 inches as shown in Fig. 3 (Anderson, Palmer, and Rowson, 1962). Rowson (1963) has given similar discussions of the observations of Cygnus A and two

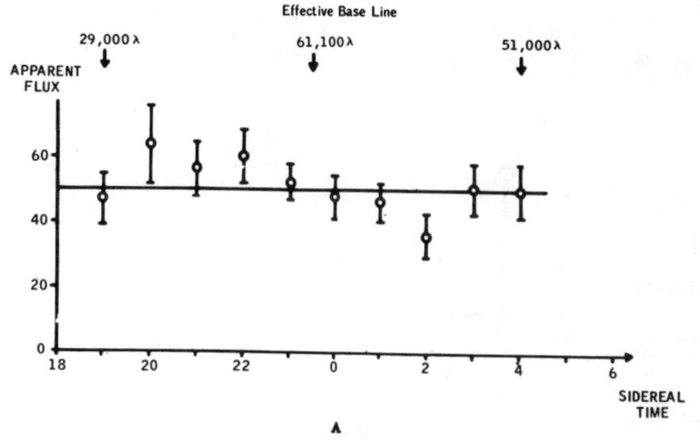

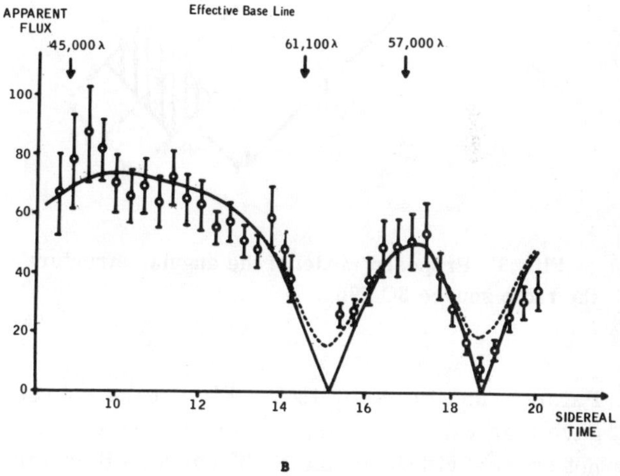

FIG. 2. Mean values of fringe amplitude observed for the quasar 3C 48 and the galaxy 3C 295 with a long base-line interferometer operating between Jodrell Bank and site 4, 115.4 km away.

other sources (3C 123, 147); so, of five sources whose structure has been studied at very long base lines, three have been found to be double.

In order to obtain still higher resolving powers, and to exploit the advantages of the tracking technique, a 25-foot telescope was constructed and operated during 1963-1964 in East Yorkshire, 132 km from Jodrell Bank (Fig. 4). The frequency of observation was changed to 408 Mc/s giving a maximum base line of λ180000. This telescope could be remotely controlled by a radio link from Jodrell Bank, and was also equipped with a small mechanical analogue computer which, when desired, caused the telescope to follow the sidereal motion of sources.

Altogether thirty-two sources were studied with this equipment, and fringe patterns were observed for sixteen of these (Anderson, Donalson.

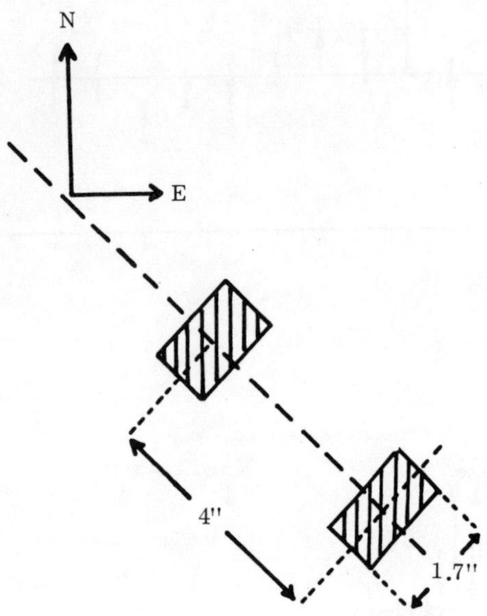

FIG. 3. Proposed model of the angular structure of the radio source 3C 295.

Palmer, and Rowson, 1965). Five of the sources gave high values of fringe visibility, which did not change with hour angle. They were therefore assumed to be unresolved and hence smaller than 0".4. An example of the fringe patterns observed for 3C 286 is shown in Fig. 5, a. Analysis of these observations is still in progress and no source models have yet been derived.

Many of the sources observed in our recent work have been identified with quasars. Approximate values for their radio angular sizes can now be derived if this data is combined with the transit data discussed in Sections II and III. Such estimates can be made for twenty-six of the thirty-one quasars reported by Sandage. The distribution of sizes is given in Table II, which shows that more than half of this group are smaller than 5", and one-third are smaller than 1". Similar estimates can be made for thirty-five of the fifty-two radio galaxies in the list of Matthews, Morgan, and Schmidt (1964). Twenty-seven of these are larger than 20", and none are smaller than that.

V. Angular scale and radio spectra

Kellerman, Long, Allen, and Moran (1962) have pointed out that there is good correlation between lists of sources having high values of surface brightness and having radio spectra which are strongly curved. Slish (1963) and Williams (1963) have suggested that the curvature at low frequencies may arise because the synchrotron radiation will be absorbed in the emitting region if the surface brightness exceeds a value of approximately 10^{10}° K. Using this theory they have predicted minimum values of the angular size of some sources. These included CTA 21 and 102, which were predicted to be smaller than 0".01. Moffet *et al.* reported that these sources were smaller than 1". In our recent work, fringe patterns were seen from both sources suggesting that they are unresolved at λ180000. This observation reduces the value of their angular size to < 0".4. Angular dimensions have been predicted for eight sources in all, and now show reasonably good agreement with the measured sizes, as may be seen in Table III.

Table II. Estimates of the angular scale of quasars and radio galaxies derived from interferometer measurements.

Angular Scale (Sec of Arc)	Size Estimates for 25 Quasars	Size Estimates for 35 Radio Galaxies
> 20	3	26
≤ 20	22	9
≤ 5	14	0
≤ 1	9	0

Table III. A comparison of the predicted and observed angular sizes for some sources which have curved radio spectra.

Source	Predicted Size (Sec of Arc)	Observed Size (Sec of Arc)	Frequency of Observation (Mc/s)
3C 48	0.4	< 0.5	408
119	0.3	< 0.4	408
147	0.2	0.6	408
295	.6	Double 1 × 1.7	159
298	.6	Double? <1×1	408
299	.2	No fringes seen	
CTA 21	.01	< 0.4	408
102	0.01	< 0.4	408

VI. Future work

The work described in this report has suggested two separate lines of development. There is the problem of extending the flux range of the

FIG. 4. The remotely controlled transportable radio telescope erected in East Yorkshire, 134 km from Jodrell Bank.

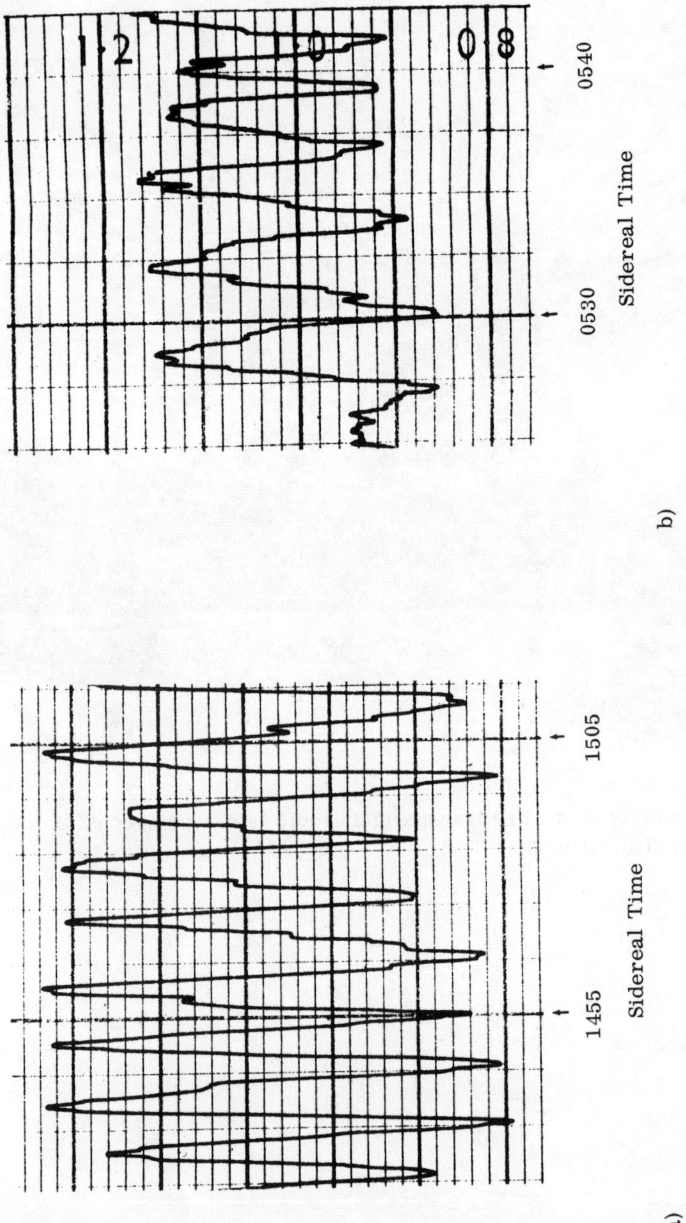

FIG. 5. Examples of the fringe patterns observed for (a) the source 3C 286 and (b) the source CTA 21, during interferometric observations at 408 Mc/s with aerials 134 km apart.

sources for which diameter measurements are made. For this work we require interferometer aerials of larger collecting area, separated by distances comparable with those discussed above. Our fully steerable and transportable Radio Telescope Mark III is being constructed for this program, at a site near Nantwich, 24 km from Jodrell Bank. With this instrument we hope to study more than 1,000 sources at several frequencies between 150 and 1420 Mc/s, that is, effective base lines d/λ between approximately 10,000 and 100,000 wavelengths.

The second problem, of resolving the smallest sources and studying their structure, requires much greater resolving power. We shall probably try to obtain this using the shorter wavelengths, in co-operative programs with other observatories. If base lines substantially longer than 100 km are required, it may be necessary to develop entirely new methods of transferring the radio noise from the aerials to the main interferometer.

REFERENCES

L. R. Allen, R. Hanbury Brown, and H. P. Palmer, 1963, *M. N.*, **125**, 57.

B. Anderson, H. P. Palmer, and B. Rowson, 1962, *Nature*, **195**, 165.

B. Anderson, W. Donaldson, H. P. Palmer, and B. Rowson, 1965, *Nature*, **205**, 375.

R. C. Jennison, and M. K. Das Gupta, 1953, *Nature*, **172**, 996.

K. I. Kellerman, R. J. Long, L. R. Allen, and M. Moran, 1962, *Nature*, **195**, 692.

T. A. Matthews, W. W. Morgan, and M. Schmidt, 1964, *Ap. J.*, **140**, 35.

B. Y. Mills, 1952, *Nature*, **170**, 1063.

R. Minkowski, 1960, *Pub. A. S. P.*, **72**, 354.

D. Morris, H. P. Palmer, and A. R. Thompson, 1957, *Observatory*, **77**, 103.

B. Rowson, 1963, *M. N.*, **125**, 177.

F. G. Smith, 1962, *Nature*, **170**, 1065.

V. I. Slish, 1963, *Nature*, **199**, 682.

P. J. S. Williams, 1963, *Nature*, **200**, 56.

NEW LIMITS TO THE ANGULAR SIZES OF SOME QUASARS

R. L. Adgie, H. Gent, and O. B. Slee*
Royal Radar Establishment, Great Malvern

and

A. D. Frost†, H. P. Palmer, and B. Rowson
University of Manchester, Nuffield Radio Astronomy Laboratories, Jodrell Bank

In previous experiments at Jodrell Bank the angular sizes of discrete radio sources have been investigated by means of long base line interferometers[1,2]. The highest resolving power used previously was obtained during observations at a wavelength $\lambda = 0\cdot73$ m with telescopes 180,000 wavelengths apart (134 km). Four quasi-stellar and one unidentified source were found to be unresolved in those observations. Their angular sizes were thus shown to be smaller than 0·4 sec of arc. In a further attempt to resolve these sources another experiment has been carried out using the Mark I 250-foot radio telescope at Jodrell Bank, and one of the 82-foot radio telescopes operated by the Royal Radar Establishment, Malvern. The separation of these telescopes is 127 km, in a direction which is close to north—south. This interferometer worked on a wavelength $\lambda = 0\cdot21$ m, so that the maximum resolving power was more than three times greater than had been obtained in the previous observations. The effective resolving power changes as sources are observed in different directions, but its maximum value is greater than 600,000 wavelengths for most sources. With this baseline the output fringe frequency produced as the radio source moves through the lobes of the interferometer pattern varies with hour angle from 0 to as much as 40 c/s. An almost identical frequency was produced continuously by a digital 'fringe speed machine' and this was subtracted from the interferometer output, so that the fringe patterns displayed on the chart recorder were normally slower than 0·5 cycles/min. A microwave link system was established between the two observatories, via two repeater stations. A new very-high-frequency phase-locking system was also developed, with equipment at each site.

During these observations, the five sources which had not been resolved previously (3C 119, 286, CTA 21 and 102) all gave clear fringe patterns over a wide range of hour angles, though for only one source did the amplitude remain approximately constant. These sources are listed in column 1 of Table I. Column 2 shows their catalogued[8] flux density, S, at $\lambda - 0\cdot21$ m. The observed fringe amplitudes were normalized by daily observations of the source 3C 147, which was found to give clear fringe patterns at all hour angles. Corrections arising from various instrumental effects have been applied to these normalized values of fringe amplitude. It was found that when the sources were observed at elevations less than 15° these corrections were frequently greater than 20 per cent, and such observations have not been used. The corrected values of fringe amplitude have been calibrated by considering the maximum value observed for each source, which is shown in column 3 of Table I. It was found that these maximum values are in an almost constant ratio to the total flux from each source, as shown in column 4, the mean value of these ratios being $14\cdot1 \pm 0\cdot9$.

A constant ratio in column 4 would be obtained if the minimum linear dimensions of all these sources were similar, and they were at comparable distances, and were partially resolved to the same extent by effective baselines between 500,000λ and 600,000λ. This is an improbable situation, and it seems more likely, as we assume, that each source was unresolved by the effective baseline at which the maximum value of fringe amplitude was observed. This means that each of

Reprinted from NATURE, 208, 275-276, 1965.

Table I. Radio sources smaller than 0·1 sec of arc in at least one dimension.

Source	Flux S at $\lambda = 0{\cdot}21$ m (flux units)*	Max. fringe amplitude, A, observed with tracking interferometer of aerial spacing 605,000λ (arbitrary units)	A/S
3C 119	8·5 ± 0·8	125 ± 10	14·7 ± 1·8
3C 286	16·1 ± 0·4	245 ± 20	15·2 ± 1·5
3C 287	7·6 ± 1·5	95 ± 5	12·5 ± 2·5
CTA 21	8·0 ± 0·8	115 ± 10	14·4 ± 2·0
CTA 102	6·6 ± 1·6	95 ± 10	14·4 ± 2·0
		Mean value	14·1 ± 0·9

*One flux unit = 10^{-26} Wm^{-2} (c/s)$^{-1}$.

Table II. Data on 3C 273 at 0.21 M

Total flux density (1962·3)	39·8 ± 2 flux units
Flux ratio component B : A (1962·9)	1·40
Therefore component A was then 16·5 and B,	23·3 flux units
Long baseline interferometer (1965·5)	
Two parts of one component gave	30 ± 2 flux units
Total power measurements (1965·6) $A + B$	46 ± 1·5 flux units
Provisional interpretation (1965·7)	

Component A is still 16·6 flux units, and component B is now 30 flux units. Flux ratio B : A now 1·84.

these sources is smaller than 0·1 sec of arc in at least one direction. The source CTA 102 did not appear to be resolved at any hour angle when its elevation was greater than 15°, so that its angular size is to be less than 0·1 sec of arc in all position angles. Each of the other sources was partially resolved at some hour angle, and the corresponding maximum angular dimensions (assuming gaussian source models) were in the range 0·12-0·16 sec of arc, as shown in column 5 of Table I.

The quasar 3C 273 was also observed and was found to give clear fringe patterns of unexpectedly large amplitude with one well-marked minimum near meridian transit. Observations of lunar occultations of this source show that it consists of two components, A and B, the centres of which are 19·5 sec apart in position angle 044°. If both these components had contributed to the fringe patterns observed at this base line, a pronounced minimum would have been observed every 10-15 min at most hour angles. As these frequent minima were not observed, it follows that only one component of the source is small enough to give fringe patterns at this base line, and that it, in turn, probably consists of two parts, which are each smaller than 0·1 sec of arc.

As only one minimum was observed, the angular separation of these parts and the position angle cannot, by inspection, be determined uniquely, though they may be obtained in due course by more detailed analyses. If it is assumed that this position angle is also 044°, the separation is of the order 0·4 sec of arc.

This interpretation of our observations, when calibrated with the factor derived from the sources discussed earlier, shows that the two parts of one component of 3C 273 are now giving a flux density not less than 30 ± 2 flux units. As may be seen in Table II, this is significantly greater than the value of 23·3 flux units derived for the brighter component, B, from the ratio given by the occultation observations of 1962·9. The total flux from this source was therefore re-measured recently, with the Mark I telescope at Jodrell Bank, and compared with the radio galaxies 3C 348 and 353. These measurements show that, on the assumption that the radio galaxies have remained constant, the value of total flux from 3C 273 in 1965·6 was 46 ± 1·5 flux units. This is compatible with the other results if it is assumed that the interferometer observations refer to the component B, and that the flux of this component has now increased

Table 3. CURRENT INTERPRETATIONS OF THE RADIO MEASUREMENTS OF THE ANGULAR AND LINEAR DIMENSIONS OF SOME QUASARS

Source	Redshift z	Remarks on individual sources	Approximate linear dimensions (parsecs)	
			If redshifts cosmological*	If sources 'local' at about 10 Mpc
3C 273	0·158	Source has two components 19·5 sec apart in position angle 044°	Separation of A and B 32,000	950
		Component B, coincident with optical quasi-stellar object, has two parts, each smaller than 0·1 sec, separation probably 0·4 sec. At $\lambda = 0{\cdot}21$ m flux of component B has increased by approx. 30 per cent in 2·7 yr	Each part smaller than 170 separation probably 700	5 20
		Component A has dimensions 5 × 1·5 sec flux probably unchanged	8,500 × 2,500	240 × 70
3C 48	0·367	At $\lambda = 0{\cdot}73$ m, elliptical, 0·4 by < 0·3 sec†‡§	1,100 × ⩽900	20 × ⩽15
3C 47	0·425	At $\lambda = 0{\cdot}21$ m, this source has two components (ref. 9) Separation 62 sec Each component ⩽10 sec, but little structure ⩽2 sec¶	 ⩾180,000 < 30,000 but > 6,000	 ⩾3,000 < 500 but > 100
3C 147	0·545	At $\lambda = 0{\cdot}73$ m, elliptical 0·6 by < 0·4 sec†‡§	2,000 × ⩽1,300	30 × ⩽20
3C 254	0·734	At $\lambda = 1{\cdot}89$ m, ≃6 sec. No structure measurements at any wave-length	≃21,000	≃300
3C 286	0·86	At $\lambda = 0{\cdot}21$ m, 0·12 by < 0·1 sec‡	420 × < 350	6 × < 5
3C 245	1·029	At $\lambda = 1{\cdot}89$ m, > 12 sec. No structure measurements at any wave-length¶	> 41,000	> 600
CTA 102	1·037	At $\lambda = 0{\cdot}21$ m, < 0·1 sec in all position angles	< 360	< 5
3C 287	1·055	At $\lambda = 0{\cdot}21$ m, 0·14 × < 0·1 sec‡	500 × < 360	7·0 × < 5
3C 9	2·012	At $\lambda = 1{\cdot}89$ m. > 10 sec. No structure measurements at any wave-length¶	> 32,000	> 450

* These linear dimensions have been calculated for model universes (ref. 4) in which the acceleration parameter $q_0 = +1$. If $q_0 = 0$, these values must be multiplied by a factor $(1 + 0{\cdot}5z)$ which for these sources lies in the range 1–2·01.

† Results (ref. 7) of detailed analysis of observations at $\lambda = 0{\cdot}73$ m.

‡ None of the available evidence suggests that the source contains fine structure within this elliptical component, but further analyses or observations at even higher resolution could conceivably reveal a more complex structure, within these overall dimensions, as in the case of 3C 273B.

§ Fringe patterns corresponding to partial resolution of this source were recorded during the observations at $\lambda = 0{\cdot}21$ m described above. They have not yet been analysed in sufficient detail to improve significantly the earlier interpretation of the angular structure of this source.

¶ A weak source at $\lambda = 0{\cdot}73$ and $\lambda = 0{\cdot}21$ m. Even if it were unresolved, the signal-to-noise ratio would be poor under optimum conditions, and the fringe pattern might not have been recognized during those observations which were attempted.

to 30 flux units, that is, by approximately 30 per cent in three years. It is not known which part of component B has increased. This conclusion may be compared with the measurements reported by Dent[3] which show that at 8,000 Mc/s, where almost all the radiation comes from component B, the flux has increased by 40 per cent in 2·5 years.

We have summarized in Table III our current interpretations, based on these and earlier measurements, of the angular dimensions of the ten quasi-stellar sources the red-shifts of which have been published[4,5]. Possible values of the linear dimensions of the radio-emitting regions of these sources are shown in columns 4 and 5. Those in column 4 have been calculated on the hypothesis that the redshifts of these sources arise from the general expansion of the Universe, and so correspond to distances greater than 600 Mpc. The values given in column 5 are calculated on the hypothesis suggested by Hoyle and Burbidge[6], that the emitting regions are at distances of order 10 Mpc, and their redshifts arise from very high intrinsic velocities of recession of the individual objects.

REFERENCES

*On leave from C. S. I. R. O., Sydney, Australia.

†On leave from University of New Hampshire.

[1] L. R. Allen, B. Anderson, R. G. Conway, H. P. Palmer, V. C. Reddish, and B. Rowson, 1962, *Mon. Not. Roy. Astro. Soc.*, **124**, 477 (1962).

[2] B. Anderson, W. Donaldson, H. P. Palmer, and B. Rowson, 1965, *Nature*, **205**, 375.

[3] W. A. Dent, 1965, *Science*, **148**, 1458.

[4] M. Schmidt, 1965, *Astrophys. J.*, **141**, 1295.

[5] I. S. Shklovsky, 1963, *Astro. Circ., U. S. S. R.*, No. 250.

[6] F. Hoyle, and G. R. Burbidge, (personal communication).

[7] B. Anderson, Ph. D. thesis, Univ. Manch. (1965).

[8] R. G. Conway, K. I. Kellerman, and R. J. Long, 1963 *Mon. Not. Roy. Astro. Soc.*, **125**, 261.

[9] M. Ryle, B. Elsmore, and Ann C. Neville, 1965, *Nature*, **207**, 1024.

RADIO SOURCES HAVING SPECTRA WITH A LOW FREQUENCY CUT-OFF

J. M. Hornby and P. J. S. Williams*

I. *Introduction*

The analysis of Conway, Kellermann and Long (1963) provided the spectra of 160 discrete sources, and for most of these the variation of flux-density with frequency between 38 Mc/s and 1,420 Mc/s could be represented by a simple power law. For a few sources, however, the flux density at low frequencies fell progressively below the value predicted from the observations at higher frequencies, and these sources were also distinguished by an exceptionally high surface brightness. More accurate values of the flux density at 38 Mc/s have since been obtained, and as a result further examples of this class of source have been established. At the same time, several of these sources have been identified with the optically compact and very luminous extra-galactic objects provisionally described as quasi-stellar. Mechanisms which might be responsible for such a cut-off in the low frequency spectrum have been suggested by Slish (1963) and Williams (1963); these mechanisms are discussed further in the present paper.

2. *Observations*

The flux densities used are taken from Conway *et al.* (1963) with the following additions:

(i) More accurate flux densities are now available at 38 Mc/s as the result of a new survey with the large moving-T synthesis instrument at Cambridge. The observations cover most of the sky between declinations $-16°$ and $+80°$ with a beam of about $0.7° \times 1.0°$ (Williams, Kenderdine & Baldwin 1966).

(ii) In order to provide additional information on the form of the spectra between 38 and 178 Mc/s, use has been made of the observations at 81.5 Mc/s (Shakeshaft *et al.* 1955) and 85 Mc/s (Mills *et al.* 1958). Correction factors of 0.85 and 0.79 respectively have been applied to the published values, the factors being determined by interpolation from a comparison of the quoted flux densities for the more intense sources which have simple power law spectra.

(iii) Flux densities at 610 Mc/s have also been taken from recent measurements at the Nuffield Radio Observatory, Jodrell Bank, which include almost all the sources in the revised 3C catalogue with $S_{178} \geq 14.5 \cdot 10^{-26}$ m.k.s. (Long, personal communication).

With these observations it is now possible to define the low frequency spectra with greater certainty. In the present paper an examination will be made of fourteen of the sources which have been found to exhibit significant curvature. The observed spectra are shown in Fig. 1 from which it can be seen that a number of sources exhibit a maximum flux density at a frequency f_c within the range of the present observations.

3. *Radio and optical properties of the sources.*

It has already been noted that the sources with the highest brightness temperature tend to have convex spectra (Kellermann *et al.* 1962); the present observations confirm this. With the possible exceptions of 3C171 and 3C212 all the spectra in Fig. 1 belong to the small proportion of sources with angular diameter less than 5" of arc, as estimated from measurements with a long baseline interferometer (Allen *et al.* 1962).

Furthermore, for those sources which have been optically identified and whose distance has been estimated from red-shift measurement, it is possible to determine the radio luminosity of the source at 178 Mc/s, P_{178}. In general, these sources turn out to be among the most powerful

* J. M. HORNBY and P. J. S. WILLIAMS, Mullard Radio Astronomy Observatory, Cavendish Laboratory, Cambridge. Reprinted from MON. NOT. R. ASTR. SOC. 131, 237-246, 1966.

Reprinted from Mon. Not. R. astr. Soc., 131, 237-246, 1966.

known, P_{178} being greater than 10^{26} watts $(c/s)^{-1}$ in all cases.

A third correlation appears to exist, several of these sources being identified with the class of compact extra-galactic objects known as quasi-stellar, which are intrinsically very bright and exhibit an excess of ultra-violet radiation.

These various properties are summarized in Table 1 where it is seen that 3C48 and 3C 147 show them all to a very marked degree. There is, however, no complete correspondence between any two of these properties and care should be taken before attempting to classify radio sources by one of these properties alone. For instance, sources that have been identified as compact optical objects with an ultra-violet excess show a wide range of radio properties; on the other hand, 3C 295, which has a convex spectrum and a high brightness temperature, has been identified with a galaxy that does not show these extra-ordinary optical properties.

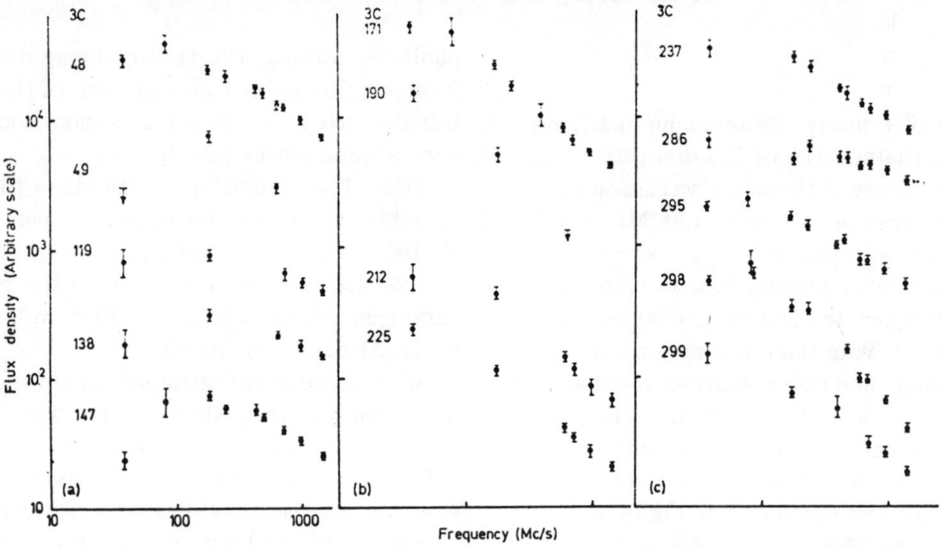

FIG. 1. Radiation spectra of fourteen radio sources showing a low frequency cut-off.

4. *Possible cut-off mechanisms.*

In order to explain a low-frequency cut-off it is necessary to postulate that at low frequencies there is either a decrease in the source's emissivity or else an increase in the absorption coefficient, within or immediately outside the source.

In the following discussion, the differential energy spectrum of the relativistic electrons is taken to be of the form:

$$N(E)\,dE = AE^{-\gamma}\,dE, \qquad (1)$$

except when we wish to discuss a cut-off in the electron energy spectrum. Such a cut-off we shall take to be the sharpest possible; that is, the electrons have the above spectrum for energies greater than E_0 while there are no relativistic electrons with energies less than E_0. The radio spectrum at high frequencies we take to vary according to the simple power law:

Table I. Sources with curved spectra at low frequencies

3C	T_{178} min. (°K)	log P_{178} $(w(c/s)^{-1}st^{-1})$	Identification
48	5.10^9	26.7	Quasi-stellar
49	3.10^8	...	...
119	1.10^9	...	...
138	2.10^9	...	...
147	6.10^9	27.3	Quasi-stellar
171	3.10^6	26.2	Galaxy
190	3.10^8	...	...
212	8.10^6	...	Quasi-stellar
225	5.10^8	...	...
237	6.10^7	...	...
286	6.10^8	...	Quasi-stellar
295	3.10^9	27.1	Galaxy
298	2.10^8	...	...
299	1.10^9	...	...

$$S(f) \propto f^{-\alpha}, \qquad (2)$$

where S(f) is the flux density at a frequency f. T(f) denotes the surface brightness temperature of the source. The conditions inside the source are taken to be uniform and when absorption cut-offs are considered, cubical geometry is assumed for the sources. The electron density (n) and magnetic field strength (B) in the source may be specified by the plasma and gyro-frequencies, f_p and f_H; the temperature (T) of these electrons will usually be about 10^4 °K, but it can be as high as $3 \cdot 10^4$ °K in the nuclei of Seyfert galaxies.

A decrease in the emissivity at low frequencies can arise in two ways.

(1) A cut-off in the energy spectrum of the electrons. Since we have no knowledge of the exact shape of a cut-off in the electron energy spectrum, we have taken the sharpest possible cut-off, defined above, and derived the corresponding cut-off in the radio-frequency spectrum. The low frequency spectral index of the radiation from a single electron is - 1/3. The radiation intensity from any electron distribution will therefore also vary as $f^{1/3}$ provided that all the electrons are radiating at a frequency much less than their critical frequency. If the calculated cut-off is more gradual than that actually observed, a cut-off in the electron energy spectrum cannot explain the observed radio spectrum.

(2) The Tsytovich effect. The refractive index for radio waves in an emitting region where electrons are present is slightly less than unity. Although the change is small, its effect on the emissivity can be large when the radiating particles are travelling at nearly the speed of light; this is because the interference between the electro-magnetic waves emitted at different points on a particle's orbit is strongly affected by the small increase in the phase velocity of the waves. A more detailed qualitative explanation has been given by Scheuer (1964).

The power radiated by a particle moving in a circular orbit in an isotropic but dispersive medium was first calculated by Tsytovich (1951). Using his results, we have calculated the power radiated per cycle per second from a system of electrons with a power law energy spectrum. This has been done for two cases, firstly when the particles move in an ionized medium and secondly when they move in a vacuum. The ratio of the radiated power in the first case to that in the second case is plotted as a function of ff_H/f_p^2 in Fig. 2 for $\gamma = 2$ and 3.

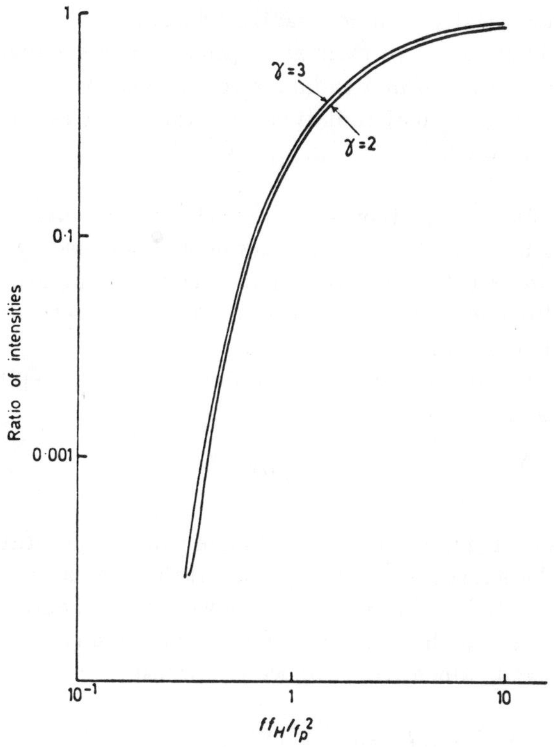

FIG. 2. Ratio of intensity of radiation from relativistic electrons with power law energy spectrum in a magneto-ionic medium to the intensity of radiation from the same electrons in vacuo, as a function of ff_H/f_p^2.

This result is only strictly valid for particles moving in circular orbits. To treat the case where the particles have an isotropic pitch angle distribution, the ratio must be interpreted as the ratio of the intensities perpendicular to the magnetic field. In the case when ϕ, the angle between the line of sight and the magnetic field, is not equal to 90° one can substitute $f_H \sin\phi$ for f_H when using Fig. 2, because in a vacuum all the essential synchrotron radiation formulae involve only $f_H \sin\phi$. The introduction of a refractive index whose magnitude is independent of f_H and ϕ will not change this result.

Four qualitative features of Fig. 2 are of interest:

(i) The cut-off is characterized by a frequency $f_p^2/f_H \sin\phi$ which we shall call for convenience the Tsytovich frequency, f_T.

(ii) The curves are almost independent of γ.

(iii) For frequencies greater than f_T the ratio of intensities increases slowly with frequency; e.g. for $f = 10f_T$ there is still a 10% reduction in intensity.

(iv) At low frequencies the ratio decreases exponentially with decreasing frequency.

There are two main absorption processes that can take place in a radio source corresponding to the two principal radiation mechanisms at radio frequencies.

(3) Synchrotron self-absorption. The electrodynamics of this process has been discussed by Le Roux (1960, 1961). There are three important features of the cut-off arising from synchrotron self-absorption.

(a) At frequencies for which the optical depth is large,

$$T(f) = 10^9 (f/f_H)^{1/2}, \qquad (3)$$

where $T(f)$ is the surface brightness temperature of the source at a frequency f. If the angular diameter (θ) of the source is known, the magnetic field strength B in the region of emission can be derived, since from (3) we can write

$$B\theta^{-4} = 1.7 \cdot 10^{-37} f^5/S^2, \qquad (4)$$

where θ is measured in seconds of arc, f in cycles per second and S in units of 10^{-26} watts meter^{-2} (c/s)$^{-1}$.

(b) For frequencies slightly below f_S, the frequency at which the optical depth is unity, $T(f)$ rapidly approaches its asymptotic value, which is proportional to $f^{1/2}$, e.g. at $f = \frac{2}{3} f_S$, the error in the spectrum (3) is less than 5%.

(c) For frequencies slightly above f_S, self-absorption is negligible, e.g. for a relativistic electron distribution with $\gamma = 3$ when $f = 2f_S$, the optical depth is 0·1.

(4) Thermal absorption. For convenience we shall discuss two extreme situations. In the first, the H II absorbing region occupies the same region of space as the radio source; in the second, it surrounds the radio source. The spectrum of the received radiation from the former type of source should vary as

$$S(f) \propto f^{-\alpha+2} (1 - \exp(-f_A^2 f^{-2})), \qquad (5)$$

where

$$f_A = 2 \cdot 10^{10}\, nL^{1/2}\, T^{-3/4}, \qquad (6)$$

L being the thickness of the absorbing region in kiloparsecs. In the second situation

$$S(f) \propto f^{-\alpha} \exp(-f_A^2 f^{-2}), \qquad (7)$$

f_A again being given by equation (6).

5. *Analysis of the observational data.*

(1) Comparison of the observed and theoretical spectra. In Fig. 3 are shown the observed points on the spectra of some of the radio sources whose low frequency spectrum is very curved, together with the spectra obtained by taking each cut-off mechanism in turn and fitting the theoretical spectra to the observed points. In two cases, 3C 49 and 147, it can be seen that the low frequency cut-off which would be associated with a cut-off in the electron energy spectrum is too gradual to explain the observations. None of the other mechanisms can be excluded for any of the other sources. However, points on the observed spectra at about 80 Mc/s could possibly exclude an energy cut-off or the Tsytovich cut-off. Such points could not exclude the rather sharper cut-offs associated with absorption, because of possible non-uniformities in the distribution of ionized hydrogen or the relativistic electrons.

In Table II, we list the parameters required to give the theoretical spectra shown in Fig. 3 together with the parameters of the other sources, whose curvature is not so marked. In calculations involving absorption by ionized hydrogen, the temperature of the gas was taken as 10^{40} K.

The values given in Table II for $B\theta^{-4}$, n/B, n^2L and $E_0^2 B$ are all upper limits, if it cannot be decided which mechanism is responsible for the cut-off in any particular source. If, on the other hand, the cut-off mechanism is known, then one of the above quantities will give a more or less exact value for the source in question, the remaining three being only upper limits.

(2) Angular diameters. In order to obtain values for the field strength on the assumption that the cut-off is caused by synchrotron self-absorption, it is necessary to determine the

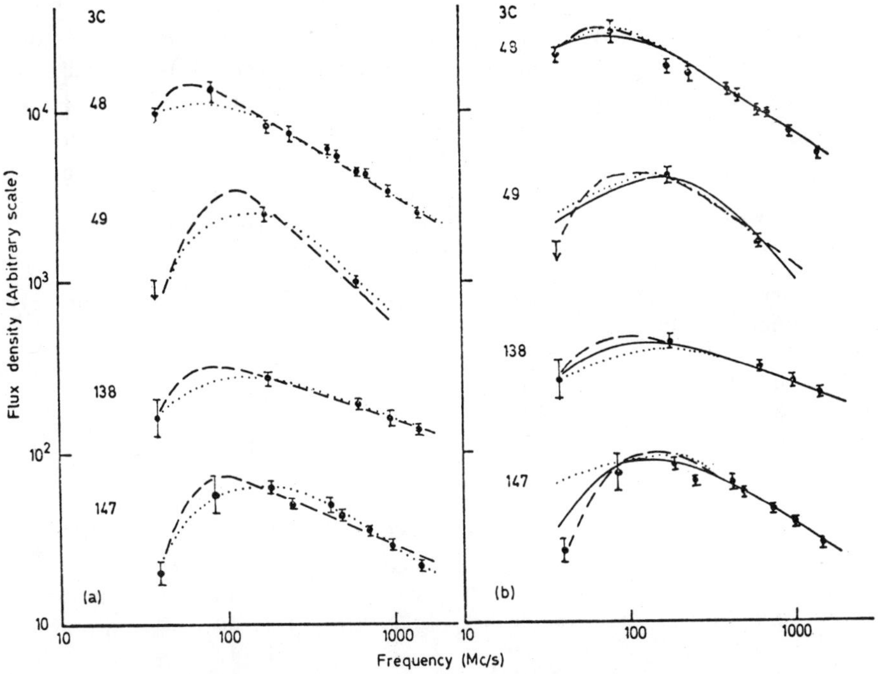

FIG. 3. *(a) Synchrotron self absorption,* — — — — ; *Tsytovich effect,* .... *(b) Thermal absorption,* ——— *(inside)* ; — — — — *(outside). Electron energy cut-off,* ....

TABLE II

Source	Synchrotron self-absorption	Thermal absorption Within	Outside	Tsytovich effect	Energy cut-off	$B_{eq}\theta^{6/7}$
	$B\theta^{-4}$	n^2L		n/B	$E_0{}^2B$	
3C	(g s^{-4})	(cm^{-6} kpc)		(g^{-1} cm^{-3})	(MeV^2 g)	
48	1·2 . 10^{-2}	9·5	2·5	8·5 . 10^5	6	10^{-4}
49	1	30	6	24 . 10^5	...	
119	1·2 . 10^{-2}	4	1·5	7 . 10^5	6	
138	1	7	3	12 . 10^5	12	
147	6·3 . 10^{-2}	20	6	21 . 10^5	...	10^{-4}
171	4 . 10^{-3}	3	1	7 . 10^5	3	10^{-4}
190	1·2 . 10^{-2}	4	1·5	8·5 . 10^5	3	
212	6·3 . 10^{-2}	7	2·5	10 . 10^5	6	
225	4 . 10^{-3}	4	1·5	8·5 . 10^5	3	
237	1·2 . 10^{-2}	7	1·5	8·5 . 10^5	6	
286	1·2 . 10^{-2}	2·5	1	5 . 10^5	6	
295	4 . 10^{-3}	3	1·5	8·5 . 10^5	6	10^{-4}
298	1·2 . 10^{-2}	9·5	3	14 . 10^5	6	
299	1·2 . 10^{-2}	2·5	1	5 . 10^5	4	

sources' angular diameters. Of the sources in Table II, Allen *et al.* (1962) found that all except 3C171 and 212 had angular diameters less than 3.3" of arc, 3C171 and 212 had visibility less than 0.3 at spacings of 32,000 and 61,000 wavelengths, while their visibility at 9,700 wavelengths was 0.2 and 0.5 respectively. We have taken upper and lower limits on these sources' angular diameter of 40" and 10". More recent work (Allen, personal communication) has shown that 3C48, 147 and 295 are still visible at a spacing of 160,000 wavelengths, so that the angular size of their components must be less than 0.3".

Hewish, Scott and Wills (1964) have examined the variations in the apparent intensity of some sources, caused by scintillations in the interplanetary medium. They conclude that of the sources in Table II, 3C48, 119 and 138 have angular diameters less than 0.3" while 3C147, 225, 286, 295 and 298 have angular diameters less than 0.8". 3C49 and 299 were not found to scintillate, implying angular diameters greater than 0.3" and 0.5". 3C190 was not studied, since it was confused by side-lobes of Cygnus A. The data on the sources' angular diameters are summarized in Table III and limits on the field strength are calculated, assuming that the cut-off was caused by synchrotron self-absorption. Redshifts are available for the sources 3C48, 147, 171 and 295 and one can therefore put limits on the physical size of the source. If the cut-off is attributed to thermal absorption within the source, one can then put limits on n, the thermal electron density. For the other sources, the upper limit on B can be combined with the upper limit on n/B to derive a not very useful upper limit on n.

Table III

	Diameter (in seconds of arc)		Field strength (gauss)		Electron number density (cm^{-3})	
3C	Upper limit	Lower limit	Upper limit	Lower limit	Upper limit	Lower limit
48	0.3	...	10^{-4}	...	...	3.0
49	3	0.3	80	$8 \cdot 10^{-3}$	10^{8}	...
119	0.3	...	10^{-4}	...	70	...
138	0.3	...	$8 \cdot 10^{-3}$	...	10^{4}	...
147	0.3	...	$5 \cdot 10^{-4}$	...	...	4.5
171	40	10	10^{4}	40	0.3	0.15
190	3	...	1	...	$21 \cdot 10^{6}$	...
212	40	10	$1.4 \cdot 10^{5}$	630	$1.4 \cdot 10^{11}$	...
225	0.8	...	$1.5 \cdot 10^{-3}$	...	$1.2 \cdot 10^{3}$	...
237	0.8	...	$5 \cdot 10^{-3}$	...	$4 \cdot 10^{3}$	...
286	0.8	...	$5 \cdot 10^{-3}$	...	$2.5 \cdot 10^{3}$	...
295	0.3	...	$3 \cdot 10^{-5}$	...	...	2
298	0.8	...	$5 \cdot 10^{-3}$	...	$7 \cdot 10^{3}$	...
299	3	0.5	1	$8 \cdot 10^{-4}$	$5 \cdot 10^{5}$	...

Two conclusions can be drawn from Table III. First, the field strength required for synchrotron self-absorption to take place in a source of angular diameter 10" or more is implausibly large, and it appears that synchrotron self-absorption can be excluded as the cut-off mechanism in 3C171 and 212. Second, in two of the sources 3C48 and 295, the upper limit on the field strength is much less than the value for the field strength deduced by equipartition arguments (cf. Table II), i.e. in these sources, the cosmic ray energy density is much greater than the magnetic energy density.

(3) Faraday rotation in sources. If the amount of Faraday rotation within a source is known, it is possible to draw some conclusions, as to whether the Tsytovich cut-off is or is not likely to occur at a higher frequency than the cut-off caused by thermal absorption within the source. We shall take as an upper limit on the rotation measure (R) associated with Faraday rotation within the source, the observed rotation measure of the radiation from the source. The fact that Morris and Berge's results (1964) indicate that this will give a rather generous upper limit on R only makes the conclusions of this section more certain. Only two sources in Table I have so far been observed to be polarized, namely 3C48 and 286. The values of R and the percentage polarization at 10.6 cm have been taken from Seielstad, Morris and Radhakrishnan (1963) and are given in Table IV.

If the electron density is negligible outside the radio source, then using the equations for f_A and f_T, one can derive the relation

$$f_A^2/f_T = 1.3 \times 10^{19} n |B_\perp| L T^{-3/2} \text{ c/s}, \qquad (8)$$

where $|B_\perp|$ is the average magnitude of the component of the field perpendicular to the line of sight, $B_\parallel$ will be the average value of the component of the field parallel to the line of sight, account being taken of sign. At infinite frequency, the depolarization of the radio source

Table IV

	P	R	$1.7 \cdot 10^{12} P^{-1} RT^{-3/2}$	f_c
3C 48	> 3.4	53	< 25 Mc/s	~ 50 Mc/s
3C 286	> 8.5	4	< 01 Mc/s	~ 50 Mc/s

is caused entirely by tangling of the magnetic field. If the percentage polarization at infinite frequency is P, then

$$\frac{B_{\|}}{|B_{\perp}|} \simeq P/100. \tag{9}$$

In general, one will only have a lower limit on P. |R| will be given by the relation (e.g. Gardner and Whiteoak 1963).

$$|R| = 8.1 \cdot 10^{8} nB\ L. \tag{10}$$

Substituting equations (9) and (10) into equation (8) one obtains

$$f_A^2/f_T = 1.7 \cdot 10^{12}\ P^{-1}\ R\ T^{-3/2}. \tag{11}$$

If the observed cut-off frequency f_c is equal to either f_A or f_T but it is not known which, then

$$f_c < 1.7 \cdot 10^{12}\ P^{-1} |R| T^{-3/2} \text{ if } f_c = f_A > f_T$$

$$f_c > 1.7 \cdot 10^{12}\ P^{-1}\ RT^{-3/2} \text{ if } f_c = f_T > f_A.$$

In other words, if the right-hand side of equation (11) is less than the observed cut-off frequency, the cut-off cannot have been caused by thermal absorption; if, on the other hand, it is greater than the cut-off frequency, then the Tsytovich effect can be excluded.

The relevant parameters for 3C48 and 286 are shown in Table IV. It is therefore clear that for these sources, it is possible to state that the observed cut-off in the radio frequency radiation spectrum is not caused by thermal absorption within the source.

(4) Energy requirements. The total energy requirements of a source of known intensity and red-shift are minimized by assuming that the magnetic energy and the energy of all the relativistic particles in the source are equal. Here we have assumed (a) that all the relativistic particles are electrons and (b) that the electron energy spectrum cuts off at a certain energy, such that the critical frequency of a particle with this energy is 10 Mc/s, when the field strength is equal to its equipartition value B_{eq}. Values for B_{eq}. are given as a function of θ, in the right-hand column of Table II. It can be seen that, in the case of 3C48 and 295, B_{eq} must be greater than $3 \cdot 10^{-4}$ gauss. This is three and ten times greater than the upper limits on B derived for these two sources (cf. Table III). It can therefore be said with certainty that for these two sources, the energy of the relativistic electrons is at least a hundred times greater than that stored in the magnetic field.

6. *Conclusions.*

The spectra of fourteen sources have been examined and limits have been obtained for the field strength and thermal electron density within the source. For two of the sources (3C171 and 212) it appears implausible that the cut-off was caused by synchrotron self-absorption and for two others (3C49 and 147) the observed cut-off is much sharper than that associated with a cut-off in the relativistic electron energy spectrum. For the two sources that are observed to be polarized (3C48 and 286) the cut-off cannot be caused by thermal absorption within the source. Further observations at 80 Mc/s may exclude the possibility that some of the observed cut-offs are caused by either the Tsytovich effect or a cut-off in the electron energy spectrum. In 3C48 and 295, the energy of the relativistic particles must be very much greater than that stored in the magnetic field.

7. *Acknowledgments*

We wish to express our gratitude to Dr. R. J. Long for allowing us to use the results of his observations at 610 Mc/s prior to their publication. This work was carried out during the tenure of D.S.I.R. Maintenance Awards.

REFERENCES

L. R. Allen, B. Anderson, R. G. Conway, H. P. Palmer, V. C. Reddish, and B. Rowson, 1962. Mon. Not. R. astr. Soc., 124, 477.

R. G. Conway, K. I. Kellermann, and R. J. Long, 1963. Mon. Not. R. astr. Soc., 125, 261.

F. F. Gardner, and J. B. Whiteoak, 1963. Nature, 197, 1162.

A. Hewish, P. F. Scott and D. Wills, 1964. Nature, 203, 1214.

K. I. Kellermann, R. J. Long, L. R. Allen, and M. Moran, 1962. Nature, 195, 692.

E. Le Roux, 1960. Annls Astrophys., 23, 1010.

E. Le Roux, 1961. Annls Astrophys., 24, 71.

B. Y. Mills, O. B. Slee, and E. R. Hill, 1958. Aust. J. Phys., 11, 360.

D. Morris, and G. L. Berge, 1964. Astrophys. J., 139, 1388.

P. A. G. Scheuer, 1965. Quasi-stellar Sources and Gravitational Collapse, University of Chicago Press.

G. A. Seielstad, D. Morris, and V. Radhakrishnan, 1963. Astrophys. J., 138, 602.

J. R. Shakeshaft, M. Ryle, J. E. Baldwin, B. Elsmore, and J. H. Thomson, 1955. Mem. R. astr. Soc., 67, 106.

V. I. Slish, 1963. Nature, 199, 682.

V. N. Tsytovich, 1951. Vest. mosk. gos. Univ., 11, 27.

P. J. S. Williams, 1963. Nature, 209, 56.

P. J. S. Williams, S. Kenderdine, and J. E. Baldwin, 1966. Mem. R. astr. Soc., in press.

Quasi-Stellar Sources: Variation in the Radio Emission of 3C 273

W. A. Dent
Radio Astronomy Observatory,
University of Michigan, Ann Arbor

Since July 1962 repeated measurements of the flux densities of some 35 nonthermal radio sources have been made at 8000 megacycles per second (wavelength, λ, 3.75 cm) with the University of Michigan's 85-foot (26-m) parabolic antenna. Since the report of the observed light variations (*1*) of the quasi-stellar object identified with the radio source 3C 273, special attention has been given to this source in an attempt to detect a variation in its radio emission. This report presents evidence for a more or less steady increase of 40 percent in the radio emission of 3C 273 at 8000 Mc/sec in the past 1000 days. Until the recent announcement of a cyclic variation in CTA 102 at 940 Mc/sec (*2*), no variations in the radio emission of extragalactic sources had been reported.

Forty-seven separate sets of measurements of the ratio of the antenna temperatures of 3C 273 to Virgo A were made with a pencil beam of 5.9 minutes of arc. Virgo A was chosen as a reference source since the position of Virgo A is only about 10 degrees north of 3C 273. Both sources were observed within 1 hour of the meridian on the same day. The measured antenna temperatures were corrected (*3*) for the finite extent of Virgo A, the differential change in antenna gain with orientation, atmospheric extinction, and the linear polarization component of the sources. The overall corrections were 3 percent for 3C 273 and 5 percent for Virgo A, and any changes in these corrections were too small to detect.

The results of the measurements are shown in the upper half of Fig. 1. Each point represents an average of at least two separate observations made on a given day. The bars are standard deviations computed from the system noise and the number of observations on the particular day. The ratio of the antenna temperature of 3C 273 to Virgo A has increased between 31 July 1962 and 14 April 1965 from 0.525 to 0.74. The measured ratio of Virgo A to Cygnus A over the same period (lower half of Fig. 1) does not show any systematic trend or variation and supports the assumption that the flux density of Virgo A did not change significantly during the period of these observations.

If a flux density for Virgo A at 8000 Mc/sec of 45.0×10^{-26} watt m^{-2} cps^{-1} (cps, cycles per second) (*4*) is adopted, then the flux density of 3C 273 has increased from 23.6 to 33.3 in units of 10^{-26} watt m^{-2} cps^{-1}, corresponding to an increase in antenna temperature of from 2.18° to 3.07°K over the above period.

The radio source 3C 273 consists of two components separated by 20 seconds of arc (*5*). Component 3C 273A, identified optically with a faint jet (*6*), has a nonthermal radio spectrum with a steep spectral index ($\alpha \approx -0.7$), while component 3C 273B, identified with a quasi-stellar object, has a nearly flat ($\alpha \approx 0$) radio spectrum. The steep spectrum of component A dominates the observed composite spectrum at frequencies below 1000 Mc/sec while the nearly flat spectrum of component B dominates above.

At 8000 Mc/sec the flux density of component B is about five times that of component A (*7*). Thus if the 8000-Mc/sec emission from the quasi-stellar component is varying, it must increase at a rate of approximately 17 percent per year compared to the much larger variation of 84 percent per year required if the increasing emission is from the jet. The fact that such a large annual variation has not been reported for this source at frequencies below 1000 Mc/sec strongly supports the assumption that the radio variation is due to the quasi-stellar component. This assumption is further supported by the fact that 3C 273B shows light variations having an irregular period of about 13 years (*8*). Continued radio observations of 3C 273 over a longer time base will establish whether its radio variation is also periodic.

Photometric observations (*9*) of 3C 273B over a 10-month interval in 1963 reveal a decrease in the optical emission in each of three colors of about 0.2 magnitude or 20 percent. This suggests that there may be a coupling between the mechanisms responsible for the optical continuum and the observed radio spectrum. If the optical continuum is synchrotron radiation then the decay of the radiating electrons ($\sim 10^{12}$ ev) to lower energies by radiation and inverse Compton losses will produce an increase in the radio emission.

In addition to 3C 273, there is some evidence for variations at 8000 Mc/sec in two other quasi-stellar sources. Figure 2 shows measurements of the ratios of the corrected antenna temperatures of three sources to the antenna temperature of Virgo A. The optically identified (*10*) quasi-stellar sources 3C 279 and 3C 345 show a net decrease of about 17 and 19 percent, respectively, in their radio emission over the past year; the data are too sparse to rule out shorter-termed variations. Observations of the quasi-stellar sources 3C 286 and 3C 147 (not shown) do not allow any variations greater than 10 percent over the past 2 years.

Like 3C 273, the observed radio spectra of both 3C 279 and 3C 345 are composite, with a steep and a flat spectral component (*4, 7*); the principle contribution to the total flux density at 8000 Mc/sec is due to the flat spectral component. Thus it appears that variable radio emission may be associated with quasi-stellar sources having flat spectra, whereas the quasi-stellar sources with steep noncomposite radio spectra like 3C 147 and 3C 286 do not show obvious or large rates of variation.

This correlation suggests that quasi-stellar sources with flat spectra are more unstable and probably younger

Reprinted from SCIENCE, 148, 1458-1459, 1965.

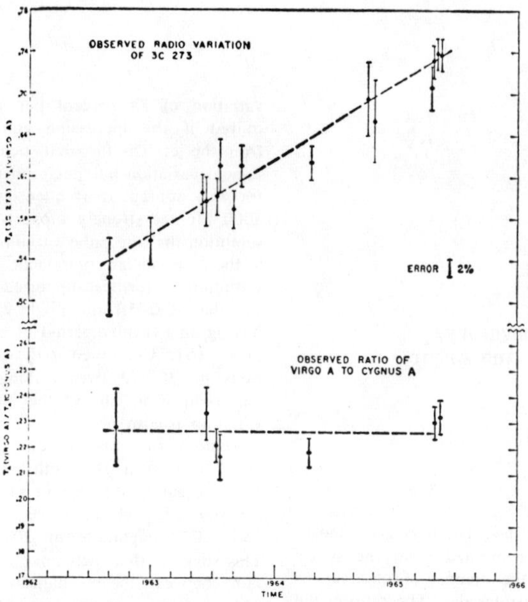

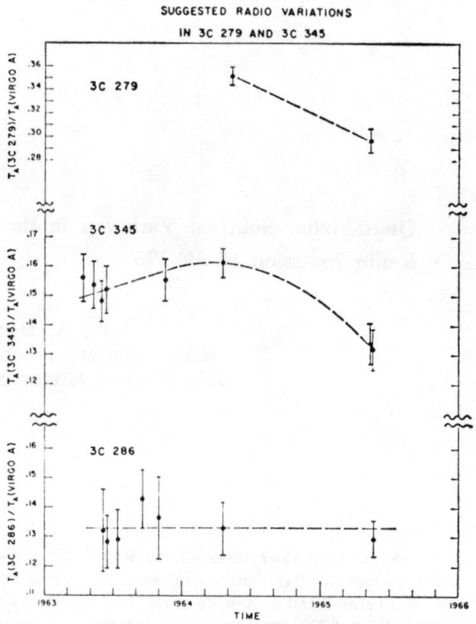

Fig. 1 (left). The ratio of the corrected antenna temperature, T_A, at 8000 Mc/sec of the quasi-stellar source 3C 273 to that of Virgo A, showing the observed 40-percent increase in the radio emission over a nearly three-year period. The measured ratio of Virgo A to Cygnus A shows no variation over the same period. Fig. 2 (right). Evidence for possible variations in the radio emissions of the quasi-stellar sources 3C 279 and 3C 345 at 8000 Mc/sec, and for a lack of variation in the quasi-stellar source 3C 286. Like 3C 273, which shows variations, both 3C 279 and 3C 345 have flat radio spectra at 8000 Mc/sec.

than the evolutionarily older sources whose initial spectra were probably flat and have since been steepened by synchrotron radiation losses and inverse Compton energy losses of the radiating electrons. The presence of both a steep and a flat component in the spectra 3C 273, 3C 279, and 3C 345 could suggest that recurrent outbursts have occurred in these sources.

The time scale of the observed radio variation of 3C 273B is short and is probably of the same order as the 13-year period of the optical variations. Since large fractional variations in the emission must occur over a time scale greater than the "light-travel" time through the source (*11*), the upper limit to the linear size of the radio component of 3C 273B is less than about 13 light years or 4 parsecs. Thus, knowledge of the angular diameter of the varying component will give an upper limit to the distance of the source.

The large red shift, $\Delta\lambda/\lambda = 0.158$, observed for 3C 273 (6) has been taken to indicate a distance of about 470 megaparsecs on the basis of Hubble's law (*12*). However, at a distance of 470 megaparsecs the linear size deduced above would subtend an angle of less than 0.002 seconds of arc, and the source would be opaque at 8000 Mc/sec owing to synchrotron self-absorption for a magnetic field greater than 10^{-5} gauss (*13*). Since the radio spectrum of 3C 273B is nearly flat ($\alpha = 0$) above at least 400 Mc/sec (*7*) and is not characteristic of self-absorption ($\alpha = 2.5$), the apparent angular diameter of 3C 273B must be greater than 0.1 second if the emission is synchrotron radiation.

Lunar occultation observations of 3C 273 at 1420 Mc/sec (*5*) show that 3C 273B consists of a central bright core component about 0.5 second of arc in diameter which contributes about 80 percent of the total flux density embedded in a weaker 7-second diameter halo. If an angular diameter of 0.5 second at 8000 Mc/sec is adopted, the upper limit of 4 parsecs to the linear size of the radio component of 3C 273B would set an upper limit to its distance at 2 megaparsecs and thus place 3C 273 within our own local group of galaxies.

It should be emphasized that this deduced distance is based on an assumed angular diameter of 3C 273B. If the angular diameter of 3C 273B is actually very much smaller than the reported measured value (*5*) or if the radio source at 8000 Mc/sec consists of a 0.002-second diameter nucleus of variable radio emission embedded in a much larger 0.5-second diameter non-varying core, then the distance to the source could still be 470 megaparsecs. However, in both of these cases the radio emission from 3C 273B could not be due to synchrotron radiation, since self-absorption effects are not observed in the radio spectrum. A similar argument applies against a cosmological interpretation of the recently reported red shift (*14*) of CTA 102, unless its radio emission is not synchrotron radiation. It should be noted that a measured angular diameter greater than a few thousands of a second of arc for the varying components of these quasi-stellar sources will exclude a cosmological interpretation of their red shifts.

References and Notes

1. H. J. Smith and D. Hoffleit, *Nature* **198**, 650 (1963).
2. G. B. Sholomitsky, Commission 27 of the International Astronomical Union, Information Bulletin No. 83, 27 February 1965.
3. W. A. Dent and F. T. Haddock, in preparation.
4. ———, *Nature* **205**, 487 (1965).
5. C. Hazard, M. B. Mackey, A. J. Shimmins, *ibid.* **197**, 1037 (1963).
6. M. Schmidt, *ibid.* **197**, 1049 (1963).
7. W. A. Dent and F. T. Haddock, in *Quasi-Stellar Sources and Gravitational Collapse*, I. Robinson, A. Schild, E. L. Shucking, Eds. (Univ. of Chicago Press, Chicago, 1965), p. 381.
8. H. J. Smith, *ibid.*, p. 221.
9. A. R. Sandage, *Astrophys. J.* **139**, 416 (1964).
10. ——— and J. D. Wyndham, *ibid.* **141**, 328 (1965).
11. J. Terrell, *Science* **145**, 918 (1964).
12. J. G. Greenstein and M. Schmidt, *Astrophys J.* **140**, 1 (1964).
13. V. I. Slish, *Nature* **199**, 682 (1963).
14. M. Schmidt, *Astrophys. J.* **141**, 1295 (1965).
15. I thank Professor F. T. Haddock for many helpful suggestions and for making the facilities of the Radio Astronomy Observatory available to me. I also thank T. V. Seling for developing the highly sensitive radiometer and J. L. Talen for the development of the digital data system. Supported by the ONR under contract Nonr-1224(16).

QUASI-STELLAR SOURCE 3C 273 B: VARIABILITY IN RADIO EMISSION

George B. Field
Astronomy Dept., University of California, Berkeley

In a recent issue of *Science* Dent (1) has reported the very important discovery of variability in the 8,000-Mc/sec radio emission of the quasi-stellar source 3C 273 B. The purpose of this note is to call attention to a numerical error in Dent's calculations which invalidates his conclusion that if the variable source emits by the synchrotron mechanism it must be much closer than the 1.5×10^9 light years assigned to it on the basis of its red-shift (2).

Dent based his discussion on an article by Slish (3), who shows that below the frequency

$$\nu = [10^{33}\ \theta^{-2}\ S_\nu\ \nu^{\alpha}]^{2/(5+2\alpha)} B^{1/(5+2\alpha)} \tag{1}$$

self-absorption of synchrotron radiation will cause a rapid drop in flux. In this expression θ is the angular diameter of the source, S_ν its flux density in MKS units, α its spectral index at high frequencies ($S'_\nu \propto \nu^{-\alpha}$), and B the magnetic field in gauss. For the case $\alpha = 0$ applicable to 3C 273 B and with $S_\nu = 3 \times 10^{-25}$ watt m^{-2} cps^{-1} (cps, cycles per second), Eq. 1 gives $\theta'' = 3 \times 10^9\ B^{1/4}\ \nu^{-5/4}$ arc seconds. If we take $B > 10^{-5}$ gauss and assume that the spectrum of the variable component extends down to 400 Mc/sec (following Dent), we find $\theta'' > 3 \times 10^{-3}$ sec. This is 1/30 of the value derived by Dent. The error must be Dent's, as I derived Eq. 1 independently as a check. At the cosmological distance of 3C 273, this corresponds to a lower limit of 23 light years for the diameter of the source, a value which is just marginally compatible with the observed variability. It should be noted that this argument is based on the assumption that the source is variable at 400 Mc/sec also. The lower limit on θ based on the 8000-Mc/sec data alone (with $\alpha = 0$) is 6×10^{-5} sec, or 0.5 light years (whereas Dent found that the source would be optically thick at 8000 Mc/sec for $\theta < 2 \times 10^{-3}$ sec). We conclude that the 8000-Mc/sec data alone do not justify Dent's conclusion that the variable component cannot be due to a synchrotron source at great distance. Even the further assumption that the variable component is still strong at 400 Mc/sec is marginally compatible with such a source. Obviously any further information which can be obtained on the variability at lower frequencies will bracket the angular diameter better and permit conclusions as to the nature of the source.

A model which appears compatible with the present data consists of about 50 sources, each contributing to the flux and varying independently with periods of approximately 10 years. If each source is about 10 light years in radius, and has a magnetic field of 1.2×10^{-2} gauss, the source can be variable down to 400 Mc/sec. Standard formulas for synchrotron emission would require about 1.5×10^{52} erg of relativistic electrons in each source. Such amounts of energy (8×10^{-3} $M_0 c^2$) are marginally obtainable from explosions of massive stars (several hundred solar masses) according to the calculations of Colgate and White (4). The proposal of Field (5), that quasars represent the earliest phase in the life of a galaxy, when supernovae occur at a rate of about five per year, is therefore pertinent.

REFERENCES

[1] **W. A. Dent, Science 148, 1458 (1965).**
[2] **M. Schmidt, Nature 197, 1040 (1963).**
[3] **V. I. Slish, ibid. 199, 682 (1963).**
[4] **S. A. Colgate and R. H. White, U. S. At. Energy Comm. Rept. UCRL 7777 (Lawrence Radiation Laboratory, University of California, 1964).**
[5] **G. B. Field, Astrophys. J. 140, 1434 (1964).**

Reprinted from SCIENCE, 150, 78, 1966.

3C 273: VARIATIONS IN ITS 3.4-mm FLUX*

E. E. Epstein
Aerospace Corp., El Segundo, California

J. P. Oliver
University of California, Los Angeles, California

R. A. Schorn
Jet Propulsion Laboratory, Pasadena, California

We previously reported observations of 3C 273 at 3.4 mm (88.4 GHz) which suggested a flux variation by a factor of about 2 or 3 in only a few months' time (Epstein 1965*a*). To verify this suggestion we have continued observing 3C 273. We used the dual-beam technique and equipment described previously. An illustration of the 15-foot antenna with which we made our observations appears in Gary, Stacey, and Drake (1965). Nighttime atmospheric attenuation corrections were computed with the aid of equation (6) of Shimabukuro (1966) and radiosonde data; daytime corrections were computed with the aid of observations of the Sun by using the "average temperature" method (Epstein 1965*b*). The attenutation correction factors ranged from 11.15 to 1.38; the uncertainties in the flux values resulting from these correction factors are $\lesssim 6$ per cent. Corrections for antenna pointing errors amounted to only a few per cent.

Figure 1 presents our results. The flux increase observed in the spring of 1966 appears to confirm the previous suggestion that flux variations by a factor of 2 or so occur in a few months.

We note that, unlike the situation at longer wavelengths, where it is possible to make flux measurements relative to a standard source (cf. Dent 1965), the absence of strong point sources at 3 mm means that our measurements are necessarily absolute. However, we have used our equipment almost daily for a number of observing programs, including

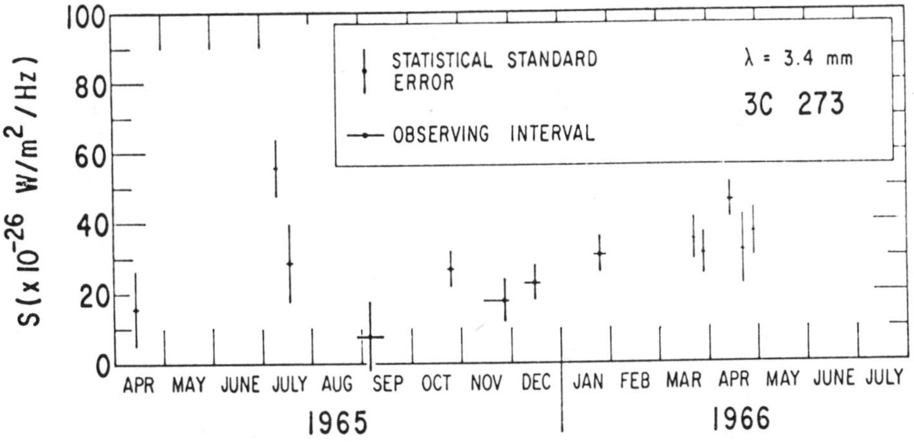

FIG. 1.—3.4 mm flux, S, of 3C 273 versus time. *Horizontal bars* indicate the time intervals over which the observations were made. *Vertical bars* indicate the statistical standard errors; these error values are consistent with those predicted on the basis of radiometer noise. The total estimated errors (including systematic calibration errors) are about 15 per cent greater. Each point represents about 20 hours of observing time.

* This work was supported by U.S. Air Force contract AF 04(695)-669. Schorn's portion was also supported in part by the Jet Propulsion Laboratory, California Institute of Technology, under contract NAS 7-100 sponsored by the National Aeronautics and Space Administration.

Reprinted from Ap. J., 145, 367-368, 1966.

precision measurements of Venus from August, 1965, to April, 1966. The results of these programs are inconsistent with the presence of any significant systematic errors in the 3C 273 data. We also have repeated "blank" sky observation tests (Epstein 1965*a*), including tests within several degrees of the Sun (Epstein 1966), with very satisfactory results. In addition we have continued to plot histograms to verify that the scatter in each set of data is no larger than that expected from radiometer noise.

Comparisons of the 3.4-mm variations with those at optical and other radio wavelengths should be extremely valuable in testing the numerous models of quasi-stellar radio sources.

REFERENCES

Dent, W. A. 1965, *Science*, **148**, 1458.
Epstein, E. E. 1965*a*, *Ap. J.*, **142**, 1285.
———. 1965*b*, *A.J.*, **70**, 721.
———. 1966, *Science*, **151**, 445.
Gary, B., Stacey, J., and Drake, F. D. 1965, *Ap. J. Suppl.*, **12**, 239.
Shimabukuro, F. I. 1966, *IEEE Trans. on Antennas and Propagation*, AP-14, 228.

OBSERVATIONS OF 3C 273 AND 3C 279 AT 1 mm

F. J. Low
National Radio Astronomy Observatory, Green Bank, West Virginia

Two quasi-stellar radio sources, 3C 273 and 3C 279, have been observed at a wavelength of 1 mm. The 200-inch telescope at Palomar Observatory was used with a thermal detection radiometer (Low 1965) developed at the National Radio Astronomy Observatory and at Texas Instruments, Inc. The absolute calibration adopted in this work is based on the results of lunar and planetary observations at 1 mm discussed by Low and Davidson (1965).

The Moon, Mars, Jupiter, 3C 144, and 3C 273 were observed in January, 1965, using an unpolarized feed horn producing a Gaussian beam on the sky of half-power width equal to 60". In June, 1965, observations of the Moon, Mars, Saturn, 3C 273, 3C 279, and M17 were obtained with the single horn and a special chopper which produced two beams separated in right ascension by 190". A simple on-off observing procedure was used in January with a 15-sec integration time. In June each source was moved alternately between the two beams again using a 15-sec integration. A maximum error of 5" was found in positioning the source in the center of the beam. Four factors contributed to the improved signal-to-noise ratio in June: (1) the measured precipitable water vapor in June was one-fourth that in January, (2) the use of two beams increased the signal, (3) almost perfect cancellation of sky-background fluctuations was produced by the dual beams, and (4) extraneous noise in the radiometer was eliminated prior to the June run.

Table 1 gives the date, flux density, probable error computed from statistics, and the number of independent observations. The method of absolute calibration can introduce a systematic error of ± 15 per cent in the flux density of a point source. Because of the unknown angular distribution of flux in 3C 144 (the Crab Nebula) and M17 (the Omega Nebula) the observations at 1 mm give uncertain limits on their total fluxes.

Table 1. Observational Data for 3C 273 and 3C 279

Object	Date	Flux Density $[W/m^2/(c/s)]$	P.E. $[W/m^2/(c/s)]$	No. Obs.
3C 273	Jan. 22	300×10^{-26}	$\pm 100 \times 10^{-26}$	120
3C 273	June 7	22	± 5	104
3C 279	June 6	3	± 6	66

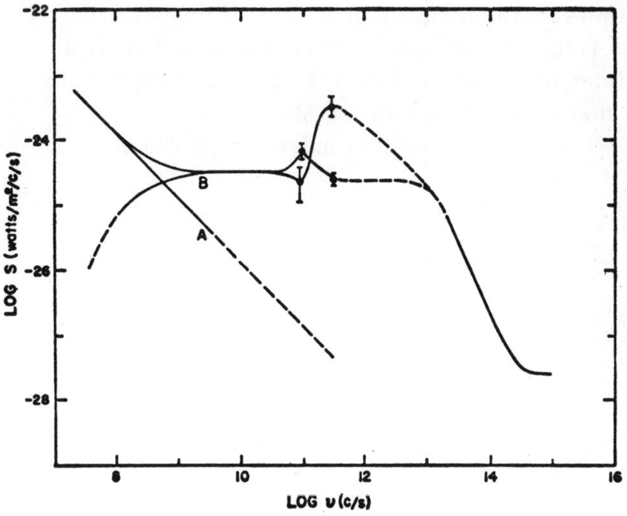

FIG. 1. The flux density of 3C 273A and B as a function of frequency. The curve is drawn through both the 1-mm and 3.4-mm values observed in January and April, 1965, and through the corresponding values obtained in June and and July, 1965.

Fig. 1 shows the present knowledge concerning the increasingly complex spectrum of 3C 273. The observations at 1 mm and the recently reported data of Epstein (1965a) have added to the spectrum published by Low and Johnson (1965). Dent (1965) has reported a small but significant increase in flux at 3.75 cm over a 3-year interval. The apparent decrease in the 1-mm flux by a factor of 10 in a 5-month interval occurred

Reprinted from Ap. J., 142, 1287-1289, 1965.

simultaneously with the increase at 3.4 mm reported by Epstein (1965a). Observations of the flux at 2.2 μ (Johnson 1964) were begun in March, 1964. The flux quoted by Low and Johnson (1965) has not varied at 2.2 μ by more than ± 20 per cent over a 15-month interval. Thus the wavelength dependence of the slowly varying component is not yet established, whereas the rapid, large-scale variations appear to be confined to the millimeter wavelengths. The release of energy at 1 mm, the subsequent apparent shift of the peak in the spectrum to lower frequencies, and the observed increase in the spectral index of the almost flat microwave spectrum suggest the sudden injection of high-energy particles that decay in energy on an exceedingly short time scale.

If further observations confirm these results and a unique physical interpretation can be made, it may become possible to place unambiguous limits on the distance of 3C 273. Flux densities as large as 300 units combined with light-travel times of months will be difficult to reconcile with a distance as large as 400 Mpc.

The positive spectral index for 3C 279 at frequencies above 1.5 Gc/s (Dent and Haddock 1965), the possible decrease in flux with time at 3.75 cm (Dent 1965), and the large flux at 3.4 mm (Epstein 1965b) suggest a behavior not unlike 3C 273. The observation at 1 mm on June 6 implies a flux significantly less than the value $(30 \pm 10) \times 10^{-26}$ W/m^2/(c/s) at 3.4 mm. This also is similar to the spectrum of 3C 273 as measured in June.

The use of the Hale Telescope at 1 mm was made possible by the co-operation of the staff at the Mount Wilson and Palomar Observatories.

REFERENCES

[1] W. A. Dent, 1965, Science, 148, 1458.

[2] W. A. Dent, and F. T. Haddock, 1965, Nature, 205, 487.

[3] E. E. Epstein, 1965a, Ap. J., 142, 1282.

[4] E. E. Epstein, 1965b, ibid., p. 1285.

[5] H. L. Johnson, 1964, Ap. J., 139, 1022.

[6] F. J. Low, 1965, Proc. I.E.E.E., 53, 516.

[7] F. J. Low, and A. W. Davidson, 1965, Ap. J., 142, 1278.

[8] F. J. Low, and H. L. Johnson, 1965, Ap. J., 141, 336.

A MODEL FOR VARIABLE EXTRAGALACTIC RADIO SOURCES

H. Van Der Laan

Department of Astronomy, University of Western Ontario

THE recently discovered[1] temporal variations in the millimetre and centimetre radiation from a number of quasi-stellar radio sources and one Seyfert galaxy[2-5] provide radio astronomy with a potentially rich source of information on the physics of non-thermal radio sources. It has been suggested[5,6] that the varying components are sources of synchrotron radiation which are initially optically thick but, as they expand, become optically thin at progressively longer wavelengths. Indeed, data now available[5] suggest that the variability in the high frequency part of the radio spectrum has a systematic character. The purpose of this article is to present a quantitative discussion of the evolving synchrotron self-absorption model. The time-dependence of the radio spectrum is derived, and it is shown how variations observed at one frequency may be used to predict the behaviour at any other frequency.

The attitude here is to admit the present mysterious nature of the initial event(s) giving rise to the relativistic electron flux. Rather than add to the multi-parameter speculation concerning such events, a definite model for the subsequent evolution of the radiation source is considered, which has a number of features accessible to direct observational tests. In this manner it may be possible, as additional and refined data are obtained, to approach the initial event by imposing well defined boundary conditions and other requirements.

Properties of the model. Consider a uniform and spherical cloud of radius r, expanding at the rate $\dot{r}$, filled with a flux of electrons at relativistic energies. Let these electrons have an isotropic velocity distribution and an energy distribution $N(E)\mathrm{d}E = K(t)E^{-\gamma}\mathrm{d}E$ (cm^{-3}) confined to a range $E_1(t) \leqslant E \leqslant E_2(t)$. They radiate as they are accelerated in a magnetic field of strength B gauss (G). Initially, the cloud is sufficiently compact to be optically thick at all radio frequencies. The condition that the brightness temperature of the radio source cannot exceed the equivalent temperature of the electrons chiefly responsible for the synchrotron radiation at any given frequency yields the well known relation[7,8]

$$\nu_m = CB^{1/5}\,\theta^{-4/5}\{S(\nu_t)\,[\nu_t/\nu_m]^{(\gamma-1)/2}\}^{2/5} \qquad (1)$$

where ν_m is the frequency at which the spectral curve reaches a maximum, ν_t is some frequency at which the source is transparent, $S(\nu_t)$ is the flux density at frequency ν_t, θ is the angular diameter, and C is a numerical constant dependent on the units used. For synchrotron radiation the volume emissively $p(\nu) \propto \nu^{-(\gamma-1)/2}$ while the absorption coefficient $\mu(\nu) \propto \nu^{-(\gamma+4)/4}$, and for this reason it follows that the frequency dependence of the spectral curve takes the form

$$S(\nu) \propto \frac{p(\nu)}{\mu(\nu)}[1-e^{-\tau(\nu)}]$$
$$\propto \nu^{5/2}[1-e^{-\tau(\nu)}] \qquad (2)$$

where the optical depth $\tau(\nu) = \int\mu(\nu)\mathrm{d}l \propto \nu^{-(\gamma+4)/2}$.

For frequencies $\nu \ll \nu_m$ the flux density is given by

$$S(\nu) = k_1 B^{-1/2}\,\theta^2\,\nu^{5/2}. \qquad (3a)$$

Reprinted from NATURE, 211, 1131-1133, 1966.

while at $\nu \gg \nu_m$

$$S(\nu) = k_2 K B^{(\gamma+1)/2}\,\theta^3\,\nu^{-(\gamma-1)/2} \qquad (3b)$$

where k_1 and k_2 are constants.

The parameters change as in the adiabatic expansion model first proposed by Shklovsky[9] to represent the early phase of supernova remnant evolution. Magnetic flux is conserved

$$B = B_0(r/r_0)^{-2} \qquad (4a)$$

and the relativistic gas cools adiabatically, so that for an individual particle

$$E = E_0(r/r_0)^{-1} \qquad (4b)$$

while the angular diameter, of course, varies as

$$\theta = \theta_0(r/r_0) \qquad (4c)$$

In the model, particles neither enter nor escape the reservoir after the initiating event, so that

$$\frac{\mathrm{d}}{\mathrm{d}t}\left\{r^3(t)K(t)\int_{E_1(t)}^{E_2(t)} E^{-\gamma}\,\mathrm{d}E\right\} = 0 \qquad (4d)$$

The subscript 0 denotes parameter values at a specified instant. From relations (3) and (4) it follows that for $\nu \ll \nu_m$

$$S(\nu, \rho) = S_0(\nu)\,\rho^3 \qquad (5)$$

and for $\nu \gg \nu_m$

$$S(\nu, \rho) = S_0(\nu)\,\rho^{-2\gamma} \qquad (6)$$

where $\rho = r/r_0$ is the relative radius of the source. From relations (1), (4) and (6) the dependence of ν_m, the frequency of maximum flux density, on radius is seen to be

$$\nu_m(\rho) = \nu_{m0}\,\rho^{-(4\gamma+6)/(\gamma+4)} \qquad (7)$$

From equations (7) and (5) or (6) the variation of the maximum flux density follows:

$$S_m(\rho) = S_{m0}\,\rho^{-(7\gamma+3)/(\gamma+4)} \qquad (8)$$

The evolution of the spectrum is thus seen to be very simple. The spectral curve, as given in the usual log $S(\nu)$ – log ν plane, does not change, but moves down and towards the left, that is, to lower flux densities and lower frequencies. Observing such a source at one frequency will show the flux density to increase rapidly at first, reach a maximum and then decrease more gently. Simultaneous observation at a lower frequency will show the same fractional rate of increase, but the maximum will be reached later and have a smaller value.

Secular variations take the form

$$\dot{\nu}_m/\nu_m = -\left(\frac{4\gamma+6}{\gamma+4}\right)\dot{r}/r;\quad \dot{S}_m/S_m = -\left(\frac{7\gamma+3}{\gamma+4}\right)\dot{r}/r \qquad (9)$$

and for $\nu \ll \nu_m$ $\qquad$ $\nu \gg \nu_m$

$$\dot{S}/S = 3\dot{r}/r\ ;\ \dot{S}/S = -\,2\gamma\dot{r}/r \qquad (10)$$

To calculate $S(\nu, \rho)$, the flux density at a given frequency and radius, use expressions (2), (7) and (8). It is noted that the spectral curve has no identifiable features other than the curvature associated with the turnover, which provides the point of reference. The flux density and frequency are therefore given relative to their values at the spectral maximum when $\rho = 1$. Then

$$S(\nu,\rho) = S_m\left(\frac{\nu}{\nu_m}\right)^{5/2}\left[\frac{1-\exp(-\tau)}{1-\exp(-\tau_m)}\right]$$

$$= S_{m0}\,(\nu/\nu_{m0})^{5/2}\,\rho^3\,\frac{\left[1-\exp\left\{-\tau_m\left(\frac{\nu}{\nu_{m0}}\right)^{-(\gamma+4)/2}\rho^{-(2\gamma+3)}\right\}\right]}{[1-\exp(-\tau_m)]} \quad (11)$$

where τ_m, the optical depth corresponding to the frequency at which the flux density is a maximum, is the solution of

$$e^x - \left(\frac{\gamma+4}{5}\right)x - 1 = 0 \quad (12)$$

which is the condition for expression (2) to be a maximum.

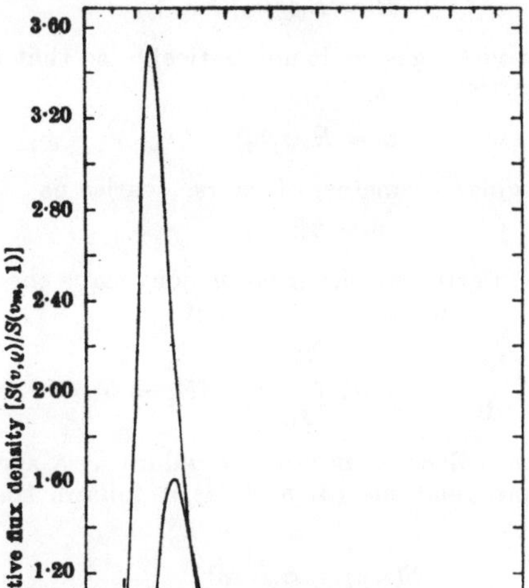

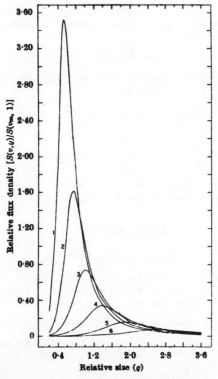

Fig. 1. The radial dependence of flux density. The frequencies of successive curves differ by a factor of two. The spectral index (α) in this case is 0·25. Values for the various curves are ν_{m0}/ν : (1) 1/2; (2) 1; (3) 2; (4) 4; (5) 8; (6) 16.

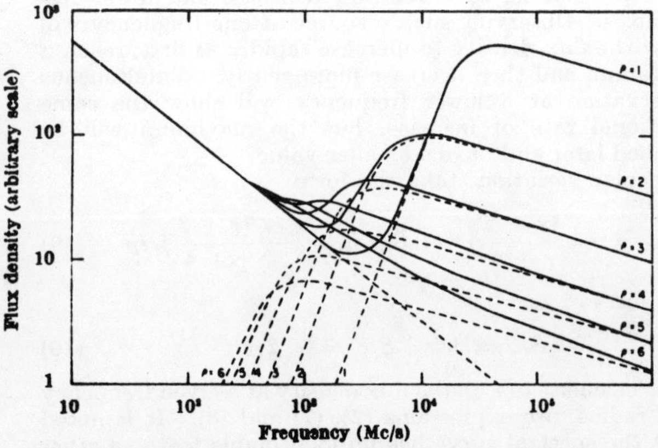

Fig. 2. The evolution of a typical variable radio source spectrum. The variable source spectrum is added to that of the constant source and the resultant is shown at successive integer values of ρ.

Flux density as a function of frequency and time. In order to obtain the flux density at any frequency as a function of time, it is necessary to specify the time dependence of the expansion velocity. Let $t_0 = r_0/\dot{r}_0$ be the "apparent age" of the source when its radius is r_0. Suppose $\rho = (t/\beta t_0)^\beta$ and define βt_0 as the "expansion age" at radius r_0. (The "true age" is not defined when the circumstances that give rise to the source are unknown.) Two cases will be referred to in particular: (*a*) constant expansion velocity, that is, $\beta = 1$; and (*b*) deceleration in the manner of a relativistic gas cloud expanding into a uniform gaseous medium[10], asymptotically $\beta = 2/5$.

The function $S(\nu,\rho)/S_{m0}$, as given in equation (11), is presented graphically in Figs. 1 and 2 for several values of ν/ν_{m0} and a range of ρ. These curves were computed using a spectral index $\alpha = (\gamma - 1)/2 = 0{\cdot}25$. If the frequency of maximum flux ν_{m0} is known and S_{m0} is obtained from measurement or from equation (11), then the vertical scale of curves such as in Fig. 1 can be specified in flux units and each curve can be assigned a definite frequency.

The time scale factor t_0 can be determined from a measurement of one observable and its rate of change. Any of the equations (9) and (10) is an expression for the apparent age when $1/t_0$ is substituted for $(\dot{r}/r)_0$. If, for example, the flux density and its rate of change are measured at a frequency $\nu_0 \ll \nu$

$$t_0 = 3S_0/\dot{S}_0 \quad (13)$$

In order to relate radius and age, let the time t' be measured from the instant $t' = 0$ when S_0, $\dot{S}_0$ and ν_{m0} are specified. Then

$$\rho = (1 + t'/\beta t_0)^\beta$$
$$\simeq 1 + t'/t_0 \text{ for } |t'| \ll \beta t_0 \quad (14)$$

The function (14) is sketched in Fig. 3 for several values of the deceleration parameter β. For times small compared with βt_0, before and after the instant $t' = 0$, the model can be compared with given sources at any frequency and time. For times separated from the instant $t' = 0$ by intervals of order βt_0 the curves in Fig. 3 serve to relate ρ and t' for a chosen β.

If the maximum of the spectral curve is at frequency ν_1 when $t' = t_1$, then this maximum reaches another frequency ν_2 at

$$t'(\nu_{m2}) \equiv t_2 = (\beta t_0 + t_1)\,(\nu_1/\nu_2)^{(\gamma+4)/\beta(4\gamma+6)} - \beta t_0 \quad (15)$$

If at $t' = t_1$ the maximum flux is $S_m(\nu_1)$, then at the time given by relation (15) the flux density maximum is at ν_2 and is given by

$$S_m(t_2) = S_m(t_1)\left(\frac{\beta t_0 + t_2}{\beta t_0 + t_1}\right)^{-\beta(7\gamma+3)/(\gamma+4)} \quad (16a)$$

or the equivalent expression

$$S_m(\nu_2) = S_m(\nu_1)(\nu_1/\nu_2)^{-(7\gamma+3)/(4\gamma+6)} \quad (16b)$$

These relations together can be applied to any source the spectrum of which is variable and known over a fraction of the expansion age. Spectral changes in the immediate future are then predictable; for later times they depend on the expansion mode. Prolonged observations can be expected to provide information about the dynamics of expansion and to yield a value for β.

The applicability of the model *überhaupt* can be tested by the requirement that the expressions for t_0 equivalent to expression (13), given by each observable and its rate of change as given in equations (9) and (10), result in values equal within experimental errors. Data now available are spread too thinly in time and wavelength to attempt as yet quantitative tests; qualitatively the model conforms to observed radio source behaviour[1,4,5].

There are ways other than synchrotron self-absorption which can give rise to a turnover in the spectral curve[11]. A magneto-ionic medium dense enough to cause significant dispersion leads an otherwise inverse power law spectrum to decrease sharply on the low frequency side of $\nu = 20\, n_e/B$ Mc/s where n_e is the free electron number density in cm^{-3} (Tsytovich effect). In that case the adiabatic expansion model without synchrotron self-absorption gives

$$\nu_m \propto n_e/B \propto \rho^{-1}$$

Qualitatively the alternatives are similar, but quantitatively they differ considerably. Compare, for example, the following ratio in the two cases:

$$\left\{\frac{\dot{S}(\nu_m)/S(\nu_m)}{\dot{\nu}_m/\nu_m}\right\} \begin{array}{ll} = (3\gamma + 1)/2 \text{ (dispersive medium)} & (17a) \\ = (7\gamma + 3)/(4\gamma + 6) \text{ (synchrotron self-absorption)} & (17b) \end{array}$$

Because for most sources $\gamma \simeq 1{\cdot}5$–$2{\cdot}5$, spectral information over a time $\sim t_0$ will enable such cases to be distinguished.

Multiple bursts. Although current spectral data are incomplete, in a few instances the spectral information is sufficient to indicate that the observed spectrum is a superposition of two or more time-varying remnants of "bursts" and a constant component with the usual inverse power law spectrum. Such is the case for *3C* 273 and *3C* 279, for example. Very homogeneous data, at closely spaced time and frequency intervals, will make it possible to separate the contributions of each component, to which the method here outlined can be applied to predict the evolution of the composite spectrum.

The presence of more than one variable component in a single radio source has interesting implications. It cannot be ruled out *a priori* that the magnetic field is an attendant result of the event that produced the relativistic particles, but it is simpler to regard the former as an ambient field, present before the event in which energy is released. If this were indeed the case, then the spectra would indicate that the events occur at spatially separate points. Repeated outbursts at the same position would result in the following sequence of events: the first burst sweeps out the surrounding magneto-ionic medium, leaving an explosion vacuum. Particles produced in a second burst at that point would travel in a vacuum and not radiate until they entered the expanding shell owing to the preceding event. The result would be a shift in the spectral curve towards higher frequencies and an intensity increase. The observed spectra show, however, that the frequency spectra rather than the energy spectra are superimposed. It is therefore probable that regions exist in the quasi-stellar radio sources where conditions favour the occurrence of many independent outbursts. The available data show separate components which become observable only 2–5 years apart[5]; the regions of explosive activity may therefore be only a fraction of a kiloparsec in extent. The cumulative effect of many outbursts is expected to be a non-varying radio source with a conventional spectrum[12], resulting when the expansion volumes overlap and the energy spectra are superposed.

Discussion. In the first instance the model is intended to account for and to predict the gross features of early radio source evolution. Departures from the predicted behaviour will occur to the extent that the complexity of the radio source exceeds that of the model. Such discrepancies will either show that the model is not related to the phenomenon observed or suggest what additional structure the model requires for a more adequate representation.

It should be noted that the higher the frequency at which the variation is detected, the shorter the time scale factor t_0. This shortens the time interval over which predicted variations can be compared and different expansion modes can be distinguished, proportionately. Millimetre and centimetre observations will yield the parameter values required to predict expected variations at longer wavelengths. Because the measurements cannot be repeated, it is highly desirable that at least two instruments are used for simultaneous observations at (nearly) equal frequencies. Systematic monitoring of the variable sources at many frequencies is necessary to acquire homogeneous data[13], so that theoretical suggestions can be tested. Optical and X-ray data are equally necessary; correlated variations at widely separated regions of the electromagnetic spectrum could shed much light on the nature of these sources. Such correlations are expected for plausible parameter ratios, where some radio photons, by inverse Compton interaction with the relativistic electrons, are converted to hard radiation.

Homogeneous radio data will make it possible to obtain reliable life times and linear diameters of the varying components. This information may then enable radio interferometry with baselines of several thousand kilometres to settle the controversial question of quasi-stellar distances.

I thank Drs. K. I. Kellermann and I. I. K. Pauliny-Toth for informative discussions and a preprint of their paper. I also thank Dr. D. S. Heeschen, director of the National Radio Astronomy Observatory, Charlottesville, for the hospitality of this institution.

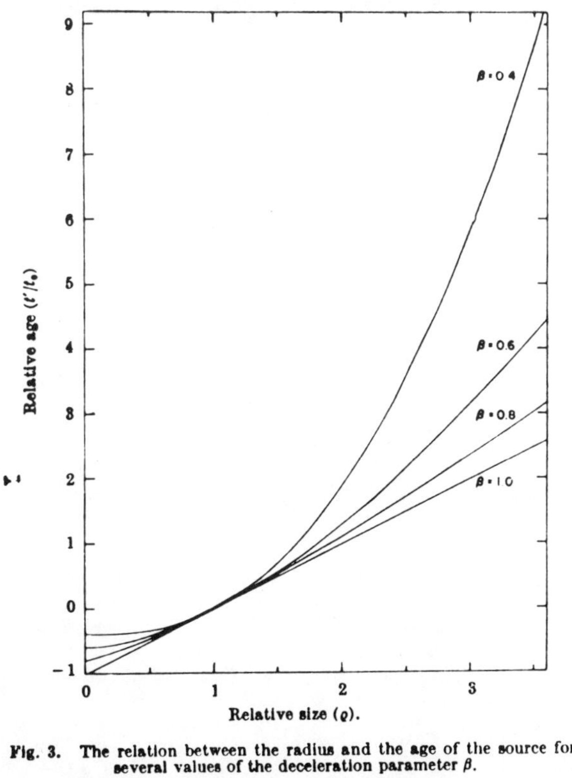

Fig. 3. The relation between the radius and the age of the source for several values of the deceleration parameter β.

[1] Dent, W. A., *Science*, **148**, 1458 (1965).
[2] Epstein, E., *Astrophys. J.*, **142**, 1282 (1965).
[3] Low, F. J., *Astrophys. J.*, **142**, 1287 (1965).
[4] Dent, W. A., *Astrophys. J.*, **144**, 843 (1966).
[5] Pauliny-Toth, I. I. K., and Kellermann, K. I., *Astrophys. J.* (in the press)
[6] Shklovsky, I. S., *Nature*, **206**, 176 (1965).
[7] Slish, V. I., *Nature*, **199**, 682 (1963).
[8] Williams, P. J. S., *Nature*, **200**, 56 (1963).
[9] Shklovsky, I. S., *Astron. Zh.*, **37**, 256 (1960).
[10] van der Laan, H., *Mon. Not. Roy. Astro. Soc.*, **126**, 535 (1963).
[11] Hornby, J. M., and Williams, P. J. S., *Mon. Not. Roy. Astro. Soc.*, **131**, 237 (1966).
[12] Kellermann, K. I., *Astrophys. J.* (in the press).
[13] Kellermann, K. I. (personal communication) has initiated efforts to observe variable radio sources systematically on an international scale.

PART 2

RADIO GALAXIES AND QUASARS, OPTICAL OBSERVATIONS AND MODELS

MODELS OF QUASI-STELLAR SOURCES

Maarten Schmidt
Mount Wilson and Palomar Observatories, Carnegie Institution of Washington,
California Institute of Technology

We shall briefly review the main properties of the quasi-stellar sources (Sec. I), investigate various interpretations of the redshifts (Sec. II), and then discuss some models proposed for these objects (Sec. III).

I. Properties of Quasi-Stellar Sources

The optical properties that distinguish quasi-stellar sources from other objects may be summarized as follows: (1) starlike objects identified with radio sources; (2) variable light; (3) large ultraviolet excess; (4) broad emission lines; and (5) large red-shifts.

(1) All the quasi-stellar objects have optical diameters less than 1"; the actual diameters are not directly known. Some of the objects have wisps or jets reaching out to 20" (3C 273), 9" (3C 48), or 3" (3C 196). These wisps contain up to 10 per cent of the total light.

(2) Light variations have amplitudes up to 0.6 mag. with a typical cycle of about 10 years. Variations at rates up to 0.05 mag. (i.e., 5 per cent) per month have been observed; the reality of more rapid variations is not established. The colors do not vary systematically with the total light. An exception is 3C 2 which has brightened several magnitudes and become much bluer over the past decade, according to Sandage.

(3) The ultraviolet is about twice as intense as that of stars with the same energy distribution in the blue and yellow parts of the spectrum. Sandage uses this property to discover quasi-stellar objects near the positions of radio sources.

(4) Broad emission lines have been found in fourteen quasi-stellars (March, 1965). No lines have been detected in 3C 196 (and perhaps one or two others still under observation), but its red-shift may be such that no strong lines happen to be present in the visual part of the spectrum. The line widths are about 20-50 Å. No absorption lines have been found yet.

(5) The red-shifts of the emission lines in quasi-stellars are up to seven times as large as those of the intrinsically brightest galaxies of the same magnitude. Only four red-shifts have been determined till now.

Property (1) constitutes the present definition of the quasi-stellar radio sources. Practically all known quasi-stellars exhibit property (3) because of discovery techniques. It remains to be seen whether all quasi-stellar sources adhere to all five properties. It may be necessary to revise the definition of quasi-stellar sources eventually if observations on further sources reveal conflicting properties. It is, for instance, not clear whether the radio emission is important for the definition of quasi-stellar sources. However, there is no proof at present of the existence of "quasi-stellar non-radio sources."

It is remarkable that there is not a single radio property that distinguishes the quasi-stellar sources uniquely from the radio galaxies (i.e., sources identified with galaxies). The most interesting radio property concerns the angular diameter of a radio component; for most of the quasi-stellar radio sources it is less than 1", while for almost all radio galaxies it is larger than 1". A glaring exception is the quasi-stellar source 3C 47, which contains no radio component much smaller than 1'. Another statistical property of the quasi-stellar sources is that their spectrum shows more often low-frequency curvature than do radio sources in general. Other radio properties of the quasi-stellar sources such as polarization, spectral index, component separation in doubles, etc., seem to be no different from those exhibited by radio galaxies.

II. Redshifts

The red-shifts can be interpreted as (1) Doppler shifts of nearby objects; (2) gravitational red-shifts; and (3) cosmological red-shifts.

(1) Doppler shifts. The low upper limit of the proper motion of 3C 273 in conjunction with the radial velocity corresponding to the red-shift allows an estimate of the minimum distance. If the tangential velocity is of the same order of magnitude as the radial velocity, this minimum distance is 10 Mpc (3×10^{25} cm). At this distance the mass of hydrogen required by the Balmer emission spectrum is 100 solar masses (2×10^{35} gm); the acceleration of such a mass to relativistic velocities would be a major problem. Also, one would expect half of the spectral shifts to be blue-shifts, but all four known are red-shifts. Also not understood would be the observed correlation between red-shift and magnitude of the quasi-stellar sources: there would be no reason why the brightest quasi-stellar (3C 273) should have the smallest red-shift.

A modification of the above has been suggested by Terrell (1964). He considers that 3C 273 may have been ejected from the central part of our Galaxy. On this basis its minimum distance becomes 200 kpc. The absence of observed blue-shifts would require that all quasi-stellars have their origin in our Galaxy. The red-shift-magnitude relation could only be understood if it really were a velocity-distance relation. This would require roughly the same absolute brightness for each of the ejected quasi-stellars, and would point to one particular time of ejection for all sources, about 10^7 or more years ago. At constant output 3C 273 would have radiated around 10^{54} ergs over this period of time, or close to 1 $M_\odot c^2$. At a reasonable efficiency of 10^{-3}, this requires the available nuclear energy from a mass of 10^3 $M_\odot$. Since the quasi-stellars move at velocities not much less than c, the kinetic energy of each is close to 10^3 $M_\odot c^2$, or around 10^{57} ergs. Since there must exist many more than the fifty or so quasi-stellars now known, the total kinetic energy in all quasi-stellars would amount to perhaps 10^{60} ergs. If it were at all possible to accelerate objects of a 1000 $M_\odot$ to relativistic velocities, the energy problem would be very severe, especially if we consider that the required 10^{60} ergs would have to be produced as recently as 10^7 years ago.

(2) Gravitational red-shifts. If the observed red-shifts were gravitational, then the emission lines would have to originate in a thin shell, the thickness of which is limited by the variation of gravitational potential corresponding to the observed widths of the emission lines. The electron density in this shell is very low, because forbidden lines are seen in the spectra. An analysis of the spectra of 3C 48 and 3C 273 by Greenstein and Schmidt (1964) gives electron densities of 3×10^4 (or less) and 3×10^6 cm^{-3}, respectively. The emission in, say, Hβ from ionized hydrogen at electron temperatures between 10^{40} and 3×10^{40} K is around 10^{-25} N_e^2 ergs per cm^3 per sec, where N_e is the electron density.

Thus, for a given mass of the object, we can compute the radius R (from the gravitational redshift), the emitting volume (from the line width), the total line emission (from N_e), and hence the distance (from the Hβ flux density at Earth). Table 1 shows radii and distances computed for a large range of masses. The upper part applies to 3C 48 if the thickness of the emitting shell is taken as 0.016 R. Limits

Table 1. Mass, radius, and distance of 3C 48 if its red-shift were a gravitational red-shift

log M/$M_\odot$.	log R	log r	r
Thickness of Emitting Shell 0.016 R			
0 ...	5.6	6.2	14 km
2 ...	7.6	9.2	...
4 ...	9.6	12.2	...
6 ...	11.6	15.2	100 A.U.
8 ...	13.6	18.2	0.5 pc
10 ...	15.6	21.2	0.5 kpc
12 ...	17.6	24.2	0.5 Mpc
14 ...	19.6	27.2	500 Mpc
Thickness of Emitting Shell 0.00001 R			
13 ...	18.6	24.0	0.3 Mpc
15 ...	20.6	27.0	300 Mpc

imposed by gravitational perturbations on the planetary system, the angular size of the objects, and gravitational perturbations on our local part of the Galaxy exclude masses less than 10^{11} $M_\odot$ at distances less than 15 kpc. For a mass of 10^{14} $M_\odot$ the cosmological red-shift would already be half the total red-shift; hence, on the basis of gravitational red-shifts only extragalactic objects in the mass range 10^{11} to 10^{14} $M_\odot$ seem to be allowed.

The upper part of Table 1 is actually a very conservative compilation. It can be shown that the line emission must originate in not more than 1 scale height in the objects' atmosphere. In order to estimate the scale height, we need the temperature; it seems safe to take as an upper limit for the temperature that corresponding to the line widths. The scale height turns out to be only 10^{-5} R; the lower part of Table 1 is based on this thickness of the emitting shell. With this revision we get a minimum mass of 2×10^{13} $M_\odot$ at a distance of at least 2 Mpc; these minima are required to avoid perturbing the Local Group of galaxies for which the total mass from the virial theorem is found to be not more than 2×10^{12} $M_\odot$ (cf. Humason and Wahlquist 1955).

This minimum mass of 2×10^{13} $M_\odot$ exceeds the largest known mass of a galaxy by a factor of at least 10. Whether such a mass could have a stable configuration within a radius of 20 pc is extremely doubtful. Also, the hypothesis of gravitational red-shifts provides no understanding of the observed red-shift-magnitude relation of the quasi-stellar sources.

(3) Cosmological red-shifts. At first, cosmological red-shifts seemed perhaps less attractive for the quasi-stellar sources because the large distances led to optical luminosities 4 to 50 times larger than those of the brightest galaxies. Also, it seemed objectionable to have light fluctuations in extragalactic objects. Cosmological red-shifts, however, do explain the red-shift-magnitude relation; they lead to radio luminosities that are not abnormal, about equal to those of the strongest radio galaxies.

As we have seen above, there are important objections to interpretation of the redshifts as relativistic velocities of local objects, or as gravitational red-shifts of supergalactic masses outside our Local Group. Hence, it is most likely that the red-shifts are indeed cosmological red-shifts. On this basis some of the intrinsic properties of the four quasi-stellar radio sources with known red-shifts are given in Table 2. The

Table 2. Data for Quasi-stellar radio sources with known red-shifts.

Source	z	M_ν	L (Radio)† (erg/sec)	"Size" (Radio) (kpc)
3C 47	0.425	-23	1.5×10^{44}	250
3C 48	.367	-25	4.7×10^{44}	< 3
3C 147	.545	-25	2.0×10^{45}	10
3C 273	0.158	-26	3.1×10^{44}	40

*From Schmidt and Matthews (1964).

†Computed by integration between emitted frequencies of 10^7 c/s and 10^{11} c/s, except for the following lower limits of observed frequency: 10^8 c/s for 3C 48, 5×10^7 c/s for 3C 147, 10^9 c/s for component B of 3C 273.

total optical luminosities depend on the unobserved ultraviolet part of the energy distribution. They range probably from about 10^{45} to over 10^{46} ergs per sec.

III. Models of Quasi-Stellar Sources

A model of the quasi-stellar sources is likely to consist of three parts: a nucleus providing the energy and part of the optical continuum radiation, an H II region generating the emission lines and part of the optical continuum, and a larger halo producing the radio emission.

The H II region is observed in the emission lines. The density of the gas may be estimated from the relative strengths of forbidden lines. For densities below 3×10^4 cm^{-3} the strengths of all observed emission lines vary as the square of the electron density, N_e. For higher densities collisional de-excitation becomes important and the intensity increases only as N_e. The line that is affected at the lowest density is the λ 3727 line of [O II]. From its absence, and from the presence of the λ 5007 line of [O III] in 3C 273, Greenstein and Schmidt (1964) derived a density of 3×10^6 cm^{-3}. In 3C 48 the intensity ratios of the lines are about "normal," so it can only be stated that the maximum density is 3×10^4 cm^{-3}.

For a given electron temperature, these densities determine the emission in $H\beta$ per cm^3; for $T_e = 2 \times 10^4$ °K, it is about 10^{-25} N_e^2 erg sec^{-1} cm^{-3}. With the known total emission in $H\beta$, we at once compute the emitting volumes as about 2×10^{55} cm^3 for 3C 273, 2×10^{58} cm^3 for 3C 48. The corresponding radii are 1 pc and 10 pc, respectively. The observed light variations suggest that the radius of the source of optical continuum is not more than 0.3 pc (1 light-year) for 3C 273, and not more than 1 pc for 3C 48. Greenstein and Schmidt (1964) have argued on this basis that the H II region is larger than the optical continuum source.

A complication arises from the fact that the optical depth for electron scattering to the center is about 6 for 3C 273 (0.5 for 3C 48). Hence a photon emitted at the center would be scattered every half-year on the average. It would emerge after about 36 scatterings, 18 years later. The dispersion in this emersion time would be around 3 years, or just the light travel time of radius R. Hence, the light variations from a small source with a multiple-scattering atmosphere of radius R cannot be faster than those from a distributed light source with radius R. One solution to our dilemma would be to make the H II region just as small as the continuum source, but this would lead, especially for 3C 48, to densities higher than permitted by the observed spectrum. A more probable solution is that the gas has a patchy or filamentary distribution, the density in the filaments being those mentioned above, and the densities between the filaments being at least an order of magnitude lower. I do not believe that this is an artificial solution: Many planetary nebulae and the Crab Nebula show highly irregular or filamentary structure. The optical depth will decrease as the over-all area, so that for 3C 273 a total volume of 10 pc radius filled with 2×10^{55} cm^3 filaments of density 3×10^6 cm^{-3} will account for the line emission, yet be essentially transparent for electron scattering. If the optical continuum nucleus is very small, one might even speculate that part of the light variations are due to the interception of light by filaments moving across the nucleus as seen from the Earth. The constancy of the colors of the quasi-stellars is consistent with this possibility.

Except for the introduction of a patchy structure in the H II region, the above model is essentially that of Greenstein and Schmidt (1964). They found that any radio emission from the shell would be effectively unpolarized because of excessive Faraday rotation, and that the shell would be optically thick in free-free at, say, 1000 Mc/s. These conclusions would be little affected by the introduction of a patchy structure in the shell. The non-thermal character of the radio spectra and the observed radio polarization then suggest that the radio emission must originate outside the H II region, in the halo. This suggestion is supported by the observed angular diameter of 0".5 of 3C 273B corresponding to a radius of 500 pc for the radio halo.

At the electron temperature of around 2×10^4 °K used till now, the thermal equivalent width of Hβ is about 1300 Å, i.e., the energy radiated in Hβ equals that of 1300 Å of the neighboring continuum generated by bound-free and free-free processes. The observed equivalent widths are only 86 Å in 3C 273, 28 Å in 3C 48. Thus the H II region contributes a thermal continuum that is about 7 per cent of the total continuum in 3C 273 (2 per cent in 3C 48). The remainder of the optical continuum is attributed to the nucleus which we assign a radius of less than 1 light-year so as to permit the observed light variations.

Oke (1965) has carried out a spectrophotometric study of 3C 273. He finds that the energy distribution is remarkably flat below 5000 Å; at longer wavelengths it turns up. Oke considers the latter as being synchrotron radiation, but his main thesis is that the flat continuum in the blue and violet is due to bound-free, free-free, and 2-photon emission from a hot gas for which he adopts a temperature of 1.6×10^5 °K. At this high temperature the gas would produce the Balmer emission spectrum and the entire continuum; hence, there would be no need for an extra source of optical continuum. This high temperature has been chosen such that the thermal equivalent width of Hβ in terms of the continuum is just that found in 3C 273. However, the work by Pengelly (1964) on line emission from ionized hydrogen, together with Oke's evaluation of the continuum, suggests that the equivalent width of Hβ is roughly constant with temperature, at a value of about 1300 Å. If this is correct, then the ratio of line emission to continuum contains no information about the temperature, at least in the range 10^{40} to 10^5 °K, but only about the fraction of non-thermal continuum present in the source. It is not necessarily an advantage to explain the entire continuum as produced by the same gas that yields the emission lines. The time variation in the continuum suggests a radius of not more than 1 light-year. At this size, Oke's model would require an electron density of 10^8 cm^{-3}, a value quite unlikely on the basis of the observed spectrum of 3C 273.

Field (1964) has proposed that a quasi-stellar source represents the birth of a spherical galaxy. He considers the fate of a gas cloud with sufficiently little angular momentum to allow collapse to densities at which the formation of stars may take place at high rate. Field derives a radius of the cloud (shortly after star formation starts) of not less than 100 pc if the total mass is 10^{11} $M_{\odot}$. He then accounts for the luminosity in terms of starlight, or in supernovae. In this connection it is of interest that Sandage has pointed out that the colors of quasi-stellars are similar to those of bright spiral-arm segments in M33, which probably represent freshly formed stars. The observed light variations would have to be attributed to the supernovae in Field's model. Colgate and Cameron also had suggested that the light of the quasi-stellars could be due to supernovae. A most important question is whether the light-curves of the quasi-stellar sources admit interpretation in terms of a number of independent outbursts. It seems to me that the slow cyclic variations seen in Smith's light-curve for 3C 273, and in Sandage's curve for 3C 48 would not allow this, but more careful analyses should be carried out.

The total output of a quasi-stellar source is about 10^{54} ergs per year. The absence of secular light variation suggests a lifetime of at least 1000 years. If, on the other hand, observed features such as wisps, jets, and distant radio components of a quasi-stellar source are all part of one single major event, then the age may be around 10^6 years, the total energy output some 10^{60} ergs on the assumption of constant luminosity. The production of this energy would require the available nuclear energy from about 10^8 $M_{\odot}$ hence implies the presence of 10^9 $M_{\odot}$ or more. No "ordinary" galaxy has been observed around the stellar image of a quasi-stellar source yet, but the observational problem of detecting a faint distributed source around a much brighter point source is severe.

The relation between quasi-stellar radio sources, radio galaxies, and normal giant galaxies is still very unclear. The ratio of the numbers per unit volume may be estimated at about $10^{-8}:10^{-4}:1$. If all normal galaxies during their evolution had passed through these radio stages, then the quasi-stellars would have a lifetime of 100 years, the radio galaxies of 10^6 years. The quasi-stellar sources almost certainly have a lifetime of at least 1000 years, hence at most 10 per cent of all galaxies can have passed through the quasi-stellar stage.

The similarity in radio properties of radio galaxies and quasi-stellar sources does suggest a relation. If each radio galaxy passes through a quasi-stellar stage, then this will last 10^{-4} of the radio galaxy's radio life. For lifetimes of quasi-stellar sources of 10^3 to 10^6 years, this then implies lifetimes of 10^7 to 10^{10} years for the radio galaxies. Ages as long as 10^9 to 10^{10} years would tend to make the energy problem of the radio galaxies even more severe than it is at present. This suggests that either only a small fraction of all radio galaxies passes through the quasi-stellar stage, or a lifetime of the quasi-stellar phenomenon considerably less than 10^6 years.

REFERENCES

[1] **G. B. Field, 1964, Ap. J., 140, 1434.**

[2] **J. L. Greenstein, and M. Schmidt, 1964, Ap. J., 140, 1.**

[3] **M. L. Humason, and H. D. Wahlquist, 1955, A. J., 60, 254.**

[4] **J. B. Oke, 1965, Ap. J., 141, 6.**

[5] **R. M. Pengelly, 1964, M.N., 127, 145.**

[6] **M. Schmidt, and T. A. Mathews, 1964, Ap. J., 139, 781.**

[7] **J. Terrell, 1964, Science, 145, 918.**

THE SUPERGIANT GALAXIES

W. W. MORGAN AND JANET ROUNTREE LESH
Yerkes Observatory, University of Chicago

1. In a recent paper (Matthews, Morgan, and Schmidt 1964), a description was given of the optical forms of identified extragalactic radio sources. Of the "strong" sources identified, approximately one-half are associated with galaxies having the following characteristics: (*a*) they are located in clusters, of which they are outstandingly the brightest and largest members; (*b*) they are centrally located in their clusters; (*c*) they are never highly flattened in shape; and (*d*) they are of a characteristic appearance, having bright, elliptical-like nuclei, surrounded by an extended amorphous envelope; the nuclei may be double or multiple in nature. These supergiant galaxies have been given the form-type class of cD in Morgan's classification.

2. In the same paper, a description was given of a group of galaxy clusters from Abell's *Catalogue* (1958). The characteristics of this group appear to be similar to those just described for the radio sources—with the single exception that most of them are not known to be radio sources. Of the twenty-six clusters in Abell's *Catalogue* of richness greater than 1 and distance less than 5, ten clusters contain cD galaxies; of these ten, only two are known to be radio sources. We therefore have the following situation: In the Abell *Catalogue* of galaxy clusters, an appreciable fraction of the clusters have a characteristic appearance of being dominated by a centrally placed cD galaxy—or multiple galaxy; of this group of clusters, a small fraction are known to be radio sources.

3. These supergiant galaxies are of considerable interest—both as a class and individually. Their "main bodies" are considerably larger in size than those of such giant systems as the Andromeda Nebula (M31), and they are much higher in luminosity. Figure 1 is a composite photograph of the inner part of the cluster Abell 2199, taken from the Yerkes copy of the National Geographic Society–Palomar Observatory Sky Survey; the insert is a photograph of M31 taken by the late E. E. Barnard with a 6-inch refractor. The absolute scales of the two photographs are the same; the diameter of the "main body" of M31 illustrated is 100′, which corresponds to a true size of 24000 pc. The sizes of the main bodies of M31 and of the two cluster members to the lower left are similar; the supergiant galaxy NGC 6166 (= 3C 338) in the upper part of the illustration is much larger and more luminous. If the sizes of the photographic "main bodies" are compared, it seems likely that these cD galaxies will turn out to be the largest known single structures.

4. A survey has been carried out of all galaxy clusters in Abell's *Catalogue* having richness greater than 2 on his scale; there are eighty-five such clusters; these richest clusters all belong to distance groups 4, 5, and 6. The classification of the cluster forms was carried out on the Yerkes copy of the National Geographic Society–Palomar Observatory Sky Survey. Of the eighty-five clusters examined, twenty-two—or about one-fourth—were found to have dominating, centrally placed, cD galaxies. The twenty-two clusters are listed in Table 1. The significance of the three classes in the table is as

*Reprinted from Astrophysical Journal **145**, 1364 (1965).

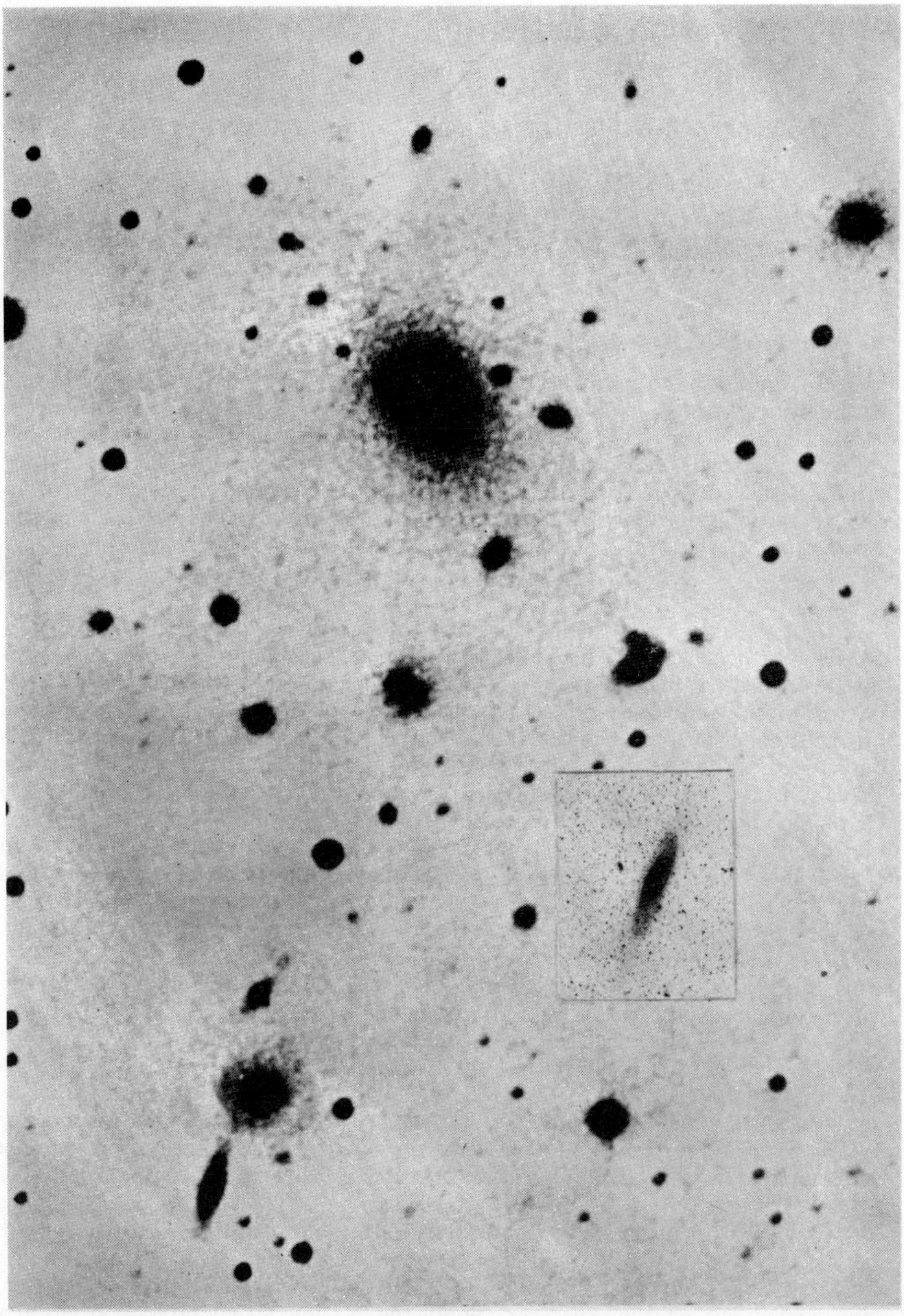

FIG. 1.—Central region of the cluster Abell 2199. The supergiant galaxy is NGC 6166 = 3C 338. The insert is M31, reduced to the corresponding linear scale of the cluster. NGC 6166 can be considered a prototype for the cD galaxies. The main body of M31 as illustrated is around 24000 pc.
Cluster photograph from National Geographic Society–Palomar Observatory Sky Survey; M31 photograph from 6-inch refractor plate by E. E. Barnard.

follows: (1) highest class examples; (2) examples of somewhat lower weight; (3) distant clusters where the nature of the central galaxy has been assumed by analogy with nearer clusters. We label these clusters containing cD galaxies as "cD clusters."

5. The cD clusters listed in Table 1 increase to about thirty the probable members of this class; nine other clusters of richness 2 have been published earlier (Matthews, Morgan, and Schmidt 1964). We therefore now have available a sufficient number of specimens for detailed investigation of their nature. Among numerous questions that must be asked are the following: (1) Why are some cD clusters radio sources, while some are not? (2) Are there significant structural differences between the two categories? (3) What

TABLE 1

RICHEST GALAXY CLUSTERS CONTAINING cD GALAXIES

Abell	Class	Notes
22	1	Two cD galaxies in common envelope. Nuclear part of cluster elongated.
42	1	Two cD galaxies in common envelope. Circular, compact cluster nucleus.
222	3	Distant, very elongated cluster in shape of integral sign. Another, similar cluster nearby.
586	1	Tremendous central galaxy. Very rich, compact cluster nucleus. Rich star field.
655	2	cD galaxy not centrally located.
750	3	Distant. Double nucleus.
795	3	Distant.
868	1	Circular, rich cluster nucleus. In rich star field.
910	2	Borderline case.
963	1	Outstanding in brightness. Probably double nucleus.
1146	1	Compact, rich cluster nucleus.
1413	1?	Very large and luminous central galaxy. This could be largest of all cD galaxies, but the possibility of its being a foreground object should be checked from redshift.
1437	2	
1689	1?	Rather distant. Three separated, brilliant nuclear galaxies. Concentrated, rich cluster nucleus.
1704	3	Distant.
1918	2	Very luminous central galaxy.
1961	3	Distant. Foreground ellipticals?
2204	1	cD galaxy triple on blue plate. In halation from bright star.
2317	3	Distant. cD galaxy probably triple. Concentrated cluster nucleus.
2554	2	cD galaxy not at center of cluster.
2645	3	Distant. Nuclear galaxy double.
2670	1	Type example. Also in list published earlier.

is the significance of the subclusterings around the cD galaxies; and are there numbers of fainter galaxies included within the volume of the cD galaxies? And, perhaps most interesting of all: (4) Are we observing "old" or "young" structures in the cD galaxy clusters?

We are indebted to Dr. H. W. Babcock, director of the Mount Wilson–Palomar Observatories for permission to reproduce Figure 1 from the National Geographic Society–Palomar Observatory Sky Survey glass copy at Yerkes. This work was supported by a contract with the Office of Naval Research.

REFERENCES

Abell, G. O. 1958, *Ap. J. Suppl.*, **3**, 211.

Matthews, T. A., Morgan, W. W., and Schmidt, M. 1964, *Ap. J.*, **140**, 35 (reprinted in 1965 in *Quasi-stellar Sources and Gravitational Collapse*, ed. I. Robinson, A. Schild, and E. L. Schucking [Chicago: University of Chicago Press], p. 105).

PHOTOELECTRIC SPECTROPHOTOMETRY OF QUASI-STELLAR SOURCES

J. B. Oke
Mount Wilson and Palomar Observatories,
Carnegie Institution of Washington, California Institute of Technology

Introduction

Early in 1963 a program was begun to measure absolute spectral energy distributions of the star-like objects associated with radio sources such as 3C 273 and 3C 48 to discover something about their nature. The importance of these objects was emphasized with the discovery that they had very large red-shifts (Schmidt 1963; Oke 1963; Greenstein and Matthews 1963). The program to measure absolute spectral energy distributions was expanded to include other fainter sources. The observations of 3C 273 were easily obtained with the photoelectric spectrum scanner then available; the results have been published by Oke (1965a). Since even in 3C 48, at magnitude 16.3, observations became difficult with the 200-inch telescope, a new prime-focus scanner was designed and built. It was anticipated that this new instrument, with suitable offset guiding devices, could provide absolute spectral energy distributions of objects as faint as 19 mag. in a reasonable amount of observing time, and this expectation has been met. This new scanner has now been in operation for two years, and in this paper absolute energy distributions of six quasi-stellar sources are presented and discussed.

It should be emphasized that the results given here are not definitive. The aim was rather to make a survey of as many quasi-stellar sources as possible with sufficient accuracy and spectral resolution to indicate the characteristics of the absolute spectral fluxes. Consequently the mean error of each observed point is moderately large, and a resolution of only 50-200 Å has been used in most cases. More accurate data or higher-resolution data can be obtained by expenditure of sufficient observing time. In this paper only limited data are given for points to the red of λ 6000, but such observations are being made at the present time.

The objects measured are so faint that the background sky brightness becomes important. It is worthwhile discussing briefly the prime-focus scanner and the technique used for subtracting the sky background. The instrument is a conventional all-reflection system in which the dispersive power is provided by a 600 groove/mm plane grating, blazed at λ 3750 in the second order. The instrument is used in slitless form, usually with round entrance apertures of diameter ranging from 7" to 20", depending on the seeing. The wavelength can be set with an accuracy of 1 Å and the resolution for a stellar image of 1" is approximately 2 Å in the second order. The exit slit can be set for band passes from 2 Å to 170 Å in the second order. Observations from λ 3200 to λ 5840 are made in the first or second order with a blue-sensitive Ascop 541 B photomultiplier tube with an S17 cathode. Infrared observations are made in the first order with an RCA 7102 photomultiplier which has an S1 semi-transparent cathode. A tri-alkali S20-cathode tube for the spectral range from λ 5000 to λ 8000 is planned but not yet in use. All observations are made with pulse-counting equipment with a resolution time of approximately 1 μsec.

The instrument is provided with two entrance apertures separated by 40" which produce two separate spectra at the exit slit. Light in both spectra reaches the photomultiplier-tube cathode through a single Fabry lens. A light chopper just behind the exit slit with a frequency of 15 c/s allows each separate spectrum to enter the photocell only half the time. The light chopper also provides a reference signal by intercepting the light beam going from a small lamp to a photo-

Reprinted from Astrophysical Journal.
145, 668-683.

TABLE 1

OBSERVED QUASI-STELLAR SOURCES

Object	m_V	$B-V$	$U-B$	$1+z$
3C 48..........	16.3	0.40	−0.59	1.367
3C 286.........	17.3	.26	− .91	1.848
3C 245.........	17.3	.46	− .82	2.029
CTA 102.......	17.3	.42	− .79	2.037
3C 446.........	18.4	.44	− .90	(2.40)
3C 9...........	18.2	0.23	−0.74	3.012

diode. This reference signal controls the direction of a reversible pulse counter in such a way that when light from one spectrum or aperture enters the photomultiplier the counter counts upward; when light enters from the other aperture it counts downward. If a star is placed in one aperture, the resulting net count is that from the star alone. Numerous precautions must be taken. (a) Since the two apertures may not be quite identical, two measurements must be made with the star in each hole; the resulting counts are averaged to eliminate any asymmetry. (b) Since even a synchronous motor does not run quite uniformly, the reference signal is used only to initiate a count. The accurate internal timer in the counter itself does the actual timing. With this technique the count time in each aperture remains accurately the same. A small deadtime interval exists at the end of each half-cycle. (c) The integration time must be controlled so that an integral number of cycles is included in an observation. In practice the system works extremely well even when the background sky brightness is much larger than the signal itself. Since the error of an observation depends on the total star-plus-sky count, rather than on the net star count, some provision must be made to record total counts, and in some cases it may be preferable to employ two "up" counters and switch the input into these with the synchronizing signal.

The Observations

The list of quasi-stellar sources for which observations are presented is given in Table 1. The table includes the source number, apparent visual magnitude, the B - V and U - B colors, and 1 + z where $z = \Delta\lambda/\lambda_0$. All the observations discussed below were made with the prime-focus scanner described above. To place the observations on an absolute-flux basis, standard stars with known absolute fluxes were also observed. The list of standards and the reduction techniques have been described by Oke (1964, 1965c). The errors in the absolute-flux system are in almost all cases smaller than the errors in the observations to be described.

The absolute fluxes are expressed in magnitudes AB which are defined by the equation

$$AB = -2.5 \log f_\nu - 48.55, \qquad (1)$$

where f_ν is the observed absolute flux in units of ergs sec^{-1} cm^{-2} $(c/s)^{-1}$. The constant is obtained from absolute flux data given by Code (1960) for a star of apparent visual magnitude 0.00. Values of AB are listed in Table 2 together with the corresponding values of $1/\lambda$, where the wavelength λ is measured in microns and the band width, $\Delta\lambda$, in Ångstroms. Also listed in the table are the numbers of observations, n, which were averaged to yield the tabulated magnitudes. A single observation always consists of two measurements with the object placed in each of the two entrance apertures. The fluxes and reciprocal wavelengths in Table 2 all refer to their observed values at Earth with no attention being paid to any redshift effects. The data are plotted in Figures 1 - 6.

Estimates of the error in the magnitudes AB can be made by (a) comparing observations made on different nights and (b) determining the statistical error on the basis of the total photon count. As a rule these two estimates of errors are in good agreement, indicating that the errors are produced largely by the statistical fluctuations of the limited number of photons counted. For the blue observations with $1/\lambda > 1.7$ the mean errors in magnitudes for individual tabulated points are as follows: 3C 48, 0.03 mag.; 3C 286, 0.03 mag.; 3C 245, 0.05 mag.; CTA 102, 0.03 mag.; 3C 446, 0.07 mag.; 3C 9, 0.06 mag. These errors are only approximate and are larger for points where only one observation was made. In the case of the infrared measures of 3C 286, the mean errors are 0.08 mag. in the red and increase further in the infrared. For CTA 102 the mean errors in the red and infrared are approximately 0.15 mag.

For observations made below λ 6000 no particular effort was made to detect emission lines since

TABLE 2

OBSERVED ABSOLUTE FLUXES, $-2.5 \log f_\nu - 48.55$

$1/\lambda$	AB	$\Delta\lambda$	n	$1/\lambda$	AB	$\Delta\lambda$	n	$1/\lambda$	AB	$\Delta\lambda$	n
					3C 48						
1.418....	15.62	50	1	1.778....	16.21	50	2	2.235....	16.40	50	3
1.439....	16.03	50	1	1.794....	16.29	50	2	2.260....	16.49	50	3
1.449....	15.23	50	1	1.810....	16.24	50	2	2.286....	16.47	50	3
1.460....	14.11	50	1	1.827....	16.29	50	2	2.312....	16.53	50	3
1.471....	14.59	50	1	1.844....	16.21	50	2	2.339....	16.55	50	3
1.481....	15.76	50	1	1.861....	16.21	50	2	2.367....	16.65	50	3
1.493....	14.96	50	1	1.878....	16.27	50	2	2.395....	16.74	50	3
1.504....	15.10	50	1	1.896....	16.23	50	2	2.424....	16.69	50	3
1.515....	15.77	50	1	1.914....	16.31	50	2	2.454....	16.61	50	3
1.527....	15.72	50	1	1.933....	16.37	50	2	2.484....	16.62	50	3
1.538....	15.19	50	1	1.952....	16.20	50	2	2.516....	16.59	50	3
1.550....	15.29	50	1	1.971....	16.24	50	2	2.548....	16.62	50	3
1.562....	15.59	50	1	1.990....	16.41	50	2	2.581....	16.58	50	3
1.575....	16.12	50	1	2.010....	16.38	50	2	2.614....	16.31	50	3
1.587....	15.95	50	1	2.031....	16.39	50	2	2.649....	16.63	50	2
1.660....	16.09	50	1	2.052....	16.41	50	2	2.685....	16.69	50	2
1.674....	16.02	50	1	2.073....	16.42	50	2	2.721....	16.78	50	2
1.688....	16.13	50	1	2.094....	16.32	50	2	2.759....	16.86	50	2
1.702....	16.13	50	1	2.116....	16.30	50	2	2.797....	16.85	50	2
1.717....	16.02	50	2	2.139....	16.22	50	2	2.837....	16.90	50	2
1.732....	16.17	50	2	2.162....	16.39	50	2	2.878....	16.85	50	2
1.747....	16.14	50	2	2.186....	16.37	50	3	2.920....	16.86	50	2
1.762....	16.21	50	2	2.210....	16.44	50	3				
					3C 286						
1.091....	16.25	200	2	1.900....	17.09	200	2	2.750....	17.23	100	5
1.113....	16.74	200	2	1.900....	17.08	200	5	2.850....	17.09	100	5
1.136....	17.25	200	2	2.000....	17.20	200	5	2.950....	17.31	100	5
1.195....	17.09	200	2	2.090....	17.23	200	5	3.030....	17.30	100	5
1.285....	17.14	200	2	2.190....	17.18	100	5	2.794....	17.34	50	8
1.473....	17.18	200	2	2.240....	17.22	100	5	2.834....	17.11	50	8
1.651....	17.11	200	2	2.350....	17.31	100	5	2.875....	17.27	50	8
1.712....	17.24	200	2	2.400....	17.38	100	5	1.916....	17.19	50	3
1.712....	17.03	200	5	2.478....	17.38	100	5	1.934....	16.91	50	3
1.800....	17.21	200	2	2.589....	17.24	100	5	1.953....	17.21	50	3
1.818....	17.15	200	5	2.700....	17.26	100	5				
					3C 245						
1.712....	16.95	100	2	2.240....	17.70	50	2	2.750....	17.57	50	2
1.818....	17.17	100	2	2.350....	17.58	50	2	2.800....	17.68	50	1
1.900....	17.17	100	2	2.400....	17.56	50	2	2.850....	17.53	50	2
2.000....	17.22	100	2	2.478....	17.73	50	2	2.900....	17.62	50	1
2.090....	17.28	100	2	2.589....	17.34	50	2	2.950....	17.53	50	2
2.190....	17.63	50	2	2.700....	17.65	50	2	3.030....	17.35	50	1

TABLE 2—*Continued*

1/λ	AB	Δλ	n	1/λ	AB	Δλ	n	1/λ	AB	Δλ	n
					CTA 102						
0.980....	16.23	200	2	1.809....	17.28	200	4	2.478....	·17.95	100	4
1.148....	16.83	200	2	1.835....	17.36	200	1	2.545....	17.66	100	1
1.195....	16.55	200	2	1.900....	17.44	200	4	2.581....	17.56	100	4
1.285....	17.05	200	2	2.000....	17.49	200	4	2.610....	17.58	100	1
1.325....	17.15	200	3	2.090....	17.57	200	4	2.700....	17.81	100	4
1.408....	17.09	200	3	2.190....	17.78	100	4	2.763....	17.88	100	4
1.554....	17.21	200	3	2.242....	17.77	100	4	2.853....	17.94	100	4
1.651....	17.22	200	3	2.294....	17.78	100	1	2.950....	17.84	100	4
1.712....	17.24	200	4	2.350....	17.78	100	4	3.030....	17.81	100	4
1.754....	17.14	200	4	2.400....	17.88	100	4				
					3C 446						
1.695....	17.83	200	2	2.410....	18.62	100	2	2.797....	18.52	50	1
1.754....	18.03	200	2	2.469....	18.48	100	2	2.817....	18.70	50	1
1.818....	18.18	200	2	2.532....	18.30	100	2	2.837....	18.68	50	1
1.887....	18.23	200	2	2.516....	18.29	50	1	2.878....	18.69	50	1
1.961....	18.37	200	2	2.548....	18.35	50	1	2.899....	18.41	50	1
2.020....	18.28	200	2	2.581....	18.28	50	1	2.920....	18.87	50	1
2.041....	18.37	200	2	2.597....	18.44	50	1	1.639....	16.86	100	2
2.062....	18.46	200	2	2.614....	18.15	50	1	1.667....	17.89	100	2
2.105....	18.37	100	2	2.649....	17.79	50	1	1.695....	17.79	100	2
2.151....	18.43	100	2	2.667....	17.77	50	1	1.724....	17.86	100	2
2.198....	18.31	100	2	2.685....	17.40	50	1	1.754....	17.91	100	2
2.247....	18.44	100	2	2.721....	18.12	50	1	1.786....	18.02	100	2
2.299....	18.52	100	2	2.740....	18.41	50	1				
2.353....	18.57	100	2	2.759....	18.81	50	1				
					3C 9						
1.712....	17.93	200	2	2.478....	18.72	100	3	2.676....	17.99	25	1
1.739....	18.13	200	2	2.589....	18.65	100	3	2.694....	17.61	25	1
1.800....	18.05	200	2	2.655....	17.99	100	3	2.712....	17.38	25	1
1.900....	18.17	200	2	2.728....	17.63	100	3	2.731....	17.23	25	1
2.000....	18.08	200	2	2.804....	18.61	100	3	2.750....	17.74	25	1
2.090....	18.09	200	2	2.885....	18.85	100	3	2.769....	18.06	25	1
2.098....	18.19	100	3	2.971....	19.08	100	3	2.788....	18.14	25	1
2.143....	17.70	100	3	2.589....	18.26	25	1	2.807....	18.85	25	1
2.190....	18.34	100	3	2.606....	18.28	25	1	2.827....	18.91	25	1
2.242....	18.48	100	3	2.623....	18.54	25	1	2.847....	18.67	25	1
2.350....	18.40	100	3	2.641....	18.02	25	1				
2.100....	18.57	100	3	2.658....	17.99	25	1				

the red-shifts were already known from the work of Schmidt (1965) and Greenstein and Matthews (1963) and the emphasis, in this study, was on the continuous spectrum. The only exceptions are 3C 286, where a special set of observations was made to find a second emission line so that the redshift could be determined (Oke 1965b), and 3C 9 where an attempt was made to find the λ 1909 line of C III]. It is apparent, however, that the present observations do show a number of emission lines. When infrared observations were being made, an effort was made to verify the redshifts given by Schmidt by searching for expected strong emission lines such as Hβ and λ 5007 of [O III]. In those cases where a reasonable measure of the emission-line strength could be made, this has been done and the equivalent widths, in terms of the continuum, are given in Table 3. The accuracy is moderately high when the observations give an indication of the line profile, but may be underestimates when only one measurement centered on the line has been made. Using data below for the continuum, these equivalent widths have been converted to absolute energies j ergs sec^{-1} cm^{-2} at Earth and absolute emitted energies J ergs sec^{-1}. The emission lines are discussed in more detail below.

3C 48: A series of observations in the far red (tabulated but not plotted) show the [Ar IV] lines λ 4711 plus λ 4740, Hβ, and λ 4959 plus λ 5007 of [O III]. The ratio of [O III] to Hβ is 2.4 while Hβ and [Ar IV] have comparable strengths equivalent to 110 Å of continuum. In addition to the easily recognized emission lines in the blue, there are two broad regions of emission centered near 1/λ = 2.20 and 2.60. These may be unresolved emission lines which could include lines of He I, [Ar IV], [Ar V], O III, [Na VI], the [Ne V] lines λ 3345 and λ 3425, and λ 3203 of He II.

3C 286: In addition to the previously detected

lines λ 1909 of C III] (Schmidt 1962) and λ 2798 of Mg II (Oke 1965b), measurements in the infrared show very high fluxes in the positions of Hβ and λ 4959 plus λ 5007 of [O III]. Since no point has been measured to the red of these two wavelengths, it is conceivable but unlikely that the increase in flux observed is due to an abrupt rise in the continuum rather than to the emission lines.

3C 245: The λ 1909 line found by Schmidt (1965) has been observed but no measurement at the position of the other observed line λ 2798 has been made.

CTA 102: Both λ 1909 and λ 2798 found by Schmidt are observed and are very intense and very broad. The half-widths correspond to velocities of approximately 10000 km/sec. Infrared measures show high fluxes in the positions of Hδ and λ 5007 of [O III], and if these lines are also broad the equivalent widths in Table 3 are too small by a factor of 2. No measures have been made at Hβ and Hγ.

3C 446: No redshift has been determined with certainty. There is a very strong and broad line at $1/\lambda = 2.68$, some evidence for a weaker line at $1/\lambda = 2.53$, and possibly a weak line at $1/\lambda = 2.20$. Extra measurements near $1/\lambda = 1.7$ show an

TABLE 3

EMISSION-LINE INTENSITIES

Object and Line (λ_0)	E.W. (Å)	$j \times 10^{15}$ (ergs cm^{-2} sec^{-1})	$J \times 10^{-42}$ (ergs sec^{-1})
3C 48:			
2798	15	23	3.3
4861	110	120	17
5007	264	280	40
3C 286:			
1909	10	11	8
2798	15	8	6
4861	86	17	13
5007	250	47	36
3C 245:			
1909	18	12	13
CTA 102:			
1909	82	42	48
2798	112	43	50
4101	98*	26*	30*
5007	144*	30*	35*
3C 446:			
1550	230	63	130
1640(?)	50	13	26
1909	20	4.4	9
3C 9:			
1216	460	116	500
1550	98	24	100

* Probably underestimated by a factor of 2.

TABLE 4

CONTINUUM ABSOLUTE FLUXES

$1/\lambda$	$\log f_\nu$	$1/\lambda_0$	$\log F_{\nu_0}$	$1/\lambda$	$\log f_\nu$	$1/\lambda_0$	$\log F_{\nu_0}$	$1/\lambda$	$\log f_\nu$	$1/\lambda_0$	$\log F_{\nu_0}$
3C 48				3C 286				CTA 102			
1.6	−25.836	2.187	30.187	1.2	−26.272	2.217	30.344	1.2	−26.212	2.444	30.541
1.8	−25.912	2.461	30.111	1.4	−26.284	2.587	30.336	1.4	−26.272	2.852	30.481
2.0	−25.980	2.734	30.043	1.6	−26.292	2.957	30.328	1.6	−26.332	3.259	30.421
2.2	−26.044	3.007	29.979	1.8	−26.300	3.326	30.320	1.8	−26.392	3.666	30.361
2.4	−26.092	3.281	29.931	2.0	−26.308	3.696	30.312	2.0	−26.452	4.074	30.301
2.6	−26.136	3.554	29.887	2.2	−26.316	4.066	30.304	2.2	−26.508	4.481	30.245
2.8	−26.168	3.827	29.855	2.4	−26.324	4.435	30.296	2.4	−26.572	4.889	30.181
2.9	−26.180	3.964	29.843	2.6	−26.332	4.805	30.288	2.6	−26.588	5.296	30.165
				2.8	−26.336	5.174	30.284	2.8	−26.584	5.703	30.169
				3.0	−26.344	5.544	30.276	3.0	−26.552	6.111	30.201
3C 245				3C 446				3C 9			
1.7	−26.220	3.449	30.530	1.7	−26.644	4.08	30.298	1.7	−26.612	5.120	30.545
1.8	−26.264	3.652	30.486	1.8	−26.680	4.32	30.262	1.8	−26.644	5.421	30.513
2.0	−26.352	4.058	30.398	2.0	−26.748	4.80	30.194	2.0	−26.708	6.024	30.449
2.2	−26.428	4.464	30.322	2.2	−26.804	5.28	30.138	2.2	−26.776	6.626	30.381
2.4	−26.480	4.869	30.270	2.4	−26.856	5.76	30.086	2.4	−26.844	7.229	30.313
2.6	−26.492	5.275	30.258	2.6	−26.884	6.24	30.058	2.6	−26.908	7.831	30.249
2.8	−26.460	5.681	30.290	2.8	−26.904	6.72	30.038	2.8	−26.972	8.433	30.185
3.0	−26.420	6.087	30.330	3.0	−26.908	7.20	30.034	3.0	−27.040	9.036	30.117

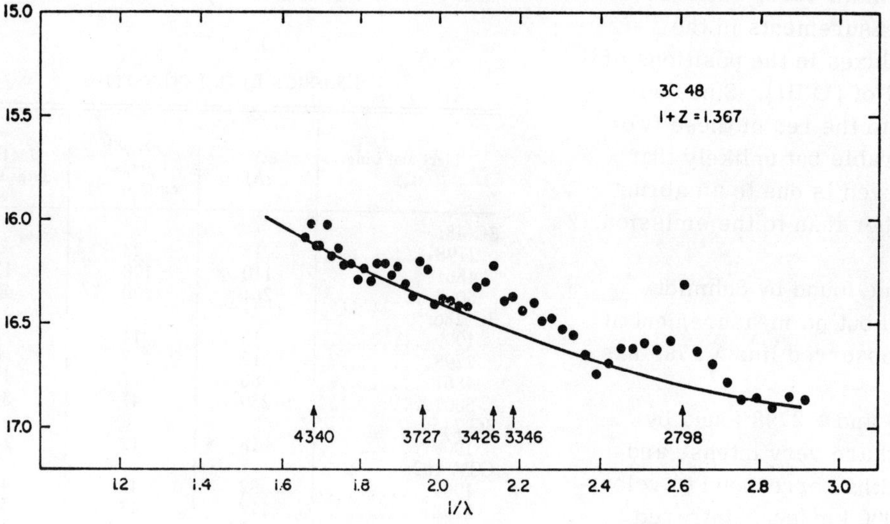

FIG. 1. Observed absolute fluxes, -2.5 log f_ν -48.55, plotted against 1/λ, where λ is the wavelength in microns, for 3C 48. The red-shifted locations of various emission lines are indicated.

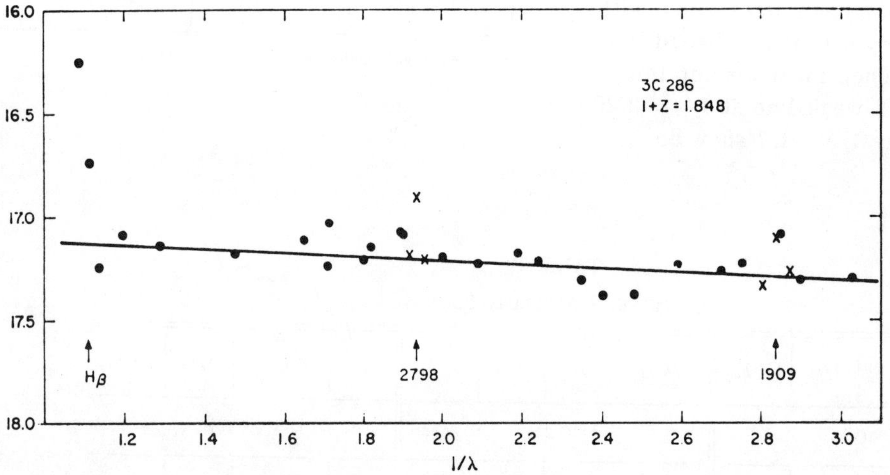

FIG. 2. —Same as Fig. 1 but for 3C 286. Crosses indicate special measures made to detect and measure the two emission lines λ 2798 and λ 1909.

increase in the flux toward the red, but no measures far enough to the red have yet been made to determine whether the flux continues to rise or to decrease again. If the strongest line is λ 1550 of C IV, the other two possible lines would be λ 1640 of He II and λ 1909 of C III]. The possible line at 1/λ = 1.7 could be due to λ 2424 of [Ne IV], λ 2470 of [O II], and λ 2506 of [Mg VII]. For this case 1 + z = 2.40. Identification of the strongest line with Ly-α would place λ 1550 at 1/λ = 2.10 where no line is seen, and λ 1909 at 1/λ = 1.71 where a line may exist. The half-width of the strong line is 8000 km/sec. In the present analysis of the continuum 1 + z = 2.40 is adopted. The red-shift of this object has also been discussed by Burbidge (1965).

3C 9: Both Ly-α and λ 1550, identified by Schmidt (1965) are observed. Lyman-α is strong, has a half-width of 9000 km/sec, and falls off more sharply on the violet than on the red side. Measurements at the position of λ 1909 have not revealed a line.

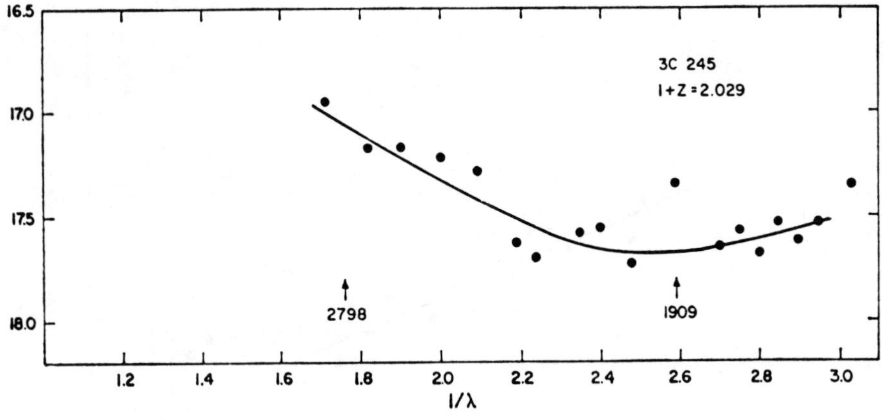

FIG. 3. --Same as Fig. 1 but for 3C 245

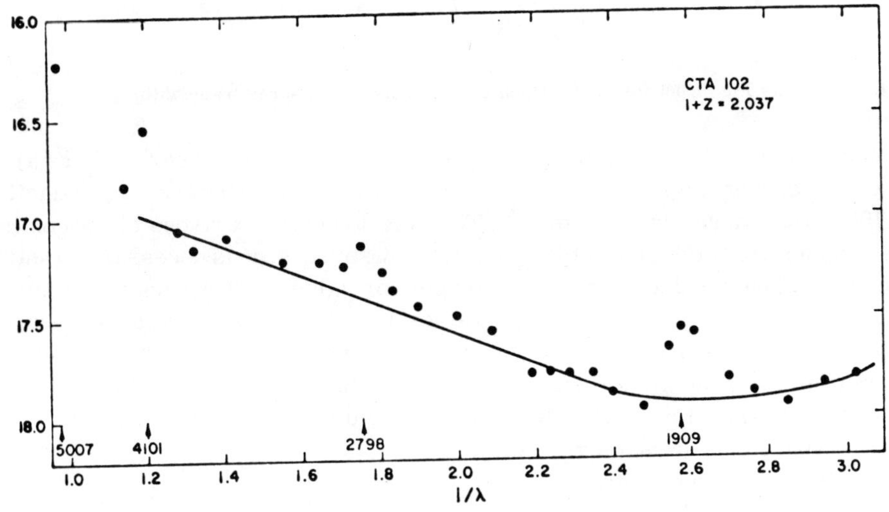

FIG. 4. --Same as Fig. 1 but for CTA 102.

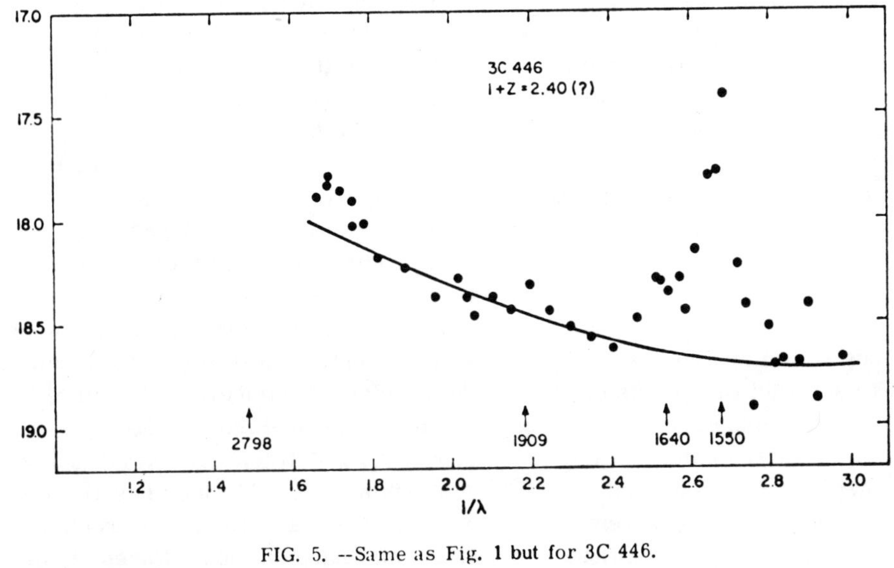

FIG. 5. --Same as Fig. 1 but for 3C 446.

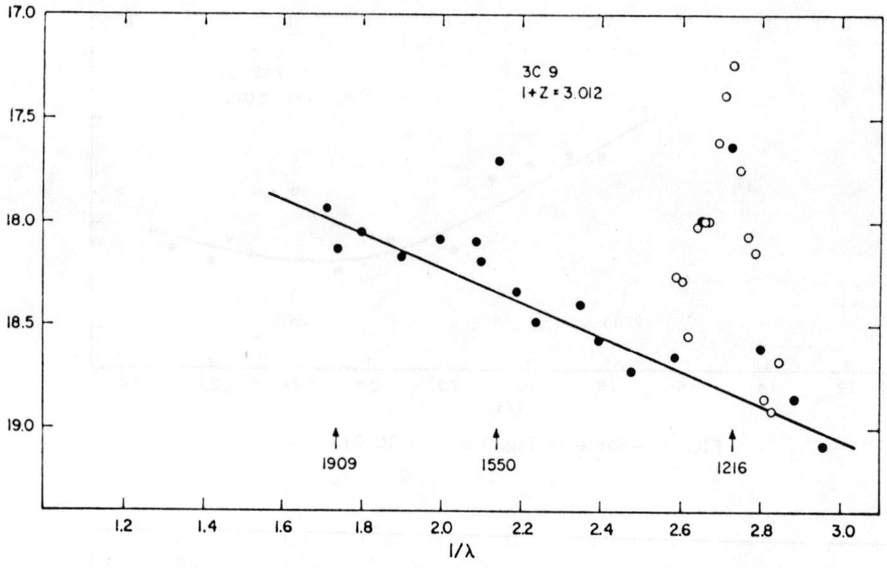

FIG. 6. --Same as Fig. 1 but for 3C 9. Open circles denote measures made with a 25 Å band pass.

It has been suggested on the basis of microphotometer tracings of photographic spectra (Gunn and Peterson 1965) that there is a drop of the continuum as one crosses from the red to the violet side of Ly-α in 3C 9. The absolute flux measurements shown in Figure 6 do not confirm this drop, if the continuum shown is adopted. The absence of a drop in intensity modifies their conclusions concerning the amount of neutral hydrogen in intergalactic space only in making their result an upper limit. Absence of a drop has been interpreted by Bahcall and Salpeter (1965) as indicating a temperature for the intergalactic hydrogen between 2×10^5 ° and 2×10^6 ° K.

There does appear to be an absorption feature between $1/\lambda = 2.4$ and 2.5 in 3C 286 that shows up on all observations made on different nights. It corresponds to rest wavelengths λ_0 between λ 2160 and λ 2260. It could also be a gap between regions of broad, blended emission lines.

The Continuous Spectrum

With the observations shown in Figures 1 - 6 it is clearly possible to subtract the effects of the strong emission lines, leaving only the continuous spectrum, which is the chief contributor to the total observed flux. The curves drawn in the figures are intended to represent these continuum fluxes. It is assumed that the quasi-stellar sources observed here are not reddened since they all have large galactic latitudes. To make direct comparisons of the continua of the different sources, it is necessary to shift all wavelengths or values of $1/\lambda$ back to the laboratory values $1/\lambda_0$. The observed absolute fluxes, f_ν are converted into absolute emitted fluxes, $F_{\nu 0}$, assuming a Hubble constant of 100 (km/sec) Mpc^{-1} and a world model with $q_0 = +1$. The relation is

$$F_{\nu 0} = 1.07 \times 10^{57} f_\nu \frac{z^2}{1+z} \text{ ergs sec}^{-1} \text{ (c/s)}^{-1}. \quad (2)$$

Had we chosen a world model with $q_0 = 0$, there would have been a further factor $(1 + \frac{1}{2}z)^2$ multiplying the right-hand side of the equation. In Table 4 are listed $1/\lambda$, $\log f_\nu$, $1/\lambda_0$, and $\log F\nu_0$; the last two quantities are plotted against each other in Figure 7 for the six sources. In this figure no attempt has been made to match absolute emitted fluxes since these depend on the cosmological model and are, in any case, not constant from one source to another. The curves have been shifted vertically to group objects with similar continuum fluxes and are numbered by the 3C or CTA numbers. The curve for 3C 48 has been extended into the red by using fluxes reported by Matthews and Sandage (1963). The continuum of 3C 273 which has already been discussed by Oke (1965a), is replotted in the diagram. The continuum fluxes, f_ν or F_{ν_0}, have

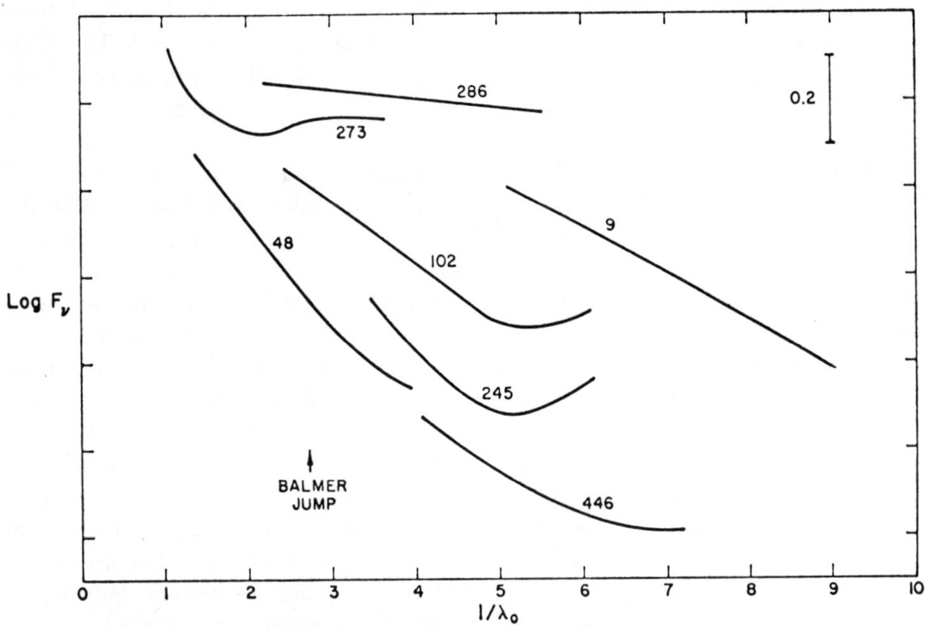

FIG. 7. Continuum absolute emitted fluxes, log $F\nu_0$, plotted against $1/\lambda_0$; λ_0 and ν_0 refer to rest wave lengths and frequencies respectively. The curves have been displaced vertically by arbitrary amounts.

been used to convert the emission-line equivalent widths of Table 3 into absolute values that are also given in the table.

Discussion of the Continuum

There are a number of ways by which a continuous spectrum can be produced, and we will discuss three of these in some detail. A continuous spectrum is produced by the surface layers of stars and since it has been suggested that a quasi-stellar source may have a massive star at the center (Fowler 1964) it is of interest to discuss the possibility. The observed energy distributions correspond to black-body or stellar radiators whose effective temperatures are in the neighborhood of 15,000°K or less. At such temperatures the predominant spectral features are the Balmer jump and Balmer lines provided the chemical composition includes at least a reasonable amount of hydrogen; there is no evidence for a Balmer jump in absorption in any of the quasi-stellar sources or evidence of any hydrogen absorption lines. It could be argued that the hydrogen lines have been filled in with emission, but there are usually other absorption features, such as the H- and K-line of Ca II and the G-band, where this would not happen. The only stars that can produce smooth continua are very hot stars, and the observed energy distributions do not fit such stars. The only possible way of producing an appropriate continuum would be to combine very hot and rather cool stars, and even in this case it is difficult to see how a spectrum could be produced without absorption lines. We will therefore not pursue this possibility further.

Synchrotron radiation is a second source of smooth continuous spectra. The spectrum of a radiating electron has been given by Oort and Walraven (1956). If the distribution of electron energies E in an assembly of electrons is given by $N(E)\,dE = k\,E^{-\beta}dE$, then it can easily be shown (Woltjer 1958) that the radiated flux is given by

$$S_\nu = 1.171 \times 10^{-22} k \nu^{(1-\beta)/2} L^{(\beta-1)/2} H_\perp^{(\beta+1)/2} \int_{\alpha 2}^{\alpha 1} \alpha^{(\beta-3)/2} F(\alpha)\, d\alpha, \qquad (3)$$

where k and L are constants, $H_\perp$ is the magnetic-field component perpendicular to the velocity vector of the electron, β is a constant, and $F(\alpha)$ is a function tabulated by Oort and Walraven. We also have $\alpha = \nu/\nu_c = \nu/LH\,E^2$, and the integral limits α_1 and α_2 correspond to the lower and upper limits E_1 and E_2 of the electron energies in the distribution N(E). It is usually adequate to set $E_1 = 0$ or $\alpha_1 = \infty$. If we treat α_2 and β simply as free parameters, it is possible to produce a wide variety of spectra S_ν. A flat spectrum, S_ν = const., is achieved if $\beta = 1$, so that the $\nu^{(1-\beta)/2}$ term is unity, and by choosing a very small value of α_2 corresponding to a very large energy E_2. For $\beta = 1$, as E_2 is decreased, S_ν becomes steeper with larger fluxes at lower frequencies. For $\beta = 2$, S_ν increases toward lower frequencies even when E_2 is very large. The function S_ν is smooth, and very roughly log S_ν is linear with $1/\lambda_0$ for values of $1/\lambda_0$ between 1 and 10. Any smoothly varying optically observed energy spectrum that is flat or increases in intensity toward lower frequencies can be fitted with a suitable choice $\beta \geq 1$ and some value of E_2 or α_2.

A third source of continuous radiation is bound-free and free-free transitions. These can occur for any constituent element, but the only one of importance if the usual cosmic abundances prevail is hydrogen. The equations for the spectrum have been given in convenient form by Seaton (1960). Computations have been carried out using these formulae and Gaunt factors published by Berger (1956) for temperatures from 8,000° to 1,000,000°K. A slight complication arises since two-photon emission may also be important, and its radiation occurs in just the spectral region which is now obserbed. Calculations of this have also been made. If the density in the gas is greater than 2×10^4 electrons/cm^3, then two-photon emission is not important; evidence from 3C 48 and 3C 273 (Oke 1965a; Greenstein and Schmidt 1964) suggests that such densities do prevail. If we neglect two-photon emission, the radiative flux below the Balmer discontinuity goes approximately as $\exp(-h\nu/kT_e)$ if the electron temperature T_e is constant. In terms of the coordinates used in Fig. 7 the hydrogen-emission spectrum is almost a straight line; at low temperatures the flux decreases toward higher frequencies, while at temperatures greater than 100,000 °K the spectrum becomes almost flat. Preliminary attempts to fit the observed spectra with hydrogen-radiating spectra indicated that two-photon emission should not be included, and in the following discussion it is neglected.

We now discuss the observed fluxes shown in Fig. 7 in terms of the two mechanisms just described. Since both mechanisms produce spectra that are smooth, it is immediately evident that the spectra of 3C 273, CTA 102, 3C 446, and 3C 245 cannot be explained by a bound-free plus free-free hydrogen model with a single electron temperature, or by a synchrotron model where the electron-energy spectrum is a simple law of the form $N(E)dE = k\,E^{-\beta}dE$. It is therefore appropriate to divide the spectra of these objects into two parts. Particular synchrotron-emission cases, where there are two quite different electron-energy distributions, have been discussed by Greenstein (1964). For 3C 273, CTA 102, and 3C 245 we can simply subtract a constant flux F_ν indicated by the moderately flat blue part of the spectrum; this is a questionable procedure for 3C 245. The residual red part of the spectrum of each object has a very steep slope in the observed spectral region. In the case of 3C 446 a spectrum with a slope given by the fluxes from $1/\lambda_0$ = 6-7 can be subtracted. For 3C 286 and 3C 9 there is no evidence that two radiating components are present although the possibility cannot be ruled out. The spectrum of 3C 48 is quite steep, but there is an indication that it is beginning to flatten in the violet.

Consider, first, the observed fluxes for 3C 48 and CTA 102. Below the Balmer jump, marked on Fig. 7, it would be possible to fit 3C 48 and the red component of CTA 102 by hydrogen bound-free and free-free radiation with an electron temperature in the neighborhood of 10,000°K. But at such a temperature there would be a large Balmer jump which is completely excluded by the observation. In these cases we are left with a source such as synchrotron radiation, and Matthews and Sandage (1963) have shown that for 3C 48 a synchrotron spectrum can be fitted accurately. The red component of 3C 273 cannot be fitted with a hydrogen spectrum of any reasonable temperature and can also be attributed to synchrotron radiation (Oke 1965a). Likewise the red components of 3C 446 and 3C 245 are almost certainly of the same character.

We now consider the blue parts of 3C 273, CTA 102, 3C 446, 3C 245, and the whole spectrum of 3C 286. These are characteristically almost flat with the possible exception of 3C 245. It has already been indicated that a synchrotron model described by equation (3) can explain such spectra if $\beta = 1$ and E_2 is taken extremely large. Shklovskii (1965) has presented arguments which indicate that the flat spectra in the visual and ultra-violet may be produced by photons which have been shifted from the radio region by the inverse Compton effect. The fact that the observed flat spectra extend far into the ultraviolet makes this suggestion very attractive. We can also explain these flat spectra by free-free and bound-free transitions of hydrogen as has been proposed for 3C 273 by Oke (1965a). 3C 286 would require T_e = 400,000°K, CTA 102 and 3C 245 would require as high or higher temperatures, and 3C 446 a temperature of approximately 150,000°K. At the present time there does not appear to be any way to decide between the various possibilities, although it might be done by measuring the optical polarization at different wavelengths.

The spectrum of 3C 9 can be explained by either mechanism. If it is produced by a hydrogen gas, the electron temperature would be 50,000°K and the fit of the observed and computed fluxes is excellent. In this case there should be a large Balmer jump at λ 11000. It would be possible but extremely difficult to observe. A simple synchrotron model also can be selected that will produce the correct spectrum.

Discussion of the Emission Lines

The equivalent width of Hβ in 3C 48 is 110 Å. Adopting an electron temperature of approximately 16,000°K (Greenstein and Schmidt 1964), we estimate from computations of Pengelly (1964) and continuum values of γ given by Oke (1965a) that the equivalent width relative to bound-free plus free-free hydrogen radiation should be approximately 2,000 Å. (The ratio given by Oke is lower than this since two-photon emission was also included in the continuum.) On this basis only 5 per cent of the continuous radiation at Hβ is produced by hydrogen. On the violet side of the Balmer jump, however, 30 per cent of the radiation should be from hydrogen and a Balmer jump of 0.3 mag. should be observed. The confluence of the Balmer emission lines above the Balmer jump will wipe out any actual discontinuity, but the excess radiation should be observed. The observed continuum does not appear to contain such a contribution, although it cannot be ruled out since the shape of the non-thermal continuum is unknown. If T_e = 40,000°K, the Balmer jump would be only 0.1 mag., which is in better agreement with the observed continuum fluxes. The strength of Hβ in 3C 286 is nearly the same as that in 3C 273. Since the continua in these objects are similar, the discussion of 3C 273 by Oke (1965a) applies also to 3C 286.

The absolute strength of Ly-α and its strength relative to 1,000 Å of bound-free plus free-free continuum at λ 1216 is given in Table 5 for various electron temperatures and for Case B, assuming that no Ly-α photons are destroyed in the nebula They have been computed from formulae and tables given by Seaton (1959, 1960). $I_{Ly-\alpha}$ is in units of ergs cm^{-3} sec^{-1}. The continuum energy is also given in the table in terms of γ_{1216} computed from formulae given by Seaton (1960).

For 3C 9 the observed equivalent width of Ly-α is 460 Å. If we assume that the continuum is all produced by hydrogen, which as we have seen requires T_e = 50,000°K, the predicted equivalent width is 5,000 Å, or a factor 10 larger than observed. To force agreement with observations an electron temperature of 200,000°K or more would be required, and in this case the continuum would be nearly flat, contrary to observations. We must conclude that either the continuum is mostly non-thermal or most of the Ly-α photons have been destroyed. From an observational point of view there is an indication that Ly-α is asymmetric, and this could be caused by absorption on the violet side of the line. Since the measured position of the line center is correct (Schmidt 1965), it is unlikely

TABLE 5

COMPUTED EMISSION-LINE INTENSITIES

T_e (°K)	$(10^{24} I_{Ly-\alpha})/N_e^2$	$10^{14}\times\gamma_{1216}$	$I_{Ly-\alpha}/I(1000\ Å)$	$10^{21}\times F(T_e)$	$10^{21}\times H(T_e)$
10000	4.23	0.014	2240	0.16	0.15
14000	3.21	0.090	265	1.93	1.09
20000	2.37	0.329	53.3	11.7	4.49
30000	1.66	0.817	15.0	43.5	10.3
40000	1.27	1.22	7.68	82.5	20.7
50000	1.02	1.52	4.97	116	26.7
70000	0.74	1.88	2.90	166	34.8
100000	0.50	2.12	1.75	208	38.6
200000	0.24	2.26	0.81	...	...
400000	0.11	2.11	0.40	...	...

that the strength of Ly-α has been underestimated by as much as a factor of 2.

In the following discussion it will be assumed that photons of both Ly-α and λ 1550 of C IV are not destroyed; this assumption will be justified. The C IV line λ 1550 is probably produced by collisional excitation followed by radiative de-excitation. The probability per second of a collisional excitation is given by

$$D_{12} = \frac{8.63 \times 10^{-6} N_e}{T_e^{1/2}} \frac{\Omega(12)}{g_1} 10^{-E\theta_e} \qquad (4)$$

(Seaton 1960), where $\theta_e = 5040/T_e$, E is the excitation potential of the upper level relative to the lower one, and Ω/g_1 is the collision parameter. The values of Ω/g_1 used in this paper are taken from the work of Osterbrock (1963). If every collision is followed by radiative de-excitation and if the photon produced escapes, the strength of the λ 1550 line is given by

$$I_{1550} = 8.63 \times 10^{-6} \frac{\Omega(12)}{g_1} X_3 \frac{N_e^2 10^{-E\theta_e}}{T_e^{1/2}} \; h\nu \text{ ergs cm}^{-3}\text{ sec}^{-1}$$

$$= 1.66 \times 10^{-16} X_3 \frac{N_e^2 10^{-7.96\theta_e}}{T_e^{1/2}} \equiv X_3 N_e^2 F(T_e) \qquad (5)$$

$X_3 N_e$ is the number density of C^{+++} atoms. For the C III] line λ 1909, similar assumptions yield

$$I_{1909} = 2.69 \times 10^{-17} X_2 \frac{N_e^2 10^{-6.46\theta_e}}{T_e^{1/2}} \equiv X_2 N_e^2 H(T_e). \qquad (6)$$

Both $F(T_e)$ are given in Table 5. For $T_e \geq 20{,}000°$K and assuming $X_2 = X_3$, the ratio $I_{1550}/I_{1909} = 4$, reflecting largely the difference in the collision cross-sections. The fact that the λ 1909 line has not yet been observed in 3C 9 means that its equivalent width is less than 20 Å and as a consequence $X_2 \geq X_3$. Also because of the large ionization potential of C^{+++} (64.2 eV) and the very small transition probability for radiative ionization from the ground level, it is reasonable to assume that a substantial fraction of all carbon atoms are in the form of C^{+++}. Taking the solar ratio for the relative abundance of carbon to hydrogen we find $X_3 = 5 \times 10^{-4}$ and from the line intensities and data in Table 5 the electron temperature is 13,000°K. A still reasonable value of $X_3 = 10^{-5}$ leads to a temperature of 30,000°K. While smaller values of X_3 and larger values of T_e are possible, we adopt an electron temperature of 20,000°K.

We now discuss the question of whether λ 1550 or Ly-α photons are lost before they escape from the nebula. The photons of the resonance line λ 1550 of C IV can be lost by (a) collisional de-excitation from the second level or (b) excitation from the upper level to a still higher level followed by a return to the ground level. For a two-level atom we have

$$N_1(B_{12}u_\nu + D_{12}) = N_2(B_{21}u_\nu + A_{21} + D_{21}), \qquad (7)$$

where N_1 and N_2 are the populations of the lower and upper levels, respectively, B_{12}, B_{21}, and A_{21} are the Einstein probability coefficients, and D_{12} and D_{21} are the collisional probabilities. We will assume for 3C 9 a radius of 1 pc, $N_e = 10^6$, and $T_e = 20{,}000°$K; this is consistent with the observed strength of Ly-α. The total emitted energy from 3C 9 in the ultraviolet is 10^{20} ergs $\text{sec}^{-1}(\text{c/s})^{-1}$. This corresponds to a flux at the surface of 3×10^{-8} erg cm^{-2} $\text{sec}^{-1}(\text{c/s})^{-1}$ and an energy density $u_\nu = 10^{-18}$ erg $\text{cm}^{-3}(\text{c/s})^{-1}$. In the center of the line, u_ν would be only slightly higher, since we need consider only the small volume of nebula which has a small Doppler shift. We also find $B_{12} = 6.0 \times 10^{19}$, $B_{21} = 2.0 \times 10^{19}$, and $A_{21} = 9.6 \times 10^{8}$. We compute D_{12} from equation (4), which gives 2.3×10^{-3}. We can also obtain D_{21} from D_{12}. In equation (7) only the $B_{12}u_\nu$ and A_{21} terms are significant, and we find that $N_2/N_1 = 6 \times 10^{-8}$. We can write the cross-section $a_\nu \approx a/\Delta\nu_D$, where $\Delta\nu_D$ is the appropriate half-width of the a_ν profile and which we choose to be 10^{12} c/s. This gives $a_\nu = 2.6 \times 10^{-14}$. The actual emission lines have much larger half-widths than this, but we will assume that the smaller value is appropriate and that once a photon has traveled a small fraction, say 10 per cent, of the radius of the nebula, largescale Doppler motions such as radial expansion will provide sufficient Doppler shift so that the photon readily escapes from the nebula. For $N_1 = 10$, the optical depth for transitions from the N_2 level is of the order of 10^{-4} and we can assume that further excitation from the upper level is unimportant.

The ratio of radiative to collisional de-excitation is given by

$$\frac{A_{21}}{D_{21}} = \frac{2.91 \times 10^{6} a T_e^{1/2}}{N_e \lambda^2 [\Omega(12)/g_1]}, \qquad (8)$$

where λ is the wavelength and the other factors have been defined. For $N_e = 10^6$ and $T_e = 20{,}000°$K, this ratio is 4×10^{10}. The optical depth at the center of the λ 1550 line for a path distance of 0.1

pc is 8×10^4. The number of scatterings, Q, as a function of the optical depth has been computed by Osterbrock (1962); for this τ, $Q = 10^6$. Since this number is much smaller than the ratio A_{21}/D_{21}, collisional de-excitation is not important. Our assumption that practically all λ 1550 photons eventually escape from the nebula is valid.

We next consider the destruction of Ly-α photons by (a) collisional de-excitation and (b) radiative excitation from the n = 2 level. Two-photon emission has already been discussed and shown to be unimportant. Radiative excitation from the n = 2 level in a nebula optically thick to Lyman radiation will eventually lead to the release of a Ly-α photon and no Ly-α photons are destroyed. As in the case of the λ 1550 line, the ratio A_{21}/D_{21} for Ly-α will be approximately 4×10^{10} which must be compared with the number of scatterings Q. The number of neutral hydrogen atoms can be calculated approximately from the relation.

$$N_1(H) \int_{\nu_0}^{\infty} B_{1f}\, u_\nu d\nu = N_e^2\, a_A, \tag{9}$$

where $N_1(H)$ is the number density of hydrogen atoms in the ground level; $N_e^2\, a_A$, the number of recombinations per second per cubic centimeter onto all levels; and ν_0, the frequency of the Lyman limit. The energy density u_ν was obtained by extrapolating the observed absolute flux shown in Fig. 7 linearly beyond the Lyman limit and allowing for the size of the nebula. Since most ionizations are produced by photons with frequencies just above ν_0, the extrapolation should not be seriously in error. For $N_e = 10^6$ and a radius for the nebula of 1 pc, we find $N_1(H) = 0.2$; $N_1(H)$ goes as $N_e^{2/3}$. The optical depth in the center of the line for a path length of 0.1 pc is roughly 600, and the number of scatterings will be much smaller than the ratio A_{21}/D_{21}. Therefore all Ly-α photons should escape from the nebula. One further relevant calculation is the ratio of the number of quanta capable of ionizing hydrogen to the number of Ly-α quanta; they should be nearly identical. The result is 3.45×10^{55} ionizing photons per second for the nebula. The number of Ly-a photons is 3.0 $\times\ 10^{55}$ per second, and the ratio is practially unity as required.

A few general comments can be made about the emission-line intensities measured in the other quasi-stellar sources. The [O III] lines, λ 4959 + λ 5007, have a range in intensity of at most a factor of 2 in the three objects where it has been measured. The ratio of [O III] to Hβ ranges from the normal value of 3 in planetary nebulae down to 0.25 in 3C 273 (Oke 1965a). The C IV line at λ 1550 has the same intensity in 3C 9 and 3C 446, in spite of the great apparent strength in 3C 446. the strength of λ 1909 of C III has a range of a factor of 6. The ratio of λ 1550/λ 1909, which has a lower limit of 5 in 3C9, is 14 in 3C 446. This suggests that the abundance of C^{++} is less than that of C^{+++} and probably variable due to the difference in ionization. The other line which should be highly variable due to ionization effects is the Mg II line λ 2798. The observed spread is a factor 15. In the two objects CTA 102 and 3C 286 where both λ 2798 and λ 1909 have been measured, the intensity ratio is close to unity even though the absolute intensities are very different in the two objects; if this is generally true it suggests that similar ionization conditions exist for these two ions.

The prime-focus scanner and associated electronics were built with funds provided by the National Aeronautics and Space Administration through contract NsG-426.

REFERENCES

J. N. Bahcall, and E. E. Salpeter, 1965, Ap. J., 142, 1677.

J. M. Berger, 1956, Ap. J., 124, 550.

E. M. Burbidge, 1965, Ap. J., 142, 1674.

A. D. Code, 1960, Stars and Stellar Systems, Vol. 6: Stellar Atmospheres, ed. J. L. Greenstein (Chicago: University of Chicago Press), chap. ii.

W. A. Fowler, 1964, Rev. Mod. Phys., 36, 545.

J. L. Greenstein, 1964, Ap. J., 140, 666.

J. L. Greenstein, and T. A. Matthews, 1963, Nature, 197, 1041.

J. L. Greenstein, and M. Schmidt, 1964, Ap. J.. 140, 1.

J. E. Gunn, and B. A. Peterson, 1965, Ap. J., 142, 1633.

T. A. Matthews, and A. R. Sandage, 1963, Ap. J., 138, 30.

J. B. Oke, 1963, Nature, 197, 1040.

_____, 1964, Ap. J., 140, 689.

_____, 1965a, ibid., 141, 6.

_____, 1965b, ibid., 142, 810.

_____, 1965c, Ann. Rev. Astr. and Ap., Vol. 3, ed. L. Goldberg (Palo Alto, Calif.: Annual Reviews, Inc.), p. 23.

J. H. Oort, and Th. Walraven, 1956, B.A.N., 12, 285.

D. E. Osterbrock, 1962, Ap. J., 135, 195.

_____, 1963, Planet. Space Sci., 11, 621.

R. M. Pengelly, 1964, M.N., 127, 145.

M. Schmidt, 1962, Ap. J., 136, 684.

_____, 1963, Nature, 197, 1040.

_____, 1965, Ap. J., 141, 1295.

M. J. Seaton, 1959, M.N., 119, 81.

_____, 1960, Rept. Progr. Phys., Vol. 23, ed. A. C. Strickland (London: Physical Society), p. 313.

I. S. Shklovskii, 1965, Astr. Zh., 42, 893.

L. Woltjer, 1958, B.A.N., 14, 39.

THE SPECTRUM OF 3C 273

F. L. Low
The National Radio Astronomy Observatory, Green Bank, West Virgina

H. L. Johnson
Lunar and Planetary Laboratory, University of Arizona, Tucson, Arizona

Because of the great interest in the quasi-stellar radio sources (quasars), we have continued our efforts to measure the brightness of 3C 273 in the infrared spectral region. Further observations at 2.2 μ (magnitude K) have confirmed the original measures (Johnson 1964a), and we report here on measures of this object made through the 7.5-14-μ atmospheric window.

Our first effort to measure 3C 273 at 10μ was made in June, 1964, with the 28-inch telescope of the Catalina Station. We obtained a positive result, with a rather large probable error. More recently (December, 1964), using the 82-inch telescope of the McDonald Observatory, we have confirmed the 28-inch result. The observational data, on the N-magnitude system (Low and Johnson 1964) are:

28-inch telescope; N = +2.6 ± 0.5 (p.e.) mag.,

82-inch telescope; N = +2.4 ± 0.4 (p.e.) mag.

The multicolor data which we have so far obtained for 3C 273 are listed in Table 1; these are mostly new data not previously published. The absolute fluxes given in Table 2 were computed by the same procedure applied before (Johnson 1964a), which requires the assumption that the solar spectral energy distribution is similar to that of other stars which have been observed. The solar colors deduced by Johnson (1964b) were used with the exception of an improved value of +1.50 for V - N, based on additional unpublished data. Taking V for the Sun to be -26.73, and using the known absolute solar spectrum (Allen 1963; Saiedy 1960), we then compute the flux density for a zero magnitude star at each wavelength. We find that, for a star for which N = 0.00, the monochromatic flux density at the effective wavelength of our filter is 1.87×10^{-16} watt/cm^2/μ. From the measured spectral response of our system we get 9.25 μ as the effective wavelength and 2.4 μ as the effective width for stars, whereas for 3C 273 we get 10.4 μ for the effective wavelength and 1.1×10^{13} c/s for the effective band width. We are, therefore, able to convert the relative measures of 3C 273 on the N-magnitude system to absolute flux densities at 10.4 μ. An independent procedure would be to start with the flux

Table I. The observational data.

	U-V	*B-V*	*V-R*	*V-I*	*V-K*	*V-N*
3C 273	-0.68	+0.20	+0.24	+0.74	+3.60	+10.4

Table II. The absolute fluxes.

	U	*B*	*V*	*R*	*I*	*K*	*N*
3C 273	2.3	2.6	2.8	2.4	2.8	12.7	420×10^{-28} watt/m^2/(c/s)

Reprinted from the Astro Physical Journal 141, 336 (1965).

value 2.0×10^{-16} watt/cm^2/μ for a N = 0.00 star derived from the absolute measures of Wildey and Murray (1964). Although this flux is not much different from our calculated value of 1.87×10^{-16} watt/cm^2/μ, their effective wavelength is appreciably larger than ours and would lead to a higher flux for 3C 273.

From our data, and the radio lunar occultation data (Hazard, Mackey, and Shimmins 1963; von Hoerner 1964), we plot in Fig. 1 the spectrum of 3C 273. Curve A applies to the invisible radio source at the end of the jet; curve B, to the star-like object; and curve C, which fits the BVR data for 3C 273, is the observed spectral energy distribution of an A7 V star (Johnson 1964b). The error bars for the K and N points do not designate probable errors; instead, they represent the 90 per cent confidence intervals (Elandt 1961). This means that the probability is 90 per cent that the true values lies within the intervals designated. The limits of the 90 per cent confidence intervals are for N, 0.8 and 7.6×10^{-26} watt/m^2/(c/s); and for K, 0.95 and 1.58×10^{-27} watt/m^2/(c/s).

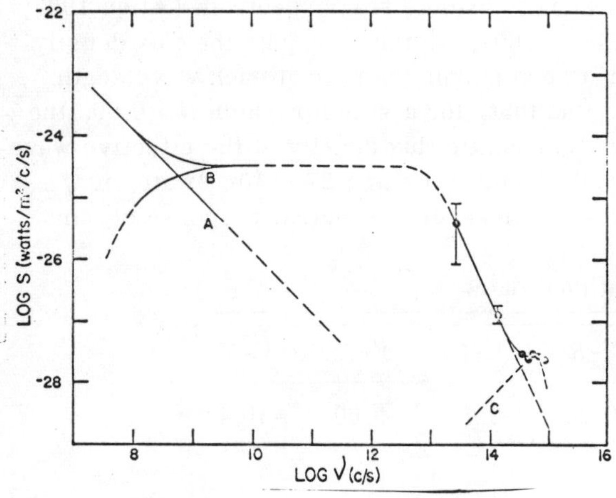

FIG. 1. The spectrum of 3C 273

3C273 is the first radio source for which there are sufficient optical, infrared, and radio data to construct a spectrum of the type shown in Fig. 1. It is apparent from Fig. 1 that the spectrum consists of at least three parts: the two well-known radio components A and B and an optical component superimposed on the high frequency cutoff of B. The dashed line extension of curve B at high frequencies can be subtracted from the total to yield values for the optical spectrum. The curve C is included only for reference and is not intended to imply that the optical component is necessarily of stellar origin. This would still leave the ultraviolet excess to be explained as yet a fourth component.

The N measures were made with a dual-beam photometer using a cooled interference filter and a low-temperature germanium bolometer (Low 1961) operated at 2° K. The estimated signal on the detector from the 28-inch telescope was only 3×10^{-14} watt.

We gratefully acknowledge the able assistance of Mr. D. Steinmetz in taking much of the data. This research was supported in part by the Office of Naval Research and the National Science Foundation.

REFERENCES

C. W. Allen, 1963, Astrophysical Quantities (London: Athlone Press).

R. C. Elandt, 1961, Method of Least Squares and Principles of the Theory of Observations (London: Pergamon Press), p. 95.

C. Hazard, M. B. Mackey, and A. J. Shimmins, 1963, Nature, 197, 1037.

S. von Hoerner, 1964, private communication.

H. L. Johnson, 1964a, Ap. J., 139, 1022.

__________. 1964b, Bull. Tonantzintla and Tacubaya Obs., 3, 305.

F. J. Low, 1961, J. Opt. Soc. Amer., 51, 1300.

F. J. Low, and H. L. Johnson, 1964, Ap. J., 139, 1130.

F. Saiedy, 1960, M. N., 121, 483.

R. L. Wildey, and B. C. Murray, 1964, Ap. J., 139, 435.

THE CORRELATION OF COLORS WITH REDSHIFTS FOR QSS LEADING TO A SMOOTHED MEAN ENERGY DISTRIBUTION AND NEW VALUES FOR THE *K*-CORRECTION

ALLAN SANDAGE
Mount Wilson and Palomar Observatories, Carnegie Institution of Washington,
California Institute of Technology

ABSTRACT

Following earlier work by McCrea, Kardashev and Komberg, and Barnes, the colors of forty-three quasi-stellar radio sources (QSS) are found to correlate statistically with redshift. The empirical functions $B - V = f(z)$ and $U - B = g(z)$ permit an energy distribution function $F(\lambda)$ to be derived such that the redshifting of this distribution through the U, B, and V measuring bands reproduces the correlations. This function is the monochromatic flux distribution smoothed by the broad bands of the measuring system. Relative to the distribution $F(\lambda) \propto \lambda^{-1}$, a depression exists near $\lambda_0 = 2100$ Å and a peak occurs near $\lambda_0 = 2800$ Å. These features produce the statistical color change with redshift.

The K-terms to be applied to measured magnitudes for the effects of redshift are relatively small but the color corrections, $K_U - K_B$ and $K_B - K_V$, vary systematically with $\Delta\lambda/\lambda_0$ as required. The redshift varies across the $U - B$ and $B - V$ diagrams in a regular fashion—a circumstance which permits statistical predictions of the redshift using $U - B$ and $B - V$ values alone.

Previously unpublished two-color photoelectric measurements for twenty-two QSS candidates are given in the Appendix.

I. INTRODUCTION

Three papers have recently appeared concerning the intrinsic color properties of quasi-stellar sources (QSS). McCrea (1966) has shown that the sources can be classified by the sign of the curvature of segments of the log $F(\nu)$ energy distribution as derived from UBV measurements, and further that this classification divides the QSS into different redshift groups. Objects whose spectra are concave downward have small redshifts, while those that are concave upward have larger redshifts. He suggested that this property might be useful as a rough predictor of the redshift.

Kardashev and Komberg (1966) independently discovered the same phenomenon. They showed the relation quantitatively by plotting an index of the curvature against redshift for sixteen QSS available in March, 1966. A remarkable systematic relation was found, and these authors suggested that the cause is an intrinsic energy distribution with discontinuities being shifted through the U, B, and V measuring bands.

Barnes (1966) independently discovered a segment of the same phenomenon, but came upon it in a different way. He plotted the observed $B - V$ and $U - B$ colors against redshift and found a correlation with $U - B$, but not with $B - V$. However, the scarcity of the data masked the $B - V$ correlation (which indeed does exist), and Barnes could not discuss the cause of the correlation because of the inadequacy of the $B - V$ material.

Of the three papers, only Kardashev and Komberg's emphasized (1) that the shape of $F(\nu)$ governs the correlations, and (2) that information on this function can be obtained from the observational data. Maarten Schmidt reached the same conclusions from McCrea's data as discussed at the Miami conference on cosmology in December, 1965.

These highly suggestive papers have stimulated the present extension of the results. Redshift and color data are now available for forty-three sources.[1] The correlations have been repeated, and the consequences concerning the energy distribution function and the $B - V$, $U - B$ diagram are discussed.

[1] The closing date for data was April 13, 1966.

Reprinted from the Astrophysical Journal, Volume 146.

II. CORRELATIONS OF COLOR WITH REDSHIFT

The available data for thirty-seven of the forty-three sources are given in Table 1, ordered by redshift value and divided into five groups for later analysis in § V. Data for the other six are not listed because the redshifts were unpublished at the time of writing, but they are plotted in the diagrams. I am indebted to Drs. C. R. Lynds and G. R. Burbidge for the communication of the results for these six objects. The tabulated redshifts are taken either from the literature or from preprints and are due principally to four research groups: M. Schmidt, E. M. Burbidge, C. R. Lynds, and W. K. Ford and Vera Rubin. The color data are entirely photoelectric and are summarized elsewhere (Sandage 1965, Table 2), except for seven sources that are given in the Appendix.

Following Kardashev and Komberg, the tabulated functions α_1 and α_2 are the spectral

TABLE 1

REDSHIFTS, COLORS, AND SPECTRAL INDICES FOR THIRTY-FIVE QSS AND TWO QSG

Object	z	$B-V$	$U-B$	α_1	α_2	$\alpha_1-\alpha_2$
3C 273..........	0.158	+0.21	−0.85	−0.48	−0.28	−0.20
1217+02........	0.240	− .02	−0.87	+0.46	−0.19	+0.65
3C 249.1........	0.311	− .02	−0.77	+0.46	−0.65	+1.11
3C 277.1........	0.320	− .17	−0.78	+1.07	−0.60	+1.67
3C 48...........	0.367	+ .42	−0.58	−1.33	−1.51	+0.18
3C 351..........	0.371	+ .13	−0.75	−0.15	−0.74	+0.59
3C 47...........	0.425	+ .05	−0.65	+0.18	−1.19	+1.37
3C 279..........	0.536	+ .26	−0.56	−0.68	−1.60	+0.92
3C 147..........	0.545	+ .35	−0.59	−1.05	−1.46	+0.41
3C 334..........	0.555	+ .12	−0.59	−0.11	−1.46	+1.35
3C 345..........	0.594	+ .29	−0.50	−0.80	−1.87	+1.07
MSH 03−19....	0.614	+ .11	−0.65	−0.07	−1.19	+1.12
3C 263..........	0.652	+ .18	−0.56	−0.35	−1.60	+1.25
3C 380..........	0.691	+ .24	−0.59	−0.60	−1.46	+0.86
3C 254..........	0.734	+ .15	−0.49	−0.23	−1.92	+1.69
3C 286..........	0.846	+ .22	−0.84	−0.52	−0.33	−0.19
1252+11........	0.870	+ .35	−0.75	−1.05	−0.74	−0.31
3C 196..........	0.872	+ .60	−0.46	−2.07	−2.06	−0.01
0922+14........	0.900	+ .54	−0.52	−1.82	−1.78	−0.04
0957+00........	0.908	+ .47	−0.71	−1.54	−0.92	−0.62
MSH 14−121....	0.942	+ .44	−0.76	−1.41	−0.69	−0.72
3C 245..........	1.029	+ .45	−0.83	−1.46	−0.37	−1.09
CTA 102........	1.038	+ .42	−0.79	−1.33	−0.55	−0.78
3C 287..........	1.054	+ .63	−0.65	−2.19	−1.19	−1.00
3C 204..........	1.112	+ .55	−0.99	−1.86	+0.35	−2.21
3C 208..........	1.112	+ .34	−1.00	−1.01	+0.40	−1.41
BSO 1..........	1.241	+ .31	−0.78	−0.88	−0.60	−0.28
3C 181..........	1.382	+ .43	−1.02	−1.37	+0.49	−1.86
3C 446..........	1.403	+ .44	−0.90	−1.41	−0.05	−1.36
3C 270.1........	1.519	+ .19	−0.61	−0.39	−1.37	+0.98
3C 454..........	1.756	+ .12	−0.95	−0.11	+0.17	−0.28
3C 432..........	1.804	+ .22	−0.79	−0.52	−0.55	+0.03
PHL 938........	1.93	+ .32	−0.88	−0.93	−0.15	−0.78
3C 191..........	1.946	+ .25	−0.84	−0.64	−0.33	+0.31
3C 9............	2.012	+ .24	−0.76	−0.60	−0.69	+0.09
0106+01........	2.107	+ .15	−0.70	−0.23	−0.96	+0.73
1116+12........	2.118	+0.14	−0.76	−0.19	−0.69	+0.50

indices between the effective frequencies of the V and the B bands for α_1, and between the B and the U bands for α_2, i.e.,

$$\alpha \equiv \frac{\Delta \log F(\nu)}{\Delta \log \nu}. \quad (1)$$

This is the same definition used by radio astronomers.

Using the transformations that convert V, B, and U magnitudes into absolute fluxes (Matthews and Sandage 1963), it is easily shown that these optical indices are related to the colors by

$$\alpha_1 = 0.38 - 4.08\,(B - V)\,, \quad (2)$$

$$\alpha_2 = -4.14 - 4.54\,(U - B)\,, \quad (3)$$

where we have adopted the effective frequencies $\log \bar{\nu} = 14.735$ for V, $\log \bar{\nu} = 14.833$ for B, and $\log \bar{\nu} = 14.921$ for U.

Figure 1 shows the correlation between redshift and the colors obtained from Table 1. The solid lines represent the best estimate of the correlations as drawn by eye from 31 points which were available at the start of the investigation. The other 12 points were added after the calculations were completed, and indeed these later points serve as a confirmation of the systematic trends found from the first 31 values. The dotted curves, discussed later, represent a prediction from an energy distribution curve derived from the entire material.

Although there is considerable scatter in Figure 1, it is clear that most of the QSS do follow a systematic trend of color with redshift. The effect is almost certainly real because the shape of the correlation is similar in $B - V$ and in $U - B$, and the minimum in $B - V$ is shifted along the abscissa by a factor of 1.25 in $1 + \Delta\lambda/\lambda_0$. This was the first clue that a common energy distribution curve with a peak and a valley could explain the data because this factor is just the value required from the ratio of the effective wavelengths of $\bar{\lambda}_V = 5515$ Å, $\bar{\lambda}_B = 4408$ Å, and $\bar{\lambda}_U = 3593$ Å, giving $\bar{\lambda}_V/\bar{\lambda}_B = 1.25$ and $\bar{\lambda}_B/\bar{\lambda}_U = 1.23$.

The correlation is even more striking in Figure 2, which follows Kardashev and Komberg's procedure of plotting the curvature index $\alpha_1 - \alpha_2$ against redshift. Again, the solid curve was drawn as the best empirical fit to the first 31 points, and the dotted curve is from a reverse calculation from an adopted energy distribution curve now to be discussed.

III. THE ENERGY DISTRIBUTION CURVE

Figure 1 contains enough information to reconstruct an energy distribution curve which, when redshifted through the band passes of the *UBV* system, will reproduce the observed correlations. This follows because $B - V$ and $U - B$ are related to the first derivatives of the energy distribution function, and the function itself can be found by direct numerical integration. We cannot, of course, recover more information than is implicitly contained in the broad-band colors, and this restricts the accuracy of the reconstructed $F(\lambda)$ because the band widths of the filters are 600 Å for U and 1000 Å for B and V.

What we can do, however, is to obtain the true function as smoothed by the 1000-Å filters. Consistent use of effective wavelengths, both in constructing $F(\lambda)$ and in reversing the calculation to predict values of $B - V$ and $U - B$ is legal as regards the calculated colors, but caution must be used in any astrophysical interpretation of the derived $F(\lambda)$ because the true monochromatic function will be considerably more lumpy than our derived relation. Recognizing this limitation, we proceed as follows.

We first need the absolute calibration of the *UBV* system, and this was adopted from

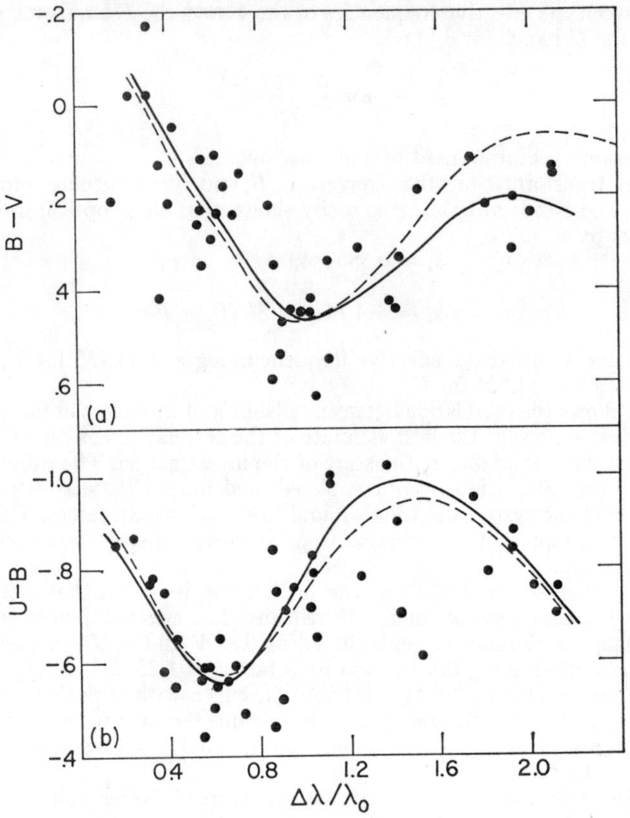

FIG. 1.—The variation of $B - V$ and $U - B$ with redshift for forty-three quasi-stellar sources. The solid line is an empirical fit while the dotted curve is calculated from the flux distribution of Fig. 3.

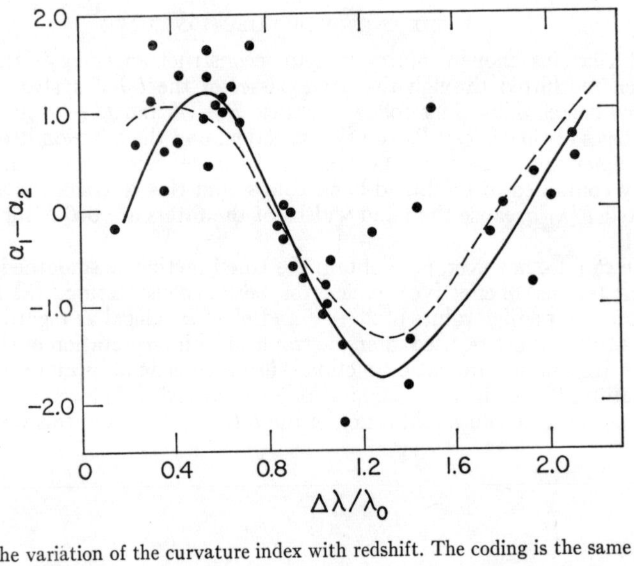

FIG. 2.—The variation of the curvature index with redshift. The coding is the same as in Fig. 1

an earlier study (Matthews and Sandage 1963). The zero points given there require that

$$U - B = 2.5 \log \frac{F_B(\nu)}{F_U(\nu)} - 0.910\,, \tag{4}$$

$$B - V = 2.5 \log \frac{F_V(\nu)}{F_B(\nu)} + 0.091\,, \tag{5}$$

where the fluxes are expressed per unit frequency interval. It was more convenient to work in flux per unit wavelength interval, and for this the constants in equations (4) and (5) must be changed by 2.5 log $(\lambda_i/\lambda_j)^2$, giving

$$U - B = 2.5 \log \frac{F_B(\lambda)}{F_U(\lambda)} - 0.47\,, \tag{6}$$

$$B - V = 2.5 \log \frac{F_V(\lambda)}{F_B(\lambda)} + 0.58\,, \tag{7}$$

as the fundamental equations of the problem. The values of the constants are of vital importance because any errors here will introduce progressively large errors in the derived $F(\lambda)$ function as the integration proceeds. The constants are no better than the absolute calibration of the UBV system which rests ultimately on the absolute energy distribution of Vega and the Sun, and these are believed to be known to better than 10 per cent over most of the wavelength range. We do have a check on the ratio of the constants because equations (6) and (7) each produce an independent $F(\lambda)$ function. If one of the constants is wrong but the other correct, these two determinations would disagree; but, in fact, they do agree to within about 10 per cent.

Adopting effective wavelengths with the values previously quoted, the numerical integration was accomplished as follows.

1. Values of $B - V$ and $U - B$ were read from Figure 1 at various values of the redshift z.

2. The flux ratios corresponding to these colors were calculated from equations (6) and (7).

3. The proper (unshifted) wavelengths corresponding to the chosen redshifts were calculated from

$$\lambda_0 = \frac{\lambda_{\text{eff}}}{1 + z} \tag{8}$$

for U, B, and V. Consequently, for every redshift, the flux ratios and the proper effective wavelengths are known. We need now only adopt an arbitrary starting value for $F(\lambda_0)$ at some selected λ_0 and obtain F at other discrete proper λ-values by an obvious stepwise procedure. A graphical method was used for interpolation of the $F(\lambda)$ in the various built-up segments of the developing distribution-curve. As previously mentioned, two independent integrations were made—one based on equation (6) and the other on equation (7). The two functions differed only slightly, and Figure 3 shows the compromise curve. Table 2 lists the result from $\lambda_0 = 1100$ Å to $\lambda_0 = 5600$ Å.

The extent to which this mean curve reproduces the correlations was tested by applying equations (6) and (7) in reverse to predict the colors. The dotted lines in Figures 1 and 2 and the final two columns of Table 3 show the result. The agreement between observed and calculated values is good except for the slight departure of the $B - V$ curves at large redshifts. This could have been corrected by a second compromise solution for $F(\lambda)$, but a recalculation was not considered worthwhile with the available data.

The dotted curve in Figure 3 is the only distribution which will produce no change of

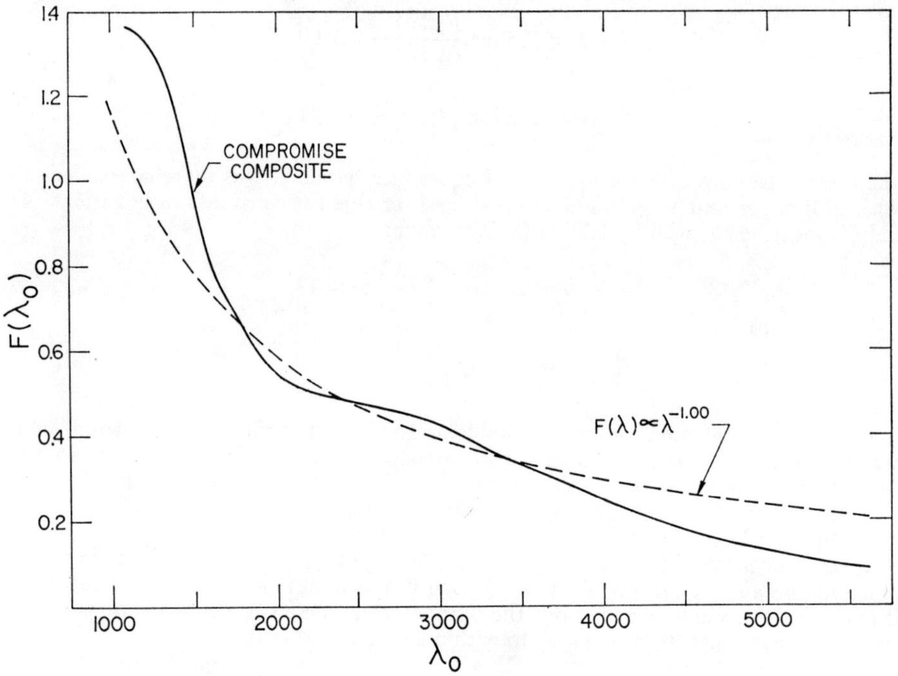

FIG. 3.—The energy distribution per unit wavelength interval obtained as a compromise from a numerical integration of the empirical curves of Figs. 1, *a* and *b*. The dotted distribution is that unique curve which gives no change of *V*, *B*, or *U* magnitude with redshift.

TABLE 2

THE COMPROMISE COMPOSITE ENERGY-CURVE

λ (Å)	$F(\lambda)$	log ν	log $F(\nu)$	λ (Å)	$F(\lambda)$	log ν	log $F(\nu)$
1100..........	1.365	15.436	0.218	2800..........	0.450	15.030	0.548
1200..........	1.345	15.398	.287	3000..........	.422	15.000	.580
1300..........	1.262	15.364	.329	3200..........	.380	14.972	.590
1400..........	1.101	15.330	.334	3400..........	.344	14.946	.600
1500..........	0.960	15.301	.334	3600..........	.311	14.921	.605
1600..........	0.820	15.274	.322	3800..........	.280	14.897	.607
1700..........	0.722	15.246	.319	4000..........	.248	14.875	.599
1800..........	0.648	15.221	.322	4200..........	.218	14.854	.586
1900..........	0.582	15.199	.322	4400..........	.188	14.834	.562
2000..........	0.548	15.176	.341	4600..........	.164	14.814	.540
2100..........	0.519	15.154	.360	4800..........	.148	14.796	.533
2200..........	0.500	15.135	.384	5000..........	.130	14.778	.512
2300..........	0.492	15.115	.418	5200..........	.112	14.761	.482
2400..........	0.482	15.097	.443	5400..........	.100	14.745	.465
2600..........	0.465	15.062	0.497	5600..........	0.088	14.729	0.441

magnitude with redshift and, like all power-law distributions, it produces no change of color with redshift. This is easy to show by considering the K-correction as in § IV. The dip of $F(\lambda)$ below the dotted curve near $\lambda_0 \simeq 2100$ Å and the rise above the curve near $\lambda_0 \simeq 2800$ Å are the features which produce the correlations of Figures 1 and 2 as they are redshifted in and out of the measuring bands. It should again be emphasized that these features must, in reality, have greater contrast relative to the dotted curve than is shown in Figure 3 because of the smoothing action of the broad-band filters.

Figure 4 shows the derived energy distribution expressed per unit frequency interval as found from Figure 3 by multiplying $F(\lambda)$ by λ^2 at each wavelength. The results are given in Table 2. This diagram is in a form where one can understand McCrea's discovery that ordering the QSS by spectral curvature tends to order them by redshift. The following construction is helpful. Cut a rectangular window in a piece of paper (or

TABLE 3

OBSERVED AND CALCULATED CORRELATION OF $B-V$ AND $U-B$ WITH REDSHIFT

z	Observed		Calculated	
	$B-V$	$U-B$	$B-V$	$U-B$
0.0	(−0.20)	(−1.02)	−0.18	−1.06
0.1	(− .14)	(−0.95)	− .11	−0.92
0.2	− .07	−0.82	− .10	−0.85
0.4	+ .06	−0.65	+ .11	−0.66
0.6	+ .22	−0.56	+ .25	−0.58
0.8	+ .37	−0.63	+ .41	−0.62
1.0	+ .46	−0.79	+ .46	−0.75
1.2	+ .43	−0.93	+ .42	−0.87
1.4	+ .36	−0.99	+ .36	−0.99
1.6	+ .28	−0.97	+ .20	−0.93
1.8	+ .22	−0.90	+ .12	−0.90
2.0	+ .20	−0.81	+ .06	−0.79
2.2	+0.24	−0.69	+ .05	−0.66
2.4	....	....	+ .07	....
2.6	....	....	+ .17	....
2.8	....	....	+ .27	....
3.0	....	....	+0.37	....

a card) of width equal to the arrow labeled log ν_U/ν_V in the upper right of Figure 4. Make the height of the rectangle about 3 inches. The width defines the observed frequency interval between the V and U effective frequencies for any QSS, regardless of its redshift. Place the right-hand edge of this window at the appropriate arrow along the abscissa for any desired redshift. The segment of the spectrum which will be observed in the UBV system will now show through the window. As the window is progressively shifted toward larger redshifts, the curvature of the observed segments changes in the same systematic manner as the segments in McCrea's Figure 1. This, then, provides a ready explanation of McCrea's discovery.

An even more striking confirmation of the fact that most QSS statistically follow the log $F(\nu)$ curve of Figure 4 can be seen by plotting each of the thirty-seven objects separately on transparent paper in a log $F(\nu_0)$, log ν_0 diagram, using the three observed optical flux values obtained from V, B, and U measurements and the equations of Matthews and Sandage (1963). Shifting each partial segment along the ordinate of Figure 4 shows the remarkably systematic agreement of most QSS with the adopted distribution function.

IV. THE K-TERM

Figures 3 and 4 now permit the K-correction terms to be computed in a more accurate way than has previously been possible. The definition of K and the procedure have been described elsewhere (Humason, Mayall, and Sandage 1956, Appendix B; Sandage 1965, Appendix). The results for K_V, K_B, and K_U are given in Table 4 as computed from Figure 3 (or Table 2), again using the effective wavelengths $\bar{\lambda}_V = 5515$ Å, $\bar{\lambda}_B = 4408$ Å,

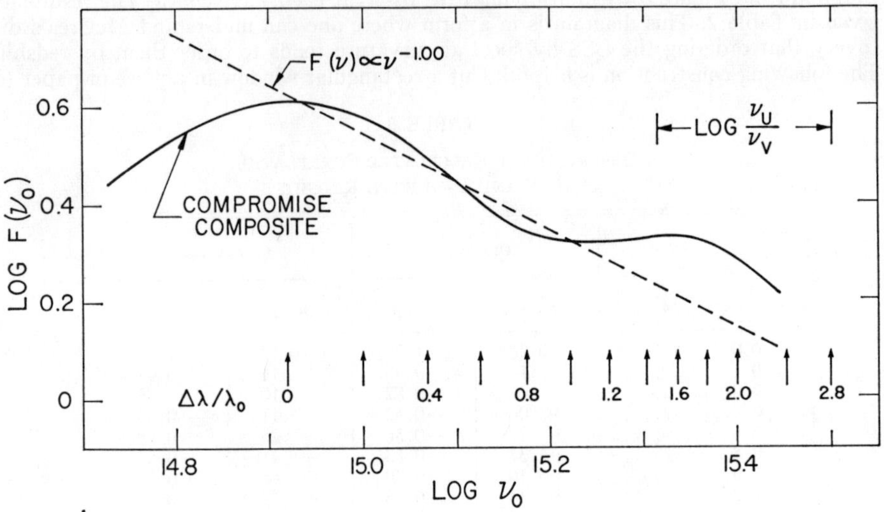

FIG. 4.—The energy distribution per unit frequency interval obtained from the compromise curve of Fig. 3. The frequency interval shown in the upper right and the arrows along the abscissa marked with redshift values are useful in understanding McCrea's result, as explained in the text.

TABLE 4

THE K*-TERM NORMALIZED TO $z = 1.0$

z	2.5 log (1+z)	K_V	K_B	K_U	K_B-K_V	K_U-K_B	($B-V$) Calc.	($U-B$) Calc.
0.0..........	0.00	(−1.00)	(−0.35)	(−0.05)	+0.65	+0.30	−0.19	−1.05
0.2..........	0.20	−0.57	.00	+ .09	+ .57	+ .09	− .11	−0.84
0.4..........	0.36	−0.22	+ .13	+ .03	+ .35	− .10	+ .11	−0.65
0.6..........	0.51	−0.09	+ .13	+ .05	+ .22	− .18	+ .24	−0.57
0.8..........	0.64	+0.01	+ .06	− .08	+ .05	− .14	+ .41	−0.61
1.0..........	0.75	0.00	.00	.00	.00	.00	+ .46	−0.75
1.2..........	0.86	−0.06	− .02	+ .10	+ .04	+ .12	+ .41	−0.87
1.4..........	0.95	−0.11	− .01	+ .22	+ .10	+ .23	+ .36	−0.98
1.6..........	1.04	−0.14	+ .12	+ .30	+ .26	+ .18	+ .20	−0.93
1.8..........	1.12	−0.14	+ .21	+ .36	+ .35	+ .15	+ .11	−0.90
2.0..........	1.19	−0.10	+ .31	+ .35	+ .41	+ .04	+ .05	−0.79
2.2..........	1.26	−0.03	+ .38	+0.29	+ .41	−0.09	+ .05	−0.66
2.4..........	1.33	+0.03	+ .42	....	+ .39	....	+ .07	....
2.6..........	1.39	+0.13	+ .42	....	+ .29	....	+ .17	....
2.8..........	1.45	+0.19	+ .38	....	+ .19	....	+ .27	....
3.0..........	1.51	+0.23	+0.33	....	+0.10	....	+0.36	....

* K is to be added to observed magnitude to obtain what would have been observed if the effects of redshift were zero. If the sign of K is negative, the corrected magnitude will be brighter than the observed value.

and $\bar{\lambda}_U = 3593$ Å. The tabulated values include both the selective term, 2.5 log $F(z)/F(0)$, and the non-selective band-width term, 2.5 log $(1 + z)$. For fluxes expressed in energy per unit wavelength interval, the selective term makes the source appear brighter under redshift as the high intensity blue portion of the curve is shifted into the bands. However, the non-selective band-width term works in the opposite direction. It is obvious that the two effects just cancel for all redshifts if $F(\lambda) \propto \lambda^{-1.00}$ because $\lambda_z = \lambda_0(1 + z)$.

The values in Table 4 have been normalized to a redshift of $z = 1.0$ so as to fall in the middle range of the observational data. A negative value of K means that the magnitude of the redshifted source (i.e., the observed magnitude) is too faint compared to what it would have been at a redshift $z = 1.0$; i.e., the observed magnitude must be made brighter to correct it to $z = 1.0$. Therefore, the K-corrections are to be added,

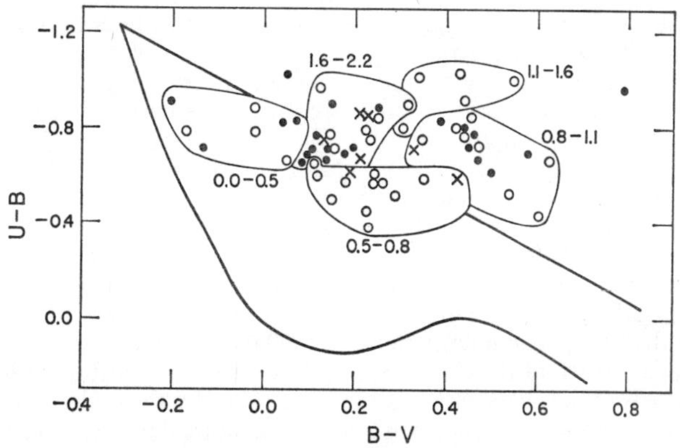

FIG. 5.—The domains of redshift intervals in the two-color diagram. *Open circles*, QSS of known redshift where the z-values agree with the domain assignment. *Crosses*, QSS where the measured redshifts violate the assignment. *Closed circles*, QSS of unknown redshift. Note the progressive increase of redshift as one proceeds counterclockwise in a spiral fashion around the domains. The black-body line is shown for comparison.

with their tabulated sign, to observed magnitudes to obtain corrected values. The magnitude corrections are very small except near $z = 0$. This agrees with earlier conclusions (Sandage 1965, Appendix) based on an approximate energy distribution curve which was suggested by the 3C 48 data.

Table 4 also lists the $K_B - K_V$ and $K_U - K_B$ color corrections. It is the variation of these quantities with redshift that produces the correlations of Figure 1. A check of Table 4 is therefore provided by applying $K_B - K_V$ and $K_U - K_B$ values to the observed colors at $z = 1.0$. The comparison is satisfactory.

V. THE $U - B$, $B - V$ DIAGRAM

The correlations of Figure 1, *a* and *b*, require that the redshift must vary systematically across the two-color diagram. The mean $B - V = f(U - B)$ relation is given by the values in Table 3, and these define a spiral in the two-color diagram turning counterclockwise as the redshift increases. Figure 5 with the observational data plotted shows that this is indeed the systematic trend. Five regions are blocked out in the diagram corresponding to the five redshift ranges marked along the boundaries of each area. Sources for which redshifts are unknown at the moment are shown as closed circles. Sources which have known redshifts are shown as open circles or crosses. The open circles represent QSS whose redshifts conform to the values assigned to the areas. The crosses represent those cases which violate the assignment. Only seven of the forty-three cases lie in the wrong area, while thirty-six follow the systematics. This shows that, at least statistically, there is a correlation between redshift and position in the two-color diagram. It appears that the correlation has some predictive value. Of course Figures 1, 2, and 5 are uniquely related—each shows the same phenomenon but in a different representation.

VI. CONCLUSION

The remarkable fact which emerges from these data is that quasi-stellar sources taken as a group are similar enough so as to produce the systematics shown. The cause of the valley and the peak at $\lambda_0 \simeq 2800$ Å is not yet understood. Kardashev and Komberg suggest that interstellar or intergalactic absorption is responsible, but it is also possible that the cause is in the source function of the radiation itself or, as Lynds has privately suggested, the effect might be due to the intense emission lines of Mg II ($\lambda_0 = 2800$ Å) and C III ($\lambda_0 = 1909$ Å). High-resolution spectrum scans of a representative sample of QSS chosen from the objects which show small scatter in Figures 1 and 2 should ultimately lead to a much better $F(\nu)$ curve than we have been able to derive here.

Note added in proof, August 8, 1966.—The recent outburst of QSS 3C 446, with its brightening by 3.2 mag. and its large change of the $(U - B)$-color (Sandage 1966), emphasizes again that the correlations shown in Figures 1 and 2 can only be statistical. Any individual QSS can disobey the general trend at any given time. Source 3C 446 obeyed the relations before outburst, having $B - V = +0.44$, $U - B = -0.90$ with $z = 1.403$. After outburst, $B - V = 0.54$, and $U - B = -0.51$. This $(U - B)$-value disobeys Figure 1, *b*. Analysis of the color data (Sandage, Westphal, and Strittmatter 1966) shows that the predominant cause of the large change in $U - B$ was the decrease in the equivalent width of the C IV (1550 Å) line during the outburst phase. This is quite direct evidence that the shape of Figure 3 is controlled by the underlying continuum radiation *as modified by the strong emission lines* and smoothed by the large band width of the U, B, and V filters. In view of this, it seems probable that the peak near $\lambda_0 = 2800$ Å in Figure 3 is indeed due to the Mg II line as suggested by Lynds. The argument of this paper is in no way changed by this deeper understanding of the correlations because we have not assumed in the preceding sections that we have derived the *continuum* flux, but only a smoothed distribution which produces Figure 1, *a* and *b*, statistically. A detailed discussion of the effect of the lines, superposed on an assumed continuum, is given by Strittmatter and Burbidge (1967) and shows that the correlations of Figure 1 may indeed be less smooth than drawn, and that the compromise continuum of Figure 3 may be even lumpier than we have derived. But it is clear that the QSS are sufficiently similar in their continuum distribution and in the strength of the emission lines relative to this continuum that the statistical correlations of Figures 1 and 2 do exist and do not degenerate into a complete scatter diagram.

It is a special pleasure to thank Drs. Kardashev and Komberg for sending a reprint of their remarkable work at an early date and for an accompanying letter describing the main points of their discovery. I am also grateful to Maarten Schmidt for an initial conversation about McCrea's work, and to C. R. Lynds for a helpful discussion on the first draft of this paper.

APPENDIX

NEW PHOTOMETRY OF QSS CANDIDATES

Table A1 gives new photoelectric measurements of twenty-two candidates for QSS. The identifications have been made by the large number of workers listed in the references to the table and by P. Véron (unpublished). The observations were made with the 200-inch reflector in

October 1965 and January 1966 with a conventional 1P21 three-color photometer, relative to standards on the Johnson-Morgan system. The mean errors should all be less than ±0.04 mag.

The most interesting objects are those from the Haro-Luyten *Catalogue* (1962) that have been identified by Scheuer and Wills (1966) with radio sources observed in the 4C survey. The color distribution of these eight objects follows the same pattern as other QSS, and there is every reason to believe that they are correctly identified and conform to the group characteristics of quasi-stellar sources in general.

Table A1, together with the summary published previously (Sandage 1965, Table 2), lists a total of sixty-six sources, most of which have proved to be QSS upon spectrographic identification. There are, however, a few in the earlier list which are not now bona fide identifications. The object 3C 217 is in a confused radio field, and there are two very blue objects near the several available radio positions. Each object has QSS colors, but the identification is quite uncertain at

TABLE A1

MAGNITUDES AND COLORS OF TWENTY-TWO QSS NOT PREVIOUSLY REPORTED

Object	Date (U.T.)	V	$B-V$	$U-B$	Chart
3C 2	Oct. 19, 1965	19.35	0.79	−0.96	1
3C 57	Jan. 14, 1966	16.40	.14	−0.73	2
3C 232*	Jan. 13, 1966	15.78	.10	−0.68	3
3C 268.4	Jan. 13, 1966	18.42	.58	−0.69	4
3C 288.1	Jan. 13, 1966	18.12	.39	−0.82	4
3C 309.1	Jan. 13, 1966	16.78	.46	−0.77	4
3C 345	Oct. 19, 1965	15.96	.29	−0.50	5
3C 351	Oct. 19, 1965	15.28	.13	−0.75	6
3C 380	Oct. 19, 1965	16.81	.24	−0.59	5
3C 432	Oct. 19, 1965	17.96	.22	−0.79	4
3C 454	Oct. 19, 1965	18.40	.12	−0.95	4
3C 454.3	Oct. 21, 1965	16.10	.47	−0.66	7
MSH 03−19	Oct. 19, 1965	16.24	.11	−0.65	8
MSH 14−121	Jan. 13, 1966	17.37	.44	−0.76	9, 10 (object *a*)
PHL 658	Oct. 19, 1965	16.40	.11	−0.70	11
PHL 923	Oct. 19, 1965	17.33	.20	−0.70	11
PHL 1078	Oct. 19, 1965	18.25	.04	−0.81	11
PHL 1093	Oct. 19, 1965	17.07	.05	−1.02	11
PHL 1305	Oct. 19, 1965	16.96	.07	−0.82	11
PHL 1377	Oct. 19, 1965	16.46	.15	−0.89	11
PHL 3740	Oct. 19, 1965	18.61	.09	−0.65	11
PHL 6638	Oct. 19, 1965	17.72	0.18	−0.69	11

* 3C 232 was identified by C. M. Wade with Tonantzintla 469. Tonantzintla 470, in same field, has $V = 16.40$, $B - V = 0.07$, $U - B = 0.13$.

REFERENCES FOR CHARTS

1. Sandage, A., Véron, P., and Wyndham, J. D. 1965, *Ap. J.*, **142**, 1307.
2. Bolton, J. G., Shimmins, A. J., Ekers, J., Kinman, T. D., Lamla, E., and Wirtanen, C. A. 1966, *Ap. J.*, **144**, 1229.
3. Iriarte, B., and Chavira, E. 1957, *Bol. Obs. Tonantzintla y Tacubaya*, No. 16, p. 3.
4. Wyndham, J. D. 1966, *Ap. J.*, **144**, 459.
5. Sandage, A., and Wyndham, J. D. 1965, *Ap. J.*, **141**, 328.
6. Lynds, C. R., Stockton, A. N., and Livingston, W. C. 1965, *Ap. J.*, **142**, 1667.
7. Sandage, A. 1966, *Ap. J.*, **144**, 1238.
8. Matthews, T. A. 1964, *Carnegie Institution Year Book*, p. 44.
9. Hazard, C., Mackey, M. G., and Nicholson, W. 1964, *Nature*, **202**, 227.
10. Hazard, C. 1964, *Quasi-stellar Sources and Gravitational Collapse*, ed. I Robinson, A. E. Schild, and E. L. Schucking (Chicago: University of Chicago Press), chap. xi.
11. Scheuer, P. A. G., and Wills, D. 1966, *Ap. J.*, **143**, 274.

present. Through the courtesy of M. Schmidt, it is now known that the identification of 3C 247 suggested by Sandage, Véron, and Wyndham (1965) is incorrect. The object marked on the published chart is a star. Its color suggests that it is an extreme subdwarf, characteristic of the main-sequence stars of halo globular clusters.

REFERENCES

Barnes, R. 1966, *Ap. J.*, **146**, 285.
Haro, G., and Luyten, W. J. 1962, *Bol. Obs. Tonantzintla y Tacubaya*, **3**, No. 37, p. 17.
Humason, M. L., Mayall, N. U., and Sandage, A. R. 1956, *A.J.*, **61**, 97.
Kardashev, N. S., and Komberg, B. V. 1966, *Astr. Circ. USSR*, No. 357.
Matthews, T. A., and Sandage, A. R. 1963, *Ap. J.*, **138**, 30.
McCrea, W. H. 1966, *Pub. A.S.P.*, **78**, 49.
Sandage, A. 1965, *Ap. J.*, **141**, 1560.
———. 1966, *I.A.U. Circular*, No. 1961.
Sandage, A., Westphal, J. A., and Strittmatter, P. A. 1966, *Ap. J.*, **146**, 322.
Sandage, A., Véron, P., and Wyndham, J. D. 1965, *Ap. J.*, **142**, 1307.
Scheuer, P. A. G., and Wills, D. 1966, *Ap. J.*, **143**, 274.
Strittmatter, P. A., and Burbidge, G. R. 1967, *Ap. J.* (in press).

INTENSITY VARIATIONS OF QUASI-STELLAR SOURCES IN OPTICAL WAVELENGTHS

Allan Sandage

Mount Wilson and Palomar Observatories, Carnegie Institution of Washington, California Institute of Technology

Routine photoelectric photometry of quasi-stellar sources (QSS's) has been in progress with the 200-inch reflector since 1963. The principal emphasis has been the confirmation of peculiar colors as an aid to positive optical identification of radio sources. However, as time permits, repeated observations of previously identified sources have been made to monitor optical variations. We report here observations of those QSS's where more than one observation exists. The QSS's which have been followed most extensively are 3C 48, 3C 196, and 3C 273. The new data extend the time coverage for these three sources to January, 1966, two years beyond that reported previously (Sandage 1964). In addition to these three objects, data for 3C 9, 3C 43, 3C 47, 3C 216, 3C 245, 3C 286, 3C 287, and 3C 454.3 are given. The QSS's for which the evidence for intensity fluctuations is beyond doubt are listed in Table 1. Three sources with more than one measurement but where variations have not yet been detected with certainty are listed in Table 2.

The magnitudes and colors are on the *UBV* system as measured with a conventional 1P21 refrigerated photomultiplier relative to Johnson-Morgan standards. The mean error of the magnitudes is of the order of ± 0.03 mag., although occasional values can be in error by more than this value would imply because of the fairly rapid techniques of survey photometry which were employed. Differences between various readings of up to 0.10 mag. are therefore not necessarily proof of real variation of these faint objects.

The observations are spaced rather widely in time and no positive statement can yet be made about the existence of very short-term variations of the order of hours or a few days. But, as previously reported (Sandage 1964), several runs of 3 hours' duration were made on 3C 48 with no evidence of variations on this time scale.

Remarks on the individual objects follow.

3C 47: Although observed only twice, the variation of $\Delta B = 0.21$ mag. is beyond doubt because all three independently measured magnitudes show the same variation; i.e., $B - V$ and $U - B$ remain nearly constant.

3C 48: Remained relatively quiescent during the 16-month period, hovering near $V = 16.29$. The light-curve from 1960 through January, 1966, is shown in Figure 1.

3C 196: The reality of the variation is beyond doubt. The source has gradually brightened in all three wavelengths. An apparent sudden burst in December, 1963, appears to be real, as two independent concordant observations were made on December 12, 1963.

3C 216: The observation of May 10, 1964, appears to be correct. The measurement is of high accuracy with mean errors of $\epsilon_V = \pm 0.022$ mag., $\epsilon_B = \pm 0.018$ mag., and $\epsilon_U = \pm 0.034$ mag., which are small compared with the amplitude of the change between December 12, 1963, and January 10, 1965.

3C 245: The variation is not beyond doubt because of its smallness.

3C 273: This source has remained remarkably constant over the 2-year interval of these new observations. Since photoelectric observations began on February, 1963, the variation has only been $\Delta B = 0.38$ mag., gradually decreasing during the first year, and remaining steady thereafter. This is in contrast to the large variations observed in earlier years by Smith and Hoffleit (1963*a*, *b*) and by Smith (1964), but, of course, these earlier variations are on a considerably longer time scale.

Reprinted from Astrophysical Journal, 144, 1234-1238.

TABLE 1

PHOTOELECTRIC DATA FOR

3C47, 3C48, 3C196, 3C216, 3C245, and 3C273

Date	JD 2437+	V	B	U	B−V	U−B	Remarks
				3C47			
Dec. 12, 1963	1375.8	18.12	18.15	17.51	0.03	−0.64	
Oct. 5, 1964	1673.8	18.32	18.36	17.69	0.04	−0.67	
				3C48			
Oct. 3, 1964	1671.8	16.27	16.76	16.21	0.49	−0.55	
Oct. 5, 1964	1673.8	16.28	16.75	16.20	0.47	−0.55	
Jan. 8, 1965	1768.8	16.23	16.66	16.12	0.43	−0.54	
Jan. 9, 1965	1769.8	16.28	16.68	16.14	0.40	−0.54	
Jan. 10, 1965	1770.8	16.30	16.70	16.17	0.40	−0.53	
Feb. 15, 1965	1806.8	16.27	16.61	16.03	0.34	−0.58	
Oct. 19, 1965	2052.8	16.30	16.70	16.17	0.40	−0.53	
Oct. 20, 1965	2053.8	16.33	16.74	16.22	0.41	−0.52	
Jan. 13, 1966	2138.8	16.29	16.73	16.21	0.44	−0.52	
Jan. 14, 1966	2139.8	16.32	16.74	16.21	0.42	−0.53	
Jan. 15, 1966	2140.8	16.33	16.73	16.18	0.40	−0.55	
				3C196			
Apr. 1, 1962	755.8	17.72	18.38	17.94	0.66	−0.44	
May 16, 1963	1165.8	17.69	18.31	17.99	0.62	−0.32	
Dec. 12, 1963	1375.8	17.45	18.05	17.56	0.60	−0.49	Mean of 2
Feb. 14, 1964	1439.8	17.60	18.09	17.64	0.49	−0.45	
Mar. 9, 1964	1524.8	17.54	18.12	17.63	0.58	−0.49	
May 10, 1964	1525.8	17.50	18.02	17.57	0.52	−0.45	
Jan. 10, 1965	1770.8	17.42	18.00	17.53	0.58	−0.47	
Jan. 13, 1966	2138.8	17.40	17.90	17.44	0.50	−0.46	
Jan. 14, 1966	2139.8	17.37	17.96	17.37	0.59	−0.59	
				3C216			
Dec. 12, 1963	1375.8	18.48	18.97	18.37	0.49	−0.60	Mean of 2
May 10, 1964	1525.8	18.75	19.34	18.95	0.59	−0.39	$\varepsilon_v = \pm 0.022; \varepsilon_B = \pm 0.018; \varepsilon_u = \pm 0.034$
Jan. 10, 1965	1770.8	18.28	18.78	18.19	0.50	−0.59	
				3C245			
Dec. 12, 1963	1375.8	17.29	17.75	16.93	0.46	−0.82	
Feb. 14, 1964	1439.8	17.25	17.67	16.81	0.42	−0.86	
Jan. 14, 1966	2139.8	17.20	17.63	16.81	0.43	−0.82	
				3C273			
Jan. 20, 1964	1414.8	12.90	13.11	12.26	0.21	−0.85	
Feb. 14, 1964	1439.8	12.90	13.11	12.22	0.21	−0.89	
Mar. 9, 1964	1463.8	12.90	13.11	12.21	0.21	−0.90	Mean of 2
May 9, 1964	1524.8	12.88	13.09	12.20	0.21	−0.89	Mean of 2
June 7, 1964	1553.8	12.94	13.13	12.25	0.19	−0.88	Mean of 2 100"
Jan. 10, 1965	1770.8	12.94	13.12	12.26	0.18	−0.86	
Jan. 27, 1965	1787.8	12.92	13.10	12.16	0.18	−0.94	60"
Jan. 29, 1965	1789.8	12.93	13.07	12.11	0.14	−0.96	60"
Mar. 27, 1965	1846.8	12.87	13.06	12.19	0.19	−0.87	
Jan. 13, 1966	2138.8	12.90	13.09	12.18	0.19	−0.91	
Jan. 14, 1966	2139.8	12.90	13.10	12.19	0.20	−0.91	
Jan. 15, 1966	2140.8	12.92	13.11	12.20	0.19	−0.91	

TABLE 2

PHOTOELECTRIC DATA FOR NON-VARIABLE QSS'S
3C 9, 3C 286, AND 3C 287

Date	JD 2437+	V	B	U	$B-V$	$U-B$	Remarks
			3C 9				
Dec. 12, 1963	1375.8	18.21	18.44	17.70	0.23	−0.74	
Oct. 5, 1964	1673.8	18.22	18.48	17.69	0.26	−0.79	
			3C 286				
May 15, 1963	164.8	17.30	17.54	16.76	0.24	−0.78	
Feb. 14, 1964	1439.8	17.29	17.47	16.61	.18	− .86	
May 10, 1964	1525.8	17.31	17.52	17.14:	.21	− .38:	
Jan. 15, 1966	2140.8	17.29	17.56	16.71	0.27	−0.85	
			3C 287				
Feb. 14, 1964	1439.8	17.67	18.22	17.53	0.55	−0.69	Color change suspicious
May 10, 1964	1525.8	17.68	18.40	17.79	0.72	−0.61	Does it vary?

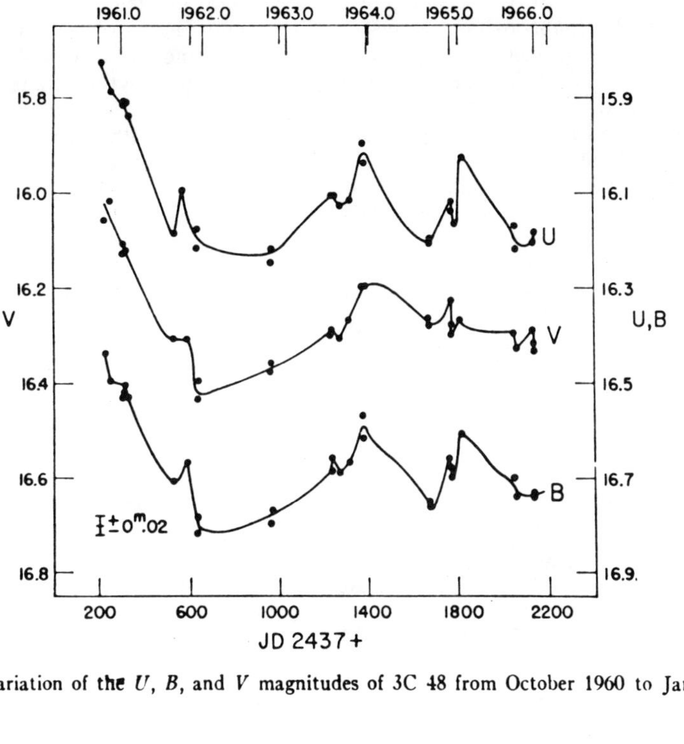

FIG. 1.—Variation of the U, B, and V magnitudes of 3C 48 from October 1960 to January 1966

In addition to these photoelectric measurements which indicate variation in at least five sources and probably six, photographic evidence exists for intensity changes in at least four other QSS's.

1. Variations in 3C 345 have been especially well followed by Goldsmith and Kinman (1965). Here the blue magnitude changed by 0.4 mag. in an interval of 18 days. This is the best documented case of a rapid variation on a short-time scale.

2. 3C 2 changed by $\Delta B \simeq 1.5$ mag. between September 27/28, 1954, and August 20/21, 1963, as discussed elsewhere (Sandage, Véron, and Wyndham 1965).

3. 3C 43, also described in this reference, changed in blue magnitudes by 0.9 mag. between September 27/28, 1954, and September 28/29, 1964. An illustration of the change is shown in Figure 2 (*bottom*). 3C 43 is the object at the tip of the arrow. The left picture is copied with 25× enlargement from the original Sky Survey blue paper print; the right print is copied from a new plate taken September 28/29, 1964. The image quality of the left and right prints is not the same because one was made from a paper print and the other from a plate, but the large change of magnitude of 3C 43 is evident by differential comparison of the images on both prints with stars A, B, and C.

4. An equally striking case is 3C 454.3, identified in Figure 2 (*top*). Again, there is a decrease in brightness between the Sky Survey plate of August 1/2, 1954, and a new Schmidt plate of September 24/25, 1965. Photoelectric photometry of stars 1, 2, and 4 on October 20/21, 1965, provides a skeleton sequence from which estimates of the magnitude of the QSS's on various dates can be made. The magnitudes of the sequence stars are given in the caption to Figure 2. Three plates exist of 3C 454.3. These are (1) the original Sky Survey plate of August 1/2, 1954, (2) the plate by Véron on September 24/25, 1965, and (3) a special plate by Luyten on October 29/30, 1965. A color equation exists between the photoelectric B system and the photographic plates because the latter were all taken on 103aO emulsion with no filter—a combination that admits much radiation below λ 3900 Å. The equation is especially severe for QSS objects because they are intense in the ultraviolet and give a falsely bright image on the plates for their B magnitude on the Johnson-Morgan system. The size of the correction is 0.7 mag. as determined by a comparison of the photoelectric measurements of 3C 454.3, obtained on October 20/21, 1965, with the estimated m_{pg} magnitude on Luyten's plate taken only 9 days later. With this correction applied, 3C 454.3 was $B = 15.7$ on August 1/2, 1954, 16.75 on September 24/25, 1965, and 16.55 on October 29/30, 1965. The photoelectric values are $V = 16.09$, $B - V = 0.47$, and $U - B = -0.66$ on October 20/21, 1965. This well-verified case gives an amplitude of $\Delta B = 1.05$ mag. for the QSS.

The results discussed here for ten QSS's again indicate that many, perhaps most QSS's are variable, but it is clear that very much work must be done on a more systematic basis before we know if the variations are entirely random or have some systematic character. A photographic survey program toward this end has recently been started at Palomar using the 48-inch Schmidt telescope.

REFERENCES

Goldsmith, D. W., and Kinman, T. D. 1965, *Ap. J.*, **142**, 1693.
Sandage, A. 1964, *Ap. J.*, **139**, 416.
Sandage, A., Véron, P., and Wyndham, J. D. 1965. *Ap. J.*, **142**, 1307.
Smith, H. J. 1964, *Quasi-stellar Sources and Gravitational Collapse*, ed. I. Robinson, A. E. Schild, and E. L. Schucking (Chicago: University of Chicago Press), chap. xvi.
Smith, H. J., and Hoffleit, D. 1963*a*, *Nature*, **198**, 650.
———. 1963*b*, *A.J.*, **68**, 292.

THE CHANGE OF INTENSITY, COLOR, LINE STRENGTH, AND LINE POSITION IN THE QSS 3C 446 DURING THE 1966 OUTBURST

Allan Sandage, J. A. Westphal, and P. A. Strittmatter
Mount Wilson and Palomar Observatories, Carnegie Institution of Washington, California Institute of Technology

During the course of routine observations of quasi-stellar radio sources, 3C 446 was found to be abnormally bright on June 24.4 U.T., 1966 (Sandage 1966). A further observation on July 12.4 U.T., 1966, confirmed this result, showing that the object had undergone an outburst of at least 3.2 mag. sometime between October 5, 1964, and June 24, 1966. This time interval can be narrowed appreciably because a visual estimate of the V magnitude by Schmidt on September 23 and 25, 1965, gave V not brighter than 18 on those dates. The event is of particular importance as it enables us to make several tests concerning the intensity and position of the emission-line features.

TABLE 1

PHOTOELECTRIC DATA ON THE OUTBURST IN 3C 446

Date (U.T.)	V	$B-V$	$U-B$
Oct. 3.2, 1964........	18.42	0.45	−0.93
Oct. 5.2, 1964........	18.36	.43	− .88
June 24.4, 1966.......	15.14	.57	− .51
July 12.4, 1966.......	15.28	0.52	−0.52

Table 1 gives the photoelectric data. The large change in the $U - B$ color is well above the probable error ($\approx \pm 0.03$ of the determination; $B - V$ has also changed but by a smaller, though still significant, amount. We adopt the following mean values in the subsequent discussion. On October 3–5, 1964: $V = 18.39$, $B - V = 0.44$, $U - B = -0.90$; June and July, 1966: $V = 15.21$, $B - V = 0.54$, $U - B = -0.51$. The increase of 0.14 mag in the V magnitude between the June and July observations suggests that the outburst was already then in the decay phase.

To determine the extent of the change in the emission-line intensities, three well-widened spectrograms were obtained with the Hale 200-inch reflector on July 13.45 U.T., 1966, July 14.44 U.T., 1966, and July 15.40 U.T., 1966, at a dispersion of 190 Å mm. The emission lines of C IV (1550) and C III (1909), originally found by Burbidge (1965) and by Schmidt (1966) to be of strong to moderate intensity before the outburst,

Reprinted from Astrophysical Journal
146, 322-326.

were not clearly visible, indicating that the contrast between these lines and the continuum had drastically decreased. There was only a faint suggestion to the eye of the C IV (1550) line near $\lambda = 3729$ Å; no trace of the C III (1909) line could be found.

In order to obtain more precise identification and positions of the lines and to measure their intensities, the plates were microphotometered, rectified to intensities using a calibrated wedge, and analyzed by the cross-correlation filtering technique (Westphal 1966) with three different filter widths. Figure 1 shows the result for the plate taken on July 15.40 U.T., 1966. The He comparison spectrum is shown below the filtered QSS spectrum. A feature at $\lambda_{obs} = 3732$ Å, visible well above the plate noise, is present on all three spectrograms, and we are convinced that it is the C IV (1550) line seen by Burbidge (1965), Schmidt (1966), and Oke (1966). No trace of the C III (1909) line at λ 4585 was recovered on any of the three plates, and we conclude that it has been submerged in the increased continuum radiation of the outburst.

The equivalent width (E.W.) of the C IV line has been measured on all three filtergrams with the result that E.W. = 9.2, 9.3, 9.7 Å on the plates obtained at July 13.45, 14.44, and 15.40 U.T., respectively. These values are quite sensitive to the adopted

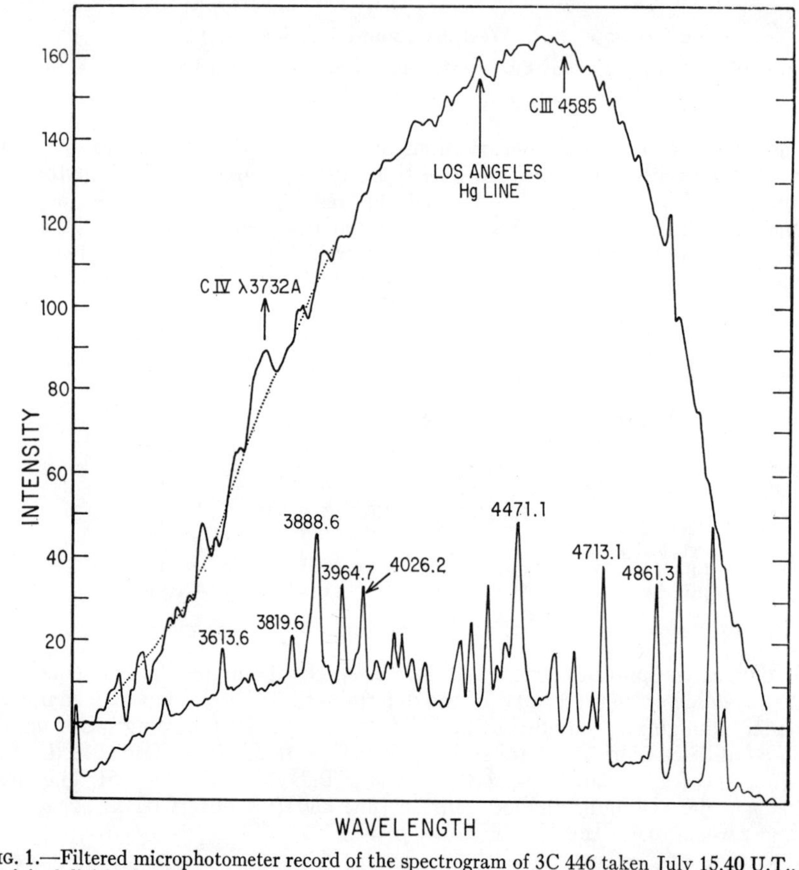

FIG. 1.—Filtered microphotometer record of the spectrogram of 3C 446 taken July 15.40 U.T., 1966. The original digitized output was smoothed with a Gaussian correlation function with a total dispersion width of 13 Å. The He comparison lines are shown below the QSS record. The C IV line at λ 3732 is easily seen. No trace of the C III line expected at λ 4585 is seen.

continuum level (see Fig. 1), the resultant uncertainty in equivalent width being perhaps as large as 50 per cent. The equivalent width of 230 Å for C IV (1550) measured in 3C 446 by Oke (1966) before the outburst is much larger than our value. The considerable change in the contrast between line and continuum is beyond doubt and suggests strongly that each is produced in a different region of the object and by unrelated mechanisms.

The digital technique used to filter the microphotometer outputs is ideally suited for the measurement of the wavelength of any desired feature. The He comparison lines are used as a calibration of the wavelength scale in digital units. We sampled the spectrogram at spacings of 1.58 Å which corresponds to every 3 μ on the plate. The computer output permits the determination of the point of maximum correlation on the filtergrams to within one digital unit, or closer if interpolation is desired. This technique was used to determine the wavelength of the maximum correlation of the smoothing filter with the C IV feature, giving λ 3736.4, λ 3726.0, and λ 3733.4 for the wavelength of the C IV line on each of the three plates. The mean value is λ 3731.9 ± 2.8 (m.e.). This should be compared with Burbidge's (1965) value of λ 3727 Å, and Schmidt's (1966) value of λ 3729 Å. In view of the line width of at least 50 Å, we do not consider the differences to be significant and conclude that there is no evidence for line shifts. The point is of some importance in view of Gold's discussion as summarized by Burbidge and Burbidge (1966).

Finally, there is the question of the colors. We wish to test the hypothesis that the lines have remained essentially constant in absolute intensity, and that the color change is due to the increase in the equivalent width as the continuum intensity changes. We therefore examine the following possibilities: (*a*) the continuum has increased in intensity by the same amount at all wavelengths and the observed change of color is due solely to the decreased contrast of the lines; (*b*) the change in color is due to a combination of the effect of the lines and of a change of slope of the continuum energy distribution again keeping the absolute strength of the lines constant. We now show that hypothesis (*a*) is probably incapable of explaining the observed color changes while (*b*) gives a self-consistent picture.

Hypothesis (*a*).—The new observations during outburst give $B - V = +0.54$, $U - B = -0.51$. These colors are consistent with a power-law spectrum $F(\nu) \propto \nu^{-1.8}$ as shown either by Table A4 of Matthews and Sandage (1963, Appendix A) or by direct calculation using Appendix B of the same reference. Since the measured E.W. of C IV (1550) is only about 10 Å at this phase of the outburst, the energy distribution is nearly pure continuum, uncontaminated by the lines. We now ask what the E.W. of the C IV (1550) line at λ 3729 Å and C III (1909) at λ 4585 Å would have to have been before the outburst to produce the observed color change of $\Delta(B - V) = 0.10$ mag, $\Delta(U - B) = 0.39$ mag, if this were the only mechanism operating. The λ 4585 line affects only B, and λ 3729 affects only U. Hence, 0.10 mag must be the contamination factor of the λ 4585 line in B, while 0.49 mag must be the factor in U due to the λ 3729 line to give $\Delta(U - B) = 0.39$ if hypothesis (*a*) is correct. The E.W. before outburst of the two relevant lines required to cause the observed color change can be calculated from the data given by Strittmatter and Burbidge (1966). They are E.W. = 120 Å for λ 4585, and E.W. = 380 Å for λ 3729. The E.W. after outburst in the June–July period should then be these values reduced by the outburst intensity of 3.2 mag (a factor of 19), which gives E.W. (now) = 6 Å for λ 4585, and E.W. (now) = 20 Å for λ 3729, again if hypothesis (*a*) is correct. These four values (before and after outburst) are all high in view of (1) Oke's (1966) measurement of 20 Å for the λ 4585 line and 230 Å for the λ 3729 line before outburst, and (2) our measurement of only E.W. (now) = 10 Å for λ 3729 and our failure to detect the line at λ 4585. Furthermore, the required intensity ratio of $\frac{380}{110}$ for the lines before outburst is entirely inconsistent with Oke's (1966) ratio of $\frac{230}{20}$. We feel therefore that hypothesis (*a*) should be rejected. However, we note that Oke's (1966) photoelectric scans of 3C 446 show several additional features, one of which is identified with

He II (1640). If all these features are real, the combined equivalent widths under the U and B filters would be more nearly capable of explaining the observed color changes under hypothesis (*a*). In view of the uncertainty, however, we shall provisionally discount this possibility.

Hypothesis (*b*).—Here we allow the continuum to change in slope and predict the equivalent width of the C IV (1550) line at λ 3729 which is required to account for the additional color change. Again we require the total color changes to be $\Delta(B - V) = 0.10$ and $\Delta(U - B) = 0.39$. To simplify, we adopt Oke's pre-outburst value of E.W. = 20 Å for C III (1909). Since the line effectively disappeared in June–July, and hence contributed nothing to the colors at that time, the total change in $B - V$ due to the line change is 0.02 mag (cf. Strittmatter and Burbidge 1966). Hence the remainder of the observed change, namely, $\Delta(B - V) = 0.08$ mag must be caused by the change in the slope of the continuum which we henceforth assume to be of a power-law form with index $n(F_\nu \propto \nu^{-n})$. Referral to Table A4 of Matthews and Sandage (1963) shows that $\Delta(B - V)/\Delta n = 0.24$, giving a change in n of ≈ 0.33 to 1.47 if the color change in $B - V$ is to be explained. The corresponding change in $U - B$ is likewise 0.08 mag.

The change in $U - B$ due to the C IV (1550) line is then calculated as follows. The observed $U - B$ color after outburst was -0.51 minus the effect of the 10 Å equivalent width of the λ 3729 line. This contamination effect is about 0.01 mag giving a continuum color after outburst of $U - B \approx -0.50$ which is to be compared with the observed color of $U - B = -0.90$ before outburst. We have seen that 0.08 mag can be attributed to the change in continuum slope, leaving a difference of 0.32 mag to be explained as the effect on the U magnitude of the C IV line. From the work of Strittmatter and Burbidge (1967) we obtain a required equivalent width of C IV (1550) before outburst of 250 Å. This is in good agreement with Oke's (1966) measured value of 230 A.

Furthermore, we can predict the equivalent width which the C IV (1550) line should have had in June–July, 1966, if its absolute intensity remained constant. The intensity increase in the V magnitude is a factor 19 (3.2 mag). The change in slope of the continuum, characterized by a change in exponent from $n = 1.8$ to $n = 1.47$, means that the intensity increase at λ 3729 is reduced from 19 to 16.7, giving a predicted equivalent width of $\frac{250}{16.7} = 15$ Å. Although this is not in perfect agreement with the measured value of about 10 Å, we believe the difference is not significant in view of the errors involved both in the estimate of E.W. = 250 Å, which is based on the somewhat uncertain observed colors, and in the adopted position of the continuum in Figure 1.

We conclude that the observed change in color is consistent with the hypothesis (i) that the continuum slope became steeper, and the associated colors therefore somewhat redder, during the outburst, and (ii) that the lines remained essentially constant in absolute intensity, changing in equivalent width from about 250 Å to 15 Å for C IV (1550) at λ 3729 and from 20 Å to about 1 Å for C III (1909) at λ 4585. Detailed spectrum scans will provide a clear test of these conclusions.

The participation of one of us (P. A. S.) was made possible, in part, by a grant from the National Science Foundation.

REFERENCES

Burbidge, E. M. 1965, *Ap. J.*, **142**, 1674.
Burbidge, E. M., and Burbidge, G. R. 1966, *Ap. J.*, **143**, 271.

Matthews, T. A., and Sandage, A. R. 1963, *Ap. J.*, **138**, 30.
Oke, J. B. 1966, *Ap. J.*, **145**, 668.
Sandage, A. 1966, *I.A.U. Circ. No. 1961.*
Schmidt, M. 1966, *Ap. J.*, **144**, 443.
Strittmatter, P. A., and Burbidge, G. R. 1967, *Ap. J.* (in press).
Westphal, J. A. 1966, *Ap. J.*, **142**, 1661.

THE EXISTENCE OF A MAJOR NEW CONSTITUENT OF THE UNIVERSE: THE QUASI-STELLAR GALAXIES

ALLAN SANDAGE
Mount Wilson and Palomar Observatories
Carnegie Institution of Washington, California Institute of Technology

ABSTRACT

Photometric, number count, and spectrographic evidence is presented to show that most of the blue, starlike objects fainter than $m_{pg} = 16^m$ found in color surveys of high-latitude fields are extragalactic and represent an entirely new class of objects. Members of the class called here quasi-stellar galaxies (QSG) resemble the quasi-stellar radio sources (QSS) in many optical properties, but they are radio-quiet. The QSG brighter than $m_{pg} = 19^m$ are 10^3 times more numerous per square degree than the QSS that are brighter than 9 flux units. The surface density of QSG is about 4 objects per square degree to $m_{pg} = 19^m$.

The evidence is developed in three parts: (1) Photoelectric photometry shows that a fundamental change occurs in the color distribution of high-latitude blue objects at about $V = 14.^m5$. Brighter than this, the objects fall near the luminosity class V line of the $U - B$, $B - V$ diagram. Fainter than this, 80 per cent of the objects lie in the peculiar region known to be occupied by the quasi-stellar radio sources. (2) The observed integral-count-curve, log $N(m)$, for objects in the Haro-Luyten catalogue undergoes a profound change of slope between $m_{pg} = 12^m$ and $m_{pg} = 15^m$, steepening and reaching a constant slope for m_{pg} fainter than 16^m. This magnitude interval is the same as that in which the color distribution changes, as discussed above. The slope fainter than 16^m is $d \log N(m)/dm = 0.383$. It is shown that this is the expected value from the theory of cosmological number counts for uniformly distributed objects with large redshifts. (3) Spectra of five of the faint blue objects are similar to spectra of quasi-stellar radio sources. Intense, sharp emission lines of forbidden [O III], [O II], and [Ne III], together with very broad (35 Å wide) lines of Hβ, Hγ, Hδ, Hε, and [Ne V] are present in two of the five. Two broad emission lines are present in another at λ 3473 and λ 4279, identified as C IV (1550) and C III (1909). The other two objects have featureless spectra with only a blue continuum showing. The redshifts ($\Delta\lambda/\lambda_0$) for the three objects with lines are 0.0877, 0.1307, and 1.2410. The position of the objects in the redshift–apparent-magnitude diagram shows each of the three to be superluminous.

The space density of the quasi-stellar galaxies is estimated to be about 5×10^{-80} QSG/cm^3, which is to be compared with the space density of normal galaxies of about 1×10^{-75} galaxies/cm^3. The ratio, per unit volume, of QSG to QSS is estimated to be 500, which gives a lifetime of the QSG phase as 5×10^8 years if the lifetime of the radio source is 10^6 years.

The objects would seem to be of major importance in the solution of the cosmological problem. They can be found at great distances because of their high luminosity. QSG at $B = 22^m$ are estimated to have a mean redshift of $\Delta\lambda/\lambda_0 \simeq 5$ for a model universe of $q_0 = +1$. At these redshifts, we are sampling the universe in depth to 0.63 of the distance to the horizon (for $q_0 = +1$), and are looking back in time more than 0.9 of the way to the "creation event" in an evolutionary model. Study of the $[m, z]$- and log $N(m)$-curves using the QSG should eventually provide a crucial test of various cosmological models. But even more important, comparative study of the quasi-stellar galaxies and the intimately connected quasi-stellar radio sources is expected to shed light on the evolutionary processes of the violent events that characterize the two classes.

I. INTRODUCTION

A systematic search for quasi-stellar radio sources (hereinafter called "QSS") has been in progress at the Mount Wilson and Palomar Observatories for the past several years. Following the discovery of an excess in the ultraviolet radiation of 3C 48, 3C 196, and 3C 286 (Matthews and Sandage 1963), the search was begun with the 100-inch reflector using the two-color photographic method previously described (Ryle and Sandage 1964).

When the telescope was set at a catalogued radio position, nearly every ultraviolet object that occurred near the plate center proved to be a correctly identified QSS, as shown by later optical position determinations (Veron 1965) and photoelectric measurements. However, a curious circumstance developed as many survey plates accumulated. Objects were found on the plates that imitated the ultraviolet excess of the true QSS but

Reprinted from Astrophysical Journal
141, 1560-1578.

that did not occur at the radio positions. The first four such images were found on plates centered on the Cambridge 3C R positions of 3C 194, 3C 205, 3C 225, and 3C 280. Consultation with Ryle concerning the possibility that the radio positions could be in error showed that the four objects could not possibly be identified with the four catalogued radio sources because they were displaced from the radio positions by amounts far in excess of the radio probable errors. The nature of these so-called interlopers remained obscure and no further work was done on them for several months.

Many more interlopers were found when the two-color search was transferred to the wide-field 48-inch Schmidt telescope. The observed frequency of about 3 per square degree to a limiting magnitude $B \simeq 18.^m5$ was low enough so as not to impair the discovery program for the QSS, because the chance of finding a random interloper within a radio error square of 30″ by 30″ was only one in 5000. But the frequency was high enough to suggest that they might be the same type of objects found by Luyten (1953, 1954, 1955, 1956*a*), Iriarte and Chavira (1957), Chavira (1958), and Haro and Luyten (1962) in high galactic latitudes, which occur with a frequency of 4 objects per square degree to $B \simeq 19.^m0$.

This identification was adopted as a working hypothesis until photometry of the first four interlopers showed that the $U - B$, $B - V$ indices, given in Table 1, were far different from the run of colors tabulated in the Haro-Luyten (HL) catalogue, so the identification with the HL objects remained ambiguous. The colors closely resemble those for U Gem and SS Cyg eruptive variables (Walker 1957; Grant and Abt 1959; Mumford 1964; Eggen and Sandage 1965), and because eruptive variables are known to occur among the blue-halo objects (Luyten and Haro 1959; Haro and Chavira 1960; Haro and Luyten 1960), the objects in Table 1 were tentatively identified as members of this class. However, evidence accumulated in the past several months shows that this cannot be true. The interlopers and the HL objects are one and the same, but most of them are not stars.

TABLE 1

PHOTOELECTRIC COLORS FOR THE FIRST FOUR INTERLOPERS

Object	V	$B-V$	$U-B$
Near 3C 194	17.72	0.45	−0.83
Near 3C 205	18.13	.46	−0.77
Near 3C 225	19.25	.23	−0.81
Near 3C 280	19.48	0.23	−1.16

The majority of the blue objects in high galactic latitudes appear to be superluminous galaxies with very large redshifts. They appear to be optically similar to quasi-stellar radio sources, but are radio-quiet to a flux level of 10^{-25} W/m² Hz at 178 MHz. Their average absolute blue magnitude is of the order of $\langle M_B \rangle \simeq -25^m \pm 2^m$ (with $H = 75$ km/sec 10^6 pc). The average redshift at $B = 22.^m0$ is estimated to be $\Delta\lambda/\lambda_0 \simeq 5$, and may be higher. At this redshift, the metric distance reached by members of the class is about 63 per cent of the way to the horizon of the universe (if $q_0 = +1$). To this magnitude limit we are looking back into an evolutionary universe ($q_0 = +1$) about 90 per cent of the way in time to the "creation event" ($R \approx 0$). Because they are radio-quiet to the limit of existing surveys, the name "QSG," denoting quasi-stellar galaxies, might be appropriate for the class.

The evidence for these conclusions is given in the following sections, presented in the order in which the clues and final confirmation unfolded between February and May of this year. The large redshifts were predicted on the basis of §§ II and III. Confirmation came from spectral observations made with the 200-inch telescope between April 23 and May 6.

II. EVIDENCE FROM THE COLORS

We consider in this section the colors of two classes of objects: (*a*) the true quasi-stellar radio sources identified to April 15, 1965, and (*b*) the blue objects with $9^m < B < 19^m$ found in surveys at high galactic latitudes.

a) *QSS*.—Identification and photoelectric confirmation of forty-four quasi-stellar radio sources are now available. The data, taken from the literature (Matthews and Sandage 1963; Ryle and Sandage 1964; Sandage and Wyndham 1965; Sandage, Veron, and Wyndham 1965; Bolton, Clarke, Sandage, and Veron 1965), are collected in Table 2. In all but a few cases, the identifications are confirmed by excellent agreement between the radio positions and the new optical positions measured by Veron.

TABLE 2

PHOTOELECTRIC COLORS AND MAGNITUDES FOR QSS AVAILABLE TO APRIL 15, 1965

Object	*b*	*V*	*B−V*	*U−B*	Object	*b*	*V*	*B−V*	*U−B*
3C 9......	−47°	18.21	+0.23	−0.74	3C 263........	+50	16.32	+0.18	−0.56
47......	−41	18.1	+ .05	−0.65	270.1......	+81	18.61	+ .19	− .61
48......	−29	16.2	+ .42	−0.58	273........	+64	12.8	+ .21	− .85
93......	−38	18.1	+ .34	−0.50	275.1......	+79	19.00	+ .23	− .43
138*.....	−11	18.84	+ .53	−0.16	277.1......	+60	17.93	− .17	− .78
147†.....	+10	17.8	+ .65	−0.37	279	+57	17.75	+ .26	− .56
181......	+15	18.92	+ .43	−1.02	280.1......	+77	19.44	− .13	− .70
186......	+26	17.60	+ .45	−0.71	281........	+69	17.02	+ .13	− .59
190......	+22	17.46	− .20	−0.90	286........	+81	17.30	+ .22	− .84
191......	+21	18.4	+ .25	−0.84	287........	+81	17.67	+ .63	− .65
196......	+33	17.6	+ .60	−0.43	298........	+61	16.79	+ .33	− .70
204......	+36	18.21	+ .55	−0.99	334........	+41	16.41	+ .12	− .59
207......	+30	18.15	+ .43	−0.42	336........	+42	17.47	+ .44	− .79
208......	+33	17.42	+ .34	−1.00	446........	−49	18.39	+ .44	− .90
215......	+37	18.27	+ .21	−0.66	CTA 102......	−38	17.32	+ .42	− .79
216......	+43	18.3	+ .50	−0.60	MSH 13-011...	+55	17.68	+ .14	− .66
217......	+43	18.50	+ .25	−0.86	0106+01......	−61	18.39	+ .15	− .70
245......	+56	17.25	+ .45	−0.83	0922+14......	+41	17.96	+ .54	− .52
247......	+62	18.82	+ .52	−0.25	0957+00......	+40	17.57	+ .47	− .71
249.1....	+38	15.72	− .02	−0.77	1116+12......	+64	19.25	+ .14	− .76
254......	+66	17.98	+ .15	−0.49	1217+02......	+64	16.53	− .02	− .87
261......	+73	18.24	+0.24	−0.56	1252+11......	+75	16.64	+0.35	−0.75

* 3C 138, corrected by $E(B-V) = 0.30$, $E(U-B) = 0.22$, gives $V_0 = 17.9$, $(B-V)_0 = +0.23$, $(U-B)_0 = -0.38$.
† 3C 147, corrected by $E(B-V) = 0.30$, $E(U-B) = 0.22$, gives $V_0 = 16.9$, $(B-V)_0 = +0.35$, $(U-B)_0 = -0.59$.

The distribution of colors is shown in Figure 1. Closed circles are the measured colors with no correction for galactic reddening. The two open circles represent 3C 138 and 3C 147, where reddening corrections of $E(B-V) = 0.^m30$ and $E(U-B) = 0.^m22$ have been applied. The correction for 3C 147 was determined directly by measurement of *UBV* for neighboring galactic stars. The same correction was arbitrarily applied to 3C 138—justified by the equal latitudes of the two QSS on either side of the galactic plane and by the similarity of galactic longitude. The five open triangles are for the N-type galaxies 3C 109, 3C 212, 3C 234, 3C 287.1, and 3C 459, which were measured in the general identification program for radio sources.

The two straight lines represent the black-body relation and the distribution

$$F(\nu)d\nu = Ae^{-C\nu}\,d\nu\,, \qquad (1)$$

where A and C are constants. The $U - B$ and $B - V$ colors were computed by relations given elsewhere (Matthews and Sandage 1963).

The distinguishing ultraviolet excess of the QSS relative to normal stars is evident. About half of the QSS fall in the region below the black-body line, known to be occupied by the white dwarfs (Johnson and Morgan 1953; Eggen and Greenstein 1965). The other half fall above the black-body line in a region occupied by old novae, SS Cyg, U Gem, and Z And–type variables, as first described by Walker (1957).

b) High-latitude blue objects.—Many surveys for blue stars in high galactic latitudes have been made in the past 40 years by a number of people. The earliest are the two studies by Malmquist (1927, 1936), which covered an area of about 150 square degrees near the north galactic pole to $m_{pg} \simeq 15^m$. Eighteen stars with negative colors were found in a sample which contained nearly 8200 stars. Following this pioneer work, surveys were

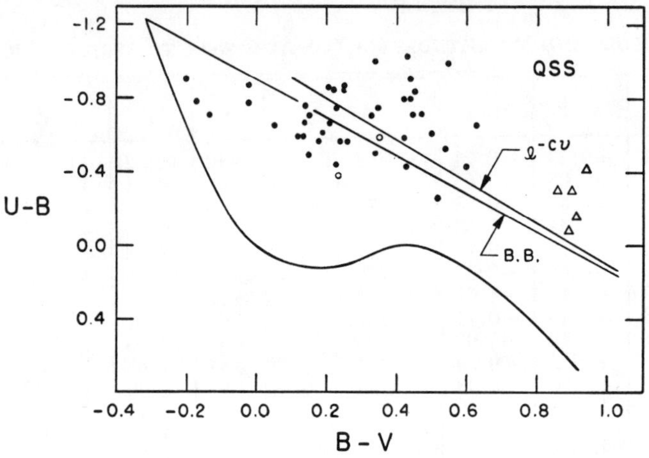

FIG. 1.—The distribution of color indices for 44 quasi-stellar radio sources (*circles*) plus 5 N-type galaxies. The normal luminosity class V line is shown together with lines for black bodies and for objects that radiate with a distribution shown by eq. (1) of the text.

made by Humason and Zwicky (1947), Luyten (many papers from 1952 to the present), Iriarte and Chavira (1957), Chavira (1958), Cowley (1958), Feige (1958), Slettebak and Stock (1959), and Haro and Luyten (1962).

Most of the detailed spectroscopic and photometric work on these objects has been confined to stars brighter than $B = 15^m$. Spectroscopic work by Humason (Humason and Zwicky (1947), Greenstein (1956, 1961), Slettebak (Slettebak, Bahner, and Stock 1961), Klemola (1962), and Berger (1963) had shown that the blue objects brighter than $B \simeq 15$ are stars of a variety of types including white dwarfs, hot subdwarfs, horizontal-branch stars, and composite stars. Photometry by Iriarte (1958, 1959), Slettebak, Bahner, and Stock (1961), Klemola (1962), Harris (unpublished), and Eggen and Sandage (1965) indicated that the spectral classifications are consistent with the colors. Furthermore, statistical parallax determinations giving $\langle M_B \rangle \simeq +3$ for stars brighter than $B \simeq 15^m$ by Luyten (1956*b*, 1959, 1960, 1962*a*, *b*), and Klemola (1962) are consistent with assigning many of the stars to the globular-cluster-like halo population.

But study of the photometric data already in the literature gives a hint that something unexpected occurs for $B > 15^m$. Klemola's discussion (1962, Fig. 2) of the available data, especially those of Iriarte (1959), suggested that a transition in the color distribution occurs at about $B = 14.^m5$, but the data were too scanty for definite conclusions.

In February of this year, following the clue of Table 1, Veron and I made a special color survey on two plates taken with the Palomar 48-inch Schmidt to see if abnormal colors are a universal property of the interlopers. Veron marked thirty-one objects that had large negative $U - B$ indices. Photoelectric photometry was obtained for twenty-one of these objects at the end of February. The unexpected result (Sandage and Veron 1965) was that fifteen of the twenty-one interlopers had colors that imitated the known QSS, five had colors close to the luminosity class V line, and one appears to be an F subdwarf. These data, together with those discussed by Klemola, suggested that the majority of blue objects fainter than $B \approx 15$ are fundamentally different from those brighter than this limit.

Figure 2 shows the results, divided into two groups at apparent visual magnitude $V = 14.^m5$. The data are from Feige (1958), Iriarte (1959), Slettebak, Bahner, and Stock (1961), Klemola (1962), Harris (unpublished observations of the Humason-Zwicky stars made available by Greenstein), Eggen and Sandage (1965), Sandage and Veron (1965), and Table 1 of this paper. The sharp division between Figure 2, *a* and *b*, is clear. Stars

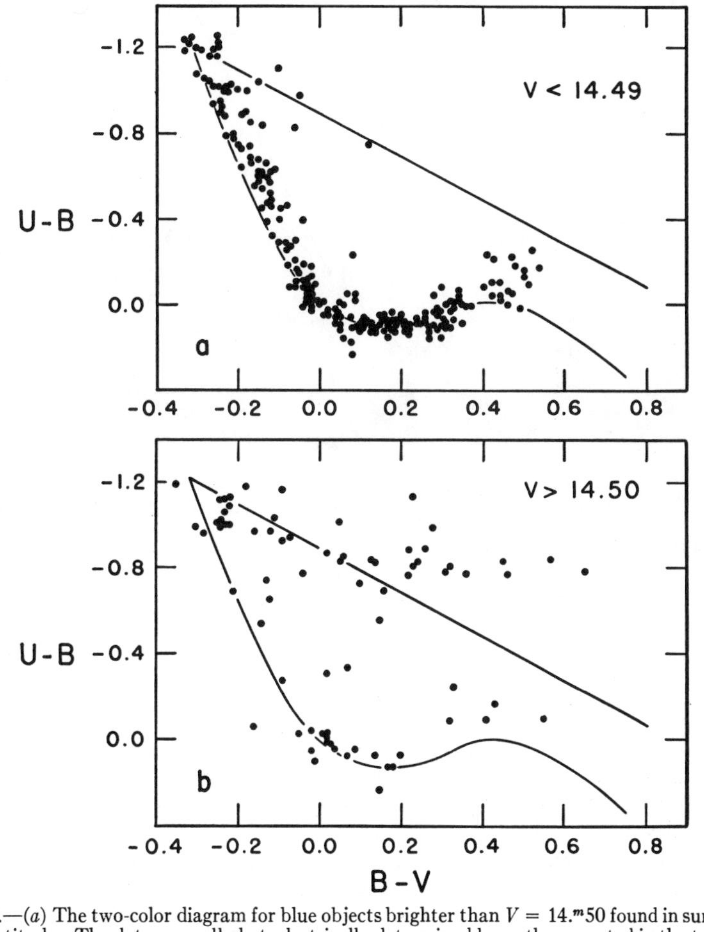

FIG. 2.—(*a*) The two-color diagram for blue objects brighter than $V = 14.^m50$ found in surveys at high galactic latitudes. The data were all photoelectrically determined by authors quoted in the text. (*b*) Same as *a* for objects fainter than $V = 14.^m50$.

in Figure 2, *a*, follow the luminosity class III–V line, while most objects in Figure 2, *b*, imitate the distribution of QSS in Figure 1. The results suggest that we are beginning to run out of galactic halo stars at about $B \simeq 15^m$ and are starting to pick up objects with QSS colors fainter than this limit. But are members of this new class extragalactic? Evidence that this is the case came from analysis of the integral count, $N(m)$, from the catalogue of Haro and Luyten (1962).

III. EVIDENCE FROM THE COUNTS

The HL catalogue lists 8746 blue objects found in 2000 square degrees of the south galactic polar region. The care with which the survey was conducted insures adequate

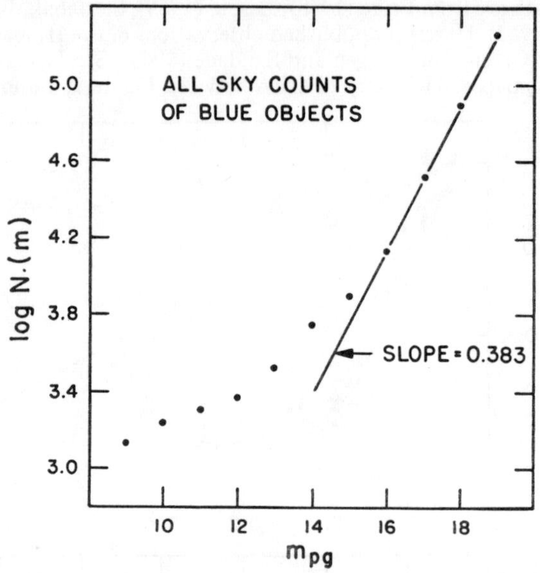

FIG. 3.—The integral-count–apparent-magnitude relation for objects in the Haro-Luyten catalogue with tabulated $U - V$ values smaller than $-0.^m8$. The counts have been normalized to give all-sky values. $N(m)$ is the number of objects in 41253 square degrees brighter than apparent magnitude m. Note the sharp change of slope near $m_{pg} = 15^m$.

completeness to $B = 19.^m0$ within the imposed color limits. Figure 3 shows log $N(m)$ versus m_{pg} for objects in the catalogue with tabulated $U - V$ values smaller than $-0.^m8$. The counts are from the HL Table V, but with log $N(m)$ increased by 1.314 to convert the tabulated numbers to all-sky values in 41253 square degrees.

The striking feature of Figure 3 is the sharp change of slope near $m_{pg} = 15$. Brighter than $m_{pg} = 13$, the slope is $d \log N(m)/dm = 0.070$. From $m_{pg} = 16^m$ to the limit of the data, the slope is remarkably constant with a value $d \log N(m)/dm = 0.383$.

Figure 3 *could* be explained by a single class of galactic stars (of constant mean M_B) only if the spatial density $\Delta(r)$ drops precipitously from $m_{pg} = 9$ to $m_{pg} = 12$, undergoing a transition in the gradient $d\Delta(r)/dr$ from magnitude 12 to magnitude 16, and then leveling off to a constant gradient beyond $m_{pg} = 16$. In the distance interval corresponding to apparent magnitudes 9–12, the slope of $d \log N(m)/dm = 0.070$ requires that

$$\frac{d \log \Delta(r)}{dm} = -0.530, \tag{2}$$

which means

$$\frac{d \log \Delta(r)}{dr} = -\frac{115}{r} \tag{3}$$

or

$$\Delta(r) = \text{const.}\, r^{-2.65} \qquad \text{for } m_{pg} < 12^m , \tag{4}$$

where r is the distance from the Sun in the direction of the Haro-Luyten survey. Between $m_{pg} = 12$ and $m_{pg} = 16$, the gradient would have to become less steep, reaching a constant value for distances corresponding to $m_{pg} = 16$, such that

$$\Delta(r) = \text{const.}\, r^{-1.085} , \qquad m_{pg} > 16^m . \tag{5}$$

Fainter than $m_{pg} = 19$, the density gradient would presumably have to steepen again to prevent the mass of the halo from becoming infinite. A change of density gradient of this type is unknown for any class of object in the Galaxy.

The possibility that all the blue objects from $19 \geq B > 9$ are stars of a single type cannot be excluded on the basis of Figure 3 alone, but, if this were true, the change in color shown in Figure 2, *a* and *b*, at the point of slope transition of the log $N(m)$-curve would not be explained. Of course, two main types of galactic stars could be postulated—the bright class with $\langle M_B \rangle \simeq +3$, as required by the proper-motion calibration (Luyten 1956*b*, 1959, 1960, 1962*a*, *b;* Klemola 1962), and the faint class with a density decline as $r^{-1.085}$, but this second group would then have to be almost entirely SS Cyg–U Gem-type variables or old novae to explain the colors of Figure 2, *b*. This possibility seemed so remote that a second explanation was sought in terms of QSG.

The principal question now to be answered is: if the objects fainter than $B \simeq 16^m$ are predominantly extragalactic, why does the slope, $d \log N(m)/dm$, have a value of 0.383 rather than 0.60, which applies for normal galaxies at this magnitude level? The answer is: the objects are not of normal absolute luminosity but are brighter. They are so distant for apparent magnitudes fainter than 19^m that the predicted cosmological bending of the log $N(m)$-curve (Sandage 1961*a* [hereinafter called "Paper I"], Fig. 4; 1963, eqs. [26]–[33]) has taken place at these bright apparent magnitudes. The theoretical slope at any desired m can be predicted once the mean absolute luminosity is fixed, and the result is that $d \log N(m)/dm = 0.40$ *if we assign* $\langle M_B \rangle$ *the value that applies to the quasi-stellar radio sources of known redshift*. The argument is as follows.

The important theoretical point is that the log $N(m)$-curve is obtained from theory by a series of transformations from the volume $V(r)$ contained within coordinate distance r. The number count $N(r)$ is proportional to $V(r)$ for uniformly distributed sources. Now $N(r)$ can be uniquely transformed into $N(z, q_0)$ because the metric distance is given once z and q_0 are known (see Sandage 1961*a*, eqs. [50] and [10′]). The fundamental point is that $N(z, q_0)$ is unique—it depends only on the geometry and not upon the objects being counted. However, if we wish to transform $N(z, q_0)$ into the observable $N(m, q_0)$ relation, the shape of the function at any m does depend on a physical parameter of the objects; namely, the mean absolute luminosity, or, more explicitly, on the zero point of Hubble's expansion law, which gives m as a function of z.

These considerations show that the slope of the $N(m)$ relation has a particular value at a particular redshift; but, if we are counting two classes of objects that differ by ΔM in mean absolute magnitude, this given slope will occur at *different* apparent magnitudes, separated by $\Delta m = \Delta M$.

Numerical values of $N(m)$ have been tabulated (Sandage 1961*a*,[1] Tables 3 and 4) for

[1] It has been pointed out by E. L. Schücking and by van Albada (1963) that an error exists in Fig. 3 and Table 3 of my paper (1961*a*) due to choice of the incorrect value of the double-valued arc sin P beyond arc sin $P = 1$. The correction procedure on p. 372 (eq. [41]) of the paper happens to be valid, but for the wrong reasons. Therefore, Table 4 and Fig. 4 are correct, but the discussion about the antipole is

the special case of average field galaxies, which obey a Hubble relation

$$m_i - K_i = 5 \log z + 22.516 \qquad \text{for } z \ll 1\,. \quad (6)$$

Here m_i is the observed heterochromatic magnitude in band pass i, and K_i is the "bolometric" correction due to redshifting the spectrum through the ith filter band (see Humason, Mayall, and Sandage 1956, Appendix B; Sandage 1965, for the theory of the K-correction). For a different class of objects where the constant in equation (6) has a

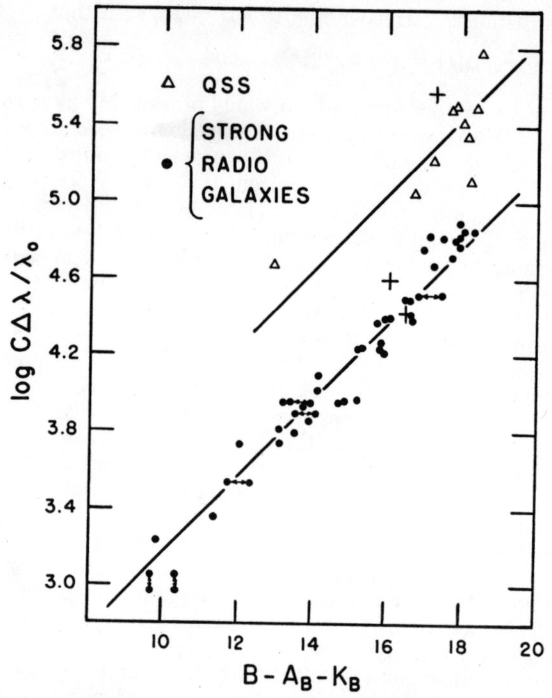

FIG. 4.—The redshift-apparent-magnitude relation for strong radio galaxies (*circles*) and for quasi-stellar radio sources (*triangles*). Arrows connect dumbbell galaxies for which the magnitude of the bright component and of the two components combined is given. Correction for K-dimming has been applied to the galaxies. No K-correction has been applied for the quasi-stellar sources. The three crosses are discussed in § IV.

different value C, Tables 3 and 4 of Paper I can still be used to obtain the shape of $N(m)$, providing that the argument of the tables is changed by $22.516 - C$ magnitudes. When this is done, the tables will be entered at the same value of z for the two objects. It is important, therefore, to know the $[m, z]$ relation for the objects under discussion.

Figure 4 shows the Hubble diagram for strong radio galaxies and for quasi-stellar sources as determined from unpublished photoelectric photometry (Sandage 1965), and from redshifts by Schmidt (1965*a*), with a few by Minkowski and by Greenstein. The radio galaxies are shown as closed circles, while the quasi-stellar sources are open tri-

erroneous. The caption on Table 5 should read *equator* rather than *antipole*. But again, all numerical values of the critical Table 3, supplemented where necessary by Table 4, can be used without change as they are unaffected by the conceptual mistake.

angles. Plotted[2] is log $c\Delta\lambda/\lambda_0$ versus the photoelectric B magnitude corrected for aperture effect, galactic absorption A_B, and heterochromatic K-dimming discussed elsewhere (Sandage 1965). Redshifts for the quasi-stellar sources are due entirely to Schmidt, except for 3C 48, which was announced by Greenstein and Matthews (1963). The spectroscopic and photometric data on the ten quasi-stellar sources are summarized in Table 3.

The K-correction has been applied in Figure 4 to the galaxies but not to the QSS because it is shown in the Appendix that, if the energy distribution of quasi-stellars is of the form of equation (1), the K-dimming is very small even for large redshifts. This is because the high-energy part of the spectrum is being shifted into the visual region, in contrast to the situation for normal galaxies where the ultraviolet energy is very small. Equation (1) has the further property that the anomalous $U - B$ index will continue to be abnormal for redshifts as high as $z = 3$ (see Table A2 of the Appendix), which explains why 3C 9, with $z = 2.012$, still has the peculiar observed color indices. This situation, ideal from the discovery standpoint, will probably change, however, for redshifts greater than $z = 3$ when the Lyman continuum will be shifted to $\lambda \geq 3600$ Å. If a large Lyman jump

TABLE 3

DATA FOR THE QSS SHOWN IN FIGURE 4

Object	z	log cz	V	$B-V$	$U-B$	Redshift Ref.*
3C 9......	2.012	5.781	18.21	+0.23	−0.74	(5)
3C 47.....	0.425	5.106	18.1	+ .05	− .65	(3)
3C 48.....	0.367	5.042	16.2	+ .42	− .58	(2)
3C 147....	0.545	5.214	16.9	+ .35	− .59	(3)
3C 245....	1.029	5.490	17.25	+ .45	− .83	(5)
3C 254....	0.734	5.343	17.98	+ .15	− .49	(5)
3C 273....	0.158	4.676	12.9	+ .21	− .85	(1)
3C 286....	0.846	5.404	17.30	+ .22	− .84	(4)
3C 287....	1.054	5.500	17.67	+ .63	− .65	(5)
CTA 102..	1.038	5.493	17.32	+0.42	−0.79	(5)

* (1) Schmidt 1963; (2) Greenstein and Matthews 1963; (3) Schmidt and Matthews 1964; (4) Schmidt 1962; (5) Schmidt 1965.

is present, the observed U magnitude will be depressed, as explained in the Appendix, and the two-color technique may no longer be useful for survey purposes. This is expected to occur near $B \simeq 21^m$, which is well within the limit of the 200-inch telescope.

Figure 4 shows that the ten QSS show a relatively small scatter about a mean $[m, z]$ line whose equation is

$$m_B = 5 \log z + 18.186 . \quad (7)$$

The data are not yet extensive enough to detect a second-order term, and a straight line ($q_0 = +1$) is adopted, which is accurate enough for the present purpose. (Future work will, of course, aim to find q_0 by using data of this kind.)

The difference in the constant between equations (6) and (7) is $4.^m33$, which is the difference in the mean absolute magnitudes of QSS and the field galaxies in the redshift catalogue (Humason, Mayall, and Sandage 1956). Table 3 of Paper I has been entered with this shift of argument and the predicted log $N(m)$-curves for $q_0 = 0$, and $q_0 = +1$ are shown in Figure 5. Here the $q_0 = 0$ curve has been normalized to fit the observed

[2] It should be noted in passing that multiplication of $\Delta\lambda/\lambda_0$ by the velocity of light is a convention which should probably now be abandoned. The function $c\Delta\lambda/\lambda$ is not the "velocity of expansion" and the relativistic Doppler equation must not be applied to the ordinate in Fig. 4. The fundamental parameter in the theory of expanding spaces is $\Delta\lambda/\lambda_0$ itself, which of course becomes larger than 1, approaching ∞ as the horizon is reached.

counts in the HL catalogue, but the normalization is slightly arbitrary because our present knowledge does not yet tell us what percentage of real stars exist at $B = 19^m$. However, the problem at hand, namely, the value of the slope of log $N(m)$, is independent of the normalization. The mean slope between $m_{pg} = 16$ and $m_{pg} = 19$ for the two cases are $d \log N(m)/dm = 0.403$ for $q_0 = 0$, and $d \log N(m)/dm = 0.360$ for $q_0 = +1$, values that nicely bracket the observed slope of 0.383 and that are very different from 0.60 expected for extragalactic objects with negligible redshifts.

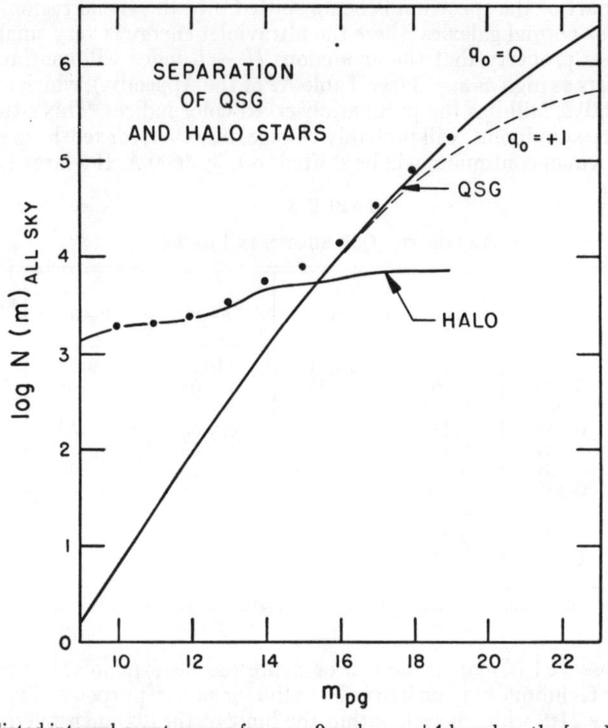

FIG. 5.—Predicted integral-count-curves for $q_0 = 0$ and $q_0 = +1$ based on the $[m, z]$ relation for QSS of Fig. 4, fitted to the data of Fig. 3. The normalization has been arbitrarily made for the $q_0 = 0$ curve at $m_{pg} = 19$. The curve marked *halo* is obtained by subtraction of the QSG-curve for $q_0 = 0$ from the observed points. The $N(m)$-values are the number of objects brighter than m_{pg} in 41253 square degrees.

The present analysis should not be carried too far at the moment. For example, no cosmological conclusions on the value of q_0 should be inferred from Figure 5, even though the log $N(m)$-values differ significantly between $q_0 = +1$ and $q_0 = 0$. We must have observations of the QSG at bright magnitudes to properly normalize the log $N(m)$-curves of Figure 5. Equally important, Figure 4 must be better defined because a small change in the constant in equation (7) is equivalent in its effect to the differences between $N(m)$-curves for the various q_0-values in Figure 5.

Figure 5 shows the subtraction of the cosmological $N(m)$-curve (for $q_0 = 0$) from that of the observed HL points. The subtraction separates the true halo stars from the (hypothetical to this point in the discussion) blue galaxies, and shows that the two $N(m)$-curves cross near $m_{pg} = 15$, which is just where the color distribution changes. Table 4 gives the ratio of stars to QSG based on this subtraction, but it should be emphasized that these numbers show only the trend to be expected. The numbers themselves are quite uncertain because the subtraction of two large numbers is involved. All that can safely be generalized is that Table 4 qualitatively explains the color change in Figure 2, *a* and *b*, but the exact ratios must await the results of more detailed surveys.

The results of §§ II and III were established before redshifts were obtained. They constituted a prediction that the radio-quiet blue objects fainter than $B \simeq 15$ that have peculiar colors should have large redshifts. To prove this it was necessary to test for the redshifts, and observations to this end were undertaken April 23–27 and May 2–6, 1965.

IV. EVIDENCE FROM THE REDSHIFTS

Objects for the test were chosen from Iriarte's photometric list (1959) of Tonantzintla blue stars, from Veron's list (Sandage and Veron 1965) for which photometry had been completed in February, and from the interlopers of Table 1. Maarten Schmidt generously offered to observe a few of the candidates during his Palomar run in the last week in April to insure against the possibility of cloudy weather during my scheduled run in the first week of May. The insurance paid off, because the May 2–6 period was abnormally stormy on Palomar Mountain, characterized by mountain fog blown by high winds from the valley, making the seeing images large during the periods of clearing. However,

TABLE 4

PREDICTED RATIO OF HALO STARS TO QSG FROM FIGURE 5

Survey Limit (*B* Mag.)	$N(m)$ Stars/ $N(m)$ QSG	Survey Limit (*B* Mag.)	$N(m)$ Stars/ $N(m)$ QSG
12	26.3	16	0.60
13	9.6	17	.24
14	4.1	18	.10
15	1.5	19	0.04

spectra were obtained of several of the brighter objects of the candidate list during this time.

A total of ten spectrograms of six objects were taken during the two runs with the 200-inch prime-focus spectrograph, using the F/1 camera at a dispersion of 400 Å/mm. The plates were baked Eastman Kodak IIa-D, covering the wavelength range 3180–6300 Å, or IIa-F, which extends the spectral region to 6700 Å. Schmidt obtained four spectra of three faint objects, one of which proved to be crucial, while I got six spectra of three additional objects, two of which gave positive confirmation. The results are as follows:

BSO 16.—This is one of the blue stellar objects found in the special photographic survey carried out by Veron in February. The photoelectric photometry gave $V = 16.17$, $B - V = -0.25$, and $U - B = -1.01$. One spectrogram by Schmidt shows conclusively that the object is a star of small velocity with Hβ, Hγ, Hδ, and Hϵ in absorption. No other features are clearly visible. The result is expected because the colors are unlike those of QSS, but lie close to the luminosity class III–IV line in Figure 2, *a*, for normal stars.

Near 3C 194.—This was the first of the interlopers, whose photometric properties are given in Table 1. One spectrogram by Schmidt is continuous, with no obvious emission or absorption features. It is similar in this respect to several other QSS, such as 3C 196 (Schmidt 1965*c*). The continuous spectrum extends far into the ultraviolet, as expected from the $U - B$ color, and looks similar in all respects to the continuum of other quasi-stellar radio sources studied to date. However, the absence of lines precludes use of this object as a test of the hypothesis.

BSO 1.—Two spectra by Schmidt show the presence of two very broad emission lines. Visual measurement for wavelength with a microscope gave values of λ 3473 and λ 4279 Å for the two lines. The redward line is exceedingly difficult to see, but analysis of the plates using Westphal's (1965) cross-correlation technique definitely proves that the line is present with a wavelength of λ 4284 Å. The value is well determined by the cross-correlation method if the line is symmetrical. On both plates the blueward line is easy to see and is bisected by a sharp absorption feature, flanked by the broad emission. We interpret this to mean self-absorption. The identification of the lines is not completely certain because the wavelength ratio of 1.233 is given by four possible pairs of coincidences in lists of candidate lines prepared by Schmidt (1965*b*) and by Bowen (private communication). The four cases are (1) Mg II (λ 2798) and [Ne V] (λ 3426), ratio 1.224; (2) O III (λ 3133) and [Ne III] (λ 3869), ratio 1.235; (3) [Ne V] (λ 3346) and Hδ (λ 4102), ratio 1.226; and (4) C IV (λ 1550) and C III (λ 1909), ratio 1.232. Cases (1), (2), and (3) can be excluded because of the absence of other lines that should be present, such as the Balmer series and λ 3727 of [O II]. Furthermore, the wavelength ratios for cases (1) and (3) are outside the measuring errors. Case (2) can be further excluded because the [Ne V] lines (λλ 3346, 3426) should appear, but they do not. Case (4) appears to be the most probable for the following reasons: (1) The λ 1550 line has been previously observed in the quasi-stellar sources 3C 9 and 3C 287, while λ 1909 appears in 3C 254, 3C 245, CTA 102, and 3C 287 (Schmidt 1965*b*). (2) The observed wavelength ratio agrees almost exactly with the laboratory value. (3) The self-absorption of the observed λ 3473 line is consistent with the identification as the ground-state doublet λ 1550, because this is the resonance line of C IV, being the same sort of transition as the D-lines of sodium and the H- and K-lines of singly ionized calcium. The one objection to the identification is the apparent absence of Mg II (2798), which should appear at λ 6269. Schmidt took a special IIa-F plate to attempt to find the line. There may be faint features in that region on both plates, but neither visual inspection nor Westphal's cross-correlation technique has definitely confirmed its existence. The value of the redshift of BSO 1 is, therefore, not beyond all question, but what *is* established is that a radio-quiet object, resembling the QSS in optical properties, exists whose redshift is certainly not zero.

The redshift $\Delta\lambda/\lambda_0 = 1.241$ has been adopted as highly probable for the discussion. The photometry gives $V = 16.98$, $B - V = 0.31$, and $U - B = -0.78$, which places BSO 1 among the brightest QSS in Figure 4. The object is shown as a cross in this diagram at $B = 17.^m29$.

BSO 105.—One spectrogram shows a continuum extending far into the ultraviolet with no obvious emission or absorption features. It closely resembles the object near 3C 194 and has the same bearing on the problem—neither confirming nor denying the hypothesis.

Ton 730.[3]—Three spectra are available. Two were taken in very poor seeing (10″ and 5″ for the seeing disk), and one was obtained under more normal conditions. Photometry by Iriarte (1959) gives $V = 15.91$, $B - V = 0.57$, $U - B = -0.84$—colors that resemble those of QSS. The spectrum shows sharp, strong lines of forbidden [O III] (λλ 5007, 4959), and very broad hydrogen lines of Hβ, Hγ, Hδ, and Hε.

The lines of [O II] (λ 3727) and [Ne V] (λλ 3426, 3346) are also present, but with lower intensity and with smaller contrast with the continuum. The hydrogen lines are about 35 Å wide and resemble in this respect all quasi-stellar radio sources known to date. Ton 730 is, however, radio-quiet to 10 flux units at 178 MHz. It does not appear in the 3C catalogue nor in the summary catalogue of Howard and Maran (1965).

[3] It is of some interest that, after the observing list had been prepared for the May 2–6 Palomar run, a letter, dated April 20, 1965, was received from Iriarte in which he pointed out that the colors of Ton 730 were like those of some of the QSS and that the object was worthy of special attention. A finding chart was inclosed, and I wish to record my indebtedness to him for his letter and to suggest that he had obviously made a tentative connection between certain of the blue objects and the quasi-stellar sources.

The redshift is unambiguous with a value $\Delta\lambda/\lambda_0 = 0.0877 \pm 0.0004$ (m.e.), which shows that the object is extragalactic. The position of Ton 730 in Figure 4, shown by the cross at $B = 15.48$, is fainter than any of the ten QSS with known redshifts. It falls near the line for the strong radio galaxies rather than near the mean QSS line. However, it should be remembered that this fainter line is almost the *upper envelope* of the $[m, z]$ relation for all known normal galaxies, including the brightest optical galaxies in the great clusters (Sandage 1965). Thus, Ton 730 is an extremely bright object as galaxies go ($M_B = -22.^m2$ if $H = 75$ km/sec 10^6 pc). Furthermore, comparison with "normal" QSS is dangerous at the present time because the sample of QSS is much too small to know if the dispersion in absolute magnitude of the quasi-stellars is great enough to contain Ton 730 within the scatter.

Although the results were encouraging, they did not completely verify the hypothesis of stellar-appearing galaxies as the main constituent of the faint blue objects in the poles, for the following reason. The optical image of Ton 730 is slightly fuzzy and elongated on 48-inch Schmidt plates. On very close inspection, this object would not be identified as a star in surveys. The image looks like an extremely compact galaxy on long-exposure plates, although to the eye at the 200-inch telescope it is stellar, and it appears almost stellar on short exposures with the Schmidt. Our present interpretation is that the fuzziness is caused by material blown off from the central core as in the case of 3C 48 and 3C 196, with their attendant small nebulae (Matthews and Sandage 1963). But this circumstance appeared to exclude Ton 730 as a *proof* of the existence of stellar-appearing galaxies. In any case, Ton 730 is exceedingly peculiar and is unlike any non-radio object found to date. Its spectrum is similar to the QSS 3C 249.1 (Schmidt, private communication). Its colors imitate the QSS and it is undoubtedly related to the general class of objects under discussion.

Ton 256.—Iriarte (1959) gives $V = 15.91$, $B - V = 0.57$, $U - B = -0.84$. The object looks completely stellar on the sky survey plates and has the peculiar colors of the QSS. The two spectrograms obtained show intense, sharp lines of [O III] (λλ 5007, 4959), [Ne III] (λ 3869), [O II] (λ 3727), and broad emission lines of Hβ, Hγ, Hδ, Hε, and [Ne V] (λλ 3426, 3346). The spectrum is similar to that of Ton 730 except that the emission lines are more intense relative to the continuum. The best spectrogram is reproduced in Figure 6, where a convenient fiducial wavelength reference is provided by the bright night-sky line of [O I] at λ 5577. All spectral features are shifted by $\Delta\lambda/\lambda_0 = 0.1307 \pm 0.0002$ (m.e.), which is a very large shift for this apparent magnitude for normal galaxies. The object is radio-quiet to the limit of the 3C catalogue.

Ton 256 has been placed in Figure 4 as a cross at $B = 16.^m49$. It is brighter than the radio-galaxy line and has an absolute magnitude similar to the quasi-stellar radio source 3C 47. This object, together with BSO 1, with confirmatory evidence from Ton 730, appears to provide substantial evidence for the hypothesis of the existence of a new class of extragalactic objects—the radio-quiet QSG.

V. THE SURFACE AND SPATIAL DENSITIES OF QSG

Surface density.—If most of the objects fainter than $m_{pg} = 16^m$ in the HL catalogue are QSG, the average surface density at any apparent magnitude can be found from Figure 5. Reading from the curves and reducing to numbers per square degree gives the values in Table 5, where the numbers for $q_0 = +1$ have been renormalized to give the observed surface density at $m_{pg} = 19^m$.

The K-correction will begin to be appreciable for the fainter entries in the table because the redshifts are expected to be about $z \geq 2$ at $m_{pg} - K = 20^m$, and larger than 5 at $m_{pg} - K = 22^m$. At these redshifts the problems involving the Lyman continuum will be paramount, making the observed magnitudes much fainter than would otherwise be the case. Furthermore, the expected disappearance of the ultraviolet excess at redshifts greater than 3, due to the presence of the Lyman jump under the U filter, means

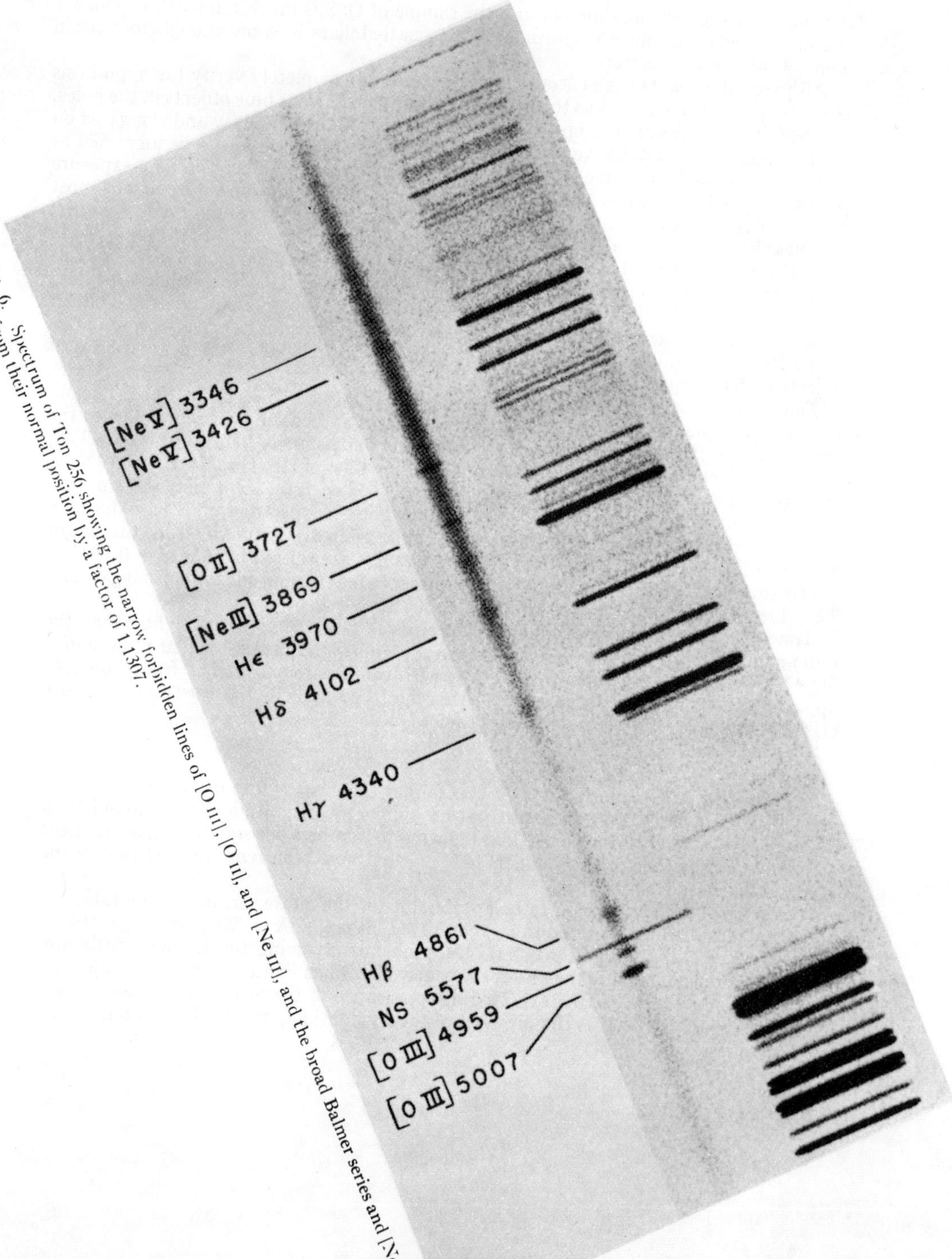

FIG. 6. Spectrum of Ton 256 showing the narrow forbidden lines of [O III], [O II], and [Ne III], and the broad Balmer series and [Ne V] lines, redshifted from their normal position by a factor of 1.1307.

that new search techniques must be developed before objects of this type can be discovered at the faintest light levels. The most likely bet appears to be long-exposure, two-color plates in the red and near-infrared (to compensate for the redshift), but criteria must be found for picking up the abnormal colors of QSG at these long wavelengths.

Table 5 also shows the necessity of using instruments of wider field than that possessed by the 200-inch prime focus. Even neglecting the K-term at $m_{pg} = 23$, there are only 4.7 expected quasi-stellar galaxies in the 0.077 square-degree field for the $q_0 = 0$ model, and only 1.6 QSG per plate for $q_0 = +1$. Nevertheless, with proper instrumentation and new techniques, the possibility of choosing between model universes on the basis of counts now seems to exist, and every effort toward this goal will be expended.

It is of some interest to note how many bright QSG there should be at various apparent magnitudes. Figure 5 predicts that 6 of these objects should exist brighter than $m_{pg} = 10^m$, 23 brighter than 11^m, 83 brighter than 12^m, and 316 brighter than 13^m over the entire celestial sphere—numbers that can eventually be checked.

Spatial density.—The volume density of QSG can be estimated approximately by combining the data of Figures 4 and 5 at bright magnitudes where the effects of spatial curvature and redshift are negligible. Figure 5 shows that there should be 83 QSG over the entire sky brighter than $m_{pg} = 12^m$. The mean redshift at this magnitude, obtained from Figure 4, is $z = 0.058$, which corresponds to a distance of 2.3×10^8 pc, or a volume of 5.2×10^{25} pc^3 (assuming a Hubble constant of 75 km/sec Mpc). There is, then, an average of one QSG every 6.2×10^{23} pc^3, which gives a density of 5.4×10^{-80} QSG/cm^3.

TABLE 5

PREDICTED NUMBER OF QSG BRIGHTER THAN m PER SQUARE DEGREE AND PER 200-INCH PRIME-FOCUS PLATE

$m_{pg} - K$	$q_0 = 0$		$q_0 = +1$	
	$N(m)$	$N(m)/12.9$	$N(m)$	$N(m)/12.9$
18	1.8	0.14	2.1	0.16
19	4.2	0.32	4.2	0.32
20	8.7	0.67	7.3	0.56
21	17.8	1.4	11.6	0.90
22	33.1	2.6	16.0	1.2
23	60.3	4.7	20.2	1.6

Compare this with the density of galaxies whose mean luminosity is the same as those contained in the redshift catalogue (Humason, Mayall, and Sandage 1956); i.e., galaxies that obey $m_p - K_p = 5 \log z + 23.616$. This equation, together with the integral count relation of Holmberg (1957) given by $\log N(m) = 0.6m_{pg} - 8.87$ for $m_{pg} < 13^m$, gives a spatial density of one galaxy every 1.0×10^{75} cm^3, or 1.0×10^{-75} gal/cm^3. The ratio of the number of average galaxies to QSG is 1.9×10^4. This ratio would be higher by a factor of 8 if the mean absolute magnitude of the galaxies was taken to be $1.^m5$ fainter—justified by noting that the redshift catalogue is biased toward the brighter, more easily observed galaxies. Reasonable limits to the ratio of numbers per unit volume of normal optical galaxies to QSG would then be 2×10^4 to 2×10^5.

Therefore, the QSG contribute negligible mass density to the universe. Even supposing they had a very high mass of $10^{14}\ \mathfrak{M}_\odot$, the average density due to QSG would be only 10^{-32} gm/cm^3, which is about 10 times smaller than that usually assumed for normal galaxies (Hubble 1926; Whitford 1954; Oort 1958).

An important ratio is the number of QSG to QSS per unit volume because *if* a QSS evolves into a QSG, as seems likely, the ratio gives the relative lifetimes. There are 329 non-galactic radio sources in the 3C revised catalogue (Bennett 1962), which covers slightly more than half the sky to 9 flux units. The current optical identification program shows that 30 per cent of the 3C R sources are QSS, giving a surface density of 5×10^{-3} QSS/square degree. This is to be compared with 4 QSG/square degree to $m_{pg} = 19$, assuming that most of the faint Haro-Luyten objects are extragalactic, giving a ratio of surface densities of 800. An estimate of the volume-density ratio can be obtained if we know the radio volume surveyed for QSS to 9 radio flux units and the optical volume corresponding to $m_{pg} = 19^m$. A simple calculation can be made only if the radio and optical absolute luminosities of the QSS and QSG show small dispersion. Schmidt's (1965*b*) data for the nine QSS with known redshifts suggest that this is the case, and we shall adopt it. Reducing his monochromatic radio powers to 178 MHz with a spectral index of -0.7 and adopting a Hubble constant of 75 gives an average absolute power of 2.7×10^{28} W(Hz)$^{-1}$ for the known QSS, a value which permits the distance of a source with flux density 9×10^{-26} W m^2 Hz at 178 MHz to be found from the standard equations (Sandage 1961*a*, eq. [50]). Similarly, adopting the QSS line in Figure 4 to represent the QSG relation, we can find the redshift at $m_{pg} = 19$ and, hence, the coordinate distance surveyed for such sources, using the same value of H. Fortuitously, the distances of the radio and optical surveys to the stated limits turn out to be nearly equal, so the volumes surveyed will be about the same. The actual distance ratio is (optical/radio) = 1.16, which gives a volume ratio of 1.56, with the optical volume being larger. Consequently, the number ratio per unit volume of QSG to QSS is $800/1.56 \simeq 500$, which must be the ratio of the lifetimes if all QSG go through the radio phase.

Evidence has been presented elsewhere (Burbidge, Burbidge, and Sandage 1963) that the lifetime of strong radio sources is of the order of 10^6 years. This then requires that the quasi-stellar galaxies can exist in their observed state for only 5×10^8 years. If this rough estimate is substantially correct, it raises the possibility that we may be witnessing galaxies in the process of formation, because 5×10^8 years is the order of the collapse time of the halo into a disk of a normal galaxy (Eggen, Lynden-Bell, and Sandage 1962). Such a process would agree with the model of Field (1964), arrived at from entirely different grounds. If there is no way out of the short lifetime for QSG, then the conclusion is forced that the QSG phenomenon takes place at times later than the "creation event" of an evolutionary universe, because QSG are present at all distances, spanning nearly 10^{10} years in light-travel time. This fact requires that the objects either periodically flare to their superluminosity from an older, more stable state, or are continuously being born.

VI. COSMOLOGY

Schmidt's discovery (1965*b*) of exceedingly large redshifts for several quasi-stellar objects emphasizes again the importance of the QSS, and now the QSG, in testing different world models. The most sensitive test is still the deviation from linearity of the $[m, z]$ relation (Sandage 1961*a*, Fig. 1). Equations (23), (24), and (25) of Paper I show that very large differences exist in $m - K$ at a given z between the $q_0 = +1$, 0, and -1 cases for redshifts larger than 2, which is now within the observable range. Relative to a straight line ($q_0 = +1$), the $q_0 = 0$ curve should be fainter by $1.^m50$ and the steady-state case of $q_0 = -1$ should be fainter by $2.^m37$ at $z = 2$. For $z = 5$ the deviations become $2.^m25$ and $3.^m89$, respectively.

If the K-corrections can be found empirically by spectrum scanning or multicolor photometry, these theoretical differences should be easy to check by increasing the number of data points in Figure 4. Furthermore, the problem of the evolutionary change in absolute optical luminosity appears to be simpler for QSG than for giant ellipticals (Sandage 1961*b*), because the lifetimes of the QSG are short compared with the interval of cosmic time into which we are gazing; hence averages over a sufficiently large sample of objects should cancel out the evolutionary luminosity decline of a single object. The only remaining problem concerns the cosmic dispersion in luminosity of the sources themselves—a problem which should soon be solved by increasing the sample of both QSS and QSG in Figure 4. As long as the intrinsic spread is smaller than about 2^m, differences between the q_0 models as large as predicted above can be found.

A large fraction of the observable universe can be sampled using the QSG. The equation for the co-moving coordinate distance is given by

$$r = \frac{(2q_0 - 1)^{1/2}}{q_0^2(1+z)} \{q_0 z + (q_0 - 1)[(1 + 2q_0 z)^{1/2} - 1]\} \tag{8}$$

(see eq. [50] and [10′] of Paper I, noting the misprint in eq. [50]). For the simple case of $q_0 = +1$, this reduces to $r = z/(1+z)$. The horizon in this model occurs at $r = 1$. The volume contained within r is

$$V(r) = 2\,R_0^3\,[\sin^{-1} r - r(1 - r^2)^{1/2}] \tag{9}$$

(Sandage 1963, eq. [26]). Therefore, to a redshift of $z = 5$, $r = 0.833$ and the volume $V(r) = 1.048 R_0^3$. The total volume contained from the observer to the horizon at $r = 1$ is πR_0^3. We therefore sample one-third of the total space available using objects with $z = 5$. The metric distance along a geodesic from the observer to the object is $R_0 \sin^{-1} r = 0.986 R_0$ for $z = 5$, which should be compared with the maximum distance of $(\pi/2)R_0$ which can be observed in the model, showing that we reach 63 per cent of the way to the horizon with objects of this redshift.

The cosmic time into the past that we are looking, given by equations shown elsewhere (Sandage 1961*b*), is $\tau/t_0 \simeq 0.92$ for $q_0 = +1$, $z = 5$, where t_0 is the proper time between the present and the occurrence of the singularity at $R_0 = 0$.

These numbers show that a significant fraction of the total universe (in the closed case) can be sampled and that any deviation from regularity in the distribution of QSG at $m_{pg} \simeq 19$ would be of cosmic significance. Counts for the surface distribution should therefore be able to give us information about the largest-scale clustering tendencies of objects in space.

It is a pleasure to thank W. C. Miller for specially baking the plates used to obtain the spectrograms, James Westphal for applying his cross-correlation technique to the spectra of BSO 1, and Helen Czaplicki for tying the manuscript under great pressure. Dr. I. S. Bowen was helpful in discussions of expected ultraviolet lines and in comments on the spectrograms. Very special thanks go to Maarten Schmidt, whose help in obtaining the spectroscopic evidence was paramount and whose communication of the results on his spectacular redshifts before publication allowed the prediction of § III to be made at an early date. Finally, Philippe Veron's enthusiastic, hard work on all phases of the QSS identifications and on the BSO survey made parts of this development possible.

APPENDIX

THE *K*-EFFECT FOR QUASI-STELLAR SOURCES

The change of apparent magnitude due to redshifting the continuous spectrum through fixed measuring bands must be known for any class of object before the proper redshift–apparent-magnitude relation can be constructed from observational data. This change in magnitude is called the *K*-correction and has been discussed in recent times by several authors (Humason, Mayall, and Sandage 1956; Sandage 1955). The correction is now determined with moderate accuracy for E and S0 galaxies with known $I(\lambda)$ distribution (Table 1 of Sandage 1965), but the problem for quasi-stellar sources is not yet properly solved because of lack of knowledge of (1) $I(\lambda)$ to $\lambda \simeq 500$ Å, and (2) the variation of $I(\lambda)$ from source to source.

Detailed knowledge of $I(\lambda)$ exists for only 3C 273 (Oke 1965) from $\lambda_0 \simeq 2800$ to $\lambda_0 \simeq 9350$, and 3C 48 from $\lambda_0 \simeq 2640$ to $\lambda_0 \simeq 7220$ due to photometry by Baum. Figure 7 shows log $F(\nu)$ in absolute units of W m^{-2} $(Hz)^{-1}$ versus ν for 3C 48, taken from Table 6 of Matthews and Sandage (1963), where the results of Baum's work are tabulated. For both 3C 48 and 3C 273, the energy distribution per unit frequency interval follows equation (1) of the text with moderate accuracy; 3C 48 follows the equation exactly as Figure 7 shows.

Equation (1) is the form for optically thin bremsstrahlung when $C = h/kT$. The following calculations were made assuming equation (1) is correct, although Figure 1 shows beyond doubt that individual QSS deviate on both sides of the two-color relation.

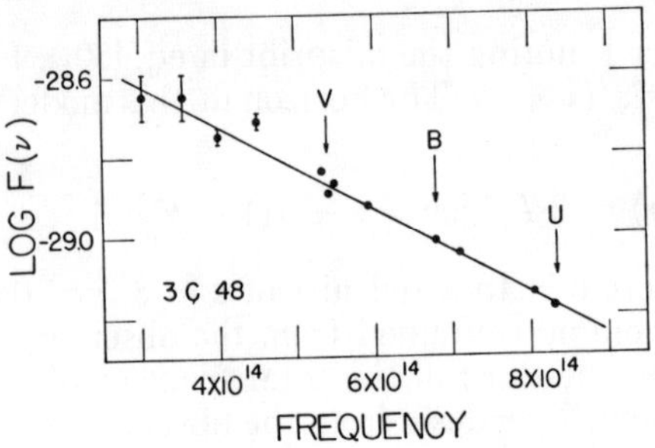

FIG. 7.—The continuous spectral distribution for 3C 48 from the measurements by W. A. Baum and by the author. The ordinate is the observed flux in W m^{-2} $(Hz)^{-1}$. Effective wavelengths of the U, B, and V filters are shown.

TABLE A1

THEORETICAL COLORS FOR RADIATORS OBEYING $F(\nu)$ = CONST. EXP $(-h\nu/kT)$

T (° K)	$U-B$	$B-V$
∞	−0.909	+0.096
100000	− .832	+ .176
30000	− .648	+ .339
20000	− .518	+ .459
13000	− .303	+ .649
10000	−0.121	+0.810

Expected $U - B$ and $B - V$ values for various parameter temperatures T were computed by the method of Matthews and Sandage (1963, Appendix A) and are given in Table A1, from which the line in Figure 1 was drawn.

The assumed function has the property that, upon redshifting the spectrum by $\lambda/\lambda_0 = 1 + z$, or $\nu/\nu_0 = (1 + z)^{-1}$, the function retains its form, but the distribution then imitates a source of lower-parameter temperature given by $T_z = T_0/(1 + z)$. The color values merely walk down the $\exp(-C\nu)$-curve of Figure 1 but retain the ultraviolet excess relative to luminosity class V stars for all redshifts. This is an important property because the true quasi-stellar sources are known to possess abnormal $U - B$ values even for extreme redshifts, as proved by 3C 9 whose $1 + z = 3.012$ (Schmidt 1965*b*).

The selective part of the K-correction was computed by quadrature for the cases $T_0 = 100000°$ K and $T_0 = 30000°$ K, using the v sensitivity function previously tabulated (Matthews and Sandage 1963). It was found that, instead of using the complete function, a simpler computation using effective wavelengths gave answers that agreed to within a few per cent with the more exact values. K_B- and K_U-values were calculated by this shorter method. Table A2 gives the results where the tabulated K_i-values represent the total change in magnitude due to redshifting in the band pass i. The non-selective term Δ mag. $= 2.5 \log (1 + z)$, due to the change in effective band width, has been included in the tabular values. A negative sign means the object becomes brighter due to redshift than it would if the measuring bands were moved along the spectrum by $1 + z$ and were increased in band width by $1 + z$ to keep up with the redshift. The K-values with their tabulated signs reversed should be applied to all observations made at fixed wavelengths so as to place the observed magnitudes on a "bolometric" scale.

The important feature revealed by Table A2 is that the K-corrections are small, even going negative, for redshifts as large as $z = 3$ if $T_0 > 50000°$ K. This is the probable value of the parameter temperatures (for $z = 0$) for many QSS, showing that, to first approximation, no K-

TABLE A2

K-CORRECTIONS FOR VARIOUS PARAMETER TEMPERATURES AND REDSHIFTS FOR RADIATORS OBEYING $F(\lambda)d\lambda$ = CONST. $\lambda^{-2}\, d\lambda$ EXP $(-hc/k\lambda T)$*

z	$T_0=\infty$			$T_0=100000°$ K			$T_0=50000°$ K			$T_0=30000°$ K		
	K_U	K_B	K_V	K_U	K_B	K_V	K_U	K_B	K_V	K_U	K_B	K_V
0.0	0.00	0.00	0.00	0.00	0.00	0.00	0.00	0.00	0.00	0.00	0.00	0.00
0.5	−0.44	−0.44	−0.44	− .22	− .26	− .30	0.00	−0.08	−0.15	+0.28	+0.15	+0.03
1.0	−0.76	−0.76	−0.76	− .31	− .40	− .46	+0.13	−0.04	−0.17	+0.70	+0.44	+0.19
1.5	−1.00	−1.00	−1.00	− .33	− .46	− .57	+0.33	+0.07	−0.13	+1.18	+0.78	+0.42
2.0	−1.19	−1.19	−1.19	− .33	− .47	− .63	+0.55	+0.22	−0.06	+1.71	+1.17	+0.69
2.5	−1.36	−1.36	−1.36	− .27	− .47	− .66	+0.82	+0.40	+0.06	+2.23	+1.60	+1.00
3.0	−1.50	−1.50	−1.50	− .20	− .45	− .66	+1.11	+0.61	+0.20	+2.83	+2.04	+1.32
3.5	−1.63	−1.63	−1.63	−0.11	− .40	− .65	+1.41	+0.84	+0.35	+3.45	+2.49	+1.67
4.0	−1.75	−1.75	−1.75	......	− .33	− .61	......	+1.09	+0.53	......	+2.98	+2.03
4.5	−1.85	−1.85	−1.85	......	− .26	− .58	......	+1.33	+0.71	......	+3.45	+2.40
5.0	−1.94	−1.94	−1.94	......	− .18	− .54	......	+1.59	+0.89	......	+3.98	+2.77
5.5	−2.03	−2.03	−2.03	......	− .09	− .49	......	+1.90	+1.09	......	+4.49	+3.13
6.0	−2.11	−2.11	−2.11	......	+0.01	−0.43	......	+2.20	+1.30	......	+5.02	+3.56

* The K-corrections contain the selective term

$$\Delta \text{ mag.} = 2.5\left[\log \frac{\int_{\lambda_1}^{\lambda_2} S(\lambda) I(\lambda)_0 d\lambda}{\int_{\lambda_1}^{\lambda_2} S(\lambda) I(\lambda)_z d\lambda}\right]$$

plus the non-selective term Δ mag. $= 2.5 \log (1 + z)$.

correction need be applied to the QSS data of Figure 4 in the range of redshifts there encountered. The K-corrections will become large near $m_{pg} \simeq 22^m$, where the redshifts are expected to be near $z = 4$. Here special problems in the survey for ultraviolet objects will be encountered because the change in $B - V$ and $U - B$ due to redshift will be large. The objects cannot then be easily distinguished from normal stars. A most important factor not considered in Table A2 is the effect of the Lyman continuum at $\lambda = 912$ Å as it is shifted into the observable range. A redshift of $z \simeq 3$ will bring $\lambda = 912$ Å into the center of the U filter at $\langle\lambda\rangle = 3600$ Å and will cause K_U to be much larger than given in Table A2. This is expected to occur near $m_{pg} \simeq 21^m$ and may seriously hamper searches for QSG by color methods. Detailed energy-distribution-curves for QSG, which will be determined in due course, will determine the seriousness of the problem.

Finally, it should be noted that the expected change in color due to redshift is given by $\Delta(U - B) = K_U - K_B$ and $\Delta(B - V) = K_B - K_V$, which can be found for any z from Table A2.

REFERENCES

Albada, G. B. van. 1963, *B.A.N.*, **17**, 127.
Bennett, A. S. 1962, *Mem. R. Astr. Soc.*, Vol. **68**, Part 5.
Berger, J. 1963, *Pub. A.S.P.*, **75**, 393.
Bolton, J., Clarke, M., Sandage, A., and Veron, P. 1965, *Ap. J.* (in press).
Burbidge, G. R., Burbidge, E. M., and Sandage, A. R. 1963, *Rev. Mod. Phys.*, **35**, 947.
Chavira, E. 1958, *Bull. Obs. Ton.*, No. 17, p. 15.
Cowley, C. R. 1958, *A.J.*, **63**, 484.
Eggen, O. J., and Greenstein, J. L. 1965, *Ap. J.*, **141**, 83.
Eggen, O. J., Lynden-Bell, D., and Sandage, A. 1962, *Ap. J.*, **136**, 748.
Eggen, O. J., and Sandage, A. R. 1965, *Ap. J.*, **141**, 821.
Feige, J. 1958, *Ap. J.*, **128**, 267.
Field, G. 1964, *Ap. J.*, **140**, 1434.
Grant, G., and Abt, H. A. 1959, *Ap. J.*, **129**, 323.
Greenstein, J. L. 1956, *Proc. 3d Berkeley Symposium*, **3** (Berkeley: University of California Press), 11.
———. 1961, *Stellar Atmospheres* (Chicago: University of Chicago Press), p. 676.
Greenstein, J. L., and Matthews, T. A. 1963, *Nature*, **197**, 1041.
Haro, G., and Chavira, E. 1960, *Bull. Obs. Ton.*, No. 19, p. 11.
Haro, G., and Luyten, W. J. 1960, *Bull. Obs. Ton.*, No. 19, p. 17.
———. 1962, *ibid.*, **3**, 37.
Holmberg, E. 1957, *Medd. Lunds Obs.*, Ser. II, No. 136.
Howard, W. E., III, and Maran, S. P. 1965, *Ap. J. Suppl.*, **10**, 1.
Hubble, E. P. 1926, *Ap. J.*, **64**, 321.
Humason, M. L., Mayall, N. U., and Sandage, A. R. 1956, *A.J.* **61**, 97.
Humason, M. L., and Zwicky, F. 1947, *Ap. J.*, **105**, 85.
Iriarte, B. 1958, *Ap. J.*, **127**, 507.
———. 1959, *Lowell Obs. Bull.*, **4**, 130.
Iriarte, B., and Chavira, E. 1957, *Bull. Obs. Ton.*, No. 16, p. 3.
Johnson, H. L., and Morgan, W. W. 1953, *Ap. J.*, **117**, 313.
Klemola, A. R. 1962, *A.J.*, **67**, 740.
Luyten, W. J. 1953, *A.J.*, **58**, 75.
———. 1954, *ibid.*, **59**, 224.
———. 1955, *ibid.*, **60**, 429.
———. 1956*a*, *ibid.*, **61**, 261.
———. 1956*b*, *A Search for Faint Blue Stars* (Pub. Obs. Univ. of Minn., Paper 7).
———. 1959, *ibid.*, Paper 17.
———. 1960, *ibid.*, Paper 21.
———. 1962*a*, *ibid.*, Paper 29.
———. 1962*b*, *ibid.*, Paper 30.
Luyten, W. J., and Haro, G. 1959, *Pub. A.S.P.*, **71**, 469.
Malmquist, K. G. 1927, *Medd. Lunds Obs.*, Ser. II, No. 37.
———. 1936, *Stockholm Obs. Ann.*, Vol. **12**, No. 7.
Matthews, T. A., and Sandage, A. R. 1963, *Ap. J.*, **138**, 30.
Mumford, G. S. 1964, *Ap. J.*, **139**, 476.
Oke, J. B. 1965, *Ap. J.*, **141**, 6.
Oort, J. H. 1958, *La Structure et l'évolution de l'univers* (Brussels: Solvay Conference), p. 163.
Ryle, M., and Sandage, A. 1964, *Ap. J.*, **139**, 419.
Sandage, A. 1961*a*, *Ap. J.*, **133**, 355.
———. 1961*b*, *ibid.*, **134**, 916.
———. 1963, *Robertson Mem. Vol.* (Soc. Indust. Applied Math., Philadelphia), p. 43.
———. 1965, *Proc. Padua Conference on Cosmology*, ed. L. Rosino (Italy).
Sandage, A., and Veron, P. 1965, *Ap. J.* (in press).
Sandage, A., Veron, P., and Wyndham, J. D. 1965, *Ap. J.* (in press).
Sandage, A., and Wyndham, J. D. 1965, *Ap. J.*, **141**, 328.
Schmidt, M. 1962, *Ap. J.*, **136**, 684.
———. 1963, *Nature*, **197**, 1040.
———. 1965*a*, *Ap. J.*, **141**, 1.
———. 1965*b*, *ibid.*, **141** (in press).
———. 1965*c*, Paper presented at Second Texas Symposium on Relativistic Astrophysics, Austin, Texas.
Schmidt, M., and Matthews, T. A. 1964, *Ap. J.*, **139**, 781.
Slettebak, A., Bahner, K., and Stock, J. 1961, *Ap. J.*, **134**, 195.
Slettebak, A., and Stock, J. 1959, *Astr. Abhandlungen Hamburger Sternwarte*, Vol. **5**, No. 5.
Veron, P. 1965, *Ap. J.*, **141**, 332.
Walker, J. F. 1957, I.A.U. Symposium No. 3 (Cambridge: Cambridge University Press), p. 46.
Westphal, J. A. 1965, *Ap. J.* (in press).
Whitford, A. E. 1954, *A.J.*, **59**, 194.

THE NATURE OF THE FAINTER HARO-LUYTEN OBJECTS*

T. D. KINMAN
Lick Observatory, University of California, Mount Hamilton, California
Received July 26, 1965

ABSTRACT

A survey for faint RR Lyrae stars near the north galactic pole shows that the space density $\rho(z)$ per kpc^3 varies with height z (in kpc) above the galactic plane according to the relation

$$\log \rho(z) = 0.70 - 0.072z .$$

The inferred integral count of the blue horizontal-branch stars for $m_{pg} > 16$ is greater than the contribution of galactic stars to the Haro-Luyten objects of the same magnitude postulated by Sandage in 1965. For $15 < B < 17$, the number of RR Lyrae stars is about one-third of the number of blue Tonantzintla stars observed by Iriarte and Chavira in the same area of the sky. Low-dispersion spectra of an unbiased sample of twelve Tonantzintla stars ($15 < B < 16$) showed that all were galactic objects (seven white dwarfs, four horizontal-branch stars, and one hot subdwarf), whereas, according to Sandage, one-half should have been quasi-stellar galaxies.

If one-half of the Haro-Luyten blue objects with $15 < B < 16$ are white dwarfs which have a z distribution like the stars of the galactic disk, their integral count (together with that assumed for the horizontal-branch stars) is equal to 75 per cent of the integral count of the Haro-Luyten objects at $B = 16.5$ and shows a similar increase with magnitude. The majority of faint Haro-Luyten objects are therefore likely to be galactic stars and not quasi-stellar galaxies as Sandage suggests. It therefore is premature to infer the cosmological properties of the fainter Haro-Luyten objects from the slope of their integral count-curve.

A faint and quasi-stellar galaxy near the position of Haro-Luyten 293 has an emission spectrum and blue continuum similar to that observed in the Irr 1 galaxies NGC 1569 and NGC 2366 and in the blue galaxy Haro 6. These galaxies appear to have rather similar size and structure with bright compact regions which would appear stellar at distances greater than about 100 Mpc. Although they are less bright ($-14 < M_{pg} < -17.5$) than the quasi-stellar galaxies observed by Sandage (and their Balmer lines are narrower and their level of excitation is lower), it is tentatively suggested that they may represent a related class of objects.

About 3–4 per cent of the Tonantzintla blue objects appear to be non-stellar. The number of blue extragalactic objects which appear stellar is unknown, but it is unlikely to exceed 20 per cent of the fainter Haro-Luyten objects and may well be less. It is suggested that precise proper motions could be used to recognize the white dwarfs which are likely to constitute a large fraction of the fainter Haro-Luyten objects.

I. INTRODUCTION

Sandage (1965) has shown that a sample of the blue stars at high galactic latitudes shows a transition in their color distribution at $V = 14.5$, the colors of several of the fainter stars being similar to those found for the quasi-stellar radio sources. The rate of increase with magnitude of the integral count $N(m)$ of the blue objects found by Haro and Luyten (1962) shows an increase at about $m_{pg} = 15$ as shown in Figure 1. Sandage therefore suggested that most of the blue objects fainter than $m_{pg} = 16$ must be a new class of extragalactic objects (called quasi-stellar galaxies) and considered that the contribution of the galactic stars to the integral count of the blue objects must be small at faint magnitudes (*dashed line*, Fig. 1).

II. ESTIMATES OF THE SPACE DENSITIES OF GALACTIC STARS

A survey for RR Lyrae stars in 74 square degrees near the north galactic pole (see Kinman and Wirtanen 1963 for a preliminary report) shows that the space density of RR Lyrae stars ($\rho(z)$ per kpc^3) is related to the height z above the galactic plane (in kpc) by

$$\log \rho(z) = 0.70 - 0.072z . \quad (1)$$

* *Contributions from the Lick Observatory*, No. 196.

Reprinted from Astrophysical Journal
142, 1241-1248.

The corresponding whole-sky values of $N(m)$ derived from this expression are shown in Figure 1. These values of $N(m)$ refer to RR Lyrae stars with periods greater than 0.44 days and photographic amplitudes greater than 0.75 mag. The values of $N(m)$ for all classes of these stars will therefore be larger. The integral counts for the RR Lyrae stars are markedly different from those predicted for the blue halo stars by Sandage. Since the RR Lyrae counts refer to a region close to the north galactic pole and the blue star counts refer to a region between the south galactic pole and latitude −36°, a more direct comparison is desirable.

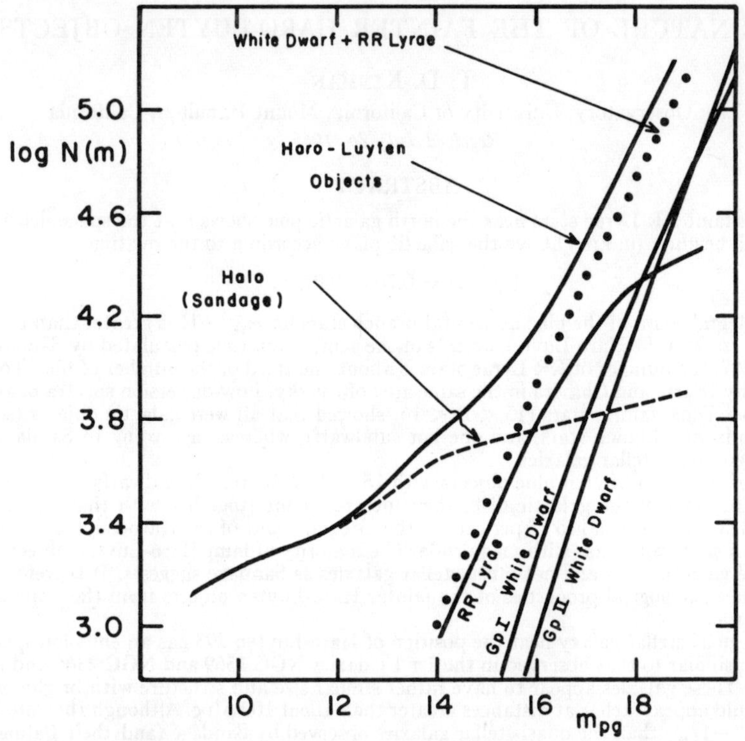

FIG. 1.—The whole sky integral count–apparent magnitude relation for the Haro-Luyten objects after Sandage (1965). The dashed line is the integral count for halo stars predicted by Sandage. The curves marked RR Lyrae, White dwarf Gp 1, and White dwarf Gp 2 are integral counts for these classes of objects derived as described in the text. The total integral count for the RR Lyrae stars and the White dwarfs is shown by a dotted line.

The observed magnitude distribution of the RR Lyrae stars is shown in Figure 2 together with that of the blue stars in the same region of the sky (Iriarte and Chavira 1957; Chavira 1959). The B magnitudes of the blue stars were redetermined from astrophotometer measures of astrograph plates using the same photoelectric sequences used for the RR Lyrae stars. The difference $(B - m_{pg})$ between these new magnitudes and the original Tonantzintla magnitudes are shown as a function of the Tonantzintla magnitude by the filled circles in Figure 3. The crosses are the differences between Iriarte's (1959) photoelectric B magnitudes and the Tonantzintla magnitudes and show a very similar effect indicating significant systematic errors in the Tonantzintla magnitudes. Figure 2 shows that for $15 < B < 17$ there are about a third as many RR Lyrae stars as blue stars in this area of the sky. Only four of the RR Lyrae stars were identified as being included as blue stars. It may be expected, however, that there should exist with

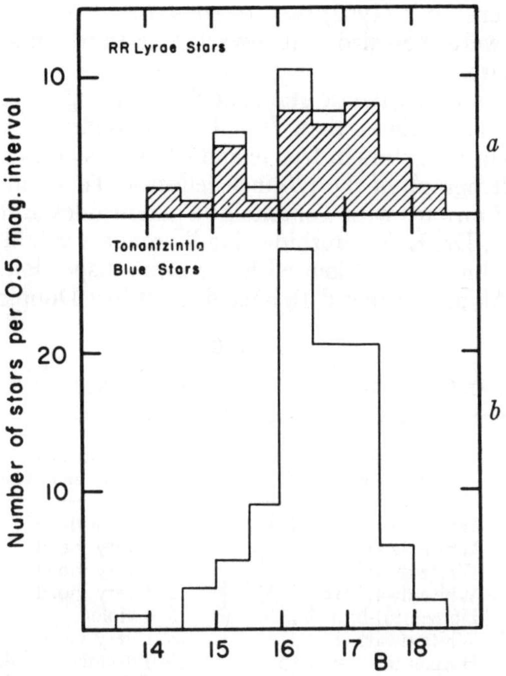

FIG. 2.—(*a*) The distribution of RR Lyrae variables as a function of *B* magnitude in a 74-square-degree area of the sky near the north galactic pole. *Hatched area:* variables with photographic amplitude > 0.75 mag.; *open area:* variables with lower amplitudes. (*b*) The distribution with magnitude of the Tonantzintla blue stars in the same area of the sky.

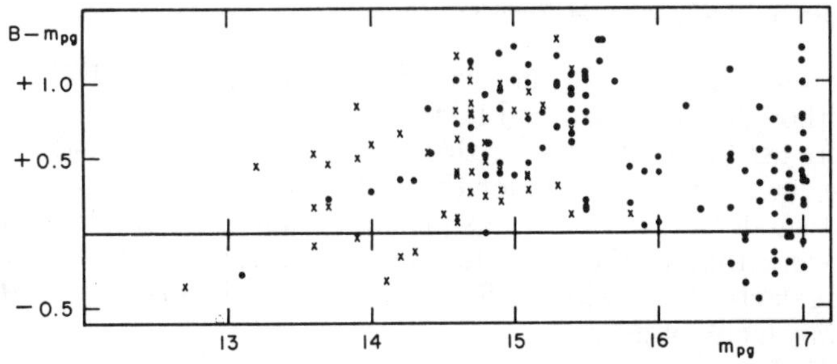

FIG. 3.—The differences ($B - m_{pg}$) between Lick and Tonantzintla magnitudes for the blue stars in the RR Lyrae search area (*filled circles*) as a function of Tonantzintla magnitude. The differences between Iriarte's (1959) photoelectric magnitudes and the Tonantzintla magnitudes for stars in the same list are shown by crosses.

the RR Lyrae stars a comparable number of blue horizontal-branch stars of about the same absolute magnitude. These blue horizontal-branch stars should therefore make a significant contribution to the blue star count in this magnitude range. This would be compatible with Greenstein's (1962) spectroscopic result that at $B = 15$ and $B - V < -0.1$, the blue stars were one-third white dwarfs, one-third hot subdwarfs, and one-third horizontal-branch stars.

It therefore appears that galactic stars should make up a significant fraction of the blue stars at magnitudes fainter than $B = 15$. To confirm this result, low-dispersion spectra (200 Å/mm) of nine stars in Iriarte and Chavira's (1957) list were obtained with the prime-focus spectrograph of the 120-inch reflector. To avoid observational selection all stars from No. 804 onward with Tonantzintla magnitudes in the range 14.5–15.5 were observed. In addition, Dr. E. M. Burbidge kindly made available spectra of three other stars from this list which were taken with the same dispersion. These spectra are described in Table 1. All stars showed Balmer lines of low Doppler displacement and no

TABLE 1

SPECTRAL CLASSIFICATION OF OBSERVED TONANTZINTLA STARS

Tonantzintla No.	Spectral Classification	Tonantzintla Magnitude and Color	Corrected Magnitude*
229†......	White dwarf (DC–DA)	15.2 very violet	15.9
237†......	White dwarf (DC)	15.3 very violet	16.0
797†......	White dwarf (DA)	15.2 very violet	15.9
804.......	White dwarf (DC–DA)	15.2 very violet	15.9
805.......	Horizontal-branch (A0)	15.4 violet	16.1
807.......	White dwarf (DC–DA)	15.1 very violet	15.8
809.......	Horizontal-branch (A0)	15.0 decidedly violet	15.7
811.......	Horizontal-branch (A0)	15.1 decidedly violet	15.8
812.......	Horizontal-branch (A3)	14.5 decidedly violet	15.0
813.......	White dwarf (DA)	15.5 violet	16.2
816.......	White dwarf (DA)	14.8 very violet	15.2
817.......	sdB (like Feige 95)	15.0 decidedly violet	15.7

* Using corrections to Tonantzintla magnitudes given by Fig. 3.
† Spectra taken by Dr. E. M. Burbidge.

emission lines were observed. In the case of Ton 237 (classified DC), the Balmer lines were visible although very broad and faint. Since prints of white-dwarf spectra (Greenstein 1961) were used as standards for this classification, the white-dwarf spectral types are necessarily rough but adequate for the present purpose.

The present small but, as far as was possible, unbiased, sample of blue star spectra indicates therefore that in the corrected magnitude range $15.0 < B < 16.2$, the blue objects are all galactic stars, whereas Sandage (his Fig. 5) predicted that about one-half should be extragalactic objects.

Eggen and Greenstein (1965) divided the white dwarfs into two groups. Group 1 consists mainly of sharp-lined DA stars while the stars of group 2 show mainly non-DA spectra. Our sample of blue stars indicates that at $15 < B < 16$, each of these groups accounts for about a quarter of the blue stars. We take $M_B = +11.3$ and $+12.8$ for the absolute magnitudes of the two groups, respectively, and assume that the variation of the space density with height z above the galactic plane is given by

$$\rho(z) = \rho(0)\, e^{-z/\beta} . \qquad (2)$$

Allen (1963) gives $\beta = 280$ pc for the white dwarfs. The total number of blue stars observed by Haro and Luyten ($15 < m_{pg} < 15.9$) was 280, so that the expected number of white dwarfs of each group with this magnitude is 1442 for the whole sky. The integral counts $N(m)$ for each group were computed from this information and are shown in Figure 1. Not only is the slope of the log $N(m)$ versus m_{pg} relation for the white dwarfs similar to that of the blue stars for $m_{pg} > 16$, but the total number of white dwarfs accounts for about two-thirds of the total number of blue stars in the Haro-Luyten Survey. The number density of white dwarfs in the solar neighborhood is predicted to be 0.96×10^{-3} pc^{-3} and 3.80×10^{-3} pc^{-3} for group 1 and group 2, respectively, which is in good agreement with the value of 4.5×10^{-3} for white dwarfs of spectral classes B, A, and F given by Allen (1963). While the above calculations are extremely rough, they do suggest that galactic stars may account in large part for the form of integral count curve of the Haro-Luyten objects.

The color distribution of the stars with $V > 14.5$ is less easy to understand because of the significant number of stars found by Sandage for which the $U - B$ color is more negative than for the black-body relation. Old novae and the U Gem and SS Cyg stars have such colors, as Sandage notes, but it is difficult to estimate their frequency in the absence of suitable surveys. The general consistency in the relation between the Tonantzintla magnitudes and the re-determined magnitudes makes it unlikely, however, that a large percentage of the blue stars could be eruptive variables of large amplitude.

III. EXTRAGALACTIC HARO-LUYTEN OBJECTS

The above results suggest that Sandage has overestimated the frequency of quasi-stellar galaxies, but they do not affect the important discovery of their existence. In an attempt to rediscover some of the Haro-Luyten objects (for which only rough coordinates are given), the blue-ultraviolet double-exposure method of Ryle and Sandage (1964) was used at the prime focus of the 120-inch reflector. Near the position of HL 293 ($m_{pg} = 16.7$), two blue objects were found (A and B in Fig. 4). The fainter (B) image is definitely non-stellar and is rather similar to that of Ton 730 with a faint jet projecting out about 5″ (northward) from the central image. This non-stellar object is fainter relative to the nearby stars on plates taken with the 120-inch reflector (f/8.3) and with the 20-inch Astrograph (f/7) than on the Palomar Sky Survey plates as might be expected for an extended object. A very rough estimate of the B magnitude is 17.7 on the Sky Survey plates and about a half-magnitude fainter on the 120-inch plates. No obvious change in appearance or magnitude is apparent between the images of this object on the 1965 and 1949 plates taken with the Astrograph. The fuzzy core, which is seen visually and with short-exposure plates at the prime focus of the 120-inch reflector, has a diameter of 1″ to 2″.

Two spectra in the blue were obtained of this object with the prime-focus spectrograph of the 120-inch reflector (200 Å/mm, 100 and 165 min on Kodak baked IIa-O emulsion). These showed a faint blue continuum and the emission lines of [O III] $\lambda\lambda$ 5007, 4959, and 4363; [O II] λ 3727; [Ne III] λ 3869; the Balmer lines $H\beta$, $H\gamma$, $H\delta$, $H\epsilon$, and $H\zeta$. It is possible that the [S II] line at λ 4068 is also present. The $H\alpha$ line only was found on a single red spectrum (360 Å/mm, 150 min on Kodak 103a-F emulsion). The absence of [Ne V] shows that the object is of lower excitation than either Ton 256 or Ton 730 observed by Sandage. Also the Balmer lines are much narrower, and their Doppler width cannot exceed 100–200 km/sec; the blue continuum may also be less prominent relative to the emission spectrum, but this is difficult to judge from spectra taken with different spectrographs. The redshift ($c[d\lambda/\lambda]$) for HL 293B is 1550 km/sec which makes it unlikely to be a galactic object. Corrected for galactic rotation, the redshift is 1700 km/sec. The distance is therefore 22.6 Mpc ($H = 75$ km/sec 10^6 pc) and $M_B \sim -14$ or about 7–8 mag. fainter than Ton 256 and 730. The diameter of the central condensation is thus only 100–200 pc and the faint jet is perhaps 500 pc long. There is no radio source listed close to the position of HL 293B in the *General Catalogue* of Howard and Maran (1965).

The spectrum of HL 203B is not like that of a typical Irr 1 (Magellanic Cloud–type irregular galaxy) such as NGC 4449 that has a relatively low excitation emission spectrum, but corresponds more closely to the spectra of the irregular galaxies NGC 1569 and NGC 2366. Mayall (1935) showed that these galaxies have a moderately high excitation spectrum (with [Ne III]) and a lineless blue continuum which is particularly strong in NGC 1569. Both of these galaxies have an apparent integrated m_{pg} of about 12 and integrated $B - V$ colors corrected for interstellar extinction of 0.4–0.5 (G. and A. de Vaucouleurs 1964). Their redshifts are small—119 km/sec for NGC 1569 and 229 km/sec for NGC 2366 (Humason, Mayall, and Sandage 1956). Both galaxies have a bar structure (which is brighter in the case of NGC 1569) that has a concentrated bright region at one end. It is of interest to note that Baade (1963) classified NGC 2366 as an Sc rather than an Irr 1 galaxy presumably because of the faint extensions to its bar structure.

According to Holmberg (1950), NGC 2366 is a member of the M81 group (modulus $M - m = 27.1$; Sandage 1954) and has been resolved into H II regions and stars brighter than $M_{pg} = -4.1$ (Sandage 1961). Its absolute magnitude (M_{pg}) is therefore about -15 and its bar is about 3 kpc long. The concentrated part of this galaxy must, however, be only a few hundred parsecs in diameter. It is therefore somewhat larger and brighter than HL 293B.

A number of blue galaxies that are bright in the ultraviolet and which have emission spectra have been found by Haro (1956). Their positions on the two-color diagram are not dissimilar to those of the quasi-stellar galaxies and radio sources (Hiltner and Iriarte 1958). They also appear to have bluer colors than the Seyfert galaxies according to Haro. Münch (1958) examined a number of them spectroscopically and concluded that they showed a wide range in structure and luminosity and that some were quite compact. One of the most compact of these galaxies (Haro 6) has a spectrum which is similar to HL 293B (see Fig. 5) as was judged from Haro's observations and a spectrum taken by Mayall with the Crossley nebular spectrograph. The redshift of Haro 6 is 1927 km/sec (Mayall and de Vaucouleurs 1962), and on the Palomar Sky Survey (blue) it has an elongated image of length 15″ and apparent magnitude $m_{pg} \sim 14.5$ at a rough estimate. The distance of Haro 6 is therefore ~26 Mpc, and its diameter ~2 kpc and its integrated magnitude (M_{pg}) ~ -17.5. Haro 6 is therefore similar in size but somewhat brighter than both HL 293B and NGC 2366. In these latter objects, the Hβ line does not show appreciable widening, but in Haro 6 this line appears to have a width of perhaps 750 km/sec although this is difficult to judge on the low-dispersion Crossley spectrum.

Two of the blue objects (Ton 151 and Ton 698) in the RR Lyrae survey region have elongated non-stellar images that are similar in appearance to that of Haro 6. In another sample of Tonantzintla blue objects, three more non-stellar objects were found. Such non-stellar objects could therefore be relatively common and comprise perhaps 3 or 4 per cent of the Tonantzintla objects. Since, as we have seen, some of these objects have bright concentrated regions with diameters of only a few hundred parsecs, they may be expected to appear stellar or nearly so at distances greater than 100 Mpc. It will be necessary, however, to observe a larger sample of these objects before we can derive a luminosity function and local space density for them and from this deduce their expected frequency at faint magnitudes.

The quasi-stellar galaxies observed by Sandage (BSO 1, Ton 730, and Ton 256) are much brighter ($M_{pg} \sim -21$ or brighter) than the irregular galaxies discussed above. Sandage's galaxies also show higher excitation emission lines (e.g., [Ne V]) and have Balmer lines which are much broader (Doppler width ~2000 km/sec). It is suggested as a hypothesis, however, that the same physical process could be responsible for both classes of objects. In this case the range in the absolute magnitude and other parameters might correspond either to a range in the scale and intensity of the phenomenon in galaxies of different size or might represent different aspects of the phenomenon at different stages of evolution. In the latter case we might picture the objects as consisting of a luminous compact core in which there is considerable mass motion, and in which the

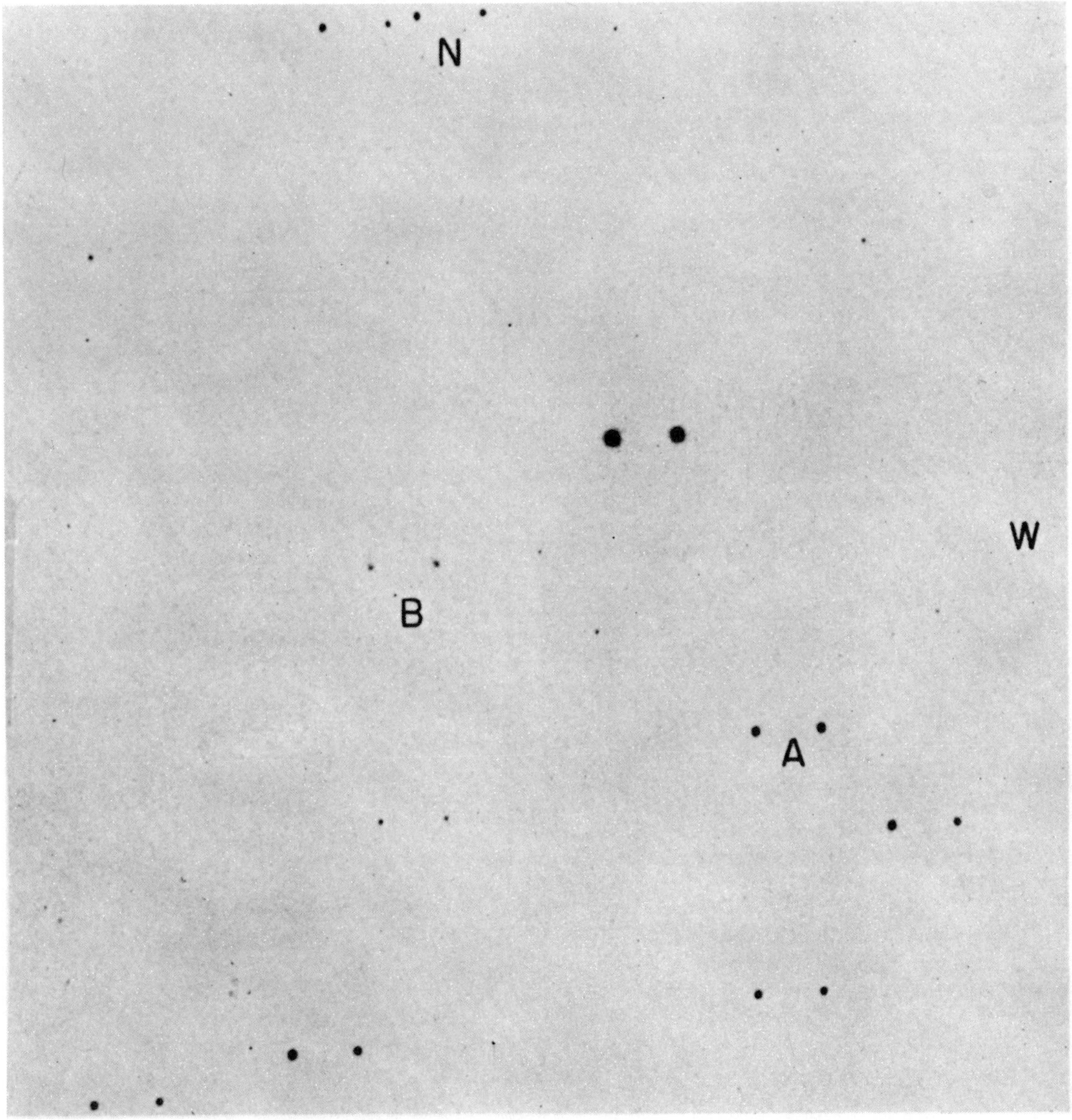

FIG. 4.—The two blue objects found near the position of Haro-Luyten object No. 293. The ultraviolet image is 20″ west of the blue image. Object *B* is non-stellar on the original plate and has a faint jet extending northward 5″ from the object. Blue exposure 10 min (Schott GG13 filter), ultraviolet exposure 20 min (Corning 9863 filter) on Kodak 103a-O emulsion. Taken with the 120-inch reflector diaphragmed to 72-inch aperture.

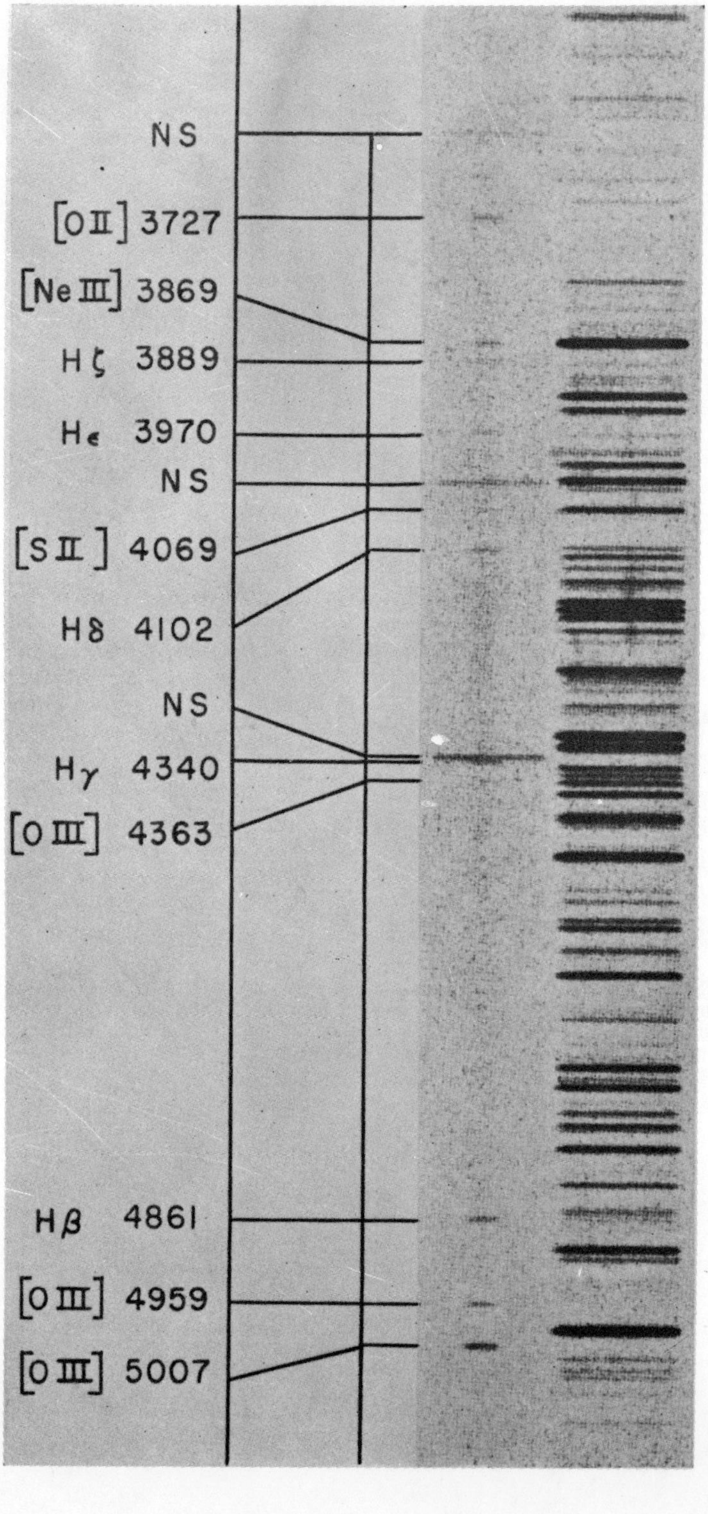

FIG. 5.—Spectrum of HL 293B showing forbidden lines of [O III], [O II], and [Ne III] and the Balmer lines Hβ–Hζ. The presence of [S II] is uncertain

Balmer lines are produced, surrounded by a tenuous outer region in which the forbidden lines arise. The fainter galaxies might then represent a later stage in the evolution in which the luminosity and mass motion in the core had decayed. The number of objects of this type known at present is, however, so small that this relationship suggested by the absolute magnitude, excitation, and line width could well be fortuitous. A survey for more objects of this general class and a more precise study of them is therefore highly desirable.

The number of blue extragalactic objects with stellar images is not known, but it seems unlikely from our estimates of the numbers of galactic stars that they can comprise more than about 20 per cent of the Haro-Luyten objects with $B > 16$ and their number may well be less. One way in which the nearby blue galactic stars could be eliminated from the fainter Haro-Luyten objects is by the determination of precise proper motions. These fainter Haro-Luyten objects, as we have seen, are likely to contain a high proportion of white dwarfs which (for $B \leq 20$) will mostly lie within 500 pc of the Sun. At this distance a tangential velocity of 24 km/sec corresponds to an annual proper motion of $0\overset{''}{.}01$. The estimated probable error in the annual proper motion of a single star derived from plates taken 20 years apart with the Lick Carnegie Astrograph (against a reference frame of 50 galaxies) is estimated to be $0\overset{''}{.}005$ (Vasilevskis 1957). It should therefore be possible to identify a large fraction of the white dwarfs in the blue-object population by means of their proper motions.

IV. CONCLUSIONS

The observed space densities of the RR Lyrae stars and the spectral classification of a sample of the Tonantzintla stars suggest that a large fraction of the fainter Haro-Luyten blue objects are galactic stars of the disk and halo and not quasi-stellar galaxies as Sandage has suggested. It is therefore premature to infer the cosmological properties of the fainter Haro-Luyten objects from the slope of their integral count versus apparent magnitude relation.

The irregular galaxies NGC 1569 and NGC 2366 have similar spectra and dimensions to those found for a Haro-Luyten object (293B) and the blue galaxy Haro 6. These objects have bright concentrated regions which would appear stellar or nearly so at great distances. It is possible that these objects are related to the brighter and more highly excited objects found by Sandage.

About 3–4 per cent of the Tonantzintla blue objects appear to be non-stellar. The number of extragalactic objects which would appear stellar is unknown but is unlikely to exceed 20 per cent of the Haro-Luyten objects fainter than $B = 16$ and may be less. It is suggested that white dwarfs could be distinguished from other faint blue objects by their proper motions.

It is planned to publish a more detailed account of the RR Lyrae survey at the north galactic pole. Acknowledgment is made to the National Science Foundation for a grant (NSF GP-907) in support of this work.

REFERENCES

Allen, C. W. 1963, *Astrophysical Quantities* (London: Athlone Press), p. 237.
Baade, W. 1963, *Evolution of Stars and Galaxies*, ed. C. Payne-Gaposchkin (Cambridge, Mass.: Harvard University Press), p. 225.
Chavira, E. 1958, *Bol. Obs. Tonantzintla y Tacubaya*, No. 17, p. 15.
Eggen, O. J., and Greenstein, J. L. 1965, *Ap. J.*, **141**, 83.
Greenstein, J. L. 1961, *Stellar Atmospheres* (Chicago: University of Chicago Press), p. 676.
———. 1962, *Proc. 11th Gen. Assembly I.A.U.*, Vol. **11B**, (New York: Academic Press), p. 311.
Haro, G. 1956, *Bol. Obs. Tonantzintla y Tacubaya*, No. 14, p. 8.
Haro, G., and Luyten, W. J. 1962, *Bol. Obs. Tonantzintla y Tacubaya*, No. 22, p. 37.
Hiltner, W. A., and Iriarte, B. 1958, *Ap. J.*, **128**, 443.

Holmberg, E. 1950, *Medd. Lunds. Astr. Obs.*, Ser. II, No. 128.
Howard, W. E., III, and Maran, S. P. 1965, *Ap. J. Suppl.*, **10**, 1.
Humason, M. L., Mayall, N. U., and Sandage, A. R. 1956, *A.J.*, **61**, 97.
Iriarte, B. 1959, *Lowell Obs. Bull.*, **4**, 130.
Iriarte, B., and Chavira, E. 1957, *Bol. Obs. Tonantzintla y Tacubaya*, No. 16, p. 3.
Kinman, T. D., and Wirtanen, C. A. 1963, *Ap. J.*, **137**, 698.
Mayall, N. U. 1935, *Pub. A.S.P.*, **47**, 320.
Mayall, N. U., and Vaucouleurs, A. de. 1962, *A.J.*, **67**, 363.
Münch, G. 1958, *Sky and Telescope*, **17**, 231.
Ryle, M., and Sandage, A. 1964, *Ap. J.*, **139**, 419.
Sandage, A. 1954, *A.J.*, **59**, 180.
———. 1961, *The Hubble Atlas of Galaxies* (Washington, D.C.: Carnegie Inst. Pub. No. 618), p. 39.
———. 1965, *Ap. J.*, **141**, 1560.
Vasilevskis, S. 1957, *A.J.*, **62**, 113.
Vaucoulers, G. and A. de. 1964, *Reference Catalogue of Bright Galaxies* (Austin: University of Texas Press).

REMARKS ON EVOLUTION OF GALAXIES

Thornton Page
Van Vleck Observatory, Wesleyan University

In discussion of the quasi-stellar objects (QSO's) I would like to emphasize their place in the general evolution of galaxies. This was touched upon by Hanbury Brown when he defined the QSO as an event, and is dramatically illustrated by the optical brightening of 3C 2 reported by Alan Sandage. The question whether QSO's are to be defined in terms of their radio characteristics or their optical characteristics (or in terms of present, imperfect theories of their energy output) is thus reduced to the level of: "Which comes first, the chicken or the egg?" It also leads to a set of optical observations not mentioned among the many topics discussed at this Symposium: the "compact galaxies" announced by F. Zwicky at the I.A.U. meetings in Hamburg in August of 1964.

If QSO's represent a brief phase in the life of a galaxy, it is reasonable to expect other forms of similar nature that precede and follow in the evolutionary pattern or time sequence of events and Zwicky claims to have observed many of these on Palomar photographs, recognizing them a "fuzzy starlike images," several of which show large redshifts. He claims that there is about one per square degree in the unobscured regions of the sky brighter than 17th or 18th magnitude. This adds another practical basis for selecting objects for optical and radio observation and may fit an evolutionary sequence from normal galaxies (extended optical images of types Sc to E0) through compact galaxies and starlike images of large UV excess to QSO's with strong radio emission and small radio angular diameter. The next stage may be an invisible collapsed mass of density larger than the Schwarzschild limit.

I suspect that there are other evolutionary tracks, possibly characterized by different total mass, or different angular momentum in the initial stage, accounting for the variety of galaxy forms, many of which show no strong radio emission, and that a complete theory of the evolution of galaxies will soon be able to account for most of them.

QUASI-STELLAR RADIO SOURCES AS SPHERICAL GALAXIES IN THE PROCESS OF FORMATION

GEORGE B. FIELD
Princeton University Observatory

ABSTRACT

The properties of a galaxy forming from an intergalactic gas cloud are calculated. If the angular momentum is large enough to prevent formation of a "massive object," but small enough to prevent formation of a gaseous disk, the gas contracts to a small radius and high density before forming stars. When star formation commences, it proceeds rapidly to completion, resulting in the simultaneous formation of a large number of main-sequence stars, with masses up to 100 suns. The resulting luminosity, 7×10^{46} erg/sec for a total mass of 10^{11} suns, may be adequate to explain the quasi-stellar sources. Supernovae may be the cause of the light variations in 3C 48, although 3C 273 appears to be too young for supernovae to occur. The kinematical properties of the galaxy at this early phase do not disagree with the data for 3C 48 and 3C 273, and ages of about 3×10^6 and 3×10^5 years are indicated for these objects. It is expected that they will evolve into normal ellipticals with little flattening or gas. A constant rate of formation of quasi-stellar sources over 10^{10} years would have resulted in the creation of $\sim 10^7$ spherical galaxies. As this is ~ 10 per cent of the observed number, the present rate of creation of spherical galaxies would be a significant fraction of the average rate over 10^{10} years, if the present interpretation is correct.

I. INTRODUCTION

The quasi-stellar radio sources are the brightest objects known; Matthews and Sandage (1963) estimate that the luminosity of 3C 48 is 2×10^{46} erg/sec. One therefore asks if there is a phase in the evolution of a galaxy characterized by such a luminosity. Shklovsky (1960) pointed out that, very early in the life of a galaxy, the gas density and therefore the rate of star formation would be high. He estimated an absolute magnitude of -23 due to supernovae alone.

In this paper we study the earliest phases in the life of a galaxy, following Shklovsky's suggestion. We find that the radiation emitted by the young stars in such a system is of the order required in the quasi-stellar sources. Furthermore, the size of the galaxy at these phases is found to be in accord with the present observed upper limits on the size.

II. CONTRACTION OF A GAS CLOUD

We assume that an intergalactic gas cloud of sufficiently high density to be stable against the universal expansion is formed by some unspecified process, perhaps thermal instability (Gold and Hoyle 1959; Field 1964). An initial density $\rho_0 = 10^{-28}$ gm cm^{-3} is sufficient. If the mass is typical for a galaxy, say 10^{11} suns, the initial radius R_0 of an approximately spherical cloud is 0.2 Mpc. If the cloud is to start to contract to galactic dimensions, $2T + \mathfrak{M} + \Omega$ must be initially negative. (T is kinetic, $\mathfrak{M}$ is magnetic, and Ω is gravitational energy.) For simplicity we take ρ_0 to be uniform, in which case the value of $\Omega(R_0)$ is -3×10^{57} erg. In order for contraction to begin, the equatorial rotation velocity (v_0) must be less than 40 km/sec, the turbulent velocity must be less than 30 km/sec, the temperature must be less than 20000° K, and the magnetic-field strength must be less than 2×10^{-7} gauss.

Once the contraction starts, it will continue until $2T + \mathfrak{M}$ becomes comparable with $-\Omega$; $-\Omega$ increases like R^{-1} (assuming that the cloud remains uniform). Hoyle (1953) has shown that the temperature does not rise appreciably until the density becomes so

Reprinted from Astrophysical Journal,
140, 1434-1444.

large that the optical depth at the relevant wavelengths approaches unity. Let us denote by R_T the radius at which this occurs. Then as long as $R > R_T$, the thermal energy is independent of R; only after $R = R_T$ can it rise adiabatically like R^{-2}. It follows that the thermal energy can halt the contraction only at an $R \leq R_T$. The amount of turbulent energy tends to be less than or equal to the amount of thermal energy, owing to the rapid increase in dissipation when the Mach number exceeds unity; we may therefore ignore its effects for $R > R_T$. Magnetic energy rises like R^{-1}, and can therefore never grow to equal $-\Omega$ if it starts out smaller. Rotational energy increases like R^{-2}, equaling $-\Omega$ when the radius approaches $R_R = (v_0/40)^2 R_0$ if we assume solid-body rotation.

Let us suppose for a moment that star formation becomes important at a critical radius R_S, in the sense that the condensation of gas into stars and protostars of small cross-section increases rapidly at that phase. The behavior of the contraction is quite different, depending on whether R_R or R_S is the larger, if R_T is smaller than both.

If R_R is larger, the contraction is halted by the centrifugal forces associated with the rotation. Since these forces do not act parallel to the axis of rotation, however, the contraction continues in that direction, forming a disk. The virial theorem states that the energy of the disk is $\frac{1}{2}\,\Omega(\text{disk})$. The latter is $<\frac{1}{2}\,\Omega(R_R)$, since the disk is more compact than the sphere of the same radius. Also, $\Omega(R_R) \ll \Omega(R_0)$, since $R_R \ll R_0$. Therefore, a large amount of energy must be dissipated. This is not difficult to do, since the system is still mainly in the form of gas, by hypothesis. The result is that a spiral galaxy is formed, as described by Mestel (1963).

If R_S is larger, stars form before centrifugal forces become important. The radius continues to decrease through R_S, the gas going more and more into stars. Again centrifugal forces halt the contraction when R_R is approached, but this time very little gas is left to dissipate the energy, as required if a disk is to be formed. The stars simply continue in their orbits without dissipation, and the system begins to expand. The maximum radius of expansion, R_M, is given by the criterion that $-\Omega(R_M) = E$, the energy dissipated by the gas during the contraction phase. Since the latter will usually be only a small fraction of $T(R_S) \simeq -\Omega(R_S)$, R_M will considerably exceed R_S. The ensuing behavior, as the stellar system pulsates between R_R and R_M, poses an interesting problem in stellar dynamics. How does the pulsation energy become randomized among the stellar orbits? Because of the relative lack of flattening and of gas in the final system, we may tentatively identify the result with an elliptical galaxy.

At least in the extreme case $R_R \ll R_S$, there is nothing to keep the gas which remains at any stage from contracting to a high density, while in the opposite extreme, centrifugal forces prevent a build-up of high gas density. We therefore expect that star formation will go rapidly to completion when R_S is largest, and that the release of nuclear energy by stars will be most rapid in this case. This circumstance, together with the indication below that R_S is in fact only of the order of 100 pc, suggests that the properties of this case be studied more carefully, for comparison with those of the quasi-stellar sources.

The remaining case, when R_T is largest, corresponds to a significant increase in internal pressure at radii $>R_R$ or R_S. This results in the formation of a single "massive object" of the type discussed by Hoyle and Fowler (1963). The evolution of such objects to the point where rotation and formation of condensations become important, and their possible connection with the quasi-stellar sources, have been considered by Hoyle and Fowler (1964).

It is useful to have R, ρ, V (radial velocity of outer boundary), and T (kinetic energy), as functions of the time during the contraction phase. If, in accordance with the previous discussion, we assume that turbulent, thermal, rotational, and magnetic energies are considerably less than $-\Omega$ as long as $R > R_R$, the motion is approximately a free fall, with the release of gravitational energy mostly going into the kinetic energy of radial motion. Hunter (1962) has shown that an initially uniform sphere remains uniform under such motion, the radial coordinate of each gas element decreasing in proportion to that of the outer boundary, which obeys the relation

$$R = R_0\left[\frac{3\pi}{4}\left(\frac{t_0 - t}{t_0}\right)\right]^{2/3}, \tag{1}$$

where

$$t_0 = \left(\frac{3\pi}{32G\rho_0}\right)^{1/2} = 6.6 \times 10^9 \text{ years} \tag{2}$$

is the time required for $R \to 0$ (if centrifugal force did not intervene). Equation (1) is approximate, holding only when $R \ll R_0$. The density is

$$\rho = \frac{1}{6\pi G(t_0 - t)^2}, \tag{3}$$

and the velocity is

$$V = -\tfrac{2}{3}\left(\frac{3\pi}{4}\right)^{2/3}\frac{R_0}{t_0}\left(\frac{t_0 - t}{t_0}\right)^{-1/3}. \tag{4}$$

Since at internal points r, $v = Vr/R$, the kinetic energy is readily found to be

$$T \doteq \tfrac{3}{5}\frac{GM^2}{R} = -\Omega(R). \tag{5}$$

This expression reflects the fact that the total energy of the system is small compared to T if there is little dissipation and if $R \ll R_0$.

III. STAR FORMATION

In order to estimate the value of R_S (or the corresponding value of $t_0 - t$, denoted by t_*), we must have recourse to a theory of star formation. However, understanding of this problem is so limited that it does not appear possible to do this deductively. Instead, we adopt the empirical approach developed by Schmidt (1959, 1963*b*), who showed in his first paper that several observations related to the conversion of gas into stars in the solar neighborhood can be understood if the rate of star formation depends primarily on the density of available gas. He adopted a law of the form

$$\frac{d\rho_S}{dt} = K_n \rho_G{}^n, \tag{6}$$

where ρ_S and ρ_G are the densities of stellar and gaseous matter, respectively, and n and K_n are positive constants. It was found that $n = 2$ and $K_2 = 10^6 \text{ cm}^3 \text{ gm}^{-1} \text{ sec}^{-1}$ fitted the observations. In his later paper, Schmidt developed evidence that the effective value of n varies with the stellar mass concerned, being 1, 3, and 5 for 1, 10, and 100 solar masses, respectively. This conclusion was based on the fact that the formation rate of bright stars relative to that of faint ones seems to have been higher during the early history of the Galaxy (when the gas density was high) that it is at present.

For simplicity we shall use a single mean value of n in what follows. If we use equations (3) and (6) together with the requirement that $(\rho_S + \rho_G)R^3$ is constant, we find a differential equation in t for the quantity ρ_G/ρ. Its solution is

$$\frac{\rho_G}{\rho} = 1 - \frac{\rho_S}{\rho} = \left[1 + \left(\frac{t_*}{t_0 - t}\right)^{2n-3}\right]^{-1/(n-1)}, \tag{7}$$

where

$$t_* = \left(\frac{n-1}{2n-3}K_n\right)^{1/(2n-3)}(6\pi G)^{-(n-1)/(2n-3)}. \tag{8}$$

Equations (7) and (8) are valid for $n > \frac{3}{2}$, and indicate that gas is converted into stars as the cloud contracts, the process being about half complete when $t_0 - t \simeq t_*$. When $(t_0 - t)/t_* \ll 1$, $\rho_G/\rho \to 0$ like $[(t_0 - t)/t_*]^{(2n-3)/(n-1)}$, so that the depletion of gas is almost complete if $R_R \ll R_S$. The important point is that taking $R_S = R(t_*)$ is a meaningful procedure if star formation indeed proceeds according to equation (6), with $n > \frac{3}{2}$.

For definiteness we have adopted $n = 2$ and $K_2 = 10^6$, as in the solar neighborhood. This gives

$$\frac{\rho_G}{\rho} = \left(1 + \frac{t_*}{t_0 - t}\right)^{-1}; \qquad t_* = \frac{K_2}{6\pi G} = 2.9 \times 10^4 \text{years}. \tag{9}$$

This value of t_* is very uncertain for several reasons. First, the physical significance of equation (6) is not yet clear, and therefore we are not sure that it is legitimate to apply it to regions other than the solar neighborhood. Second, our choice $n = 2$ describes only the behavior of moderately bright stars. We shall see below that it is the very bright stars which are of interest for the quasi-stellar sources, so that we might have chosen an even larger value of n. (However, this would only have accentuated the peaking of star formation in the neighborhood of t_*, and hence strengthened the use of the concept of R_S.) Finally, the value of K_2 probably depends on the scale and velocity of the turbulence and the strength of the magnetic field, so that the appropriate value of K_2 may differ considerably from that in the solar neighborhood.

These uncertainties notwithstanding, we proceed on the basis of equation (9). From equations (1)–(5) we then find that $R_S = R(t_*) = 110$ pc, $\rho(t_*) = 10^{-18}$ gm cm^{-3}, $V(t_*) = -2600$ km/sec, and $T(t_*) = 5 \times 10^{60}$ erg. The small values of t_* and R_S, and the correspondingly large values of ρ, V, and T, are due to the fact that the cloud must contract to a small radius and high density if the rather weak process of star formation (small K_2) is to keep pace with the contraction. Since R_S is very small, R_R must be even smaller, indicating that v_0 must be <1 km/sec.

The behavior of the density is shown in Figure 1. The approach to R_R (arbitrarily taken as 1 pc in the figure) is indicated, as is the approach to R_M (fixed at 3 kpc by Fish's [1964] value of -1.5×10^{59} erg for the energy of an elliptical galaxy of this mass).

IV. ENERGY SOURCES

One possible energy source for the quasi-stellar sources is the kinetic energy of the gas, plotted in Figure 2 from equations (1), (5), and (9). Notice that it reaches a peak of 2.5×10^{60} erg at $t_0 - t = 10^4$ years, decreasing thereafter because of the conversion of gas into stars. It is evident that a moderate fraction of this energy could be converted into radiation without modifying the dynamics significantly. Ginzburg (1961) has suggested that fast electrons trapped in the increasing magnetic field of the cloud might be accelerated by the betatron effect and subsequently radiate synchrotron emission. It is easily shown that bremsstrahlung losses in the surrounding gas ($\sigma = 2.4 \times 10^{-26}$ cm^2) effectively limit this process to the phases earlier than $t_0 - t = \sigma c/4\pi G\, m_H = 1.6 \times 10^7$ years, when the density is still low enough so that losses do not outweigh the acceleration. $R = 5000$ pc and $T = 7 \times 10^{58}$ erg at this phase. Since this value of R is considerably larger than that of the quasi-stellar optical objects, it appears impossible to form the quasi-stellar objects themselves in this manner. On the other hand, it may be possible in this way to interpret the jet (or wisps) observed in 3C 273 and 3C 48, as they are considerably larger. Since they may emit a total energy of the order of 10^{57} erg, a few per cent of the available kinetic energy would be needed.

The fast electrons which are accelerated may be part of an intergalactic population, or may be secondaries due to nuclear collisions of an intergalactic population of cosmic rays with the gas in the cloud. If the electrons are present throughout the contraction, their total energy increases as R^{-1} as the cloud contracts. Since $-\Omega$ increases in the same way, the ratio would be constant, and a few per cent of $-\Omega(R_0)$, corresponding to

$\sim 2 \times 10^{-17}$ erg cm^{-3} would be needed. This is $\frac{1}{200}$ of the energy density of fast electrons in the Milky Way.

If the electrons are secondary, one finds that an initial energy density of 10^{-16} erg cm^{-3} in 10-GeV protons would be converted into 10^{57} erg of 12-GeV electrons at $t_0 - t = 2 \times 10^7$ years. (We assume acceleration of the protons to 400 GeV and a yield of 7 secondary electrons per collision, following Pollack and Fazio, 1963.) In this case, the intergalactic cosmic-ray energy would have to be 10^{-4} that in the Milky Way.

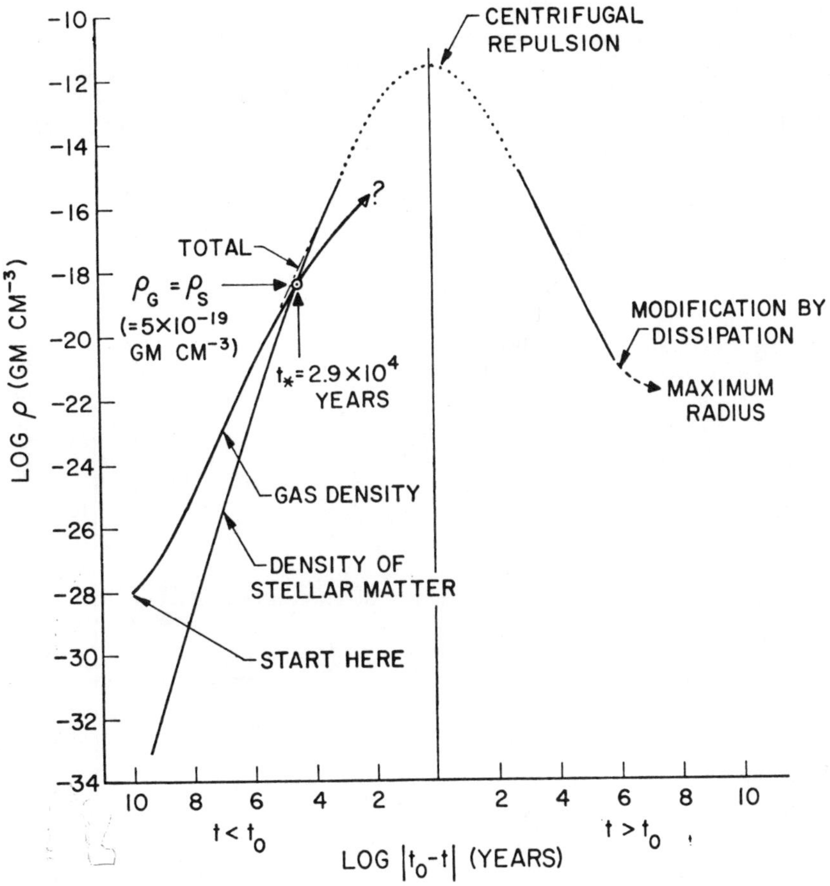

FIG. 1.—Variation of density during contraction and expansion phases. Star formation is 50 per cent complete at $t = t_0 - t_*$. The course of the curve for $t \simeq t_0$ depends on the angular momentum. Dissipation limits the expansion.

Nuclear energy available from the hydrogen in the contracting cloud will be released both in the form of radiation from normal stars and in the form of kinetic energy and relativistic particle energy associated with explosions of massive stars (Type II supernovae). At least some of this energy will be available when the radius is of the order of R_S, and is therefore of interest for the interpretation of the quasi-stellar objects themselves.

We calculate approximate values of the luminosities associated with normal stars and with supernovae by assuming that they have characteristics like those of galactic Population I; this assumption omits possible effects of low metal abundance. Restricting attention to the main sequence only (thereby underestimating the luminosity), we may use a relation of the form

$$\frac{L}{M}=\frac{\int_{M_L}^{M_U} L(M)\,dN(M)}{\int_{0.1}^{100} M\,dN(M)} \tag{10}$$

for the mean luminosity-to-mass ratio of main-sequence stars (in solar units). M_L and M_U are lower and upper mass limits (functions of the time) which correspond to the lower turn-off from the main sequence associated with Helmholtz contraction and to the upper turn-off associated with stellar evolution. The limits on the mass integral are taken arbitrarily, but the results depend only on the lower limit to the -0.35 power. The data on $L(M)$ given in Allen (1963, p. 203) can be represented by

$$L = kM^q\,, \tag{11}$$

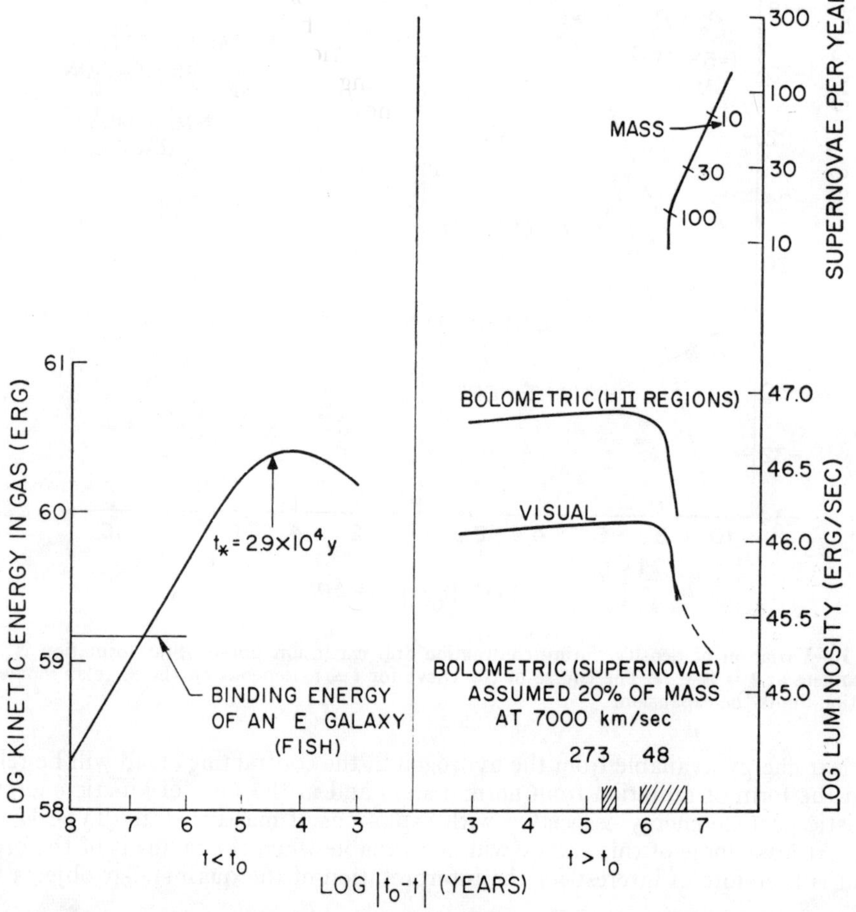

FIG. 2.—Energy sources in the young galaxy. Kinetic energy in the gas due to gravitational acceleration plotted at left. Observed binding energy for 10^{11} suns is 6 per cent of maximum kinetic energy, suggesting either small dissipation or underestimate of t_*. Luminosities of normal stars and supernovae at right. Acceptable ages for 3C 48 and 3C 273 from kinematical data.

with $k = 3.3$ and $q = 2.84$ for the bolometric luminosity, and $k = 2.5$ and $q = 2.48$ for the visual luminosity.

$$dN(M) \sim M^{-2.35}\, dM \tag{12}$$

was taken from Salpeter (1955). M_L was computed from Chandrasekhar's (1939) formula and the stellar radii given in Allen; the results were 1, 3, 10, and 30 solar masses at 5, 1, 0.2, and 0.01×10^6 years, respectively. M_U was taken from Limber's (1960) compilation of model calculations to be 20, 30, 60, and 100 solar masses at 6, 4, 3, and 2×10^6 years, respectively. These evolution times are 30 per cent less than Limber's because of a correction introduced to bring them into line with calculations since 1960. (While this paper was in preparation, Dr. R. Stothers kindly informed the author that such a correction is unnecessary. The correct times are therefore 1.4 times those given.) Masses greater than 100 were assumed to be absent (Schwarzschild and Härm 1959). Notice that we have implicitly assumed that the protostars are all formed at the same moment, $t = t_0 - t_*$. This seems legitimate since the actual spread of times of formation is only $\sim 10^5$ years, much smaller than the evolution time of the most massive stars ($>10^6$ years). The luminosity is then a function of the time since formation, or $t - (t_0 - t_*)$. The results are shown in Figure 2 for $t > t_0$ (expansion phase). There is little change in L from $t - t_0 = 10^3$–10^6 years, since little light is added by the stars of small mass which are contracting to the main sequence during this period. The peak value of the bolometric luminosity is 7×10^{46} erg/sec, corresponding to $L(\text{bol})/M = 370$. The typical star is therefore an early B star. The visual luminosity is about 20 per cent of the total (B.C. = 1.8^m).

At 2×10^6 years (or 3×10^6 years without the correction) the luminosity drops rapidly owing to the evolution of stars near 100 solar masses, and supernovae commence at about the same time. Figure 2 indicates the number of supernovae per year of various masses calculated from equation (12) and the main-sequence lifetimes, assuming that all massive stars become supernovae. If we assign an energy per supernova equal to $10^{50} M$ erg, corresponding to a shell whose mass is 20 per cent of that of the star, ejected at 7000 km/sec, we obtain the bolometric luminosity due to supernovae given in Figure 2. The energy estimate agrees with Hoyle and Fowler's (1960) estimate of 3×10^{51} erg of nuclear energy in a star of 30 solar masses. It is conservative, however, in comparison to Colgate and White's (1963) estimate of up to 3×10^{54} erg, based on the energy released by the gravitational collapse of the stellar core.

The curves beyond $t - t_0 = 10^7$ years, when masses less than 18 $M_\odot$ are involved, are uncertain because they depend on the lowest mass which will become a supernova. Stothers (1963) demonstrates that the observed supernova frequency indicates that most stars above 18 $M_\odot$ become supernovae. He points out that the actual limit may be only 4 solar masses, owing to the effects of neutrino emission, in which case 2–3×10^{-2} would be expected annually in the Milky Way. According to Payne-Gaposchkin (1957), such a high frequency is not necessarily inconsistent with the absence of historical galactic Type II supernovae, because of the effects of interstellar extinction. If this is the case, supernova activity might extend to 3×10^7 years, with frequencies up to 200/year.

V. COMPARISON WITH QUASI-STELLAR SOURCES

It is of interest to compare the data for 3C 48 and 3C 273 with the radius and velocity-curves of Figure 3. We shall assume that $t > t_0$, so that the objects are expanding; supernova explosions may then possibly play a role in the light variations. Greenstein and Matthews (1963) place an upper limit of 2500 pc on the radius of 3C 48, while Schmidt (1963*a*) states that the radius of 3C 273 is less than 500 pc. Therefore, $t - t_0$ is less than 6×10^6 years and 3×10^5 years, respectively. On the other hand, they also give widths for the emission lines which permit one to estimate upper limits on the expansion velocity (since random velocities may also contribute to the observed widths).

These upper limits, 750 km/sec for 3C 48 and 1500 km/sec for 3C 273, imply that $t - t_0$ is greater than 10^6 and 2×10^5 years, respectively. Therefore there are non-vanishing ranges of $t - t_0$ which are consistent with the data for both sources. We note that there is sufficient time in each case for the jets (1.5×10^5 light-years long) to have been formed by ejection with speeds less than c at about the epoch of star formation.

According to Greenstein (1964), the visual light of 3C 48 is 10^{45} erg/sec, and of 3C 273, 4×10^{45} erg/sec. According to Figure 2, the computed visual luminosity of the model is somewhat greater. He estimates that the bolometric light is $>4 \times 10^{45}$ erg/sec and

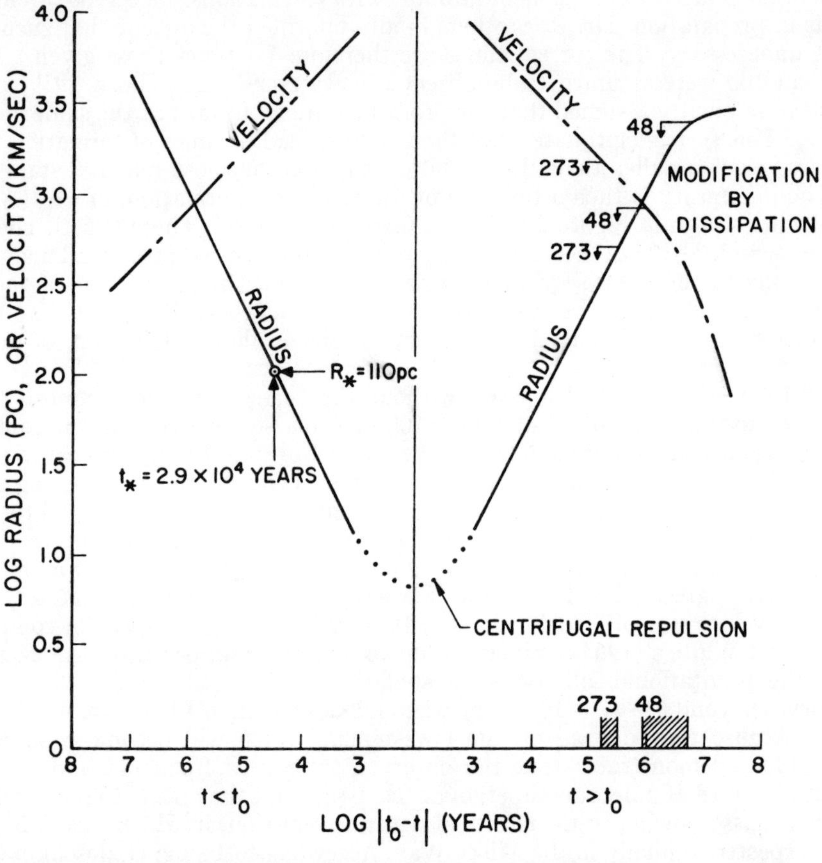

FIG. 3.—Kinematics of contraction and expansion, assuming free-fall conditions with small rotation and dissipation. Rotation modifies curves as R approaches R_R (small), and dissipation modifies curves at R_M (large). Actual value of R_M from observed binding energy of ellipticals. Acceptable ages for 3C 48 and 3C 273 from upper limits on radius and expansion velocity.

$>10^{46}$ erg/sec for the two cases; Figure 2 indicates that even these values can be interpreted as due to normal stars. We make no attempt here to interpret the observed spectra, however (see Greenstein and Schmidt [1964]).

A major problem for the interpretation proposed here is the observed light variation. For example, Sandage (1964) found that 3C 48 faded 0.4^m in one year, then recovered 0.2^m during the next two years. It would seem that this variability requires a source of light less than a few light-years across if the light-curve is periodic. On the other hand, if the light-curve has a random character, it may be possible to explain it by the superposition of several independent sources spread over a larger region. Since the young galaxy in our model is much larger than is required for a single source, we tentatively adopt the latter interpretation. We note that 3C 48 ($t - t_0 = 1$–6×10^6 years) may be in a phase when supernovae are active. If $t - t_0$ were about 3×10^6 years so that supernovae had recently commenced, the expected rate would be ~10/year, and their mean bolometric luminosity, $\sim 6 \times 10^{45}$ erg/sec. Could they possibly account for the variability?

To investigate this question, we first assume that the light-curve of each supernova is approximately a step function 5 years in duration. Second, we assume that all of its energy is converted to visible light, giving a luminosity of 8×10^{43} erg/sec. (The luminosity follows from the assumed energy and duration.) If the explosions occur at

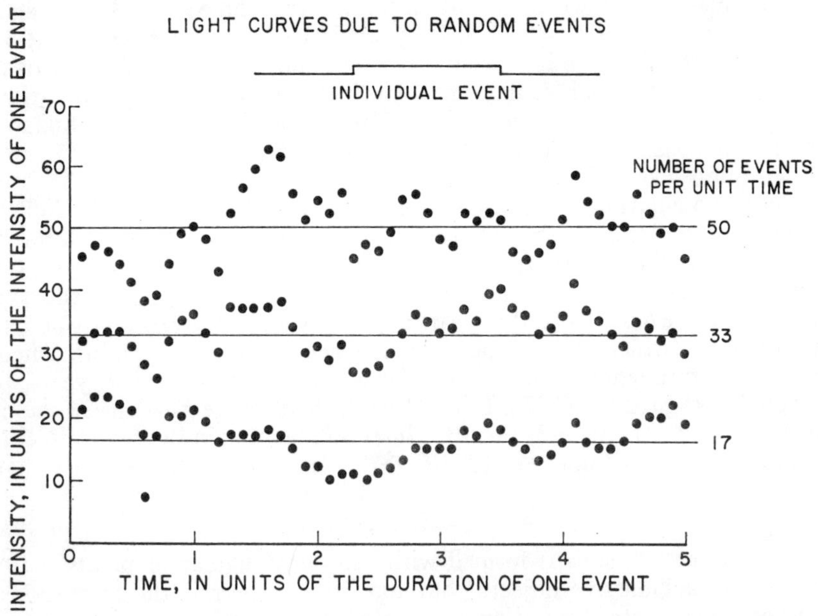

FIG. 4.—Light-curves calculated by random superposition of many samples of light-curve at top. Note adherence to rules that mean intensity is N times individual intensity, while rms variation is $N^{-1/2}$ times the mean. Typical periods in variation appear to be somewhat less than duration of single event. For interpretation of 3C 48 light-curve using supernovae see text.

random, we may calculate the observed light-curve by superposing individual light-curves with phases taken from a random-number table. It is easy to show that the mean number of supernovae (N) present at any time is just the mean number which commence during a period equal to the duration of the light-curve and that the mean luminosity is N times the individual value. The variability occurs largely on a time scale equal to the duration of each light-curve, and amounts (in rms value) to $N^{-1/2}$ in units of the mean total luminosity. Our assumptions therefore yield a light-curve like that for $N = 50$ in Figure 4, where the unit of time is 5 years, the mean luminosity is 4×10^{45} erg/sec, and the rms variability is 0.15^m. These data agree roughly with the observed values, if normal stars make only a minor contribution. Moreover, the observed light-curve does not appear to be radically different from the calculated one (between 1.6 and 2.6 time units, say).

Now the adopted parameters for the individual light-curves seem to be very different from those for typical observed Type II supernovae. The latter light-curves last only 10^{-2} times as long and contain only 10^{-4} times the energy. This may be explained by the fact that the supernova energy is largely in the motion of the shell, and the light observed in the usual case corresponds only to the tiny fraction emitted during the brief phase when the shell is dense enough to emit efficiently. Heiles (1964) has shown that the energy imparted by the supernova shell to the surrounding interstellar gas is ultimately radiated, but on a time scale of 10^6 years or so under normal conditions. Colgate and Cameron (1963) calculated that the energy of the shell is emitted optically in a few days if the density is $\sim 10^8$ times normal. It appears that if the interstellar density should be $\sim 10^5$ times normal, we might obtain the observed time scale. According to our model, the gas density at $t - t_0 = -3 \times 10^6$ years is 10^{-22} gm cm^{-3}, and by $+3 \times 10^6$ years, most of this should have been converted into stars, leaving a density of perhaps 10^{-24} gm cm^{-3}, which is not far from normal. The interpretation of the light variations of 3C 48 by supernovae would therefore require improbably high concentrations of interstellar matter in regions of star formation—up to 10^5 times the average.

Even more severe difficulties arise in the case of 3C 273, which has undergone two well-defined oscillations in light between 1930 and 1958 (Smith and Hoffleit 1963). The sinusoidal character of these oscillations already argues against our interpretation. Moreover, the upper limit on its radius implies an age less than 3×10^5 years, far too early for significant stellar evolution to have occurred, and therefore, for supernovae to have commenced. The only way out appears to be a revision of our estimate of t_* up to 3×10^6 years or more (which may be permissible since K_2 is so uncertain). Then supernovae would occur throughout the whole expansion phase out to $\sim 10^7$ years, including the early phase assigned to 3C 273. In that event, the average gas density could conceivably be as high as 10^{-20} in 3C 273, and 10^{-22} in 3C 48, tending in the right direction to shorten the radiation time of supernovae.

VI. SUMMARY

If a cloud of 10^{11} suns is formed with sufficient angular momentum to prevent the formation of a "massive object," but too little to permit the formation of a disk, a period of vigorous star formation is expected. Crude estimates of the phase of the commencement of star formation yield radii and velocities in the vicinity of those of the quasi-stellar sources. Furthermore, the available gravitational- and nuclear-energy sources may be sufficient to account for the luminosities of these objects. Gravitational energy may be responsible, via acceleration of electrons and subsequent synchrotron radiation, for the formation of the jets. Normal stars, which extend to large masses and correspondingly high luminosities during these early phases, may provide much of the optical light, either directly or by degradation of stellar ultraviolet in H II regions. Finally, if conditions are right, supernovae may account for the optical variability and perhaps also a synchrotron component of the emission. The crucial point is the suddenness of star formation associated with the rapid rise of gas density in the contraction of a slowly rotating system.

If one adopts a time scale of 10^6 years for the quasi-stellar phase of a galaxy, and assumes that there may be 10^3 sources in the observable universe, it follows that 10^7 galaxies are former quasi-stellar sources, if the rate of formation has been constant over 10^{10} years. What would be the characteristics of these galaxies? It seems reasonable that at least some of the normal radio galaxies are among them, so that observations bearing on the ages of these objects are important. The older ones should appear among the normal ellipticals, and because the angular momentum must be very low for the quasi-stellar phenomenon, they should appear particularly among the spherical galaxies. If there are of the order of 10^8 spherical galaxies (de Vaucouleurs 1959), the quasi-stellar phenomenon may be quite common among them, particularly if the rate of galaxy formation has been larger in the past. Again, age estimates for ellipticals would be useful.

The initial rotation velocity required to permit contraction inside 100 pc is 1 km/sec. It is not possible to say on the basis of our limited knowledge of the intergalactic gas whether this low value would be expected to occur frequently enough among intergalactic clouds to account for the number of quasi-stellar sources.

It is a pleasure to thank Drs. L. Spitzer, M. Schwarzschild, D. G. Wentzel, and N. Woolf, of the Princeton University Observatory, for helpful comments. Drs. J. L. Greenstein, M. Schmidt, and J. B. Oke, of Mount Wilson and Palomar Observatories, also made several important suggestions.

REFERENCES

Allen, C. W. 1963, *Astrophysical Quantities* (London: Athlone Press).
Chandrasekhar, S. 1939, *Stellar Structure* (Chicago: University of Chicago Press).
Colgate, S. A., and Cameron, A. G. W. 1963, *Nature*, **200**, 870.
Colgate, S. A., and White, R. H. 1963, *Bull. Amer. Phys. Soc.*, **8**, 306.
Field, G. B. 1964, to be published.
Fish, R. A. 1964, *Ap. J.*, **139**, 284.
Ginzburg, V. L. 1961, *Soviet Astr.—A.J.*, **5**, 282.
Gold, T., and Hoyle, F. 1959, *Paris Symposium on Radio Astronomy*, ed. R. N. Bracewell (Stanford, Calif.: Stanford University Press), Paper 104, p. 583.
Greenstein, J. L. 1964, lectures at the Institute for Advanced Study.
Greenstein, J. L., and Matthews, T. A. 1963, *Nature*, **197**, 1041.
Greenstein, J. L., and Schmidt, M. 1964, *Ap. J.*, **140**, 1.
Heiles, C. 1964, *Ap. J.*, **140**, 470.
Hoyle, F. 1953, *Ap. J.*, **118**, 513.
Hoyle, F., and Fowler, W. A. 1960, *Ap. J.*, **132**, 565.
———. 1963, *Nature*, **197**, 533.
———. 1964, California Institute of Technology (Preprint).
Hunter, C. 1962, *Ap. J.*, **136**, 594.
Limber, D. N. 1960, *Ap. J.*, **131**, 168.
Matthews, T. A., and Sandage, A. R. 1963, *Ap. J.*, **138**, 30.
Mestel, L. 1963, *M.N.*, **126**, 553.
Payne-Gaposchkin, C. 1957, *The Galactic Novae* (Amsterdam: North-Holland Publishing Co.).
Pollack, J. B., and Fazio, G. G. 1963, *Phys. Rev.*, **131**, 2684.
Salpeter, E. E. 1955, *Ap. J.*, **121**, 161.
Sandage, A. R. 1964, *Ap. J.*, **139**, 416.
Schmidt, M. 1959, *Ap. J.*, **129**, 243.
———. 1963*a*, *Nature*, **197**, 1040.
———. 1963*b*, *Ap. J.*, **137**, 758.
Schwarzschild, M., and Härm, R. 1959, *Ap. J.*, **129**, 637.
Shklovsky, I. S. 1960, *Soviet Astr.—A.J.*, **4**, 885.
Smith, H. J., and Hoffleit, D. 1963, *Nature*, **198**, 650.
Stothers, R. 1963, *Ap. J.*, **138**, 1085.
Vaucouleurs, G. de. 1959, *Hdb. d. Phys.*, **53**. 275.

A MODEL OF QUASI-STELLAR RADIO SOURCES

P. A. Sturrock
Institute for Plasma Research,
Stanford University, Stanford, California

The nature of quasi-stellar objects[1], originally identified by association with strong radio sources, remains an open question. It was remarked earlier[2] that the explosions which occur in quasi-stellar objects and those which occur in radio galaxies are similar not only to each other, but also to solar flares.[3] The proposal that the explosions of galaxies and of quasars, leading to radio clouds, could be interpreted as "galactic flares" could not at that time be developed in more detail because the mechanism of solar flares was not adequately understood. The model for the high-energy phase of solar flares, described in the preceding article[4], now makes it possible to construct a more detailed model of quasi-stellar objects.

One may argue very simply in favour of the explosions being flares. Ultra-relativistic electrons must have been accelerated by an electric field. A sudden spontaneous conversion of energy from one form to another is an instability. The kind of instability which generates electrical fields is a plasma instability. If there is to be more acceleration than heating, the gas density must be low. As a consequence, the energy released by the instability will be mainly magnetic. This is precisely what is generally believed to happen in the main and high-energy phases of solar flares.

Hoyle and Fowler[5] have argued that the energy released in some galactic explosions is so great (10^{59} ergs or more) that it is probably derived from the nuclear and gravitational energy of a compact body of gas of mass $10^5 M_\odot$ - $10^8 M_\odot$ and small size, generically related to the nucleus of a Seyfert galaxy. If the mass is supported by gas and radiation pressure, it is prone to instabilities, including some of a relativistic nature[6], but these do not necessarily release energy of the required magnitude.[7] Moreover, a non-electromagnetic model cannot, of itself, yield particle acceleration.

These difficulties are overcome if it is assumed that the mass is supported principally by magnetic stresses. The stored magnetic energy will then be comparable with the gravitational binding energy so that the model contains enough energy in such a form that it can readily be converted into high-energy particles. The assumption that the magnetic field is closed presents certain difficulties, not the least of which is that of explaining the origin of the field. The assumption that the magnetic field is open circumvents these difficulties. The magnetic field must now be assumed to be primeval.

The nucleus of a quasar is in this model coupled magnetically to inter-galactic space. If there is a sharp drop in density at the surface of the central mass, there will be a field reversal, constituting a "sheet pinch," at the "ecliptic plane" of the system, as shown in Fig. 1. The sheet pinch will terminate in a circular Y-type neutral line outside which there will be a ring-shaped trap for accreting gas. The field lines coupled directly to the nucleus constitute a pair of "funnels" for accreting gas. Photographic evidence that such rings and funnels actually occur is provided by plates 143, 146 and other plates of Arp's *Atlas of Peculiar Galaxies*.[8] It is now tempting to re-interpret the observed gas flux of both M82[9] and NGC 1275[10] as accretion rather than explosion. (Dr. George Field has independently suggested, in private conversation, that the gas flux of M82 may be an influx.)

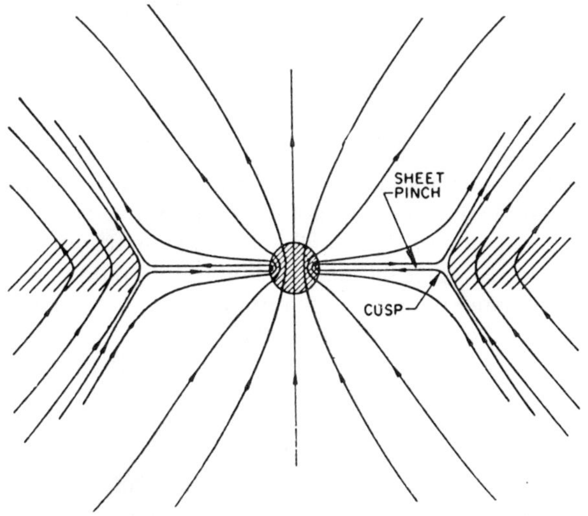

FIG. 1. Field configuration associated with quasar (twist not shown)

Reprinted from NATURE, 211, 697-700, 1966.

On balancing the magnetic energy and gravitational binding energy of a simple model, not too far from spherical,[2] a mass-magnetic-flux relation is obtained of the form:

$$\Phi \approx 10^{-3} \text{ M} \tag{1}$$

where Φ is in gauss cm^2, M in grams, and the exact coefficient depends on the geometry. It is noteworthy that a typical sample of radio clouds, discussed by Maltby, Matthews and Moffet,[11] are calculated to have fluxes in the range 10^{34} - 10^{38} gauss cm^2, in excellent agreement with equation (1) and the Hoyle-Fowler hypothesis[5] that the explosions originate in galactic nuclei, for which $10^4 M_\odot$ - $10^8 M_\odot$ is a reasonable mass range.

It appears that a magnetic field can offer only gross neutral stability. However, magnetohydrodynamic turbulence has an effective γ of 3/2 and so can provide gross stability. For a certain simple model, one finds that the oscillation frequency for radial pulsations ω may be related to the mean density ρ, the radius R and the Schwarzschild radius $R_S (=2GMc^{-2})$ by:

$$\omega^2 \approx \pi G\rho(A^2 - 0.55|R_S/R|) \tag{2}$$

where A is the Alfvén number, that is, the ratio of the turbulent velocity v_t to the Alfvén velocity $v_A [= B(4\pi\rho)^{-1/2}]$.

The above and related equations make it possible to estimate parameters for the nucleus of the quasar 3C 273, if we interpret the observed 13 year periodicity[12] as a radial pulsation and the dispersion in gas velocity of "half-width" $10^{8.3}$ cm sec^{-1}, inferred from emission line widths[13], as the turbulent velocity v_t. With the choice $A = 10^{-1}$, we then obtain $M = 10^{42.3}$ g $= 10^9 M_\odot$, $R = 10^{16.2}$ cm (= 5 light days), $\rho = 10^{7.1}$ g cm^{-3}, $\Phi = 10^{39.3}$ gauss cm^2, $B = 10^{6.4}$ gauss, $v_A = 10^{9.3}$ cm sec^{-1}. The escape velocity is approximately v_A.

As was observed earlier,[2] the likelihood of anomalous density gradients provides for the occurrence of flare flashes,[14] which may explain the observed flashes in both optical and radio emission. It is also possible that magnetohydrodynamic instabilities may occur which could produce sudden changes in radius, either increasing or decreasing the optical luminosity.

We may now seek, in the present model, the analogue of the high-energy phase of solar flares described in the preceding article. The site for this phenomenon will be the sheet pinch which lies in the "ecliptic plane," and the mechanism will be the tearing-mode instability. The result will be a "pocket" of turbulent magnetic field containing high-energy particles being ejected by the tension of the magnetic field, as indicated in Fig. 2. If this occurs over only a small range of longitude, it will constitute a jet, and it is proposed that the jet of 3C 273 and the jet of M 87 (ref. 15) are manifestations of this mechanism. The fact that the synchrotron radiation of the jet of 3C 273 is detectable optically but not in the radio wavelengths indicates that the spectrum is somewhat flat, implying that acceleration is in progress.

The hypothesis that radiation from the jet of 3C 273 is the result of continuing acceleration enables one to estimate the strength of the magnetic field at the tip of the jet. The rate of emission is about $10^{44.3}$ ergs sec^{-1} and the cross-sectional area of the jet is about $10^{43.9}$ cm^2. We make a conservative estimate of the field strength by assuming that the shock front at the tip of the jet is relativistic, advancing at the speed of light. On equating the rate of radiation to the rate of "dissolution" of magnetic energy, one obtains the estimate $B = 10^{\bar{5}.6}$ gauss at the tip of the jet. Because the tip of the jet is at least $10^{23.2}$ cm from the nucleus, the magnetic flux through each hemisphere amounts to $10^{42.8}$ gauss cm^2. If we now refer to the mass-magnetic-flux relation, we see that this magnetic flux must be associated with a mass of $10^{45.8}$ g, that is, $10^{12.5} M_\odot$ Our model will not permit the visible nucleus 3C 273B to have so great a mass. We are therefore led to the conclusion that surrounding the visible object 3C 273 is a much larger mass of gas which is invisible, presumably because it is fully ionized. We may note with some satisfaction that the presence of a mass of this magnitude will make it possible for the quasar to evolve into a galaxy as massive as M87, or conceivably into a cluster of galaxies.

It is interesting to enquire further into the likely fate of the jet of 3C 273. It will continue as a jet

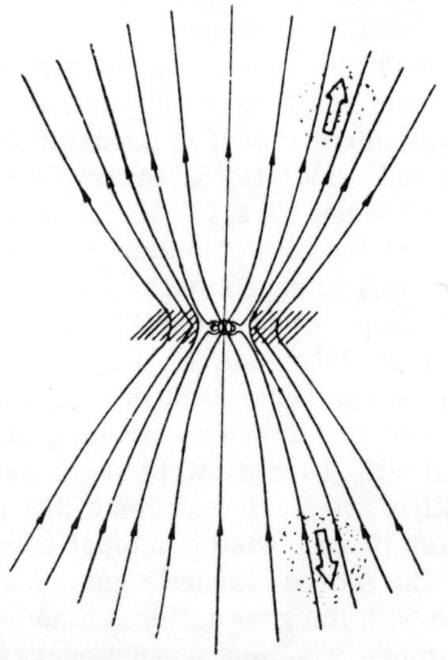

FIG. 2. Field configuration associated with quasar following flare, leading to jet formation

until it arrives at the ring-shaped neutral line. At this point, the supply of magnetic energy will cease, and there will be no further acceleration. The cloud of electrons and ions, which have relativistic energies, will now divide into two clouds, about equal in size, one moving "northwards" and the other "southwards." This is indicated in Fig. 3.

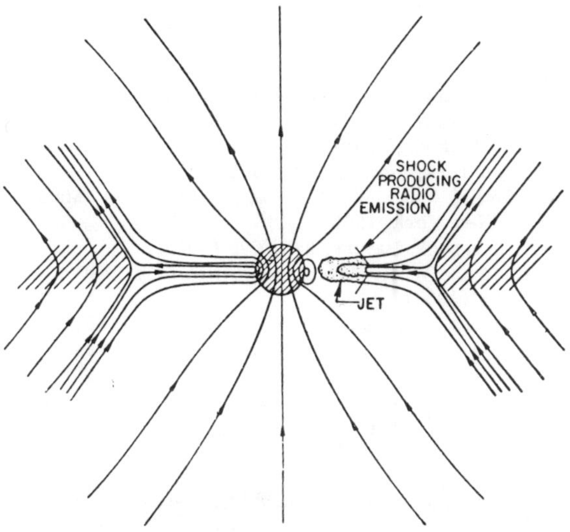

FIG. 3. Later stage, showing division of jet into two radio clouds (twist and effect of twist not shown)

It is notable that a single explosion eventually produces two similar radio clouds at equal distances from the nucleus. In the case of 3C 273, the nucleus will not be on the centre line of the double radio source when it forms (although it may seem so to other observers). However if, as seems quite possible, the flare were to extend over all longitudes, the end result would be two "smoke ring" radio clouds diametrically opposite with respect to the nucleus. We see that this model offers a simple explanation of the most striking characteristic of radio galaxies--their double structure.

Two possible interpretations of the radio source 3C 273A present themselves. One is that the electrons responsible for the radio emission have been accelerated in a shock front at the tip of the jet... the same mechanism as is responsible for a Type II solar radio burst, although the electron energies would be greater so that we are seeing synchrotron radiation rather than radiation from plasma oscillations. The other alternative is that the jet has just arrived at the cusp so that we are seeing radiation from the first batch of electrons to move out of the acceleration region of the sheet pinch.

As was noted earlier, the funnelling action of the magnetic field has the result that intergalactic gas streams directly into the nucleus. Although the Jeans criterion is open to criticism,[16] it will be used to provide an estimate of the radius of the sphere of intergalactic gas which has condensed to form the quasar. If in the formula $R_c = a(4\pi G\rho_0)^{-1/2}$, where a is the speed of sound and ρ_0 the density, we adopt the current estimate[17] $\rho_0 \approx 10^{-28}$ g cm^{-3}, we find the required temperature to be $T_0 = 19^{6\cdot2}$ °K; this may seem high, but it has recently been shown that the density of neutral hydrogen in intergalactic space is exceedingly low.[18] The corresponding value of R_c is $10^{24\cdot4}$ cm (800 kiloparsecs). This estimate enables us to calculate the intensity of the primeval magnetic field. As the flux of $10^{42\cdot8}$ gauss cm^2 originally threaded a sphere a radius $10^{24\cdot4}$ cm, the primeval magnetic field is of intensity $10^{7\cdot5}$ gauss. The primeval magnetic pressure was therefore comparable with the gas pressure. However, as Chandrasekhar[19] has noted, this would not have inhibited the gravitational condensation.

It was noted previously that the funnelling action of the magnetic field permits a high accretion rate leading to rapid evolution of the quasar. If we identify the optical radiation from the nucleus of 3C 273, 10^{46} ergs sec^{-1}, as being due to accretion, the gas influx will be $10^{27\cdot2}$ g sec^{-1} or more, so that the evolution time (the time for the nucleus to double its mass) is 10^{15} sec ($10^{7\cdot5}$ y) or less. The inequality quoted in the previous sentence is based on the expectation that the inflowing gas will give up the bulk of its energy at a shock front (or series of shock fronts) at which the temperature is likely to be much higher than that of the visible surface of the nucleus. If the shock reheats the gas to a temperature comparable with the original (intergalactic) temperature, the radiation would be primarily in the X-ray band so that the X-ray output from the nucleus of 3C 273 may well exceed the visible radiation.

The main "galactic" mass of gas should also be a source of X-rays. Whether or not it is more intense than the X-rays originating near the nucleus depends on the size of the galactic mass (and weakly on its temperature). If the temperature is about 10^6 °K, X-ray emission from the galactic mass will exceed 10^{46} ergs sec^{-1} if the radius of the mass is less than $10^{21\cdot9}$ cm, that is, 2·5 kpc.

The present model suggests answers to many questions concerning quasars. In addition to those already mentioned, it is possible to explain why the formation of a small heavy object has not been inhibited by angular momentum: the "open" magnetic field configuration provides a mechanism for retaining a low angular velocity during condensation by transferring angular momentum from the inner regions to the outer regions. One may also note that, with the high magnetic field quoted, the inverse Compton effect will be negligible in comparison with synchrotron radiation, so avoiding a difficulty which has led Hoyle, Burbidge and Sargent[20] to suggest that quasars are not at cosmological distances. Further, one may note that the explosion mechanism suggested for quasars will be applicable also to radio galaxies, if one assumes that the magnetic field of the nucleus of such a galaxy is similar to that which we associate with a quasar. Such a galaxy would indeed evolve from each quasar, according to the present model.

This chain of thought now offers an explanation of the recent remarkable observations of Byram, Chubb and Friedman[21], that certain radio galaxies emit X-rays at a power 10-100 times greater than the combined optical and radio output. This X-ray emission may be ascribed to an influx of intergalactic gas into the nucleus.

In view of the high magnetic field strength of the nucleus of 3C 273 (which is expected to be typical of quasars), the great mass of plasma surrounding the nucleus, and the various plasma phenomena which may arise, analysis of the radiation (including radio emission) from the nucleus appears a complex problem. One will need to bear in mind the possibility of proton synchrotron radiation and also of coherent radiation processes, as has recently been suggested by Ginzburg and Ozernoy.[22]

Finally, we may suspect from the fact that solar flares produce high-energy protons ("solar cosmic rays") of energies ($\sim$ 1 beV) much greater than that of electrons responsible for the radio emission (a few meV), that galactic flares should accelerate some ions to energies which are much greater than the energy of electrons in, say, the jet of 3C 273. The jet must contain electrons with energies of about $10^{12 \cdot 3}$ eV in order to give optical radiation in a magnetic field of strength $10^{5 \cdot 6}$ gauss. These electrons are decelerated by synchrotron radiation at a rate 10^6 eV sec^{-1}. Thus the stochastic acceleration process[23], which arises from the electric-field spectrum produced by the instability, has a maximum acceleration rate of at least this value. Protons are subject to the same acceleration, without experiencing significant deceleration. Any protons which have been accelerated at the maximum rate for the life-time of the jet (which must be at least $10^{12 \cdot 4}$ sec, as its length is 10^{23} cm) will have attained an energy of at least $10^{18 \cdot 5}$ eV. Multiply charged ions may have achieved a greater energy. Thus, it is possible that galactic flares are responsible for the high-energy component of cosmic rays.

This work was supported by the Air Force Office of Scientific Research, and the National Aeronautics and Space Administration.

REFERENCES

[1] Quasi-stellar Sources and Gravitational Collapse, edit. by Robinson, I., Schild, A., and Schucking, E. L. (University of Chicago Press, 1965).

[2] P. A. Sturrock, Nature, 205, 861 (1965).

[3] H. J. Smith, and E. V. P. Smith, Solar Flares (Macmillan Co., New York, 1963).

[4] P. A. Sturrock, Nature, 216. 695 (1966).

[5] F. Hoyle and W. A. Fowler, Mon. Not. Roy. Astro. Soc, 125, 169 (1963).

[6] S. Chandrasekhar, Astrophys. J., 140, 417 (1964).

[7] W. A. Fowler, Proc. Belfer Science Forum (Yeshiva University Press, 1965).

[8] H. Arp, Atlas of Peculiar Galaxies (California Institute of Technology, 1966).

[9] C. R. Lynds and A. R. Sandage, Astrophys. J., 137, 1005 (1963).

[10] E. M. Burbidge and G. R. Burbidge, Astrophys. J., 142, 1351 (1965).

[11] P. Maltby, T. A. Matthews, and A. T. Moffet, Astrophys. J., 137, 153 (1963).

[12] H. J. Smith, in Quasi-stellar Sources and Gravitational Collapse, edit. by I. Robinson, A. Schild, and E. L. Schucking, 221 (University of Chicago Press, 1965).

[13] J. L., Greenstein, and M. Schmidt, Astrophys. J., 140, 1 (1964).

[14] P. A. Sturrock, and B. Coppi, Astrophys. J., 143, 3 (1966).

[15] W. Baade, Astrophys. J., 123, 550 (1956).

[16] D. Layzer, Ann. Rev. Astron. Astrophys., 2, 341 (1964).

[17] D. W. Sciama, Quart. J. Roy. Astron. Soc., 5, 196 (1964).

[18] J. E. Gunn, and B. A. Peterson, Astrophys. J., 142, 1633 (1966).

[19] S. Chandrasekhar, Hydrodynamic and Hydromagnetic Stability, 589 (Oxford, 1961).

[20] F. Hoyle, G. R. Burbidge, and W. L. W. Sargent, Nature 209, 751 (1966).

[21] E. T. Byram, T. A. Chubb, and H. Friedman, Science 152, 66 (1966).

[22] V. L. Ginzburg, and L. M. Ozernoy, Astrophys. J., 144, 599 (1966).

[23] P. A. Sturrock, Phys. Rev., 141, 186 (1966).

A QUASAR MODEL BASED ON RELAXATION OSCILLATIONS IN SUPERMASSIVE STARS*

William A. Fowler
California Institute of Technology, Pasadena, California

The work described in this paper constitutes a return to the early point of view of Hoyle and Fowler (1963a, b) that supermassive stars can meet the energy requirements in radio sources, specifically in the quasars. General relativity leads to dynamic instability in non-rotating massive stars but the first result is relaxation oscillations energized by hydrogen burning rather than catastrophic collapse at least for masses not exceeding 10^6 $M_\odot$. It is noted that the introduction of rotation raises this limit by several orders of magnitude and that a forthcoming paper will treat this matter. It is emphasized that the exotic forms of energy emission observed in the quasars can serve to damp the relaxation oscillations in such a way that stable pulsations result. It is thus suggested that quasars consist of pulsating supermassive stars, energized by nuclear reactions, with radio and optical emissions from extended surrounding regions which the star excites with ultraviolet radiation and relativistic particles. With the exhaustion of nuclear energy, gravitational energy may become available and the evolution of a quasar into an extended radio source becomes possible.

*Supported in part by the Office of Naval Reserach [Nonr-220(47)] and the National Aeronautics and Space Administration [NGR-05-002-028]. (A talk presented at the Belfer Science Forum of Yeshiva University in New York on November 16, 1964.)

The Optical Identification of Radio Sources

Since World War II a revolution has ocurred in astronomy and astrophysics. The heroes of this revolution are the radio astronomers who have not only detected radio waves from extragalactic sources but have devised techniques for pinpointing the location of these sources on the celestial sphere. In the forefront of these efforts have been radio astronomers throughout the world—in Australia, England, the Netherlands, the Soviet Union and the United States.

The break-through in the observation and identification of radio sources has been made possible by the development and construction in several locations throughout the world of radio telescopes capable of making position determinations to better than ten seconds in angle. As one example, radio astronomers of the California Institute of Technology, using the funds furnished by the Office of Naval Research, have constructed at the Owens Valley Radio Observatory in Bishop, California, an interferometer consisting of two 90-foot dishes which can be separated on railroad tracks up to distances of 1,000 meters. Since the radio waves studied are the order of 30 centimeters in wave length, each dish is capable of an angular resolution near 10^{-3}, in observations taken with great care, while the interferometer arrangement yields a limiting resolution near 3×10^{-5}

The precise determination of the position of a radio object makes possible an accurate comparison with the position of optical objects visible through large conventional telescopes which have very high angular resolution because of the short wave length of visible light. The ultimate objective is to make an "identification" of the radio source with an optical object. Radio astronomers at the California Institute of Technology have the unique advantage of being able to cooperate with staff members of the Mount Wilson and Palomar Observatories in using the 200-inch Hale Telescope on Mount Palomar for making position comparisons and identifications.

Early identifications, made before great precision had been reached in the radio observations, indicated that in some cases radio sources seemed to be associated with pairs of galaxies in close proximity and perhaps even in collision. This led naturally to the assumption that the energy freed in such a collision might be the source of the radio energy. It is now believed that collision energy is inadequate in this regard but more importantly the great majority of the more precise identifications for radio sources outside of our Galaxy, the Milky Way, are with single, isolated galaxies and not with pairs of galaxies.

Until recently there has been no way to determine directly the distance to the radio objects of interest although red shift measurements have been made on the 21-cm atomic hydrogen line from nearby objects and are being rapidly extended to more distant objects. On the other hand the distance to the galaxy can be calculated if optical red shift measurements have been made and if the red shift is assumed to be proportional to distance in accordance with Hubble's Law.

The Energy Requirements of the Radio Sources

Identification with an optically red shifted galaxy thus makes it possible to determine the absolute luminosity of radio sources from the measured apparent luminosity, that is, the radio flux at the earth in erg cm^{-2} sec^{-1} can be translated into the total rate of energy emission in erg sec^{-1} at the source with the additional assumption of isotropic emission. (The possibility that the radio waves are directed at the earth is rightly given scant attention.) The results are staggering. More than 50 radio sources have been listed by Matthews, Morgan and Schmidt (1964) with luminosities exceeding 10^{38} erg sec^{-1} and ranging up to 2×10^{45} erg sec^{-1}, the value for 3C 295 (the 295th object in the third Cambridge University catalogue of radio sources). The optical luminosity of the sun is 4×10^{33} erg sec^{-1} and that from the Galaxy is approximately 10^{44} erg sec^{-1}. Thus 3C 295 has a radio luminosity almost 10^{12} times that of the optical emission from the sun and more than ten times that from the Galaxy.

The total amounts of energy required to sustain these luminosities can be calculated in several ways. It is reasonable to assume that the minimum age to be assigned to the sources is that given by dividing the observed dimensions by the velocity of light. Actually the linear growth of the sources might well have taken place at considerably smaller velocities. Even so the ages fall in the range 10^5 to 10^6 years or 10^{13} seconds in order of magnitude and thus the cumulative emissions are at least as high as 2×10^{58} ergs.

Another method of determining the total energy involved in the radio sources is based on the assumption that the radio emission is synchrotron radiation from high energy electrons spiraling in a magnetic field extending throughout the object. This process is thought to be the most efficient for the generation of radio waves and accounts qualitatively at least for the polarization observed in many of the sources. The synchrotron theory implies that energy is stored in the radio objects in the form of magnetic field energy and relativistic electron energy. The magnetic field energy is proportional to the mean square of the field intensity (B) and to the volume. The total energy of the electrons is proportional to the rate at which they emit energy, the radio luminosity, divided by the three halves power of the field intensity and the square root of the characteristic radio frequency emitted. Thus for a given observed volume, luminosity and radio emission spectrum the total energy is equal to a term proportional to B^2 plus one proportional to $B^{-3/2}$. The field intensity is, of course, unknown but even so the total energy exhibits a minimum as a function of B and this minimum can be readily determined. The values for the minimum stored energy even exceed those of the minimum cumulative energies. In the case of the Hercules A source the minimum stored energy, if the only high energy particles are electrons, is approximately 10^{60} erg. The theory does not explicitly indicate the method by which the electrons are accelerated to high energy but it is reasonable to assume, as is the case in the cosmic radiation, that the nuclear component (mostly protons) of the neutral medium or plasma must have considerably greater total energy content than do the electrons. Upon taking this factor into account the stored energy in Hercules A, for example, is almost 10^{61} erg. Because of their greater mass the protons do not take part in the synchrotron emission.

In coming to a realistic estimate of the energy requirements in radio objects there remains the knotty problem concerning the efficiency with which the energy generated has been converted into relativistic particles and magnetic fields. Acceleration mechanisms employed in terrestrial laboratories are notoriously inefficient but this may well be due to the very small scale, astrophysically speaking, within which such mechanisms must operate. However, it is estimated that even in solar flares not more than a few per cent of the energy released is in the form of relativistic particles, the main energy release occurring in mass motions and electromagnetic radiation.

On the above basis, the figure 2×10^{62} ergs is frequently quoted as a representative value of the energy requirement in the larger radio sources and for the purposes of argument this figure will be accepted as the maximum value in what follows. Suggestions have been made which modify the simple synchrotron model in such a way as to reduce the energy requirements. The magnetic field can be imagined to have a "clumpy" structure such that the effective emitting volume, where the field is highest, is much smaller than the overall volume observed. The magnetic field energy is proportional to the emitting volume not the overall volume. The emission may come from groups of electrons radiating coherently and thus much more efficiently. Detailed studies of modifications along these lines will be necessary before the energy problem can be considered to be solved.

It will be noted that there is considerable disparity in the two estimates which it is possible to make for the energy requirements in the extended radio sources. On the one hand the cumulative emissions range up to 2×10^{58} ergs while the stored energies on the synchrotron model have been estimated to be as high as 2×10^{62} ergs.

Supermassive Stars

The immensity of $a \times 10^{62}$ ergs, can best be appreciated by a comparison with the equivalent rest mass energy of a single star, for example, the sun. The mass of the sun is 2×10^{33} grams and the square of the velocity of light is $(3 \times 10^{12})^2 \sim 10^{21}$ ergs per gram. Thus Einstein's relation between energy and mass.

$$E = Mc^2 \tag{1}$$

become numerically

$$E \approx 2 \times 10^{54} \, (M/M_\odot) \text{ erg} \tag{2}$$

where $M/M_\odot$ is the stellar mass expressed in units of the solar mass. We see then that the energy stored on the synchrotron theory in particles and magnetic fields in the invisible radio objects requires the original production of energy of the order of that obtained by the complete annihilation of the mass of one-hundred million suns, 10^8 $M_\odot$. The problem can be taken in a quite literal sense on the grounds that the conversion of mass is the fundamental mechanism for the production of energy. On this basis the problem reduces to how, when and where did the conversion take place.

Before proceeding it is advisable to write Einstein's relation in a form more directly applicable to the problem under consideration as follows

$$\Delta E = (M_0 - M)\, c^2 = 2 \times 10^{54} \, (M_0 - M)/M_\odot \text{erg} \tag{3}$$

where ΔE is the energy made available from a system of particles with total rest mass M_0 when by some mechanism the mass, measured through gravitational or inertial effects by an external observer, has been reduced to M. The quantity ΔE is the energy store available for transformation at varying efficiencies into the various observable forms—gamma ray, x-ray, optical, radio, neutrino and high energy particle emission.

In principle it is possible for M to decrease to zero but not to negative values and so the maximum available energy is indeed $M_0 c^2$. One mechanism by which this can occur is through the annihilation of equal amounts of matter and antimatter. No detailed theory has been advanced showing how annihilation can take place in radio objects, the main problem having to do with the assembly of matter and antimatter in sufficient quantities on a time scale at most equal to that associated with the assumed explosive origin of these objects. Although annihilation will not be discussed further in this paper, it may prove to be the ultimate solution to the problem.

The success of the idea of nuclear energy generation in stars led quite naturally to the extension of this idea to the radio sources. Hoyle and Fowler (1963a) investigated the possibility that a mass of the order of $10^8 M_\odot$ has condensed into a single star in which the energy generation takes place. On this point of view, using the standard theory of stellar structure in Newtonian hydrostatic equilibrium, one immediately obtains optical luminosities of the order of 10^{46} erg sec^{-1} and lifetimes for nuclear energy generation of the order of 10^6 to 10^7 years so that the overall energy release is approximately 10^{60} ergs.

There is, of course, a basic limitation inherent in thermonuclear energy generation. The conversion of hydrogen into helium involves the transformation of only 0.7 per cent of the rest mass into energy and further nuclear burning leading to the most tightly bound nuclear species near iron brings this figure only to slightly less than one per cent. Thus M_0 - M in equation (3) is at most equal to $0.01\ M_0$ and the complete nuclear conversion of 10^8 solar masses of hydrogen into iron group elements leads to the release of 2×10^{60} ergs. In general

$$\Delta E_{\text{nucl}} < 2 \times 10^{52} M_0/M_\odot \text{erg} \tag{4}$$

Equation (4) is expressed in terms of an upper limit for the following reason. In the observed stars with masses ranging approximately from 1 to 100 $M_\odot$ the conversion never seems to reach completion before steady mass loss or supernova explosion terminates the life of the star. Thus it is clear that the nuclear generation of 2×10^{62} ergs, the maximum value discussed above, involves at least 10^{10} $M_\odot$. This figure corresponds to the entire mass of a medium size galaxy! On the other hand, the nuclear generation of 2×10^{58} ergs, the minimum value discussed above, involves the order of $10^6 M_\odot$. This figure corresponds to the mass of the larger globular clusters of stars in the halo of the Galaxy and in other galaxies. If globular clusters are involved in the energy production, it need not necessarily take place at the center of the galaxy.

In the massive galaxies associated with the strong radio sources there seemed to be no observational evidence for the abnormal heavy element concentration which would presumably follow from the nuclear conversion of $10^{10} M_\odot$. If the larger energy requirements are accepted, it can be argued that nuclear energy might prove inadequate and so Hoyle and Fowler (1963b) turned to another possibility, gravitational energy. On classical Newtonian theory the gravitational binding energy of a system of rest mass M_0 with maximum radius R is approximately given by

$$\Omega \approx \frac{GM_0^2}{R} \tag{5}$$

This expression does not include a numerical coefficient of the order of unity which depends upon the distribution of matter in the star. If no energy is stored in the system which remains as "cold" gas or "dust" then Ω becomes ΔE, the energy freed by the system on condensing from the dispersed state in which the gravitational interaction can be neglected. If equation (5) is written

$$\Delta E_{\text{grav}} = \Omega \approx \frac{GM_0}{Rc^2} M_0 c^2 \tag{6}$$

it will be seen that the dimensionless quantity GM_0/Rc^2 is just the fraction of the rest mass energy made available. Classical Newtonian theory places no limitation on GM_0/Rc^2 but the theory of general relativity limits it to the order of unity in the approximation here under consideration. Thus

$$\Delta E_{\text{grav}} < M_0 c^2 < 2 \times 10^{54} M_0/M_\odot \text{erg} \tag{7}$$

in agreement with the statement made previously that M could not become negative.

In what way can use be made of the release of gravitational energy? We assume that in some way this energy is removed from the collapsing core and is either absorbed in the outer envelope or is completely lost by the star. In either case the hydrostatic balance in the envelope is destroyed and the envelope material is ejected with high velocity. The energy loss from the core may occur through photon or neutrino emission. Another possibility exists if the massive star is in rotation. After the exhaustion of nuclear energy the star will contract with the contraction of the core being much more rapid than that of the envelope. It is reasonable to suppose that the angular momentum of the core will be conserved once it has contracted away from the envelope and that eventually the core will become unstable to fission into two bodies rotating about each other as in a binary star. Such a system loses rotational energy by radiating gravitational waves.

All emission mechanisms suffer from the limiting effect of the gravitational red shift. In order for gravitational energy to be released from the core it is necessary that the core contract or that $(GM_0/Rc^2)_{core}$ increase. But the red shift in radiation is just proportional to this dimensionless quantity in first order. Radiation arrives at a distant observer with less energy than that calculated by a local observer where the radiation is emitted. This is true for all forms of energy transfer, by particles as well as radiation. Thus the rate of any form of energy loss by the core is greatly reduced as $(GM_0/Rc^2)_{core}$ increases and, as a result, the energy loss is not complete as implied in equation (6) where it was assumed that no internal energy of motion or radiation remained in the star during contraction. As a matter of fact even the most optimistic caluclations have not revealed mechanisms whereby a contracting massive star can transfer more than a few per cent of the gravitational energy of its core to the outer envelope. The gravitational release of energy may be somewhat more efficient than nuclear release but not by a large factor. Thus the release of 2×10^{62} ergs must, on just about any grounds, involve a mass of the order of 10^{10} $M_\odot$. The large elliptical galaxies associated with radio sources have total masses estimated at $10^{12} M_\odot$. Thus if 2×10^{62} ergs is indeed the correct value for the energy requirement in radio galaxies, then of the order of one per cent of the mass of the galaxy has been involved in the generation of this energy.

Quasars

It has been noted previously that Hoyle and Fowler (1963a) had obtained optical luminosities of the order of 10^{46} erg $\sec^{-1}$ for a massive star of $10^8 M_\odot$ in hydrostatic equilibrium and in fact it was found that the luminosity is just proportional to the mass for $M > 10^3 M_\odot$. These large optical luminosities did not seem to have any immediate connection with the extended radio sources since the problem concerning the transformation of the optically emitted energy into high energy electrons and magnetic fields remained unsolved.

However, at the same time that these calculations were being made, an observational discovery of great significance was made in Pasadena by Schmidt (1963) and was quickly confirmed by Oke (1963) and by Greenstein and Matthews (1963). It has been known for some time that certain of the radio sources were located in coincidence with star-like objects which apparently had diameters too small to be resolved by optical telescopes and which showed on photographic plates as diffraction images characteristic of the telescope. These objects were called "radio stars."

The Pasadena group pioneered in the use of the 200-inch Hale Telescope on Mount Palomar to investigate the spectroscopy of these "radio stars." For several years their investigations of four of these objects led nowhere; they were unable to understand the peculiar emission lines of the spectra which the telescope revealed. There the matter rested until Schmidt began studying the spectrum of a fifth object catalogued by Cambridge University radio astronomers as 3C 273. This time the Gordian knot was cut. Several of the emission lines from 3C 273 formed a simple harmonic pattern, with separation and intensity decreasing toward the ultraviolet. The lines obviously belonged to a series of the type expected from hydrogen or any other atom that had been stripped of all electrons but one. Schmidt soon concluded that no atom gave the observed wave lengths. If he assumed, however, that the spectrum lines had been shifted toward the red by 16 per cent, the observed wave length agreed with those of hydrogen. Shortly thereafter Oke found the Hα-line in exactly the position predicted by the red-shift hypothesis and Greenstein and Matthews found an even greater red shift or 37 per cent in 3C 48 when they properly identified the lines observed as corresponding to well-known lines from the elements oxygen, neon and magnesium.

Greenstein and Schmidt (1964) soon showed that the red shifts could not be gravitational red shifts associated with large masses confined to regions of very small radius. The masses involved are found to be quite large but the radii of the emitting regions are so great that the gravitational red shift is negligible. They suggested that their "quasi-stellar" objects or "quasars" are extragalactic and that the red shifts arise from the general cosmological expansion of the universe. With this interpretation they were then able to determine the luminosity distance for the objects and to convert the observed apparent luminosities into absolute luminosities. The calculations indicated that the quasars have optical luminosities of the order of 10^{46} ergs sec^{-1} or more than one-hundred times the optical luminosity of our Galaxy. The quasars may or may not be located in galaxies, but if they are, they outshine the surrounding galaxy so that it is lost in the diffraction pattern of the quasar image.

The optical luminosities of the quasars are very high, $\sim 10^{46}$ ergs sec^{-1}, but there is no convincing evidence that these objects have lifetimes in excess of 10^5 to 10^6 years. Thus the cumulative optical emission is the order of 10^{59} ergs which is well within the nuclear resources of a star with $M = 10^8$ $M_\odot$. Only seven per cent of the hydrogen of such a massive star need be converted into helium to release this amount of energy. Because of the small volume the stored energies required are small.

It is now well established that the quasars exhibit variability in optical luminosity (Smith and Hoffleit 1963; Matthews and Sandage 1963; Sandage 1964; Sharov and Efremov 1963; Geyer 1964). In addition to luminous flashes with durations of the order of days or weeks, there is some evidence for cyclic variations with periods of the order of ten years. It is generally agreed that the occurrence of the cyclic variations is crucial to the question whether the primary radiation object is a single coherent massive star (10^4 - 10^8 $M_\odot$) as originally proposed by Hoyle and Fowler (1963a, 1963b) or a system of smaller stars (1-10^2 $M_\odot$) as discussed by numerous authors (Burbidge 1961; Hoyle and Fowler 1965; Gold, Axford and Ray 1965; Woltjer 1964; Ulam and Walden 1964; Field 1964). It is difficult on the basis of collisions or supernova outbursts in a system of many stellar objects to explain variations which exhibit a fairly regular periodicity. Thus, without prejudice to the problem of the reality of the cyclic variations since only additional and more precise observations will settle this matter, the possibility is investigated in what follows that such variations can arise from nonlinear relaxation oscillations in a single massive star. The star is taken to have no rotation and to be spherically symmetric with all physical parameters depending only on the radial variable. Rotation or other mechanisms which destroy the spherical symmetry change the behavior of the star markedly and will be mentioned briefly at the end of the paper.

Relaxation Oscillations in Non-Rotating Super-Massive Stars

The rapid generation of nuclear energy during the general relativistic collapse of a massive star is considered to be the triggering agent for the relaxation oscillations. From the standpoint of the model discussed in Hoyle and Fowler (1965) it is necessary to assume that the early fragmentation in the original gas cloud resulted in the formation of stars small enough (< 10 $M_\odot$) that significant nuclear evolution (consumption of hydrogen) did not occur in the time scale ($\sim 3 \times 10^6$ years) in which stellar collisions reduced the system of stars once again to a single gaseous object. Thus the starting point is a massive star, say $M \sim 10^6$ $M_\odot$, with a characteristic dimension of 10^{17} cm, central temperature of the order of 10^5 °K, with pressure support due almost entirely to radiation and a structure closely approximating that of a polytrope of index $n = 3$ (Fowler and Hoyle 1963, 1964). The composition is the same as that of the original gas cloud, for example, $X = 0.75$, $Y = 0.22$ and $Z = 0.03$. After some exhaustion of hydrogen a representative composition for purposes of computation will be taken to be $X = 0.50$, $Y = 0.47$, $Z = 0.03$ with Z primarily made up of CNO-nuclei.

General relativistic considerations lead to dynamic instability in non-rotating massive stars when the radius falls below a certain critical value (Iben 1963; Fowler 1964a, b; Chandrasekhar 1964a, b, c; McVittie 1964; Gratton 1964; Zel'dovich 1964). In what follows we make use of the Post-Newtonian approximation in the notation of Fowler (1964a). The significant results are illustrated in Figure 1 where the energy content of the star exclusive of the rest mass energy of the constituent particles is presented as a function of the radius and central temperature. The heavy solid curve represents the equilibrium binding energy in solar rest-mass energy equivalent units given by the Post-Newtonian approximation. The decrease to a minimum at a certain outer radius and central temperature followed by a

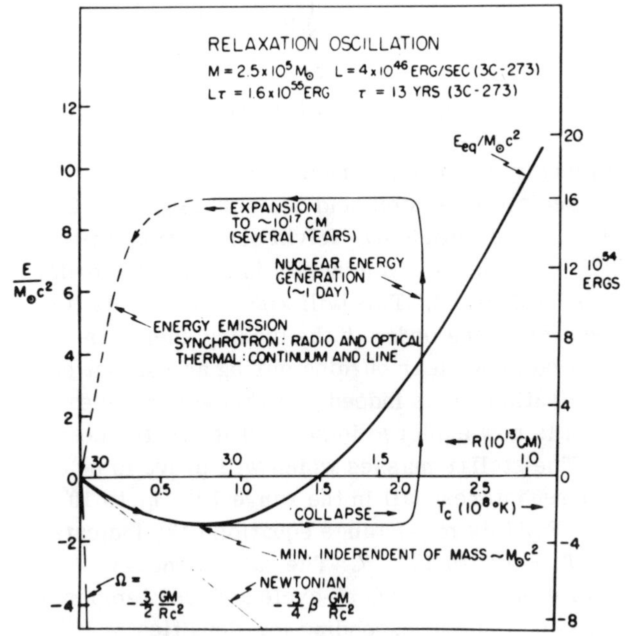

FIG. 1. The internal energy of a non-rotating massive star (2.5×10^5 M) in excess of the rest mass energy is shown as a function of radius and central temperature. The heavy curve shows the energy required for hydrostatic equilibrium when general relativistic considerations are taken into account. This curve deviates quadratically from the linear Newtonian term and reaches a minimum with absolute value the order of $M_\odot c^2$ at $R \sim 4 \times 10^{13}$ cm, $T_c \sim 0.7 \times 10^8$ °K. This minimum is reached before nuclear energy generation begins in the interior. A general relativistic collapse occurs which is stopped and reversed by hydrogen burning through the CNO bi-cycle in a time of approximately one day near $R \sim 10^{13}$ cm, $T_c \sim 2 \times 10^8$ °K. A radial shock wave is initiated and the resulting expansion extends to a radius of approximately 10^{17} cm in a time scale of the order of several years. Damping of the expansion occurs through radio and optical synchrotron emission and by non-equilibrium continuum and line emission. The overall process can best be described as a relaxation oscillation. The case illustrated employs the luminosity and period observed for the quasar in 3C 273B.

rise into the unbound region is in marked contrast to the linear decrease exhibited by the Newtonian term, $-3\beta GM/4Rc^2$. To the left of the minimum in the "classical" range an adiabatic perturbation toward smaller radii leads to more energy than that required for equilibrium and thus to more pressure than that necessary for hydrostatic equilibrium. Thus the contraction is opposed as will clearly also be the case for a perturbing expansion and thus the system is inherently stable. The same argument used to the right of the minimum (Fowler 1964b) indicates that a contraction leads to less pressure than that needed for hydrostatic equilibrium while an expansion leads to more so that the system is dynamically unstable to adiabatic perturbations.

The equilibrium energy in the Post-Newtonian approximation can be written for n = 3 as

$$\frac{E_{eq}}{Mc^2} = -\frac{3}{8}\beta\left(\frac{2GM}{Rc^2}\right) + \frac{3}{16}\left(\frac{3}{\pi}\right)^{1/2} R_3 \left(\frac{2GM}{Rc^2}\right)^2 + \cdots \tag{8}$$

where $R_3 = 6.897$ is one of the constants of integration for the polytropic equation for n = 3. From Eddington's quartic equation the ratio of gas pressure to total pressure is given for small values by Fowler and Hoyle (1964) as

$$\beta \approx 6\,(\Gamma_1 - 4/3) \approx \frac{4.28}{\mu}\left(\frac{M_\odot}{M}\right)^{1/2} \ll 1 \tag{9}$$

where μ is the mean molecular weight and $\Gamma_1 = d\ln p/d\ln\rho$ with p, the pressure and ρ, the density. For $M = 10^4$ to $10^8\ M_\odot$, $\beta \sim 10^{-1}$ to 10^{-3}. In massive stars the main pressure support is that due to radiation, β is small, the Newtonian term in (8) is small and the Post-Newtonian term becomes significant for small values of $R_g/R = 2GM/Rc^2$ of the order of β. The limiting gravitational radius or Schwarzchild radius is designated by $R_g = 2GM/c^2 = 3 \times 10^5\ (M/M_\odot)$ cm.

The critical radius can be determined by setting the derivative of (8) equal to zero in which case

$$\frac{R_c}{R_g} = \left(\frac{3}{\pi}\right)^{1/2}\frac{R_3}{\beta} = \frac{6.74}{\beta} \sim \left(\frac{M}{M_\odot}\right)^{1/2} \sim 10^2 \text{ to } 10^4 \tag{10}$$

Numerically it is found that

$$R_c = 3 \times 10^5\ (M/M_\odot)^{3/2} \text{ cm} \sim 10 \text{ to } 10^7 \text{ light seconds} \tag{11}$$

At this point and in what follows it will be assumed that the gross internal structure of the star is insensitive in first order to static or dynamic changes in E/Mc^2 of the order of β. Thus the equations relating central temperature and density with the radius for a polytrope with n = 3 can be freely employed. In fact the second order term in equation (8) is correctly derived by using first order terms where required. The critical central temperature and density are thus

$$T_c = \frac{Rc^2}{2\,\mu a^{1/2}\,G^{3/2}\,M} = \frac{1.25\times 10^{13}}{\mu}\left(\frac{M_\odot}{M}\right)\,{}^\circ K \qquad (12)$$

$$\sim 10^9 \text{ to } 10^5 \;{}^\circ K \text{ for } M \sim 10^4 \text{ to } 10^8\, M_\odot$$

$$\rho_c = \frac{2.54\times 10^{17}}{\mu^3}\left(\frac{M_\odot}{M}\right)^{7/2} \text{gm cm}^{-3} \qquad (13)$$

$$\sim 10^4 \text{ to } 10^{-10} \text{ gm cm}^{-3} \text{ for } M \sim 10^4 \text{ to } 10^8\, M_\odot$$

Moreover it is possible to show (Fowler 1964a) that equation (8) can be expanded in increasing powers of T with only the linear and quadratic terms retained in the Post-Newtonian approximation.

The minimum equilibrium energy at R_c, T_c, ρ_c turns out to be independent of the stellar mass and is given by

$$E_{eq}^{min} = -\frac{3M_3\,R^2\,c^2}{4\,\mu^2\,R_3\,a^{1/2}\,G^{3/2}} \qquad (14)$$

where $M_3 = 2.018$ is the second constant of integration for the polytropic equation for n = 3. Equation (14) can be rewritten by introducing R $R^4/a = (15/\pi^2)(\hbar^3 c^3/M_u^4)$ and after evaluation of numerical factors becomes

$$E_{eq}^{min} = -\frac{0.27}{\mu^2}\left(\frac{\hbar c}{GM_u^2}\right)^{3/2} M_u c^2 \sim -10^{57} \times 10^{-24}\, c^2 \qquad (15)$$

where M_u is the atomic mass unit. It is well known that the dimensionless gravitational interaction constant $GM_u^2/\hbar c$ is very small, of order 10^{-38}. The corresponding fine structure constant in electromagnetism is $e^2/\hbar c = 1/137$. As indicated in equation (15) this leads to

$$E_{eq}^{min} \sim -M_\odot c^2 \sim -2\times 10^{54} \text{ ergs} \qquad (16)$$

which indicates that the maximum binding energy of a non-rotating massive star is of the order of one solar mass-energy equivalent. More precisely

$$E_{eq}^{min} = -\frac{0.51}{\mu^2}\,M_\odot c^2 = \frac{0.91\times 10^{54}}{\mu^2} \text{ erg} \qquad (17)$$

In a first appraisal of the problem it is interesting to consider equation (16) in relation to the luminosity of the quasars which is of order $L \sim 10^{46}$ erg sec^{-1} (Greenstein and Schmidt 1964, and Oke 1965). The Helmholtz-Kelvin contraction time, E_{eq}^{min}/L, with this luminosity is thus of the order of several years. This will also be the cycle time if energy of the order of the binding energy is supplied by nuclear burning during an oscillation or pulsation. It is indeed just this general idea which is now explored in somewhat greater detail.

The stellar masses which will prove to be of greatest interest fall in the range $10^5\ M_\odot$ to $10^6\ M_\odot$. For this mass range equation (12) indicates that $T_c \sim 10^8$ to 10^7 °K. The rate of energy generation by the CNO bi-cycle at these temperatures and the corresponding low densities is considerably less than that required to maintain a luminosity of the order of 10^{46} erg sec^{-1}. Thus when the star reaches the minimum energy in Figure 1, collapse will commence and will continue until temperatures are reached at which nuclear energy generation becomes adequate for stability. Collapse will in fact be halted when, over the appropriate time interval, the nuclear energy generation matches that required to supply $E_{eq} - E_{eq}^{min}$. Collapse will be reversed to expansion in approximately the same time interval so that the nuclear processes overshoot and deliver in each pulse or cycle the following amount of energy

$$E_{cyc} = 2\,(E_{eq} - E_{eq}^{min}) = 2\,|E_{eq}^{min}|\,(x_n - 1)^2 = \frac{1.82\times 10^{54}}{\mu^2}\,(x_n - 1)^2 \text{ erg} \qquad (18)$$

where $x_n = T_n/T_c$ and T_n is the temperature at which the nuclear burning takes place while T_c is the critical temperature for the minimum in E_{eq}. The factor of 2 also follows directly from the conservation of momentum during the stopping and reversal of the collapse. The quadratic dependence on the temperature term in x_n follows directly from the fact that in equation (8) for E_{eq} only linear and quadratic terms in $R^{-1} \propto T$ are retained. In $E_{eq} - E_{eq}^{min}$ only the quadratic term remains. To determine T_n requires knowledge of the time scale for collapse and of the rate of energy generation in the CNO bi-cycle at elevated temperatures.

The time scale for nuclear burning during general relativistic collapse can be calculated with sufficient accuracy using the post-Newtonian approximation for the acceleration which can be written as follows:

$$\ddot{r}(1+\cdots\cdot) = -\frac{dp}{dr}\frac{(1+v^2/c^2-2GM_r/rc^2)}{(\rho+p/c^2)} - \frac{GM_r}{r^2}\left(1+\frac{4\pi pr^3}{M_rc^2}\right) \quad \text{(all r)}$$

$$\approx \frac{1}{\rho_0}\frac{dp}{dr}\left(1-\frac{u}{\rho_0 c^2}-\frac{p}{\rho_0 c^2}+\cdots\cdot\right) - \frac{4}{3}\pi G\rho_0 r\left(1+\frac{u}{\rho_0 c^2}+\frac{3p}{\rho_0 c^2}+\cdots\cdot\right)$$

$$\text{(post-Newtonian, center)} \tag{19}$$

$$\approx \frac{1}{\rho_0}\frac{dp}{dr} - \frac{4}{3}\pi G\rho_0 r$$

$$\text{(Newtonian, center)}$$

In equation (19), r is the Lagrangian radial coordinate, p is the pressure, u is the internal energy density, ρ_0 is the mass-density, ρ is the mass-energy density in mass units and M_r is the mass-energy interior to r. Relativistic terms have not been explicitly indicated on the left-hand side of equation (19). The first form on the right-hand side is relativistically exact, the second form is the post-Newtonian approximation at the center of the star, while the third form is the customary Newtonian approximation at the center. In this Newtonian approximation at hydrostatic equilibrium, $\ddot{r} = 0$. Thus to first order $\ddot{r}$ is just the difference of the first post-Newtonian terms on the right-hand side and the post-Newtonian terms on the left-hand side are not needed and have not been explicitly presented. It will be noted that the difference in the post-Newtonian terms is proportional to $(2u+4p)/\rho_0 c^2 \sim 10p/\rho_0 c^2$ for $u \sim 3p$ as is the case in massive stars where radiation pressure dominates.

Standard methods of integration applied to equation (19) with $\ddot{r} = \dot{r} = 0$ at the initial critical conditions lead, for small $(p/\rho_0 c^2)_c \approx (aT^4/3\rho_0 c^2)_c$, to

$$\frac{\dot{T}}{T} \approx -\frac{\dot{r}}{r} \approx \left(\frac{40\pi aG}{9c^2}\right)^{1/2} T_c^2\, x(x^2-2x\ln x-1)^{1/2} \tag{20}$$

$$\approx \left(\frac{40\pi aG}{27c^2}\right)^{1/2} T_c^2\, x(x-1)^{3/2} \quad \text{for } 1<x<2 \tag{21}$$

$$\sim \left(\frac{aG}{c^2}\right)^{1/2} T_c^2\, x^2\,(x-1)^{1/2} \quad \text{for } 2<x<10 \tag{22}$$

where $x = T/T_c$. Solving for the e-folding time in T or r one finds numerically for equation (22), which is the case of primary interest, that

$$\tau_{gc} \equiv T/\dot{T} = \frac{dt}{d\ln T} \sim \frac{1.3\times10^5}{T_8^2\,(x-1)^{1/2}}\ \text{sec} \tag{23}$$

Thus the e-folding time in temperature or radius for general relativistic gravitational collapse (gc) at hydrogen burning temperatures, $T_8 \sim 2$, in massive stars for which $T_{8c} \sim 1$ is somewhat less than one day. This is considerably greater than the classical free fall time which is $\tau_{ff} = (8\pi G\rho/3)^{-1/2} = 1340\,\rho^{-1/2}$ sec $\sim 10^3$ sec. However it is the shortness of τ_{gc} relative to the overall period of order 10 years which illustrates the extreme non-linearity of the oscillations under consideration.

In the discussion in Hoyle and Fowler (1965) of the behavior of CNO-burning of hydrogen in massive stars it was noted that the proton capture reaction by nuclei such as N^{13} proceed at a rate comparable to the beta-decay of N^{13} and that alpha-particle reactions lead to some transmutation of the CNO-nuclei into heavier nuclei. However these are not serious effects in the cases of primary interest in this paper and it is sufficiently accurate to make the assumption that all CNO-nuclei actively participate and remain as catalysts, mostly as N^{14} ($\sim 0.9Z$), and that the rate of energy generation is primarily determined by the $N^{14}(p,\gamma)$ reaction for which Hebbard and Bailey (1963) give the empirical parameters, $S_0 = 2.75 \pm 0.50$ keV-barns and $\langle dS/dE\rangle \approx 0$. This leads to a slight modification of the results of Caughlan and Fowler (1962). When expressed as a power law in temperature near $T_8 \sim 2$ the nuclear energy generation rate for the HCNO-burning is

$$\epsilon \approx 3.7\times10^{12}\,\rho XZT_8{}^8 \quad \text{erg gm}^{-1}\ \text{sec}^{-1}$$
$$\approx 5.6\times10^{10}\,\rho T_8{}^8 \quad X=0.05,\ Z=0.03 \tag{24}$$

In a massive star with polytropic index n = 3, Fowler and Hoyle (1964) express the density in their equation (B120) as

$$\rho \approx 130 \left(\frac{M_\odot}{M}\right)^{1/2} T_8{}^3 \quad \text{gm cm}^{-3} \tag{25}$$

so that

$$\epsilon \approx 7.3 \times 10^{12} \left(\frac{M_\odot}{M}\right)^{1/2} T_8^{11} \quad \text{erg gm}^{-1}\ \text{sec}^{-1} \tag{26}$$

Fowler and Hoyle (1964) also give the energy generation averaged over the star and using their equation (C84) one finds

$$\overline{\epsilon} \approx 4.4 \times 10^{11} \left(\frac{M_\odot}{M}\right)^{1/2} T_8^{11} \quad \text{erg gm}^{-1}\ \text{sec}^{-1} \tag{27}$$

This is still a quantity effectively representative of the central region of the stellar interior. As noted above, equation (27) must not be used when the burning becomes rapid enough that beta-decay processes limit the rate of energy generation. The limit comes when the time for the conversion of four protons into helium is just the sum of the mean lifetimes for proton capture by O^{14} (100 sec) and O^{15} (180 sec) and is given by

$$\begin{aligned} \overline{\epsilon} &\sim \left(\frac{4}{14.5} \times \frac{6.0 \times 10^{18}}{280}\right) Z \sim 5.9 \times 10^{15}\ Z \quad \text{erg gm sec}^{-1} \\ &\sim 1.8 \times 10^{14} \qquad Z = 0.03 \end{aligned} \tag{28}$$

The limiting temperature given by combining equations (27) and (28) is

$$\begin{aligned} T_8 &< 1.7 \left(\frac{M}{M_\odot}\right)^{1/22} \\ &< \sim 3 \qquad \text{for } M = 10^5 \text{ to } 10^6\ M_\odot \end{aligned} \tag{29}$$

The simplest procedure is now to equate $\overline{\epsilon}$ multiplied by the stellar mass and by the effective time for nuclear burning to ϵ_{cyc} given by equation (18). The effective time can be estimated as follows. The quantity $\overline{\epsilon}\tau_{gc}$ varies approximately as T^9. The e-folding time for T^9 is 1/9 that for T. However the velocity is reduced from its initial value to zero and is then reversed by the energy generation so that the effective time for nuclear burning during each cycle is (4/9) τ_{gc}. Thus

$$\epsilon_{cyc} = \frac{4}{9}\overline{\epsilon}\,\tau_{gc}\, M \tag{30}$$

Equations (12), (18), (23), (27) and (30) can be combined to yield

$$\frac{M}{M_\odot} \approx 1.0 \times 10^5\ \frac{(x_n)^{18/17}}{(x_n - 1)^{5/17}} \sim 10^5\,(x_n)^{13/17} \tag{31}$$

where $x_n = T_n/T_c$ as before. T_c is the critical temperature at which collapse begins and T_n is now the temperature at which the hydrogen burning generates ϵ_{cyc} in the available time determined by the reversal of the collapse. Equations (18) and (31) then yield (for $x_n \sim 3$ as found below)

$$\epsilon_{cyc} \sim 10^{41} \left(\frac{M}{M_\odot}\right)^{34/13} \quad \text{erg} \tag{32}$$

In this equation $\mu = 0.73$ has been used corresponding to X = 0.50, Y = 0.47, Z = 0.03.

We have now arrived at the nuclear energy generated in the pulse which triggers each relaxation oscillation or cycle. This must be equal to the total luminosity L for all forms of radiation multiplied by the cycle period τ_{cyc} so that

$$L\,\tau_{cyc} \sim 10^{41} \left(\frac{M}{M_\odot}\right)^{34/13} \quad \text{erg} \tag{33}$$

or

$$\frac{M}{M_\odot} \sim 2 \times 10^{-16}\,(L\,\tau_{cyc})^{13/34} \tag{34}$$

In the case of the quasar 3C 273 the observations indicate $L \sim 4 \times 10^{46}$ erg sec^{-1} (Oke 1965) and $\tau_{cyc} \sim 13$ yr $\sim 4 \times 10^8$ sec (Smith and Hoffleit 1963) so that $L\,\tau_{cyc} \sim 1.6 \times 10^{55}$ erg and

$$\frac{M}{M_\odot} \sim 2.5 \times 10^5 \tag{35}$$

Corresponding to this value for $M/M_\odot$ it is found from equation (12) that $T_{8c} \sim 0.7$ and from equation (31) that $x_n \sim 3$ so that $T_{8n} \sim 2$. These values are illustrated in Figure 1. At $T_{8c} \sim 0.7$ the stellar radius is $\sim 4 \times 10^{13}$ cm while at $T_{8n} \sim 2$ the radius is $\sim 1.3 \times 10^{13}$ cm. The Schwarzschild limiting radius is $\sim 8 \times 10^{10}$ cm.

It will be noted in equation (32) that ϵ_{cyc}

varies as a fairly high power, ~ 2.6, of the stellar mass. Thus ϵ_{cyc} rapidly approaches the total nuclear energy content of the star. At this point the nuclear energy would suffice for only one pulse of energy generation. These considerations lead to the conclusion that relaxation oscillations without serious overshooting due to excessive energy generation can only take place in non-rotating stars with mass not exceeding $\sim 10^6\ M_\odot$. It may prove significant that this is the order of magnitude of the mass of the larger globular clusters.

The above discussion has treated the relaxation oscillations almost solely in terms of energy considerations. Damping and stabilizing mechanisms will be discussed in the sequel but an important point in this connection should be noted at this time. The time spent during the oscillation with $R < R_c$ when the star is dynamically unstable is very short compared to the overall period being of the order of one day. Thus practically the entire oscillation occurs with $R > R_c$ during which the star is dynamically stable as in the classical, non-relativistic case. The maximum radius reached during the oscillation will be the order of the ratio of 10 years to 1 day multiplied by R_c or approximately 10^{17} cm. The classical period for linear oscillations is not strictly applicable to the very large amplitude oscillations under discussion. This period is given by $\Pi \sim (\beta\bar{\rho}G)^{-1/2}$ and is the order of 10 years for the mean density in a star with $M \sim 2.5 \times 10^5\ M_\odot$ (3C-273) at an intermediate stage between the minimum ($R_n \sim 10^{13}$ cm) and maximum ($R_{max} \sim 10^{17}$ cm) excursions of the relaxation oscillations.

Equation (34) yields the mass of a non-rotating central stellar object having properties consistent with the product of the luminosity and period of a variable quasar such as 3C 273 if the period is that for relaxation oscillations in the star maintained by energy generation through HCNO-burning. The actual period will be determined by other considerations to be discussed later. At this point we turn our attention to the cumulative emissions from 3C 273.

The mass given by equation (35) can only satisfy the cumulative energy requirement for 3C 273 if the lifetime is relatively short. The nuclear energy resources for $M = 2 \times 10^5\ M_\odot$ are at most 4×10^{57} erg from equation (4). For a luminosity equal to 4×10^{46} erg sec^{-1} this yields a lifetime of 10^{11} sec or 3,000 years. The number of relaxation oscillations is approximately 250. Greenstein and Schmidt (1964) discuss a model for the quasar in 3C-273 with lifetime equal to 10^3 years. They emphasize that on this basis the quasar is considerably younger than the associated objects in 3C 273, namely the radio halo surrounding the quasar and the optical jet. The quasar has then to be taken as a later event unassociated with the origin of the older, large scale components of 3C 273.

It is beyond the scope of this paper to discuss in detail the modifications to the discussion presented here which are necessary if the value obtained for the mass from an equation such as (34) is to be substantially increased. Suffice it to note that the introduction of rotation leads to a substantial increase in the mass in which the nuclear burning can take place without excessive overshooting during the relaxation oscillations. This matter will be treated by the author in a forthcoming paper (Fowler 1965a). The present paper will be concluded with a brief enumeration of considerations connected with the energizing and damping of relaxation oscillations in such a way that stable pulsations are possible. We lean heavily on the quasar models discussed by Greenstein and Schmidt (1964) and by Oke (1965) in this enumeration. These considerations are illustrated schematically in Figure 2.

(1) Ledoux (1941) has shown that the relative radial displacement at the center of a pulsating massive star in which $\gamma \sim 4/3$ can be comparable in magnitude to that throughout the star and particularly at the surface. This means that nuclear energy generation at the center is extremely effective in triggering pulsations in the massive stars under discussion in this paper.

(2) Ledoux (1941) and Schwarzschild and Härm (1959) have emphasized that the problem of stability in massive stars depends critically on the mechanism of heat leakage in the envelope which serves to damp the oscillations energized in the core. They show that pulsational instability is to be expected for stellar masses above a critical value of the order of $\sim 10^2\ M_\odot$, if the only processes of heat transfer and loss are ordinary convection and radiation. This can be understood on the basis that the radiative luminosity is proportional to R^2T^4 which in turn is proportional to R^{-2} so that at the large radii occurring during the expansion, $L_{rad} \propto R^{-2}$ is

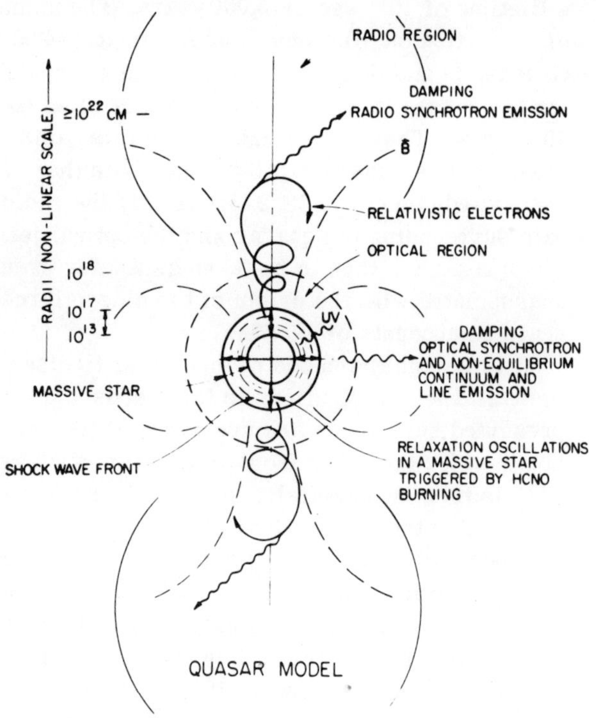

FIG. 2. Schematic model for a quasar. Large amplitude relaxation oscillations between radii of 10^{13} and 10^{17} cm are energized in a massive star by HCNO burning at a temperature near 2×10^8 °K. Shock waves transmit the energy to the tenuous outer envelope from which relativistic particles are ejected into the region surrounding the star. An associated dipole magnetic field channels the relativistic particles into two large scale regions ($\geq 10^{22}$ cm) in which radio synchrotron emission occurs. Optical synchrotron radiation is emitted from the region immediately surrounding the star ($\sim 10^{18}$ cm). Non-equilibrium continuum and line emission are also stimulated in this region by ultraviolet radiation from the star. It is this region which is visible and not the star itself.

ineffective as a damping mechanism.

(3) What is required are damping mechanisms which are effective at large radii and low surface densities. It is therefore suggested that the extraordinary modes of energy emission evidenced by the quasars, namely, radio synchrotron emission, optical synchrotron emission as well as non-equilibrium continum and line emission serve as the damping agents in stabilizing the pulsations. It is the overall rate of these emissions relative to the nuclear energy generation per pulse which determines the period of the oscillations. As discussed previously these emissions take place predominantly while $R > R_c$ during which the star is dynamically stable.

(4) As noted in the discussion of relaxation oscillations the nuclear energy generation takes place in the period of the order of a day which is very short compared to the observed overall periods of approximately 10 years. This nuclear pulse will lead to the propagation of a radial shock wave outward from the center of the star. From the work of Ôno, Sakashita and Ohyama (1961), Ohyama (1963) and Colgate and White (1964) it is known that such a shock wave will reach relativistic velocities in the tenuous outer envelope of the star and will there generate relativistic particles which are then ejected into the region surrounding the star. This high energy process becomes an especially effective damping agent during the latter stages of expansion when surface densities are low. It is generally believed that shock wave acceleration results in the production of relativistic particles with total energies comparable to that for non-relativistic particles. This seems to be required by the quasar observations.

(5) The ejection of relativistic particles leads to the formation of the region with dimensions of the order of 10^{18} cm in which an optical synchrotron continuum can be generated in the presence of an associated magnetic field. This region is in fact relatively transparent to high energy particles which can leak out to form a much more extended region with dimensions of the order of 10^{22} cm or even greater in which radio synchrotron emission takes place. The reader is referred to Greenstein and Schmidt (1964) for detailed description of the regions under discussion.

(6) If the overall magnetic field has dipole structure then the ejection of the relativistic particles will tend to occur parallel to the dipole axis and to result in the formation of a two-component radio source as is frequently observed. For the field strengths required by synchrotron theory, the Larmor radii of the relativistic particles are quite small compared to the dimensions of the radio sources. On this picture the line of centers of the radio components would lie along the axis of rotation even for an inclined magnetic dipole. For the line of centers to be perpendicular to the axis of rotation it is necessary to consider other possibilities such as the fission mechanism discussed by Hoyle and

Fowler (1965) and Fowler (1964a).

(7) From the original work of Hoyle and Fowler (1963a) on supermassive stars the surface temperature is estimated to be the order of 10^5 °K during the hydrogen burning stage in the interior. Intense ultraviolet emission at this temperature will amply suffice to excite non-equilibrium continuum and line emission from the 10^{18} cm region in which optical synchrotron radiation is also generated. At the same time the high opacity presented to the ultraviolet radiations would make observation of the embedded supermassive star impossible.

(8) Upon the exhaustion of nuclear energy resources, gravitational collapse occurs in a non-rotating massive star. For a rotating star collapse can also occur if mechanisms for the transfer of angular momentum are effective. In the case of collapse, gravitational energy becomes available and the evolution of a quasar into an extended radio source may become possible as discussed by Fowler (1964a).

Conclusion

The work described in this paper constitutes a return to the early point of view of Hoyle and Fowler (1963a, b) that supermassive stars can meet the energy requirements in radio sources, specifically in the quasars. General relativity leads to dynamic instability in non-rotating massive stars but the result is relaxation oscillations energized by hydrogen burning rather than catastrophic collapse at least for masses not exceeding 10^6 $M_\odot$. It is noted that the introduction of rotation raises this limit by several orders of magnitude and that a forthcoming paper will treat this matter. It is emphasized that the exotic forms of energy emission observed in the quasars can serve to damp the relaxation oscillations in such a way that stable pulsations result. It is thus suggested that quasars consist of pulsating supermassive stars, energized by nuclear reactions, with radio and optical emissions from extended surrounding regions which the star excites with ultraviolet radiation and relativistic particles. With the exhaustion of nuclear energy, gravitational energy may become available and the evolution of a quasar into an extended radio source becomes possible.

Acknowledgments

The introductory material in this paper has been taken with minor modifications from Fowler (1965b). Permission of the American Philosophical Society to use this material is gratefully acknowledged. The author is indebted for illuminating discussions with J. Bardeen, R. F. Christy, F. Hoyle and J. B. Oke.

REFERENCES

G. R. Burbidge, Nature, 190, 1053.

G. R. Caughlan and W. A. Fowler, 1962, Ap. J., 136, 453.

S. Chandrasekhar, 1964a, Phys. Rev. Letters, 12, 114.

_____ 1964b, ibid., 12, 437.

_____ 1964c, Ap. J., 140, 417.

S. A. Colgate, and R. H. White, 1964, Report No. UCRL-7777, Livermore Radiation Laboratory, University of California.

G. B. Field, 1964, Ap. J., 140, 1434.

W. A. Fowler, 1964a, Rev. Mod. Phys., 36, 545.

_____ 1964b, ibid., 36, 1104.

_____ 1965a, Proceedings of the National Academy of Sciences, (Washington, April 27, 1964) to be published.

_____ 1965b, Proceedings of the American Philosophical Society (Philadelphia, April 24, 1964) to be published.

W. A. Fowler, and F. Hoyle, 1963, Herstmonceux Bulletin, 67, E302.

_____ 1964, Ap. J. Suppl. Series, 91, 201.

E. H. Geyer, 1964, Z. Astrophys., 60, 112.

T. Gold, W. I. Axford, and E. C. Ray, 1965, Quasi-Stellar Sources and Gravitational Collapse (Chicago: The University of Chicago Press) p. 93.

L. Gratton, 1964, Internal report of the Astrophysical Laboratory of the University of Rome and the 4th Section of the Center of Astrophysics of the Italian National Research Council, Frascati, July 1964. Presented at the CONFERENCE ON COSMOLOGY, Padua, Italy, September 1964.

J. L. Greenstein, and T. A. Matthews, 1963, Nature, 197, 1041.

J. L. Greenstein, and M. Schmidt, 1964, Apl. J., 140, 1.

D. F. Hebbard, and G. M. Bailey, 1963, Nuclear Phys., 49, 666.

F. Hoyle, and W. A. Fowler, 1963a, Mon. Not. RAS, 125, 169.

_____ 1963b, Nature, 197, 533.

_____ 1965, Quasi-Stellar Sources and Gravitational Collapse (Chicago: The University of Chicago Press) p. 17.

I. Iben, Jr., 1963, Ap. J., 138, 1090.
P. Ledoux, 1941, Ap. J., 94, 537.
T. A. Matthews, W. W. Morgan, and M. Schmidt, 1964, Ap. J., 140, 35.
T. A. Matthews, and A. R. Sandage, 1963, Ap. J., 138, 30.
G. C. McVittie, 1964, Ap. J., 140, 401.
N. Ohyama, 1963, Prog. Teoret, Phys., 30, 170.
J. B. Oke, Nature 197, 1040.
______ 1965, Apr. J., 141, 6.
Y. Ōno, S. Sakashita, and N. Ohyama, 1961, Prog. Theoret, Phys. Suppl. No. 20.
A. R. Sandage, 1964, Ap. J., 139, 416.
M. Schmidt, Nature, 197, 1040.
M. Schwarzschild and R. Härm, 1959, Ap. J., 129, 637.
A. S. Sharov, and Yu. N. Efremov, 1963, Information Bulletin on Variable Stars, Number 23, Commission 27 of I.A.U.
H. J. Smith, and D. Hoffleit, 1963, Nature, 198, 650.
S. M. Ulam, and W. E. Walden 1964, Nature, 201, 1202.
L. Woltjer, 1964, Nature 201, 803.
Ya. B. Zel'dovich, 1964, Soviet Physics—DOKLADY, 9, 195.

SUMMARY

R. Hanbury Brown
University of Sydney

I would like to congratulate the speakers who presented papers on the astronomical observations of radio sources, and it is my job to summarize something of what they have said. Let me remind you first of the history of the discovery of the quasi-stellar sources (QSS's). Some years ago it was noticed that a few radio sources have very small angular size; the positions of some of these sources were then measured precisely, and an optical telescope was used to photograph their positions. Stellar-like images were found, but their spectra could not be interpreted. After a year or two the occultation of 3C 273 by the Moon provided an extremely precise position and another identification with a curious optical object; the redshifted lines in the optical spectrum of this object were interpreted, and it was shown, with reasonable certainty, that the object is extragalactic with an unusually high luminosity. Today we are calling these objects quasi-stellar sources. It would be nice if we could find a better name for them. In his talk Harlan Smith asked what is "fundamental" about the observed properties of these objects so that we might define what we call a QSS. I very much doubt whether any of the single properties observed so far can be used to define a QSS; so far the only fundamental thing appears to be some sort of event.

Looking first at the radio properties of these objects, we have been told that they exhibit angular structure on two different scales, a large scale corresponding to dimensions of the order of 100 kiloparsecs and a much smaller scale; some are double and some are single. In this respect they appear to be similar to the majority of radio sources; however, Dr. Palmer has shown us that the present list of about thirty suspected QSS's have significantly smaller angular sizes than the radio sources associated with the bright radio galaxies. In his list of angular sizes for the suspected QSS's, although most of the sources had very small angular sizes, it is worth noting that some did not. Palmer also told us that a few, three or four, of the QSS's have angular sizes less than 0.4 sec of arc I feel sure that many people would agree with me that it is worth a good deal of effort to push this limit further down, and I hope Palmer will consider trying to do so. The evidence on the small angular size of the QSS's receives strong support from the observation at Cambridge that the QSS's scintillate when the line of sight passes close to the Sun. Of twenty-eight sources known to scintillate we were told that fifteen are suspected QSS's.

Dr. Baldwin showed us that, while 85 per cent of radio sources have straight spectra, some show pronounced curvature or cutoff at low frequencies. This curvature is correlated with high surface brightness and small apparent angular size. Of fourteen sources showing this low-frequency cutoff, only two had angular sizes greater than 5" and four are in the list of suspected QSS's presented by Dr. Sandage.

As far as the other radio properties are concerned, QSS's have absolute magnitudes which are quite average for radio sources, and as far as we know their flux does not vary. In short, apart from a pronounced tendency to exhibit smaller angular diameters than radio galaxies, there does not seem to be anything distinctive or unusual about the radio sources associated with QSS's.

Harlan Smith has already summarized the papers dealing with the optical properties of the QSS's. However, I would like to stress two things which interested me particularly. First, there is the problem of the light variations from the QSS's. In the early days of radio astronomy the variations in flux from the strong radio sources in Cygnus and Cassiopeia were interpreted as evidence that these sources must be small, perhaps a few light-seconds across. Later we found that the variations were introduced by the Earth's ionosphere. In the case of the QSS's we should bear in mind that there are no previous observations of small intense sources of light at the enormous distance of the QSS's; we cannot be sure from any previous evidence that a distant point source, for example, an arc-lamp at 10^9 light years, whould be seen as a steady light. We must bear in mind that the fluctuations might be in the intervening medium. A second optical observation, which interested me in particular was the observation that QSS's do not appear to be associated with clusters of galaxies. At the present time this conclusion cannot be drawn safely from the evidence; there are very few well-established identifications of QSS's, and also the QSS's are themselves so bright that the associated cluster, if any, might be beyond the plate limits. There are, however, two well-identified QSS's which do not appear to be associated with clusters and it is a matter of great interest to find whether this is the general rule.

In seeking to establish the nature of QSS's one wonders how much our conclusions are affected by observational selection. At first identifications were made by searching for optical objects in the positions of radio sources of small angular size; more recently we have been looking for very blue stellar-like objects and then identifying them with radio sources. But are the blue objects which we call QSS's necessarily associated with radio sources, or are the radio sources which we call QSS's necessarily associated with blue objects? The radio sources associated with the blue objects seem fairly typical of radio sources, although they do appear to have smaller angular sizes; however, we do see similar radio sources without blue objects. Furthermore, in the surprising case of 3C 2, described by Sandage, we see the radio source without the blue object and then a few years later a blue object appears and the radio source seems to be unaffected. Also it rather looks as though we may see the blue objects without the radio sources. Sandage told us that there are blue objects scattered all over the sky at roughly 2 per square degree; are these the same objects as the QSS's?

Perhaps radio sources and blue objects are likely to appear in the same region or object, and it might be that they are not directly related. There are a host of such questions which can only be answered by further radio and optical observations, but I believe that it is primarily an extension of the optical observations which we need at present.

OBSERVED CHARACTERISTICS OF QUASARS

Harlan J. Smith
University of Texas

Appearance

As indicated by the original name, quasi-stellar source, these objects are star-like in appearance. More specifically, this means that being less—perhaps far less—than half a second of arc in apparent angular size, their principal radiant centers cannot be resolved as other than points of light by existing ground-based telescopes. So far, fewer than a dozen quasars have been examined at the highest obtainable resolution and limiting magnitude; nearly all of these show almost vanishingly faint wispy appendages or jets, of which the 20 sec-of-arc jet of 3C273 is the clearest example. In 3C48, a pair of fuzzy jets proceed in opposite directions for a total of about 12 sec of arc. While such jets contribute almost none of the total optical luminosity, in the case of 3C273 the jet is actually brighter than the central source at radio frequencies below about 1000 Mc/sec. A jet rather similar in appearance, although much smaller, protrudes from the nucleus of a nearby radio galaxy, Virgo A; in this case, strong polarization suggests that the light is optical synchrotron radiation.

The radio appearance of a typical quasar is relatively similar to the optical—a small central core with appendages. However, radio resolutions now exceed optical resolutions, as witnessed by the important recent Manchester discovery that the central radio source of 3C273 is double with component separations of only a few tenths of a second of arc, and by scintillation results from ARECIBO suggesting that the radio core of 3C273 is smaller than 0.01 sec of arc.

Spectra

While some characteristics of quasar spectra have already been considered, several of the most interesting points concern the consequences of the great redshifts and the form of the continuous energy distribution.

Apart from the sun, astronomers have been unable to obtain much detailed spectral information below the atmospheric ozone cutoff at 3000 Å (although various space telescope-spectrograph combinations are beginning to change this). In order to see the far uv spectra, the only approach available to ground-based astronomers is to examine objects having large enough redshifts to bring the otherwise unobservable uv into detectable regions. The enormous redshifts of most of the quasars bring up a host of lines hitherto unavailable to astronomical spectroscopists. Expecially conspicuous are the strong lines of Mg II 2798 Å, C III 1909 Å, C IV 1516 Å, and Lyman alpha 1216 Å. There are no clearly established surprises in the abundances of the different elements in quasars, as suggested by the appearance of their spectral lines. In particular, after the immensely abundant hydrogen are found helium, carbon, oxygen, nitrogen, neon, magnesium, and silicon—the elements one would expect to see according to all the previous experience with so-called cosmic abundances.

Ordinary stars have a continuous energy distribution in their spectra to a first order resembling the Planck spectra of blackbodies at well-defined temperatures, usually in the range of 2,000°K to 50,000°K. Thus, over the normally observed spectral region from about 3,000-10,000 Å, the stellar spectra either peak somehwere in the observed range, as is the case with the sun and the most common stars, or they slope fairly steeply up toward the blue for hot stars or toward the red for cool stars. But photoelectric spectral scans of many quasars by Oke at Mt. Wilson show that, compared with stars, most of the quasar continuua are relatively flat; several decline only slightly into the far uv, most of them decline significantly

*Reprinted from Applied Optics 5, No. 11, November 1966, pp. 1709-1718.

into the uv, and in all cases they rise into the far ir. In view of the strength of the hydrogen emission lines, it is natural to try to explain the continuum as arising from hydrogen recombination (de-ionization of the gas). Oke finds that several of the spectra could be fairly well explained by hydrogen recombination taking place at about 40,000°K to 100,000°K, but a nonthermal component is necessary in most, if not all, cases, particularly to fit the ir; optical synchrotron radiation is the favorite nonthermal candidate.

The spectral regions discussed so far (up to about 1 μ) are the traditional domain of optical astronomy. But thanks mainly to the work of Johnson and Low at Arizona, a rich body of information filling in the gap between 1 μ and 20 μ is rapidly becoming available. The work has immediate and immensely important bearing on quasars. The great rise in the ir, when coupled with the logarithmic frequency scale, indicates that quasars radiate most of their energy (approximately ten times the optical) in the ir region of tens of microns.

The ir data show good continuity with radio fluxes from quasars in the millimeter and short centimeter region. From a few centimeters on out to decameter wavelengths, quasars resemble common radio galaxies in their spectral characteristics.

Colors

Spectroscopy, while rich in information content, is painfully expensive in terms of telescope time. So few large telescopes are available to catch the faint drizzle of photons from sources perhaps billions of light years distant, and each must spend so long building up the spectral image of even a single quasar that other more rapid means of gaining some partially equivalent information are highly desirable. Such means are at hand in the commonly used techniques of broadband photometry. Originally, the eye, with its visual (yellow-green peak) frequency response, established one astronomical photometric system; the next was that of an untreated (blue-sensitive) photographic plate. The color index of a star, often now abbreviated to color, was defined as the difference of the blue and visual magnitudes. Color proved to be strongly correlated with surface temperature, as would be expected from the fact that most stars exhibit roughly Planckian radiation. By now, photometric systems have become highly sophisticated, with a wide range of spectral sensitivities, bandwidths, and color indices used to define approximate spectrophotometry of celestial objects; for highest accuracy, photoelectric photometry is employed wherever possible.

One of the most useful of these applications of color index spectrophotometry has been the two-color diagram in which the standard Johnson uv (U), blue (B), and visual (V) magnitudes are combined in the form of U-B vs B-V. In such a two-color diagram ordinary stars occupy a well-defined locus labeled Main Sequence. The general trend from upper left (hot stars) to lower right (cool stars) simply reflects the increasing reddening of the spectral energy distribution of cooler stars; the kink in the sequence arises from gross hydrogen Balmer absorption effects at around 10,000°K, the most favorable surface temperature for neutral hydrogen excitation. Objects having an excess of uv light appear above the normal stellar distribution. There exists a sprinkling of relatively faint hot stars having such uv excesses; they tend to populate a sequence below the dashed line on the diagram. Most of these objects seem to belong to the halo population of stars in our galaxy—a somewhat older collection of stars, having significantly lower metal abundances than the stars in the disk and plane of the galaxy. White dwarfs, the collapsed final luminous state of normal stars, also occupy this portion of the diagram, as do a few odd variable stars. But, compared with most stars, quasars have such strong uv excesses that they tend to lie on or above the dashed line; this materially assists in their identification.

A most important contribution has been Sandage's extensive use of the strong uv excess of quasars as a finding and diagnostic tool. Finding is carried out by photographing regions of strong radio sources, with plates looking first through a uv filter, then (after a microscopic but finite displacement of the telescope field) through a different, say, visual filter. The images side by side on the plate immediately reveal the rare cases of objects with strong uv excess; any such object found in a region identified by the radio astronomers as containing a source has a strong probability of being a quasar. Accurate photoelectric photometry, and eventually spectra, settle the matter.

In the course of this work, Sandage observed that objects of the uv excess required to correspond with quasars occurred with reasonable frequency on his plates, regardless of whether a radio source had been reported in the vicinity. Spectra of several of these objects showed the same immense redshifts and spectral line features as the quasars detected by radio astonomers; Sandage has referred to these quasars, first detected optically, as quasi-stellar galaxies. The objects are also sometimes called BSO's (blue stellar objects), partly to distinguish their differing mode of original detection and partly because the identification of star vs quasar remains somewhat uncertain until a spectrum is taken. Along with extensive earlier work on such faint blue objects by many men including Humason and Zwicky, and Luyten and Haro, Sandage's results have made it clear that, in the future, optical detection methods should be able to rival if not outdo radio methods in the discovery of quasars.

Variability, Age, and Size

It seems unlikely whether, given the other observational information now available concerning quasars, anyone would have predicted confidently the conspicuous variations in brightness that have been found. Now we have to explain not only how so much energy can be liberated from such a small region of space, but particularly (if the cosmological distances are correct) how an object that may be one of the brightest in the universe can add or subtract such a large fraction of this brightness only in a matter of months.

Through the foresight and faith of astronomers such as Pickering and Shapley at Harvard, a program of photographing the entire accessible sky several times a month with wide-angle patrol cameras began around 1880 and has been conducted with reasonable continuity until the present. While most of the quasars are too faint to be seen except on the merest sprinkling of plates, by good fortune one is bright enough to be visible on perhaps half of the old patrol plates. Thus, for 3C273 we have an eighty-year record establishing two vital points with certainty, and others with strong suggestivity.

The first point bears on the ages of quasars. The light curve shows that the brightness of at least this object has not changed systematically at a rate even as great as 10 per cent per century; it follows that quasars are probably able to remain highly luminous for times at least on the order of 10^3 years. The only other significant type of observation bearing on the time scale of quasars is more indirect. Specifically, at conventional redshift distances, the lengths of the jets are such that they require more than 10^5 years at the speed of light to reach their present separations from the nuclei; the quasar should thus have been active in at least some form that long ago. We have no direct evidence that any quasar had its present luminosity when the jets were ejected, or over the subsequent interval, but this assumption is frequently made.

The second conclusion from the light curve is that the brightness of 3C273 varies conspicuously, often by a factor of two over periods of a few years, and by about half that amount in intervals as small as a few months. Much of the power in the spectrum of the variation is concentrated in time scales around a decade. It is also possible (although not necessary!) to imagine two trains of oscillations damped both in amplitude and in period, the most conspicuous beginning in late 1929.

These variations have their most immediate application to the question of the sizes of quasars. For, if a quasar is a single luminous source, it cannot be much larger in terms of light travel time than the time scale of the fluctuations, otherwise the differential light travel time from parts of the source at different distances from us would lead to washing out any large-scale brightness fluctuations.

Every well-studied quasar shows some evidence of brightness variation. So far, the most violent case on record has recently been found by Sandage, in the form of a three-magnitude rise (fifteenfold increase) in the brightness of 3C446 in less than a year. For such work, the original Palomar Sky Survey plates from more than a decade ago provide an invaluable baseline of brightness against which the current state of each newly discovered quasar can be checked.

Equally remarkable is the fact that quasars show radio variations comparable with the optical; however they are usually only at decimeter and especially centimeter and millimeter wavelengths. It is plausible to imagine the bright optical source as a small core, but traditionally the major cosmic radio sources have been thought of as vast clouds, at least thousands of light years in extent; for these to vary rapidly is out of the question. Yet most of the quasars do vary strongly in their radio emission, moreover in a strongly frequency-dependent manner. For example, 3C273 has recently been observed to be increasing sharply in brightness at 2-cm wavelength, while simultaneously remaining constant at 6 cm, increasing somewhat less rapidly at 22 cm than at 2 cm, remaining constant at 31 cm and 40 cm, and, incidentally, remaining more or less constant in optical output during this interval! In connection with this, it might be noted that there is no clear-cut color change over the optical region during routine fluctuations of quasars, although color changes may have accompanied major discontinuities of several magnitudes in such objects as 3C2.

Another form of variation, not intrinsic to the source, but important in the study of quasars, is radio scintillation. The solar-wind flux of plasma streaming out through the solar system has, on a scale of size ranging from hundreds to thousands of kilometers, inhomogeneities sufficient to cause irregular diffraction of radio waves from celestial sources. As is the case with twinkling of optical stars and planets caused by the earth's atmosphere, Hewish has shown that the amplitude of radio-source scintillation is strongly dependent on the angular size of the source. Large sources tend to average out the phase shifts generated by a number of the plasma inhomogeneities; the strongly scintillated sources are those which are effectively points with respect to the angular size of the inhomogeneities as seen from the earth. Exploitation of this effect has permitted single large instruments such as the 1,000-ft. Arecibo dish to achieve effectively thousandfold increases in resolution, in terms of the ability to select by scintillation from the large numbers of faint radio sources those of small angular size. In turn, most of these are presumably quasars.

Number of Quasars

Beginning with one in 1960, four in 1961, and nine in 1962, the number of quasars has since continued to rise even slightly faster than the square of the year number starting with 1960 as unity; about a hundred certain or almost certain quasars are known in 1966. While making forecasts about the ultimately detectable numbers and total space population is unwise (quasars have confounded essentially every prediction made about them so far), the question can be approached from several points of view. Although quasars prove to be a smaller proportion of the BSO candidate objects than was first believed, Sandage's results still indicate that, to the nineteenth magnitude, there may be on the average as many as one quasar per square degree. With some 42,000 square degrees in the sky, at least half of them relatively unobscured by interstellar absorption in our galaxy, some ten to twenty thousand quasars should be detectable to this limit. When searches are made to much fainter limits, a higher proportion of the objects will be quasars since at this faintness we will have left behind some of even the more distant stars in our galaxy; effects of large redshift may make the detection more difficult, but even so many hundreds of thousands if not millions should, in principle, be within reach of our large telescopes. Approaching this number from the standpoint of radio detection, radio telescopes now built or being built reach flux limits at which there should be many hundreds of thousands of sources available; again, as we go fainter, the proportion of quasars--already roughly a third--should presumably increase. (The assumption that quasars discovered through optical and radio astronomy are fundamentally the same kinds of objects follows from the lack of evidence for any basic difference—a substantial range in the ratio of optical to radio luminosity would be expected, as is the case for stars and galaxies). Within the next twenty years, when large optical and perhaps radio telescopes in space permit penetration to far fainter objects than can be studied from the ground, further extrapolation of these statistics suggests that the material backdrop of the visible universe may prove to be a tenuous curtain of quasars.

Theories of Quasars

It is almost literally true that every astronomer and physicist who has worked seriously on any aspect of quasars (as well as many who have not!) has his own point of view on these enigmatic objects. Partly for historical interest, partly because still we are so uncertain where the truth lies, I have organized the principal approaches in the form of a series of basic choices that must be made consciously or unconsciously with every theory. At each of these forks in the road we consider briefly some of the reasons for preferring one branch or the other. By Sherlock Holmes' method, it is necessary only that we make the right choice each time; in this way, we cannot fail to be led inexorably to the correct answer (assuming, of course, that it cooperates with our method by lying somewhere in the range of hypotheses considered). I have tried to be reasonably impartial in presenting the reasons for choices, but readily admit at the outset to having a point of view.

False vs True

Several speculations have been advanced that quasars do not exist at all, that we are the victims of an illusion of one sort or another. The most serious of these is probably the suggestion by the Barnothys that Einstein's classical gravitational lens may at last have found its application. It seems clear that the gravitational field of a massive object of intermediate distance (a galaxy, for instance) can produce substantial brightness enhancement through focusing the light coming from an object directly in line with it but far to the rear. But curiously enough, astronomers have not yet been convinced of the reality of a single example of this effect.

The lens theory might be taken more seriously by astronomers in the present case if only one or two quasars were known. But by now so many are in hand, displaying such remarkably similar properties including small angular size, jets, and variability--all of which are difficult for the lens theory to explain without additional hypotheses--that I prefer to reserve it along with the other illusionist theories for some difficult-to-explain situation in the future, and continue to believe that quasars are really more or less what they appear to be.

Near vs Far

This seemingly simple choice contains the core of most of the problems of quasars. To nonastronomers, it must come rather as a surprise that there really is no certain way to decide the distance of any quasar. The difficulty, of course, lies in the strictly two-dimensional world of direct astronomical observation; even for nearby objects the third dimension of distance usually comes from only the faintest glimpses of parallax, such as that arising from the earth's 3×10^8 km baseline of orbital motion around the sun. A large part of the astronomy of the last fifty years has been devoted to the patient unraveling of luminosity-dependent clues based ultimately on the few directly measured nearby distances, in the attempt to establish directly observable luminosity criteria such as variability, spectral type, size, or appearance. These criteria permit, when observed in a new member of a known class, the assignment of an absolute luminosity; observation of the apparent luminosity allows inference of the distance. Whenever astronomers encounter a completely new type of object, somehow they must associate it with other objects of known distance in order to draw conclusions about the luminosity of the class and the distance of its members. But from the beginning quasars have been extraordinarily unyielding in terms of clues and associations. They are not known to occur in single galaxies or even in clusters of galaxies; indeed, the question has been raised whether they may appear preferentially in relatively empty regions. Extremely careful attempts by Luyten and by Jefferys to detect parallactic motions have completely failed. The spectra are too unusual to permit unambiguous interpretation.

The only familiar and seemingly firm stone in this slippery path is the redshift. First with the optical galaxies, and then with the systematically brighter radio galaxies, redshift has consistently correlated as perfectly as imperfect data would permit with the numerous distance criteria for galaxies that could be devised. Accordingly, it was natural in the beginning (and still seems natural to most astronomers) to accept the quasar redshifts also as bona fide examples of the Hubble relation, and to assign corresponding distances of the type which I refer to frequently in this paper as cosmological. But it should be remembered that so far this is only an assumption; the redshifts may have some local cause, in which case the Hubble relation would be irrelevant to quasars.

Terrell at Los Alamos was the first to prepare a serious case for a non-Hubble, local, interpretation of the redshifts. A few galaxies, most notably Virgo A (M87) and M82, are observed to be active in ways that have been interpreted as the shooting out of matter. Terrell suggested that our own galaxy at some time in the last few million years may have expelled the quasars at speeds ranging up to nearly the velocity of light, in which case the redshifts are simple doppler velocities of recession. The faster ones are now fainter and farther away, giving a reasonable facsimile of the Hubble relation. The absence of observed secular parallactic shifts sets a lower limit on possible distance of approximately the diameter of our galaxy; a considerable range of greater distances is possible on this picture, depending on the time since the hypothetical explosion. However, the greater the distance and the longer the time, the greater the energy burden which this interpretation must bear, since each quasar must be that much brighter for that much longer time; the energy content of our galaxy soon becomes insufficient to give so much energy to so many quasars. Terrell has continued to defend this model with great ingenuity and success in the sense that no single argument of force sufficient to destroy it has yet been adduced. But it remains difficult for most astronomers to accept, mainly on the grounds that even in the minimum case its requirements on energy output in the hypothetical explosion from our own placid galaxy already seem unreasonable, that with the exponentially increasing number of new quasars being found the problem keeps getting worse for any single galaxy to have had to spawn them all, and that the repeating past painful Copernican experiences of astronomy teach us that our surroundings have a singularly low probability of being unique or even unusual; in which case, why do we not see quasars ejected from other nearby galaxies?

Hoyle and Burbidge have argued that, in fact, the quasars are ejected from other galaxies, particularly from one or several relatively peculiar nearby radio galaxies, e.g., NGC 5128, the ejection again being at nearly the velocity of light. In this case, of course, roughly as many must start out toward us as away from us, leading to the expectation of blue shifts. Also, the observed lack of parallactic motions once more becomes a problem. But by judiciously selecting the distance and time of the explosion, Hoyle and Burbidge can make the cross motions unobservably small. Such assumptions can permit most of the quasars ejected toward our hemisphere to have had time to pass by the perpendicular line of sight to their trajectories and thus be red shifted like the rest; also, the transverse Doppler effect will tend to diminish significantly the number of blue shifts. And, finally, blue shifts are inherently a bit more difficult to observe than red, primarily because fewer good lines exist in the ir to come into the visual region than is the case with uv shifted toward the visual. Nevertheless, the lack of a single blue shift in nearly sixty quasar velocities measured mainly by the Mt. Wilson group, and by Lynds at Kitt Peak, is rather ominous for this theory. Also, it shares the anti-Copernican flavor of Terrell's original model.

Arp at Mt. Wilson has recently added fuel to the debate by reporting evidence that radio sources including radio galaxies and quasars occur by pairs on opposite sides of, and typically rather far removed from, peculiar galaxies. Earlier, Matthews at Cal Tech had adduced much evidence for radio sources occurring in pairs on opposite sides of central source galaxies; it is not yet clear to what extent Arp's radio pairs may be partly an example of this effect in relatively nearby systems. Also, the reported separations up to many degrees cover so much of the sky that questions of selection and statistical interpretation have been raised. Furthermore, some of the objects identified on this picture as ejecta are themselves apparently among the most massive objects we know of in the universe, e.g., the giant radio galaxy M87. Nor do the ejected pairs match; on one side may be an immense radio galaxy with fully developed stellar content presumably almost as old as the universe, its supposed mate on the other side being a quasar that is difficult to visualize having been active for even 1% of the age of the universe. In order to excape the energy paradoxes that he would find if his quasars had velocities appropriate to their apparent redshifts, Arp suggests that the redshifts may be gravitational or due to infall of matter. But any simple Einstein gravitational redshift seems virtually ruled out by an argument of Greenstein and Schmidt, who note that a line of a given redshift must originate at the appropriate depth or shell in the gravitational potential well, and that the thickness of such a shell in any reasonable situation cannot be sufficient to produce the observed luminous outputs of the emission lines; furthermore, all the lines of any form of excitation would have to come from essentially the same shell because the redshifts are identical. This seems most improbable. If the redshift arises from infall of matter, then at the required velocities approaching that of light serious problems of dynamics, radiation, and time scale are raised, not to mention the geometrical question of why we should always see only the appropriate part of the front of the infalling cloud sufficient to give us redshifted but not significantly broadened lines. Still other explanations for the redshifts may be possible, but such are not presently known except, again, for the basic Hubble relation, in which case the redshifts are cosmological; and Arp's pairs and groups would be largely accidental and apparent rather than real physical associations. At present, Arp has published little interpretation of these problems; at least until more details of the observations and theory are available, most astronomers remain skeptical of this approach.

If cosmological distances for quasars offered no difficulties either, this paper could be much shorter. Serious objection to acceptance of the redshifts as part of the Hubble relation, along with ordinary galaxies and radio galaxies, rests mainly on two types of argument, each relatively easy to state.

We have already remarked that the appropriate cosmological distances for quasars would require in some cases intrinsic total luminosities including ir greater than 10^{47} erg/sec, also that historical observations imply the perseverance of such luminosities for some 10^3 years (10^{10} sec) while the jets speak for times nearer 10^6 years (10^{13} sec). Thus, the observable component of energy to be accounted for may run over 10^{60} ergs, equivalent to the total conversion of 10^6 solar masses (10^{11} earth masses) to energy, or, more plausibly, to the nuclear or gravitational liberation of available energy from some 10^7-10^8 solar masses, and all in a small region of space. Thus, any cosmological interpretation of the redshifts appears to require small objects in the once forbidden range of millions to billions of solar masses; for plausibility, a detailed model of the structure, stability, and radiation of such objects must be forthcoming. The problem is serious enough so that, as we have just seen, a few workers have preferred to avoid it from the start by seeking noncosmological interpretations of the redshifts.

The second class of difficulties is best illustrated by attempts to account in detail for the radio emission in the face of rapid variation, if the distances are cosmological. Several years ago, Slish pointed out an important relation that can be paraphrased as follows. A given radio source has an accurately known observed apparent brightness. This apparent brightness might arise from a large range of combinations of object size and volume emissivity—hence absolute luminosity—and distance. Actually, however, these combinations are not unlimited. For one thing, as pointed out earlier, if substantial radio variations are detected, the linear size of the radio source should not be much greater than the light travel distance corresponding to the time scale of the variations. To take a specific example, a radio source observed to vary strongly in times of the order of one hundred days is probably not much larger than a light year in diameter, and has only a corresponding volume of a cubic light year from which to radiate. The more distant we wish the source to be, the greater must be the total luminous flux coming from that cubic light year, in order still to produce the observed apparent brightness. In turn, limits can be set on the amount of power which can radiate from that volume, at least on the normal assumption that the emission is synchrotron radiation produced effectively by monoenergetic electrons. For in this case, to increase the power radiated, one must increase the electron density or the magnetic field. But, as the density and/or the field climb, at increasingly short wavelengths the cloud becomes, by self-absorption, opaque to its own radiation. In turn, this self-absorption sets an upper limit to the power that can escape from the surface of the cloud, that is, an upper limit to the absolute luminosity and hence an upper limit to the possible distance of the cloud. The long-wave maxima observed in radio source spectra are, in fact, conventionally assumed to arise from the onset of synchrotron self-absorption; thus, the maximum surface brightness can be surmised for each source from the peak wavelength in its radio spectrum. This establishes (in conjunction with the light travel time size) a limit on the total power that can be radiated, and hence on the maximum possible distance that could correspond to the apparent brightness.

The bearing of these points on Hubble relation interpretation of the redshifts is that, if some rather uncertain reported rapid radio variations of several quasars (notably CTA 102) should prove valid, the great cosmological distances for these sources cannot be correct, unless, to be sure, there are errors in the theoretical assumptions.

The majority of astronomers who still favor the cosmological distances for quasars answer the above arguments essentially as follows—First, while admittedly drastic, the required energies for quasars are at least no worse than those for the most extreme radio galaxies. But radio galaxies show such a smooth progression of types blending into normal galaxies, so many properties in common with normal galaxies, and such a fine fit to the Hubble relation of normal galaxies, that essentially no one denies that their immense distances and luminosities are adequately derived from their observed redshifts, even though we do not yet understand the source of their activity. There is even a suggestion of a progression from relatively normal radio galaxies through specimens with increasingly concentrated small cores and weak extremities to objects (N-galaxies) rather difficult to distinguish from quasars. Radio galaxies tend to be the brightest members of clusters of galaxies. Quasars do not show this effect, but to be sure would be so far away that their own galaxy or companion galaxies, if present, might simply be too faint to have been seen as yet. In view of these points, it seems reasonable to accept that the phenomenon of ill-understood vast energy releases might apply to the entire family of objects, all at the cosmological distances indicated by their redshifts.

The technical arguments regarding possible limits on emissivities stand or fall with the validity of their assumptions. In particular, rather than a uniform source, a complex structure containing a wide range of shells or even of multiple sources having differing electron densities and energy spectra would be expected in real objects, and this offers some release from the simplest form of the Slish argument. (Indeed, the steeper radio spectrum of the jet of 3C273 than that of the central source may be considered an example of this; the difference is consistent with electrons in the jet being older and more tired than those in the core.) And much more release can probably be obtained by dropping the synchrotron mechanism, as the major source of radiation in every case, in favor of some form of plasma radiation proposed by various authors in the last several years.

On the positive side, one of the strongest supports for the cosmological interpretation is the remarkable resemblance between the Hubble diagram for ordinary galaxies and that for quasars. In both diagrams, the really significant fact is the absence of points in the upper left half. To understand this, note first that in the case of galaxies we find throughout the observable universe a relatively uniform maximum luminosity; these uniformly brightest galaxies decline smoothly in apparent magnitude with increasing redshift (distance), thereby defining the upper limit or edge of the scatter of galaxy points. The slope of this edge, as it will be called here, corresponds to the numerical value for the Hubble relation. The entire region to the right of each point on the edge is in principle filled because, at the distance corresponding to each redshift, there exists an indefinitely large range of intrinsically fainter galaxies which we of course see as having fainter apparent magnitudes. In practice, such intrinsically fainter galaxies are only relatively easy to observe if nearby and at small redshifts; consequently, the actual filling of the diagram to the right of the edge tends to occur only near the bottom. Those quasars for which redshifts and apparent magnitudes are known display a relation remarkably similar in appearance to the part of the Hubble diagram for the nearer galaxies. Thus, it seems that for quasars we are also observing an approximate maximum possible intrinsic luminosity (to be sure, some four or five magnitudes above that for ordinary galaxies) defining an upper edge of the scatter diagram; and that, as is the case with galaxies, quasars also includes a wide range of lesser luminosities. In the case of quasars, we are, so far at least, sampling a considerably smaller logarithmic range of distances than with galaxies, so the proper comparison is only with the lower left half of the galaxy diagram. Since the quasar magnitudes so far have no correction for redshift effects on the photometry (K correction) or for variability, the rough similarity of slopes is only generally relevant to the argument. The important point is that quasars define an optical Hubble relation at least as strongly as the case with a corresponding magnitude range of galaxies, the region to the upper left of the edge being essentially empty in each diagram. Absence of points in this region should be a real effect rather than observational selection, since the greater brightness of any object to the left of the edge would make it relatively conspicuous.

It is true that the corresponding diagram for the radio magnitudes of quasars shows little Hubble effect, but a similar result should be expected for the radio galaxies wherein some of the apparently brightest are among the most distant, e.g., Cygnus A. Thus the lack of a pronounced Hubble relation in the apparent radio magnitudes may well be simply a consequence of a great range of radio luminosity among the quasars. A plot of apparent magnitude vs log distance for the forty brightest stars would show a similar poor correlation for the same reason.

Another relation having strong bearing on the question of quasar distances is that of number count vs apparent brightness. In a simple Euclidian space uniformly populated with some class of objects, the total number count (N) of the objects would increase with the cube of the distance searched, but the apparent brightness (S) of each would decline with the inverse square of the distance. Accordingly, the log N vs log S relation should have the slope -1.5. Exactly this slope is, in fact, observed for the brighter, relatively nearby radio galaxies. But Veron has recently pointed out that the presently known quasars show a slope of -2.2, corresponding to much more rapid increase of number with increasing faintness and presumably distance. In view of the order-of-magnitude greater distance of the average quasar than the average radio galaxy, and the much greater likelihood of evolutionary effects being present with quasars, this result, however important and pregnant, is nevertheless quite plausible.

Unambiguous detection of intergalactic absorption lines in the optical or radio spectra of quasars can settle this overriding question of distance. The optical absorption lines so far detected seem most likely to arise in gas clouds around the quasars, and the one promising reported case of hydrogen 21-cm absorption (in 3C273, by gas in the Virgo cluster of galaxies) seems to be specious under closer examination. But serious search for such effects has only begun.

Having now summarized a number of the important arguments bearing on distance, we find that, although the case is far from clear, a number of separate points strongly suggest that we are more likely to be right by taking the cosmological rather than the local branch.

Multiple vs Single; Large vs Small

These choices are rather intertwined. For, one of the few things we know with reasonable certainty about quasars is that if they are single they are small. The observed optical and radio variations early settled the point that, if single, the principal luminous source in at least some quasars must be only of the order of a light year (10^{18} cm) in diameter; indeed, if suggestions of major variations over a few weeks should be borne out, extreme quasar cores may be as small as 10^{16}-10^{17} cm. Of course, this in no way precludes much larger tenuous gaseous halos and jets, with their own characteristic optical and radio emission, but the latter must contribute only a small share of the radiation compared with the varying component.

On the other hand, it has been admitted that direct observation presently sets an upper limit only on the order of 0.5 sec of arc on the angular size of the cores; at cosmological distances such a large angle would permit most of the sources to be many thousand light years in diameter. And at least one plausible mechanism exists whereby the sources could have almost any size up to this limit and still vary somewhat as observed. Among other theorists, Cameron, Woltjer, and Field in particular have each suggested that quasars derive their extreme luminosity from the simultaneous supernova explosions of many massive but ordinary stars in a relatively dense galaxy nucleus. In Field's picture, a young incipient-galaxy nucleus, dense with gas, condenses into millions of massive stars that run their brief course of evolution in a few hundred thousand years and over a relatively short and roughly coincident period of time become supernovae. Each supernova reaches, for example, a luminosity of 10^{10}; if, at random, ten to a hundred are at any time simultaneously active for their individual few weeks of brilliance, the average total luminosity should range from 10^{11} to 10^{12}, as observed for quasars, and the statistics of the process would generate a continuous variation in total brightness. The supernova gas-cloud expulsions would produce both large halos emitting the observed spectral lines and radio noise by virtue of the high-energy particles ejected. Cameron's similar picture visualizes the supernovae as rather closer together, in a denser gas cloud, in order for the explosion of each supernova to be swaddled in a deep blanket which is able to convert with relatively high efficiency the energy of the immense particle flux into luminous energy. In this way, even individual bursts might become bright enough to be visible as flashes in the light curves.

Regardless of the details, this general approach is certainly appealing, with a significant probability of being correct. It has the advantage (from every point of view except interest) of being essentially fully understood today, requiring no substantial additional hypotheses or basically new physics to be added to our stock. As far as I am aware, the only arguments against it are relatively weak, although not negligible.

The first is the question whether ejection of a jet such as that from 3C273 would not require action qualitatively different from the continuous random explosion of supernovae (although perhaps an explanation could be devised, for example, in terms of an extraordinarily large fluctuation blasting out a cloud of gas through confining magnetic fields). A further objection is that the only historical light curve in our possession suggests collective or pulsational rather than sum-of-random activities. Field has pointed out that similar light curves could be simulated by summing the proper number of random events of the proper duration; on the other hand, after 1929 the remarkably uniform maxima and especially minima seem improbable consequences of a random sum. Another fairly strong point against the multiple source theories is the fact that two of the quasars (3C2 and 3C446) have shown what appear to be plateau-like rises of several magnitudes in their levels of emission; probably 3C273 experienced such a change on a smaller scale in the form of a systematic 0.07 magnitude drop in average brightness around the time of the 1929 onset of major apparent oscillation. Again, the statistics of small numbers can be used to cover a wide variety of forms of behavior, but such changes seem more consistent with alterations in the physical state of an incipiently unstable single source.

If we permit quasars to be small and single, it becomes necessary to make a reasonably plausible physical model for them. Energy arguments already discussed require masses not less than about 10^5 solar masses, and probably several orders of magnitude more. Hoyle and Fowler early surmised that in this mass range there might be another stable region for starlike objects; Fowler subsequently developed this theory in considerable detail. In brief, it appears that, after making the appropriate relativistic corrections in the theory of stellar models required when the mass-radius ratio becomes so great, a region of quasi-stability probably does exist in the 10^6 solar-mass range; rotation increases the permissible mass to about 10^8. There is a strong probability that the object would tend to collapse under its own gravitational field until densities and temperatures in the interior become so great as to initiate nuclear reactions that would run away rapidly enough to provide the energy necessary to re-expand the object. It would coast out to an excessive radius too great for maintaining the nuclear reactions before gravity would stop the expansion, again forcing collapse and renewal of the cycle. Plausible conditions could give irregular oscillations of this kind having durations of years or decades, as observed in 3C273.

Once it was hoped that gravitational collapse would prove much more effective as an energy source than the traditional nuclear reactions that power the stars. This hope was based on the form of the gravitational potential energy relation, proportional to M^2/R, where M is the mass of the contracting object and R is its radius. By taking a large enough mass, with the rapid gain from its square, and by letting the radius become small, it seemed that almost any desired energy might be made available; this stands in contrast to the linear simplicity of nuclear energy sources with their approximate limit of available energy 0.01 Mc^2. But Chandrasekhar and also Fowler have shown that with gravitational collapse Mc^2 is also a theoretical upper limit on available emitted energy, and, in practice, only a much smaller fraction will normally be available. This is still a useful accretion to the store, but renders gravitational and nuclear processes more nearly comparable in practice. The principal remaining difficulty with gravitational collapse is that no one has proved conclusively that, starting from diffuse matter a concentration could reach the quasi-stable range of 10^6 to 10^8 solar masses without being disrupted while still in the unstable range of say 10^3-10^4.

Matters stand now at the point where quasars at cosmological distances make reasonable physical sense either as great assemblages of supernovae or as single supermassive stars if somehow they could be formed. Wheeler, in private communication, has raised the interesting suggestion that, if the former theory is correct, the statistics of small numbers should lead to greater variance in the light of those quasars having fewer supernovae, hence lower quasar luminosity. Within a few years there should be enough information on quasar variability to see whether those lying to the right of the edge of the scatter in the Hubble diagram do, in fact, show the more extreme variations. Meanwhile, faced in this case with a difficult decision, we will choose the road of the small, single objects--partly from belief and partly just to see where it leads.

Old vs New

For some time there has been controversy about whether the energy output from radio galaxies represents a highly active stage in the evolution of an old galaxy or the beginning of activity of a new one. So far, there is really no evidence in favor of radio galaxies being fundamentally different from normal ones; rather there is growing evidence that the ages, if not the states of evolution, of most galaxies are about the same. These points are in harmony with the view that many galaxies become radio sources, some of them probably many times, also probably more often and more strongly in rough proportion to mass. The violent activity seems in some way to be associated with the nuclei of the galaxies; indeed, almost any large galaxy taken at random, regardless of whether it is a radio source, may display an apparently dense nucleus which is the seat of at least some unusual optical activity; our own galaxy has a strong radio source at its center.

With quasars there is no corresponding evidence of comparison with familiar objects on which we can firmly ground our speculations. But several points suggest that, in contrast to ordinary radio galaxies, quasars may well be in quite an early stage of their evolution. With their stupendous output of light and perhaps matter, it is hard to see how they could have existed in anything like their present forms for even a thousandth of the minimum age of the universe (now believed to be about 1.2×10^{10} years). Also, the impression that they occupy regions of intergalactic space in which few if any well-formed giant galaxies have yet developed suggests that they might represent an early stage in such a condensation process, which in time may lead to a full population of galaxies. This impression is quite consistent with evidence, brought out in a later section, that quasars seem to become more abundant when we view a significant fraction of the distance in time back to the beginning of at least this phase of the universe--an earlier time when presumably more major condensations were still forming. For what these arguments may be worth, they seem to point toward the quasar phenomenon as not being necessarily different from that occurring in radio galaxies, but on a larger scale and perhaps taking place in virgin territory.

Nonmagnetic vs Magnetic

Large-scale magnetic fields have played a significant role in astronomy only since about 1950 when the studies of radio astronomy and cosmic rays began to show increasingly the importance of magnetic fields as part of a total energy balance and as guiding the motions of the tenuous plasmas that make up a substantial part of the universe. But until quite recently theories of quasars made use of magnetic fields only as an afterthought: a way to convert relativistic electron energies into radiation through the synchrotron process. Now among others Sturrock, and Ginzburg and his colleagues in Russia have proposed that, in fact, the magnetic field may be integral to the nature of radio galaxies and quasars.

Importance of the field in Sturrock's picture is at three levels. First, and in some ways most significant, is the fact that, if the field lines are originally even vaguely parallel and open over a large region of intergalactic space, as any gravitational condensation forms locally, pulling in the field lines as it shrinks, a pair of enormous funnels is created in the still-open set of lines. Reaching far out into space, they channel in toward the center the matter from a vast region, building the mass and gravitational field of the core. The growing gravitational field continues to pull in matter and field lines laterally as well, eventually trapping a great magnetic field density in the core. Here violent activity churns the field to the point where, as Layzer pointed out, magnetic turbulence itself becomes a major contributor to the support of the object, now a quasar, against its own gravity. This turbulence factor--the second important use of the field—may be able to raise somewhat the upper limit on mass over Fowler's models for quasars. Finally, the magnetic fields near the surface are so great that occasional escape and shrinking of major bundles of external field lines hitherto pulled down into the core would, as Sturrock has suggested may happen with solar flares, occasionally eject great masses of plasma at relativistic velocities, thus accounting for the jets of radio galaxies and quasars. The Russian views are somewhat similar, but emphasize more the nonsynchrotron plasma radiation mechanisms that become possible. A remarkably small field (of the order of 10^{-8} G) suffices to start Sturrock's mechanism. While the details are neither complete nor fully accepted, the question of the improbability of field-free space and the general plausibility of the magnetic picture, as well as the number of points to which it can speak, make it seem that the burden of proof now lies with those who would try to produce quasars and their phenomena without taking into account magnetic effects.

Fundamentally Relativistic vs Somewhat Classical

Immediately upon its introduction by Einstein in 1915, the theory of general relativity had a brief surge of application to model universes, to the behavior of light in gravitational fields, and to celestial mechanics. But in each case the observable effects gave only rather trifling corrections to a remarkably good first-order classical theory. For the next forty years general relativity advanced slowly, because of its relatively intractable character and because its seeming lack of application did not attract many physicists. It would be no exaggeration to state that the discovery of quasars has revitalized the field, only the simplest example being the case mentioned earlier, where as the mass-radius ratio becomes large, classical theory necessarily becomes a poor approximation in calculating massive quasar models.

But the excitement for relativists grows when one questions the ultimate fate of masses so large that no known physics suffices to stop their collapse under their own gravity. Fifty years ago Schwarzschild proposed the spherical singularity that now bears his name. For any given mass, the Schwarzschild singularity, or critical radius (given by $R^* = 2\,GM/c^2$), amounts to that radius at which the gravitational potential would prevent the escape of light from the surface. For the earth's mass, this radius is about a centimeter, for the sun's mass it would be several kilometers, and for a quasar of 10^9 solar masses, for example, the radius would be a quite appreciable 10^{14} cm (recall that rapid light variations have suggested quasar core radii as small as 10^{16} cm). Matter within a Schwarzschild radius is presumably forever cut off from our universe. Even if some at present not clearly imaginable force should cause sufficient expansion of such a mass against its own gravity in order to drive it outside the critical radius, we could still never observe it because of relativistic dilation of time in the strong gravitational field. At the Schwarzschild radius, this dilation, as seen from outside, is infinite; the process could never complete itself from our point of view. Needless to say, theorists have been looking closely to see under what conditions such extreme statements rigorously hold, and whether departures from symmetry or deeper physics can show a loophole. And, whether we understand the physics or not, several workers, beginning most notably with Ambartzumian, have surmised that matter may be present in some highly condensed state in the nuclei of galaxies (and now in quasars), somehow materializing to become responsible for the activity observed. Indeed, since we do not know how matter came or perhaps continues to come into the universe, it is not unreasonable to look at such objects as quasars as possible sources of new matter. One of the ideas suggested most seriously has been the possibility that the original configuration of the universe was one of supercondensation, a state thrown off at the initial general expansion of the universe, with elements of the matter in the universe escaping more or less simultaneously from the bondage of their local Schwarzschild radii. But those which lagged even a trifle in the original time frame might be effectively caught in time dilation near their critical radii as seen from the point of view of those which had already escaped significant effects of gravity; these late bloomers, still near their critical radii, would seem to progress with such agonizing slowness that even (on our time scale) twelve billion years later they would just now be seen emerging from their gravitational cocoons.

Somewhat related ideas, mainly by Hoyle and Narlikar and by Wheeler and his colleagues, have suggested that matter disappears under such intense gravitational forces, perhaps being converted into one or another new kind of radiation, and perhaps making it possible for matter to rematerialize elsewhere.

In a different, but distantly related class, is the theoreticians' worry about why the local universe is not symmetrical with regard to amounts of matter and antimatter. Since we can only create the one with equal amounts of the other, for the time being it seems reasonable to put the same limitations on creation, in which case we wonder where the antimatter may be, and what its effects might be. Since antimatter offers the only known way actually to obtain Mc^2 from matter, and since energy has been a problem throughout this discussion, it is not surprising that several suggestions have been made that isolated encounters between large masses of matter and antimatter might cause the quasars. Perhaps the most detailed and plausible of these is due to Alfven, who also gives reasons for the large-scale separation of matter and antimatter in terms of plasma currents and anticurrents that should arise in the early stages of the universe.

Most astronomers are not well qualified to judge these particular esoteric matters, but we watch with interest the proliferation of theories, waiting for the occasional opportunity to test one, and wondering whether with quasars the time has arrived at last when reasonably classical physical theory will no longer suffice even to a rough approximation.

Conclusions

Many observations and theories have been touched on in this paper, but not much justice has been done to any part of the subject or to the men working in it. Yet a longer paper might come no closer to the real heart of the matter, which is that we simply do not know enough to decide the questions.

Further progress in understanding these matters will almost certainly rest on the ability of observational astronomers to work much more effectively and extensively on what are now thought of as the faintest, even inaccessible, objects. Progress will depend upon having more of the relatively conventional large telescopes (including, if possible, several in the 10-m class) to do basic work and to be available for the ever more rapidly developing new instrumental ideas which require large and versatile light collectors to feed them. Progress will also depend upon the development of relatively lower cost, but perhaps even larger, special-purpose instruments--particularly for work in the ir, for faint interference spectroscopy, and perhaps for angular interferometry. Progress will be heavily dependent upon the development of improved detectors that give astronomers more nearly unity quantum efficiency--a vital consideration when the quanta have to be counted one by one as they trickle in from the most distant sources--and on improved data storage and processing facilities. And the final solution will probably rely upon the availability of large telescopes located somewhere beyond the earth's atmosphere for covering all spectral regions from gamma rays to the longest radio waves, so many of which are lost to ground based instruments.

As to the general conclusions reached in this paper, perhaps it should be recalled that only a few scientists--in astronomy most notably Eddington--seem to have possessed an intuitive feeling for the universe which, even in the absence of sufficient evidence, led them almost always to the correct general picture. Most of us do not fare so well, and unfortunately we no longer have Eddington to guide us at the branches. So it is quite possible or even probable that the markers which we have read may have been backwards or not even the right ones at all.

In the corresponding dilemma nearly fifty years ago concerning the galaxies, Curtis concluded: "There is a unity and an internal agreement in the features of the island universe theory which appeals very strongly to me .... I hold, therefore, to the belief that the spirals are ... island universes, like our own galaxy, ... indicating to us a greater universe into which we may penetrate to distances of ten million to a hundred million light years."

If faced with a similar need to take a summary stand I would surmise (on grounds perhaps not too different from Curtis' unity and internal agreement which appeals strongly to me) that, when the truth is known, quasars will be found to exist at the distances and luminosities given by the Hubble interpretation of their redshifts, to represent an early stage in the condensation of matter from relatively large intergalactic regions, to contain immensely massive single cores irregularly pulsating and exploding and perhaps collapsing, to require in an essential way the effects of magnetic fields both weakly on the large scale and strongly on the small scale for their proper interpretation, and to be—as we look back to the limits of distance and time--the principal objects populating the observable universe. As such, for an understanding of the origin and structure of the universe, they will become the most important objects accessible to astronomers.

REFERENCES

Most of the early work on quasars is thoroughly reported in Quasi-stellar Sources and Gravitational Collapse, I. Robinson, A. Schild, and E. Schucking, Eds. (University of Chicago Press, Chicago, 1965).

Gravitation Theory and Gravitational Collapse, by B. Harrison, K. Thorne, M. Wakano, and J. Wheeler (University of Chicago Press, Chicago, 1965).

Important recent articles include W. Fowler, Astrophys. J. 144, 180 (1965).
A. Sandage, Astrophys. J. 141, 1560 (1965).

SUMMARY REMARKS ON QUASARS

Harlan J. Smith
University of Texas

As a framework for summarizing the conference reports on a subject still as novel and uncertain as the quasi-stellar sources, a few quasi-questions may be appropriate.

1. What Shall We Call Them? The older and rather cumbersome term "quasi-stellar radio source is now at the very least suspect, thanks to Sandage's interlopers which may not all prove to be detectable radio sources. But "quasi-stellar source" is also a bit cumbersome, and is furthermore redundant since any object in space which we can detect is the source of at least something. "Quasi-stellar" by itself is an adjective, and in addition may carry a wrong connotation as to the nature of the objects if, for example, Field and others should be right that they are really galaxies albeit with very compact nuclei. Although not yet very popular with astronomers, Chiu's suggestion of "quasar" offers a commendably brief proper noun, and one which has already received wide circulation. By its very sound it implies "puzzling starlike radio source," but does not explicitly and prematurely prejudice the issue of the true nature of the objects. These arguments appear to me sufficient to warrant adopting the term.

2. How Should the Class of Quasars be Defined? Schmidt presented five criteria for membership which apply to most if not all of the quasars for which the relevant data were available at the time of the Symposium:

a. Optically-identified radio source, having a core of star-like appearance to the optical resolutions so far achieved, (although normally sporting one or two very faint jets)
b. Optical variability
c. Large ultraviolet excess
d. Broad emission lines
e. Large redshifts

3. What is Fundamental Among These Defining Characteristics? If as perhaps is already the case, we should encounter objects having most but not all of the defining properties given above, which should be accpeted as physically essential and therefore defining the class? Detectable radio emission, while calling attention to outstanding candidates, may prove to correspond to a transient--perhaps later-phase of quasar evolution, and thus not be basic. Ultraviolet excess and broad emission lines are apparently normal consequences of the nature of the sources, but also may not necessarily be fundamental. Optical variability is perhaps more likely to be integrally related to the mechanism of quasars, yet at this stage we would hardly deny admission to a non-varying candidate which satisfied all the other criteria. The unique features so far distinguishing quasars from other known astronomical objects are their immense red shifts and attendant luminosities coupled with extremely small size in comparison with galactic dimensions. Many other properties aiding in the discovery, subclassification and study of quasars will surely be found, but the essential distinguishing property of the class will probably prove to be the sustained emission of galactic or supergalactic energies (say $> 10^{44}$ ergs/sec) from a very small volume of space (say $R < 10^{20}$ cm, perhaps 10^{19} or even less).

4. How Abundant are the Quasars? Had the Sandage-Veron campaign begun only a few months later, as of the end of 1964 the number of known quasars would have increased approximately in the progression: $N = (\text{year} - 1960)^2$. By this criterion we would have had the respectable number of a hundred known quasars in a decade. However, the effective use of displaced two-color images by Sandage, and the discovery by radio astronomers that quasars subtend so small an angle as to be distinguished by detectable solar-system scintillation, have sharply accelerated the rate of discovery. Judging from the number of interesting objects now being found per square degree, it is beginning to appear that thousands of quasars may lie within reach of present-day instruments, although to be sure most of them are on the very fringe of detectability and cannot yet be studied in any detail. This is consistent with the present indication that at least the nearer quasars occur at a frequency of several per cubic gigaparsec, whence—depending on cosmology—at least hundreds to thousands should be accessible.

5. How Far Away and How Bright are the Quasars? Schmidt summarized strong arguments (including electron density, volume luminosity, and conditions placed on the emitting volume by the width of the spectral lines, coupled with lack of observed gravitational pertubations on our galaxy) which almost completely rule out gravitational red shifts from massive objects as the cause of a significant fraction of the observed red shifts. Although Terrell's suggestion of a true Doppler shift arising from ejection of matter from the center of our own galaxy cannot yet be absolutely ruled out, it seems very unlikely in view of the lack of known blue shifts from ejecta from other galaxies, and the remarkable requirements which would be placed on an exploding source in our galaxy and on its ejecta. For what it is worth, our sense of "the fitness of things" (continuity with the appearance of radio and optical jets seen from other distant extragalactic objects, energy requirements consistent with those of known radio galaxies, and the ubiquitous nature of red shifts in the extragalactic realm) encourages belief that the red shifts of the quasars are also of the Hubble variety, in which case their unprecedented absolute magnitudes follow.

6. What are the Spectra? Extrapolating from only a few cases--really only from 3C273 in any detail--quasar spectra feature a nearly flat continuum across the near ultraviolet, visual, and red, with a rise in the infrared. Hydrogen emission lines of the Balmer series, plus forbidden and permitted lines of a few ionized gases are usually present with broadenings of the order of 50 Å. Oke's interpretation of the 3C273 spectrum as arising largely from hydrogen recombination in a hot ($T_e \sim 160,000°$ K) nebula, supplemented in the red by perhaps some synchrotron radiation, can account fairly closely for the data from his photoelectric scans. In this case, however, it is not clear how the relatively rapid variations might become visible at the surface, unless much of the gas is in filaments, subcondensations, or a thin disk, with some of the continuum coming direcly or almost directly from the active core(s) of the quasar. It remains important to find whether the spectral lines fluctuate in brightness with the continuum.

7. Which of the Reported Variations are Real, and What do They Imply? Again, based on a too-small sample, we believe that quasars normally change brightness by tenths of magnitudes gradually over years; erratic variability of hundredths of a magnitude over weeks or months also seems established. Bright flashes with durations of days, suspected from the photographic histories, have not been confirmed photoelectrically. Rates of change, including both increase and decrease, have thus not been observed with certainty to exceed about 0.2 magnitude per month. After 1929, 3C273 showed quasi-periodic fluctuations of about 0.7 magnitude rather resembling an oscillation damped in period as well as amplitude, with a time scale of over a decade. 3C2, with its birth or at least sudden rise by several magnitudes over two years, offers the largest amplitude so far encountered. 3C273 and 48 may be declining systematically by as much as a tenth of a magnitude per century, but the observations are not firm on this point.

8. What is a Quasar? If the Hubble Law distance are correct, quasars in any event must be massive objects, probably in the range of 10^5 to $10^9 M_\odot$. The light variations, great luminosity, and spectral appearance can be accounted for by the multiple supernova models of Field, Cameron, and others. But if the possible semi-regularity of 3C273 should be confirmed in it and in other quasars, a single massive core seems to be required. And in fact, Fowler pointed out in considerable detail how instability with a cyclicity of the order of a decade might arise through repeated onset of strong nuclear fusion in an early phase of gravitational collapse of a very massive body.

Thus, it is still not clear whether much of the mass is concentrated in a single core object, perhaps of the order of 10^{16} to 10^{17} cm, surrounded by a substantially larger gas cloud ($\sim 10^{18}$ to 10^{19} cm) and a much larger radio shell ($\sim 10^{21}$ cm), or whether there are many bright subcondensations, effectively hypernovae, in a very massive galaxy-nucleus type of configuration, or indeed whether some quite different picture may be correct.

9. How do Quasars Relate to other More Familiar Objects? There is no clear indication that quasars are found in clusters of galaxies, or in any preferential direction or region of space. It is even uncertain whether there may be an underlying galaxy in each case, so far masked by the brillianc of the quasar, although 3C2 may give us a first clue with its faint and reddish image prior to the appearance of the bright blue one.

And, although quasars share some features with radio galaxies, we still have little clue as to whether they represent a very early or very late stage in the formation of normal galaxies, whether they are perhaps an early stage in the activity of what we would soon call a radio galaxy, or whether they have any direct evolutionary link at all with more familiar objects.

PART 3

COSMIC RAYS

COSMIC-RAY PARTICLES OF HIGHEST ENERGY

Bruno Rossi
Massachusetts Institute of Technology

All experimental data on cosmic-ray particles with energies greater than about 10^{13} eV come from the study of the giant showers produced by these particles in the atmosphere. This paper is a brief review of those aspects of air shower research that are of particular significance in connection with astrophysical problems.

I. Maximum Observed Energy

The most reliable and perhaps most important result obtained thus far is the existence of primary particles with energies up to about 10^{20} eV. The question here arises as to what is meant by "about;" i.e., how accurate are the estimates of the primary energy obtained from the analysis of showers. These estimates include two steps. The first is an evaluation of the total number of shower particles from the densities observed at widely separated points; the second is a computation of the primary energy from the total number of shower particles.

As an appropriate example, consider the largest shower reported to date. This shower was observed by John Linsley[1] of Massachusetts Institute of Technology (MIT) with a detector array 3.6 km in diameter located at Volcano Ranch near Albuquerque, New Mexico, at an altitude of 1700 m (atmospheric depth: 820 gm/cm^2. Figure 1 shows the location of the detectors, the number of particles per square meter registered by each detector,

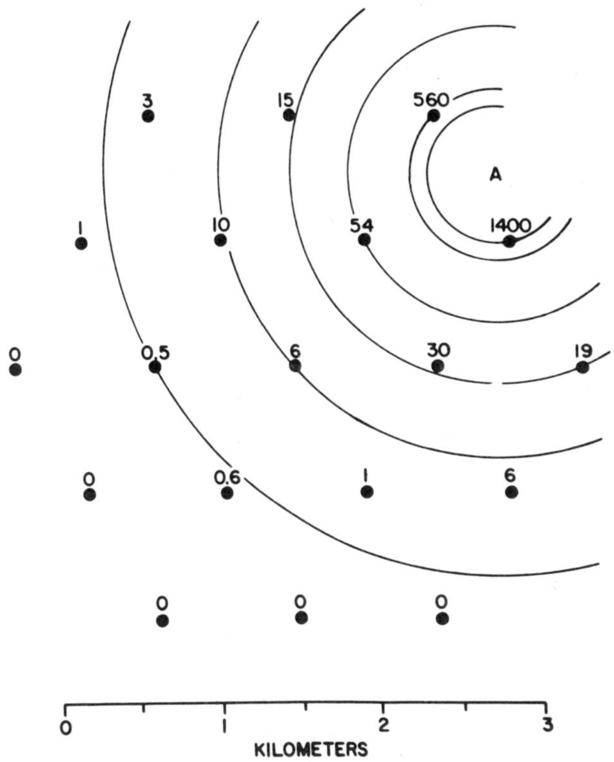

FIG. 1. Plan of the Volcano Ranch array at the time when the largest shower was recorded. The circles represent 3.3 m^2 scintillation detectors. The numbers near the circles are densities (particles per m^2) recorded by these detectors. Point A is the estimated location of the shower core. The circles are lines of constant particle density. (From Linsley 1963.)

and the point of impact of the shower axis (A), as determined from the single assumption that the particle density is a monotonically decreasing function of the distance r from the shower axis. It should be pointed out that the experimental arrangement used by Linsley provides information not only on the density of shower particles at the locations of the detectors, but also on the direction of arrival of the primary particle; this information is obtained from the times at which the shower front strikes the various detectors. For the shower in question the direction of arrival was 10° to the zenith.

In Figure 2 the circles represent the particle densities observed at various distances r from the shower axis. To these points was fitted a semi-empirical function F(r), which is supposed to represent the particle density as a function of r for all values of r.

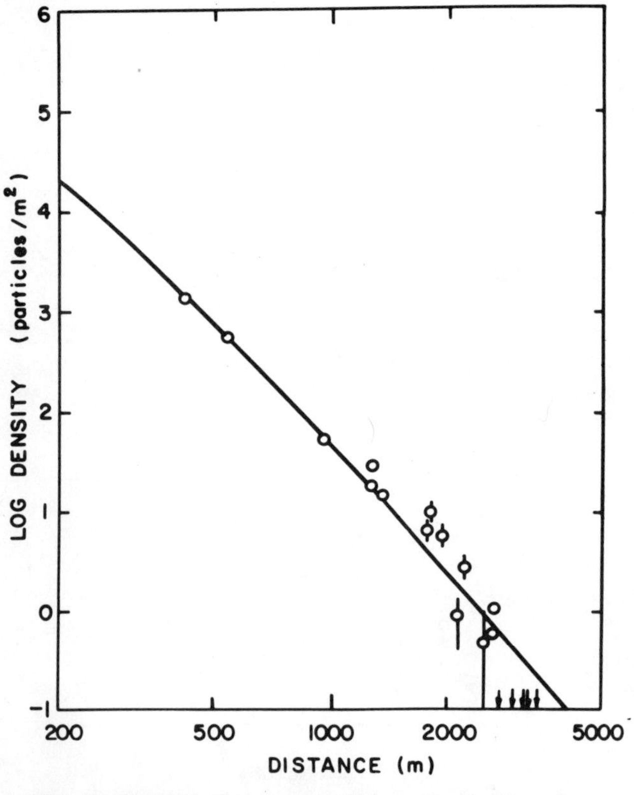

FIG. 2. Density of shower particles as a function of distance from the shower core for the event shown in Fig. 1. (From Linsley 1963.)

The total number of particles in the showers, N, is then given by

$$N = \int_0^\infty 2\pi F(r)\,dr$$

and in the present case turns out to be

$$N = 5 \times 10^{10}.$$

The uncertainty in the value of N arises from the fact that for the shower under discussion the particle density was measured only at values of r greater than 420 m. Since the number of shower particles outside the circle of 420-m radius is a minor fraction (a few per cent) of the total number of particles in the shower, the computed value of N depends critically on the assumed shape of the distribution function F(r) for $r < 420$ m.

Experimental data on the shape of the distribution function in this region are available from experiments on smaller showers (smaller showers, being more abundant, can be studied with smaller arrays, made with more closely spaced detectors). It is found that the normalized distribution function $f(r) = F(r)/N$ changes very slowly with shower size, and that, at the altitude of Volcano Ranch, its logarithmic slope $-d(\ln f)/d(\ln r)$ increases gradually with increasing N for all values of r.

The choice of the distribution function used in the analysis of the shower under discussion was based (1) on the shape of the function f(r) for showers with $N \approx 10^8$ particles (which was measured down to r = 60 m), and (2) on the dependence on N of the logarithmic slope of this function, as observed at $r > 420$ m for showers with $N \approx 10^8$, $N \approx 10^9$ and $N \approx 10^{10}$. Linsley estimates that the value of N thus computed is within a factor of 2 of the real value.

The computation of the primary energy E from the observed number of particles N involves, to some extent, the theory of the development of showers in the atmosphere, a theory which is not yet well established. However, all reasonable versions give approximately the same result at the shower maximum. Moreover, this result agrees well with an empirical estimate based on a consideration of the total energy dissipation by shower particles in the atmosphere. (The total energy dissipation is simply proportional to the number of particles at the shower maximum, times the effective width of the peak in the longitudinal distribution function, which can be evaluated with fair accuracy from a number of experimental data.) At the altitude of Volcano Ranch, showers of the sizes under consideration are near their maximum development. Thus the uncertainty in the relation between E and N is not great, certainly much smaller than the uncertainty in the value of N. The empirical method based on the energy dissipation yields

$$E = 2 \times 10^9 \, N,$$

from which one obtains the value $E = 10^{20}$ eV quoted above. The uncertainty in the estimate of E should be essentially the same as the uncertainty in the estimate of N; i.e., N should be within a factor 2 of the value quoted above.

The astrophysical significance of the existence of cosmic-ray particles in the energy range of 10^{20} eV is well known. In a magnetic field of H gauss a particle with an energy of E eV and with a charge of Z elementary charges has a radius of curvature

$$R = \frac{E}{300ZH} \text{ cm.}$$

Simple arithmetic based on this formula shows that protons of 10^{20} eV cannot be contained and accelerated by any galactic object nor by the Galaxy as a whole. For example, assuming that the average magnetic field in the galactic disk and in the galactic halo does not exceed 10^{-5} gauss, one finds for $E = 10^{20}$, $R \geq 3 \times 10^4$ light-years, a distance comparable with radius of the galactic halo. Thus, if the particle detected by Linsley is a proton, it must be of extra-galactic origin.

If the particle were a heavy nucleus, with, say, $Z = 26$, R might be as small as 1,000 light-years, a length comparable with the thickness of the galactic disk. Even under this assumption, it would be difficult to assume galactic containment, but the argument for an extragalactic origin would not be quite as strong.

These considerations point to one of the reasons why it is of vital importance to determine the nature of the cosmic-ray particles of highest energy, which brings me to the second item that I wish to discuss.

II. The Nature of Cosmic-Ray Particles of Highest Energy

An answer to the question regarding nature of the particles responsible for the largest air showers is being actively sought by various scientists, such as those at the Lebedev Institute, at the Moscow State University, at the Institute for Nuclear Research of Tokyo, at the University of Sidney, and at MIT. One method of attacking the problem, that was used both by the Tokyo group and by Linsley and Scarsi (1962)[2] at Volcano Ranch, is based on a comparison between the electron and the μ-meson components of air showers. Mu-mesons arise mainly from the decay of charged π-mesons, while electrons arise mainly from cascade multiplication of photons produced by the decay of neutral π-mesons. Charged and neutral π-mesons are produced simultaneously in the chain of nuclear interactions initiated by the arrival of the high-energy primary particle. Thus, starting from the point where the first interaction occurs, the production rate of both μ-mesons and electrons increases rapidly at first, goes through a maximum, and then decreases rapidly again as the nucleonic cascade dies out. However, since μ-mesons are much more penetrating than electronic showers, the curves representing the total numbers of μ-mesons (N_μ) and electrons (N_e) as a function of depth differ substantially. At first, they both increase at about the same rate; after the maximum, however, the electron-curve decreases rapidly while the μ-meson-curve decreases very slowly (Fig. 3).

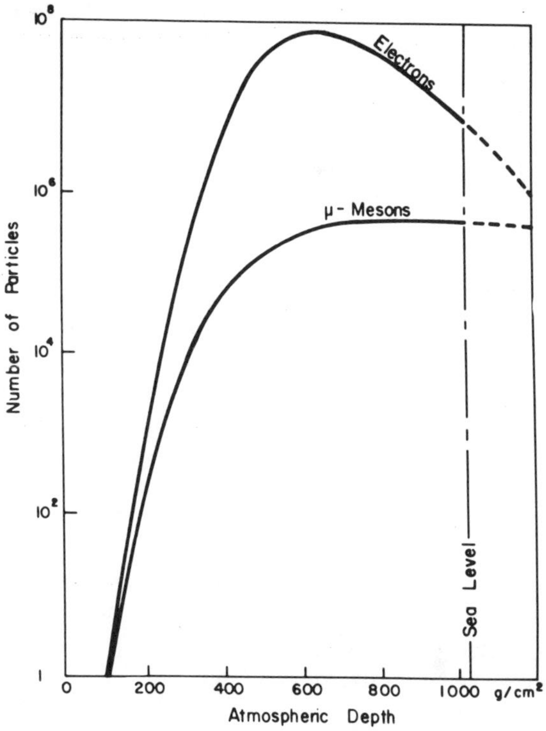

FIG. 3. Variation of the number of electrons and the number of μ-mesons with atmospheric depth, in a large shower (schematic).

Moreover, both theory and experiment indicate that, for a given depth below the starting point, the fractional number of μ-mesons, $N_\mu/(N_e + N_\mu)$, increases as the energy of the primary particle decreases. If we now consider that, for example, an iron nucleus of energy E is equivalent to 56 nucleons of energy E/56, we conclude that the shower produced by a primary iron nucleus must be richer in μ-mesons than a shower produced by a proton of the same energy (Fig. 4).

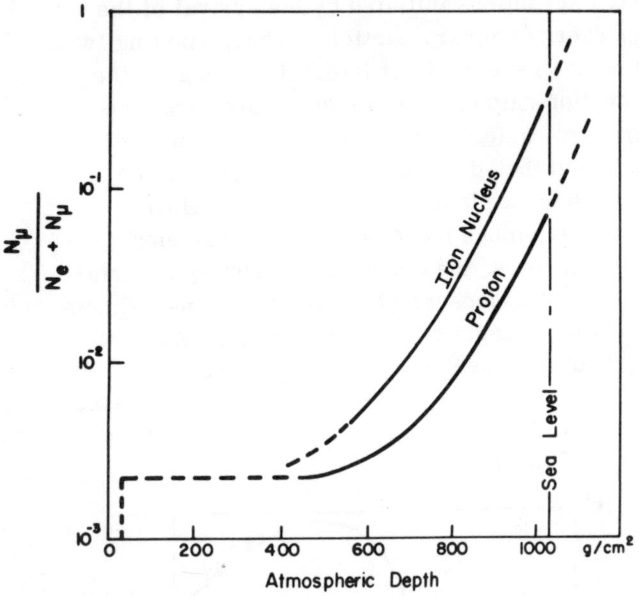

FIG. 4. Variation in the ratio between tne number of μ-mesons and the total number of particles in showers produced by a proton and by an iron nucleus of the same energy (schematic).

The experiments of the Tokyo group dealt with showers initiated by primaries in the 10^{15} - 10^{17} = eV energy range. At sea level, where the observations were made, these showers were well beyond the maximum; i.e., they were at a stage of their development where the $N_\mu/(N_e + N_\mu)$ ratio depends critically on the distance from the first interaction. Thus, one should expect fluctuations in the values of this ratio just because primary particles undergo their first interaction after traversing different thicknesses of atmosphere. The probability for the first interaction to occur between x_0 and $x_0 + dx_0$ is given by

$$\exp(-x_0/l)\ d(x_0/l),$$

where l is the mean free path. Since l is about 70 gm/cm² for protons while it is about 6 gm/cm² for iron, it is clear that the fluctuations in the depth of the first interaction, and therefore the fluctuations in the ratio $N_\mu/(N_e + N_\mu)$ due to this cause, must be much greater in the case of showers initiated by protons than in the case of showers initiated by iron nuclei.

Large fluctuations were, in fact observed; indeed showers with the same total number of particles, $N_e + N_\mu$, were found to contain a proportion of μ-mesons varying by factors near 10. It was not possible to explain these fluctuations under the assumption that the primary radiation consisted predominantly of nuclei heavier than protons. Indeed, at a given depth below the first interaction, the values of the $N_\mu/(N_e + N_\mu)$ ratio in showers produced by different heavy nuclei do not exhibit such a large spread, while the short mean free path of heavy nuclei minimizes the fluctuations related to the depth of the first interaction, as already noted. On the other hand, the experimental results could be accounted for entirely by the random distribution in the depth of the first interaction, if the primary particles were protons. The conclusion was that most cosmic-ray particles in the energy range from 10^{15} to 10^{17} eV were indeed protons; but the possibility of an appreciable admixture of heavier nuclei could not be ruled out.

The experiments of Linsley and Scarsi, dealt with showers initiated by primary particles with energies greater than 10^{17} eV. Since, at the altitude of Volcano Ranch, these showers are near their maximum development when observed in a nearly vertical direction, the $N_\mu/(N_e + N_\mu)$ ratio does not depend appreciably on the position along the shower curve; i.e., it does not depend on the depth where the first interaction occurs. Therefore, if all primary particles were of the same kind, the ratio $N_\mu/(N_e + N_\mu)$ should be a constant within the experimental errors. If, however, the primary radiation contains nuclei of different masses, then there should be a corresponding spread in the observed values of $N_\mu/(N_e + N_\mu)$.

In Figure 5 curve C represents the expected distribution of the ratio $N_\mu/(N_e + N_\mu)$ for the case that all primaries are of one kind; the spread is due entirely to experimental errors, estimated here to amount to 30 per cent. Curve B represents the corresponding distribution for the case that the primary radiation at energies greater than 10^{17} eV contains the same proportion of protons and heavier nuclei as is observed at energies smaller than 10^{12} eV.

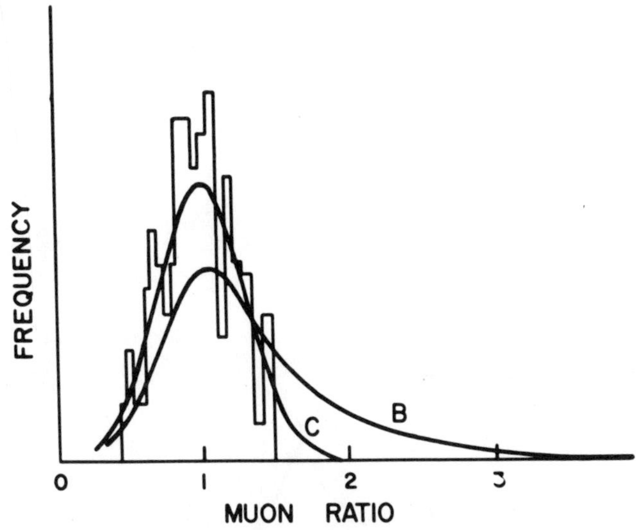

FIG. 5. Frequency distribution of the muon ratio, $N_\mu/(N_e + N_\mu)$. B: expected distribution for a primary radiation with the same mass spectrum as observed below 10^{12} eV. C: expected distribution for a primary radiation containing only one nuclear species. Experimental errors of 30 per cent are assumed. The histogram represents experimental results. The horizontal scale has been chosen so that the peaks of curves B and C correspond to the mean observed ratio, set arbitrarily equal to 1. (From Linsley and Scarsi 1962.)

The histogram represents the experimental results for nearly vertical showers obtained by Linsley and Scarsi at Volcano Ranch. The histogram agrees with curve C much better than with curve B. This means that the primary radiation at energies greater than 10^{17} eV is much "purer" than at 10^{12} eV and below. Presumably it consists almost entirely of protons. Strictly speaking, however, on the above evidence alone, it is not possible to rule out the possibility that it consists entirely of heavy nuclei, e.g., of iron nuclei.

In an attempt to distinguish between these two possibilities, Linsley and Scarsi examined a number of inclined showers, for which the effective atmospheric depth was between 1300 and 1400 gm/cm^2. We are again (as in the experiments of the Tokyo group) in the region where the $N_\mu/(N_e + N_\mu)$ ratio varies rapidly with the distance from the position of the first interaction. Thus one should expect large fluctuations if the primaries are protons, small fluctuations if the primaries are heavier nuclei.

The results of Linsley and Scarsi indicate the presence of large fluctuations and, therefore, strongly favor the view that the primaries are protons. However, this result cannot as yet be considered firmly established because of the large experimental error involved in the measurement of the $N_\mu/(N_e + N_\mu)$ ratio, particularly in the case of very inclined showers. In other words, the measured value of the $N_\mu/(N_e + N_\mu)$ ratio is a parameter only loosely related to the depth of the first interaction, and therefore does not permit an accurate determination of the mean free path of the primary particles. To some extent, the same remark applies to the experiments of the Tokyo group mentioned previously, which concerns primary particles of somewhat lower energy.

Linsley's present efforts are directed toward measuring a parameter more closely related than the $N_\mu/(N_e + N_\mu)$ ratio to the depth of the first interaction. Although no results are yet available, I wish to describe briefly the principle of the method. Consider a primary particle that undergoes its first interaction at x_0 (Fig. 6) and whose core strikes counter A. Since the shower travels through the atmosphere at nearly the speed of light, all shower particles will traverse counter A at nearly the same time. However, a counter such as B, located some distance from the core, will be traversed at different times by secondary-shower particles originating at different points along the core of the shower (which contains the

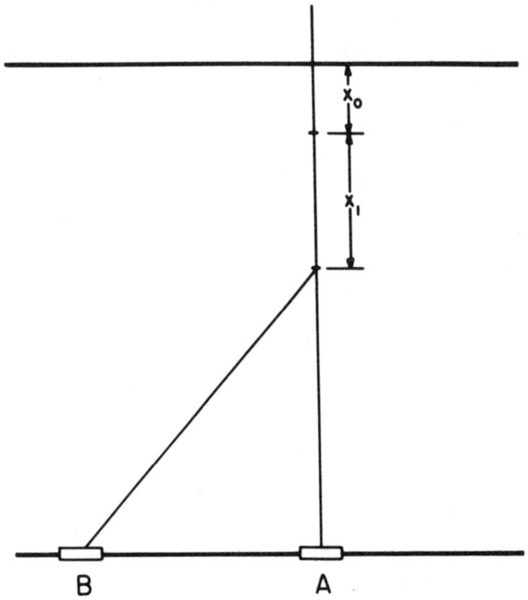

FIG. 6. Illustration of the principle of Linsley's proposed experiment.

high-energy nuclear-active component). The first particle to reach B will be presumably a μ-meson produced some distance, x_1, below the point of origin of the shower, where the production rate has reached a sufficiently high level. By measuring the time delay between the arrival times of the first shower particles at B and at A, it is possible to determine the depth $x = x_0 + x_1$. If x_1 were the same for all showers, the observed distribution in x would be identical to the distribution in x_0 because

$$\exp(-x/l)\, d(x/l) = \exp(-x_1/l)\, \exp(-x_0/l)\, d(x_0/l)$$

Actually, x_1 varies from one shower to another, but Linsley has shown by theoretical computations that the expected spread in the values of x_1 is not unduly large, and should not prevent a clear distinction between primary particles having mean free paths as different as those of protons and iron nuclei.

III. The Energy Spectrum

A third question of great astrophysical interest is the shape of the energy spectrum. If, as now appears likely, cosmic-ray particles at the low-energy end of the spectrum are mainly of galactic origin while those at the upper end of this spectrum are mainly of extragalactic origin, there should be somewhere a break indicating the transition from the galactic to the extragalactic spectrum.

The experimental data so far available are shown in Figure 7. The points corresponding to energies of 10^{15} eV or greater were obtained from the analysis of air shower data; the point at about 10^{10} eV comes from the analysis of geomagnetic effects, and the point near 10^{12} eV comes from observations with nuclear emulsion plates. The range from 10^{15} to 10^{17} eV is covered by air-shower experiments carried out by the Tokyo, MIT, and Bolivian groups at Mount Chacaltaya (5200 m altitude, 530 gm/cm atmospheric depth) and at El Alto de La Paz (4200 m altitude, 630 gm/cm^2 atmospheric depth); the range from 10^{17} to 10^{19} eV is covered by the air-shower experiments at Volcano Ranch. Between 10^{15} and 10^{18} eV the logarithmic slope of the integral energy spectrum is distinctly greater than above 10^{18} eV. It is also greater than the logarithmic slope in the energy range below 10^{15} eV. Thus it would seem that the spectrum has an inflection point somewhere between 10^{15} and 10^{18} eV. This suggests that the

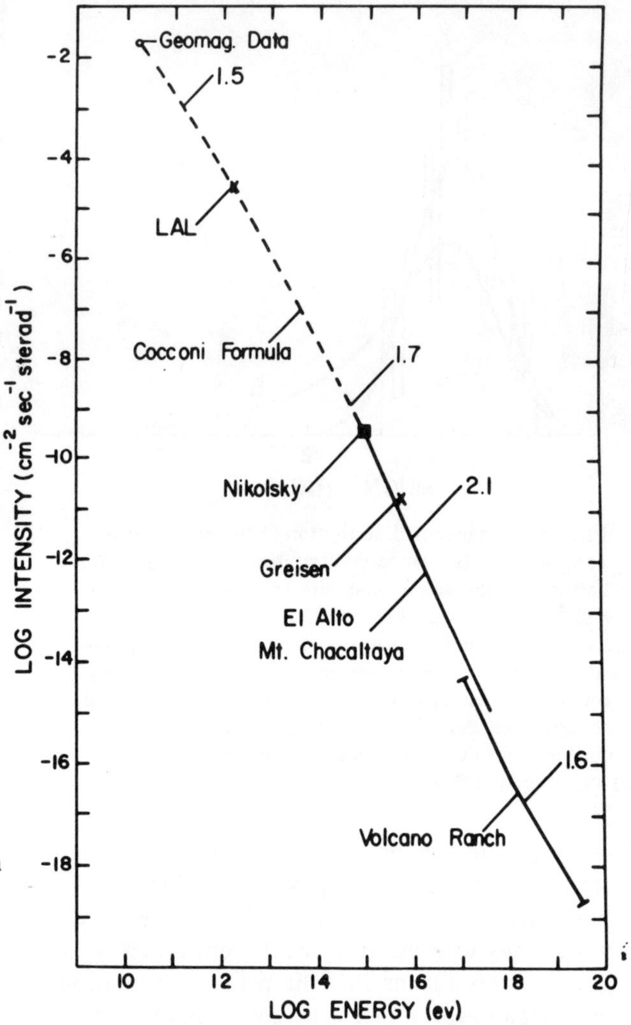

FIG. 7. The integral energy spectrum of primary cosmic rays. The numbers near the curves represent observed logarithmic slopes. (A recent analysis by G. Zatzepin of the Lebedev Institute indicates that the experimental point at 10^{12} eV is probably too high and that the spectrum has a constant logarithmic slope from about 10^{10} to about 10^{15} eV.)

spectrum of galactic cosmic rays begins to drop rapidly beyond 10^{15} or 10^{16} eV and that the extragalactic component of the radiation becomes dominant above about 10^{18} eV.

Under the assumption that the cutoff in the galactic spectrum is actually due to the inadequacy of the galactic magnetic field to contain particles with rigidity greater than some critical value, one should expect that, with increasing energy, the proton component of the galactic spectrum begins to drop before the other components do. Thus in the energy region beyond 10^{15} or 10^{16} eV, the proportion of heavy nuclei relative to protons should increase, until the extragalactic spectrum begins to dominate the picture. Extragalactic cosmic rays are presumably all protons because, as pointed out by Fugimoto and others, collisions with photons of light will effectively destroy all extragalactic nuclei of high energy. Thus, as the energy increases further, the trend will be reversed and eventually only protons will remain.

As I already mentioned, it seems that most cosmic-ray particles above 10^{17} eV are indeed protons, but I do not know of and clear evidence for the presence of a particularly large proportion of heavy nuclei in the transition region. However, the results so far obtained are not sufficiently accurate; it is hoped that experiments now under way both at Volcano Ranch, at Chacaltaya, and elsewhere will provide better information on this important question.

REFERENCES

*This work has been supported by the National Aeronautic and Space Administration and by the Atomic Energy Commission grant AT(30-1)2098.

[1] J. Linsley, 1963, Phys. Rev. Letters, **10**, 146.

[2] J. Linsley, and L. Scarsi, 1962, Phys. Rev. Letters, **9**, 123.

THE ELECTRON-POSITRON COMPONENT OF THE PRIMARY COSMIC RADIATION

Peter Meyer
Enrico Fermi Institute for Nuclear Studies and Department of Physics
University of Chicago

I. Introduction

The presence of high-energy electrons in the Galaxy was discovered more than 10 years ago when it became evident that the Galaxy is a powerful emitter of radio waves. Today this radio emission can be observed over a wide frequency region ranging from a few megacycles per second to several hundred (see, e.g., Walsh, Haddock and Schulte 1963). It was soon concluded that this radiation could not be thermal in origin because of its high intensity and the shape of its frequency spectrum. Several authors suggested at about the same time that it is synchrotron radiation produced by the helical motion of relativistic electrons around galactic magnetic fields (Alfvén and Herlofson 1950; Kiepenheuer 1950; Ginzburg 1951). This interpretation fitted the experimental data well and was generally accepted.

A direct observation of primary cosmic-ray electrons near the Earth was made much later and almost simultaneously by a cloud chamber experiment (Earl 1961) and a counter experiment (Meyer and Vogt 1961). These experiments showed that there exists a finite flux of primary electrons in the vicinity of the Earth with energies exceeding 100 MeV. Subsequently several investigations were made to study further details of the cosmic-ray electron component. The unique property of the electron component, that it can be observed within the Galaxy and specific objects through the radio emission as well as by direct observation near the Earth, makes it particularly suitable for a search of the sources which are responsible for the cosmic radiation.

II. The Origin of the Electron Component

Two distinct candidates for producing the electron component within the Galaxy have been proposed: on the one side, collision processes between energetic cosmic-ray particles and the galactic hydrogen gas; on the other side, the direct acceleration of electrons.

The collision process leads to the production of charged π-mesons which subsequently decay via μ-mesons into electrons and positrons. This process has been studied in considerable detail by a number of authors (Ginzburg and Syrovatskii 1961; Hayakawa and Okuda 1962; Jones 1963; Pollack and Fazio 1965). It has the attractive feature of lending itself to a quantitative treatment. Examples of source spectra calculated under specific assumptions for the energy spectrum of cosmic-ray protons and the cross-section for π-meson production are shown in Figures 1, a and b. This source spectrum is modified by energy-loss processes which have to be taken into account in order to

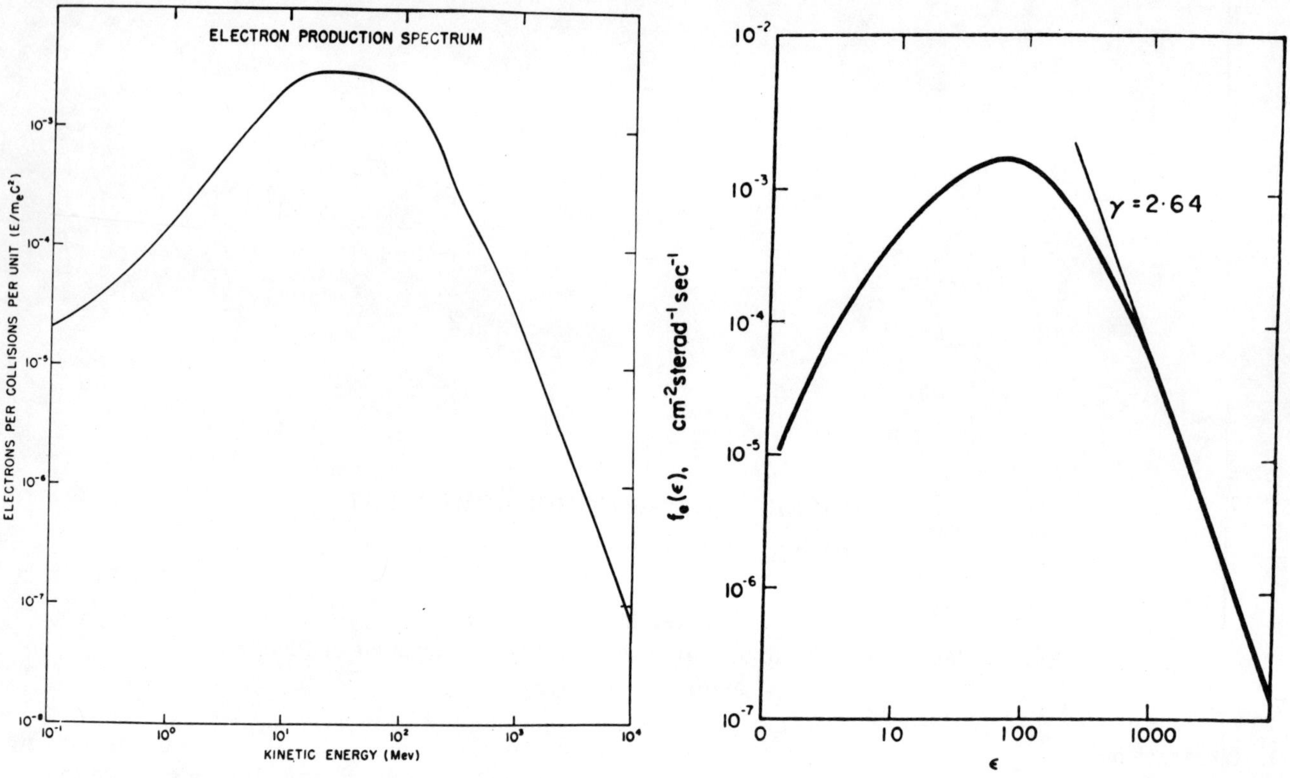

FIG. 1. The source spectrum of electrons produced in the Galaxy by proton-proton collisions (a) according to Jones (1963); (b) according to Ginzburg and Syrovatskii (1964).

arrive at an equilibrium spectrum. The important processes which lead to losses in energy are:

1. Synchrotron radiation (S)
2. Inverse Compton scattering (C)
3. Bremsstrahlung (B)
4. Ionization loss (I)
5. Leakage Loss (L).

Gould and Burbidge (1964) have calculated the rate of energy loss due to these processes in the Galaxy with the following assumptions:

Radius of the galactic halo, $R = 5 \times 10^{22}$ cm
Mean gas particle density, $\langle n \rangle = 0.03\ \text{cm}^{-3}$
Average magnetic field strength, $B = 3 \times 10^{-6}$ gauss
Electron mean free path (independent of energy), $\lambda \approx 1$ kpc.

The result of their calculation is shown in Figure 2. It is important to note that, in the energy region below 5 BeV, leakage out of the Galaxy is the dominant loss process, under the assumption chosen by the authors. This implies that the spectral shape is hardly modified by synchrotron radiation since the particles leave the Galaxy long before they lose an appreciable amount of energy through emission of synchrotron radiation. Figure 3 shows equilibrium spectra calculated by Hayakawa (1963) and by Ginzburg and Syrovatskii (1963) for collision electrons. The differences in these spectra stem mainly from different assumptions for the π-meson production cross-section and the shape of the cosmic-ray proton spectrum.

A comparison between the calculated equilibrium spectrum for collision electrons and the galactic radio emission leads to the conclusion that the electron flux is insufficient to account for the galactic radio emission (Hayakawa 1963; Ginzburg and Syrovatskii 1963) and that especially at higher energies, the electron energy spectrum is too steep to explain the frequency spectrum of the radio noise (Ginzburg and Syrovatskii 1963). Figure 4 shows, according to Ginzburg and Syrovatskii the expected radio output and frequency distribution on the basis of the calculated collision electron spectrum of Figure 3 and also a summary of the observed radio spectrum from the hemisphere around the galactic anticenter (Walsh et al. 1964). The discrepancy between the theoretical frequency spectrum and the observed spectrum is one of the reasons to search for additional sources of galactic electrons. As

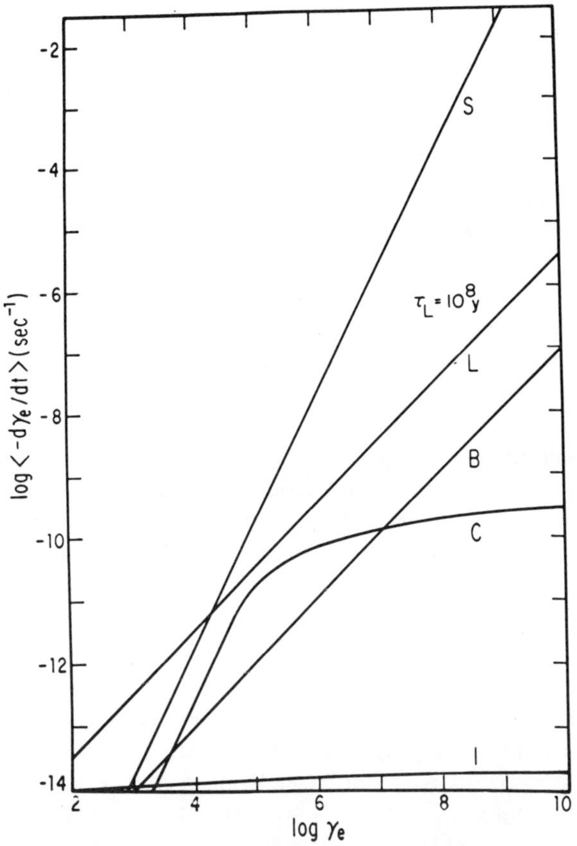

FIG. 2. The rate of energy loss of electrons in the Galaxy by various processes (Gould and Burbidge 1964). S, synchroton radiation; C, Compton scattering; B, Bremsstrahlung; I, ionization loss; L, leakage.

we shall see, the observations of electrons near the Earth also show a lower intensity than would be required to account for the radio emission.

Little quantitative information is available about directly accelerated electrons. The most conspicuous sources of such electrons within the Galaxy are the shells of supernova remnants which have been shown by optical and radio astronomical observations to contain high fluxes of relativistic electrons. The rate at which these electrons are released into the Galaxy is unknown, but one may assume that electrons of supernova origin are mainly negatively changed.

An important test on the origin of the galactic electrons is, therefore, a determination of the charge ratio of the electronic component. The collision model of the electron origin leads to a definite prediction for the electron-positron ratio. Using the cross-section and multiplicity of pion production as a function of energy, one may compute the electron-positron ratio as a function of energy. Such calculations were carried out by Hayakawa (1963; Hayakawa and Okuda 1962) and his co-workers, and more

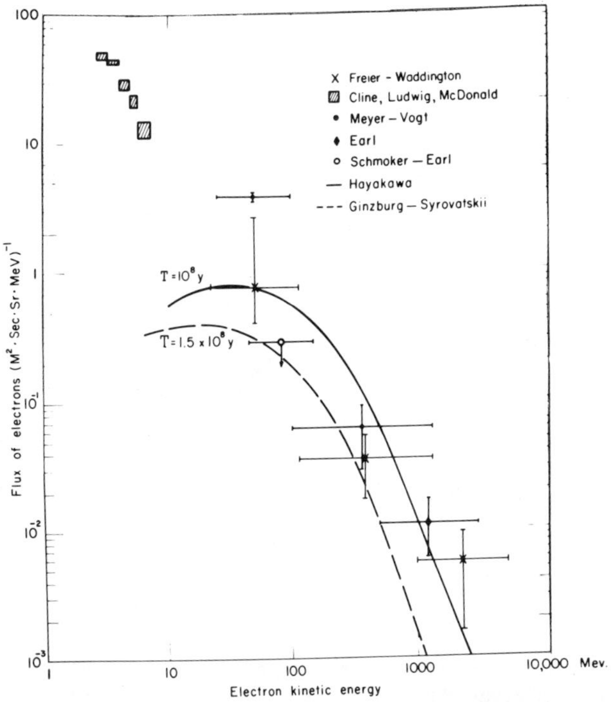

FIG. 3. Equilibrium spectra of electrons produced by collision processes in the Galaxy (Ginzburg and Syrovatskii 1964; Hayakawa 1963) and experimental results on the differential electron energy spectrum near the Earth (Earl 1961; Meyer and Vogt 1961; Cline et al., 1964; Schmocker and Earl 1965; Freier and Waddington, unpublished).

recently by Pollack and Fazio (1965). The preditions lead to a substantial positive excess in the energy region which may be covered by experiments. For example, between 100 MeV and 1 BeV, one obtains for the fraction of positrons

$$\frac{N^+}{N^+ + N^-} \approx 0.65.$$

III. The Experimental Evidence Concerning the Primary Electron Component

The flux and the energy spectrum of the primary electrons are not very well established as yet. Figure 3 shows a summary of the data which are available today. At very low energies measurements have recently been made by Cline, Ludwig, and McDonald (1964) on the Explorer 18 satellite (Ginzburg and Syrovatskii 1964). The remaining data were obtained by three different methods: (1) cloud chamber observations by Earl (1961) and J. W Schmoker and Earl (1965); (2) counter measurements by Meyer and Vogt (1961); and (3) the recent observation in emulsions by P. Freier and J. Waddington (private communication). Except for the data of Schmoker and Earl, no attempt has

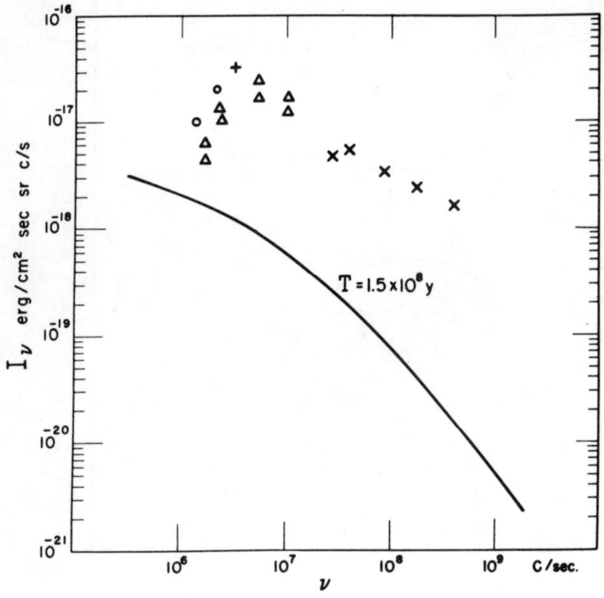

FIG. 4. The observed average cosmic radio noise frequency spectrum (Walsh et al. 1964) and the spectrum expected from collision electrons (Ginzburg and Syrovatskii 1964).

as yet been made to correct for secondary electrons produced in the atmosphere. The observation of Freier and Waddington was made at extremely high altitude (2 gm/cm^2 of residual atmosphere), and the measurements of Meyer and Vogt under 3 to 4 gm/cm^2. Except for the data below 10 MeV there are large errors in each measurement. In addition to the points shown in the differential spectrum of Figure 3 there exists a measurement of the integral flux of electrons with energies above 4.5 BeV by Agrinier et al. (1964). These authors obtain ($6.6^{+2.8}_{-1.7}$) electrons/m^2 sec^{-1} sr. More detailed measurements of the energy spectrum in the range from 100 MeV to 2.5 BeV have been made in 1964 and the results are forthcoming (L'Heureux and Meyer 1965). The new measurements will considerably improve the accuracy in the experimental determination of the electron flux and energy spectrum. It is already clear from Figures 3 and 4, however, that the flux of electrons observed near the Earth is smaller than expected from radio observations. This difference does not depend upon the nature of the source of electrons. It is unlikely that solar modulation, especially at higher energies, is sufficient to explain this difference.

The origin of the electron component may be investigated by determining its charge composition. A successful experiment in this direction was first carried out by DeShong, Hildebrand, and Meyer (1964). It was made at balloon altitude from Fort Churchill, and employs a permanent magnet for charge separation and a spark-chamber telescope for the determination of the particle trajectory before and after passing through the magnetic field. The identification of electrons is achieved by observing the electron-photon shower produced in a spark chamber which contains about 5 radiation lengths of high Z material. The results of this experiment clearly indicated a substantial excess of negative electrons. The fractions of positrons obtained under 3 and 5 gm/cm^2 of atmosphere are summarized in Table I. It should be noted that these fractions are upper limits, since the electrons produced in the atmosphere above the equipment will show a positive excess. A determination of the actual flux of primary positrons will be possible after the more accurate determination of the absolute electron flux becomes available. The electron-positron experiment has been continued in 1964 by Hartman, Hildebrand, and Meyer (1965) with improved apparatus. Although the time of exposure at altitude was very limited it was possible to confirm—with less statistical accuracy—the 1963 results and to extend the energy range of the observation to 3 BeV. The results are also shown in Table I. These data again are upper limits for the fraction of positrons in the primary radiation. The extension of the observation to 3 BeV is important since a substantial solar contribution to the electron flux at the high energies is unlikely.

The conclusions that we may draw from the observation of the charge composition of the primary electron component are the following:

1. Directly accelerated electrons are dominant in the primary electron component.
2. The relative role of electrons produced by collisions in the Galaxy is as yet undetermined but is certainly less, possibly much less than that of directly accelerated electrons.
3. Supernova remnants are a likely candidate for the production of the electron component.

IV. Summary and Conclusions

The study of primary cosmic-ray electrons has opened a new approach to the investigation of the origin of cosmic rays. Experiments on the charge ratio have shown that electrons are directly accelerated in galactic sources. Although the knowledge of the primary flux and energy spectrum of

Table I. THE FRACTION OF POSITRONS FOR VARIOUS ENERGY INTERVALS

Time	Av. Atmosp. Depth (gm/cm^2)	Energy Interval (MeV)	$N^+/(N^+ + N^-)$
July-August, 1963	4	50 - 100*	0.31 ± 0.12
		100 - 300*	.38 ± .07
		300 - 1000*	.16 ± .04
August 5, 1964	6.4	40 - 100	.79 ± .40
		100 - 300	.43 ± .21
		300 - 1000	.30 ± .14
		1000 - 3000	.33 ± .16

*The energy intervals in the 1963 experiment are not very accurate. Energy losses in the liquid Cerenkov counter used at that time will lead to an underestimate of the average electron energy.

Source: Hartman, Hildebrand, and Meyer 1965; Baldwin 1963.

electrons near the Earth is still quite incomplete, it appears that this flux is too small to account for the intensity of the galactic non-thermal radio emission, independently of what the source of the electrons might be. It is unlikely, at least at energies around 1 BeV and above, that solar modulation, during the years near solar minimum, may be made responsible for this discrepancy. The question arises whether the contribution of galactic electrons to the radio intensity is properly estimated, a question which was already raised earlier by Baldwin (1963). In the coming years it will be possible to study the solar modulation of the electron component and the changes in the energy spectrum in detail. Such observations will be necessary to more accurately determine the flux and energy spectrum of electrons in the Galaxy and to understand its relation to the radio observations. They will also be important in further studying the nature of the solar modulation mechanism. An accurate determination of the electron-positron ratio as a function of energy will be needed to clarify the contribution of electrons produced by collisions in the Galaxy. The predominantly negative charge of the electron component points strongly to a galactic rather than metagalactic origin of the electrons.

I am indebted to several authors for making available their data prior to publication.

REFERENCES

Agrinier, Koechlin, Parlier, Boella, DegliAntoni, Dilworth, Scarci, and Sironi, 1964, Phys. Rev. Letters, 13, 377.

H. Alfvén and N. Herlofson, 1950, Phys. Rev., **78**, 616.

J. E. Baldwin, 1963, Observatory, **83**, 153.

T. L. Cline, G. H. Ludwig, and F. B. McDonald, 1964, Phys. Rev. Letters, **13**, 786.

J. Deshong, R. H. Hildebrand, and P. Meyer, 1964, Phys. Rev. Letters, **12**, 3.

J. Earl, 1961, Phys. Rev. Letters, **6**, 125.

V. L. Ginzburg, and S. I. Syrovatskii, 1964, The Origin of Cosmic Rays (London: Pergamon Press).

V. L. Ginzburg, 1951, Dok. Akad. Nauk SSSR, **76**, 377.

V. L. Ginzburg, and S. I. Syrovatskii, 1961, Progr. Theoret. Phys. Suppl. (Japan), **20**, 1.

R. J. Gould, and G. R. Burbidge, 1964, I.A.U. Symposium on Astronomical Observations from Space Vehicles, Liege (preprint).

R. Hartman, R. H. Hildebrand and P. Meyer, 1965 (in press).

S. Hayakawa, 1963, Proc. Internat. Conference on Cosmic Rays, Jaipur (preprint).

S. Hayakawa and H. Okuda, 1962, Progr. Theoret. Phys. (Japan), 28, 517.

F. C. Jones, 1963, J. Geophys. Res., **68**, 4399.

R. O. Kiepenheuer, 1950, Phys. Rev., **79**, 738.

J. L'Heureux and P. Meyer, 1965 (in press).

P. Meyer and R. Vogt, 1961, Phys. Rev. Letters, **6**, 193.

J. B. Pollack and G. G. Fazio, 1965, Ap. J. (in press).

J. W. Schmocker and J. A. Earl, 1965 (in press).

D. Walsh, F. T. Haddock and H. F. Schulte, 1964, Proc. Fourth Internat. Space Sci. Symposium (Warsaw, 1963) (New York: John Wiley & Sons), p. 935.

D. C. Morton, 1964a, Nature, 201, 1308.

______. 1964b, Ap. J., 140, 460.

I. S. Shklovsky, 1964a, "On the Energy Spectrum of Relativistic Electrons in the Crab Nebula" (preprint).

______. 1964b, Astr. Tsirk., No. 304.

L. Woltjer, 1964, "X-Rays and Type I Supernova Remnants" (preprint).

DETECTION OF INTERPLANETARY 3- TO 12-MEV ELECTRONS

T. L. Cline, G. H. Ludwig, and F. B. McDonald
NASA-Goddard Space Flight Center, Greenbelt, Maryland

We report here the direct observation of interplanetary electrons of energy above 3 MeV with the IMP-1 satellite (Explorer 18).

Electrons observed in the primary interplanetary radiation in the BeV energy region by Earl (1961) and in the 200-MeV energy region by Meyer and Vogt (1961a) are believed to be of galactic origin because their energies are as high as those assumed to be necessary for their penetration into the inner solar system and because their measured intensity agrees with that which was anticipated to account for galactic radio emission. Support was lent to this hypothesis when the modulation characteristics of these particles were observed (Meyer and Vogt 1961b) to be similar to those of cosmic-ray protons and their positron-to-electron ratio was found (DeShong, Hildebrand, and Meyer 1964) to be compatible with an origin of at least half of them in meson-producing cosmic-ray interactions in the interstellar medium. We feel that the existence of an interplanetary flux of electrons lower in energy by orders of magnitude is interesting because of the possibility that these too may have a cosmic origin. If so, their study should yield entirely new information about the galactic electron sources and modulation characteristics. If they are of solar origin, there are analogous implications. We wish to demonstrate here that the flux of lower-energy electrons we observe is indeed a primary component of the interplanetary radiation, and to discuss its properties in terms of its possible origin, either galactic or solar.

The observations reported here were made with a scintillator telescope on Explorer 18, a satellite placed in an elliptic orbit with an apogee height of 193000 km. Data were taken from the launch, on November 27, 1963, until May 6, 1964, when the satellite passed into a long period in the Earth's shadow, causing failure of the detector. During this time interval the apogee moved from the sunlit side of the Earth beyond the magnetosphere (terminating at about 70,000 km) and beyond the Earth's shock front (observed with a magnetometer [Ness, Scearee, and Seek 1964] and plasma sensor [Bridge, Egidi, Lazarus, Lyon, and Jacobson 1964] at about 100,000 km) to the region behind the Earth and inside the shock front. Electron data taken only when the satellite was beyond 125,000 km are reported here; throughout the life of the instrument these data continued to be free from effects due to the trapped radiation.

The detector was developed (Bryant, Ludwig, and McDonald 1962) to study low-energy cosmic-ray protons, electrons, and light nuclei. It is composed of three scintillators: two in coincidence, measuring energy loss and total energy, and a guard counter in anticoincidence. When a table of intensity versus measured energy loss versus measured total energy is constructed from data taken at apogee, a distinct counting-rate component of mimimum-ionizing energy loss and of low apparent energy is seen. An analysis of the topology of distributions in energy loss and in total energy through this minimum-ionizing component indicates that indeed it is composed of three distinct particle groups: One group with total energies corresponding to electrons that stop within the detector, a much smaller group with a high apparent total energy equal to or exceeding the energy loss of a minimum-ionizing cosmic ray traversing the detector, and a third group with very low total energies. We believe that the latter two components are surely secondary radiations composed of, respectively, cosmic rays that avoid detection by the guard counter (by, e.g., turning into neutrals through reactions within the detector) and gamma rays made locally in the spacecraft, producing random and coherent coincidences between the energy loss and total energy detectors.

These secondary effects were eliminated to produce the left-hand side of Fig. 1, which shows the energy spectrum of electrons obtained during the first orbit (November 27-30, 1963) at a time when the observed electron intensity was at a typical minimum and when there were no measurable time variations. The right-hand side of Fig. 1 shows, for comparison, a spectrum of the difference between the first statistically significant intensity increase (January 13-16, 1964) and the immediately preceding intensity (January 9-12). No background corrections were necessary to produce the latter distribution since the electron intensity increase was unaccompanied by an increase of either secondary gamma rays or spurious cosmic rays; it was therefore possible to determine the intensity to higher energies. The nearly identical shapes of the

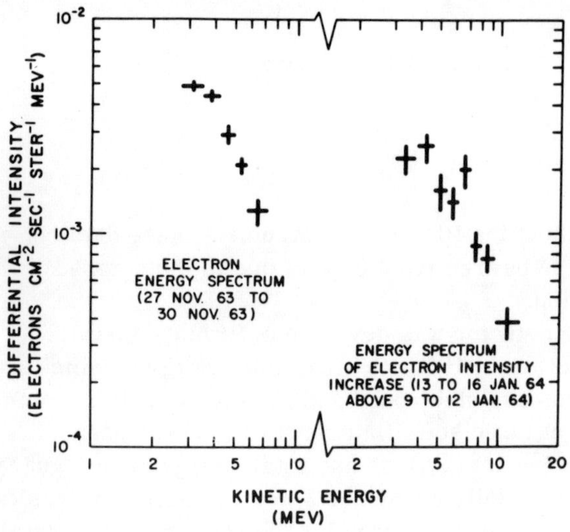

FIG. 1. Differential energy spectra of electrons observed beyond 125000 km for the Earth. The first spectrum is from the apogee of the first orbit; the second is the difference between measurements from the 13th and 12th orbits and indicates the first significant increase in intensity.

two corrected spectra suggest that the electons seen daily may have the same origin as the extra ones seen on days of increased electron flux. The integral intensity of electrons of energy between 2.7 and 7.5 MeV is 210 ± 10 electrons/m^2 sec. ster, and that of the increase between 3 and 12.5 MeV an additional 100 ± 10 electrons/m^2 sec. ster.

To demonstrate that most of the observed electrons are not of local or secondary origin at the satellite (e.g., such as knock-on or cascade-shower electrons produced in or near the detector) we consider their time variations. Figure 2 shows the counting rate of these electrons, partially corrected for slow gain drifts in the detector, plotted in the form of one-quarter-orbit averages throughout the active life of the instrument. (The gaps in the data occur at times when the satellite is within 125,000 km; the other three points per orbit are plotted so that each center one represents data taken from beyond 185,000 km.) Also shown are a comparison plot "C" of the integrated cosmic-ray flux into a scintillator with about 0.3 gm cm^{-2} shielding, and the times of a recurrent minimum in the interplanetary magnetic activity index K_p with a period of one solar rotation.

A dominant feature of the electron rate is the appearance of many statistically meaningful intensity increases, including one series apparently coincident with the recurrent K_p minimum. These electron intensity increases were not accompanied by comparable increases in the integral cosmic-ray intensity above 15 MeV: the magnitude of the electron modulation is 50 per cent on occasions, while the cosmic rays undergo modulations of less than 5 per cent. Further, following the flare of March 16, 1964, there was a solar-proton event, accompanied by Type IV solar radio emission (A. Maxwell, private communication), during which the flux of protons of energy between 15 and 75 MeV briefly increased by several orders of magnitude, while the 3- to 8-MeV electron flux rose less than 50 ± 25 per cent. (Figure 2 shows the quarter-orbit average of the total integrated cosmic-ray flux increasing at that time by about 10 per cent.) These comparisons demonstrate that, at most, an insignificant fraction of the electron modulation results from modulations of cosmic rays of energy above 15 MeV.

Modulations of protons with energies below 15 MeV, such as 27-day recurrent solar proton events similar to those observed (Bryant, Cline, Desai, and McDonald 1963) with Explorer 12, were not monitored with our apparatus; but these would be expected at the times of recurrent Forbush decreases and geomagnetic activity, rather than at the time of our repeating electron increases. Several such 1- to 10-MeV proton intensity increases were observed early in the life of the satellite by Fan, Gloeckler, and Simpson (1964), but these were about 2 weeks out of phase with our electron enhancements and appear to be accompanied by, if anything, decreases in the electron intensity and in the galactic cosmic rays.

Finally, a study of 3-hour averages of the observed intensity of these electrons indicates no variation with distance from the Earth, either during orbits of minimum intensity or during times of increased intensity; the electron rate is constant through the shock front to a distance of up to 50 per cent beyond it. Further, the satellite's passage through the wake of the moon (Ness et al. 1964) was unaccompanied by an electron intensity variation. Thus, these electrons are not secondary to cosmic rays or solar protons or due to geophysical processes.

We feel that the question of whether these primary electrons originate at the Sun or in the Galaxy cannot be definitely answered on the basis of the available data; however, the following properties of these electrons are consistent with their being galactic. First, the differential energy spectrum of this 3- to 12-MeV component fits smoothly onto a spectral plot

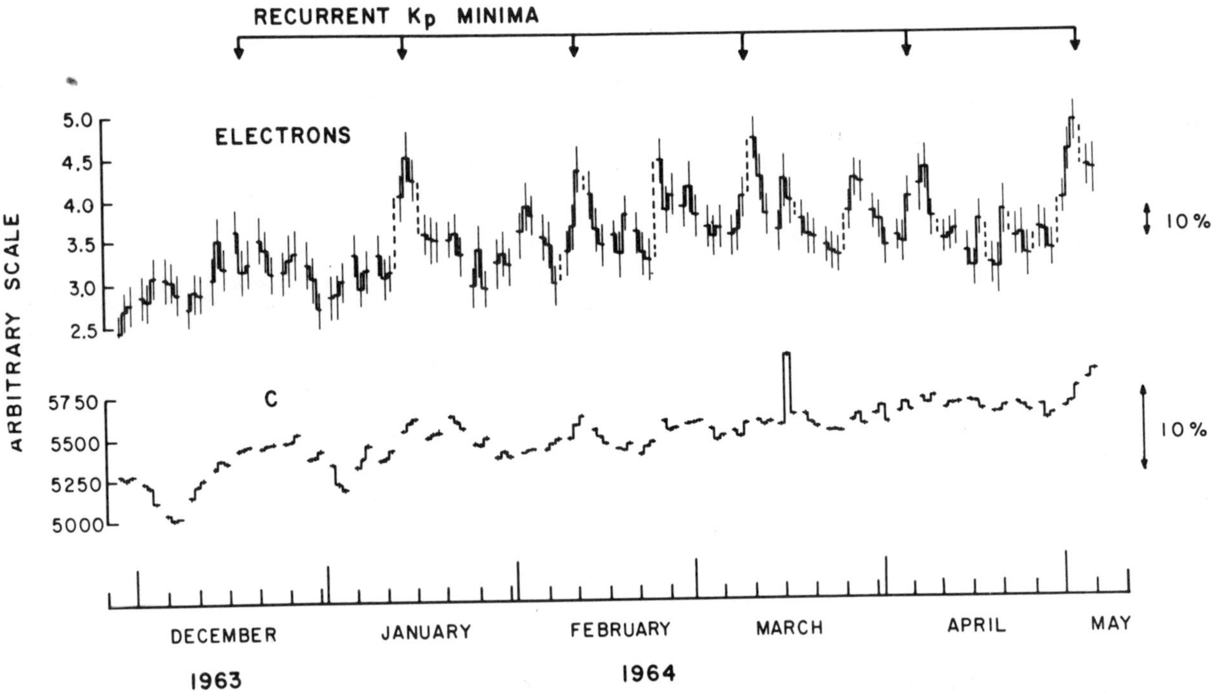

FIG. 2. Integral counting rate of electrons throughout the active life of the instrument plotted in quarter-orbit averages. The counting rate "C" of cosmic rays into a thinly shielded scintillator and the times of one recurrent minimum of the interplanetary index K_p are also shown. Recurrent Forbush decreases are seen in the cosmic rays in early December and January, and a small solar proton event occurs in March; other increases can be largely attributed to the electron mixture in the cosmic rays.

of the cosmic-ray electron intensities (Earl 1961; Meyer and Vogt 1961a, b) at much higher energies. Second, the time variations of the electrons can be compared to those of cosmic rays in that there is a strong correlation between the electron intensity increases and quiet interplanetary conditions, evidenced by K_p minima and very small sea-level cosmic-ray intensity increases. Third, there appears to be a long-term increase of electron intensity after a correction of the same order is applied for a slow, monitored drift in detector gain; if this increase is real, it is similar to the 11-year modulation of cosmic rays as solar minimum is approached. However, the fact that the differential cosmic-ray proton intensity is peaked at about 1 BV/c rigidity and negligible below 150 MV/c markedly contrasts with the fact that electrons of rigidity $\gtrsim$ 3.5 MV/c are more abundant than those of greater rigidity. Parker (1964) has recently pointed out that particles with gyroradius close to the idealized irregularity scale of the modulating medium should be deflected more than those of either extreme; thus these electrons of low rigidity might originate in the Galaxy and penetrate the solar system as easily as those of great rigidity.

In spite of the foregoing arguments for galactic origin, it is not impossible that the electrons came from the Sun. Several possibilities present themselves. For example, relativistic electrons might be generated over most of the upper surface of the solar atmosphere, in which case regions of enhanced and expanded plasma (which contain recurrent proton fluxes) would tend to contain fewer electrons while regions of quiet-time streaming would contain more, as we have observed. Further, the deceleration of the electrons in the enhanced plasma might be much greater than that in the quiet-time streaming. Alternatively, the electrons might be associated with the development of new sunspot regions, which is a characteristic of this phase of the solar cycle and appears to correlate weakly with the observed pattern of intensity increases (H. Dodson and R. Hedeman, private communication). We have not, however, found a correlation with any solar radio or optical activity.

The results we quote here are preliminary: an evaluation of the detector response, providing a more exact spectrum, and a detailed investigation of the time variations will be given elsewhere. We are happy to acknowledge the efforts of the many people who made the IMP-1 a success.

REFERENCES

H. S. Bridge, A. Egidi, A. Lazarus, E. F. Lyon, and L. Jacobson, 1964, *Space Research* 5 (Amsterdam: North-Holland Publishing Co.) (COSPAR Proceedings, Florence, 1964).

D. A. Bryant, T. L. Cline, U. D. Desai, and F. B. McDonald, 1963, *Phys. Rev. Letters*, 11, 144.

D. A. Bryant, G. H. Ludwig, and F. B. McDonald, 1962, I.R.E. *Transactions on Nuclear Science*, **NS-9**, 376.

J. A. DeShong, Jr., R. H. Hildebrand, and P. Meyer, 1964, Phys. Rev. Letters, **12**, 3.

J. A. Earl, 1961, *Phys. Rev. Letters*, **6**, 125.

C. Y. Fan, G. Gloeckler, and J. A. Simpson, 1964. Goddard *IMP Symposium*, March, 1964.

P. Meyer and R. Vogt, 1961a, *Phys. Rev. Letters*, **6**, 193. 1961b, *J. Geophys. Res.* **66**, 3950.

N. F. Ness, C. S. Scearee, and J. B. Seek, 1964, *Space Research* 5 (Amsterdam: North-Holland Publishing Co.) (COSPAR Proceedings, Florence, 1964).

E. N. Parker, 1964, *J. Geophys. Res.*, **69**, 1755.

ORIGIN OF COSMIC RAYS

G. Cocconi
CERN, Geneva

I shall limit my talk to few comments about the results presented by Professor Bruno Rossi.

The M.I.T. demonstration that charged cosmic rays exist with energies as great as 10^{19} - 10^{20} eV (the last ones reaching the Earth at the rate of one per year over a surface of 10 km^2) and that they arrive to us, as far as it can be ascertained, isotropically is a further proof that at least the most energetic band of cosmic radiation comes from outside the Galaxy. The galactic magnetic fields are in fact insufficient to bend these particles appreciably, and consequently the particles can reach us from the intergalactic spaces.

Actually, the number of 10^{20} eV particles observed by the M.I.T. experimenters is larger by a factor of about 10 than that expected by extrapolating at larger energies the data previously available. Is this an indication that the hardest cosmic rays have an energy distribution flatter than that of the less energetic ones?

While only a few years ago the evidence in favor of the extragalactic origin of the most energetic band of cosmic radiation was either ignored or dismissed by the majority of the conservative scientists, here we have faced the opposite situation. In fact, Hoyle and Burbidge were inclined to think that not only the most energetic (and consequently very rare) cosmic rays, but also those that arrive on the Earth with much smaller energies, constituting the bulk of the cosmic radiation, are of extragalactic origin. In their opinion, all the Universe could be permeated by a nearly uniform cosmic-ray flux. The extragalactic sources of high-energy particles, they argue, are powerful enough to supply all the necessary energy.

In my opinion this point of view does not take into account the fact that, for the less energetic cosmic rays, the storage factor of the galactic magnetic fields is very great, and since we know that within the Galaxy sources exist (the supernovae) which can produce these particles, the density within the Galaxy of the galactic particles is strongly enhanced. I consequently agree with those who think that the bulk of cosmic radiation is galactic, while the particles of energy above, say, 10^{15} - 10^{16} eV, are predominantly extragalactic. The importance of the magnetic storage factor in the neighborhood of the sources is illustrated by the fact that even a source as weak as the Sun can temporarily be predominant and occasionally modulate the overall cosmic-ray flux observed at the Earth.

Apart from these dissensions in the interpretation of the data, the fact remains that the observation of 10^{20} eV cosmic rays demonstrates beyond any reasonable doubt that not only electromagnetic radiations, but also matter, reach us from outside the Galaxy, where these fantastic energies and possibly even much larger ones can be produced.

PART 4

X- AND γ - RAY ASTRONOMY

RECENT OBSERVATIONS ON COSMIC X-RAYS

R. Giacconi, H. Gursky and J. R. Waters
American Science and Engineering, Inc., Cambridge, Massachusetts

B. B. Rossi, G. W. Clark, G. Garmire, M. Oda and M. Wada
Massachusetts Institute of Technology, Cambridge, Massachusetts

Introduction

The first evidence for the existence of cosmic X-rays was obtained by our group during a rocket flight in June, 1962 (Giacconi et al. 1962). We detected a very strong X-ray source in the vicinity of the galactic center and a weaker one in the direction of Cygnus. In two successive flights in 1962 and 1963 (Gursky et al. 1963) we confirmed the existence of the two sources. The stronger one appeared to be close to, but not coincident with, the galactic radio center. We assigned the Cygnus source a celestial location between +10° and +50° declination and between 20 h and 23 h R. A. In addition we detected an apparently isotropic background of 10^{-8} erg $cm^{-2} ster^{-1} sec^{-1}$.

The group at the U.S. Naval Reserach Laboratory (NRL) in April, 1963, confirmed the existence of the main source and assigned a location of $16^h 15^m$ R. A. and -15° declination (Bowyer et al. 1964a). In addition, the NRL group clearly established the existence of a source in Taurus. The lunar occultation experiment of the Crab Nebula conducted by NRL, in July, 1964 (Bowyer et al. 1964b) measured the angular dimension of the X ray source to be about 1'.

In a recent rocket flight (August, 1964) we observed a new source in the region of the galactic equator (Giacconi et al. 1964). The mechanical collimators used during this flight defined a rectangular field of view of 15° × 20° at full-width half-maximum. This field of view was sufficiently small to clearly separate the sources in Scorpio and the galactic equator but did not allow a detailed study of the structure of these sources.

In a more recent flight (October, 1964) we have observed the same region with detectors with much finer angular resolution. These measurements have revealed a complex structure of the source along the galactic equator. Two distinct sources have been detected in this region and the existence of additional ones is indicated.

In both of these flights we have also performed a measurement of the angular size of the Scorpio source by use of a novel type of modulation collimator developed by one of us (Oda 1965). The measurement yields an upper limit of about 7'. for Scorpio.

Structure of the Region in the Vicinity of the Galactic Equator

The rocket payload flown on October 26, 1964 (Fig. 1) included several detectors of different types which extended the range of observable wavelength from 0.5 to 15 Å. We have fully analyzed only the data from three banks of argon-filled Geiger counters with 0.002-inch beryllium windows. The counters were sensitive in the region from 1 to 9 Å. The effective area of each bank was about 70 cm^2. Rectangular fields of view in the form of narrow slits were defined by mechanical collimators. The disposition of these counters about the rocket spin axis is shown in Fig. 2, where the dimensions of the fields of view are also given. The spin rate of the rocket during this flight was a 2.2 rps and the yaw cone was 5.0° full width.

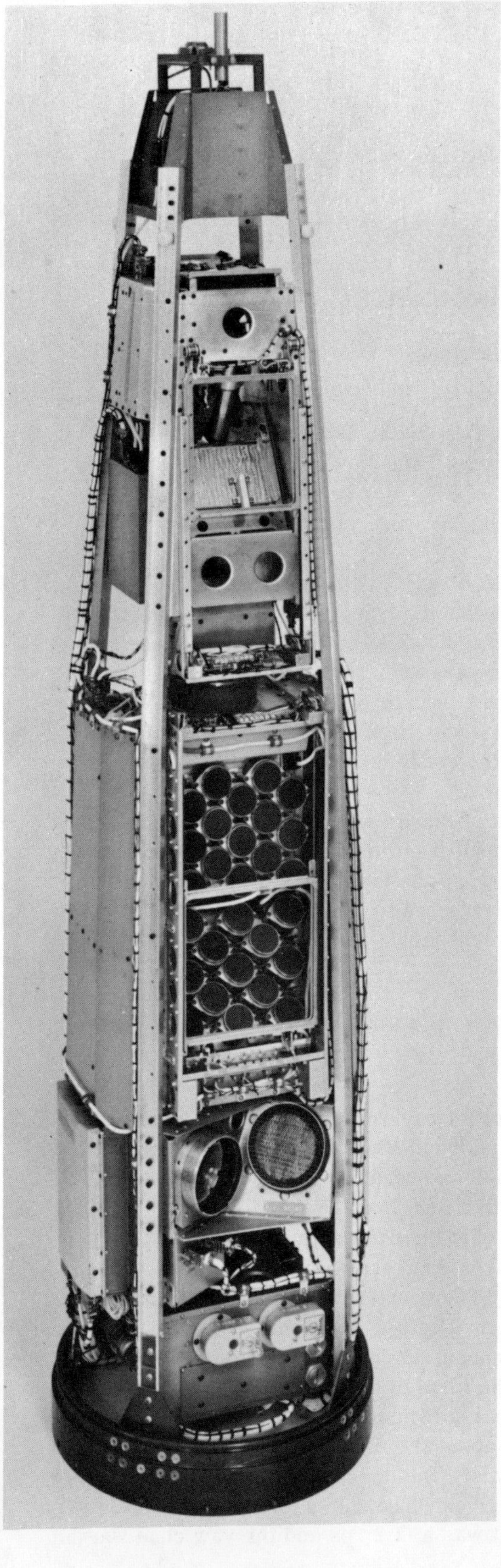

FIG. 1. Experiment payload flown by the ASE-MIT group in October, 1964, on an Aerobee rocket.

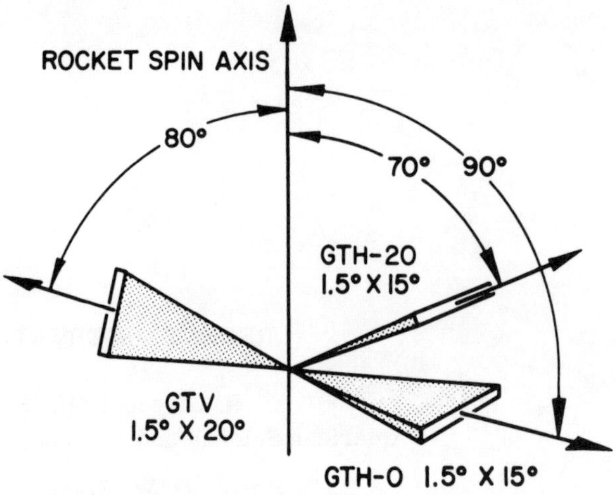

FIG. 2. Disposition of the detectors, fields of view with respect to the rocket spin axis.

Figure 3 shows the region of the sky near the galactic center scanned by the vertical slit counter (GTV) during a single rocket spin. The shadowed area represents a projection of the full field of view of the counter (3° × 40°) on the sky at a given instant of time. The direction of scan is essentially perpendicular to the galactic equator. The yaw motion of the rocket caused a ± 2.5° shift in the scan region during the flight.

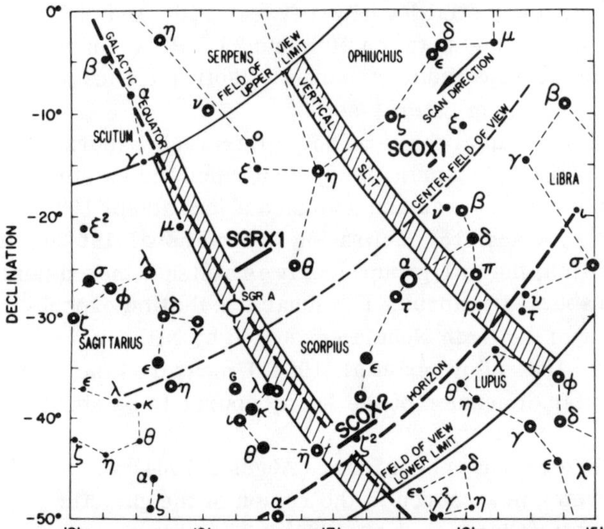

FIG. 3. Map of celestial sphere showing the region of the sky near the galactic center scanned by the vertical-slit counter (GTV).

Figure 4 shows the azimuthal distribution (rocket azimuth) of the counts received during the entire flight in about 50° azimuth. No source with intensity greater than 3 standard deviations from the background appears in the remaining interval of azimuthal angles. From these data alone one can reach a number of conclusions: (1) The angular size of Scorpio is less than 30' in the direction of scan; (2) the source along the galactic equator is either extended or composed of several point sources; (3) each of these point sources has intensity less than about $\frac{1}{10}$ Scorpio; (4) The total emission from this region is about 0.3 of Scorpio; therefore at least three point sources must exist; (5) The sources are clustered within few degrees of the galactic equator; (6) the total emission from the galactic equator as observed in the August, 1964, flight was about 0.45 Scorpio. This implies the existence of additional sources which lie outside the region explored by the vertical slit counter in the October flight.

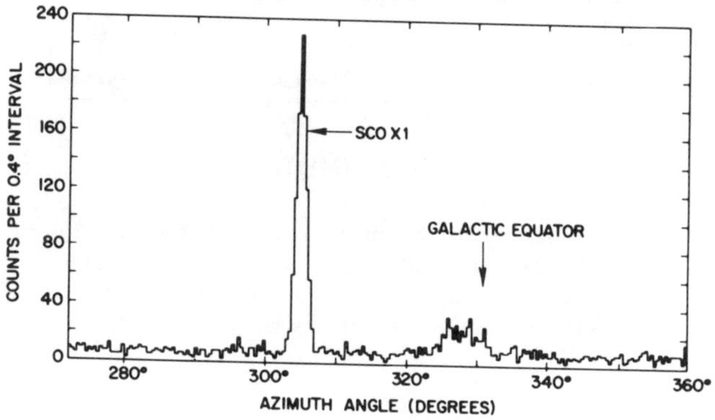

FIG. 4. Azimuthal distribution of recorded counts from the vertical slit counter flown during October, 1964 (rocket azimuth).

Figure 5 shows the regions of the sky scanned by the horizontal slit counters in the vicinity of the galactic equator. The shadowed area approximately 5° wide represents the total area scanned during a complete yaw cycle. **Figs.** 6 and 7 show the azimuthal distribution of the counts obtained in the horizontal slit counters during the entire flight in eight different portions of the yaw cone. For each of these two banks of counters the slit moves from the trace designated as "1" to the trace designated as "8" in Fig. 5 during a yaw cycle, thereby performing a scan in a direction parallel to the galactic equator.

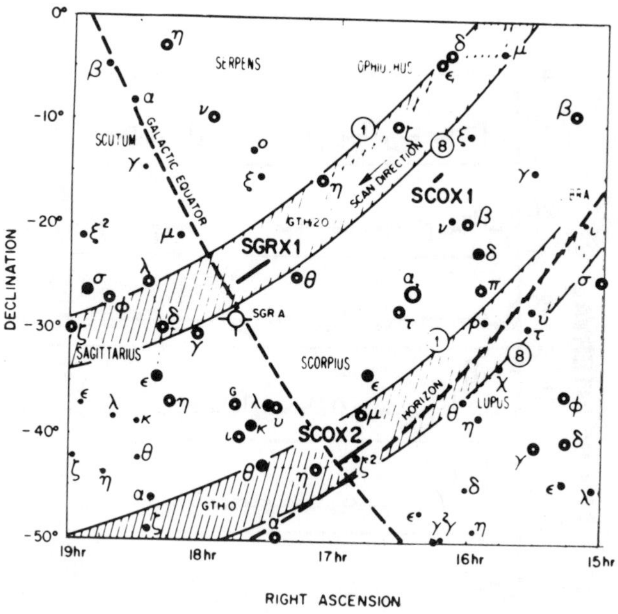

FIG. 5. Map of celestial sphere showing the region of the sky near the galactic center scanned by the horizontal slit counters (GTH-O and GTH-20). The shaded regions represent the areas scanned during the entire yaw cycle.

Figures 8 and 9 show the counting rate for the two counters averaged over a 40° interval of azimuth centered about azimuth 326° (solid dots) as a function of an angle measured along the galactic equator. The counting rate represented by the crosses in the graphs corresponds to the average level for the remainder (320°) of the azimuthal distribution. The triangular response of the detector is shown for comparison.

Figure 8 clearly shows a peak in the counting rate. The width of the peak is consistent with the measured response of the detector. We deduce the existence of a source of angular size less than $\frac{1}{2}$° in the direction of scan. The location of the source is shown in Figs. 3 and 5 and is identified as ScoX-2. This nomenclature is introduced to distinguish the new source from the one also in Scorpio at $16^h\ 15^m$ R.A., -15° declination, which will be designated as ScoX-1. (In general, we have adopted the policy of designating X-ray sources by the abbreviated name of the constellation in which they are found, the letter X, and a number designating the order of discovery.)

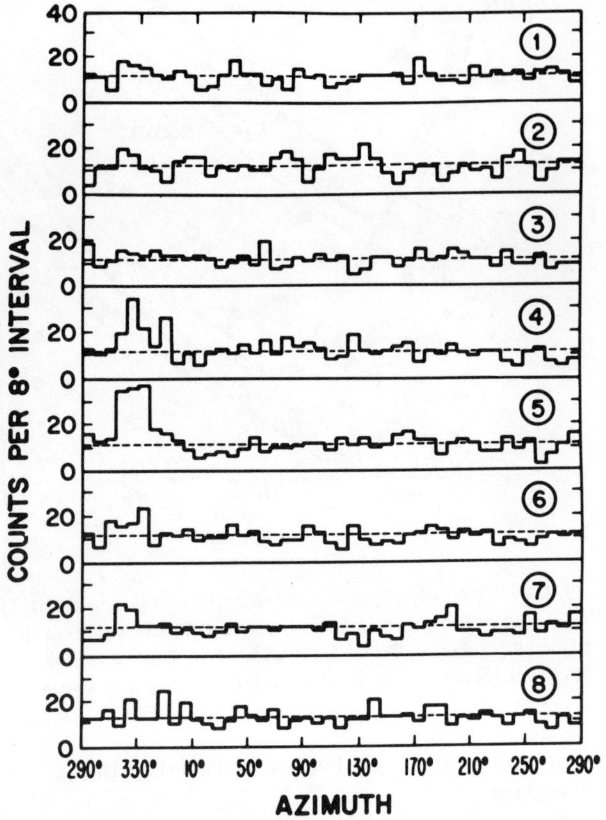

FIG. 6. Azimuthal distribution of counts for GTH-O in eight different intervals of yaw phase. The normal of the detector during the yaw phases designated as "1" and "8" traced the lines in the sky shown in Fig. 5 as "1" and "8."

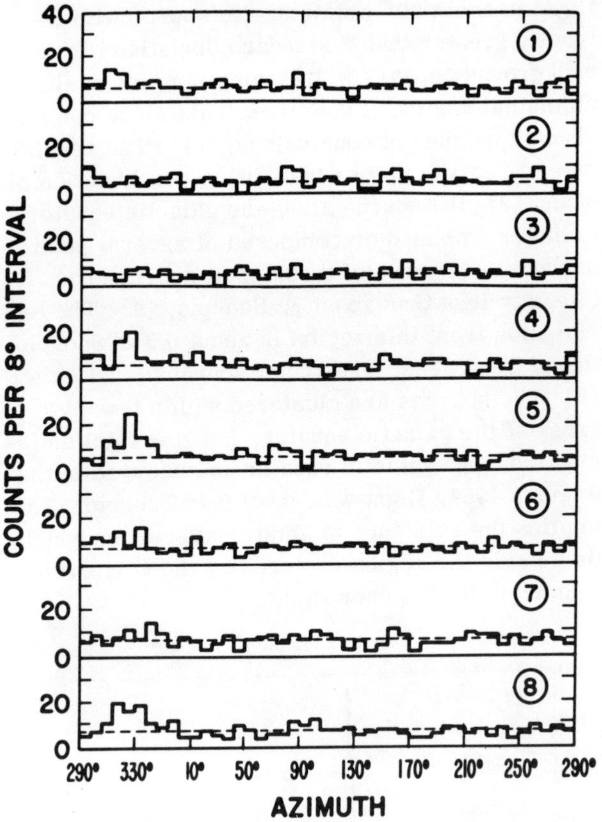

FIG. 7. Azimuthal distribution of counts for GTH-20 in eight different intervals of yaw phase. The normal of the detector during the yaw phases designated as "1" and "8" traced the lines in the sky shown in Fig. 5 and "1" and "8."

In Fig. 9 we obtain a peak in the distribution which we interpret as the presence of an X-ray source. However, the distribution indicates either an extended source or the existence of additional sources at the edge of the region explored by the counter. The approximate location of the source is shown in Figs. 3 and 5 and is designated as SgrX-1.

The source intensity is $\frac{1}{10}$ ScoX-1 for ScoX-2 and about $\frac{1}{20}$ ScoX-1 for SgrX-1. From the previous discussion on the data of the vertical-slit counter we deduce that one or more additional sources must exist within a 40° region along the galactic equator and roughly centered on Sgr A, although it is not completely clear how the sources observed in the horizontal-slit counter contribute to the sources seen in the vertical slit counter.

Data on Angular Size

The data previously discussed set an upper limit of angular size of about $\frac{1}{2}°$ for ScoX-1, ScoX-2, and SgrX-1 in the direction of scan. In the case of ScoX-1 the scan direction lies roughly along the R.A. axis and for ScoX-2 and SgrX-1 roughly along the declination axis. Much finer measurements of angular size were carried out in both the August, 1964, and October, 1964, flights by means of the modulation collimator (Oda 1965).

The principle of operation of the device is illustrated in Fig. 10. Two sets of parallel wires

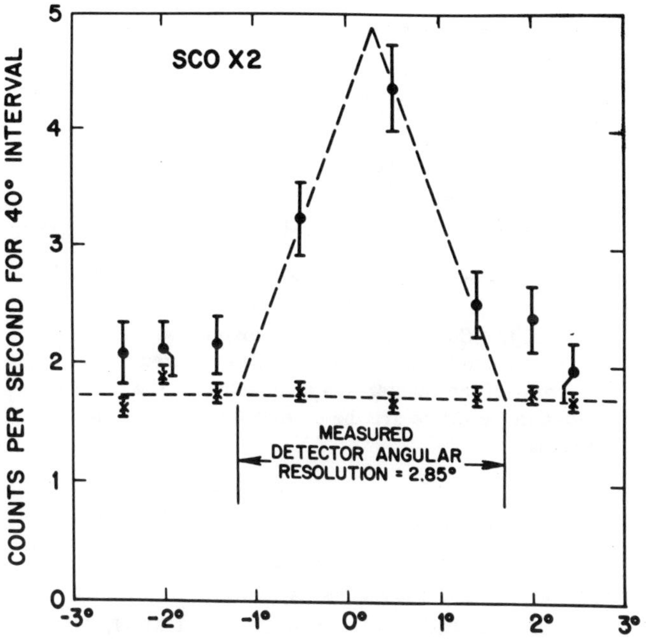

FIG. 8. Distribution of counts in GTH-O accumulated in a 40° interval centered about azimuth 326° as a function of an angle measured along an arc of the great circle passing through the source position and the center of the yaw cone. The origin corresponds to the center of the shaded region in Fig. 5.

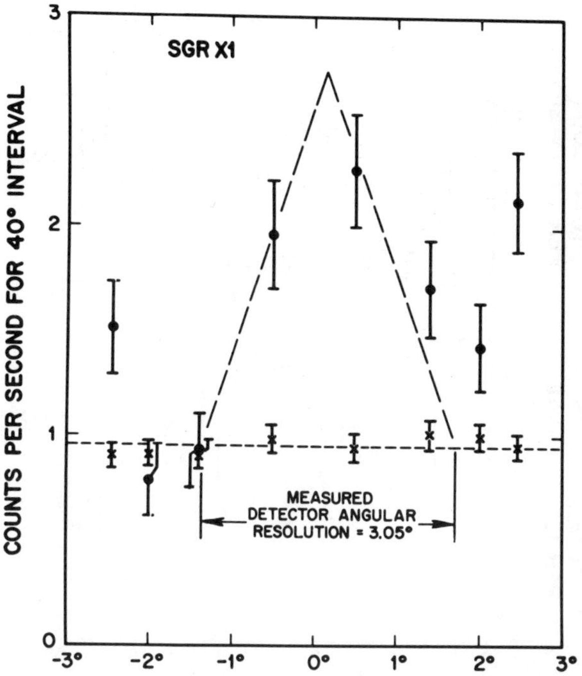

FIG. 9. Distribution of counts in GTH-20 accumulated in a 40° interval centered about azimuth 326° as a function of an angle measured along an arc of the great circle passing through the source position and the center of the yaw cone. The origin corresponds to the center of the shaded region in Fig. 5.

are placed in front of the detector with the spacing between adjacent wires approximately equal to their diameter. The angular resolution is determined by the ratio of the wire diameter to the separation of the two sets of wires. As parallel radiation moves across the field of view, the front set of wires casts a shadow on the rear set. If the shadow falls on the open spacings, the radiation is prevented from reaching the detector; if not, the radiation is transmitted. As the source moves across the field of view a series of maxima and minima of intensity

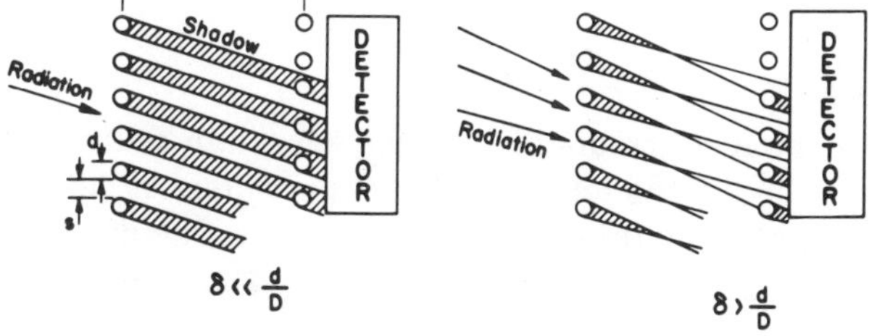

FIG. 10. Schematic diagram showing the principle of the modulation collimator.

will be observed. If, however, the source is extended, the shadow of the wires is diffuse and the modulation becomes less pronounced. When the angular divergence of the incoming beam becomes large with respect to the resolution of the device, no modulation is observed. The advantage of the device is that it permits the achievement of a fine angular resolution within a large field of view. The method is not easily applicable if several sources appear in the field of view.

In the October, 1964, flight a modulation collimator with $\frac{1}{2}°$ resolution was used. The wires were placed in a plane parallel to the spin axis. The yaw motion of the rocket produces a change in the relative elevation of the source with respect to the collimator wires and results in the modulation shown in Fig. 11. The radiation from ScoX-1 is fully modulated indicating an angular dimension much smaller than $\frac{1}{2}$ degree. The total field of view of the modulation collimator was too large to separate the various sources along the galactic equator. The partial modulation observed probably results from the strongest source that is present in that region.

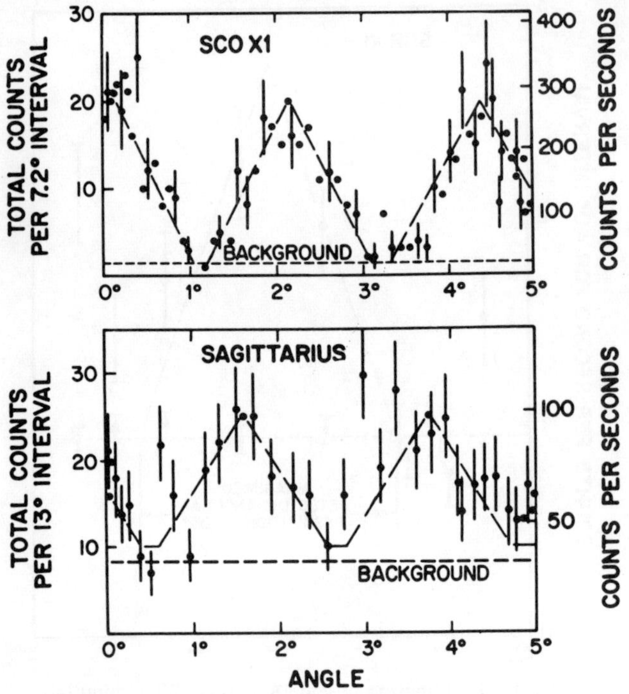

FIG. 11 Counts from ScoX-1 and Sagittarius (the group of sources along the galactic equator) obtained with a $\frac{1}{2}°$ modulation collimator plotted as a function of an angle measured along the arc of great circle through the position of the source and the center of the yaw cone.

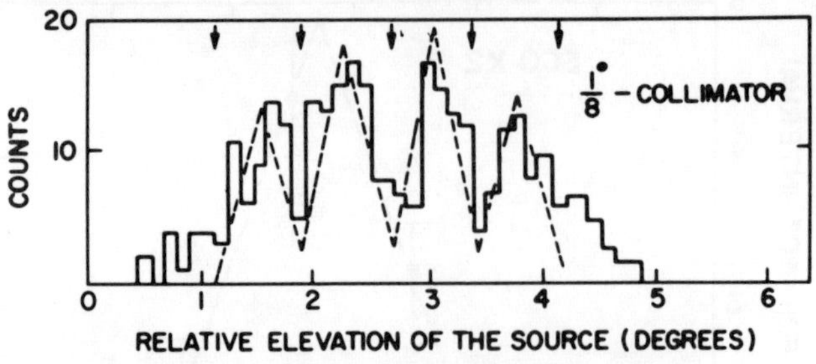

FIG. 12. Counts from ScoX-1 obtained with a $\frac{1}{8}°$ modulation collimator compared with the predicted response of the counter to a point source. The arrows shown in the figure are the predicted position of the minima.

In the August, 1964, flight two modulation collimators were flown with $\frac{1}{2}°$ and $\frac{1}{8}°$ angular resolution. The wires were mounted at an angle with respect to the spin axis so that several cycles of modulation would occur during one spin. The data on the ScoX-1 source obtained with the $\frac{1}{8}°$ collimator were summed over the entire flight and the result is shown in Fig. 12. The modulation which is observed is consistent with that from a source of angular width less than about 7'.

Discussion

A number of sources of cosmic X-radiation have been detected in this and previous experiments. Table 1 summarizes the present knowledge on location, fluxes, and angular sizes. The locations given for ScoX-2 and SgrX-1 must be taken as preliminary. With further analysis of the aspect data we hope to refine these determinations. It should be possible to achieve an area of uncertainty of 1 square degree. It should be noted that SgrX-1 is on the border of Sagittarius and Ophiuchus and may actually reside in Ophiuchus. At present, the Kepler Supernova of 1604, which lies about $7\frac{1}{2}°$ off the galactic equator, appears to be outside the region where the source is believe to be located.

Analysis of the present data with respect to spectral information is under way. In future experiments we plan to extend spectral measurements to the 20-40-Å region (Giacconi and Gursky 1965). We also plan to measure the angular size of the sources with a resolution finer than 1'. If the distance to an X-ray source is known, the determination of its angular size can aid in discriminating between various production mechanisms. The angular dimension also enters in a natural way in any relationship between the volume, distance, density, flux and total mass of a source.

The expression for the observed flux ϕ from bremsstrahlung by a "hot" plasma cloud with electrons and protons of about 10-KeV energy is

$$\phi = \frac{KNn}{4\pi R^2},$$

where R is the distance to the source in centimeters, N the total number of particles, n the particle density (cm^{-3}), and K is a constant equal to 8×10^{-24} erg cm^3 sec^{-1}. The volume of the emitting region V is

$$V = \frac{1}{6}\pi(\alpha R)^3,$$

where α is the angular diameter of the source.

Table I.

Source and Location	Flux between 2 and 8 Å	Angular Size
ScoX-1:		
$16^h\ 15^m$ R.A.	10^{-7} erg cm^{-2} sec^{-1}	< 7'
-15° decl.		
CygX:		
20^h - 23^h R.A.	1/10 ScoX-1	
+10° - +50° decl.		
TauX-1:		
Coincident with Crab Nebula	1/10 ScoX-1	= 1'
ScoX-2:		
$16^h\ 50^m$ R.A.	1/10 ScoX-1	< 30'
-41° decl.		
SgrX-1:		
$17^h 37^m$ R.A.	1/20 ScoX-1	
-24° decl.		

Using also N = nV, we derive expressions for the particle density such as

$$n = 6\left(\frac{\phi}{K^3\alpha^6 N}\right)^{1/5}, \quad n = \left(\frac{24\phi}{KR\alpha^3}\right)^{1/2}.$$

The first expression can be used to determine the particle density when α and ϕ only are known by assuming a total mass, and the second formula can be used to determine n when α, ϕ and R are known.

For ScoX-1, $\alpha < 7'$, and $\phi = 10^{-7}$ erg cm^{-2} sec^{-1}. If we assume that the total mass M is no more than 1 solar mass, then N is less than or equal to about 10^{57} and we obtain a value of 150 cm^{-3} as a lower limit for the particle density n. The corresponding value for R is 10^{21} cm, or about 300 pc.

In the case of TauX-1, which has been identified with the Crab Nebulae, $R = 5 \times 10^{21}$ cm, $\phi = 10^{-8}$ erg cm^{-2} sec^{-1}, $\alpha = 1'$ (Bowyer et al. 1964b). If we assume emission by bremsstrahlung of a "hot" plasma with particles of energy equal to 10^4 eV, the above relations require n = 450 cm^{-3}, and $N = 8 \times 10^{56}$ or $M = 1M_\odot$. These numbers are not in disagreement with presently held views on the nature of the Crab Nebula.

REFERENCES

S. Bowyer, et al. 1964a, Nature, 201, 1307.
_____. 1964b, Science, 146, 912.
R. Giacconi, et al. 1962, Phys. Rev. Letters, 9, 439.
_____. 1964, Nature 204, 981.
R. Giacconi, and H. Gursky, 1965, "Observations of X-Ray Sources outside the Solar System," Space Sci. Rev. (in press).
H. Gursky, et al. 1963, Phys. Rev. Letters, 11, 530.
M. Oda, Appl. Optics, 4, 143.

A MEASUREMENT OF THE LOCATION OF THE X-RAY SOURCE SCO X-1

H. Gursky, R. Giacconi, P. Gorenstein, J. R. Waters, M. Oda*, H. Brandt, G. Garmire, B. V. Sreekantan

American Science and Engineering, Inc.
Cambridge, Massachusetts
and
Massachusetts Institute of Technology
Cambridge, Massachusetts

We have measured the location of Sco X-1, the strong X-ray source in Scorpio, with a precision of about 1 arc min. Sco X-1 was first observed in X-rays by Giacconi, Gursky, Paolini, and Rossi (1962) during a rocket experiment in 1962 at which time its location could only be roughly estimated. A subsequent observation by Bowyer, Byram, Chubb, and Friedman (1964*a*) yielded a location with a precision of about 1°, and more recent measurements of its position have been reported by Clark, Garmire, Oda, Wada, Giacconi, Gursky, and Waters (1965) and by Fisher, Johnson, Jordan, Meyerott, and Acton (1966). The location derived from the present measurement is shown in Figure 1 and listed in Table 2 (see below).

The data which yield the present result were obtained with an instrumented payload flown on a stabilized Aerobee rocket from White Sands Missile Range, New Mexico, on March 8, 1966. The recent report by Gursky, Giacconi, Gorenstein, Waters, Oda, Bradt, Garmire, and Sreekantan (1966) that the angular size of Sco X-1 is less than 20 arc sec was based on the same data.

The Aerobee rocket was equipped with an attitude-control system that allowed pointing to a prescribed direction on the celestial sphere with a precision of a few degrees. While pointing, the rocket axes were restricted in their motion to a 1° limit cycle, and the rate of motion within the limit cycle was typically several arc minutes per second of time.

Reprinted from the Astrophysical Journal 146, 310 (1966).

The instrumentation included a group of proportional counters which detected the X-rays transmitted through two modulation collimators. It also included a camera that continuously recorded the celestial orientation of the collimators. The specifications of the instruments are listed in Table 1. Individual pulses from the X-ray counters were stretched to 0.5 msec with their amplitude preserved and telemetered to a ground receiver. The time at which the photographs were taken was recorded by telemetering the pulse that advanced the camera. The aspect photographs were recovered from the payload after the rocket flight.

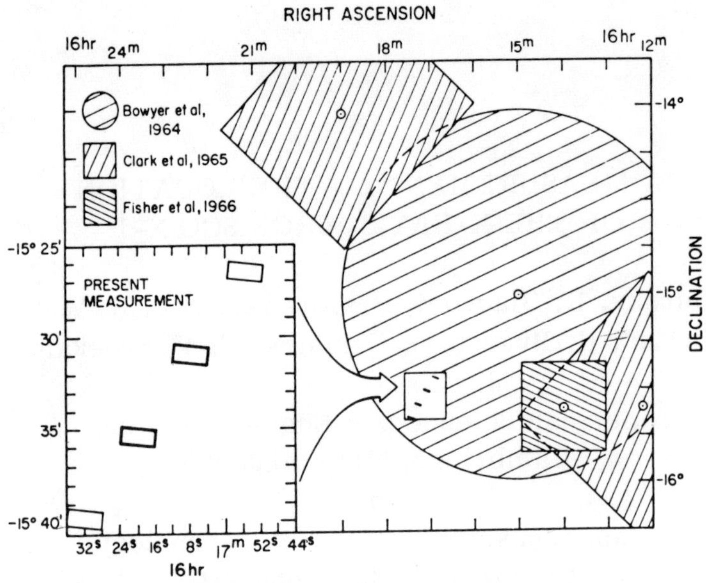

FIG. 1.—The location of Sco X-1. The results of the present experiment are shown with results from previous experiments.

TABLE 1

INSTRUMENTATION SPECIFICATIONS

Modulation collimators:	
Band width:	
Long collimator	72″
Short collimator	76″
Separation between bands:	
Long collimator	5′1″.91
Short collimator	5′16″.62
Aspect camera:	
Body	16-mm Millikan (Model DBM-3C)
Lens	f 1:1,4, 50-mm Cooke
Film	Tri-X on Estar base
X-ray detectors:	
Operating mode	Proportional region
Gas filling	Xenon at one atmosphere
Entrance window	9.0 mg/cm² beryllium
Sensitive area	120 cm² behind each collimator (5 individual detectors)
X-ray region of sensitivity	1.5 keV $< E <$ 30 keV (efficiency $>$ 20 per cent)

The angular response of the modulation collimators for X-rays consisted of a series of parallel bands separated by about four times their width (see Fig. 1 of Gursky *et al.* 1966). High angular resolution was achieved only in the direction normal to the bands. The transmission bands covered a broad field of view which allowed the observation of a given source without the necessity for particularly good pointing accuracy. The use of two modulation collimators with slightly different angular response made it possible to determine from which particular band in a given collimator the X-rays were arriving. The two modulation collimators were aligned with respect to each other to make their transmission bands nearly parallel.

The orientation of the collimator transmission bands in the sky was determined from photographs taken by the aspect camera at a 1 per sec rate. The camera lens simultaneously viewed the star field through an opening between the collimators and a diffuse-light source mounted in front of each collimator. The diffuse light was preferentially transmitted by the collimators in the same directions as were X-rays. The photographic image formed by the transmitted light was a series of bands of complex appearance due to diffraction effects. Through laboratory calibrations we established a unique correspondence between the position of these bands and the X-ray transmission directions. Thus, since star images appeared on the same photograph with the bands, the X-ray transmission directions could be directly determined with respect to the celestial sphere. There was no necessity to align the collimators with respect to the aspect camera or to maintain a fixed alignment during the flight. An example of one of the aspect photographs appears in our earlier paper (Gursky *et al.* 1966).

During the flight, Sco X-1 was observed for about 55 sec. The X-ray data taken during that portion of the flight are shown in Figure 2. A series of peaks occur in the data, each one coinciding with a transit of the source across one of the transmission bands of the collimator. The variations in the time between peaks and in the shape of the peaks for a given collimator are caused by changes in the drift rate of the rocket axes.

Figure 3 shows a plot of the coordinates of the center of the aspect photographs and the orientation of the sensitive direction (the direction normal to the X-ray transmission bands) of the modulation collimators. No peaks were observed in the counter data during the times 303–317 sec, and it is clear from the aspect data that the rocket was virtually motionless during much of that time. At $t = 325$ sec, the rocket began a programmed roll of 4°.5/sec around the central axis of the collimator. At about the same time aerodynamic effects (the rocket was beginning to re-enter the atmosphere) became noticeable and tended to force the rocket nose downward at an increasing rate. By 340 sec the X-ray source had drifted out of the field of view of the collimators.

Each time a peak in the counting rate is observed we can compute a set of narrow parallel bands in the sky separated by about 5 arc min on which the source must lie. The number of possible locations can be substantially reduced by comparing the data from the two collimators which are constructed so that they differ by about 5 per cent in the angular separation of their transmission bands. Twenty bands from the "short" collimator are separated by about the same total angle as twenty-one bands from the "long" collimator. Thus, the two collimators provide a vernier scale, which is repeated every 1°40′, the equivalent of those numbers of bands. Figure 4 shows the projection on the sky of the allowed transmission bands of the collimators. As the source traverses the field of view, peaks in the counting rate will be observed, first in one and then the other collimator. The relative times of occurrence of these peaks determine, after conversion to relative angles, which transmission band within the vernier scale corresponds to a given peak. This defines a series of bands separated by about 1°40′ on which the source must lie.

The experiment itself provided a coarse position measurement that allowed us to decide on which particular band the source lies. Five individual detectors were used behind each collimator to detect the transmitted X-rays. The upper edge of the collimators casts a shadow on the detectors. The position of the shadow was found by comparing the counting rates of the individual detectors. This, in turn, determined the angle between the X-ray source and the central axis of the collimator with a precision of better than 1°.

Given the motion of the detector axis from the aspect data we are able to follow the transit of the X-ray source from band to band throughout the flight. Once the vernier technique has established which specific band is traversed at any given time, we know which specific band is traversed at any other time. During the time when no appreciable

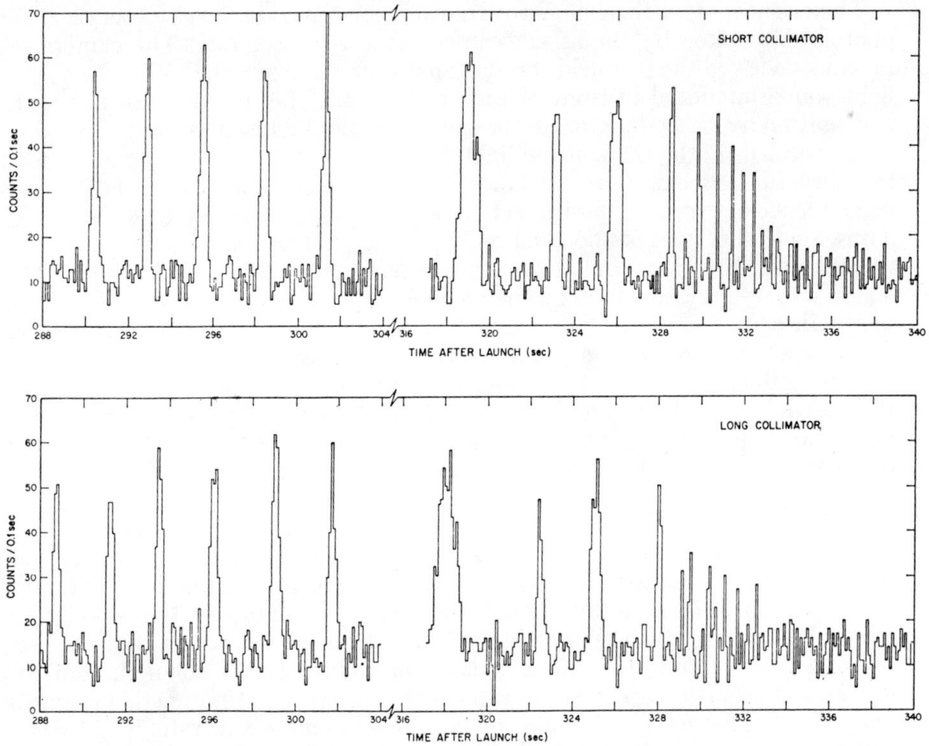

Fig. 2.—Histogram of actual counts in the energy range from 1 to 24 keV and accumulated per 0.1 sec during the time that Sco X-1 was within the field of view of the collimators.

rotation of the sensitive axis occurs, successive transits define the same line on the celestial sphere. As the sensitive direction rotates by a given angle the appearance of a maximum defines a new line tilted by that angle with respect to the previous determination. The source must be at the intersection of these two lines. The procedure is repeated for every appearance of the peak as the rotation takes place, thus resulting in a set of intersections. The average of these is taken to be the location of the source.

In practice, the experimental uncertainties in the measurement of location by the method outlined above result in the following: (1) a lack of uniqueness in the assignment of a source position; hence we list two most likely positions in Table 2; (2) an area of uncertainty around each individual position.

Our inability to assign a single source position arises from uncertainties in measuring the relative orientation of the two sets of collimator bands as recorded on the aspect photographs. One source of this uncertainty was our imprecise knowledge of the correspondence between the X-ray transmission directions and the visible light bands. The

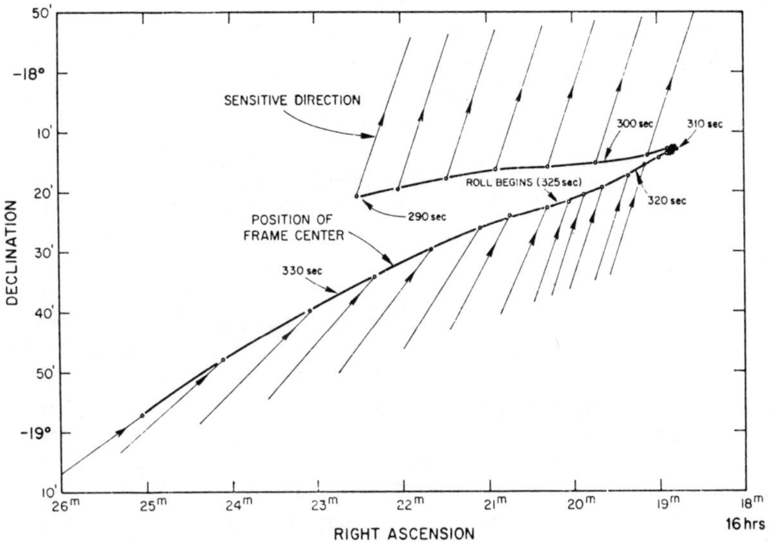

FIG. 3.—Motion of the frame center and orientation of the sensitive direction during the observation of Sco X-1. The frame center, which is defined as the geometric center of the aspect photographs, was approximately aligned with the central axes of the modulation collimators. Dots indicate 2-sec intervals from 290 to 310 sec, and 1-sec intervals from 310 to 333 sec.

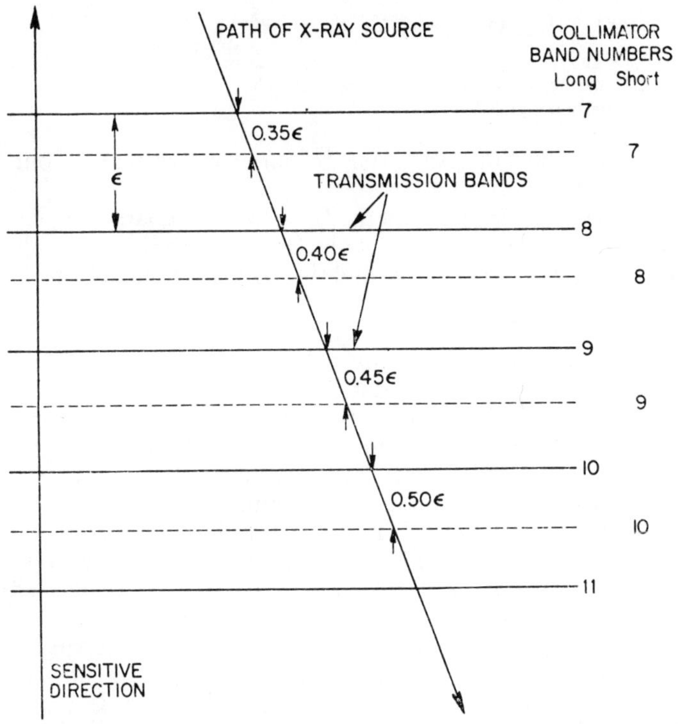

FIG. 4.—Schematic representation of several of the X-ray transmission bands from the two collimators as they appear on the aspect photographs. As successive pairs of bands are traversed, the angular separation of source transits, across correspondingly numbered bands, increases by about 0.05ϵ.

error arising from this effect proved small compared to the reading error for locating the bands on the film. Our uncertainty in determining which was the band corresponding to a measured angular separation between successive counting-rate peaks in the two collimators was about ± 1 band. Thus, four adjacent bands have to be considered as possibly containing the source, with the central two being the most probable bands.

The area of uncertainty around each individual position represents the estimated range of systematic error. While a number of uncertainties exist in determining positions on the celestial sphere from the counter data and the aspect photographs, we find that the net result of these errors is a very small scatter of the independent determinations of position from which the average is computed. The statistical error of the average is $\pm 2^s$ in right ascension and $\pm 7''$ in declination. On the other hand, there are several possible sources of systematic errors which may have magnitudes greater than the uncertainties given above. These are related to our imperfect knowledge of the properties of the modulation collimator. For example, the values of the band separation as listed in Table 1 were calculated on the basis of the measured dimensions of the collimators and are precise to $1:10^4$. We have also measured the band separations directly and find them to agree, but only with a precision of 0.3 per cent. The maximum error in position that could arise from this uncertainty is about $30''$. Accordingly, we assigned an uncertainty

TABLE 2

BEST POSITIONS OF SCO X-1 DETERMINED IN THIS EXPERIMENT*

$\alpha(1950.0)$	$\delta(1950.0)$
$16^h17^m7^s \pm 4^s$	$-15°30'54'' \pm 30''$
$16^h17^m19^s \pm 4^s$	$-15°35'20'' \pm 30''$

* The two positions are about equally probable; additional positions shown in Fig. 1 are substantially less probable.

of $\pm 4^s$ in right ascension and $\pm 30''$ in declination to our measurement as our best estimate of positional uncertainty.

As an over-all check on the accuracy of the method, we have compared the location of the X-ray source associated with the Crab Nebula (Bowyer, Byram, Chubb, and Friedman 1964*b*), also detected in this experiment, and the known position of the nebula. While the data were of much lower statistical precision than that obtained for Sco X-1, it was possible to determine the location of the X-ray source within an ambiguity of 4 bands. One of these bands passes through the visible nebula.

The location of Sco X-1 here reported is at least 30 arc min away from any of the previously reported locations. We do not consider this disagreement serious since the new location is well within the uncertainty of the measurement by Bowyer *et al.* (1964*a*) and not incompatible with the estimated errors of the result reported by Clark *et al.* (1965). It disagrees by approximately three times the experimental uncertainty quoted by Fisher *et al.* (1966) for their measurement. We interpret this discrepancy as being due to systematic errors in relating X-ray data to celestial coordinates possibly present in their measurement, but absent in the measurement reported here. It should be noted that the area of uncertainty, as defined in Figure 1, in the present measurement is between two and three orders of magnitude smaller than in the previous observations.

The significance of this measurement is that it affords the possibility of an unequivocal identification of the visible and radio counterparts of the Sco X-1 source. In a previous Letter (Gursky *et al.* 1966), we had stated that in the absence of interstellar absorption the visible image of this X-ray source should appear as a starlike object of 13th magnitude or brighter. On the basis of the earlier conclusions and of the present result, an optical identification has been made and is reported in the following companion Letter.

We are grateful to Professor Bruno Rossi, Massachusetts Institute of Technology, for many helpful discussions in connection with this experiment. We wish to acknowledge the contribution given in the computation of astronomical coordinates by G. Ouellette. Also, the contribution of Dr. G. Spada to the analysis of the data is gratefully acknowledged. This work was supported in part through funds provided by the National Aeronautics and Space Administration under contract NASW-1284 and grant NSG-386, and in part by the U.S. Atomic Energy Commission AT(30-1) 2098.

REFERENCES

Bowyer, S., Byram, E. T., Chubb, T. A., and Friedman, H. 1964*a*, *Nature*, **201**, 1307.
———. 1964*b*, *Science*, **146**, 912.
Clark, G. W., Garmire, G., Oda, M., Wada, M., Giacconi, R., Gursky, H., and Waters, J. R. 1965, *Nature*, **207**, 584.
Fisher, P. C., Johnson, H. M., Jordan, W. C., Meyerott, A. J., and Acton, L. W. 1966, *Ap. J.*, **143**, 203.
Giacconi, R., Gursky, H., Paolini, F., and Rossi, B. 1962, *Phys. Rev. Letters*, **9**, 439.
Gursky, H., Giacconi, R., Gorenstein, P., Waters, J. R., Oda, M., Bradt, H., Garmire, G., and Sreekantan, B. V. 1966, *Ap. J.*, **144**, 1249.

* On leave from the Institute for Nuclear Study, University of Tokyo, Tokyo, Japan.
† On leave from the Tata Institute of Fundamental Research, Bombay, India.

ON THE OPTICAL IDENTIFICATION OF SCO X-1

A. R. Sandage, P. Osmer, R. Giacconi, P. Gorenstein, H. Gursky, J. Waters, H. Bradt, G. Garmire, B. V. Sreekantan,* M. Oda, K. Osawa, and J. Jugaku

Lunar and Planetary Laboratory
University of Arizona, Tucson
and
Department of Space Science
Rice University, Houston

An optical search has been made of the sky surrounding the new position of Sco X-1 described in the preceding Letter (Gursky, Giacconi, Gorenstein, Waters, Oda, Bradt, Garmire, and Sreekantan 1966*a*). Preliminary results of the measurement were made available to the Tokyo Observatory and to Palomar. The search, which we believe has been successful, was based on these results and on the working hypothesis that the image should (1) appear starlike, (2) be of at least 13th apparent visual magnitude, and (3) have an ultraviolet excess relative to normal stars. The first two requirements are stated in the discussion of Gursky *et al.* (1966*b*) on the measurement of an upper limit of 20″ to the diameter of Sco X-1. The predicted lower limit on the visible magnitude was obtained by extrapolating the energy distribution from the observed range of 1–10 Å into the optical region in the assumption of a spectrum that is flat per unit frequency interval. Such a spectrum could result from either bremsstrahlung or synchrotron emission (Manley 1966) and requires, in the optical region that $B - V \simeq 0.10$ mag., and $U - B \simeq -0.91$ mag. (Matthews and Sandage 1963, Table A4 for $n = 0$).

A two-color image plate taken on Eastman 103aO emulsion with the Tokyo Observatory's 74-inch reflector on June 17/18, 1966 covered a region from about δ (1950) between $-15°$ and $-16°$ and α (1950) between 16^h15^m and 16^h and 18^m. An ultraviolet and a blue image of each star was obtained through a Hoya U2 filter ($\lambda \simeq 3600$ Å) and a

Reprinted from the Astrophysical Journal **146**, 315 (1966).

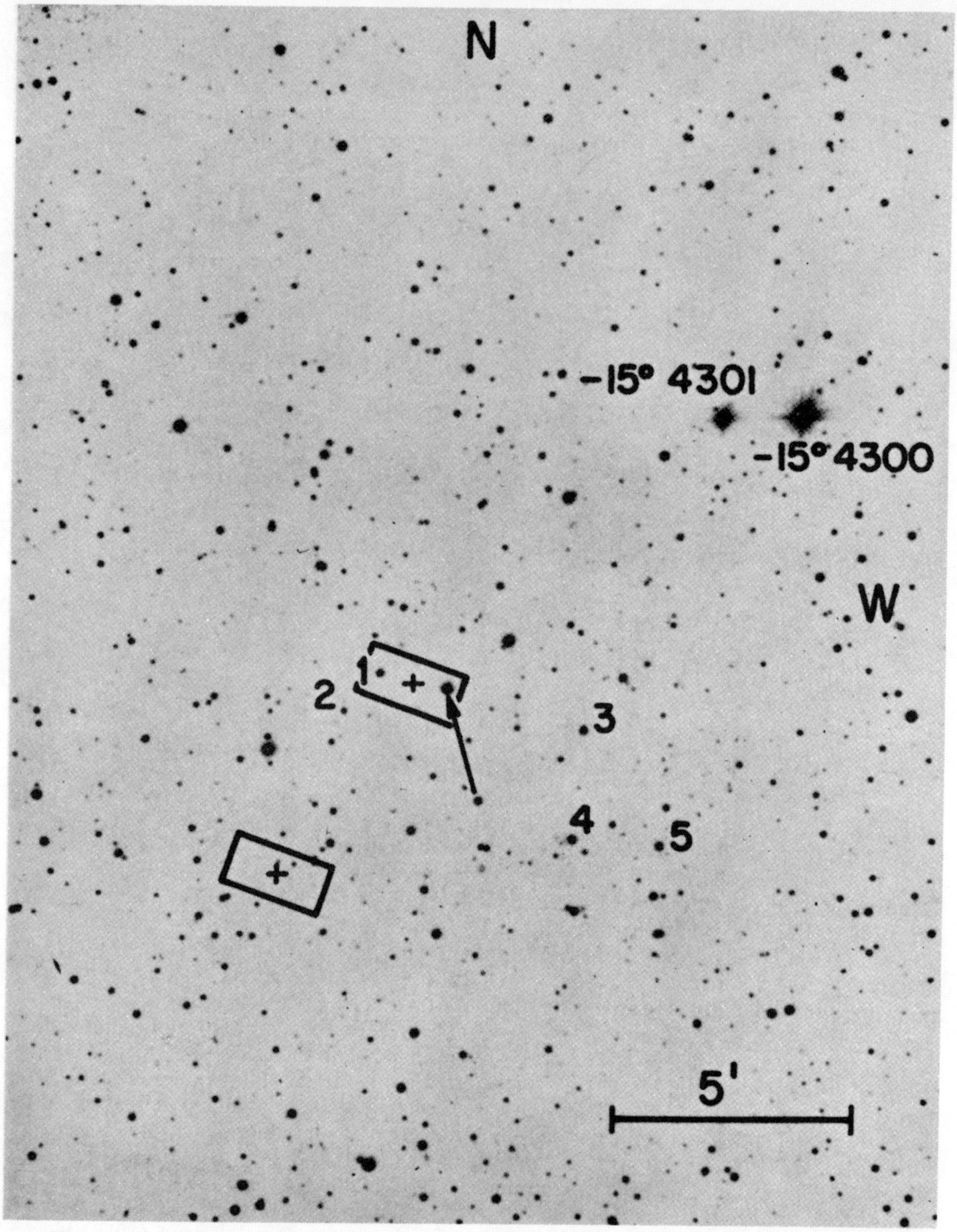

FIG. 1.—Photograph of the region containing the new X-ray position of Sco X-1, reproduced from the Palomar Sky Survey prints. The two equally probable X-ray positions are marked by crosses surrounded by a rectangle of 1 by 2 arc min. The object described in the text is marked with an arrow. The identifications of other stars for which photoelectric photometry exists are also marked.

Hoya L-39 filter ($\bar{\lambda} \simeq 4400$ Å). Immediate inspection of the plate revealed the existence of an intense ultraviolet object of $V \simeq 13$ mag. near the center of the search area and within 1 arc min of one of the two X-ray source positions quoted in the preceding Letter. Photoelectric photometry was secured of the object on the same night with the 36-inch reflector of the Tokyo Observatory, and the observation was repeated on June 22/23, 1966. Although frequently interrupted by clouds, which are prevalent in Japan during this rainy season, the observation gave $V = 12.6 \pm 0.2$ (m.e.), $B - V = +0.3 \pm 0.05$ (m.e.), and $U - B = -0.8 \pm 0.1$ (m.e.)—colors which are definitely peculiar and in the range predicted by the working hypothesis.

A spectrogram was obtained on June 18/19 at the Cassegrain focus of the 74-inch telescope with a two-prism quartz spectrograph which gives a dispersion of 90 Å/mm at 4000 Å and 150 Å/mm at Hβ. The spectrum, which was underexposed because of unfavorable sky conditions, showed a continuum with no absorption features and with faint emission at Hγ and λ 4686.

These results were communicated by cable to Giacconi, who relayed them by telephone to Palomar on June 23, P.S.T. Photoelectric observations made with the 200-inch reflector on the same night confirmed the colors measured in Japan, giving $B - V = +0.23$, $U - B = -0.88$, each with statistical measuring errors of less than ± 0.01 mag. The data further showed that the object varies. Repeated observations, made with a pulse counter and separated from each other by less than 1 min in time gave differences in the total count which were 35 times greater than the statistical $\sqrt{N}$ uncertainty. The object was monitored for 42 min on June 23 during which time its V magnitude changed irregularly from 12.44 to 12.38. The time scale of the fast flicker is of the order of 2 per cent (0.02 mag.) in several minutes. In one interval the U magnitude changed by 0.09 mag. in 8 min.

The flickering activity, together with the very peculiar colors, showed that the object has characteristics of old novae near their minimum phase. Walker's (1954, 1957) systematic survey of the photometric properties of old novae, nova-like variables, and U Gem stars showed that abnormal intensities in the ultraviolet and rapid changes in the continuum level are characteristic of the class. The object in question was tentatively identified as a member of this class on the basis of the data available by the end of June.

A second spectrogram of improved quality was obtained at the Tokyo Observatory on June 25/26, 1966. The emission features of Hβ, He II (λ4686), Hγ, Hδ, and possibly Hϵ were now clearly visible, again on a very blue continuum.

The optical position of the object was measured both at Tokyo and at Mount Wilson–Palomar with the result α (1950) = $16^h17^m4.3^s$, δ(1950) = $-15°31'13''$. This position is not definitive by astrometric standards because a refined reduction procedure using the astrographic and Yale Zone catalogues has not yet been used. Nevertheless, the position should be good to about ± 5 arc sec which is sufficient for the present purpose.

Figure 1 shows the field as reproduced from the Palomar Sky Survey prints. The two equally probable X-ray positions are shown as crosses surrounded by a rectangle of 2 by 1 arc min, corresponding to the quoted position uncertainty. The unusual blue object we have been describing is indicated by an arrow. There is no conspicuous object brighter than $V \simeq 16$ mag. in the alternate square. Furthermore, a survey of the Tokyo two-color plate shows that there is no other ultraviolet star within $\pm \frac{1}{2}°$ of the new X-ray position.

Also identified in Figure 1 are several stars whose UBV values have been measured so as to obtain an estimate of the absorption and reddening in this direction of the sky. Table 1 shows the results and indicates that $E(B - V) \simeq 0.23$ out to an apparent distance modulus of $m - M \simeq 8.7$. These reddening values, obtained by using the usual two-color diagram, indicate a rather uniform absorption in V wavelengths of 0.7 mag. over the distance interval between $(m - M)_{\text{true}}$ of 5–8 mag. (100–400 pc). Thus interstellar absorption and reddening do not adversely affect the observation of Sco X-1 in visible light.

Simultaneous photoelectric and spectroscopic observations were begun at Palomar using the 200-inch and 20-inch reflectors. Photometry with the 200-inch on July 12 gave $\bar{V} = 12.95$ with a range of $\Delta V = 0.08$ mag. in the 12-min observation interval. The colors remained relatively stable at $B - V = +0.20$, $U - B = -0.76$. These values are appreciably different from $\bar{V} = 12.39$, $B - V = +0.23$, and $U - B = -0.88$ measured two and a half weeks earlier on June 24 U.T. Continued observations on 7 successive nights with the 20-inch telescope, primarily in the B wavelength band (λλ 3800–5400 Å), are shown in Figure 2. Spectra were obtained during the times indicated by the horizontal bars on July 13–July 18.

Figure 2 shows that the object is highly unstable in its continuum radiation, resembling the behavior of the probable old nova MacRae +43° 1 which Walker (1954) discovered to have rapid optical variations. The fluctuations also resemble, but to a

TABLE 1

PHOTOMETRY OF STARS IN THE FIELD OF SCO X-1

Object	V	$B-V$	$U-B$	$E(B-V)$	$(m-M)_{app}$	Remarks
−15°4300......	8.5	0.25	+0.22	0.20	5.8	
−15°4301......	9.88	0.33	+ .27	.23	6.7	
1..............	14.97	0.94	+ .41	.28	8.4	
2..............	16.25	1.05	+ .43	.46	9.2	
3..............	14.17	1.16	+ .87	.21	6.7	
4..............	14.47	0.84	+ .33	.18	8.4	
5..............	14.46	0.83	+ .26	0.27	8.6	
Sco X-1........	12.43	0.23	− .86			June 24 U.T., 1966 selected data over a time interval of 43 min
	12.44	0.23	− .84			
	12.39	0.22	− .92	....	....	
	12.38	0.24	− .89	....	....	
	12.38	0.24	− .91	....	....	
	12.43	0.22	− .88	....	....	
	....	....	....	....	....	
	13.01	0.21	− .76	....	....	July 12 U.T., 1966 data over a time interval of 12 min
	12.97	0.20	− .73	....	....	
	12.94	0.20	− .74	....	....	
	12.95	0.19	− .76	....	....	
	12.94	0.21	− .76	....	....	
	12.93	0.21	−0.76	....	....	

lesser degree, the changes in the nova-like irregular variable AE Aquarii (Lenouvel 1957; Walker 1962). The intensity of our candidate star varied by 0.5 mag. in a 2.6-hour period on July 12 with the more rapid flicker variations which were observed on the same night at the 200-inch, presumably superposed on the decline shown in Figure 2. The variation in the general intensity level from night to night is evident, especially on the nights of July 16, 17, and 18, where the change is at least 0.9 mag.

The object has been located on old plates in the Harvard collection going back as far as 1896. Garmire and Sreekantan have inspected the Harvard material with the results shown in Figure 3 from 1935 to 1949. The variations in magnitude are real, but since no special observations were made for the photometric zero point or the scale of the comparison stars used for these estimates, the ordinate in Figure 3 is somewhat arbitrary and is not on the international m_{pg} system. Nevertheless, the data, including one point in 1896 at 12.4 mag., one in 1901 at 12.5 mag., two in 1913 at 12.5 mag. and one point in 1914 at 12.5, show that the object has been in its present state for at least 70 years with no evidence of a strong outburst in this period. However, the time coverage has not been tight enough to exclude a recurrent nova outburst with a decay time of less than a year sometime in the interval.

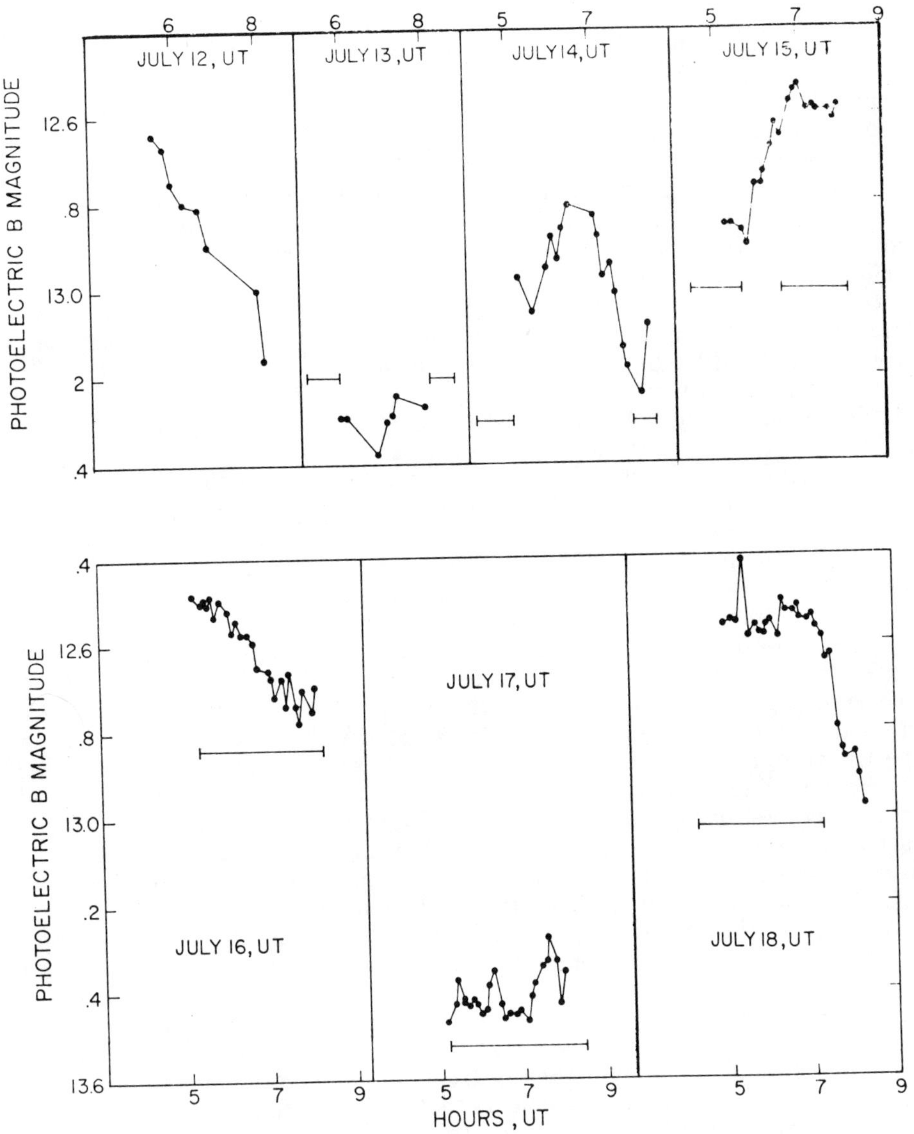

FIG. 2.—Photoelectric B magnitudes of the candidate object as obtained with the Palomar 20-inch reflector. Horizontal bars indicate the times when spectrograms were obtained with the 200-inch Hale telescope.

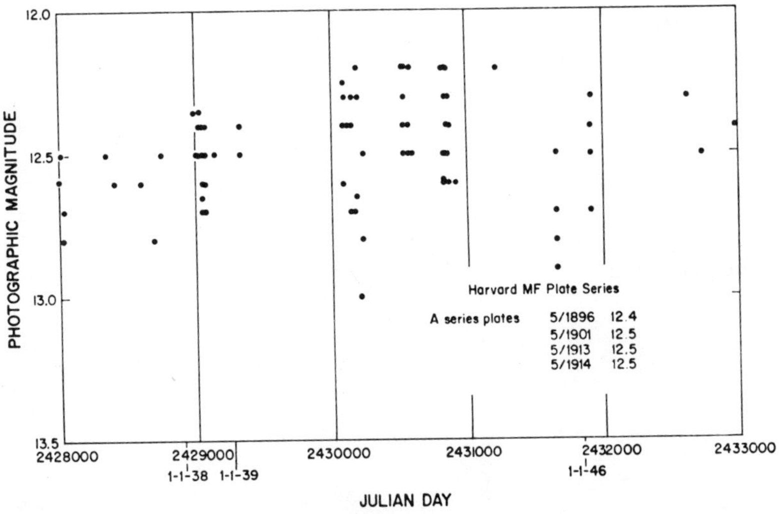

FIG. 3.—Estimates of the magnitude of the candidate object from old plates in the Harvard collection. The zero point and the scale of the photometric system are only approximate.

The spectra taken in the July 13–18 interval were obtained with the 200-inch Hale reflector at a linear dispersion of 85 Å/mm. Many of the plates were taken using the technique of a single trail along a long slit to permit the detection of short period radial velocity changes which Kraft (1963, 1964) has found to be characteristic of old novae and nova-like stars. Figure 4 shows three of the spectrograms taken on July 16, 17, and 18, respectively—nights during which the continuum intensity changes from $\bar{V} \simeq 12.6$ to $\bar{V} \simeq 13.4$ and back again to $\bar{V} \simeq 12.6$. The blank spaces in the individual spectra are time marks made by closing the dark slide of the camera for 2 min while the telescope was trailing the star along the slit.

In general, the spectra are very similar to those of old novae (Humason 1938; McLaughlin 1953; Greenstein 1961; Kraft 1964) in the types of lines present and in their strengths and widths. In particular, the hydrogen lines are in emission, He II is present, and the complex of very high excitation lines probably due to C III, N III, and possibly O II is seen near λ 4650. The interstellar K of Ca II in absorption is clearly seen but the H-line of Ca II is partially masked by Hϵ in emission.

The most striking feature of Figure 4 is the large change in the strength of the Balmer lines between the three nights. These lines are very weak on July 16 and 18 but appear in great strength relative to the continuum on July 17 when the object was very faint. Similar changes occur in D Q Her, MacRae +43° 1 and other old novae (Greenstein and Kraft 1959; Greenstein 1954) and may indicate that the low-excitation Balmer lines are formed in a different region from that in which the variable, blue continuum radiation and the high-excitation lines of He II, N III, and C III originate. This is partially borne out by the relatively small change in the equivalent width (E.W.) of the He II (λ 4686) line on the three nights compared with the large change in Hβ. Preliminary measurement of microphotometer tracings show that Hβ had E.W.'s of 0.7, 6.2, and 2.1 Å on the respective three nights while He II (4686) had E.W.'s of 3.6, 2.9 and 2.8 Å.

The most interesting feature is the broad structure between λ 4630 and λ 4655, described before as due to C III, N III, and possible O II. This structure is found in old novae such as V603 Aql (1918), DQ Her, and CP Pup, among others. On our spectrograms the intensity of the structure ranges from almost complete invisibility relative to the He II (4686) line on July 14 to twice the strength of He II on July 18.

We have looked for short-term radial velocity changes which would appear as a characteristic *S*-wave distortion in the lines on the single trailed spectra. Large changes of this type are known to be characteristic of old novae and nova-like stars (Kraft, *op. cit.*). No changes as obvious as those in DQ Her and WZ Sge (Kraft 1964) occur on our plates but suggestions of an *S*-wave are present on the July 18 spectrogram for the broad line at λ 4417 and for the H9 + He I + He II line at λ 3836 A. There are also indications from partially completed measurements that the wavelengths of other emission lines, notably He II (4686), change from night to night.

These data suggest, then, that we are possibly dealing here with an uncatalogued old nova. Many such objects must remain to be discovered because only about 150 are now known—a small number considering that the rate of outburst is about 30 novae per year in our Galaxy.

The absolute magnitudes of non-recurring novae at minimum light range between $M_B \simeq +2$ and $M_B \simeq +7$. This large spread means that the true modulus of Sco X-1 [assuming that the identification is correct and that $\bar{B}_0 = 12.1$ for the optical object as corrected for absorption by $4E\ (B - V)$] is in the range from $m - M \simeq 10$ to $m - M \simeq 5$ corresponding to distances between 100 and 1000 pc. This interval might be narrowed down by considering that the optical spectrum of Sco X-1 resembles some what that of DQ Her, CP Pup, and V603 Aql where $\bar{M}_B \simeq 5.3$ (Payne-Gaposchkin 1957, chap. 1) giving a distance of about 250 pc. The latitude of $b^{II} = +24°$ gives a height above the plane of about 100 pc which is quite characteristic of novae (Payne-Gaposch-

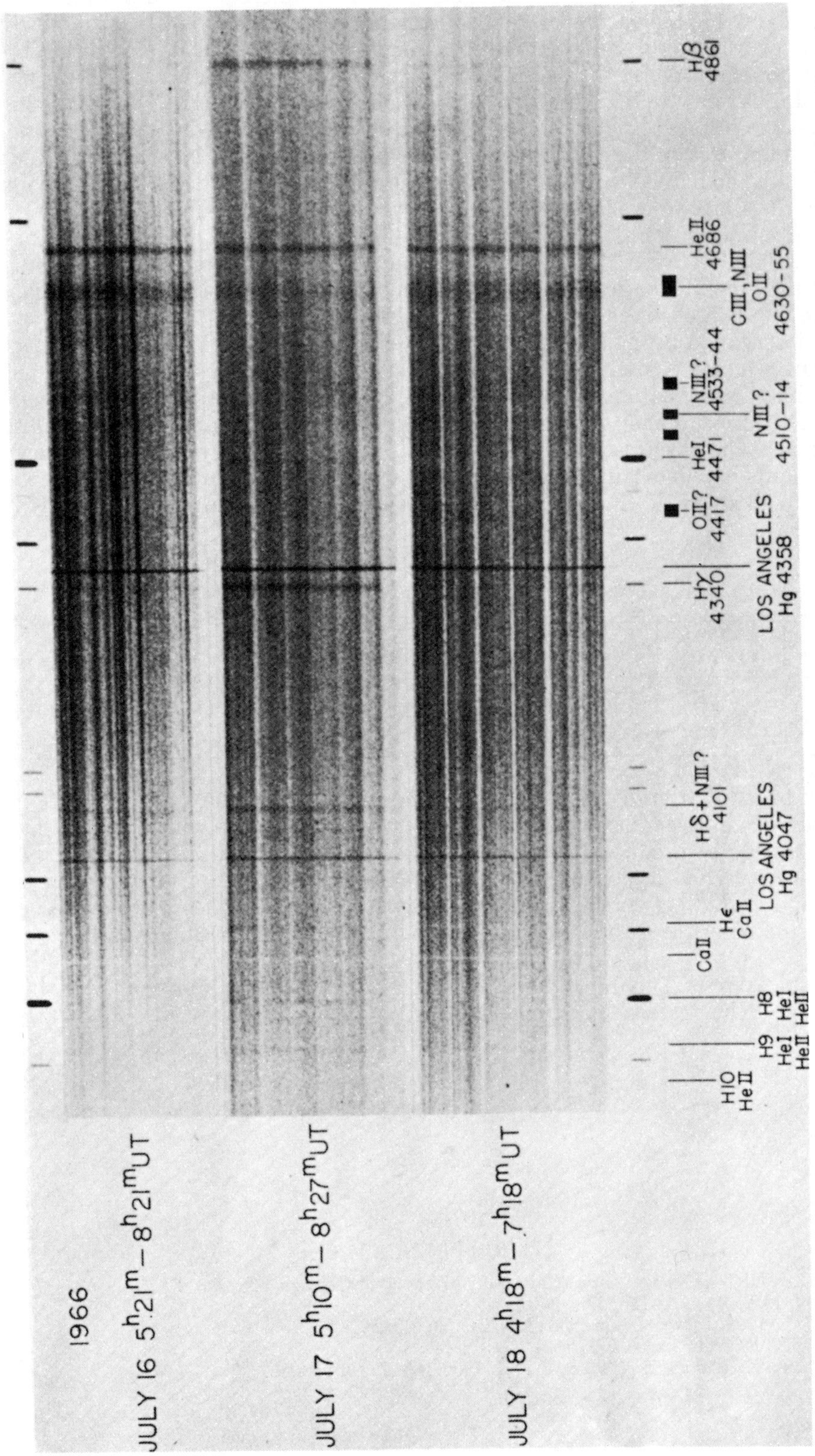

FIG. 4.—Three continuously trailed spectra in the blue spectral region. Time marks, made by closing the camera dark slide, are evident on the last two. Tentative line identifications are indicated. The comparison arc is He. The original dispersion was 85 Å mm.

deduced to lie between 1.5 and 1.65 (Harris 1961). Table 2 summarizes the values of albedo, effective temperature T_E, and brightness temperature, T_B, that we now have.

At the present epoch, the polar axis of Uranus, which lies almost in the orbital plane, is nearly perpendicular to the line from the planet to the Sun and, because of the planet's rapid rotation, we would expect a radiometric temperature close to T_E(Av). In the next 20 years, the infrared brightness temperature should rise toward T_E(Max) as the pole swings around to face the Sun. At present the planet is near perihelion; at aphelion the temperature should be 5 per cent lower because of the greater distance to the Sun.

From the above we would expect an effective temperature of about 50° K for Uranus with a peak in its spectral energy curve near 60 μ. On this basis, the fraction of the total energy radiated between 17.5 and 25 μ is about 0.1 per cent. Considering the probable error of the present 20-μ measurement, we can only conclude that radiative equilibrium with the Sun is probable, though not yet established.

The observation by Kuiper (1949) of an absorption line at 8270 Å in the spectrum of Uranus, which was attributed to pressure-induced dipole transitions in H_2 (Herzberg 1952), and the calculations of Trafton (1964) suggest that the far-infrared opacity is produced by pressure-induced dipole absorption from rotational and translational transitions in H_2. The absorption spectrum and UBVRI photometry of Uranus show that CH_4 accounts for almost all the absorbed heat from the Sun. The CH_4 level is heated by the Sun to about 85° K (Kuiper 1951) and lies under a dense clear blanket of H_2 which

TABLE 2

ALBEDOS, EFFECTIVE AND BRIGHTNESS TEMPERATURES

A	T_E (Max)	T_E (Av)	T_B (20 μ)	T_B (1.9 cm)	T_B (11.3 cm)
0.55......	74° K	52° K	55±3° K	200±35° K	130±40° K
0.75......	64° K	45° K			

is heated to 50° K from below. The 1.9-cm and 11.3-cm emission must originate in the even more dense layers below the CH_4, and it is not clear that temperatures there of 130–220° K can be produced by the greenhouse effect as just described. Observations near 60 μ are needed to establish more accurately the total energy flowing out of the planet, and high-precision observations at several radio wavelengths are needed to establish the temperature and the thermal gradient below the visible CH_4 level.

It is a pleasure to acknowledge the assistance of Arnold W. Davidson in making the observations. This research has been supported by the National Science Foundation.

FRANK J. LOW

REFERENCES

Harris, D. L. 1961, *The Solar System* (Chicago: University of Chicago Press), **3**, 272.
Herzberg, G. 1952, *Ap. J.*, **115**, 337.
Johnson, H. L. 1965, *Comm. Lunar Planetary Lab.*, **3**, No. 53.
Kellermann, K. E. 1966, *Icarus* (in press).
Kellermann, K. E., and Pauliny-Toth, I. I. K. 1966, *Ap. J.* (in press).

Kuiper, G. P. 1938, *Ap. J.*, **88**, 429.
———. 1949, *ibid.*, **109**, 540.
———. 1951, *The Atmosphere of the Earth and Planets* (Chicago: University of Chicago Press), p. 385.
Low, F. J. 1965*a*, *Ap. J.*, **142**, 806.
———. 1965*b*, *Bull. Lowell Obs.*, p. 184.
———. 1966*a*, *A. F.* (in press).
———. 1966*b*, *Ap. J.* (in press).
Low, F. J., and Johnson, H. L. 1964, *Ap. J.*, **139**, 1130.
Trafton, L. M. 1964, *Ap. J.*, **140**, 1340.

*On leave from the Tata Institute of Fundamental Research, Bombay, India.

COSMIC X-RAY SOURCES

H. Friedman
The E. O. Hulburt Center for Space Research*
U.S. Naval Research Laboratory
Washington, D. C. 20390

The first evidence for X-ray emission from the galaxy was observed by Giacconi, Gursky, Paolini, and Rossi, (1962, 1963) in June, 1962. They launched an Aerobee rocket instrumented with Geiger counters sensitive to wavelengths in a band centered at about 3 Å, from the White Sands Missile Range, and detected a strong signal from the general direction of the galactic center. The instrumentation had been designed for an attempt to observe X-rays from the Moon and was not equipped with collimation to restrict the field of view narrowly. As a result, the signal was very broad, and accurate definition of the size and position of the source was not possible. A similar experiment was repeated in October, 1962, when the galactic center was below the horizon and the strong source was not present. A third attempt, in June, 1963, verified the results of the June, 1962, flight. The conclusion drawn from these observations was that a discrete source of X-rays existed in the general direction of the galactic center.

In April 1963, Bowyer, Byram, Chubb, and Friedman (1964) launched an Aerobee equipped with a counter about ten times as sensitive as those used by Giacconi et al., and covering a wavelength range 1-8 Å. In front of the counter was a hexagonal honeycomb structure which provided collimation. The field of view was limited to about 10° at half-maximum of the transmission pattern. The rocket was deliberately given a slow spin to make it precess with a large cone angle. As a result, the circle of sky which was swept out by each roll slowly turned through the sky. Almost all of the celestial sphere above the horizon was thus scanned during the flight. At the time, the galactic center was below the horizon, but about 58 per cent of the celestial sphere was observed (Fig. 1). In this large expanse of sky, the detector found one outstanding source in Scorpius and another, about one-eighth as intense, in the direction of the Crab Nebula. No other discrete sources could be identified above the general background.

The counter was of the proportional type with an effective window area of 65 cm^2 of beryllium, 0.005 inch thick. Pulse-height discrimination was used to give rough spectral information by sorting into three channels: 1-1.5 Å, 1.5-2-5 Å, and 2.5-8 Å. From the observed distribution it appeared that the Scorpius source could be fitted to a 2×10^7 °K black-body curve.

The source in Scorpius was clearly detected on eight separate scans. Figure 2 shows the direction of each scan with intensity in counts per 0.09 second indicated along each path. The circles approximate equal intensity contours at counting rate levels of 300, 200, and 100 per second, and indicate a peak intensity of about 400 counts per second at 16^h 15^m right ascension and -15° declination. Since the Scorpius source was scanned eight times in the 1963 flight, sufficient data were available to locate the position of peak intensity with rather high accuracy. The distribution of observed counting rates versus angular displacement from the peak intensity is shown by the dots in Fig 3. The displacement of the average of the measured pattern from the computed collimator-transmission-curve for a point source is -0.2°. Bowyer et al. conclude that the true size of the Scorpius source is probably less than 0.2°

The Crab Nebula was included in the field of view of the detector on three scans, one of which passed directly over the source, the second 1.5° off, and the third 2.5° off (Fig. 4). By averaging these three scans, Bowyer et al. obtained a signal from

*Sponsored jointly by the Office of Naval Research and the National Science Foundation

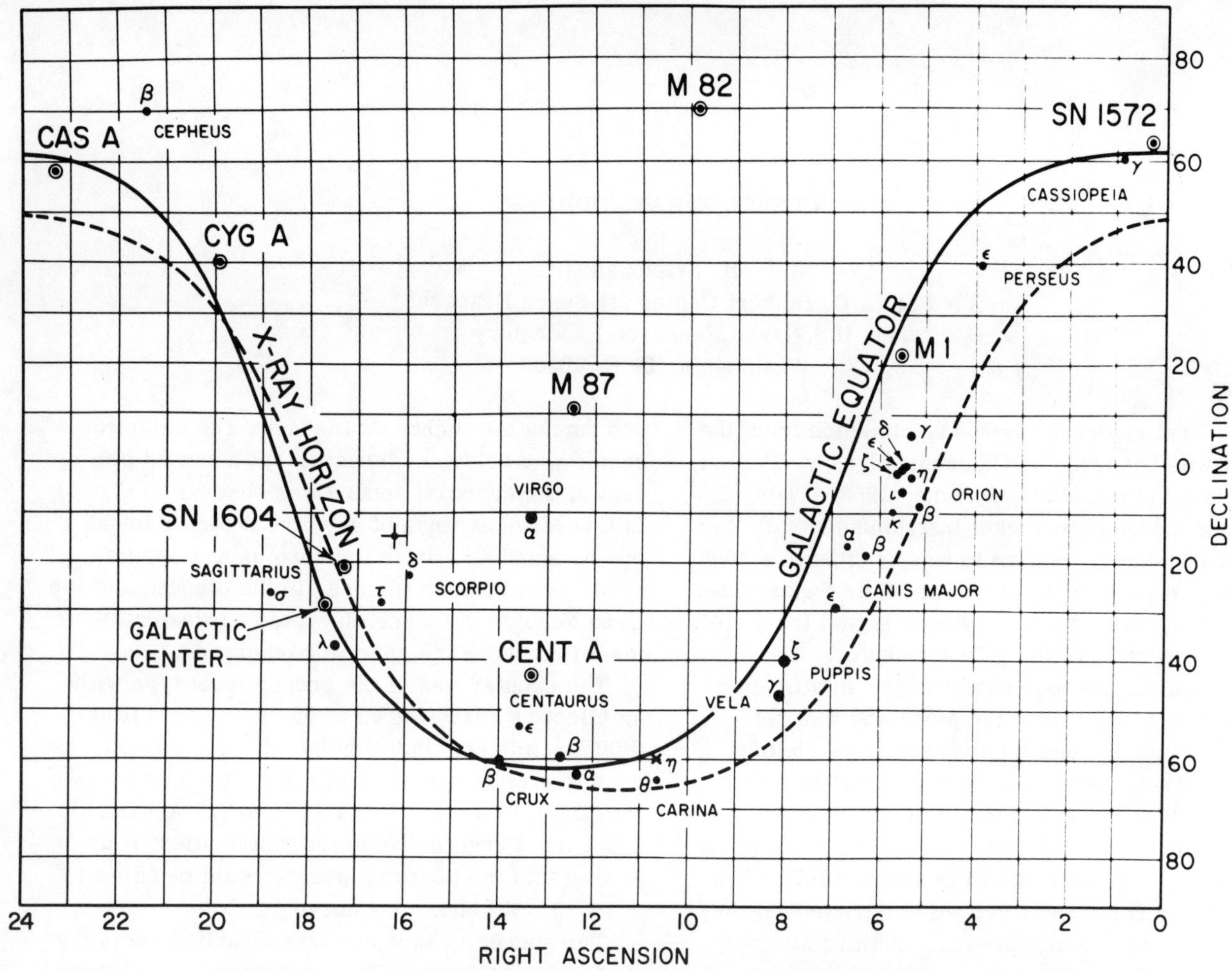

FIG. 1. X-ray horizon from peak altitude of Aerobee rocket launched April 29, 1963, at the White Sands Missile Range. All of the celestial sphere above horizon was scanned during the course of the flight. Positive X-ray responses were received from one source in Scorpio and from M1. No detectable signal was observed in scans of Cas A, Cyg A, Cent A, M87, M82, and SN 1572.

the source almost five times the standard deviation of the background. The direction of the peak response coincided with the Crab Nebula within 2°, which was as close as the statistical accuracy of the scanned profile permitted the positioning.

The intensity of X-rays from the Scorpius source is comparable to that emitted by the quiet Sun in the same wavelength range, yet the immediate neighborhood of the source is devoid of any visibly bright star, nebulosity, or radio emission. The Crab Nebula, in contrast, is a conspicuous source of radiation over a range of wavelengths from radio to visible. Its visible and radio nebula is about 6 light-years across and expanding at about 1000 km/sec. The Nebula is the debris of a supernova explosion that occurred in A.D. 1054, but no remnant of the original star can be identified.

On the basis of these early observations, it was proposed that both X-ray sources could be neutron stars, the highly compressed cores that may remain after explosion of the outer portions of supernovae. According to the theory of a neutron star (Chiu and Salpeter 1964; Morton 1964a, b), one can predict a strong X-ray source with essentially no visible or radio emission, such as required to explain the Scorpius source. A neutron star might also exist unobservable except for its X-ray emission in the center of the Crab Nebula.

Lunar Occultation Observation of the Crab Nebula

Present rocket X-ray techniques are too primitive to permit definition of the sizes of these sources

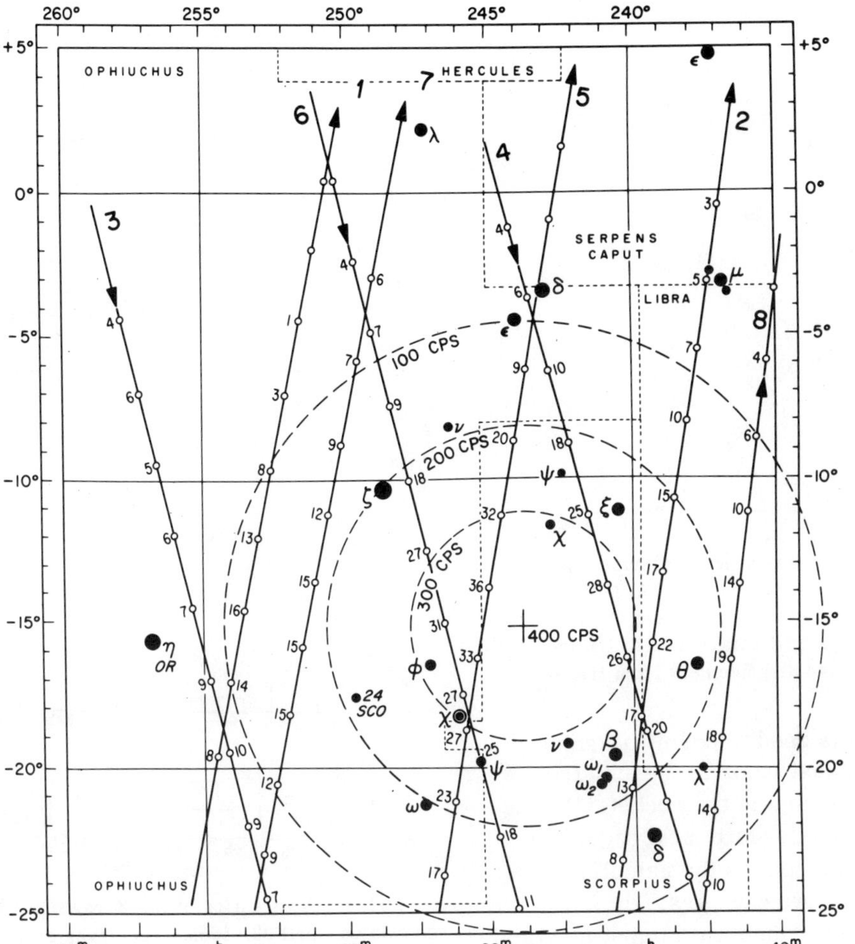

FIG. 2. Tracks of eight scans across Scorpius region. Numbers along tracks are counts per 0.09-sec interval. The dashed circles are best fits to equal intensity contours and indicate a central intensity peak of 400 c.p.s. at $\alpha = 16^h\ 15^m$, $\delta = -15°$ (Chiu and Salpeter 1964).

with sufficient resolution to support a stellar hypothesis. Bowyer et al. were able, however, to resort to the method of lunar occultation, and on July 7, 1964, at 2242 : 30 U.T. an Aerobee rocket was launched from White Sands Missile Range to observe the occultation of X-ray emission from the Crab Nebula. Lunar eclipses of the Crab Nebula are grouped at intervals of approximately 9 years. The eclipse of July 7 covered the Nebula at the rate of $\frac{1}{2}'$ (min of arc) per minute of time. Since the Nebula measures 6' across its greatest dimension, to observe a full eclipse would require 12 minutes of X-ray measurements from a rocket above the atmosphere. But the Aerobee rocket affords only 5 minutes of time above 100 km. The interval for the experiment was therefore chosen so that the eclipse of a central portion of the Nebula over a range of about 2' would be observed. The rocket was equipped with a stabilization system to orient two X-ray telescopes toward the Crab and to maintain steady pointing during the course of the occultation. If a neutron-star X-ray source exists in the center of the Crab, the occultation would be expected to produce an abrupt disappearance of X-rays. A gradual disappearance of X-ray emission would indicate that the X-ray source was an extended cloud.

Two Geiger counters of nearly identical construction were used to detect the X-rays. Each counter was in the form of a shallow rectangular box strung with five anode wires. Across the face of the box was a plastic window of Mylar, making an X-ray transmitting window of 114 cm^2 area. To provide crude wavelength discrimination, one window was 0.001 inch thick; the other was 0.00025 inch thick. Both counter volumes were connected by tubing so that they shared a common gas filling of 89.5 per cent Ne, 9.5 per cent He, and 1 per cent isobutane at atmospheric pressure.

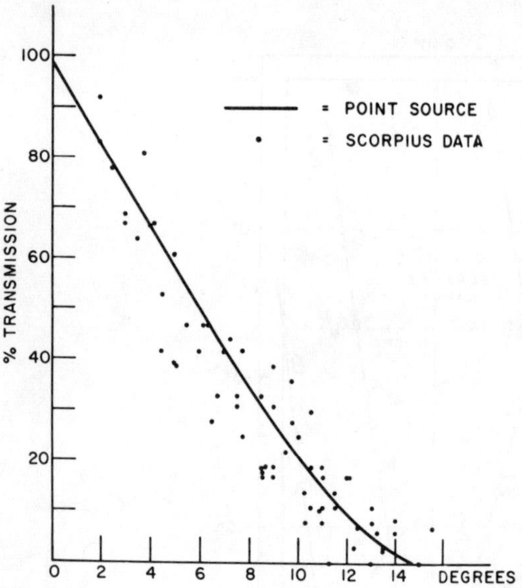

FIG. 3. Solid line is theoretical transmission-curve of honeycomb collimator used in April, 1963, Aerobee flight. The dots show the angular dependence of observed counting rates, with peak signal normalized to 100 per cent transmission. Average displacement from transmission-curve is -0.2°.

A gas flow system was used to permit flushing both tubes in series, just prior to launch, from a storage tank of gas carried in the rocket. During flight, the gas content was static at a regulated pressure. Figure 5 illustrates the calculated spectral efficiencies of the two counters. Up to a wavelength of about 3 Å both tubes had almost identical responses, but at longer wavelengths the thinner window provided increasingly greater efficiency. The two counters were mounted, one above the other, with their windows in the same plane facing outward, in a direction normal to the long axis of the rocket. Each window was covered by a honeycomb collimator, which limited the field of view. The angular response was roughly bell-shaped with a maximum cutoff angle of 14° on either side of the normal direction.

The Aerobee was stabilized by a cold gas jet-gyro referenced control system, manufactured by the Space General Corporation under contract to the U.S. National Aeronautics and Space Administration. From analysis of the sun signal, magnetometer signals, and the gyroscope data, Bowyer et al. conclude that the rocket fixed on its final orientation at 160 sec after launch and that the detectors pointed within 4° of the Crab. From 160 to 400 sec, the pointing did not deviate by more than $\frac{1}{4}$°.

Figure 6 shows the variation of X-ray counting rate through the course of the flight. At 100 sec

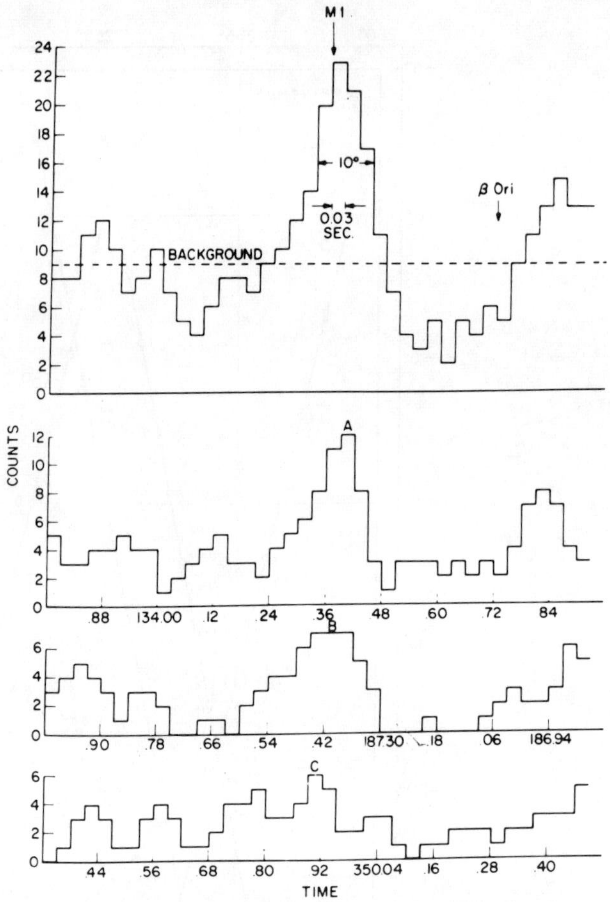

FIG. 4. Observation of the X-ray flux from the Crab Nebula on April 29, 1963. The upper graph is the sum of three separate scans A, B, and C. In scan A, the source passed directly across the center of the field of view, which was 10° in diameter; scan B was 1.5°, and scan C was 2.5° off center. A running mean of counts per 0.09 sec is plotted at intervals of 0.03 sec. The peak intensity corresponds to 52 counts/cm^2/sec.

the rocket was above the absorbing atmosphere for X-rays of wavelengths shorter than 10Å. As it rolled to acquire the Crab, the detectors scanned the Sun. The responses of the counters to the entry of the Sun into the field of view were almost perfectly mirrored as the Sun passed out of the field of view. There was no indication of counting hysteresis and, in fact, test exposures to intense sources in the laboratory had revealed no tendency of the counters to lag in recovery. Thirteen seconds elapsed between disappearance of the Sun's signal and acquisition of the final orientation to the Crab.

From 160 to 400 sec, the variation in X-ray counting rate can be attributed to the progress of the lunar occultation. Figure 7 shows the relationship between the time after launch and the

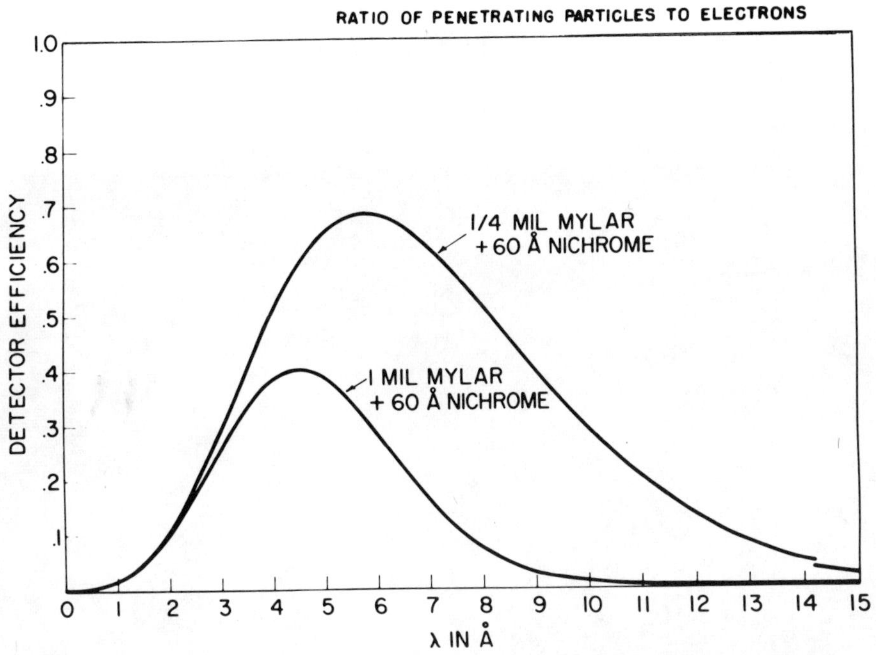

FIG. 5. Computed spectral sensitivity-curves for Mylar window counters.

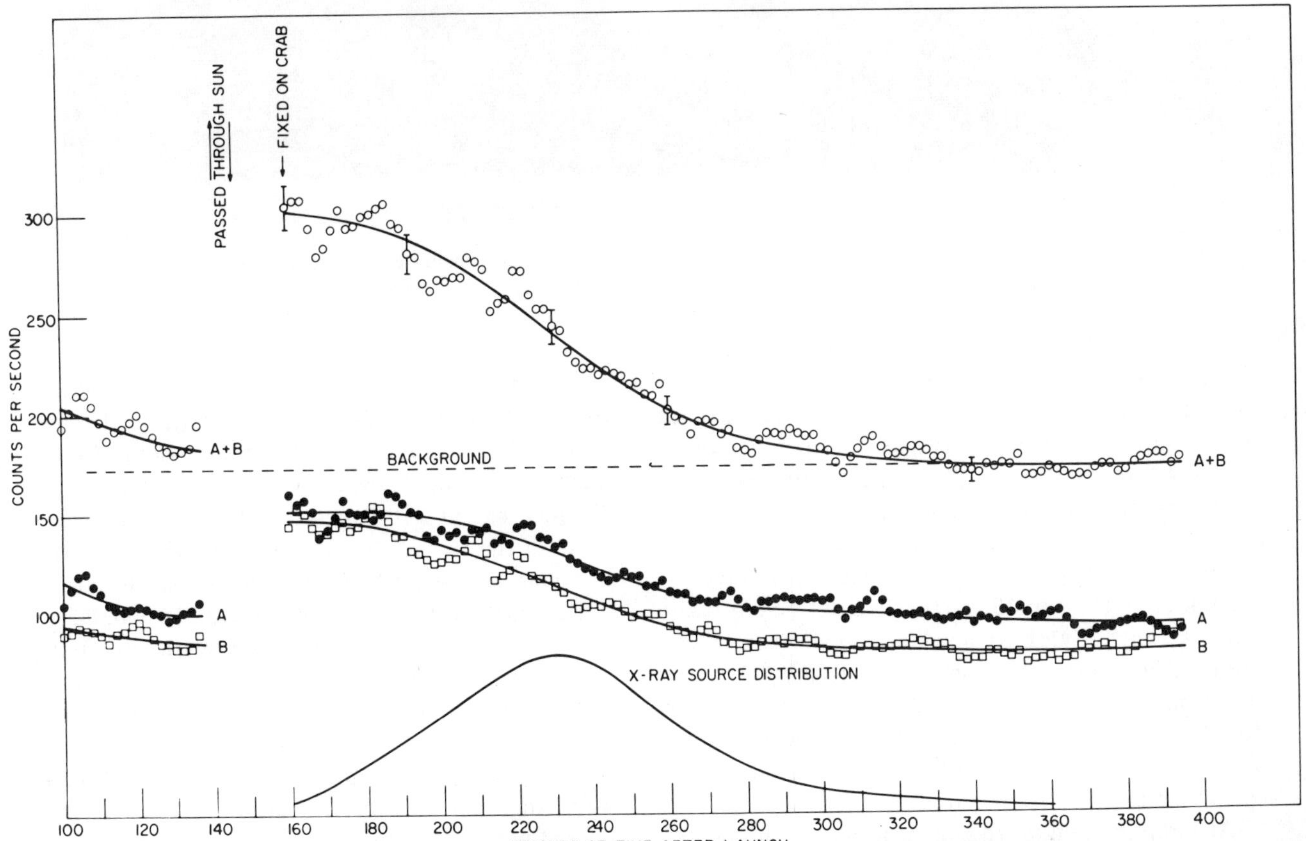

FIG. 6. Variation of the observed X-ray flux during the course of flight of July 7, 1964. The Mylar windows of counters A and B were 1 mil and $\frac{1}{4}$ mil thick, respectively. Counting rates were computed from the time required for a fixed count of 768 in each counter. A running mean is plotted at 2-sec intervals. The X-ray source distribution is the derivative of the A + B curve.

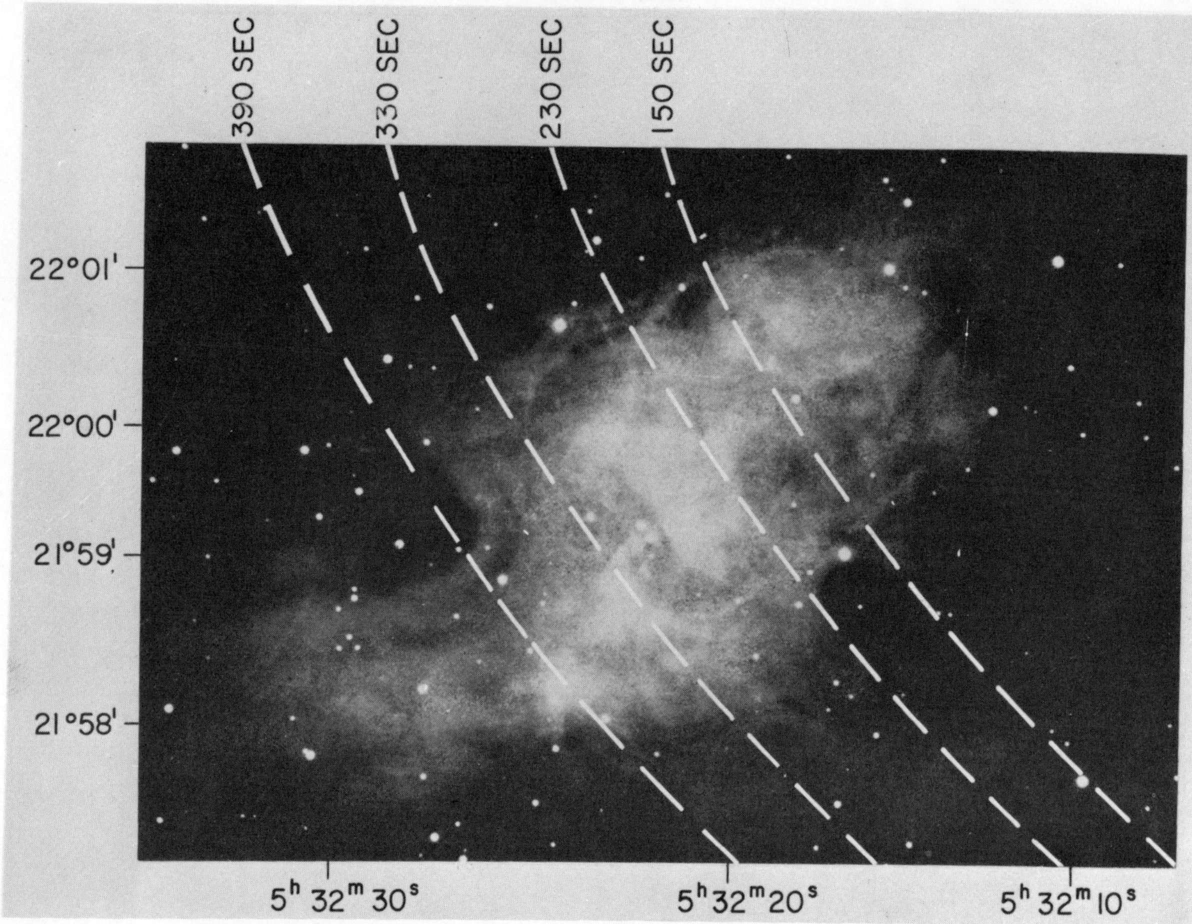

FIG. 7. Progress of the occultation of the Crab Nebula measured in seconds of time after launch of the rocket from White Sands Missile Range. The dashed curves represent the positions of the edge of the moon. A maximum rate of decrease in X-ray flux was observed at about 230 sec.

disappearance of the Crab behind the Moon. The peak of the flight was 221.4 km, which was reached at 250 sec. For all of the time interval from 160 sec (184 km) to 320 sec (199 km) during which the decrease of X-ray flux was observed, the rocket was more than five scale heights above the level of unit optical depth of the atmosphere for wavelengths shorter than 10 Å (the limit of counter A with its 1-mil Mylar window) and more than three scale heights above unit optical depth for 60 Å (the limit of counter B with its $\frac{1}{4}$-mil Mylar window). The decrease of X-ray flux during the occultation was, therefore, not related to the variation of rocket altitude.

At the beginning of the occultation, each counter exhibited a rather flat trend of counting rate versus time. This was followed by a relatively rapid decline to the background rate which had been observed in the 100 - 130 -sec interval. The curve of the sum of the two counter responses was differentiated to give the source distribution of Fig. 6. The results indicate that the total angular width of the source is about 1', which corresponds to about 1 light-year at a distance of 1,100 pc. The maximum of the X-ray flux distribution was observed at about 230 sec, when the occultation had not quite reached the center of the visible nebula. The fact that both counters responded with nearly the same counting rates would seem to indicate that the X-ray flux must be confined primarily to the region below 5 Å. However, a subsequent observation of the Crab has shown that the long-wavelength sensitivity of counter B had deteriorated before launch. The detectors had been exposed to rain while the rocket was in the launching tower for an hour before launch. Mylar is very pervious to water vapor, which destroys the long-wavelength sensitivity of the counter. Apparently the flushing of the counters before launch was not adequate to remove the trace of water-vapor contaminant.

X-Ray Surveys, 1964

Continuing the search for discrete X-ray sources, Bowyer et al. conducted two additional Aerobee surveys in 1964 from the White Sands Missile Range in New Mexico. The first of these was launched on June 16 when the zenith coordinates were $19^h\ 30^m$ right ascension, 32.4° declination. The galactic plane was mapped from the southern region of Scorpius, through Cygnus, to the northern part of Perseus. For the second flight, on November 25, the zenith right ascension was $8^h\ 48^m$, and the survey extended through Taurus to the southern portion of Puppis.

The Geiger counters were sensitive to wavelengths from 1 to 15 Å and were similar to those described in connection with the occultation observation but of larger area. Some of the counters were equipped with $\frac{1}{4}$-mil Mylar windows and some with $\frac{1}{2}$-mil Mylar windows. With the honeycomb collimator in place, the effective window aperture of each counter was 453 cm^2. By pairing two counters, mounted in the same plane parallel to the length of the rocket, an effective area for X-ray detection of about 906 cm^2 was obtained. This was about fourteen times as great as the window area used in the April, 1963, survey. Two sets of paired counters were used in each rocket and were oriented 180° apart. Because of the large area of plastic film window, a gas-flow system was required. The mixture of 89 per cent neon, 10 per cent helium, and 1 per cent isobutane was maintained at atmospheric pressure.

The counting rates of the individual counters, as well as the summed pairs, were metered by separate rate meter circuits. In the November, 1964, flight, the cosmic-ray background of one pair of counters was reduced by an anticoincidence arrangement which utilized a third pair of tray counters, with thick windows, mounted back to back with the X-ray survey counters.

On June 16, the rocket reached a peak altitude of 127 km and rolled with a period of about 8.5 sec. Its precession period was 422 sec and the precession cone angle was 144°. A solution of the orientation history was obtained to an estimated accuracy of 1.5°. The November 25 flight reached an altitude of 200 km. The roll period was 6.35 sec, precession period 236 sec, and cone angle about 140°.

In the June, 1964, survey, satisfactory data were obtained from only one pair of counters, which swept across most of the accessible sky. Much of the survey showed only a statistically varying cosmic-ray background with little difference between signals observed looking up and down. Superposed on this background were several X-ray signal peaks. The thin lines that lace the map of Fig. 8 are the traces of the counter view vector as it swept across the celestial sphere. Where the counters responded above background, the scan lines of Fig. 8 are covered by shaded strips. All of the observed X-ray sources are located relatively near the galactic plane. Those on scans 3-7 are within 5° of the galactic equator, and those on scans 1 and 10 are at galactic latitudes of 11° and 22°, respectively.

The isolated responses on scans 9 and 10 fit the position of the Scorpius source, which we derived from the April, 1963, observation. It remains the strongest source thus far observed. The telemetry record of scan 4 through Cygnus, which shows the second strongest source, is reproduced in Fig. 9.

The strong Scorpius source was observed late in the flight on passes 9 and 10 when the rocket was re-entering the absorbing atmosphere at 100.0 km and 93.4 km. On pass 11, when the rocket was at 86.7 km, no signal was detected. These three observations permit us to deduce roughly the spectral composition and the total flux out to about 10 Å. The pass at 93.4 km scanned directly across the previously determined position of the source. On the preceding and following passes, the displacement of the source relative to the center of the field of view was about 6.5°, so that the collimator transmission was about 25 per cent. The filtering effect of the air above the rocket permits one to infer qualitatively the spectral composition. At 93.4 km, with the Scorpius source at an elevation of 22.5°, unit optical thickness of the overhead air mass corresponded to $\lambda = 6$ Å. The observed flux of 6.6 counts/cm^2/sec was, therefore, confined largely to the range 1-6 Å. On the preceding pass at 100 km, the counting rate was 4.6 counts/cm^2/sec. When corrected for the collimator transmission factor of 25 per cent the flux becomes 18.7 counts/cm^2/sec. The difference in flux measured at the two altitudes may be attributed principally to radiation of wavelengths longer than 6 Å. Unit optical thickness

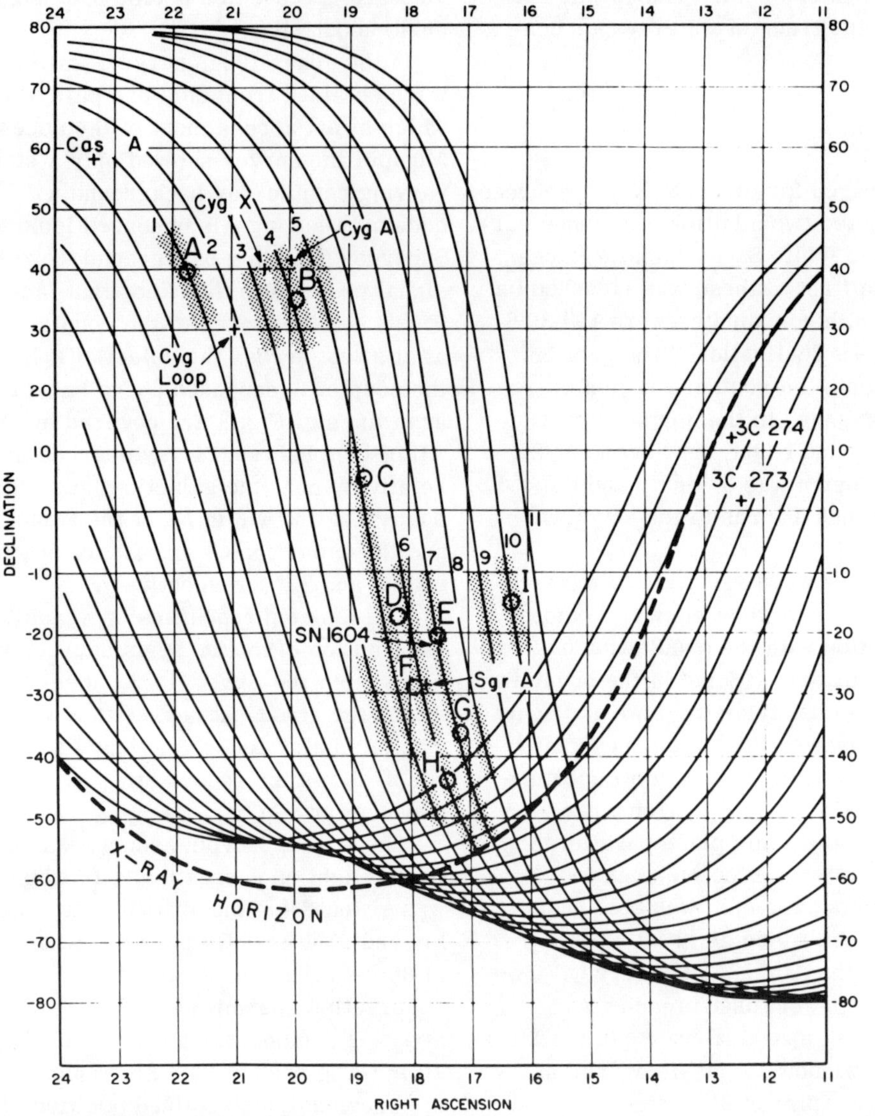

FIG. 8. Map of sky scanned by X-ray detectors in June 16, 1964, Aerobee flight. Thin lines trace path of view vector across celestial sphere on successive rolls. Shaded segments indicate portions of scan in which clearly detectable X-ray signals were observed above background. Circles are positions at which discrete sources have been identified.

corresponded to 8 Å at 100 km, and the spectrum was effectively cut off at about 10 Å. At 86.7 km, the wavelength for unit optical thickness was 4 Å. The combination of atmospheric opacity and reduced collimator transmission made the source undetectable against the background on scan 11 (Fig. 8). We conclude that approximately one-third of the observed flux from the Scorpius source falls below 6 Å and two-thirds between 6 and 10 Å. If the X-ray emission is attributed to the black-body radiation of a neutron star, the temperature of the Scorpius source, based on this spectral evidence, would be in the neighborhood of 2 or 3 million degrees.

Three of the X-ray signals, labeled A, B, and I on the map of Fig. 8, appear to come from clearly isolated discrete objects. Signal I fits the position which we identified with the strong Scorpius source in 1963. Signal B was seen on three successive passes through the Cygnus region. The telemetry record of scan 4, when the source passed close to the center of the field of view, is reproduced in Fig. 9. The signal envelope is consistent with the expected angular response from a source of small angular width compared to the collimator aperture. Although Cygnus A, the brightest extragalactic radio source in the sky, was within 6° of position B, there is no evidence of any X-ray contribution from it.

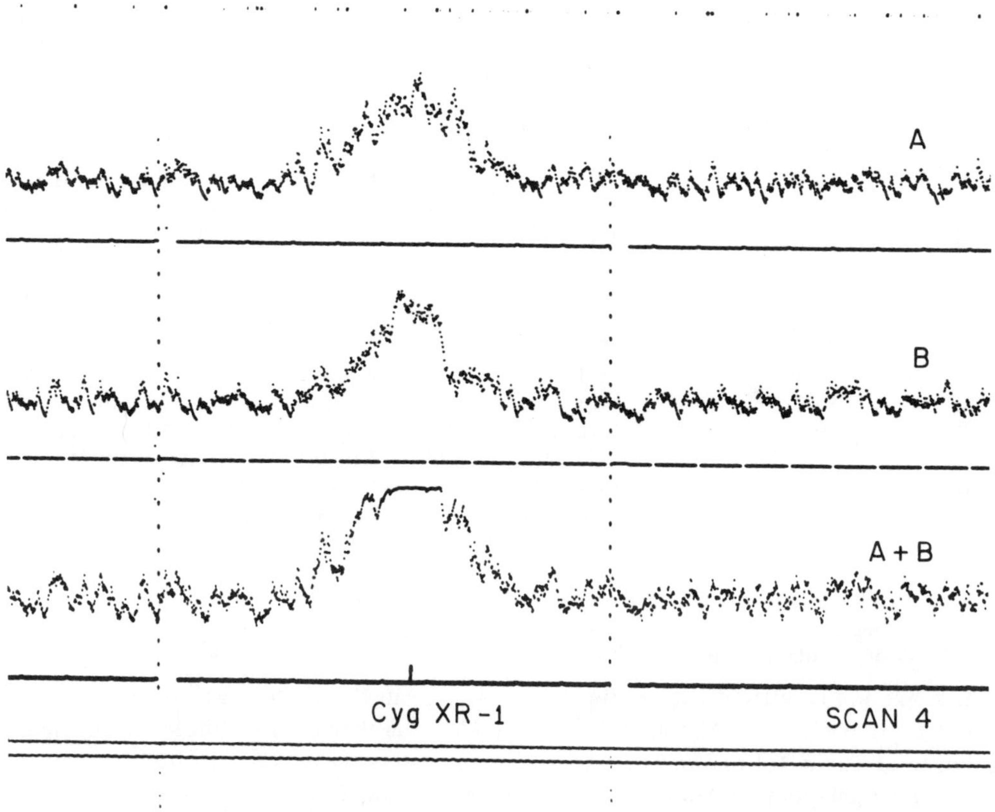

FIG. 9. Telemetry traces of Cyg XR-1 signals. Traces A and B are from individual counters, each equipped with $\frac{1}{4}$ mil Mylar window. Trace A + B is sum of signals from A and B. Dotted vertical lines are spaced 1 sec apart. Full deflection corresponds to 2,000 c.p.s.

Signal A was also discrete, and there is no outstanding optical or radio object at its position.

The portions of scans 5-7 which cover the general region near the center of the galaxy reveal a complex of emission which can be best resolved as the sum of six "point" sources labeled, C, D, E, F, G, and H. Table 1 lists the positions of the nine sources identified by the survey, the observed counting rates, and the fluxes computed for two assumed blackbody spectra. All of the listed sources were observed as signals clearly above background on each of two counters. Also included in Table 1 is the Crab Nebula. The sources are labeled "XR" for X-ray, and numbered within the various constellations according to brightness.

In the survey of November 25, satisfactory data were obtained from three of the four X-ray counters, Despite the wide expanse of sky that was searched from Perseus to Puppis, the only source observed was the Crab Nebula. The flux was 2.7 counts/cm^2/sec through $\frac{1}{4}$-mil Mylar and 1.6 counts/cm^2/sec through $\frac{1}{2}$-mil Mylar.

Discussion

The surveys completed thus far cover about 70 per cent of the celestial sphere. All of the X-ray sources observed lie rather close to the galactic plane and within ± 90° of the galactic center. This distribution resembles that of galactic novae and suggests that all the X-ray sources thus far observed may be associated with supernova remnants in the galaxy. Tau XR-1 was located by the lunar occultation observation of July 7, 1964, within 1' of the center of the Crab Nebula. Oph XR-1 fits the position of the 1604 Kepler supernova within the 1.5° accuracy of the observation.

There appear to be at least two types of X-ray "stars," those associated with radio and visible sources, such as Tau XR-1, Oph XR-1, and possibly Sgr XR-2, and those like Sco XR-1, Cyg XR-1,

Table I. X-ray sources.

Source	R.A. (1950)	Dec. (1950)	Flux* (Counts/cm2/sec)	Flux* 10^{-8} erg cm^2 sec †	‡	Remarks
Tau XR-1	$05^h\ 31.5^m$	22.0°	2.7	5.5	1.1	Within 1' of optical center of nebula
Sco XR-1	16 15	-15.2	18.7	38.0	7.9	Previous measurement (Bower *et al.* 1964) 12×10^{-8} erg/cm2/sec†
Sco XR-2	17 8	-36.4	1.4	2.9	0.6	
Sco XR-3	17 23	-44.3	1.1	2.3	0.5	
Oph XR-1	17 32	-20.7	1.3	2.7	0.6	1.1° from SN 1604
Sgr XR-1	17 55	-29.2	1.6	3.3	0.7	2.7° from galactic center
Sgr XR-2	18 10	-17.1	1.5	3.0	0.6	2.0° from M17
Ser XR-1	18 45	5.3	0.7	1.5	0.3	
Cyg XR-1	19 53	34.6	3.6	7.3	1.5	
Cyg XR-2	21 43	38.8	0.8	1.7	0.4	

*Uncorrected for atmospheric absorption. Measured 1/4-mil Mylar window
†Computed for 2×10^7 deg K black body, 1.5- 8 A
‡Computed for 5×10^6 deg K black body, 1.5- 8 A

and Cyg XR-2, which are not identified with radio or optical sources at their positions. Although the occultation observation of Tau XR-1 showed that its diameter was about 1 light-year, little is known of the dimensions of the remaining X-ray stars, other than that the true size of Sco XR-1 is probably less than 0.2°.

Since both the radio spectrum and optical emissions from the amorphous mass of the Crab have been identified as synchrotron radiation, can the observation of X-ray emission from an extended source also be attributed to synchrotron radiation? Woltjer (1964) has estimated the synchrotron flux in the far ultraviolet by assuming it to be the source of excitation of the filamentary shell. A simple extrapolation of the short-wavelength end of the optical spectrum through Woltjer's ultraviolet estimate to the X-ray region falls within an order of magnitude of the observed X-ray flux. Shklovsky (1964a) has also pointed out that, with a slight modification in the value of the interstellar absorption of light from the Crab Nebula, it is possible to fit the optical observations of synchrotron emission to the X-ray flux at about 3 Å with a common spectral index of -1.2. The spectral evidence derived from the November 25 flight can be fitted to a bremsstrahlung spectrum at about 10^7 °K, a black body at about 5×10^6 °K, and a synchrotron spectrum with an index of -1.1. More precise spectral data are urgently needed, but the evidence seems to favor the synchrotron process.

The Crab Nebula appears much larger at meter wavelengths than in the visible range. Most of the optical emission is concentrated within a region measuring about 2' in diameter, about 2-3 times smaller than the radio size. The diameter of the X-ray emitting volume is less than half that of the optical. Shklovsky (1964a) suggests that the small size of the X-ray nebula may be related to the lifetimes of highly relativistic electrons responsible for generation of X-ray synchrotron radiation. The lifetime varies inversely with the $\frac{3}{2}$ power of the magnetic field multiplied by the square root of the frequency. Estimates of the field strength range from 10^{-4} to 10^{-3} gauss, and it is most likely that the higher figure is reached near the center of the nebula. An electron radiating in the X-ray spectrum, $\sim 10^{18}$ c/s, in a field of 10^{-3} gauss would have a half-life of about 0.8 years. Shklovsky proposes that these high-energy electrons, E $\sim 10^{13}$ eV, are accelerated in a small central region of the nebula. Traveling at nearly the speed of light, they could get no farther than about 1 light-year from the region of acceleration. This would limit the size of the X-ray emitting region to roughly the observed value.

Shklovsky (1964b) and M Oda (private communication) have called attention to the close coincidence between the position which we established for the Scorpius X-ray source and the position proposed by Hanbury Brown, Davies, and Hazard (1960) for the center of a radio emitting shell associated with the radio object referred to as "the Spur." Radio frequency isophotes at 38 and 158 Mc/s have shown a spur of relatively intense emission which emerges from the galactic plane near the Sun and curves about an apparent center at 15^h, -20°. Considering the approximate nature of this position fix (about 10°), it agrees quite well with the position of $16^h\ 15^m$, -15°, for the X-ray source in Scorpius. Hanbury Brown et al., proposed that the Spur may be an object similar to the Cygnus loop which is believed to be the remnant of a Type II supernova. The Cygnus loop emission has a distribution consistent with a shell source about 10 pc thick and 40 pc in diameter at a distance of 770 pc. By analogy, the Spur would be at a distance of about 50 pc, and Shklovsky suggests that the age of this near-supernova may be 50,000 to 100,000 years. If the intensity of X-ray emission is simply related to the age, the Scorpius source may be 100 times less intense than the Crab, but its close proximity would still make it appear brighter. However, any visible nebula of synchrotron radiation associated with it would have too low a surface brightness to be detectable against the background sky. The same would be true of its radio synchrotron emission. Assuming that the Scorpius supernova remnant contains an X-ray emitting region about 1 light-year in diameter, its angular width at 50 pc would be about one-third degree. The scan data of the April, 1963, flight indicated only that the diameter did not exceed 0.2°. However, it should be possible in future observations with narrower mechanical collimation to establish the dimensions to 1' or better.

If Oph XR-1 is truly associated with Sn1604, it may be meaningful to compare it with Tau XR-1, since both are presumably Type I supernovae. Recent distance estimates place the Crab at 1.5 kpc and the Kepler supernova perhaps as far as 9 kpc. Distance alone should make the Crab approximately 50 times as bright, but its X-ray flux is only twice as bright. The Crab, however, is 550 years older, and the weakness of its X-ray flux may be attributed to aging at the rate of about 5 per cent per year. If the Kepler supernova is only 4 kpc distant, the rate would be only 1 per cent per year.

In view of the theoretical predictions of X-ray emission from the region of the galactic center (Gould and Burbidge 1963; Hayakawa and Matsuoka 1963), it is interesting to note that Sgr XR-1 is indeed close to the direction of the galactic center. We have located the X-ray source about 2.7° from Sgr A. The displacement appears to exceed the estimated positional uncertainty, but the difference is not so great as to rule out positively the possibility of a coincidence between the X-ray and radio source.

REFERENCES

S. Bowyer, E. T. Byram, T. A. Chubb, and H. Friedman, 1964, Nature, 201, 1307.

H. -Y. Chiu, and E. E. Salpeter, 1964, Phys. Rev. Letters, 12, 413.

R. Giacconi, H. Gursky, F. R. Paolini, and B. B. Rossi, 1962, Phys. Rev. Letters, 9, 439.

________. 1963, ibid., 11, 530.

R. J. Gould, and G. R. Burbidge, 1963, Ap. J., 138, 969.

R. Hanbury Brown, R. D. Davies, and C. Hazard, 1960, Observatory, 80, 191.

S. Hayakawa, and M. Matsuoka, 1963, Prog. Theoret. Phys. 29, 612.

D. C. Morton, 1964a, Nature, 201, 1308.

________. 1964b, Astr. Tsirk., No. 304.

L. Woltjer, 1964, "X-Rays and Type I Supernova Remnants" (preprint).

COSMIC X-RAY SOURCES, GALACTIC AND EXTRAGALACTIC

E. T. Byram, T. A. Chubb, H. Friedman

For the past few years, the X-ray survey program of the Naval Research Laboratory has employed large-area Geiger counters carried aloft in unstabilized rockets. The spin and precession of the rocket permit the detectors to scan slices of the sky and thus to map a grid whose fineness depends on the rate of roll and precession. Relatively fast roll and precession produce a more detailed map at the expense of poorer counting statistics in each traverse of a source. In the flight that produced the present results, the roll period was 15 seconds, as compared to 8 seconds and 4 seconds on previous flights. As a result, the detection sensitivity for a single transit was improved, but the scans were fewer and more loosely spaced. Further improvement in the detector sensitivity was attained from the rejection of cosmic-ray background count, through the use of anti-coincidence counters. With the enhancement of signal-to-noise ratio, it was possible to detect sources approximately four times weaker than in earlier surveys. For the first time, signals have been detected from the directions of individual radio galaxies. The identifications, however, are still uncertain to the extent of about 1.5 deg, which is the positional accuracy derived from the aspect solution and the location of peak responses in the telemetry record.

Previous surveys have identified approximately a dozen sources. Naval Research Laboratory (NRL) observations[1] in 1964 produced a list of ten sources, six of which were grouped in the general vicinity of the galactic center, two in the Cygnus region, and one in Scorpius. The tenth source was identified with the Crab Nebula. Surveys by Fisher et al.[2] of the Lockheed Corporation and by Clark et al.[3] have confirmed and refined some of the positions, disputed others, and identified some new positions. The NRL survey of April 1965 again covered the galactic center region. It is clear that this region is more complex than previously indicated and that higher instrumental resolution is essential to isolate the individual sources. In this report we discuss only the observations of the Cygnus region, Cassiopeia, and the scan of M-87 and neighboring regions at high galactic latitude.

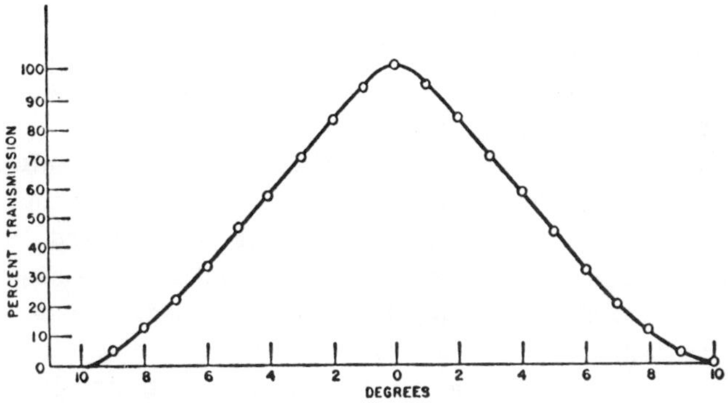

FIG. 1. Angular-transmission curve of hexagonal honeycomb collimator in rocket pitch direction.

The honeycomb baffle used to define the field of view of the Geiger counter in the April 1965 flight was identical with that used in June 1964. Figure 1 shows the angular transmission pattern in the pitch direction. The width at half-maximum was 8 deg. In theory, a perfect hexagonal honeycomb should give a triangular transmission curve with small wings at the edge of the field of view. The rounding of the peak is a result of slight imperfections and irregularities over the large collimator area. The results reported here were obtained from two counters, mounted one above the other in the same plane of the instrument section, parallel to the long axis of the rocket. The center of the field of view was normal to the roll direction. Each counter provided an effective window area of 453 cm^2. One was equipped with a window of Mylar 1/4 mil (0.0064 mm) thick; the window of the second counter was 1 mil (0.0254 mm) thick. The counting gas mixture was argon-isobutane at atmospheric pressure, and the gas path from window to wall was 2.54 cm. In the 1964 counters, the gas

Reprinted from Science **152**, 66 (1966).

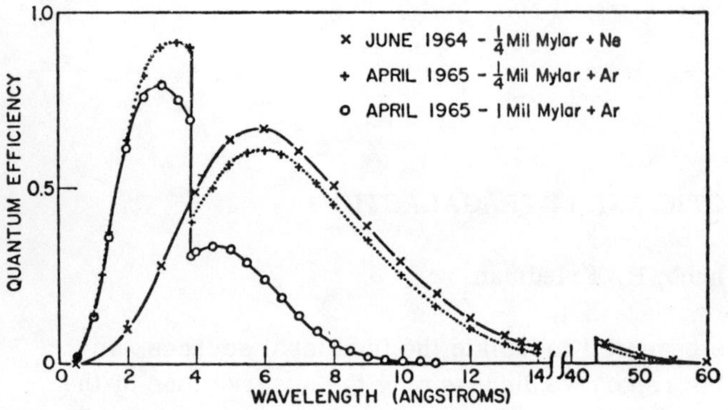

FIG. 2. Spectral-efficiency curves for Geiger counters used in June 1964 and April 1965.

was neon-isobutane. Figure 2 shows the counting efficiencies plotted against wavelength. The use of argon rather than neon greatly enhances the response below 3.9 Å. Each counter was backed by an identical, flat, box-shaped counter which served as an anticoincidence shield against cosmic rays, with rejection efficiency of about 85 percent. The combined counting rates of the two counters were prescaled by a factor of 32, and the time of arrival of every 32nd count was recorded. Pulses from individual counters were also presented directly to rate meters, whose integrated signals were separately recorded.

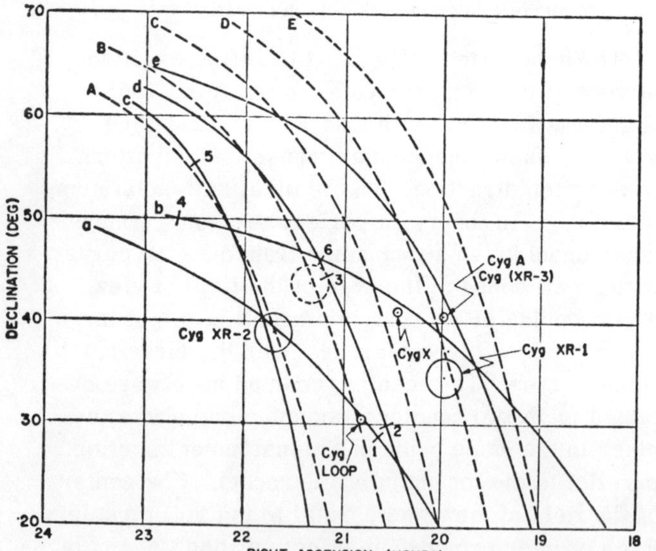

FIG. 3. Scans of Cygnus region. Dashed lines, scans A, B, C, D, and E, are tracks of June 1964 survey. Numbered positions correspond to signals shown in Figs. 6 and 7.

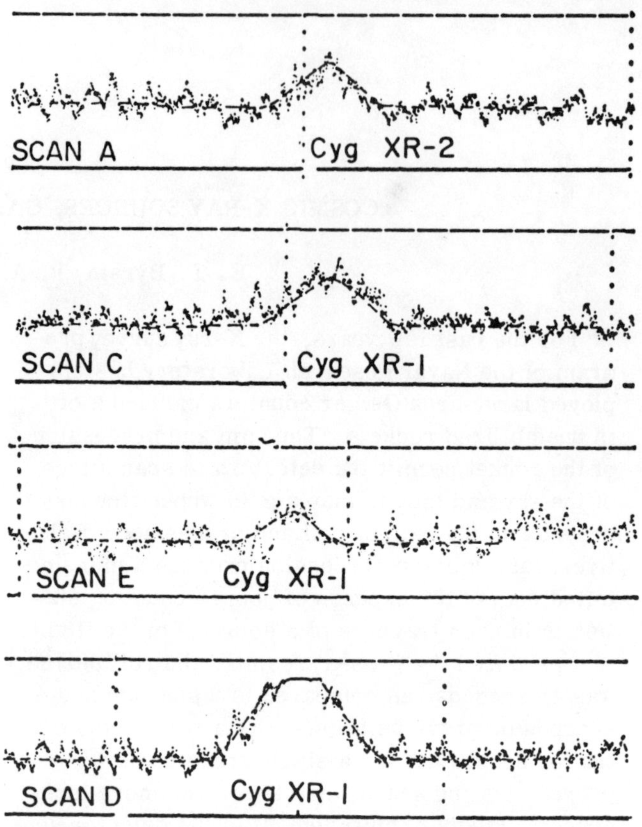

FIG. 4. Telemetry records of scans of Cygnus region in June 1964. Output of rate meter covers 0 to 2,000 count sec^{-1}. Scans are labeled to match Fig. 3. Signal on scan D is saturated.

Figure 3 is a map of scans across the Cygnus region. The dashed curves are the paths of the scans of the June 1964 flight; the solid curves are the tracks of the April 1965 scans. Figure 4 shows the telemetry records of the rate-meter signals observed in the 1964 survey. Full scale was 2,000 counts per second. The signal on scan D was saturated. The relative strengths of the three signals on scans C, D, and E indicated that the source had passed nearly through the center of the field on scan D but was displaced by 1 deg in pitch toward greater right ascension. The position was, within an estimated uncertainty of 1.5 deg, at the center of the error circle on the diagram. The computed counting rate for the source, after correction for displacement from the center of the field of view, was 3.6 counts cm^{-2} sec^{-1}. Cygnus XR-2 was observed on scan A but was not detectable on adjacent scans. Its true position was therefore taken to be on the path of scan A and the observed counting rate was 0.8 count cm^{-2} sec^{-1}.

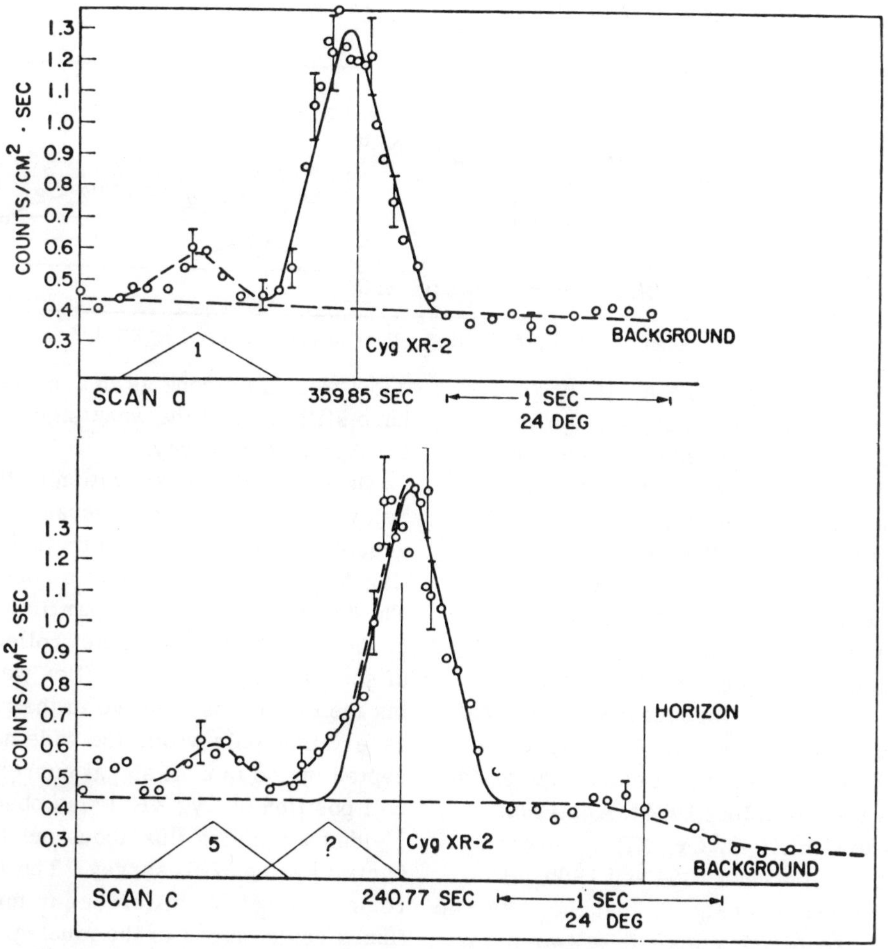

FIG. 5. Scans of Cyg XR-2 in April 1965. Numbered positions identified on Fig. 3. Solid-line envelopes match transmission pattern for a discrete source, Fig. 1. Original data are time intervals for every 32 counts. Plotted points are averages of counting rates computed for 92 counts spaced every 32 counts. Each point, therefore, retains 2/3 of its information from preceding and following points.

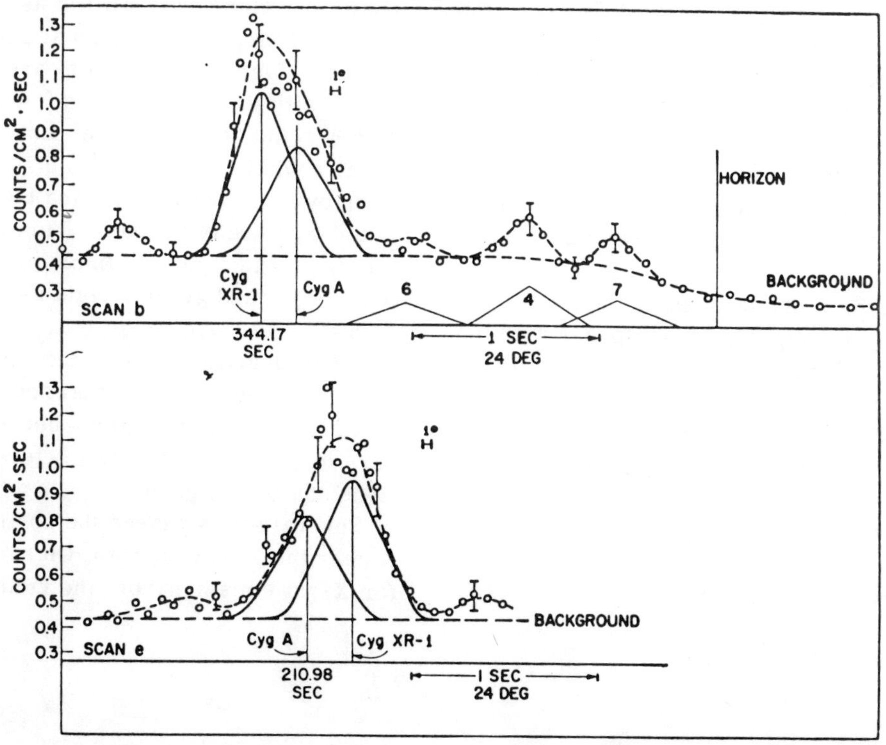

FIG. 6. Scans of Cygnus XR-1 and Cygnus A, April 1965. Solid curves match transmission pattern for a discrete source, Fig. 1. Positions 4 and 6 are shown on Fig. 3.

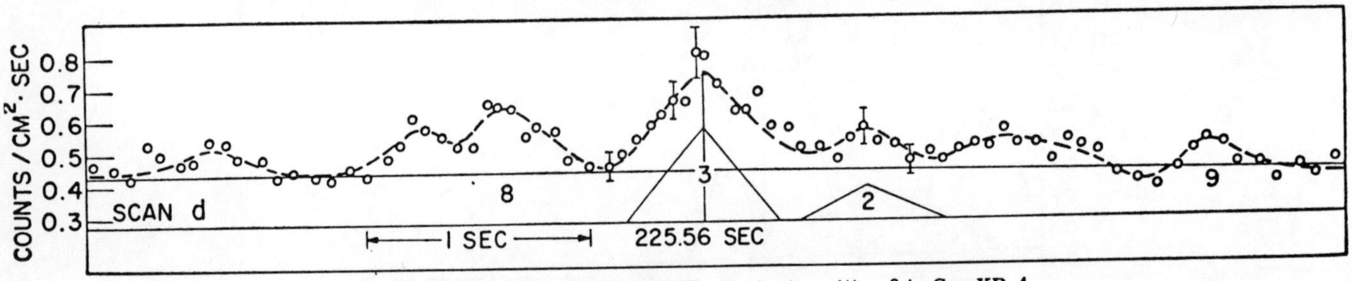

FIG. 7. Scan d (Fig. 3), April 1965. Peak at position 3 is Cyg XR-4.

Figures 5, 6, and 7 show the April 1965 observations of the same region. In Fig. 5, the vertical lines, marked Cyg XR-2, on scans a and c show the times of closest approach to the positions of June 1964. The peaks of the two signals are within the 1.5-deg error circle of the 1964 position. The fluxes were 0.9 and 1.0 count cm^{-2} sec^{-1} on scans a and c, respectively. After correcting for distance from the center of the collimator field, the averaged flux was 0.97 count cm^{-2} sec^{-1}. Without considering, for the moment, any difference in sensitivity of the 1965 counters from those of the 1964 flight, we conclude that the Cyg XR-2 source had remained essentially unchanged from June 1964 to April 1965.

In contrast to Cyg XR-2, Cyg XR-1 showed a remarkable change between 1964 and 1965. It is clear from scans b and e (Fig. 6) that the observed signal was an unresolved composite of two major sources, with just a suggestion of a weak contribution from still another component on scan b. The vertical lines marked Cyg XR-1 give the times of closest approach to the positon of that source in the 1964 survey. The lines marked Cyg A give the times of closest pointing toward that radio galaxy. Correcting for pitch displacement, the computed counting rates for Cyg XR-1 were 0.94 and 0.87 count cm^{-2} sec^{-1}, respectively, and the average was 0.91 count cm^{-2} sec^{-1}. Again, without regard for any difference in counter efficiency, the new fluxes for Cyg XR-1 are only one-fourth as great as in 1964.

From Fig. 6, it can be seen that a satisfactory fit of the second component corresponds with the position of Cyg A. The counting rates are 0.45 and 0.40 count cm^{-2} sec^{-1} on scans b and e, respectively (Fig. 6). Following our notation of X-ray sources according to brightness, Cyg A would be Cyg XR-3 in the NRL list.

The detectability of the signal from Cyg A in the April 1965 flight was made possible by the near disappearance of the flux from Cyg XR-1. If Cyg XR-1 were as strong as in June 1964, it would have still masked the weak signal from Cyg A in the April 1965 survey.

Only one other observation of the Cyg region is relevant to the present discussion. The Lockheed group surveyed the region including Cyg XR-1 and Cyg A in October 1964 with a fan-beam collimator. The length of the fan was parallel to the long axis of the rocket. As the rocket rolled, the positions of sources in the roll direction were determined, but the elevations relative to the roll axis were ascertained only within the wide band of sky accepted by the fan. A signal corresponding to the roll position of Cyg XR-1 was observed and yielded a counting flux about one-fifth of that observed by the NRL survey. The Lockheed group remarked that the difference in measured count fluxes could be due to the much greater bandwidth of the NRL detectors. No signal was observed from Cyg A; Cyg XR-2 was outside the limits of the fan beam.

Scan d, Fig. 7, clearly shows a source at position 3 which we shall list as Cyg XR-4. Another source is indicated at positions 1 and 2 near the Cygnus loop but appears not to be coincident with it. Signals 1 and 2 were studied to see if they could be fringing responses to Cyg XR-1, but it would be necessary to displace Cyg XR-1 in pitch by more than 1 deg, and the roll discrepancy would be 2 deg on trace d. Finally, signal 1 is stronger than signal 2, implying a true source position toward greater right ascension. Signals 4, 5, 7, and 8 are equal to or greater than 3 standard deviations (σ) of the background. It is clear that several additional sources are indicated in the Cygnus region, but it will be necessary to scan in finer detail before reliable positions can be assigned.

Several scans passed the position of Cassiopeia A near the X-ray horizon. On each scan, evidence for X-ray emission from the vicinity of Cas A was

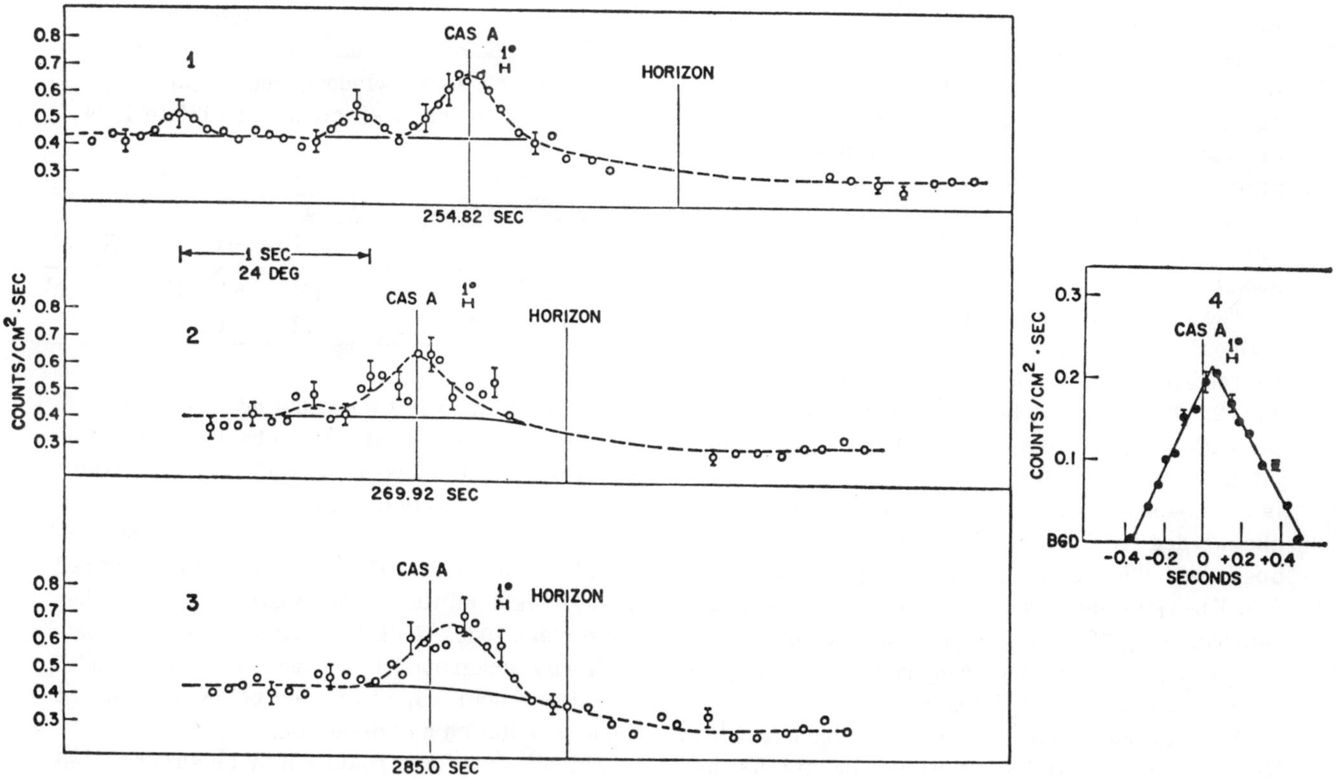

FIG. 8. Scans of region near Cassiopeia A, April 1965. Scans 1, 2, 3 each passed within 3 deg of Cas A in pitch direction. Curve 4 is the sum of scans 1, 2, and 3 plotted relative to the direction of closest approach to Cas A.

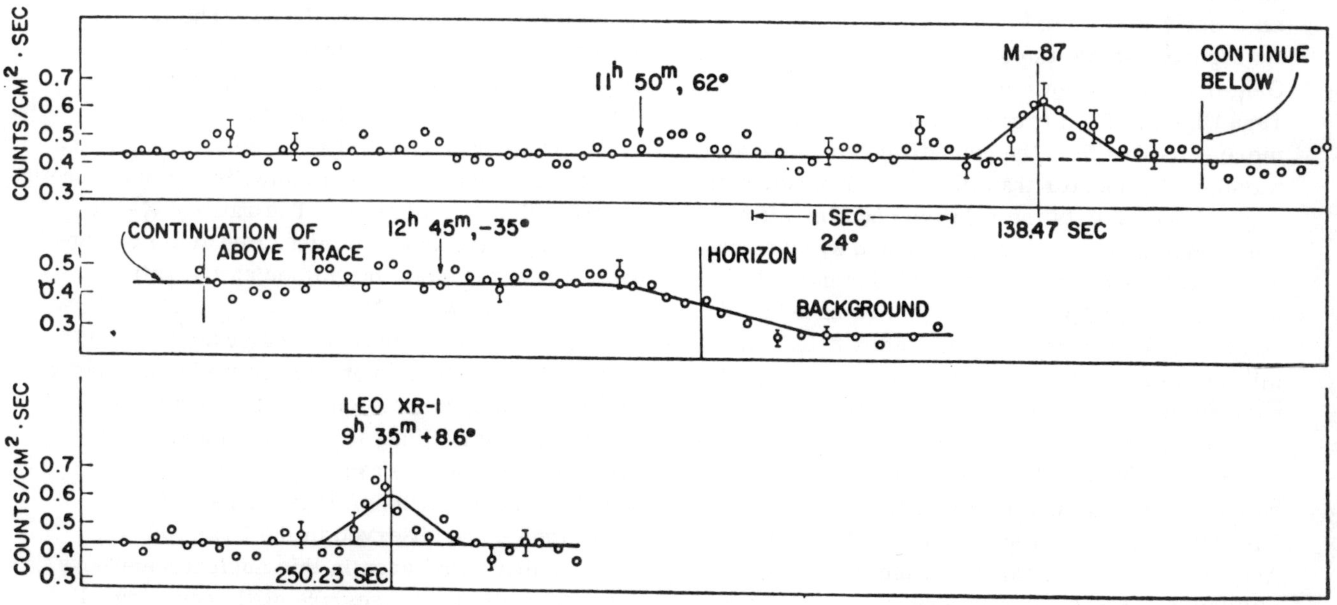

FIG. 9. Scans through M-87 and Leo XR-1.

observed. Figure 8 shows the signals along three scans that passed within 3 deg of Cas A. Curve 4 was obtained by adding the observed fluxes at corresponding angular displacements from the direction to Cas A. The resultant plot indicates a source within 1 deg of roll in the direction of closest approach to Cas A. Signals observed at pitch angles larger than 3 deg get progressively weaker, indicating that the source is indeed closer to Cas A than the 3-deg displacements of the closest scans.

The celestial sphere was loosely scanned at high galactic latitudes. Only two sources were observed above the 3-σ level. These are indicated in Fig. 9. The stronger is very close to M-87, which lay within 1 deg of the scan track. The maximum pitch-angle uncertainty is $\pm$ 3 deg, since the source is not observed on two adjacent scans separated about 6 deg from the track on which the signal was observed. The second source, which we designate Leo XR-1, is displaced 2 to 4 deg from the nearest radio galaxy, 3C-222. The indicated coordinates are for a position on the scan track; the uncertainty in pitch may be (+6 deg -9 deg).

As in previous surveys, a diffuse background of X-rays was observed. This background is distinguished from cosmic-ray background by subtracting the counting rate when the instruments are pointing downward from that observed when they are pointing above the horizon. The average X-ray background flux measured this way was 0.75 count cm^{-2} sec^{-1} $steradian^{-1}$.

Figure 2 shows the counting efficiencies of the Geiger counters flown in the June 1964 and April 1965 flights. The argon-filled counters are much more sensitive than the neon-filled counters at wavelengths shorter than the argon-K absorption edge. Table 1 lists the relative counting efficiencies based on exposures to assumed synchrotron and free-free X-ray spectra. All the data plots shown here for April 1965 flight are based on the combined responses of two counters fitted with 1/4-mil and 1-mil Mylar windows, respectively. Comparison of the responses of these two counters provides a "two-color photometry" of the X-ray sources. Here it suffices to mention that the ratios indicate a spectrum as hot or hotter than a 50-million-degree bremsstrahlung distribution. Accordingly, we conclude that the ratio of counter efficiencies appropriate to comparison of the ratio of 1965 to 1964 counting rates is unity, or slightly greater.

Table 1. Relative efficiencies of counting equipment used in April 1965 (combination of two counters, 1/4- and 1-mil-thick Mylar windows, argon-isobutane) and June 1964 (single counter 1/4-mil-thick Mylar window, neon-isobutane). Figures are ratios of efficiency of 1965 to 1964 equipment.

Synchrotron index			Free-free (deg K)		
0.5	1.0	1.5	10^8	5×10^7	3×10^7
1.1	0.9	0.7	1.2	1.0	0.9

Tables 2 and 3 list the counting rates from several sources and the diffuse background. Comparison is made between the radio, optical, and X-ray emission rates. The ratios of X-ray to radio fluxes from the discrete sources Cygnus A and M-87 and of the diffuse background compared to the integrated flux of radio galaxies are of the same magnitude. This comparison suggests that the X-ray background is composed of a multitude of discrete sources, at present below the limit of X-ray resolution and detection.

The observational results may be summarized as follows:

1) Comparison of two surveys of the Cygnus region, about 1 year apart, shows that Cyg XR-1 has decreased in brightness about 75 percent.

2) X-ray sources were observed in the directions of Cas A, Cyg A, and M-87. The X-ray emission rates for these three sources were between 1 and 2 orders of magnitude greater than the combined radio and optical emissions.

3) The unresolved X-ray background showed a degree of random variation indicative of many unresolved discrete sources. The ratio of X-ray background brightness to integrated radio brightness was of the same magnitude as the ratios observed for Cyg A and M-87.

Cygnus XR-1 is the first clear example of an X-ray variable. It cannot be specified how rapidly the variation occurred, only that it occurred between the observations in June 1964 and April 1965. At present one can only speculate about possible interpretations. It is well known that light ripples are observed in the Crab Nebula, which indicate that even in this ancient supernova very energetic processes are still occurring. If the X-radiation is synchrotron in origin, the lifetime of the high-energy electrons may be of the

TABLE 2. X-RAY FLUXES AND EMISSION RATES.

Source	Count (cm^{-2} sec^{-1})	Flux (1 to 10 Å) (10^{-8} erg cm^{-2} sec^{-1}) Free-free, 5×10^7 deg K	Flux Synchrotron $\alpha = 1$	Distance (parsecs)	Emission rate (erg sec^{-1}) X-ray (1 to 10 Å) Free-free, 5×10^7 deg K	Emission rate X-ray Synchrotron $\alpha = 1$	Emission rate Radio	Emission rate Optical	Ratio of x-ray (synch) to radio
Cyg A	0.4	0.5	0.4	220×10^6	3×10^{46}	2×10^{46}	4.4×10^{44}	4×10^{44}	45
M-87	.2	.2	.2	11×10^6	3×10^{43}	3×10^{43}	3×10^{41}		100
Cas A	.3	.4	.3	3.4×10^3	5×10^{36}	4×10^{36}	2.6×10^{35}		15
Tau A	2.7*	3.2	2.6	1.1×10^3	4.5×10^{36}	3.6×10^{36}	8×10^{33}	10^{36}	450
Cyg XR-1	3.6*	4.3	3.6						
	0.9	1.1	1.0						
Cyg XR-2	.8*	1.0	.8						
	1.0	1.2	1.1						

*June 1964. All other data refer to April 1965.

order of a year and the source may undergo a large variation due to a statistical lapse in generation of relativistic electrons. Perhaps Cyg XR-1 was in a relative "flare" condition at the time of observation in 1964. Some neutron-star models of an X-ray source indicate a cooling time of the order of a year. Although a careful analysis of observations of the remaining nine sources isolated in 1964 has not yet been completed, a preliminary examination shows no very marked variations comparable to the decay of Cyg XR-1. The need to monitor the fluxes of X-ray sources at frequent intervals is clearly indicated.

Although Cyg A is the brightest extragalactic radio source and was discovered in 1946, 5 years passed before Baade and Minkowski[4] identified it with a weak optical galaxy about 700 million light-years distant. The central portion of the galaxy appeared to consist of two bright condensations separated by about 2 seconds of arc, and it was proposed that these were two galaxies in collision. Subsequent studies indicated that the radio emission came from two vast regions whose centers lie about 100 seconds apart. The radio flux is 4×10^{46} erg sec^{-1}, about four times as great as the optical flux; and even at its great distance, Cyg A appears as bright as the sun on meter wavelengths. The radio evidence indicates that Cyg A is a synchrotron source and that the original idea of colliding galaxies is not valid. With the synchrotron hypothesis, however, it became necessary to explain how the billion-volt electrons, needed to produce the radio emission, were accelerated and to account for the total energy contained in electrons and magnetic fields. Because the synchrotron process is inefficient, the total energy content must be in excess of 10^{60} erg, and perhaps as great as 10^{62} egr.

TABLE 3.
BACKGROUND FLUXES.

Diffuse X-ray (1 to 10 Å)
7.9 count cm^{-2} sec^{-1} steradian^{-1}
9×10^{-8} erg cm^{-2} sec^{-1} steradian^{-1}
(synchroton a= 1)
Integral of radio galaxies
(100 to 10^4 MC sec^{-1})
1.5×10^{-9} erg cm^{-2} sec^{-1} steradian^{-1}
Ratio of X-ray flux to radio flux
60

In the thermonuclear burning of stars, the efficiency of conversion of mass to energy is of the order of 1 percent, and the burning of 1 sun produces about 10^{52} erg. It would, therefore, take the nuclear conversion of 10^{10} suns, or the entire mass of a medium-sized galaxy, to produce the energy content of Cyg A. Attempts to explain the energy in terms of gravitational collapse of a superstar have thus far failed. The present X-ray observation, which indicates that X-ray emission is more than an order of magnitude greater than radio plus optical emission, correspondingly increases the difficulty of explaining the total content of radio galaxies.

M-87 (Virgo A) is an elliptical galaxy, one of the brightest in the Virgo cluster, at a distance of

11 megaparsecs. It is about 5 minutes of arc in angular size, and the brightness is highly concentrated toward the center. Its total mass may be about 10^{12} solar masses, 10 times the mass of our galaxy. A luminous jet, 20 seconds long (about 1000 parsecs) bursts from its center, and its light is highly polarized. It provided Shklovsky[5] with the first evidence for the role of synchrotron radiation in radio galaxies. The radio power of M-87 is about 1000 times weaker than that of Cyg A, but it ranks immediately behind Cyg A and Cen A in flux received at the earth.

The detection of X-rays from the direction of Cas A provides evidence for a second supernova X-ray source in addition to the previously identified Crab Nebula. Cas A is believed to be a Type II supernova, whereas the Crab is Type I. It is estimated from the expansion velocity that the supernova explosion took place in the year 1702, ± 14. Taking the distance to Cas A as 3.4 kiloparsecs, its X-ray power is about equal to that of the Crab.

Several weaker sources were discovered in the April 1965 survey, and it is clear that many more are indicated by signals near the 2-σ background level. A modest increase in sensitivity and resolution should suffice to reveal these sources clearly. The great majority of X-ray sources are still unidentified with optical or radio objects, and the identifications proposed here for Cyg A, Cas A, and M-87 should be accepted with some caution because of the 1.5-deg uncertainty in positions. However, the circumstantial evidence for these identifications is strong. In the history of radio astronomy, Cyg A was the first discrete source detected; the Crab was the first radio source identified with an optical object; and M-87 was the first radio source identified with an optical galaxy. Cas A is the brightest radio source in the sky. It seems more than fortuitous that the first X-ray sources that can be associated with radio sources, within the present uncertainty of the observations, should fit four of the most spectacular radio sources. At the same time, the large number of X-ray sources that do not fit radio supernova remnants or radio galaxies implies the existence of a new type of celestial object observable only in the X-ray spectrum.

REFERENCES

[1]S. Bowyer, E. T. Byram, T. A. Chubb, H. Friedman, Science 146, 912 (1964); 147, 394 (1965).

[2]P. C. Fisher, H. M. Johnson, W. C. Jordan, A. J. Meyerott, L. W. Acton, Astrophys. J. 143, 203 (1966).

[3]G. W. Clark, G. Garmire, M. Oda, M. Wada, R. Giacconi, H. Gursky, J. Waters, Nature 207, 584 (1965).

[4]W. Baade and R. Minkowski, Astrophys. J. 119, 206 (1954).

[5]I. S. Shklovsky, Astron. J. 32, 215 (1955).

[6]Sponsored jointly by ONR and NSF.

NIGHT SKY X-RAY SOURCES

Philip C. Fisher, Willard C. Jordan, and Arthur J. Meyerott
Lockheed Missiles and Space Company,
Research Laboratories, Palo Alto, California

On October 1, 1964, a survey of night sky X-ray sources lying at low galactic latitudes was performed from a rocket launched at White Sands, New Mexico. As data reduction is not complete, this contribution is in the nature of a progress report and is primarily concerned with a description of the experiment and a preliminary appraisal of the results obtained. Some spectral information for the X-ray source in Scorpius will be given and the location of a number of sources in the general direction of the galactic center will be noted.

At the time the observing program was selected, only two X-ray sources appeared to have been reliably located. The positions were for the bright source in Scorpius observed by Giacconi and his associates (Giacconi, Gursky, Paolini, and Rossi 1962; Gursky, Giacconi, Paolini and Rossi, 1963; hereafter cited as "G^2PR"), and Bowyer, Byram, Chubb, and Friedman (1964), and a source presumed by the latter to be the Crab Nebula. Several other source positions had been tentatively identified. These included (a) a source in the range of galactic longitudes encompassing Cygnus and noted by G^2PR (1962, 1963), (b) a source in Cepheus suggested by data from the two rocket flights of Fisher and Meyerott (1964), and (c) a source in the vicinity of Cygnus suggested by data (unpublished) from one rocket flight of Fisher and Meyerott.

An attitude control system (ACS) caused the rocket to maneuver so that X ray and optical star sensors scanned slowly near the galactic equator in the manner shown in Fig. 1. The detectors had three differently oriented rectangular-shaped fields of view as indicated near the left of the figure. The scan rate was nearly constant during each scan, and these rates are given in the figure. In addition to the scan along the galactic equator, Fig. 1 indicates that three observations of the Scorpius and one additional observation of the Cygnus regions were planned. Because of a malfunction in a rocket component, the flight was terminated several seconds prior to the planned second scan of Cygnus. Difficulties with the flight's X-ray detectors resulted in only one counter providing data during scan 1.

The X-rays were detected by sealed gas proportional counters having either beryllium or aluminum windows. The detector sensitivity was highest at about 6 keV and fell off rapidly at higher and lower energies, the low-energy cutoff being about 1 keV. Pulses from the detectors were amplified, analyzed for amplitude, and sorted into several energy channels by the rocket instrumentation.

By reduction of data from the various X-ray detectors and optical star sensors, positions of X-ray sources relative to the observed stars can be obtained. In order to locate a source accurately in two celestial coordinates, it must be observed by more than one counter-collimator system. Because only one counter-collimator system provided good data during scan 1, only one angular coordinate can be determined accurately for the sources observed on this scan. Thus the position along the scan is well determined while the position perpendicular to the scan is limited by the 26° full width at half-maximum of the field of view.

The counting rate in the energy channel spanning the nominal range of 4-8 keV was observed to increase from its background value of about 75 counts/sec to rates in the range of 200 to 400 counts/sec several times during the first scan. The first occasion was when Cygnus was being traversed. For this source the counting rate versus time profile exhibited a well-defined triangular shape compatible with a single source having an angular extent small compared to the collimator resolution. At the same time, the counting rate in the 8-12-keV channel rose a detectable amount above its background level. The other regions of enhanced counting rates were grouped in longitude within about

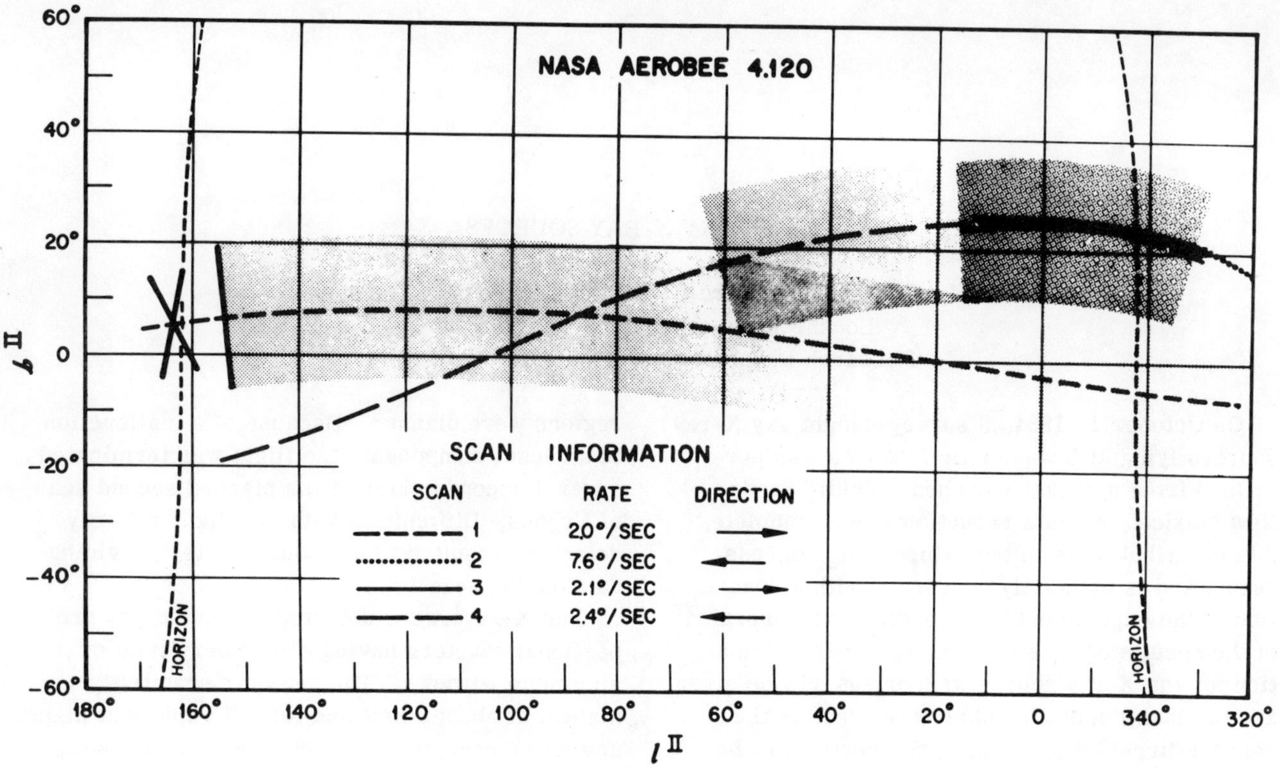

FIG. 1. The shaded areas in this galactic coordinate plot show the various swaths of sky scanned for X-ray emission. To the left, at the beginning of scan 1, the three different fields of view are schematically shown. Except for the incomplete last scan, the center of each detector's field of view moved along the indicated scan lines as the rocket rotated slowly about an axis nearly perpendicular to the galactic equator.

± 20° of the galactic center and were not completely resolved from each other. However, the count rate profile for 4-8-keV photons shows six distinct peaks, each of which could have the proper triangular shape. Four of these peaks lay on one side of the galactic center, and two on the other side. One of the strongest sources observed was about 1° from the galactic center. While the latter position is apparently not associated with the strong peak nearby, the peak's presence makes it difficult to say whether there is any radiation from the galactic center itself.

As noted above, for the Cygnus source only the angular distance along the scan path can be obtained accurately from the October 1 flight data. The longitude of the source is in the range of $70° < l^{II} < 74°$. Previously unpublished data of Fisher and Meyerott may be of use in further specifying the source's location. The highest counting rate observed on either of their two earlier flights (excluding the rates obtained when observing near the earth's horizon) was associated with the region $60° < l^{II} < 100°$ and $-5° < b^{II} < 12°$ which was scanned on one flight only. The narrow dimension of the line-shaped fields of view of the first flight's detectors were nearly orthogonal to those of Aerobee 4.120 for scans of the Cygnus region. Therefore the latitude of the source may be specified as $-5° < b^{II} < 12°$, assuming the earlier data were indeed related to this source. This approximate position agrees quite well with the NRL location of Cyg XR 1 (Bowyer et al. 1964). The Cepheus source has been proven to be spurious.

The bright source in Scorpius was observed through each of three counter-collimator systems on each of the last three scans, and was found to be within 1° of the position given by Bowyer et al. (1964). Ultimately, a position of this source will probably be established to an accuracy better than 30' from these data.

Little can be said about the spectral distribution of the sources observed on Scan 1. However, the spectral data on the bright Scorpius source are believed to be good. The peak counting rates observed (above background) during the scan 3 transit in the 2-4, 4-8, 8-12, and 12-20-keV

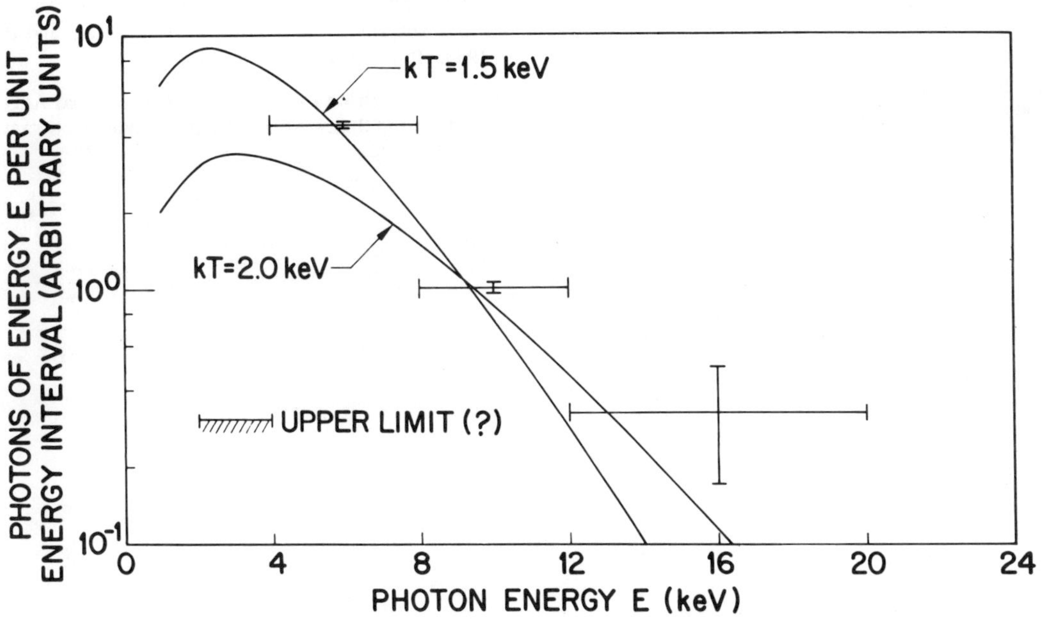

FIG. 2. Four-point spectrum obtained for the bright Scorpius source. Lack of sensitivity in the lowest energy interval has resulted in only an upper limit to that interval's flux. The error bars give only statistical errors in counts and so do not include errors in the detector's response function.

channels were < 30, 3000, 400, and ~100 counts/sec, respectively. The results of one of the first attempts at understanding the Scorpius spectrum are shown in Fig. 2. The counts observed in each of the various energy intervals were divided by the product of detection efficiency and energy width and the resultant set of numbers arbitrarily normalized to unity in the 8-12-keV interval. This procedure gives equal weight to the detector efficiency at each photon energy. For comparison purposes, two Planck distributions characterized by kT values of 1.5 and 2.0 keV have been included in the figure, k being the Boltzmann constant. A temperature of the order of 16×10^6 °K, corresponding to a kT of 1.5 keV, crudely describes the three higher-energy data points. Preliminary attempts to fit the three higher-energy data points to a power law indicate the photon flux falls off as E^{-2} or E^{-3}, where E is the photon energy. The upper limit obtained from the lack of data in a fourth-energy interval (E < 4 keV) of the same detector indicates a shortage of low-energy photons. A question mark has been placed beside the upper limit in the figure to indicate concern over the significance of the limit. Although an allowance for all the various experimental errors has not yet been made, the apparent shortage of low-energy photons is sufficient to render both the Planck and power-law distributions inadequate as descriptions of the Scorpius energy spectrum.

Failure to fit the data to a spectrum controlled by a single parameter has led to attempts to fit the data to spectra controlled by two parameters. The three higher-energy data points could be fitted to a bremsstrahlung distribution provided the maximum electron energy were of the order of 20 keV and there were more low-energy than high-energy electrons. Successful two-parameter fits have also been made to distributions consisting of either an emission line or an absorption edge near the center of the 4-8-keV range superimposed on a Planck distribution having a temperature of the order of 40×10^6 °K. Because satisfactory agreement could undoubtedly be found for many other combinations, one must conclude that the data are inadequate to specify accurately the spectral distribution of the source.

REFERENCES

S. Bowyer, E. T. Byram, T. A. Chubb, and H. Friedman, 1964, Nature, 201, 1307.

______, 1965, Science, 147, 394.

P. C. Fisher, and A. J. Meyerott, 1964, Apr. J., 139, 123; 140, 821.

R. Giacconi, H. Gursky, F. R. Paolini, and B. B. Rossi, 1962, Phys. Rev. Letters, 9, 439.

H. Gursky, R. Giacconi, F. R. Paolini, and B. Rossi, 1963, Phys. Rev. Letters, 11, 530.

This work was supported by the National Aeronautics and Space Administration under Contract NAS3-909.

OBSERVATIONAL WORK ON COSMIC GAMMA RAYS*

George W. Clark

Department of Physics and Laboratory for Nuclear Science
Massachusetts Institute of Technology
Cambridge, Massachusetts

In recent years a variety of efforts have been made at Massachusetts Institute of Technology to observe cosmic photons in the range of energies from 10^4 eV to over 10^{15} eV. I would like to review four of these efforts starting from the bottom and working up in energy.

Naïve power-law extrapolations of the measured flux densities of X-rays near 4 keV from the recently discovered cosmic X-ray sources predict intensities above 15 keV which should be detectable at the highest attainable balloon altitudes. In contrast, the black-body spectra expected from the surfaces of neutron stars, which came into vogue one year ago as possible sources of the newly discovered cosmic X-rays, should cut off sharply below 10 keV. It therefore appears that balloon measurements of X-ray spectra in the 15-60-keV range can provide a test of the neutron-star hypothesis. Furthermore, the general prospects of balloon X-ray observations are attractive for a variety of practical reasons. These considerations led to a recent successful attempt to observe cosmic X-rays in a balloon experiment (Clark 1965).

For this observation, I prepared a scintillation detector consisting of a thin crystal of Na I(Tl) optically coupled to a 5-inch-high gain photomultiplier. The pulses were analyzed and recorded in five pulse-height channels from 9 to over 62 keV. In front of the detector was a frame of brass slats which defined a field of view that was 32° wide by 110° long. The package was oriented with the axis of the detector inclined at 35° from the zenith, and the 110° direction was in a vertical plane. During the flight the package was rotated about a vertical axis by a rotator and the instantaneous azimuth of the detector axis was determined by interpolating between periodic readings of a magnetometer.

The balloon was launched at dawn on July 20, 1964, and reached its maximum altitude of 133,000 feet 3 hours before the meridian transit of the Crab Nebula. During the next 80 min, while the balloon remained above a pressure altitude of 3.9 mb and the package rotated about a vertical axis, the diurnal motion of the Crab Nebula carried it from 35° to 18° in zenith angle and from 96° to 118° in azimuth.

Figure 1 shows the observed counting rates in the five pulse-height channels plotted as a function of the azimuth of the detector axis measured relative to the Crab Nebula. A peak in the counting rates in the three middle channels is clearly evident near zero relative azimuth when the Crab Nebula was within the field of view. Since the Crab Nebula is the only known source of lower energy X-rays in this field of view, it is likely to be also the source of the higher-energy radiation observed here. The small discrepancy between the apparent position of the peaks and zero relative azimuth can be accounted for as systematic error in the magnetic-field measurement.

I calculated the relative counting rates in the five pulse-height channels expected for various hypothetical incident X-ray spectra, taking into account atmospheric absorption along the slant direction to the Crab Nebula, and the efficiency and energy resolution of the scintillation detector. Figure 2 shows the results for three trial spectra.

*This work has been supported in part by the National Aeronautics and Space Administration under Contract and Grant NsG-386, in part by the U. S. Atomic Energy Commission under Contract AT(30-1) 2098, and in part by the U. S. Air Force Office of Scientific Research under grant AF-AFOSR 546-64.

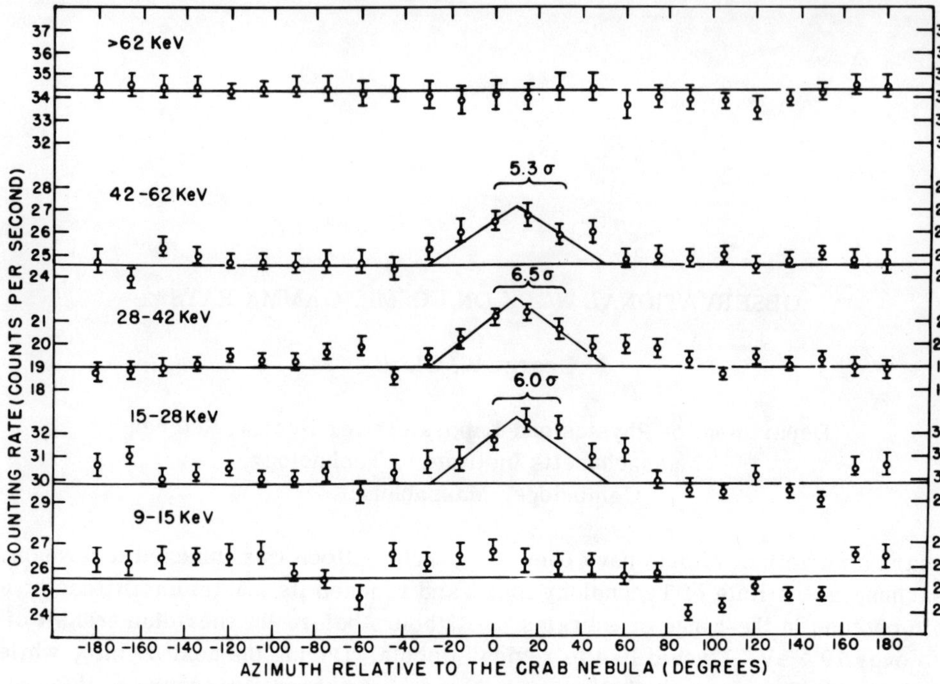

FIG. 1. The observed counting rates in the five pulse-height channels plotted against the azimuth of the detector axis measured with respect to the azimuth of the Crab Nebula.

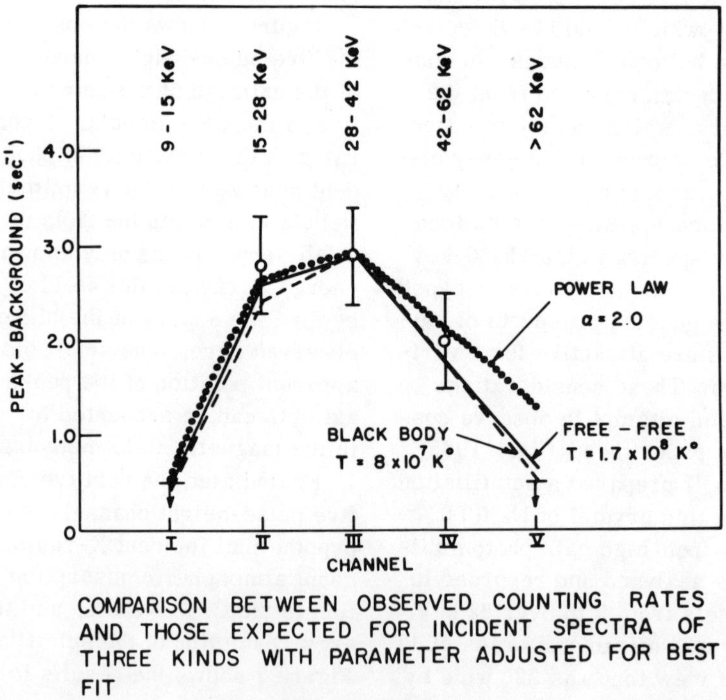

FIG. 2. Expected and observed relative counting rates in the five channels for various assumed incident spectra. The expected rates are normalized to the observed value in Channel III.

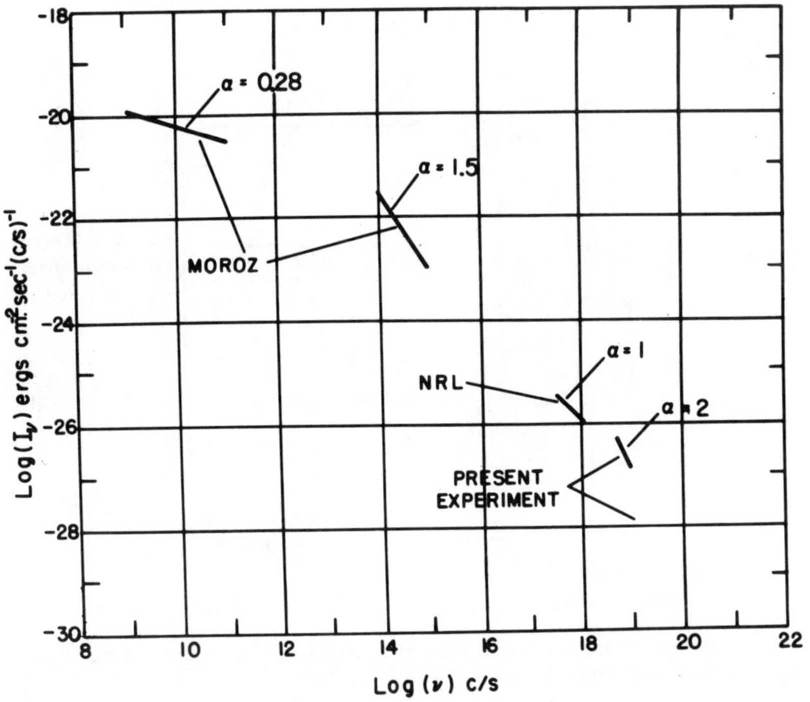

FIG. 3. Summary of observational data on the electromagnetic spectrum of the Crab Nebula.

The values are all normalized to the observed rate in Channel III. The temperature of 80×10^6 ° K required to fit a black-body spectrum to the data is far higher than the surface temperatures predicted for neutron stars. This observation is therefore strong new evidence against the simple neutron-star hypothesis for the X-ray source in the Crab Nebula.

The data from direct observations now available on the electromagnetic spectrum of the Crab Nebula are summarized in Fig. 3. The radio region is well established. The accuracy of the optical and UV results suffers from substantial uncertainties in the corrections for attenuation by interstellar dust. The data near 10^{18} c/s was calculated from the most recent results of the NRL rocket experiments. The results of the present experiment are indicated by a short piece of a power law with spectral index 2 fitted to the data.

The second experiment that I would like to describe is one carried out in February, 1964, by J. Overbeck (1964) to test whether the SCO-X1 source is a cloud of energetic electrons producing X-rays by inverse Compton scattering of starlight. M. Oda and I had pointed out previously that such a cloud, lying at a distance of about 50 light-yr, would have to have a total energy in relativistic electrons about equal to that released by a supernova in order to give the observed X-ray intensity (Clark and Oda 1963). Subsequently, P. Morrison (private communication) suggested that an ancient explosion in the galactic nucleus may have generated the required electron cloud which is presently moving out along the galactic axis where we observe it as SCO-X1. In any case, if such a cloud is the source of the observed X-rays, it must also be a source of bremsstrahlung gamma rays produced in collision between the electrons and the ambient gas. Overbeck searched for these gamma rays in a balloon experiment at 115,000 feet, in which he periodically interposed a lead shield between a large gamma-ray detector and SCO-X1. The apparatus is shown schematically in Fig. 4. He obtained negative results which appear to eliminate the inverse Compton effect as a mechanism for the X-ray production.

To see how the implications of Overbeck's experiment work out quantitatively, we note first that the average energy of the recoil photons produced in Compton collisions between electrons of energy E and isotropic photons of energy ϵ_0 is approximately

$$\bar{\epsilon} = (E/mc^2)^2 \epsilon_0 \qquad (1)$$

Taking ϵ_0 to be 2 eV for the average photon energy

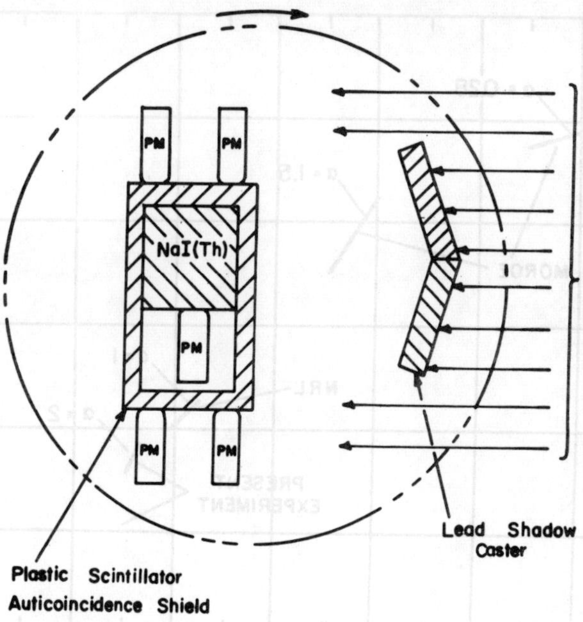

FIG. 4. Schematic diagram of the apparatus used in a search for gamma rays from the X-ray source in Scorpio.

of starlight, and $\bar{\epsilon}$ = 4 keV for the energy of the observed X-rays we find E ~ 25 MeV. The number intensity of recoil photons from a cloud with N electrons at a distance R and with a photon density n_s is

$$j_x = \frac{\sigma_\tau cNn_s}{4\pi R^2}, \qquad (2)$$

where σ_τ is the Thomson cross-section. These same electrons will suffer radiative collisions with the ambient protons with density n_p and give rise to an intensity of bremsstrahlung gamma rays which is about

$$j_\gamma = \frac{3\sigma_\tau}{137} \frac{cNn_p}{4\pi R^2} \ln 3 \qquad (3)$$

over a wide spectrum up to about 25 MeV. Combining equations (2) and (3) we can express the expected intensity of gamma rays in terms of the X-ray intensity by the relation

$$j_\gamma = \frac{n_p}{n_s} \frac{3 \ln 3}{137} j_x . \qquad (4)$$

By determining an upper limit on j_γ, one can place an upper limit on n_p in terms of the starlight density:

$$\bar{n}_p = \frac{137}{3 \ln 3} \frac{j_\gamma}{j_x} n_s . \qquad (5)$$

This limit can hardly be less than the density of electrons, for otherwise the cloud would not be electrically neutral. Thus

$$\bar{n}_p > n_e . \qquad (6)$$

But the density of electrons is approximately

$$n_e = \frac{2N}{(R\,\theta)^3}, \qquad (7)$$

where θ is the observed angular diameter; combining expressions (2), (5), (6), and (7) one finds

$$R > \frac{24\pi \ln 3}{137} \frac{j_x^2}{c\sigma_\tau n_s^2 j_\gamma \theta^3} \qquad (8)$$

Overbeck found $j_\gamma < 0.01$ in the energy range from 3.5 to 7 MeV and $j_\gamma < 0.01$ from 7 to 28 MeV (99 per cent confidence). The recent investigation by Oda, Clark, Carmire, Wada, Giacconi, Gursky, and Waters (1965) of the SCO-X1 source with the modulation collimator demonstrated that $\theta < 0.002$ radians. Rocket measurements give $j_x \sim 20\ cm^{-2}$ sec (Gursky, Giacconi, Paolini, and Rossi 1963). With these values in equation (8) one finds

$$R > 1 \times 10^{26}\ \text{cm}.$$

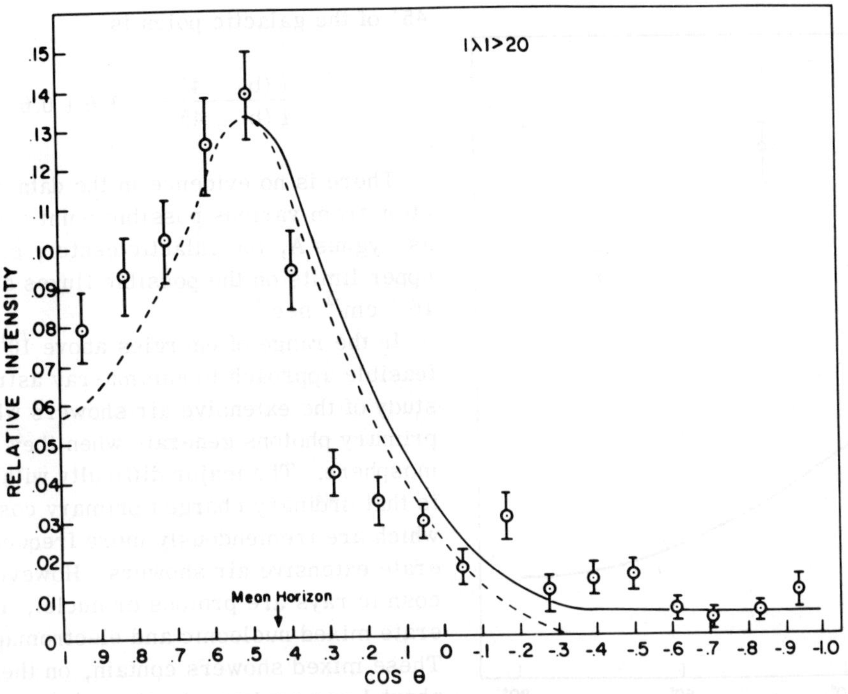

FIG. 5. Counting rate for gamma-ray events from the Explorer XI satellite plotted as a function of the angle of the detector axis from the horizon.

This puts the hypothetical electron cloud needed to produce the SCO-X1 X-rays by the inverse Compton process far outside the Galaxy and probably beyond belief.

An exhaustive treatment of the data obtained in the energy range above 100 MeV from the Explorer XI gamma-ray astronomy satellite has recently appeared (Agogino, Clark, Garmire, Helmken, Higbie, and Kraushaar 1965). This experiment was prepared several years ago by W. Kraushaar and myself, and was launched in April, 1961. It has since enlisted the efforts of several associates and students at Massachusetts Institute of Technology.

The final results from Explorer XI are essentially the same as those reported earlier in preliminary form (Kraushaar and Clark 1962). However, they are now more adequately supported by a careful laboratory study of the angular and energy response of a duplicate instrument that was carried out with tagged gamma rays at the Cal Tech synchrotron, and by a thorough analysis of the dependence of the observed counting rate in orbit on those physical conditions which should affect differently the part which is due to true cosmic gamma rays, and any part which may be due to false events caused by charged primaries. The analysis was accomplished by means of a simple and highly flexible statistical method of Monte Carlo simulation that permits one to evaluate the effective exposure time of the instrument for any given range of the parameters that describe the situation in orbit.

We designed the instrument so that a gamma ray arriving from the forward direction gave a characteristic signature of pulses which we call a gamma-ray event. Other forms of radiation could only very rarely forge this signature.

Ground tests on the duplicate instrument demonstrated that the response function representing the effective area for detecting gamma rays versus energy began at 50 MeV and rose to a plateau value of 7 cm^2 effective area at about 250 MeV. The angular response fell by a factor of 100 between 0° and 60°. The analysis of data obtained in orbit showed a maximum counting rate for gamma-ray events in the horizon direction due to albedo gamma rays, and a plateau in this rate at angles greater than 63° above the horizon, as shown in Fig. 5. Whereas the albedo rate showed a variation with geomagnetic latitude as expected, the plateau rate was constant within the poor statistics.

The total number of gamma-ray events recorded

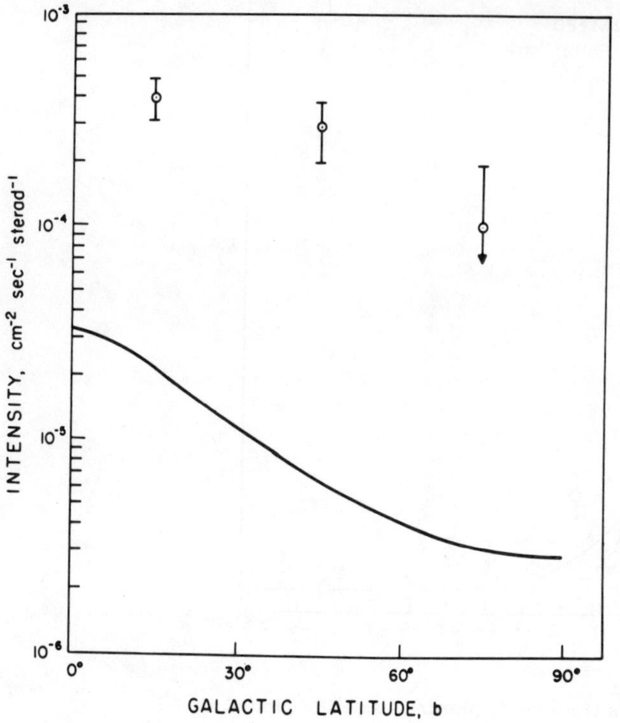

FIG. 6. Comparison between the observed intensities of gamma-ray events at different galactic latitudes (circles) and the intensity expected from interactions between cosmic rays and the interstellar hydrogen in the Galaxy (dark line).

when the axis of the detector was more than 63° above the horizon was 31. Taking into account the total exposure time, and assuming that the detected radiation is gamma rays with a differential energy spectrum of the form $W^{-\gamma}$ with $\gamma = 1.7$, we find an average integral intensity for $W > 100$ MeV of

$$I = 2 \times 10^{-4}\ \text{cm}^{-2}\ \text{sec}^{-1}\ \text{sterad}^{-1}.$$

In the absence of more definite proof as to the nature of the radiation that causes the gamma-ray events, this value must be considered only an upper limit to the true integral intensity. It is, in fact, ten times higher than the rate expected from the interaction of cosmic rays with interstellar matter in the Galaxy, assuming the distribution of atomic hydrogen determined from radio observations and also assuming a density of cosmic rays throughout the galaxy equal to the local density. The ratio of the average intensity within 45° of the galactic plane to the average intensity within 45° of the galactic poles is

$$\frac{I\,(b > 45°)}{I\,(b < 45°)} = 1.6 \pm 0.6\ .$$

There is no evidence in the data of gamma radiation from various possible point sources such as Cygnus A, the galactic center, etc. Typical upper limits on the possible fluxes are about 10^{-3} cm^{-2} sec^{-1}.

In the range of energies above 10^{14} eV the only feasible approach to gamma-ray astronomy is the study of the extensive air showers which energetic primary photons generate when they enter the atmosphere. The major difficulty with this approach is that ordinary charged primary cosmic rays, which are tremendously more frequent, also generate extensive air showers. However, ordinary cosmic rays are protons or nuclei, and they generate mixed nucleonic and electromagnetic showers. These mixed showers contain, on the average, about 1 per cent penetrating mu-mesons near their cores. In contrast, showers initiated by primary gamma-ray photons are nearly pure electromagnetic showers and contain only about 0.01 per cent of penetrating particles that arise from photomeson production. The problem therefore comes down to searching for "low-mu" showers, i.e., extensive air showers with unusually few penetrating particles. This can be done with an air-shower detector array operated in anticoincidence with a penetrating particle detector of very large area.

A group from Massachusetts Institute of Technology, the University of Tokyo, the University of Michigan, and the St. Andres University in La Paz, Bolivia, has set up an experiment called the Bolivian Air Shower Joint Experiment (BASJE) whose main purpose has been to search for gamma-ray air showers (Suga, Escobar, Murakami, Domingo, Toyoda, Clark, and La Pointe 1963). The air-shower detector consists presently of twenty 1 m^2 scintillation detectors in an array 600 m in diameter set out around the Laboratory of Cosmic Physics at an altitude of 17,000 feet on Mount Chacaltaya near La Paz. The penetrating particle detector is a 60 m^2 scintillation detector covered with 200 tons of concrete and galena (PbS). The shielding ensures the complete absorption of all electromagnetic radiation that strikes the 60 m^2 area. The detectors underneath are sensitive enough to measure a single particle

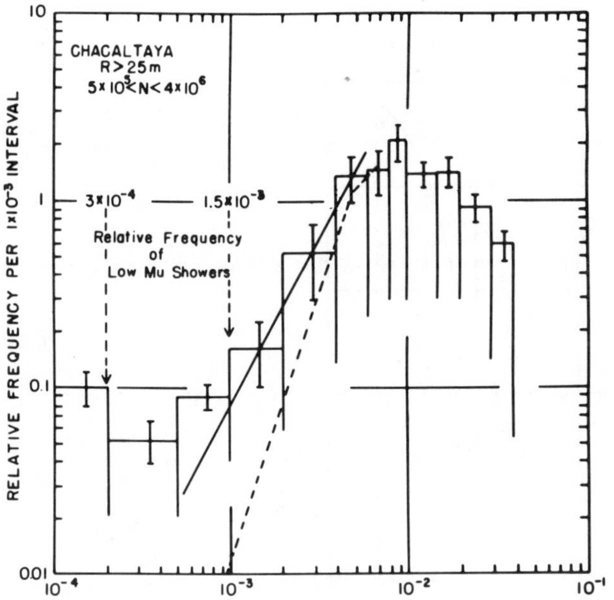

FIG. 7. Relative frequencies of extensive air showers with various proportions of penetrating particles as observed at 17000 feet altitude.

traversing any place in the 60 m^2 area. The air-shower array gives data from which the arrival direction can be determined within $\Delta\ \theta \sim 3°$, the core location within Δ R $\sim$ 2 m, and the size within Δ N/N $\sim$.1.

We have studied the distribution in penetrating particle content of a large number of showers with sizes in the range from 5×10^5 to 4×10^6 corresponding roughly to primary energies from 10^{15} eV to 10^{16} eV. We selected showers whose cores were far enough from the 60 m^2 detector so that local fluctuations in structure near the core did not affect the measurements. We also required that the expected total number of particles striking the 60 m^2 detector on top of the shielding exceed 1,400. The observed distribution in penetrating particle content (Fig.) shows that the average value was about 10^{-2}, which, for a shower of minimum acceptable size, corresponds to 14 observed penetrating particles in the 60 m detector. The distribution at low values gives evidence for the existence of a distinct class of "low-mu" showers whose penetrating particle content is less than 10^{-3} as expected for pure electromagnetic showers. The relative proportion of these "low-mu" **showers** among all showers is between 1.5×10^{-3} and 3×10^{-4}, depending on where the dividing line is placed. Firkowski, Gawin, Zawadski, and Maze (1963) have also searched for low-mu showers in an experiment at sea level, and they have reported positive results at 10^{16} eV.

On the basis of the evidence presently available one cannot prove that the nearly pure electromagnetic showers observed in this experiment are caused by primary gamma rays. One cannot exclude the possibility that they arise in rare nuclear interactions of primary protons in which nearly all of the primary energy is transferred to a neutral pi-meson which decays into two gamma rays. The clearest evidence in favor of a gamma-ray origin for some of the events would be an observation of some form of anisotropy—perhaps a concentrated source or a tendency to cluster near the Milky Way where we can expect a higher rate of production. In Fig. 8 we have plotted the arrival directions as dots on a celestial map. We have also compared their distribution in galactic latitude with the one expected on the basis of the exposure time. In neither case is there significant evidence of anisotropy.

If one assumes that the primaries of the observed low-mu showers are photons, then the relative proportion of photons among all cosmic ray primaries in the energy range from 10^{15} to 10^{16} eV is between 5×10^{-4} and 1×10^{-4}, when account is taken of the different rates of development for pure electromagnetic and mixed showers.

An interesting prospect for the future of the BASJE is the conclusion which may be drawn from data now being recorded on the relative proportion of "low-mu" showers above primary energies of 10^{16} eV. Photons of this energy can arise from neutral pi-mesons produced in photonuclear interactions between protons over 10^{17} eV and starlight. If the abundance of protons at this energy is universal, then the metagalactice production of $> 10^{16}$ eV photons by this process may give an appreciable flux at the Earth. Conversely, if the BASJE observations place a low upper limit on the relative proportion of "low-mu" showers above 10^{16} eV, then it may be possible to place significant upper limits on the metagalactic density of cosmic rays over 10^{17} eV. In any such analysis the attenuation of the gamma rays by collision with metagalactic infrared and microwave photons will have to be taken into account.

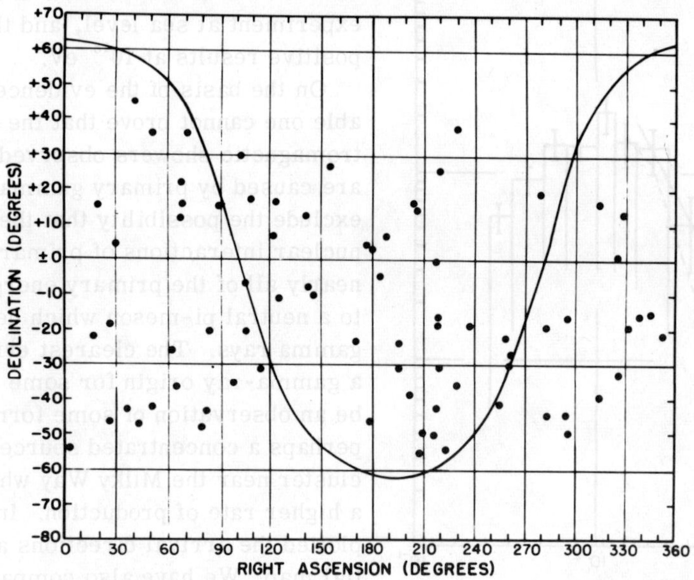

FIG. 8. Celestial arrival directions of showers which were in the lowest 10^{-3} fraction in penetrating particle content. The concentration of events between declinations -55° and +30° is the result of atmospheric absorption which limits the effective field of view to zenith angles of less than about 45°.

REFERENCES

M. Agogino, G. W. Clark, G. Garmire, H. Helmken, P. Higbie, and W. Kraushaar, 1965, Ap. J., 000, 000.

G. W. Clark, 1965, Phys. Rev. Letters, 14, 91.

G. Clark, and M. Oda, 1963, International Conference on Cosmic Rays, Jaipur.

R. Firkowski, J. Gawin, A. Zawadski, and R. Maze, 1963, Nuovo Cimento, 26, 1422.

H. Gursky, R. Giacconi, F. Paolini, and B. Rossi, 1963, Phys. Rev. Letters, 11, 530.

W. L. Kraushaar, and G. W. Clark, 1962, Phys. Rev. Letters, 8, 106.

M. Oda, G. Clark, G. Garmire, M. Wada, R. Giacconi, H. Gursky, and J. Waters, 1965, Nature, (in press).

J. Overbeck, 1964, unpublished Ph. D. thesis, Massachusetts Institute of Technology.

K. Suga, I. Escobar, K. Murakami, V. Domingo, Y. Toyoda, G. Clark, and M. La Pointe, 1963 Internat. Conf. Cosmic Rays, Proc., 4, 9.

EXTENDED SOURCE OF ENERGETIC COSMIC X RAYS

Elihu Boldt, Frank B. McDonald, Guenter Riegler, and Peter Serlemitsos
National Aeronautics and Space Administration, Goddard Space Flight Center, Greenbelt, Maryland

We report here on evidence for an extended source of x rays in a region near the direction of the north galactic pole. This result is based upon data collected during two recent balloon flights launched from Holloman Air Force Base in New Mexico.

The x-ray telescope was suspended vertically so that the detector axis had celestial coordinates given by a right ascension equal to the sidereal time at the local meridian and a declination given by the latitude of the balloon. The stability of the attitude was established to within $\frac{1}{2}$ deg by a camera that continuously photographed the horizon on infrared sensitive film. During the major portion of the flight of 6 December 1965 the detector axis was well out of the galactic plane, riding at a declination of +31 deg, and passed near the north galactic pole. The flight of 13 January 1966 carried the telescope across the galactic plane at a declination of +34 deg, intercepting the x-ray source Cygnus XR-1. The residual atmosphere during the ceiling coverage of these flights was 2.6-2.8 g/cm^2.

Figure 1 shows the main elements of the x-ray telescope. The central x-ray detection element is a cesium-iodide (thallium activated) scintillation crystal that is used to measure x rays in the interval 20-100 keV partitioned by 64 channels. The plastic scintillator (CH) serves as an anticoincidence detector of charged particles and also as the innermost section of a three-element graded x-ray shield consisting of tin on the outside and copper sandwiched between. The light from the cesium iodide is distinguished from that of the plastic scintillator by its longer decay time. The entrance port is a krypton gas proportional counter that serves as an anticoincidence detector and is also used to measure x rays in the approximate interval 10-30 keV, again partitioned by 64 channels. The transmission of the gas counter exceeds the 50% level at 25 keV (K x ray for Sn^{119m}). At this energy, the proportional counter exhibits a resolution of about 20% full width at half-maximum (FWHM) for both the primary and escape peaks. The resolution for the cesium-iodide crystal is about 30% FWHM at 74 keV (K x ray of Bi^{207}).

A structured response pattern was achieved

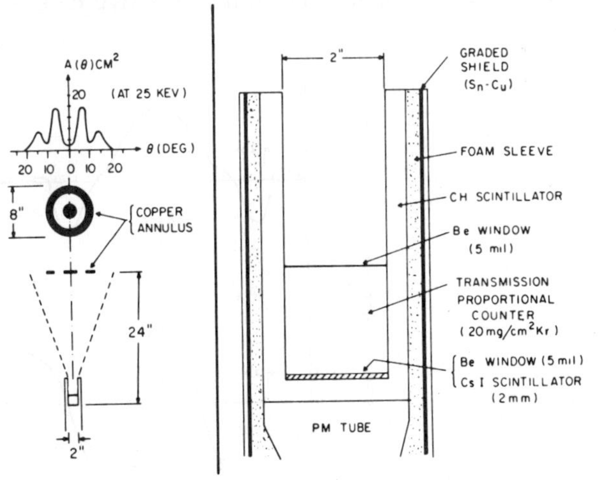

FIG. 1. (a) The x-ray spectrometer telescope utilizes the occultation scheme outlined; the detector appears at the bottom. (b) The essential features of the detector.

Reprinted from Phys. Rev. Letters **17**, 447 (1966).

through the use of a copper occultation disk and annulus placed well above the detector [see Fig. 1(a)]. The effective detection cross section (A) as a function of the zenith angle (θ) is shown in Fig. 1(a). This was obtained by measuring the counting rate due to a Sn^{119m} source placed at positions corresponding to several values of θ for a fixed radial distance of 12 ft.

At the ceiling altitude, the minimum counting rate encountered was $\approx 1\ \sec^{-1}$ for the proportional counter and $\approx 5\ \sec^{-1}$ for the cesium-iodide crystal. For a reference base line we note that the absolute minimum counting rate occurred shortly after the launch, at about 10 000-ft altitude, and was $\approx 0.3\ \sec^{-1}$ for the proportional counter and $\approx 1\ \sec^{-1}$ for the cesium-iodide crystal.

A stroboscopic location of Cygnus XR-1 is shown in Fig. 2(a). The observed counting rate (dN/dt) of the proportional counter is here folded with the response function (F) of the detector as follows:

$$\langle \text{folded rate} \rangle \equiv \frac{[\int_t^{t+\Delta t} (dN/dt) F dt]}{(\int_t^{t+\Delta t} F dt)}. \tag{1}$$

The response function (F) at any time (t) is determined by the zenith angle (θ) between the hypothetical celestial object under consideration and the actual axis of the telescope at that time. It is proportional to the effective detection cross section $A(\theta)$ depicted in Fig. 1(a). When, in fact, no localized sources contribute to the observed count, then the folded counting rate (1) exhibits the necessary feature that, independent of the assumed source location, its expectation value remains constant at a level equal to the average counting rate. As a measure of correlation then, we take the difference between the folded counting rate and the average counting rate for the same interval of data, viz.,

$$(\text{Strobe function}) \equiv \frac{\int_t^{t+\Delta t} (dN/dt) F dt}{\int_t^{t+\Delta t} F dt} - \frac{\int_t^{t+\Delta t} (dN/dt) dt}{\Delta t}. \tag{2}$$

In Fig. 2(a) we strobe in declination for sources at a fixed right ascension of 299 deg. The maximum correlation occurs at a declination of +34 deg, which is close to that reported[1] for Cygnus XR-1. The negative correlation for hypothetical sources that are about 9 deg away from Cygnus XR-1 in declination is indicative that the observed counting rate structure followed the response pattern of the occultation scheme. This could have masked any contribution to the counting rate from Cygnus-A,

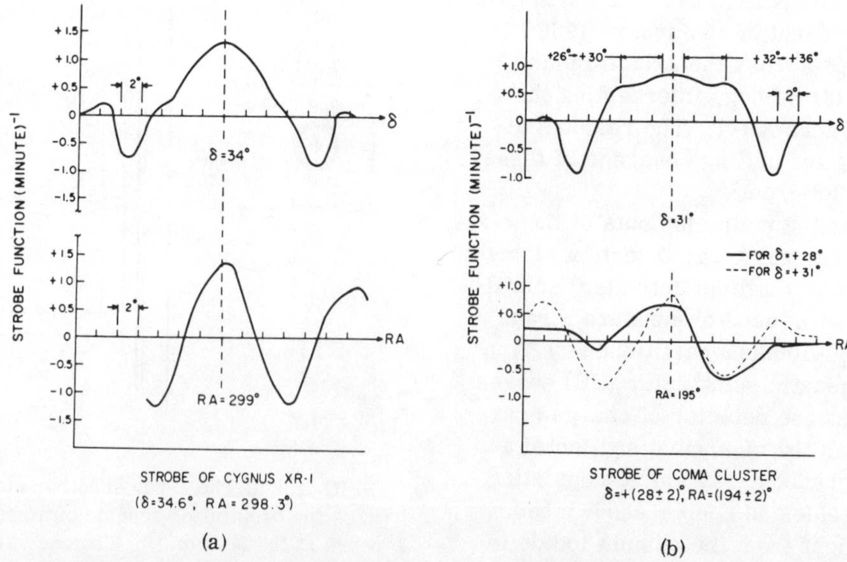

FIG. 2. (a) The strobe function [Eq. (2)] for the proportional-counter data of 13 January 1966 for the vicinity of Cygnus XR-1 (Ref. 1). (b) The strobe function for the proportional-counter data of 6 December 1965 for the vicinity of the Coma cluster (Ref. 2).

which is at about the same right ascension considered here, but is about 7 deg higher in declination. It does permit us to state that any flux from Cygnus-A above 20 keV is smaller than that from Cygnus XR-1 during the time of this observation (~19 h U.T., 13 January 1966).

Also shown in Fig. 2(a) is a strobe in right ascension for hypothetical sources at a declination of +34 deg. This was restricted in right ascension to those hypothetical sources that would have passed both into and out of the telescope's field of view during the ceiling coverage of the flight. The pronounced lobes in this pattern are characteristic of a scan in right ascension for a source that passes near the telescope axis. The principal maximum occurs at about 299 deg, close to the right ascension reported[1] for Cygnus XR-1.

The net spectrum from Cygnus XR-1 collected during a 30-min period of maximum yield is shown in Figs. 3(a) and 3(b). The response function for an object located at Cygnus XR-1 was used to establish the effective exposure during this period. The exposure for a 30-min interval of partial occultation bracketing this period of maximum yield was also obtained. By subtracting the occultation exposure value from the maximum yield value we obtain a net exposure which for the CsI crystal turns out to be 27×10^3 cm^2 sec and for the proportional counter is 22×10^3 cm^2 sec. The net spectrum shown here corresponds to this net exposure. The spectrum given by the CsI crystal was used to infer the magnitude expected for the peak interval indicated for the proportional counter; that is shown as a dotted region on Fig. 3(b). In the interval of spectral overlap for the two detectors, near 25 keV, we have referred the flux to the top of the atmosphere and obtained an incident primary flux value of $\sim 10^{-2}$ (cm^2 sec keV)$^{-1}$.

Figure 2(b) shows a stroboscopic analysis of the proportional-counter counting rate for the flight of 6 December 1965. The strobe in declination is at a fixed right ascension of 195 deg. Since the mean declination of the detector axis during the ceiling coverage of this flight was +31 deg, this procedure examines a region of the sky in the immediate vicinity of the north galactic pole. This pattern is different from that obtained for Cygnus XR-1 in that it exhibits a broad structured central peak with shoulders symmetrically situated about

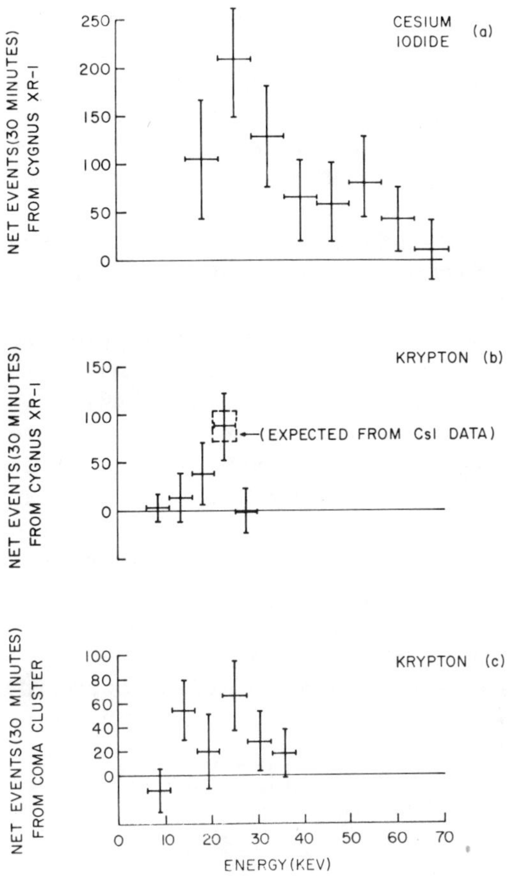

FIG. 3. (a) The net spectrum (background subtracted but with no corrections applied for atmospheric absorption or variations of detector efficiency) from Cygnus XR-1 detected by the CsI scintillator during a 30-min interval on 13 January 1966. (b) The net spectrum detected by the proportional counter (Kr gas) at the same time. (c) The net spectrum from the Coma cluster detected by the proportional counter during a 30-min interval on 6 December 1965.

5 deg on either side of +31 deg. This indicates radiation coming from a declination of +26 deg and/or +36 deg; the ambiguity here is an inherent feature of this scheme. That the correlation does not dip at the axis of symmetry indicates the passage of a source region near the detector declination of +31 deg. We interpret these features as evidence for an extended x-ray source at a mean declination of +28 or +34 deg and with an angular size of 4-5 deg. It is well established[2] that the Coma cluster of galaxies is an extended region of about that size located at a mean declination of +28 deg.

The next step in this procedure is to consider a fixed declination of +28 deg and strobe in right ascension. This is shown on Fig. 2(b) by a solid curve that exhibits a broad peak at a mean right ascension of about 195 deg, consistent with the Coma cluster.[2] If we strobe at +31 deg, which is the declination of the telescope axis, then we obtain the dotted curve shown on Fig. 2(b). We note that here we pick up a multilobed pattern somewhat similar to that obtained for Cygnus XR-1, and we interpret this as evidence that an edge of the Coma cluster source extends to a declination near +31 deg.

The net spectrum from the Coma cluster detected by the proportional counter during a 30-min interval of maximum yield is shown in Fig. 3(c). This corresponds to a net exposure value of 27×10^3 cm^2 sec and exhibits a positive net count of 169 events with an expected statistical standard deviation of 55 counts. A malfunction of the pulse-height analysis for the CsI scintillator channel during the flight of 6 December 1965 prevents an examination of the spectrum much above 30 keV.

The expected total number of events recorded by the proportional counter during a total exposure to the Coma cluster of about two hours was evaluated as 474 ± 113. For comparison, the expected total number of events recorded by the proportional counter during a similar exposure to Cygnus XR-1 was evaluated as 809 ± 121. Since the exposure values for these two sources were essentially the same, this allows us to estimate that the flux from the Coma cluster is comparable to that from Cygnus XR-1 for energies in the vicinity of 25 keV, vix., $\approx 10^{-2}$ (cm^2 sec keV)$^{-1}$ at the top of the atmosphere.

It is a pleasure to acknowledge valuable discussions with Dr. V. K. Balasubrahmanyan and the technical contributions of Mr. David Clark, Mr. Frank Clese, and Mr. Henry Doong.

[1]S. Bowyer, E. T. Byram, T. A. Chubb, and H. Friedman, Science 147, 394 (1965).
[2]G. O. Abell, Ann. Rev. Astron. Astrophys. 3, 1 (1965).

See Appendix A.

EVIDENCE FOR A SOURCE OF PRIMARY GAMMA RAYS*

J. G. Duthie, R. Cobb, and J. Stewart†
Department of Physics and Astronomy, The University of Rochester, Rochester, New York

Since Morrison[1] suggested the possibility of gamma-ray astronomy in 1958, there has been a growing interest in the field. Many theoretical reasons for expecting a measurable primary γ-ray flux at the top of the earth's atmosphere have been discussed. These are summarized in four excellent reviews of the field[2-5] which have been published in the past two years. To date, several experiments have been reported[6-13] in which a variety of instruments have been used to search for point sources. None of these experiments yielded any definite evidence for the existence of localized source intensities. Cobb, Duthie, and Stewart[11] have set upper limits of 5×10^{-5} cm^{-2} sec^{-1} from the Crab Nebula and a few times 10^{-4} cm^{-2} sec^{-1} from three other celestial objects. Frye and Smith[12] have also set upper limits of a few times 10^{-4} cm^{-2} sec^{-1} from a variety of celestial objects. Similar results were reported by Kraushaar et al.[6] In this Letter we wish to report the existence of an anomalously high count of gamma rays from the direction of the constellation Cygnus. This high count was associated with an energy spectrum which appears to differ significantly from the spectrum of secondary γ rays generated by cosmic rays interacting in the atmosphere above the balloon-borne detection system.

The present detection system is similar to that described elsewhere[11] except for a change in the location of the anticoincidence counter. A scintillation-and-Cherenkov telescope was used as the trigger for the detection of gamma rays converting in a $\frac{1}{16}$-in. lead radiator placed between two spark chambers. The system is estimated to become very inefficient at detecting gamma rays with energies less than 50 MeV. The conversion efficiency for vertically incident γ rays approaches 19% at high energies. The area solid angle factor was 25.8 cm^2 sr. The two spark chambers were used to identify the γ rays and to determine the direction of the incident photon. The data reported here rep-

Reprinted from Physical Rev. Letters **17**, 263 (1966).

resents observations made during the first $9\frac{1}{2}$ h of a balloon flight at 3-mm residual atmospheric pressure. The balloon was launched from Palestine, Texas on 23 October 1965 and reached altitude at 19-h 44-min U.T. The flight remained at a constant altitude except for a short period at sunset, when the balloon descended for about one hour corresponding to an average increase of 15% in residual atmospheric pressure. Assuming that all detected γ rays were of secondary origin, the effect of this small increase in pressure can be eliminated by weighting the events, where indicated, in proportion to the reciprocal of the pressure at which they occur; a residual pressure of 3 mm is considered normal. This linear normalization is well justified on the basis of experimental results, both with counter telescopes and the instrument used here over similar ranges of atmospheric pressures.

As in our previous report,[11] we observe both "singles" and "pairs." "Singles" are those events in which only one track is observed in the lower and none in the upper spark chamber, whereas events in which two tracks are identified in the lower chamber and none in the upper are classified as "pairs." There were 395 "single" events within the geometry of the counter telescope during the $9\frac{1}{2}$ h, but none of the results on our observation of singles will be presented here.

A total of 488 "pairs" were detected within our geometry during the first $9\frac{1}{2}$ hours of flight. Assuming a detection efficiency of 14%, this represents a flux of $(3.94 \pm 0.18) \times 10^{-3}$ cm^{-2} sec^{-1} sr^{-1} at a residual pressure of 3 mm of mercury.[14] The histogram in Fig. 1 shows the

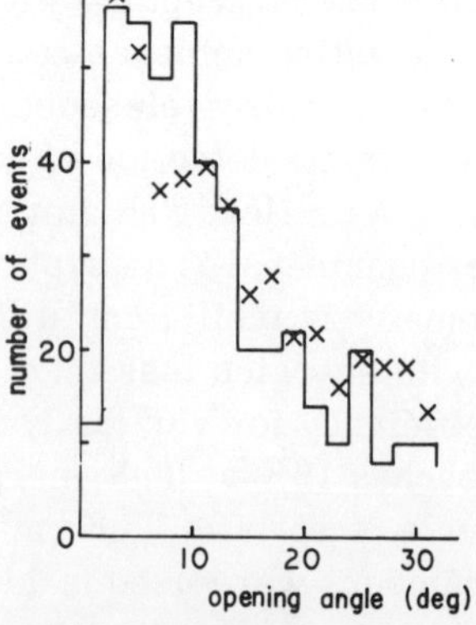

FIG. 1. Opening-angle distribution of positron-electron pairs observed during the flight. The crosses indicate the result of a Monte Carlo calculation.

distribution in the observed opening angles of the "pairs"; the opening angles are predominantly a result of multiple Coulomb scattering. The distribution has a broad maximum extending from 2 to 10°, then falls off slowly. The apparent deficiency of events with opening angles less than 2° may be experimental insofar as many such tight pairs could be classified as "singles" and would thus be absent from the present sample. The crosses represent the results of a Monte Carlo calculation[15] of the response of a similar system to the spectrum of secondary γ rays produced in the atmosphere as given by Svensson.[16] The calculated distribution involved 2588 events in the region of opening angles shown, but in the figure it has been normalized so that the number of events with opening angles between 2 and 32° is equal to the number of corresponding events in the experimental distribution. The two distributions are well matched, supporting the belief that the majority of the events are γ rays generated in the atmosphere above the detector by the nuclear interaction of cosmic rays.

The uncertainty in the direction of a gamma ray will be of the order of $\theta/2$, where θ is the observed opening angle. The fact that Fig. 1 shows many events with large θ indicates the difficulty in using the system for experiments requiring high directional resolution for the incident γ rays.

An IBM 7074 computer was used to determine the arrival direction of each γ ray and to tally the pressure-weighted number of events in cells 3° wide in right ascension (R.A.) and 3° wide in declination (dec). An examination of this tally indicated an interesting region in the vicinity of 20-h R.A. The number of events in each cell was small, however, and to improve the statistics, we have to accumulate data in larger cells. Since the detector efficiency varies with zenith angle, there will be a variation of counting rate with declination. We elected, therefore, to assembly the data into cells 9° wide R.A. and 30° wide dec, each cell centered on 31.5° N, the approximate dec lination of the zenith throughout the flight. Each cell then had an equal exposure. Furthermore, the collection of the data into the larger cells more realistically reflects the large uncertainty in the arrival direction of those events with large θ. Table I indicates the results of this analysis for pairs whose opening angles were less than 30°. The largest number of events

Table I. The weighted number of events for 16 cells on the sky, each cell being centered on 31.5° dec. The cells are 30° wide dec and 9° wide R.A. The first column indicates the right ascension at the center of each cell.

R.A. (h)	Number
16.1	19.55
16.7	20.08
17.3	10.73
17.9	16.35
18.5	23.45
19.1	13.79
19.7	31.80
20.3	20.35
20.9	12.27
21.5	16.67
22.1	13.89
22.7	27.12
23.3	20.79
23.9	24.37
24.5	15.62
25.1	16.68

was found to lie at about 20 h, where the total in the cell was 31.8 as opposed to an average rate of 19.5 per cell. With the exception of this cell, the data are well fitted to a Poisson distribution with a mean of 19. The probability of getting a single cell with an excess greater than 2.9 standard deviations (s.d.) is less than 1/300, and we have 16 cells in all. Shifting the boundaries of the 9° wide strips by whole 3° subcells did not significantly change the value of the maximum number of events per large cell. The investigation of events whose bisectors fell outside the declination limits of Table I showed no anomalously high number of events at any right ascension. The excess count in itself is interesting, but not really statistically significant. What is important, however, is that 2.9-s.d. excess comes at the time when the galactic plane transited and at the same time as another independent feature of the data showed an anomaly.

An approximately square region of the sky, about 18 by 18° and centered on 20-h R.A. and 35° dec, was selected for further analysis. This region was chosen on the basis that the anomalous counting rate indicated in Table I seemed to be due to events originating from a point somewhat north of the zenith direction at 20-h R.A. The opening-angle distribution of all events whose bisectors fell inside this

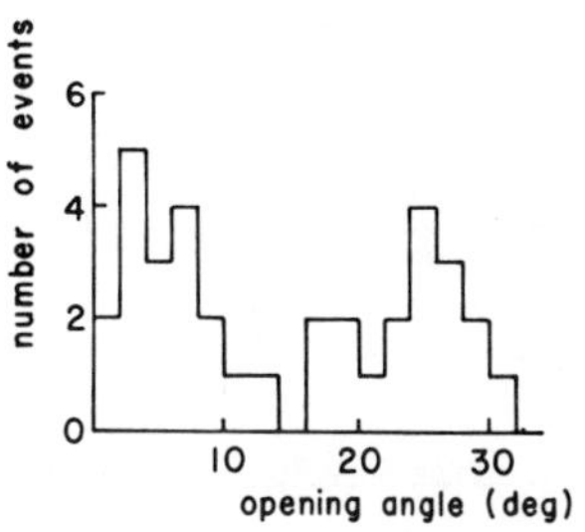

FIG. 2. Opening-angle distribution of positron-electron pairs whose bisectors fall in an 18 by 18° region of the sky centered on 20-h R.A. and 35° dec.

selected region was obtained. The results are shown in Fig. 2. The distribution is indeed quite different from the one shown on Fig. 1. A χ^2 test of the hypothesis that the data on Fig. 2 represented the same distribution as that on Fig. 1 had a confidence level of 0.3 %. The two distributions are therefore significantly different.

It remains to be established if the opening-angle distribution depends on the declination of the region. If this distribution is a function of the angle between the γ ray and the vertical axis of the detector, then the distribution for a region of the sky may depend on its declination. Thus all events in an 18° wide band of sky between 26 and 44° dec were selected. Of these events, those which fell in the 18 by 18° region of the sky centered on 20-h R.A. were eliminated and an opening-angle distribution of the remaining events was obtained. The resulting distribution was consistent with that of Fig. 1 and again different from that of Fig. 2; a χ^2 test comparing the distribution of Fig. 2 with the distribution for events from the above band gave a 1% confidence level for the hypothesis of identical distributions.

We have thus two pieces of information indicating an anomalous region of the sky at 20-h R.A. The extra counts, according to Fig. 2, are associated with opening angles greater than 14°. This indicates that the excess count is attributed to γ rays which typically are softer than those produced in the atmosphere. Fluctuations in the point of conversion in the lead radiator, disparity in the distribution of energy between the electron and positron, and effects of multiple Coulomb scattering make us unwilling to make a strong statement as to the energy of the photons. However, the suggestion that the excess lies in the events with opening angles greater than 14° allows us to attempt

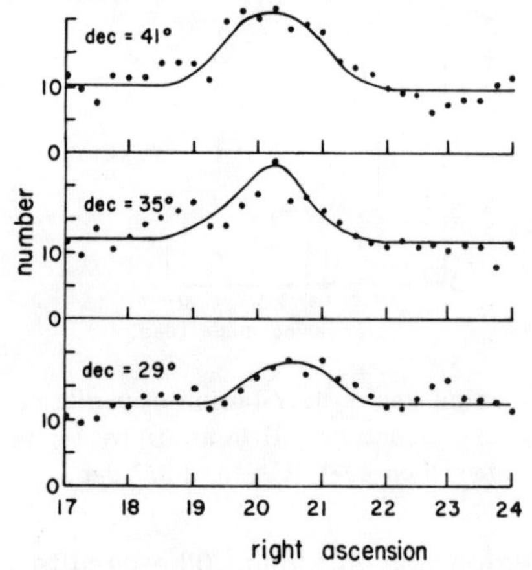

FIG. 3. "Fixed declination scans" for three declinations indicated. The scans were made using pairs having opening angles between 14 and 38°. The solid line is a visual fit to the data. The reader is cautioned not to misinterpret the meaning of "fixed declination scans" and is referred to the text for a fuller explanation.

a better method of locating the direction of the source of the anomalously high count.

Figure 3 shows the result of what we call a "fixed declination scan" in which a direction in space is chosen, and a search for the pressure-weighted number of gamma rays which can be associated with that direction is made. The criterion for associating a gamma ray with a chosen direction involves the difference between $\frac{1}{2}\theta$ and φ, where θ is the opening angle of the pair, and φ is the angle between the direction of interest and the direction of the bisector of the observed tracks in the chamber. If $\frac{1}{2}\theta > \varphi$ for any event, then we associate that event with the direction of interest. Such searches were made at three-degree intervals right ascension on many lines of constant declination. Figure 3 illustrates the results of such scans for pairs with opening angles greater than 14° and less than 38°. The lower limit was chosen since our examination of the data indicated that the opening-angle distribution in the range $\theta < 14°$ showed no anomalous behavior. Indeed a declination scan for such events showed no region of great interest. Figure 3 shows the source to be at about 20.25-h R.A. and of 35° N dec. The mean value of the background on this representation was 12.4, and the source count from the source direction is 23.5. The statistical significance of this is not simply calculable, since the counts associated with two directions less than 38° apart are correlated; the number of directions in the sky over which an event is counted depends on θ. The result of correlation can be seen in Fig. 3. However, these scans aid in determining a most likely direction of the anomalously high count.

In addition to "fixed declination scans," one can perform "fixed right ascension scans." These are even harder to analyze as the instrumental sensitivity as a function of zenith angle and pair opening angle influences the results strongly. In Fig. 4 we show the average of several fixed right ascension scans far from 20-h R.A. as well as the data at 20 h 15 min. Again, the source is shown to stand out. We thus identify the source direction as 35° N dec, 20-h 15-min R.A. The uncertainty in the direction of the source is 6° in all directions, this being the distance such that the excess count falls to about half the peak value.

The reader is cautioned against over interpretation of Figs. 3 and 4. The method of analysis involves strong correlation between adjacent points in these scans. As a result, one might be misled into interpreting the diagrams as a reflection of the angular response of the detector to a source transiting the telescope. No such claim is made by the authors. These diagrams have been used only in an attempt to locate the anomaly rather than to establish its existence.

The flux from the source was calculated from the relation

$$F = N/\eta \int A dt.$$

Here N, the number of γ rays from the source, is taken from Table I as 12.8, the excess above background in the source cell. $\int A dt$ is the time integral of the detector area for a source at 35° dec and has the value 6×10^5 cm^2 sec. η is the detection efficiency and depends most strongly on the conversion efficiency, which

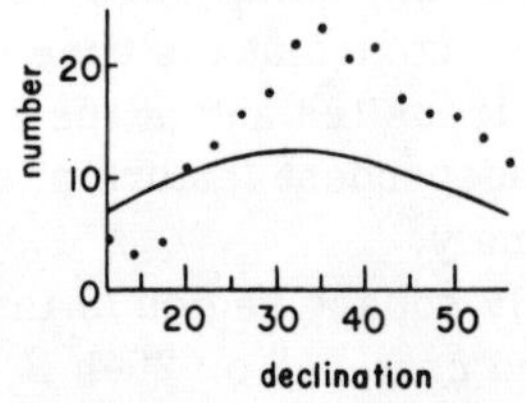

FIG. 4. "Fixed right ascension scan" for 20-h 15-min R.A. The result of the scan is indicated by dots. The solid line shows the average of several fixed R.A. scans. Each of these scans was at least $1\frac{1}{2}$-h R.A. from both 20-h 15-min R.A. and the zenith direction at the times for the beginning and end of the recording of data used in this report.

is a stronger function of energy below 200 MeV. We have taken η equal to 14%, the conversion efficiency at 75 MeV, and obtained a value of $(1.5 \pm 0.8) \times 10^{-4}$ γ's cm^{-2} sec^{-1},[17] where the error is purely statistical. This value is subject to systematic errors, the most important reflecting our lack of knowledge of the energy spectrum.

We have considered other possible explanations for our observations. Instrumental effects, and the fact that the small altitude variation at sunset occurred at approximately the time the suspected source transited, were investigated. There was no evidence of changes of instrumental sensitivity during the flight, and it is hard to imagine such a process giving both an excess count from a single region and a variation in the detected spectrum. We believe that we have correctly weighted the data for the pressure variation. Furthermore, if we have underestimated the pressure correction, then a much larger region of the sky would have shown an increased intensity. It is also difficult to reconcile the anomalous opening-angle spectrum with the slight pressure variation.

We wish to acknowledge the steady help and encouragement given to this work by Professor M. F. Kaplon. We would like to thank Professor M. P. Savedoff and members of the Cosmic-Ray Group at the University of Rochester for many helpful discussions. Considerable engineering assistance was received from Mr. R. Majka and Mr. R. Hawrylak. The further assistance of Mr. Majka in the construction and performance of this experiment was essential to its completion. Mrs. A. M. Endler gave invaluable assistance during the examination of the film.

---

*Research supported in part by the U. S. Air Force Office of Scientific Research and the National Science Foundation.

†National Aeronautics and Space Administration Predoctoral Trainee.

[1]P. Morrison, Nuovo Cimento 7, 858 (1958).

[2]S. Hayakawa, H. Okuda, Y. Tanaka, and Y. Yamamoto, Progr. Theoret. Phys. (Kyoto) Suppl. 30, 153 (1964).

[3]V. L. Ginzburg and S. I. Syrovatskii, Usp. Fiz. Nauk (USSR) 84, 201 (1964) [translation: Soviet Phys. Usp. 7, 696 (1965)].

[4]R. J. Gould, G. R. Burbidge, Ann. Astrophys. Suppl. 28, 171 (1965).

[5]G. Garmire and W. Kraushaar, Space Sci. Rev. 4, 123 (1965).

[6]W. Kraushaar, G. W. Clark, G. Garmire, H. Helmken, P. Higbie, and M. Agogino, Astrophys. J. 141, 845 (1965).

[7]T. L. Cline, Phys. Rev. Letters 7, 109 (1961).

[8]W. L. Kraushaar and G. W. Clark, Phys. Rev. Letters 8, 106 (1962).

[9]J. G. Duthie, E. M. Hafner, M. F. Kaplon, and G. G. Fazio, Phys. Rev. Letters 10, 364 (1963).

[10]E. M. Hafner, J. G. Duthie, M. F. Kaplon, and G. H. Share, in Proceedings of International Conference on Cosmic Rays, Jaipur, India (1963), edited by R. R. Daniel et al. (Commercial Printing Press, Ltd., Bombay, India, 1964-1965).

[11]R. Cobb, J. G. Duthie, and J. Stewart, Phys. Rev. Letters 15, 507 (1965).

[12]G. M. Frye and L. H. Smith, Bull. Am. Phys. Soc. 10, 705 (1965).

[13]C. E. Fichtel and D. A. Kniffen, J. Geophys. Res. 70, 4227 (1965).

[14]This is in agreement with the calculated and experimental secondary flux obtained by Share (private communications). We selected a value of 14% with a knowledge of calculations done by G. H. Share and Dr. B. R. Dennis (private communication) on the detection efficiency of comparable systems.

[15]Share, Ref. 14.

[16]G. Svensson, Arkiv Fysik 13, 347 (1958).

[17]Hafner et al. (Ref. 10) reported an excess count of γ rays detected by a simple counter telescope from this region of the sky. On that occasion a flux of 2×10^{-3} cm^{-2} sec^{-1} was reported. Later experiments with the same detector (Ref. 15) failed to confirm that result and set an upper limit of 10^{-3} cm^{-2} sec^{-1} from any point source in that region. The present results indicate a source intensity an order of magnitude smaller than the 1962 observations.

GALACTIC X-RAY SOURCES*

G. R. Burbidge, R. J. Gould, and W. H. Tucker
University of California, San Diego, La Jolla, California

During the past three years the existence of discrete x-ray sources outside the solar system has been demonstrated quite conclusively by the National Research Laboratory group,[1] by the Massachusetts Institute of Technology group,[2] and by Fisher and Meyerott.[3] About 10 such sources have been discovered so far, and a general isotropic background x-ray flux has also been reported. The discrete sources appear to have a spatial distribution showing a concentration toward the plane of the galaxy, indicating that the sources are probable galactic and at characteristic distances of ~1-10 kpc (kiloparsec). One of the sources appears to be in the direction of the galactic center. The most intense of the x-ray sources is that in the constellation Scorpius, from which a flux at the earth of about $F_X \simeq 10^{-7}$ erg/cm^2 sec is detected; the Scorpius source has not been identified with any optical or radio emitter. The one x-ray source which has been identified with any certainty is the Crab nebula for which $F_X \simeq 10^{-8}$ erg/cm^2 sec; the fluxes from most of the other x-ray sources are roughly the same as that from the Crab. Although the nature of all of these x-ray sources is not known, we feel that recent theoretical work on the interpretation of the observations allows one to reject many of the mechanisms proposed for the x-ray production. There are still problems connected with the mechanisms thought to be acceptable, and these will be discussed briefly here. A more complete discussion of these problems, along with the details of some associated calculations, will be published elsewhere.

There are essentially four possible origins or mechanisms for the x-ray production: (1) neutron stars, (2) Compton scattering, (3) bremsstrahlung, and (4) synchrotron emission. When x-ray sources were first discovered the possibility that they were neutron stars was discussed at length, but it appears that the recent work of Bahcall and Wolf[4] and others on the cooling (now thought to be very rapid) of neutron stars has shown that they are unlikely to be emitting sufficient x rays to explain the observations. Compton scattering of high-energy electrons by low-energy (~eV) photons, producing high-energy photons, has not been discussed extensively as a mechanism for x-ray production in discrete sources. It has been suggested[5] that the x rays from the Crab are due to Compton scattering of the (radio to optical) synchrotron electrons by the associated synchrotron photons. However, the intensity of this Compton-synchrotron radiation can be shown to be far too small (by a factor ~10^{-6}) to account for the observed x-ray flux.[6] This effect can easily be worked out in some detail, but its unimportance can be seen readily if one estimates the probability of a Compton scattering of a photon before escaping from the nebula. One might think that Compton scattering might produce a large x-ray flux from quasistellar radio sources in which the photon density and

Reprinted from Physical Rev. Letters **14**, 289 (1965).

the high-energy electron density are large. Again, however, simple calculations indicate a completely negligible and unobservable x-ray flux from this process. Consequently, we are led to rule out Compton scattering as an x-ray production mechanism in discrete sources. This leaves only the synchrotron and bremsstrahlung processes as possible x-ray sources.

First we consider the possibility that the x rays are synchrotron radiation.[7] We assume for the moment that the x-ray energy flux $F_X = 10^{-8}$ erg/cm^2 sec comes from a source at a galactic distance $r = 10$ kpc; the x-ray luminosity of the source is then $L_X(\propto r^2) = 1 \times 10^{38}$ erg/sec. Further, we assume for simplicity that the x-ray flux is at an effective wavelength 3 Å and frequency $\nu = 10^{18}$ cps, which is the characteristic synchrotron frequency $\nu_L \gamma_e^2$ emitted by electrons of energy $E_e = \gamma_e mc^2$; ν_L is the Larmor frequency. For a magnetic field $H = 10^{-4}$ G (the assumed value in the Crab nebula), the electron energy required is $E_e(\propto H^{-1/2}) = 3 \times 10^{13}$ eV. For such a high-energy electron the lifetime against energy loss by synchrotron emission is only $E_e(dE_e/dt)^{-1} = \tau_e(\propto H^{-3/2}) = 30$ yr. The total energy in these electrons necessary to produce the luminosity L_X is $E_t(\propto r^2 H^{-3/2}) = 1 \times 10^{47}$ erg. We note that (1) the electron energies required to produce synchrotron x rays are extremely high, (2) their lifetime is very short, and (3) the total electron energy involved is comparable to the energy released in a supernova outburst. Actually, the energy E_t quoted above is really the minimum energy of the highly relativistic electrons, since it includes only the synchrotron electrons producing x rays. The contribution of the lower energy extension of the electron spectrum to the total energy would increase the value of the total energy by an amount depending on the index of the spectrum and the low-energy cutoff. For the case[8] of the Crab nebula the extension of the x-ray spectrum (which has an index of about 1.1) to the visible leads to a total electron energy which is not excessively large (~10^{48} erg). However, it is very significant that the lifetime of the high-energy electrons is appreciably less than the age of the Crab nebula and other supernova remnants, because it would mean that the high-energy electrons would have to be continuously or at least periodically produced. If they are spasmodically produced or accelerated one might expect to observe variations in the x-ray intensity over time scales ≤ 10 yr. It is also significant that for the Crab there can at present be no continuous production via nuclear collisions and π-μ decay, since the associated flux of high-energy photons from the decay of π^0 mesons would be too high.[8]

In the case of the Crab it is well known that it is emitting optical and radio synchrotron radiation as well as x rays. We shall now compare the x ray and radio observations[9] of the galactic center region. In Fig. 1 these observations are plotted together with the extensions of power-law spectra derived for indices within limits (-0.72 ± 0.05) such as to fit the radio data. The fact that the x-ray point lies so close to this extrapolation over a factor of 10^{10} in frequency of the radio data is remarkable. While it suggests that a single mechanism is operating, the ratio of the half-lives (τ_r/τ_X) of the electrons giving rise to radiation in the two different spectral regions is $(\nu_r/\nu_X)^{-1/2}$. It should be added that if there is a continuous synchrotron spectrum over this range, detection of radiation in the optical and infrared regions will not be possible because of obscuration by interstellar matter on the one hand and thermal radiation from stars on the other.

It is easily shown that for reasonable values of the spectral index ($\alpha \geq 0.5$) of the x-ray sources, the extension of the spectra to lower frequencies would mean that they would be easily detectable as radio sources and might also be seen as faint optical nebulae. If such identifi-

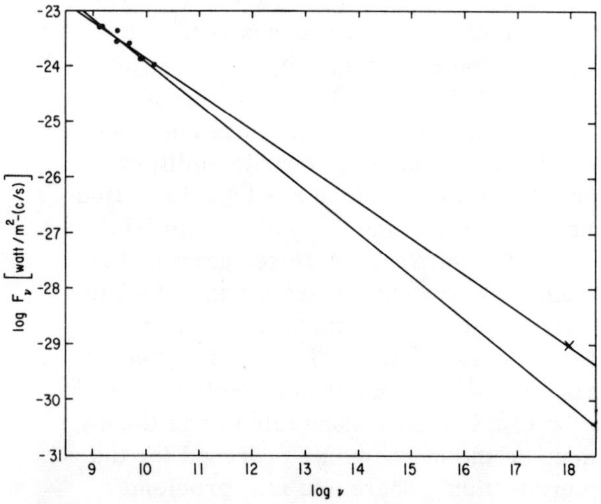

FIG. 1. The observed radiation spectrum from the galactic center. Dots denote the radio observations; an x denotes the x-ray point, determined from an energy flux 10^{-8} erg/cm^2 sec and band width $\Delta\nu/\nu = 1$ at $\nu = 10^{18}$ cps.

cations cannot be made, and if the synchrotron mechanism is responsible, there must be a sharp cutoff in the energy spectrum of the electrons, and this low-energy cutoff must be at energies greater than about 10^{12} eV for assumed magnetic fields of order 10^{-4} G. It is very difficult to see how fluxes of electrons with energies $\gtrsim 10^{12}$ eV can be produced without giving rise to large low-energy fluxes, either in an acceleration process, or by deceleration by the synchrotron process itself. The only model which would seem to be possible would be direct injection already at the highest energies from a source in the remnant and then direct escape. The source must be continuously active if the flux is to be maintained. It can also be deduced that the x-ray flux would then arise in a very small volume with an upper limit of perhaps one parsec in extent.

Because of the difficulties associated with such models, we believe that it is worthwhile considering an alternative model in which it is supposed that an outburst gives rise to a small, very hot cloud which continues to emit hard radiation as part of the thermal bremsstrahlung. We discuss now the properties associated with such a model. Two of us suggested[10] earlier that the x rays from the source at the galactic center resulted from bremsstrahlung. At the time we envisaged bremsstrahlung production by nonthermal electrons. However, as was first pointed out by Rossi,[11] about 10^5 times as much energy would be lost by these electrons in inelastic atomic collisions, so that if the x-ray luminosity of the source at the galactic center is 10^{38} erg/sec, about 10^{43} erg/sec must be supplied. This energy rate is excessively large on a galactic time scale (10^{10} yr), although perhaps it may be supplied during shorter times. One is led to consider conditions where bremsstrahlung x-ray production is more efficient; these are realized in a high-temperature ($T \sim 10^7$ °K) and low-density gas where the bremsstrahlung is produced by thermal electrons and constitutes a major source of cooling or energy loss for the gas. An associated problem is that of the cooling rate of such a high-temperature gas; we have considered this question and present here the essential results, the discussion and details to be given elsewhere. In Fig. 2 we give the rate of loss Λ_e (erg/cm³ sec) of the kinetic energy of the electron gas of density n_e by various processes in the temperature range between

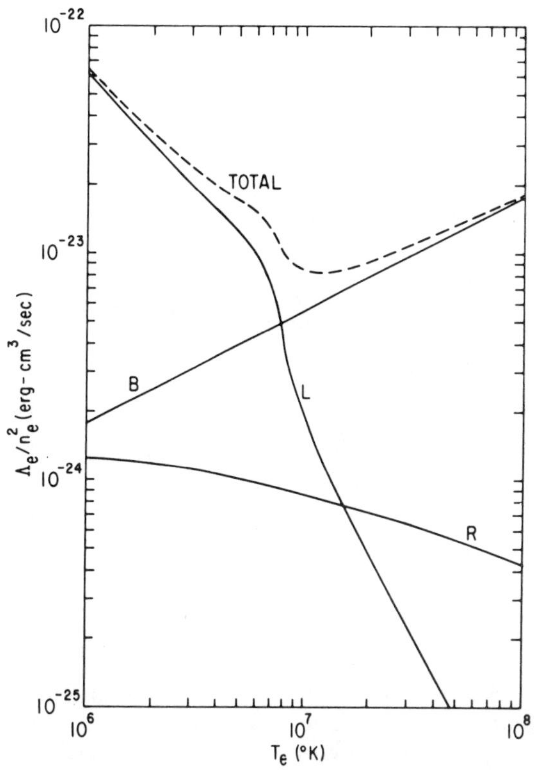

FIG. 2. Cooling rate as a function of temperature. Λ_e denotes the rate of change of the free-electron kinetic-energy density. Cooling by bremsstrahlung (B), line emission following inelastic electron collisions (L), and recombination (R) are shown. The ions of the following elements have been included: (B) H + He, (L) O + Ne, and (R) H + He + O + Ne.

10^6 and 10^8 °K.[11] It is seen that bremsstrahlung dominates the cooling at higher temperatures. In Fig. 2 we give the rate of production of x rays ($p_X = P_X/n_e^2$) in the range 1-10 keV as a function of temperature by various processes. The cooling time ($\tau_c \approx 3kT_e/n_e\Lambda_e$), density ($n_e$), and mass ($M$) of a volume ($V$) of gas required to produce the observed x-ray fluxes are of prime interest. We assume the source to be at a distance of $r = 10$ kpc and to produce an x-ray energy flux $F_X = 10^{-8}$ erg/cm² sec in the range 1-10 keV. Further, we assume the gas to be at a temperature of 10^7 °K; parameters for other values of the temperature may be determined readily from Figs. 2 and 3. Since $F_X = p_X n_e^2 V/4\pi r^2$, this choice of F_X, T, and r fixes the product $n_e^2 V$ at 4×10^{61} cm^{-3}. Then for a range $n_e = 0.1$ to 10^4 cm^{-3}, $\tau_c \sim 10^8$ to 10^3 yr, $V \sim 10^8$ to 10^{-2} pc³, and $M \sim 4\times10^5$ to 4 solar masses. The associated optical bremsstrahlung intensity is of interest and depends only

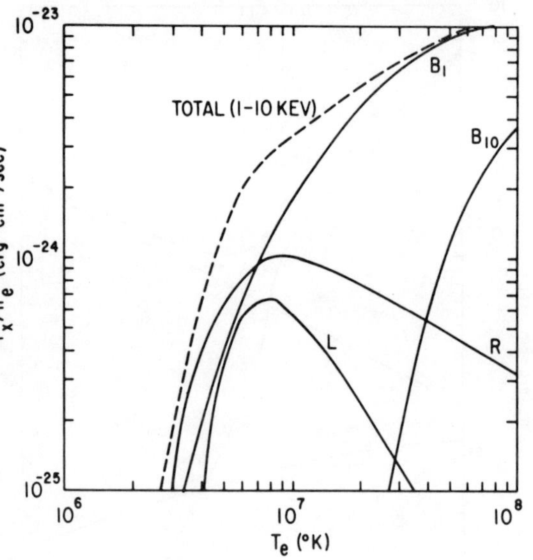

FIG. 3. X-ray production rates in the range 1-10 keV by bremsstrahlung (B_1) and recombination radiation (R) and line emission (L) (essentially only the ion Ne^{+9} contributes). The brĕmsstrahlung production rate (B_{10}) in the range 10-20 keV is also shown. The ions of the following elements have been included: (B_1, B_{10}) H + He, (L) Ne, and (R) H + He + N + O + Ne.

on the choice of T. One finds that this intensity corresponds to a twelfth-magnitude visual object which may be observable, depending on the extent of the source.

As yet it is difficult to express a preference for either the synchroton or bremsstrahlung hypothesis for the x-ray production mechanism. As we have emphasized, the energy-loss time scales for the high-energy synchrotron electrons are very short, while very high temperatures are required to produce bremsstrahlung x rays. Although the basic mechanism remains uncertain, we feel that the likely origins of the x-ray sources are supernova outbursts. For it is known that supernova remnants are synchrotron emitters in the radio range (and in the case of the Crab in the optical), and it is believed that temperatures ~10^7 °K are produced by the shock front associated with the expanding ejecta. We might remark, however, that the very high-energy (~50-keV) radiation detected from the Crab by Clark[12] requires a temperature of 2×10^8 °K, which is about an order of magnitude higher than theoretical estimates of the expected temperature. This would seem to suggest that the x rays from the Crab result from the synchrotron process.

---

*Work supported in part by the National Science Foundation and in part by the National Aeronautics and Space Administration through Contract No. NsG-357.

[1]S. Bowyer, E. T. Byram, T. A. Chubb, and H. Friedman, "Cosmic X-ray Sources" (to be published), gives references to previous work.

[2]R. Giacconi, H. Gursky, F. R. Paolini, and B. B. Rossi, "Observations of Two Sources of Cosmic X-rays in Scorpius and Sagittarius" (to be published), gives references to previous work.

[3]P. C. Fisher and A. G. Meyerott, Astrophys. J. 139, 123 (1964); 140, 821 (1964); also unpublished work.

[4]J. N. Bahcall and R. A. Wolf, "Neutron Stars" (to be published).

[5]P. Morrison, Second Texas Symposium on Relativistic Astrophysics, Austin, 1964 (to be published).

[6]One of us (R.J.G.) stated this result at the Second Texas Symposium on Relativistic Astrophysics, Austin, 1964 (to be published); and also at the Liège Symposium, Liège, Belgium, August 1964 (to be published).

[7]See also L. Woltjer, Astrophys. J. 140, 1309 (1964).

[8]R. J. Gould and G. R. Burbidge, to be published.

[9]A. Maxwell and D. Downs, Nature 204, 865 (1964).

[10]R. J. Gould and G. R. Burbidge, Astrophys. J. 138, 969 (1963).

[11]The results disagree quantitatively with C. Heiles, Astrophys. J. 140, 470 (1964).

[12]G. W. Clark, Phys. Rev. Letters 14, 91 (1965).

X-RAYS FROM THE COMA CLUSTER OF GALAXIES*

J. E. Felten, R. J. Gould, W. A. Stein, and N. J. Woolf

University of California at San Diego
and
University of Texas, Austin

Experimenters at the Goddard Space Flight Center (Boldt, McDonald, Riegler and Serlemitsos 1966) have reported an extended X-ray source of angular diameter 4-5° and spectral flux $j \approx 10^{-2}$ $(\text{cm}^2 \text{ sec kev})^{-1}$ at 25 kev. The source is centered at $\alpha \approx 13^h0^m$, $\delta \approx +28°$, and they identify it as the Coma cluster of galaxies. It is possible that this source is in fact nearer than the Coma cluster; if it is galactic, it is probably in the disk and within 100 pc, since it lies at $b^{II} \approx +88°$. Assuming, however, that the identification is correct, it implies a surprisingly large X-ray luminosity for this cluster. These X-rays might be produced in (a) individual galaxies within the cluster, or (b) the intergalactic medium. In this Letter we consider (a) briefly and show that the superposition of individual galaxies probably cannot produce the observed flux. We then resort to (b) and find that the emission is accounted for by a mass of hot gas equivalent to the familiar missing mass suggested by application of the virial theorem to this cluster, provided $T \sim 10^8$ °K.

The Coma cluster (Abell 1965) lies at a distance of 90 Mpc (for H = 75 km sec^{-1} Mpc^{-1}, which we assume throughout), so that the photon emission by the cluster at 25 kev is $J \approx 1 \times 10^{52}$ $\text{sec}^{-1} \text{kev}^{-1}$. It contains more than 800 and possibly a few thousand galaxies, brighter than M_{pv} ≈ -17; 1500 is a reasonable estimate. If the total cluster output is just the superposition of individual galaxies, it follows that the mean emission per galaxy in the Coma cluster is $\langle J_g \rangle \approx 7 \times 10^{48}$ sec^{-1} kev^{-1}. The same quantity for our own Galaxy, J_G, may be estimated fairly well from the observed flux due to the cluster of discrete sources near the galactic center: $(16 \pm 5) \times 10^{-9}$ erg cm^{-2} sec^{-1} in 20-30 kev (Giacconi et al. 1965). Assigning an appropriate distance, ≈ 10 kpc, we find $J_G \approx 5 \times 10^{44}$ sec^{-1} kev^{-1}; the average galaxy in the Coma cluster is $\gtrsim 10^4$ times brighter at 25 kev than our own!

Before attempting to interpret this surprising result, we may make another comparison. The background flux per steradian, $\delta j_B/\delta\Omega$, due to all unresolved galaxies in the universe may be estimated from

$$\delta j_B/\delta\Omega \sim (4\pi)^{-1} n_g \langle J_g \rangle R \tag{1}$$

if the mean emission per galaxy $\langle J_g \rangle$ is known. Here n_g is the mean number density of galaxies in space, often given as $\approx 1 \times 10^{-75}$ cm^{-3} (Sandage 1965); this number is appropriate if galaxies down to $M_{pg} \approx -17$ or a little brighter are counted (van den Bergh 1961, Kiang 1961). R in approximation (1) is a cosmological path length which may be taken $\sim c/2H$ for most cosmological models (Whitrow and Yallop 1964, Felten 1966). If we now assume that $\langle J_g \rangle \approx \langle J_g \rangle_{\text{Coma}}$, we find from equation (1) that the isotropic X-ray flux at 25 kev should be ~ 3 $(\text{cm}^2 \text{ sec sr kev})^{-1}$. But the observed flux (Bleeker et al. 1966; Hayakawa, Matsuoka, Ogawa and Yamashita 1966; Rothenflug, Rocchia and Koch 1966) is smaller by a factor ~ 30. Earlier workers have similarly used equation (1) to estimate the unresolved background, taking the luminosity of our Galaxy J_G as model for $\langle J_g \rangle$, and it has been shown (Oda 1966, Gould and Burbidge 1966) that J_G is too small by a factor 30-100 to account for the observed isotropic flux. Now the situation reverses, since $\langle J_g \rangle$ derived for the contents of the Coma cluster is embarrassingly large compared to the background.

How may these unexpected numbers be interpreted? It may be that $\langle J_g \rangle \not\approx \langle J_g \rangle_{\text{Coma}}$. The Coma cluster is composed mainly if not entirely of ellipticals, which are only ≈ 20 per cent of all galaxies in the universe. This correction would bring the estimate (1) within a factor of 6 of the observed value, agreement which must be regarded as satisfactory. Thus one could conclude that ellipticals as a class are powerful X-ray emitters, with mean luminosity $\sim 10^4 J_G$. It would follow that any region containing many ellipticals should be an extended X-ray source; in particular, the Virgo cluster (Abell 1965) is about half ellipticals, has ~1000 members, is six times nearer than the Coma cluster, and hence should contribute at 25 kev a flux ~10 times larger than the latter. The Virgo region has not been surveyed at this energy, but Naval Research Laboratory observations at 1 - 10 Å (Byram, Chubb and Friedman 1966) disclose a source here with count rate ≈ 0.2 cm^{-2} sec^{-1}. The NRL workers interpret this as a point source at the position of the galaxy M87. An angular extent of several degrees, however, would apparently be compatible with their data, and an alternate interpretation might be that the X rays are coming from many galaxies in the Virgo cluster. However, on the basis of an assumed spectrum $J(E) \propto E^{-2}$ and a flux of 10^{-1} $(cm^2 \text{ sec kev})^{-1}$ at 25 kev, one would expect a flux in the 1 - 10 Å region about 200 times the observed value. Only if the X-ray spectra of the galaxies were much flatter (unlike known galactic sources) would the NRL observations be consistent with the idea that emission from elliptical galaxies produces the X-ray flux at 25 kev inferred for the Virgo cluster and observed from the Coma cluster.

Taking a slightly different tack, we might suppose that $\langle J_g \rangle \neq \langle J_g \rangle_{Coma}$ because the brightness of the Coma cluster is a statistical variation, i.e., that it contains a few peculiar galaxies of exceptional X-ray luminosity, $\sim 10^6$ - 10^7 J_G, which are relatively much rarer in the universe as a whole (Note that one such object is insufficient, since the X-ray source would then not be extended). For comparison, the NRL source in Virgo, if it is indeed M87 or another galaxy in the Virgo cluster, has a 1 - 10 Å luminosity $\sim 10^5$ that of the Galaxy. Another source in Cygnus, if the NRL identification with Cyg A is correct, has luminosity $\sim 10^3$ times larger than M87. To determine whether similar sources exist in the Coma cluster, we should look for discrete radio sources there similar to M87 and Cyg A. Only three verified sources (Coma A, Coma C and 3C284; Howard and Maran 1965) lie within 5° of the cluster center. These sources, if at a distance of 90 Mpc, have absolute radio luminosities ~1/3 to 1/5 of M87. They all lie between $\delta = 27° 42'$ and $28° 24'$ and could not produce X-ray emission extending to 30° as reported by the Goddard group. It is of course possible that there exist radio-quiet galaxies which are nevertheless brighter in X-rays than M87, but until and unless the Coma X-ray source can be resolved into components this possibility must be treated with skepticism.

We turn now to a discussion of emission from the intergalactic medium within the Coma cluster. Of the many nonthermal mechanisms capable of producing X-rays therein, we may mention the Compton process and the synchrotron process. In order to obtain the observed luminosity from either of these processes, however, uncomfortably large densities of relativistic particles must be assumed. For example, it can be shown that to generate the X-rays by Compton scatterings between secondary cosmic-ray electrons produced in the cluster gas and cosmic blackbody photons, a primary cosmic-ray flux $\sim 10^3$ times the terrestrial value would be required even for a gas density $N \sim 10^{-3}$ cm^{-3}

A more promising explanation of the Coma X-ray emission is that it is due to thermal bremsstrahlung from a hot gas filling the cluster volume. Suggestive evidence for such gas comes from the well-known problem of the mass deficiency for gravitational binding (Abell 1965); the mass implied is $\sim 4 \times 10^{48}$ g. If this amount of hydrogen were spread uniformly through a sphere of angular diameter 4°, a gas temperature $T \approx 2 \times 10^8$ °K would give the emission observed at 25 kev. (For H ≈ 100 km sec^{-1} Mpc^{-1}, we would obtain instead $T \approx 1 \times 10^8$ °K.) The calculations of Woolf (1966) show that the presence of such a gas cloud is consistent with radio and optical upper limits. Clumpiness or central concentration in the gas distribution would increase the emission, which goes as $\langle N^2 \rangle$, and thus imply a lower temperature. If, however, the central concentration of the gas is as great as that reported for the galaxies (Omer, Page and Wilson 1965), then the effective angular size of the X-ray source must be ≈ 2° rather than 4°.

Because the bremsstrahlung spectrum is

relatively flat below $\hbar\omega \approx kT$, the photon flux in 1 - 10 Å from this source would not be unduly large. Nevertheless, for $T \approx 2 \times 10^8$ °K, this flux would be ~ 2 cm^{-2} sec^{-1} and should be easily detectable; if T is lower, the expected flux is larger, and for $T \lesssim 1 \times 10^8$ °K the Coma source should be among the brightest in the sky at these wavelengths. It is not clear whether this high-latitude region has been adequately scanned in 1 - 10 Å. In any case the source should be strong at $\hbar\omega \gtrsim 20$ kev, and a spectral measurement in this neighborhood would give a good determination of T.

It is interesting that if one sets

$$\tfrac{1}{2}kT \approx \tfrac{1}{2}n_H m_H v_r^{\,2}/(n_i + n_e) \approx \tfrac{1}{4}m_H v_r^{\,2},$$

where m_H is the mass of the hydrogen atom and $v_r \approx 1050$ km sec^{-1} is the rms line-of-sight velocity dispersion of the Coma cluster galaxies, one finds $T \approx 0.7 \times 10^8$ °K. This means that a temperature of this order would be expected for the gas if it has been heated by the thermalization of pre-existing mass motions in the gas with velocities comparable to the observed velocities of the galaxies.

The preconditions essential to formation of such a gas cloud are uncertain, and it is not clear whether all galaxy clusters, or even those of "compact" or "regular" form like the Coma cluster, should be X-ray sources. The latter include the Pegasus I, Cancer, Perseus, Corona and Hydra I clusters (Zwicky 1959), and these are good candidates for observation. The Virgo cluster is of rather different form, but is known to have a mass deficiency like Coma and might be a source. Koehler (1965) has argued that the small amount of neutral hydrogen he has detected in the Virgo cluster rules out the presence of ionized gas there. However, Woolf (1966) shows that most of the intergalactic matter in both the Coma and Virgo clusters may be ionized and that Koehler's argument is inadequate.

If the density of regular clusters in the universe is ~ 1 per (90 Mpc)^3, and each radiates like Coma, the expected isotropic background x-ray brightness is comparable with that observed. The spectral shape of the observed radiation, however, suggests that nonthermal processes may contribute more strongly to it (Felten and Morrison 1966).

We wish to thank E. T. Byram, R. Giacconi and P. Serlemitsos for some comments on the observational situation.

REFERENCES

G. O. Abell, 1965, Ann. Rev. Astron. Ap., 3, 1.

J. A. M. Bleeker, J. J. Burger, A. Scheepmaker, B. N. Swanenburg, and Y. Tanaka, 1966, to be published.

E. Boldt, F. B. McDonald, G. Riegler, and P. Serlemitsos, 1966, Phys. Rev. Letters, 17, 447.

E. T. Byram, T. A. Chubb, and H. Friedman, 1966, Science, 152, 66.

J. E. Felten, 1966, Ap. J., 144, 241.

J. E. Felten, and P. Morrison, 1966, Ap. J., 146. in press.

R. Giacconi, H. Gursky, J. R. Waters, B. Rossi, G. Clark, G. Garmire, M. Oda, and M. Wada, 1965, lectures at the International School of Physics "Enrico Fermi", Varenna.

R. J. Gould, and G. R. Burbidge, 1966, in Hdb. d. Phys., 46/2 (Berlin: Springer).

S. Hayakawa, M. Matsuoka, H. Ogawa, and K. Yamashita, 1966, presented to COSPAR Symposium, Vienna.

W. E. Howard, and S. P. Maran, 1965, Ap. J. Suppl. 10, 1.

T. Kiang, 1961, M. N., 122, 263.

J. A. Koehler, 1965, dissertation, Australian National University, Canberra.

M. Oda, 1966, in Proceedings of the Ninth International Conference on Cosmic Rays (London: Institute of Physics), p. 68.

G. C. Omer, T L. Page, and A. G. Wilson, 1965, A. J., 70, 440.

R. Rothenflug, R. Rocchia, and L. Koch, 1966, in Proceedings of the Ninth International Conference on Cosmic Rays (London: Institute of Physics), p. 446.

A. Sandage, 1965, Ap. J., 141, 1560.

S. van den Bergh, 1961, Zs. f. Ap., 53, 219.

G. J. Whitrow, and B. D. Yallop, 1964, M. N., 127, 301.

N. J. Woolf, 1967, Ap. J. 148, 287 (1966).

F. Zwicky, 1959, in Hdb. d. Phys., 53 (Berlin: Springer), p. 390.

*Research has been supported by the National Aeronautics and Space Administration (under Grants NsG-357 and NsG-318), the National Science Foundation, and Alfred P. Sloan Foundation Fellow, 1966-68.

SUMMARY OF X-RAYS AND γ-RAY ASTRONOMY

S. Hayakawa
Nagoya University

I summarize here the discussions starting from very high-energy γ-rays and ending up in the keV region.

1. EAS region.—Extensive air showers observed at Mount Chacaltaya provide information on the intensity of high-energy γ-rays relative to that of protons. For energies greater than 10^{16} eV, no evidence for the primary γ-ray has yet been found among 1,500 showers so far observed. This led Clark to conclude, referring to my calculation on the origin of high-energy γ-rays due to the collisions of intergalactic protons with starlight, that the intensity of intergalactic cosmic rays should be smaller than that in our neighborhood. This conclusion seems, however, to arise from misleading reference to my calculation.[1] My result gives that the ratio of γ-rays and protons expected from the process under consideration may be as small as 10^{-4} or smaller. Thus we need 15,000 showers, instead of 1,500, to say something about cosmology.

A factor of about 10 between the experimental upper limit of the γ-ray intensity and a theoretical estimate holds down to 100 MeV. The theoretical intensity is said to be modest in the sense that one cannot push it up too much because of the limit set by the positron intensity, as reported by Meyer.

2. Local sources.—The γ-ray intensities from local sources are much more difficult to predict. A curve may be drawn in such a way that it is parallel to the isotropic intensity and is presumed to be close to what we expect from the Crab Nebula. Most of the upper limits observed lie far above the curve, but the one obtained by Chudakov lies slightly below the curve. The Chudakov point gives us a strong restriction to properties of local sources.

For example, the observability of local neutrino sources, as suggested by Bahcall, would be discouraged, because such a large rate of neutrino production would have given a strong γ-ray intensity exceeding the Chudakov limit. This means that they are of common origin.

Although the γ-ray intensity is expected to be very low, it seems better to make an effort for observing local sources with good angular resolution than to find out a finite amount of widely spread γ-rays with fancy technique.

3. X-ray sources.—Much progress has been achieved in X-ray astronomy in the last year. Both ASE-MIT and NRL groups seem to have agreed to locate about ten X-ray sources, which have the following characteristic features: (a) most of them are related neither to radio sources nor to marked optical objects; (b) two are identified with remnants of SNI; (c) all are found in and near the galactic disk. The last point indicates that the X-ray sources are probably in our Galaxy. The first two points show up a rather strange feature of the spectrum of electromagnetic radiation in each X-ray source.

4. Energy spectrum of electromagnetic radiation.—The Scorpius unit of the X-ray intensity, 10^{-7} ergs cm^{-2} sec^{-1}, referred to by Giacconi, corresponds to the rate of energy generation in the X-ray region

$$10^{37}\ R^2\ \text{ergs sec}^{-1},$$

where R is the distance of a source in kpc. This in itself is a big number. The invisibility in the optical and the radio regions may indicate that the luminosities in respective regions are smaller than 10^{m} and 10^{-26} watt m^{-2}, taking account of the diffusiveness of a source as was found by Friedman with the aid of lunar occultation.

The intensities in these three spectral ranges tell us that the spectrum is as steep as or steeper than $\nu(d\nu/\nu)$ up to the X-ray region. Thereafter, the spectrum is expected to be flattened and eventually to fall off, so that it is smoothly connected with the γ-ray region. The spectrum of this shape cannot discriminate one most probably X-ray emission mechanism against many other possible ones. We can, however, say that the $\nu(d\nu/\nu)$ spectrum requires a low-energy cutoff of the spectrum of non-thermal particles, such as high-energy electrons, provided that the electromagnetic radiation is overwhelmingly non-thermal.

5. Important points for future investigations.— (a) The success in the last year was achieved mainly by narrowing the angular aperture of detectors. This should be encouraged toward a still better angular resolution, so that we shall be able to find out more sources and to get information on the size of sources. (b) The identification of the known sources by optical and radio means is exceedingly important. The present limits of 10^m and 10^{-26} watt m^{-2} will be broken through merely by a little more careful observation. Since neither new facility nor new technique seems to be needed for this, I urge astronomers here to look at these X-ray sources, as soon as they come back to their observatories. (c) The observation at lower energies is important, first of all, for the determination of a source distance in virtue of the interstellar absorption. This will become possible when the spectrum is known by the interpolation between the optical and the present X-ray regions. (d) The extension of the spectrum to higher energy is equally important, in order to determine the contributions of thermal and non-thermal components. Although experimental results of the spectrum were reported by Clark, Friedman, and Fischer, they are only preliminary and further efforts will be needed for making this conclusive.

REFERENCE

[1]A comprehensive discussion of this problem is found in Prog. Theor. Phys. Suppl., No. 30 (1964).

PART 5

NEUTRINO ASTRONOMY

SOLAR NEUTRINOS*

Raymond Davis, Jr.
Chemistry Department, Brookhaven National Laboratory,
Upton, New York 11973
Don. S. Harmer, Frank H. Neely
Nuclear Research Center, Georgia
Institute of Technology,
Atlanta, Georgia, 30232

Introduction

The prospect of studying the solar-energy-generation process directly by observing the solar-neutrino radiation has been discussed for many years. The main difficulty with this approach is that the Sun emits predominantly low-energy neutrinos, and detectors for observing low fluxes of low-energy neutrinos have not been developed. However, experimental techniques have been developed for observing neutrinos, and one can foresee that in the near future these techniques will be improved sufficiently in sensitivity to observe solar neutrinos. At present several experiments are being designed and hopefully will be operating in the next year or so. We will discuss an experiment based upon the neutrino capture reaction

$$\nu + Cl^{37} \rightarrow Ar^{37} + e^{-} . \qquad (1)$$

This reaction is the inverse of the electron-capture radioactive decay of Ar^{37}. The method depends upon exposing a large volume of a chlorine compound, removing the radioactive Ar^{37} and observing the characteristic decay in a small low-level counter. A high sensitivity for neutrino detection is achieved by using a large mass of chlorine and performing the counting measurements in a counter with a very low background. An experiment will be described that has been performed with 1,000 gal. (6.1 tons) of perchloroethylene (C_2Cl_4) that served as a pilot experiment to test the method. A detector one hundred times larger is now being built that is designed to measure the presently calculated solar-neutrino flux. The design and aims of this experiment will be described.

In planning a solar-neutrino experiment one is guided by the present calculations of solar-neutrino fluxes. These calculations have been developed in recent years, and we now have confidence that the neutrino flux can be calculated within a factor of 2. Sears (1964) in particular has calculated the solar-neutrino flux with his model using various values for the solar composition, age, luminosity, and nuclear parameters to test the errors introduced in the values of neutrino flux. Independent calculations have also been performed by Pochoda and Reeves (1964) and Cameron and Ezer (A. G. W. Cameron, private communication), and their results agree within a factor of 2 with those of Sears. It is generally agreed that the P-p chain of reactions is the dominant mechanism for the Sun. In this chain neutrinos are produced by three processes, the $H(H, e^{+}\nu)D$ reaction and the radioactive decays of Be^7 and B^8. The neutrino spectrum from these three sources may be combined to represent the gross neutrino spectrum from the Sun. Figure 1 shows this spectrum and the fluxes of neutrinos at the Earth as calculated by Sears. The neutrinos from the $H(H, e^{+}\nu)D$ reaction are below the threshold (0.816 MeV) for the capture reaction (1). However, the 0.81-MeV neutrino line from the Be^7 decay is above threshold for reaction (1), and contributes about 10 per cent of the expected capture rate in Cl^{37}. The flux of energetic neutrinos from the B^8 decay (E_{max} = 14.2 MeV) is only 1.9×10^7 cm^{-2} sec^{-1}, since the $Be^7(p, \gamma)B^8$ reaction plays only a minor role in the solar/energy-generation process. However, the cross-section of reaction (1) is so large for these energetic neutrinos that the B^8 neutrinos would be expected to produce about 90 per cent of the total signal in a detector based upon the $Cl^{37}(\nu, e^{-})Ar^{37}$ reaction.

*Research performed under the auspices of the U. S. Atomic Energy Commission.

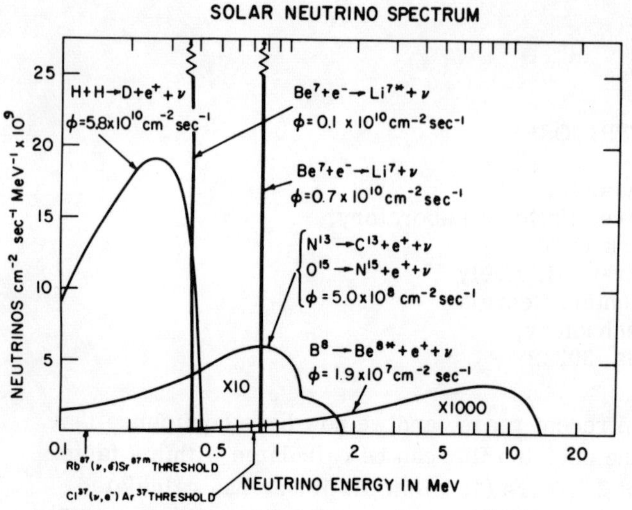

FIG. 1. Spectrum of solar neutrinos, and flux of neutrinos at Earth.

The reason that the cross-section is high for these energetic neutrinos is that neutrinos above 6.0-MeV energy will be captured to produce an excited state in Ar^{37} at 5.2 MeV. Bahcall (1964a, b; Bahcall and Barnes 1964) has pointed out that this state is the analogue state to the ground state in Cl^{37} and hence the neutrino capture to form this analogue state is a superallowed transition and therefore has a high cross-section. Bahcall's prediction has been recently verified by two separate experimental groups, at Brookhaven (Reeder, Poskanzer, and Esterlund 1964) and at McGill (Hardy and Verrall 1964). They observed the beta-decay of Ca^{37} to reveal the 5.2-MeV level in the mirror nucleus K^{37}. The fact that the cross-section for energetic neutrions is large had led us to believe we could observe the B^8 solar neutrinos by the Cl^{37}-Ar^{37} method (Bahcall 1964a, b; Bahcall and Barnes 1964; Davis 1964). It is interesting to note that the production of B^8 in the Sun depends strongly on the central temperature; therefore a measurement of the B^8 neutrino flux allows one to deduce an accurate value for the central temperature of the Sun.

Using the calculated solar-neutrino fluxes from Be^7 and B^8 and the values of the cross-section for the $Cl^{37}(\nu, e^-)Ar^{37}$ reaction, one deduces the value of the

$$\sum \phi\sigma = 3 \pm 2 \quad 10^{-35}\,\text{sec}^{-1}.$$

The error indicated reflects the uncertainties in the solar-model calculation. The error in the cross-section values are small, around 10 per cent. This sum of the products of the flux and the cross-section would correspond to an expected capture rate of 5 ± 3 per day in 600 tons of perchloroethylene. Let us now describe the pilot experiment which uses 1,000 gallons of perchloroethylene to illustrate how a neutrino detector based upon reaction (1) operates. The performance of this modest scale detector will also show how close we are to observing the presently calculated solar-neutrino flux.

The Pilot Experiment

Let us review briefly the history of the method. The advantages of the Cl^{37}-Ar^{37} method of detecting neutrinos was pointed out some years ago by Pontecorvo (1946). The experimental method was outlined and background effects were analyzed in detail by Alvarez (1949) with the view of using the technique to observe neutrinos at a nuclear reactor. We developed a detector based upon these suggestions and performed experiments near a Savannah River reactor (Davis 1957; Davis and Harmer 1959). Since a nuclear reactor is a source of antineutrinos, we found these antineutrinos would not drive the reaction. In terms of present concepts, these experiments served to test the principle of lepton conservation.

It was apparent that the Cl^{37}-Ar^{37} method with its inherent sensitivity to low-energy neutrinos would be useful to observe solar neutrinos (Davis 1964). The sensitivity of the apparatus was severely limited during the Savannah River experiments by a large background effect from cosmic-ray muons. By placing the apparatus 2,300 feet underground (1,800 meters of water equivalent) the cosmic-ray background effects were reduced to a negligibly small value. We used the limestone mine of the Pittsburgh Plate Glass Company at Barberton, Ohio. Figure 2 is a photograph of the apparatus. It consisted of two 500-gal. tanks (total of 6.1 metric tons) of perchloroethylene, C_2Cl_4, equipped with agitators, and a helium purging system. Helium was passed through the tanks in series, through condensation traps, and finally through a liquid-nitrogen-cooled charcoal trap. At the liquid-nitrogen temperatures used, argon was adsorbed on the charcoal and helium passed through without adsorption. By this simple procedure any Ar^{37} produced in the tanks could be removed and transferred to a small charcoal trap with an efficiency of 90-95 per cent. The efficiency of this process was measured by the isotope-dilution method in each experiment by introducing 0.05 cm^3 of Ar^{36} carrier at the start of each exposure and performing a mass analysis of the recovered argon. The argon so isolated was

FIG. 2. Photograph of the 1000-gal. experiment in the Barberton limestone mine, Ohio.

removed from the charcoal, purified, and counted in a small low-level counter. Pulse-height analysis was used to observe the 2.8-keV Auger electron from the electron-capture decay of Ar^{37}. The counter was operated in anticoincidence with guard-proportional and scintillation counters, a technique commonly used in low-level counting. The pulse-height spectrum is shown in Fig. 3.

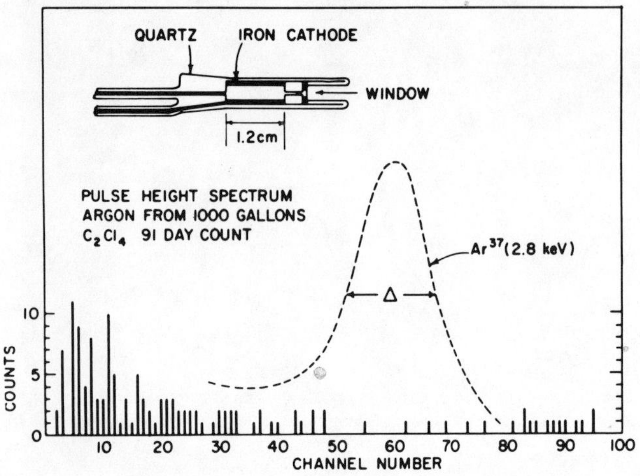

FIG. 3. Pulse-height spectrum of the argon extracted from the 1,000 gal. of perchlorethylene exposed in the Barberton mine. The resolution of the counter for Ar^{37} is indicated.

The position of the Ar^{37} Auger electron peak was determined by calibrating the counter with Fe^{55} X-radiation. The position of the Ar^{37} peak is shown, and it can be seen that no distinguishable Ar^{37} peak was observed. A safe upper limit to the neutrino-capture rate may be deduced by taking all the events in the Ar^{37} region of the spectrum, and computing the limiting solar-neutrino capture rate.

The rate so derived was found to be less than 0.3 neutrino captures per day in the 6.1 tons of C_2Cl_4. In terms of product of flux and cross-section the limit may be given as

$$\sum \phi\sigma \leq 16 \times 10^{-35}\ \text{sec}^{-1}.$$

This value may be compared to the calculated value for this product for solar neutrinos discussed earlier,

$$\sum_{\text{solar}} \phi\sigma = 3 \pm 2 \times 10^{-35}\ \text{sec}^{-1}.$$

It is apparent from a comparison of the experimental limit from this 1,000-gal. detector that we are a factor of 5 away from detecting the calculated rate.

The results of the 1,000-gal. experiment may be used to set limits on the extraterrestrial neutrino flux. Since there have been speculations on possible values of the neutrino flux it is interesting to note the limits that can be set as a function of the neutrino energy. Combining our results and the cross-sections of Bahcall (1964a) for the $Cl^{37}(\nu, e^-)Ar^{37}$ reaction, the neutrino flux limits given in Table 1 were calculated. The limit set for 1-MeV neutrinos, barely over threshold, is not very low. On the other hand, for 10-MeV neutrinos where the superallowed capture process to

TABLE 1. Limits on the Extraterrestrial Neutrino Flux

Neutrino Energy (in MeV)	Cross-Section (in cm^2)	Upper Limit to the Neutrino Flux ($cm^{-2}\ sec^{-1}$)
1. . .	5.5×10^{-46}	$<3 \times 10^{11}$
5. . .	9.3×10^{-44}	$<2 \times 10^{9}$
10. . .	2.3×10^{-42}	$<7 \times 10^{7}$
100. . .	8.3×10^{-40}	$<2 \times 10^{5}$

form the 5.2-MeV excited state in Ar^{37} is important, an upper limit of $1 \times 10^{8}\ cm^{-2}\ sec^{-1}$ can be set. This limit allows us to conclude that less than 0.2 per cent of the solar-energy cycle goes through the formation of Li^4. If Li^4 were stable, one would expect that its beta-decay to He^4 would produce energetic neutrinos which would dominate the solar-neutrino spectrum and these would have been observed by the experiment. If the central temperature of the Sun were somewhat higher than 16×10^6 °K the B^8 production would be higher. Using the flux limits tabulated, one concludes the central temperature of the Sun is below 19×10^6 °K.

The limit is given for 100-MeV neutrinos, but above a few hundred MeV the Cl^{37}-Ar^{37} radiochemical method would not be particularly sensitive. This is because the nucleon struck by the incoming neutrino with several hundred MeV energy would leave the nucleus, and Ar^{37} would not be the final product nucleus.

Plans for a Large Scale Experiment

The sensitivity of the Cl^{37}-Ar^{37} method can be improved by increasing the volume of perchloroethylene used in the detector. We are now planning an experiment using 100,000 gal. or 610 tons of perchloroethylene. A volume of liquid of this magnitude can be processed to remove argon by the same helium-sweeping method used in the pilot experiment. However, to accomplish the task in a day it is important to make the process more efficient by improving the contact between the helium gas and the liquid. This will be accomplished by the use of a pump-eductor system, and we plan to be able to remove the Ar^{37} in 10 hours with better than 90 per cent efficiency.

The main problem in scaling up the sensitivity is to keep background effects a factor of at least 50 below the calculated solar-neutrino signal. The major background effect is from cosmic radiation. Cosmic-ray muons produce protons in the liquid, and these protons produce Ar^{37} by $Cl^{37}(p,n)\,Ar^{37}$ reaction. The Ar^{37} production rate by muons has been measured at a depth of 25 m.w.e., and from the known muon intensity and cross-section for muon interaction the Ar^{37} production rate can be calculated as a function of the depth. Figure 4 shows the Ar^{37} production rate in 100,000 gal. of perchloroethylene from cosmic-ray muons at various depths. Also indicated on this plot is the Ar^{37} production rate expected from solar neutrinos, and the corresponding rates in various mines in the United States and India. It can be seen from this curve that the detector must be located at a depth of over 4,000 m.w.e., or about 4,400 feet of rock. Several mines in the United States could be used, and we are now negotiating for the use of one of these mines.

There is a background effect from fast neutrons from the rock wall. These neutrons produce Ar^{37} in the liquid by (n, p) followed by the $Cl^{37}(p,n)Ar^{37}$ reaction. Fast neutrons are produced by (α, n) reactions and spontaneous fission from the small amounts of uranium and thorium contained in the rock in the parts-per-million range. We have measured this fast-neutron background effect in two mines, and find the effect is at least a factor of 20 below the expected solar-neutrino signal. However, we are providing a water shield between the rock wall and the tank. This will be actually accomplished by flooding the tank chamber with water. Another source of background arises from small amounts of calcium and sulfur contained in the liquid. However, we find that commercial-grade perchloroethylene is free of these impurities, and background effects from this source are negligibly small.

Thus by placing the 10^5 gal. of C_2Cl_4 below 4,000 m.w.e., providing fast-neutron shielding, and insuring that the Ca and S content is low, the expected solar-neutrino signal will be clearly above the background of the detector. If Ar^{37} is observed, it may be attributed to a neutrino signal. The expected rate in the 610 tons is about 2-8 per day. With our expected sensitivity we can measure the presently calculated solar-neutrino flux to 10 per cent, or if the signal is below the presently calculated value we will be able to look for fluxes a factor of 10 lower. It would of course be important if a definite signal is observed, to test whether the neutrinos are indeed coming from the Sun. Since this method does not have directional sensitivity, one would have to look for a 7 per cent difference in flux resulting from the eccentricity of the Earth's

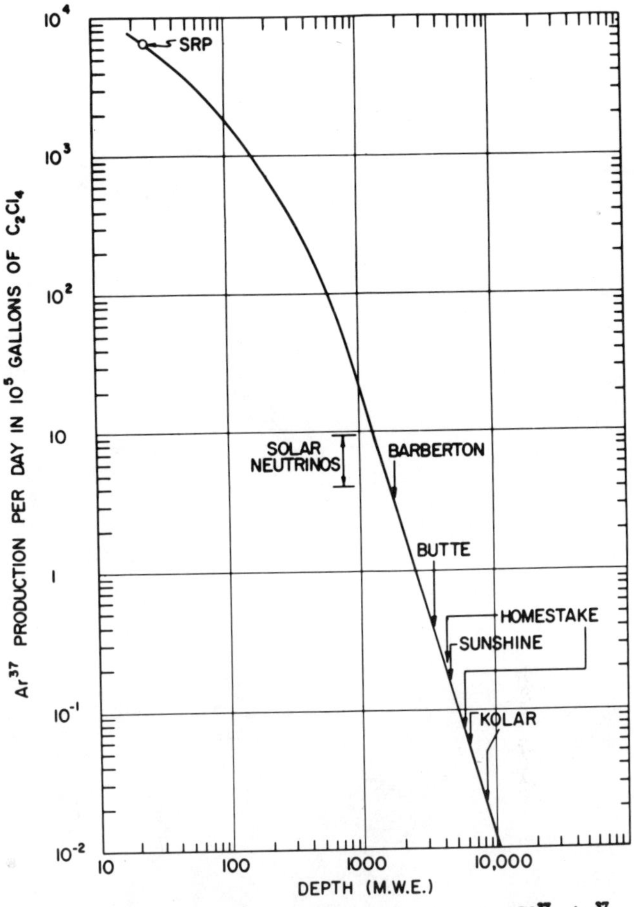

FIG. 4. The major background effect in the Cl^{37}-Ar^{37} neutrino detection method is the production of Ar^{37} by the Cl^{37} (p, n) Ar^{37} reaction in the liquid perchloroethylene by protons arising from the interactions of cosmic-ray muons. The plot shows the cosmic-ray muon produced Ar^{37} as a function of the depth in meters of water equivalent. The depths of several mines in the United States are shown that would afford sufficient cosmic-ray shielding for a solar-neutrino experiment. The Kolar mine in India is the deepest mine in the world.

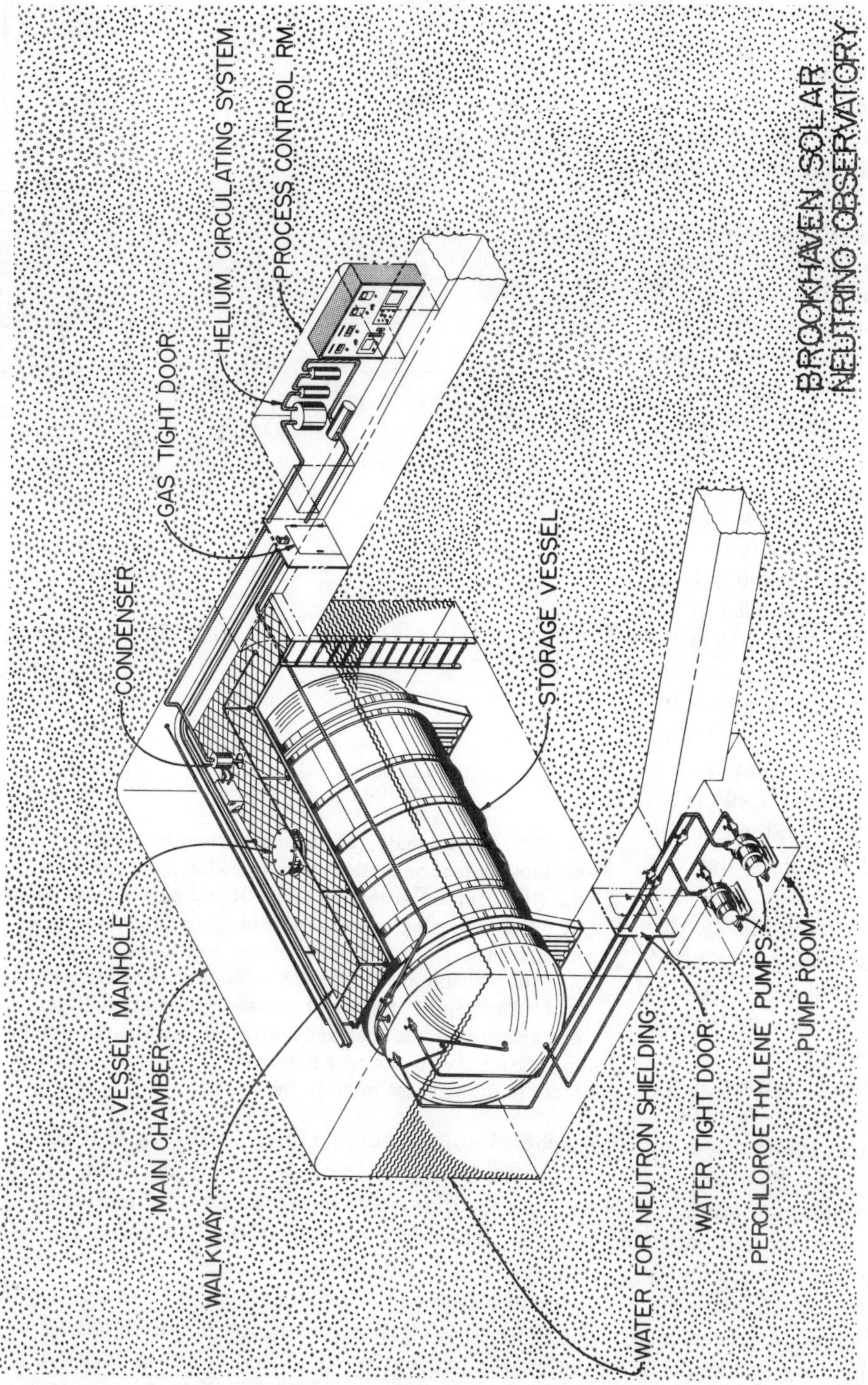

FIG. 5. Schematic layout for the Brookhaven solar-neutrino experiment.

orbit. At the levels presently calculated this would not be possible with 10^5 gal. If a higher rate is observed this test could be made.

The general arrangement of the 100,000-gal. experiment is shown in Fig. 5. Shown here is a tank 20 feet in diameter and 48 feet long in a rock cavity. Provision is made to flood the cavity for a fast-neutron shield. The equipment for purging the liquid with helium will be contained in a separate cavity indicated as the process control room. The pumps for circulating the liquid through the eductor system will be located near the base of the tank but outside of the flooded cavity. We plan to have this apparatus ready for the first experiment early in 1966.

REFERENCES

L. W. Alvarez, 1949, University of California Radiation Laboratory Rept. No. UCRL-328.

J. N. Bahcall, 1964a, Phys. Rev., 135, B137.

——. 1964b, Phys. Rev. Letters, 12, 300.

J. N. Bahcall, and C. A. Barnes, 1964, Phys. Letters 12, 48.

R. Davis, Jr., 1957, Radioisotopes in Scientific Research (Proc. 1st UNESCO Internat. Conf., Paris), 1, 728.

——. 1964, Phys. Rev. Letters, 12, 303.

R. Davis, Jr., and D. S. Harmer, 1959, Bull. Am. Phys. Soc., 4, 217.

J. C. Hardy, and R. I. Verrall, 1964, Phys. Rev. Letters, 13, 764.

P. Pochoda, and H. Reeves, 1964, Planet. and Space Sci., 12, 119.

B. Pontecorvo, 1946, Natl. Res. Council Canada Rept. No. P. D. 205.

P. L. Reeder, A. M. Poskanser, and R. A. Esterlund, 1964, Phys. Rev. Letters, 13, 767.

R. L. Sears, 1964, Ap. J., 140, 477.

COSMIC NEUTRINOS*

F. Reines
Case Institute of Technology, Cleveland, Ohio

The full title of this paper should be "A Consideration of the Experimental Possibilities for the Detection of Cosmic Neutrinos." You will recall that the neutrino is an elementary particle which has been shown to exist in four states, two associated with nuclear beta decay—or with the electron—and two associated with the mu-meson. It has spin $\frac{1}{2}$, interacts via the weak interaction, and has no detectible charge or mass. It can and does carry energy and momentum and is 100 per cent polarized, parallel or antiparallel to its linear momentum. As we are generally aware, and as Bahcall points out in his paper in this volume, the information regarding the universe and which is carried in the form of neutrinos is of great interest and is in some instances of a unique character. Indeed we are faced with the peculiar situation that the very same features which make the neutrinos so interesting from the astrophysical point of view, i.e., their weak interaction with matter, make them most difficult to observe.

At low energies—in the few MeV range—the interaction is, for most favorable targets $\sim 10^{-43}$ cm^2/nucleon. Interpreted in terms of more familiar astronomical units, this implies a mean free path of ~ 100 light years of liquid hydrogen.

As the energy is raised to the region of a GeV ($\sim 10^9$ eV) the interaction is characterized by a cross-section $\sim 10^{-38}$ cm^2/nucleon, or a mere light-day of liquid hydrogen.

If certain theoretical conjectures, notably of Glashow, are correct, the interaction with matter should rise by several orders of magnitude in the region of several hundred GeV.

The most elementary of calculations shows that the number of neutrino interactions in one metric ton per year due to all "natural" neutrinos except possibly those from our own Sun is too small to be detected.

This would be nearly the end of the story except that there is the possibility, at least in the realm of high energies, of extremely large detectors weighing many tens of thousands of tons. The situation at low energies is less hopeful—if we believe the current estimates of neutrino fluxes in the few MeV range.

Let me hasten to remark that even at low energies there is in principle no problem with obtaining a signal, assuming expense is no object. The problem in this energy range is the overwhelming background due primarily to natural radioactivity.

We recognize that there always exists the possibility of an intense source of neutrinos as from a supernova or of hitherto-unsuspected origin (Bahcall and Frautschi have suggested the Crab Nebula). Such sources could, of course, change the entire outlook but for the present, at least, the situation is in brief as described.

Figure 1 illustrates the natural sources of neutrinos graphically.

Present Limits on Natural Sources

As of now none of the sources just depicted has been observed. Indeed the neutrino has only been observed in the laboratory: near a powerful fission reactor, where the flux is 10^{13} $\bar{\nu}_e/cm^2$ sec(0-10 MeV) and at the giant electronuclear machines at Brookhaven and CERN, where the somewhat collimated neutrino beam has $\sim 10^4$ neutrinos/cm^2 sec (~ 500 MeV) at the detector.

Even under these optimum and controlled conditions the accurate observation is still not easy although the physicist is becoming accustomed to

*Work supported in part by the National Science Foundation and in part by the National Aeronautics and Space Administration through Contract No. NsG-357.

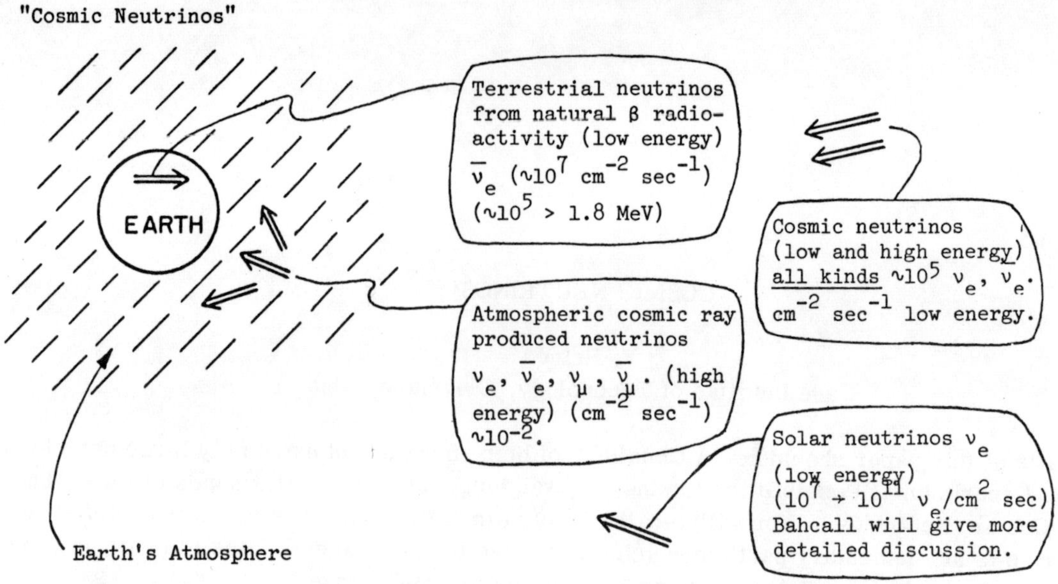

FIG. 1. Natural neutrino sources: from supernovae (low and intermediate energies); from stellar nuclear reactions (main energy sources for stars; low energies); from cosmic-ray interactions in galactic space (high energies); from unknown sources (all energies). The numbers in this figure are based on the estimates of Marx and Menyhard for terrestrial neutrinos; of Fowler, Bahcall, Sears, and Reeves for solar neutrinos; of Markov, Greisen, Zatsepin, Kuzmin, Zheleznykh, Pal, and others for atmospheric neutrinos; and of Markov, Zel'dovich, Pal, and many others for cosmic neutrinos. (Many references can be found in Markov's monograph Neutrino.)

the problem and is making progress, mostly in the development of huge scintillation detectors and most recently spark-chamber arrays.

Our present direct experimental information regarding cosmic neutrino fluxes is all of the negative kind—that is, a series of upper limits. Most of these limits have been derived with equipment which was designed for some other purpose. This may be because, with only two exceptions—the solar flux and the atmospheric neutrinos—no one has seen how to devise an experiment which can be expected to have the sensitivity and discrimination against backgrounds to yield a positive result.

To anticipate, at present the experimental limits are so great that they allow a universe which is mostly in the form of neutrinos—an intriguing notion to say the least.

Table 1 shows some direct experimental limits on natural neutrino fluxes. We could deduce limits from geophysical evidence but will not do so here. We note that since the interaction cross-sections are energy-dependent a knowledge of the spectrum must also be assumed in preparing such a table.

How close are these experimental limits to the theoretically predicted fluxes?

The solar flux at Earth due to B^8 decay in the Sun is predicted to be $(2.5 \pm 1) \times 10^7$ ν_e/cm^2 sec; thus we are within an order of magnitude or so of detecting solar neutrinos. The atmospheric cosmic-ray neutrino flux is within 2 orders of magnitude of being detected.

It is noted that the sources of neutrinos other than these two are relatively weak so that they would tend to be masked by these more local sources. To put it differently, the problem of the experimentalist who seeks cosmic neutrinos is to sort out his signals from those produced by solar and atmospheric neutrinos.

A first step in our search for cosmic neutrinos must therefore necessarily be to detect and study these "background" neutrinos.

Solar Neutrinos—ν_e

Two approaches are available in principle to attack the problem of solar neutrinos: one is to use inverse beta-decay, the other to use elastic

Table I. Flux ν/cm^2 sec in listed energy range (MeV)

Type of Neutrino	0-1.5 (GeV)	1.5-15. (GeV)	1-10 (GeV)	Reference
ν_e	-*	$< 10^{8-9}$	$<10^3$	R. Davis, Kropp, Reines (B^8 spectrum assumed)
$\bar{\nu}_e$	$< 10^{15}$	$< 10^{10}$	(Very rough)	Cowan and Reines (fission spectrum assumed elastic scattering process and inverse β decay)
ν_μ	-	-	< 1	Menon, Ramanamurthu, Sreekantan, and Miyake. (No counts in 180 m^2 days Kolar Gold Fields)
$\bar{\nu}_\mu$	-	-		

*Dash (-) indicates no limits established.

scattering of ν_e by electrons. Neither has yet been implemented, but a serious attempt is under way to make this important and unique observation of the internal condition of the Sun, and so to check current theoretical ideas in this area.

The first inverse beta-decay approach, that of Cl^{37}, has been described by Davis. He is to be congratulated on having developed such an incredibly sensitive—if somewhat massive—solar thermometer.

Another approach using inverse beta-decay was suggested last summer by R. Woods and myself. It differs from the Davis experiment in that direct counting of the neutrino interactions is proposed.

Specifically, consider the reaction

$$\nu_e + B^{11} \rightarrow C^{11} + \beta^-.$$

Observation of the product electron would give a direct determination of the ν_e energy and hence important information regarding the solar-energy cycle. A second reaction

$$\nu_e + Li^7 \rightarrow Be^7 + \beta^-.$$

which wasproposed by Woods and myself (and independently by Bahcall and, I gather in recent conversation, by Zatsepin and collaborators) has a higher cross-section than the B^{11} reaction but suffers from the restriction imposed by the weak angular correlation between the incident neutrino and its product electron. We shall return to this point shortly.

We calculate that 1 ton of ordinary lithium arranged as in the geometry to be described presently should give ~100 events/year in the energy range 6-13 MeV, a figure which is well above the background due to natural radioactivity in an achievable environment. It appears necessary to perform this experiment in a moderately deep mine to escape the background due to cosmic rays. The idea of the experiment which is planned for 1965 in a 2,000-foot-deep salt mine near Case Institute is depicted in Fig. 2.

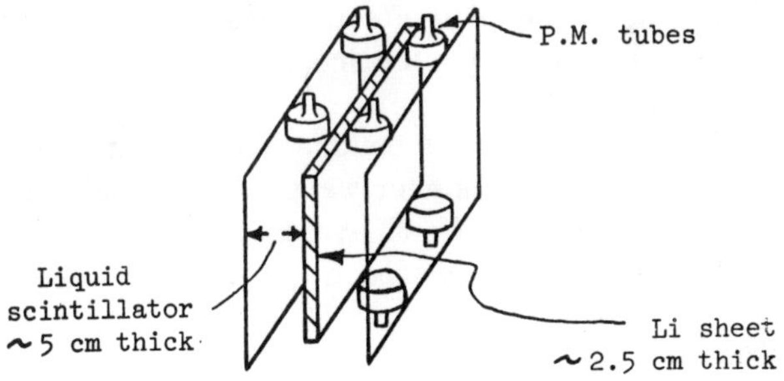

FIG. 2. Schematic of Case Institute's solar neutrino detector.

The energetic product electron is to be detected by the scintillator which surrounds the lithium sheet. The actual detector will consist of many sheets and will be surrounded by an anticoincidence to reduce charged cosmic rays to an acceptable level. Measurements and estimates of expected backgrounds indicate they will be acceptable, but ultimately the experiment itself will have to be relied upon to tell us whether this is so.

A further feature of this approach is the possibility in principle of using the angular correlation of the electron with the parent neutrino, so relating the signal to the solar source. Since, as pointed out by Bahcall and by Woods the angular correlation in Li^7 is weak, it may be necessary to use the less sensitive but more highly correlated B^{11} reaction for directional determination. The predicted mass required to determine the solar direction (within 2π !) for this geometry is $\gtrsim 10$ tons of normal Li, or 3 tons of normal B.

If the solar neutrino experiments based on inverse beta-decay confirm the current solar model, i.e., see neutrinos from the Sun, then a further step may be possible as pointed out by Kropp and myself—to study, and use the elastic scattering process $\nu_e + e^- \rightarrow \nu_e + e^-$. This process is of great interest to the theory of weak interactions and has not yet been observed, although various experimentalists are looking into the question.

Given the elastic scattering process, the angular correlation between the incident neutrino and scattered electron is quite good ($\sim 10°$) if the scattered electron is selected to have an energy not too far below that of the neutrino. This can be seen from a simple consideration of the conservation laws which yield relationships identical to those for the compton effect. Here we have a possibility for a neutrino telescope in the region of several MeV !

The detector in this instance would be a thin multiplate spark chamber weighing ~ 20 tons for a predicted signal of ~ 50 events/year, $E_{\beta^-} > 7$ MeV, a formidable prospect.

Cosmic Ray Neutrinos

Serious interest in high-energy cosmic-ray neutrinos was first shown several years ago by many groups but most notably by Pontecorvo, Markov, Zatsepin, et. al. in the U.S.S.R., by Greisen in the U. S., and by the TATA group in India. At the moment at least three groups are making plans for experimentation in this field and one experiment, that of the group* from Case Institute and the University of Witwatersrand is now in the final stages of preparation. The TATA group has long been interested in cosmic-ray neutrinos as part of their cosmic-ray program in the deep gold mines of the Kolar Gold Fields in Mysore province, India, and the University of Utah group of Keuffel is also building equipment. The TATA and Case-Wits approaches are similar in that the neutrino target is essentially external to the detector whereas the Utah approach is to have a significant target volume within the detector itself.

*The members of the group are M. Crouch, A. Hruschka, W. Kropp, H. Gurr, T. Jenkins, B. Shoffner, B. Meyer, F. Sellschop, and myself.

Two primary considerations prevail: (1) to have a detectable signal; (2) to reduce backgrounds to acceptable levels. A further consideration is to do the experiment with maximum speed and economy.

If we make conservative assumptions regarding the neutrino interaction cross-sections ($\sim 10^{-38}$ cm^2) and the neutrino flux from cosmic-ray interactions in our atmosphere we conclude that a count rate of $\sim$ 10/year can be expected in a mass of 10^3 tons. Furthermore, the background expected from cosmic rays in even a much smaller detector at sea level is many orders of magnitude greater than the signal.

We chose to achieve the necessary large mass by means of a large area detector which could view the muons produced by neutrinos in rock surrounding the detector. The background reduction is to be achieved by going very deep underground (10492 feet or 9000 meters H_2O equivalent) and relying, in addition, on the angular discrimination of the chosen arrangement. Ordinary muons are very close to the vertical, muons produced by atmospheric neutrinos are mostly horizontal (the latter due to the increased decay path in the atmosphere for more steeply inclined π, K, μ parents). The experiment is shown schematically in Fig. 3. The system has an angular discrimination both in the horizontal and vertical direction but as presently

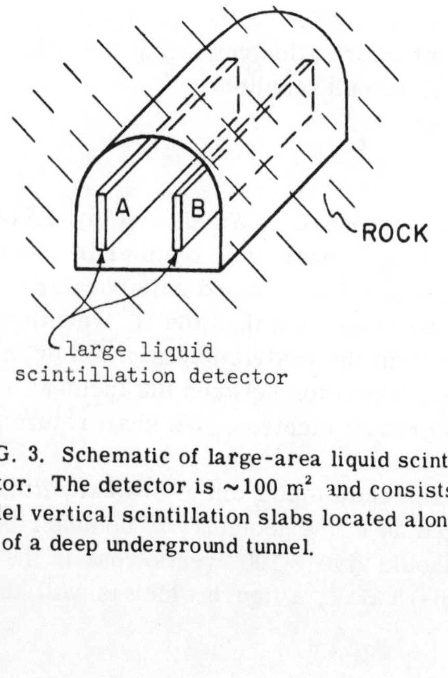

FIG. 3. Schematic of large-area liquid scintillation detector. The detector is ~ 100 m^2 and consists of two parallel vertical scintillation slabs located along the walls of a deep underground tunnel.

constituted cannot sense the parity of the muon direction. Figure 4 shows the process we hope to study. As of the present time approximately one third of the array is on the air. One sixth of the array has been operated for a few weeks.

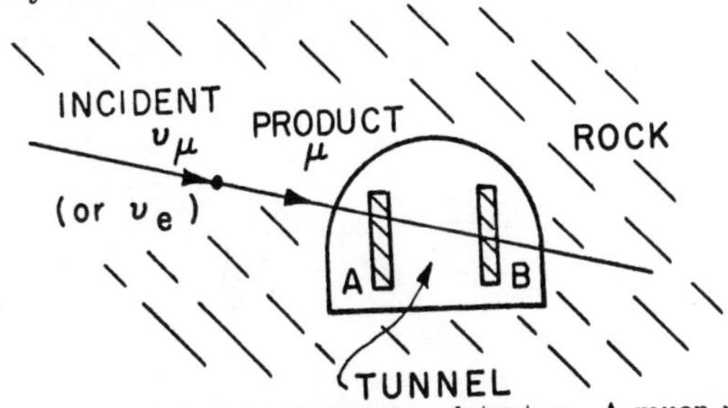

FIG. 4. End view of neutrino detector. A muon produced by a high-energy neutrino in the rock surrounding the detector will penetrate the rock and cause a coincidence in the large-area detectors A and B.

No neutrino-like signal has been seen, but this is not surprising in view of the small number of m^2 days of operation to date. One possible candidate for a penetrating cosmic-ray muon has been observed. (Incidentally, this experiment is being conducted further underground than cosmic-ray muons have been sought in the past and so should be interesting from this point of view as well.)

It is interesting that, despite the size of this detector, it has too little sensitivity, by perhaps a factor of 10^3 or more to detect an expected flux of true cosmic, that is extraterrestrial, neutrinos of high energy.

But we must take one step at a time and see whether as we increase our sensitivity, nature is as we think it is or whether some surprises might be in store for us.

LIMITS ON SOLAR NEUTRINO FLUX AND ELASTIC SCATTERING*

F. Reines and W. R. Kropp
Department of Physics, Case Institute of Technology, Cleveland, Ohio
(Received 17 February 1964)

The calculations of Bahcall, Fowler, Iben and Sears[1] on the flux at the earth of neutrinos from the decay of ^8B produced in the interior of the sun predict a value of $(2.5 \pm 1) \times 10^7\ \nu_e/\text{cm}^2\ \text{sec}$. Since the ^8B decay neutrinos have energies up to 13.7 MeV, i. e., well in excess of natural radioactivity, we are able to interpret results of an underground experiment performed in another connection[2] in terms of an upper limit on the flux of solar neutrinos providing we assume an elastic scatter of ν_e by electrons as predicted by Feynman and Gell-Mann[3] and Marshak and Sudarshan[4]:

$$\nu_e + e^- \to \nu_e + e^-.$$

Our data can also be interpreted as an upper limit on the elastic scattering process[5] assuming the flux calculations cited by Bahcall to be correct. We describe these arguments in this Letter following a discussion of the complementary relationship between the present approach and that of Davis.[1] Davis proposes to detect solar ν_e by using the inverse beta decay of ^{37}Cl,

$$^{37}\text{Cl} + \nu_e \to {}^{37}\text{Ar} + \beta^-.$$

The cross section for this reaction is predictable so that the only unknown factor is the ^8B ν_e flux, f. Therefore, given the results of the projected experiment of Davis, f can be deduced, assuming of course that the reaction which is responsible is known from other considerations – the projected Davis experiment is expected to give not f but the product of f times the cross section integrated over the relevant neutrino spectrum.

The direct counting experiment utilizing the elastic scattering process has the additional feature that it is in principle capable of giving information regarding the neutrino energy distribution, so helping to identify the responsible reaction. A negative result here in the face of a positive result from the Davis experiment could be interpreted either as due to the absence of the elastic scattering process or as an indication that the ^8B reaction is not responsible for the anticipated Davis results and does not occur in the sun as predicted.

A modest experiment was performed which enabled us to set some limits and helps assess the full-scale effort. It consisted of looking for unaccompanied counts in a 200-liter liquid scintillation detector (5×10^{28} electrons) which was surrounded by a large Cherenkov anticoincidence detector and located 2000 feet underground in a salt mine. The experimental details will be published elsewhere.[2] In a counting time of 4500 hours only three events were observed in the energy range 9 to 15 MeV, unaccompanied by pulses in the anticoincidence guard. If we ascribe these events to recoil electrons produced by the elastic scattering of solar neutrinos from the ^8B or ^4Li decays we obtain the conservative upper limits

$$f(^8\text{B}) < 10^9\ \nu_e/\text{cm}^2\ \text{sec},\ (< 2 \times 10^8\ \nu_e/\text{cm}^2\ \text{sec}).$$

The number in the brackets is the corresponding flux limit set by Bahcall and Davis. Table I summarizes the calculation of limits assuming the ^8B reaction. The results are seen to depend on the lower limit of the recoil electron energy E, which is considered. Figure 1 shows the calculated cross-section. σ_e per incident ^8B ν_e for

Table I. Elastic scattering of ^8B neutrinos.

E MeV	$\sigma_e(>E)$ ($\times 10^{45}$ cm^2)	Calculated rate[a] R_1 ($\times 10^4$ day^{-1})	Present experimental limit R_2 ($\times 10^2$ day^{-1})	R_2/R_1	Upper limit on solar ν_e flux (ν_e/cm^2 sec)
8.0	4.7	6.3			
8.5	4.2	4.6	1.6	35	9×10^8
9.0	3.1	3.4	1.6	50	1.2×10^9
9.5	2.1	2.3	1.1	50	1.2×10^9
10.0	1.4	1.5	1.1	75	

[a] $R_1 = \sigma_e(>E)Nf$, $N = 5 \times 10^{28}$ target electrons, $f = 2.5 \times 10^7\ \nu_e/\text{cm}^2$ sec.

Reprinted from Phys. Rev. Letters, 12, 457-459.

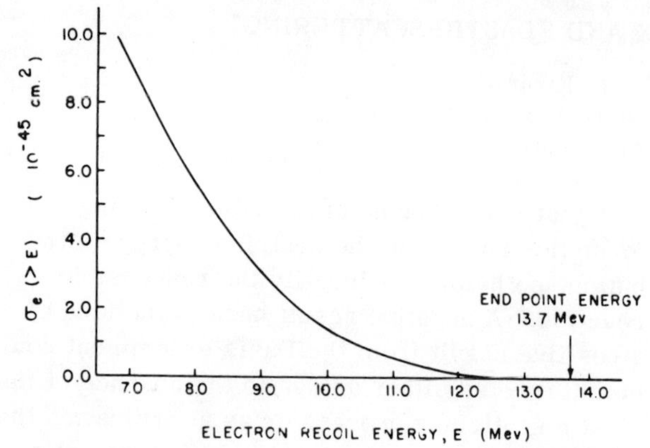

FIG. 1. Cross section, $\sigma_e(>E)$, per target electron, per incident ^{8}B neutrino, for the production of recoil electrons with energy in excess of E.

the production of recoil electrons above the energy E. On this basis, the elastic scattering cross section is seen from the table to be <35 times the expected value, a better limit by ~10 than that previously set for $\bar{\nu}_e$'s from a fission reactor.[6]

It is interesting to inquire how large a detector would be required to see the predicted ^{8}B solar neutrino and then to consider the question of backgrounds. If the flux is taken to be 2.5×10^7 ν_e/cm^2 sec and the lower limit on electron recoil energy is set at 8 MeV, then we predict a rate of 1.5 events per year in a metric ton of scintillator. Since it is within our current experience with large detectors to consider a sensitive volume of 10^4 gallons we could build a system with a predicted rate ~50/yr. The background which appears most serious is that resulting from the "pileup" of pulses due to natural radioactivity. The most troublesome natural gamma emitter is ThC″ which[7] produces a cascade of gammas of energy totalling from 2.62 to 3.96 MeV. Assuming a threefold pileup is required to give an apparent pulse >8 MeV and a resolving time of 10^{-7} sec (modest) in a detector made up of $N(=24)$ identical elements, the singles rate per element, n sec^{-1}, consistent with a total background rate, R, of 10/yr is given by

$$R = 2n^3\tau^2 \times 3.1\times10^7 N.$$

Inserting numbers we find

$$n = 88/\text{sec per element.}$$

If the resolving time is reduced a correspondingly larger value of n becomes acceptable.

As an example of attainable backgrounds[8] a scintillation detector of ~10 m^2 surface area and located 2000 feet underground in a salt mine exhibits an integral count rate above 3 MeV of only 6 sec^{-1}, an order of magnitude below the required limit per detector element estimated above.

The twofold pileup background in the complete detector array, using the integral rate above 4 MeV in the 10-m^2 test detector to make an estimate, is ~50/yr, again assuming a resolving time of 10^{-7} sec. It therefore seems reasonable that the background problem due to natural radioactivity can be reduced to acceptable levels.

A second source of background can result from the passage of cosmic rays through the detector. It is improbable that a detector located near the earth's surface can be sufficiently well shielded by a charged-particle anticoincidence detector from cosmic rays which deposit energy in the range, 8-18 MeV, of interest here. It therefore appears necessary to locate such a detector deep underground where the cosmic radiation is penetrating muons and their secondaries. If, to assess the situation, we assume a detector of 25 m^2 surface area as seen by the cosmic rays, then at a location 2000 feet underground, allowing an anticoincidence factor[9] >10^4, the residual rate due to all cosmic rays is estimated to be <300/yr. Considering only those cosmic rays which deposit between 7 and 18 MeV it appears possible that the rate would be reduced by as much as an order of magnitude or more. There always remains the possibility of going deeper underground.

We wish to thank the Morton Salt Company for continued hospitality in their Fairport Harbor Mine and Dr. M. Crouch and Dr. T. L. Jenkins for interesting discussions.

Note added in proof. – It has been called to our attention by L. Heller that Azimov and Schekhter[10] and Heller[11] have recalculated the elastic scattering process according to the conserved vector current theory and find a cross section twice as large as previously quoted. On this basis our limits on the elastic scattering process are closer by a factor of two to prediction and a detector smaller by a factor of two would suffice to detect solar neutrinos.

---

*Work supported in part by the United States Atomic Energy Commission.

[1]J. N. Bahcall, C. W. Fowler, I. Iben, Jr., and R. L. Sears, Astrophys. J. 137, 344 (1963). We are indebted to Dr. R. Davis, Jr., and Dr. J. N. Bahcall for calling our attention to the latest estimates of the ^{8}B rates as well as preprints of their publications on

the subject of solar neutrinos.

[2]A paper by the present authors is in preparation. It will contain a discussion of the various limits which can be set on nucleon stability and neutrino fluxes by a consideration of the unaccompanied counts. The earlier stages of this work are described in a paper by C. C. Giamati and F. Reines, Phys. Rev. 126, 2178 (1962).

[3]R. P. Feynman and M. Gell-Mann, Phys. Rev. 109, 193 (1958).

[4]R. E. Marshak and E. C. G. Sudarshan, Phys. Rev. 109, 1860 (1958); Proceedings of the Padua-Venice Conference on Mesons and Newly Discovered Particles, September, 1957 (Società Italiana di Fisica, Padua-Venice, 1958).

[5]Our results can also be used to set an upper limit on the product of the elastic scattering cross section averaged over the upper end of the ^4Li decay spectrum (on the unlikely assumption that it is particle stable) times the solar ^4Li-produced ν_e flux at the earth. The flux limit so obtained is $<2 \times 10^8$ ν_e/cm^2 sec. This is to be compared with the Bahcall-Davis limit of $<1 \times 10^8$ ν_e/cm^2 sec.

[6]C. L. Cowan, Jr., and F. Reines, Phys. Rev. 107, 528 (1957). This experiment was interpreted in terms of an upper limit on the neutrino magnetic moment. We here reinterpret these data in terms of the conserved vector current predictions.

[7]D. Strominger, J. M. Hollander, and G. T. Seaborg, Rev. Modern Phys. 30, 585 (1958).

[8]L. V. East, T. L. Jenkins, and F. Reines (unpublished).

[9]M. K. Moe, T. L. Jenkins, and F. Reines, Rev. Sci. Instr. 35, 370 (1964).

[10]Ya. I. Azimov and V. M. Shekhter, Zh. Eksperim. i Teor. Fiz. 41, 592 (1961) [translation: Soviet Phys.—JETP 14, 424 (1962)].

[11]L. Heller, Los Alamos Scientific Laboratory Report LAMS-3013, 1964 (unpublished).

EVIDENCE FOR HIGH-ENERGY COSMIC-RAY NEUTRINO INTERACTIONS*

F. Reines, M. F. Crouch, T. L. Jenkins, W. R. Kropp, H. S. Gurr, and G. R. Smith

Case Institute of Technology, Cleveland, Ohio

and

J. P. F. Sellschop and B. Meyer

University of the Witwatersrand, Johannesburg, Republic of South Africa
(Received 26 July 1965)

The flux of high-energy neutrinos from the decay of K, π, and μ mesons produced in the earth's atmosphere by the interaction of primary cosmic rays has been calculated by many authors.[1] In addition, there has been some conjecture[1] as to the much rarer primary flux of high-energy neutrinos originating outside the earth's atmosphere. We present here evidence[2] for the interactions of "natural" high-energy neutrinos obtained with a large area liquid scintillation detector (110 m^2) located at a depth of 3200 m (8800 meters of water equivalent, average $Z^2/A \simeq 5.0$) in a South African gold mine.

The essential idea of the present experiment[3] is to detect the energetic muons produced in neutrino interactions in a mass of rock by means of a large area detector array imbedded in it. Backgrounds are reduced by the large overburden and by utilizing the fact that the angular distribution of the residual muons from the earth's atmosphere is strongly peaked in the vertical direction at this depth. The angular distribution of the muons produced by neutrino interactions should show a slight peaking in the horizontal direction.[1]

The detector array, shown schematically in Fig. 1, consists of two parallel vertical walls made up of 36 detector elements. The array is grouped into 6 "bays" of 6 elements each. Each detector element, Fig. 2, is a rectangular box of Lucite of wall area 3.07 m^2 containing 380 liters of a mineral-oil based liquid scintillator,[4] and is viewed at each end by two 5-in. photomultiplier tubes. The array constitutes a hodoscope which gives a rough measurement of the zenith angle of a charged particle passing through it. In addition, the event is located along the detector axis by the ratio of the photomultiplier responses at the two ends. The sum of the responses then pro-

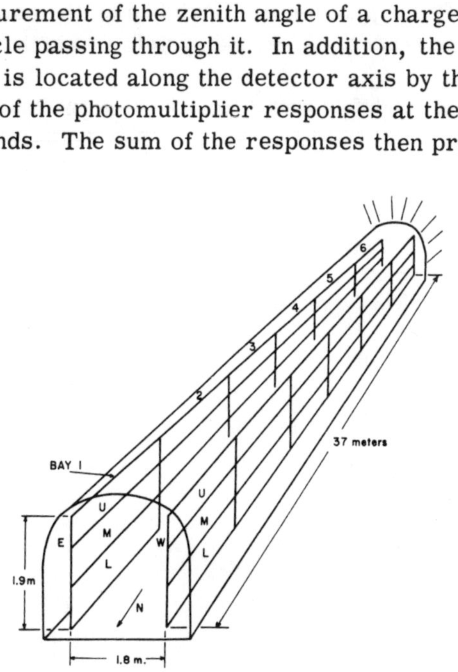

FIG. 1. Schematic of detector array.

Reprinted from Phys. Rev. Letters, 15, 429-433.

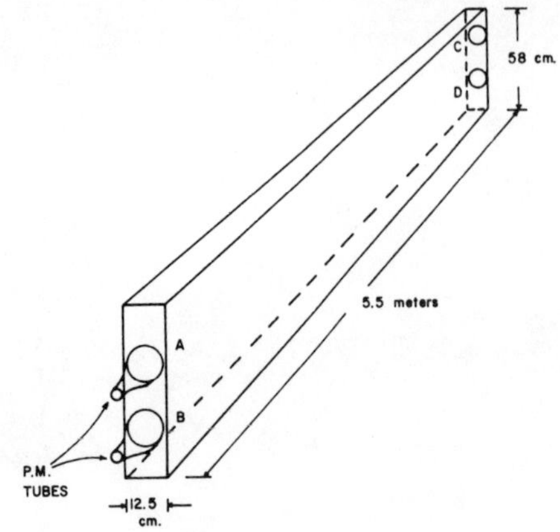

FIG. 2. Sketch of detector element.

vides a measure of the energy deposited and hence the track length in the detector. The scintillator is 20 MeV thick for minimum ionizing particles, well above energies characteristic of natural or induced radioactivity. Pulses from every photomultiplier tube were presented on two oscilloscopes, one for each side (E or W) of the array, and photographed whenever at least a fourfold coincidence of the type A_E, B_E, C_E, D_E or A_W, B_W, C_W, D_W occurred. The four tubes responsible for the coincidence, although on the same side (E or W), were not required to be located in the same detector element. A coding system consisting of signal mixers and delay lines was used to identify the photomultiplier tubes on the oscilloscope traces. A second system in which the pulse amplitudes from individual tubes were stored as charges on condensers was also employed during the phase of the experiment reported here. Calibration and system checks were accomplished by means of a light pulser placed at the center of the detector elements, an electronic pulser, and a Y^{88} source, the response to which was related to the signal from cosmic rays as seen in an identical detector element located above ground. The fourfold accidental coincidence rate is $\ll 1$/yr for energies >15 MeV. From data obtained with this system, it was possible to deduce the flux and information regarding the direction—but not sense—of the charged particles which penetrated the detector.

The system was set into operation, one bay at a time, starting in September 1964, and all six bays were completely operational by June 1965. To date we have logged 563 bay days of operation. Allowing for the solid angle seen by the detector, this corresponds to 14 200 m^2 days sr or the equivalent of all six bays for 94 days. Tables I, II, and III list events of various classes, their time of occurrence, the detector elements involved, the energy deposited, and, for events in which more than one tank gave a signal, the locations in the tanks. Figure 3 is a sample reconstruction of an eightfold event, the event of 23 February 1965.

We now present estimates of the extent to which various processes involving ordinary cosmic-ray muons can contribute to the events

Table I. Eightfold events (coincidences involving one element of each side). Run time for events was 563 bay days.

Date	Time, Greenwich (h)	Tank	Location of events (meters from north end)	Energy deposited in detector (MeV)
23 February 1965	21:48	E4*L*	2.1	29
		W4*L*	2.8	18
1 March	00:20	E5*M*	0.03	55
		W5*U*	4.9	118
17 March	18:52	E4*L*	3.6	19.4
		W4*L*	1.7	16.0
20 April	14:16	E2*M*	4.6	23.5
		W2*M*	3.4	24.5
1 June	22:37	E1*L*	5.5	18.5
		W2*L*	0.55	18.0
3 June	01:42	E4*U*	2.4	5.0
		W4*M*	3.7	18.0
1 July	15:21	E3*M*	1.3	21.0
		W3*U*	3.0	30.0

Table II. Miscellaneous multitank events. Run time for events was 563 bay days.

Date	Time, Greenwich (h)	Tank	Location of event (meters from north end)	Energy deposited in detector (MeV)
27 October 1964	19:39	E1*M*	3.6	58
		E1*L*	?	~10
		W1*U*	1.1	158
		W1*M*	2.3	116
		W1*L*	2.4	37
13 December	10:31	E2*M*	3.4	75
		E2*L*	2.1	59
22 December	11:03	E1*U*	3.7	37
		E1*M*	4.3	51
		E1*L*	4.6	16
11 February 1965	02:20	E1*U*	1.7	84
		E1*M*	1.6	78
		E1*L*	1.8	21
14 February	22:35	E4*M*	3.9	65
		E4*L*	?	?
		E5*M*	0.09	51
		W4*M*	2.9	49
		W4*L*	4.2	34
7 May	02:10	W5*M*	1.8	65
		W5*L*	1.4	97
12 June	13:40	E3*U*	3.9	60
		E3*M*	3.4	122
		E3*L*	1.7	19

listed in Table I.

(1) The angular distribution of muons at this depth can be calculated from the known depth-intensity curves.[5] Normalizing the intensity to the rate of vertical events listed in Tables II and III we find that <1 muon/yr produces a coincidence of the type listed in Table I.

(2) The energy distribution of the muons at this depth can also be determined from the depth-intensity relation. From this spectrum it is estimated the contribution to Table I events due to muons which have multiply scattered is negligible.

(3) In a similar fashion it can be shown that the number of Table I events produced by high-energy knock-on electrons or electromagnetic showers is negligible.

(4) The contribution expected from stars produced by muons is estimated from star production cross sections measured underground (at 500 meters of water equivalent)[6] to be much less than 3/yr.

(5) It is conceivable that pairs of associated high-energy muons may be responsible for events of the type listed in Table I. One argument against this interpretation is the fact that the energy deposition which would be expected from such nearly vertical particles would be much larger than observed. It is to be noted that two events of Table II (27 October and 14 February) are consistent with this multiple-muon hypothesis. This phenomenon is being studied further.

From these arguments regarding the contributions of ordinary cosmic-ray muons to the

Table III. Single-tank events with energy deposition >18 MeV.[a] Run time for events was 265 days.

Date	Time, Greenwich (h)	Tank	Energy deposited (MeV)
9 November 1964	00:48	E1*L*	45
23 November	16:05	E1*L*	48
25 November	15:09	E1*L*	63
14 February 1965	02:52	W5*M*	26
24 February	15:04	W2*M*	39
11 March	11:03	E4*L*	54
11 March	21:40	W5*M*	51
12 March	15:38	E2*L*	26

[a]Data only reduced for period to 18 March 1965.

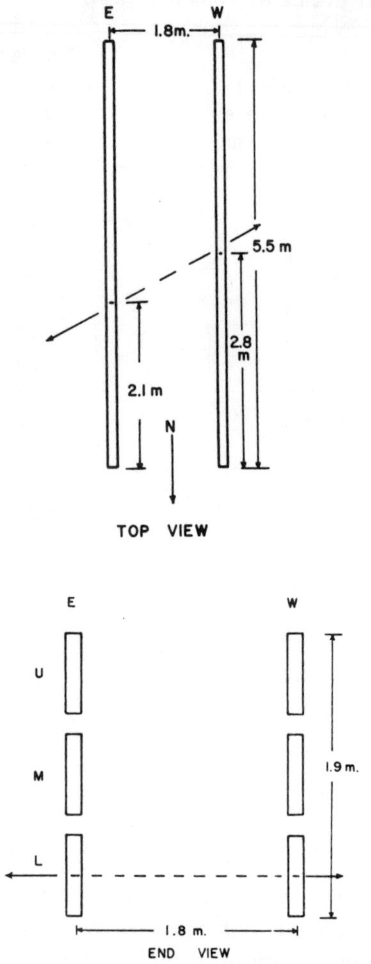

FIG. 3. Reconstruction of event of 23 February 1965.

events of Table I, it appears difficult to explain these events by known processes involving these muons. It is therefore plausible that the events of Table I are due to neutrino interactions. An estimate of the expected rate due to neutrinos produced in the earth's atmosphere is appropriate at this point.

In the following calculations we assumed a neutrino spectrum $(\nu_\mu + \bar{\nu}_\mu)$ which had the form $I_\nu = 4.8 \times 10^{-2} E^{-3.0}$ $\text{cm}^{-2}\,\text{sec}^{-1}\,\text{sr}^{-1}$ in the vertical direction above 1 BeV. The detailed angular distribution was taken into account. For elastic events,

$$\bar{\nu}_\mu + p \rightarrow n + \mu^+ \text{ and } \nu_\mu + n \rightarrow p + \mu^-,$$

produced by neutrinos above 1 BeV, we would expect 0.3 event in the period of this observation.[7] Information on the neutrino flux is less well known below 1 BeV because the associated muons are absorbed in the atmosphere. However, it is estimated that the contribution arising from these neutrinos is <0.3 event. In making this calculation, we assume that the muon receives, on the average, $\frac{1}{2}$ the neutrino energy. The expected rate of inelastic events was calculated assuming the cross section is given by

$$\sigma_{\text{in}} = 0.4 \times 10^{-38} E_\nu \text{ cm}^2 \quad (E_\nu \text{ in BeV}).$$

This cross secion is consistent with the work of the CERN bubble chamber group.[8] Inelastic events should contribute 0.8 event for neutrino energies between 1 and 10 BeV during this period. We assume here that the muon receives on the average, $\frac{1}{3}$ of the neutrino energy.

We conclude that if the seven events of Table I were all due to neutrinos, then either the neutrino flux is higher than anticipated or the interaction cross section rises more rapidly with energy above the region investigated by the accelerator groups. The combination of these effects would have to amount to a factor of five or six to remove the discrepancy between predictions and observation, setting aside the very real possibility of statistical fluctuations.

The events of Table I are consistent with isotropy in the laboratory system and show no correlation with sidereal time.

We wish to express our appreciation to Mr. A. A. Hruschka for his advice and help in designing and constructing the equipment and the laboratory facilities deep underground. Mr. Bruce Shoffner was most helpful in the design and construction of the electronics. We are grateful to the directors of the Rand Mines and their consulting engineers F. G. Hill and M. Barcza and the general manager of the East Rand Proprietary Mines and his staff for providing us with the underground laboratory.

Note added in proof.—If form factors are taken into account for elastic reactions, the muon received typically 0.8 of the neutrino energy rather than the 0.5 used in this Letter. Regarding the inelastic contribution, recent data from CERN indicate that on the average, the muon gets 0.5 of the neutrino energy rather than the 0.33 used above. If we re-evaluate the expected rates with these numbers, we obtain 0.5 (elastic >1 BeV), 1.2 (inelastic 1 to 10 BeV), or a total of 1.7 events expected during the period reported. Since the CERN data can be interpreted to imply a more rapid rise than linear in the inelastic cross section, it is conceivable that the discrepancy between observation and prediction can be explained in this way. We wish to thank Professor H. Faissner for a discussion of these points.

---

*Work supported in part by the U. S. Atomic Energy Commission.

[1]K. Greisen, Proceedings of the International Conference on Instrumentation for High-Energy Physics, Berkeley, California, September 1960 (Interscience Publishers, Inc., New York, 1961), p. 209; M. A. Markov and I. M. Zheleznykh, Nucl. Phys. 27, 385 (1961); G. T. Zatsepin and V. A. Kuzmin, Zh. Eksperim. i Teor. Fiz. 41, 1818 (1961) [translation: Soviet Phys.—JETP 14, 1294 (1962)]; R. Cowsik, Proceedings of the Eighth International Conference on Cosmic Rays, Jaipur, India, December 1963, edited by R. R. Daniels et al. (to be published).

[2]The first neutrinolike event was reported by the Case-Wits group in the Proceedings of the Informal Conference on Experimental Neutrino Physics, CERN, 1965 (to be published).

[3]Proceedings of the Eighth International Conference on Cosmic Rays, Jaipur, India, December 1963, edited by R. R. Daniels et al. (to be published).

[4]T. L. Jenkins and F. Reines, IEEE, Trans. Nucl. Sci. 11, 1 (1964).

[5]S. Miyake, private communication.

[6]L. Avan and M. Avan, Compt. Rend. 244, 450 (1957).

[7]This was calculated using the cross sections of T. D. Lee and C. N. Yang, Phys. Rev. Letters 4, 307 (1960); N. Cabbibo and R. Gatto, Nuovo Cimento 15, 304 (1960).

[8]M. M. Block et al., Phys. Letters 12, 281 (1964).

THE K. G. F. NEUTRINO EXPERIMENT*

C. V. Achar, M. G. K. Menon, V. S. Narasimham, P. V. Ramana Murthy
and B. V. Sreekantan
Tata Institute of Fundamental Research, Bombay, India

K. Hinotani and S. Miyake
Osaka City University, Osaka, Japan

and

D. R. Creed, J. L. Osborne, J. B. M. Pattison and A. W. Wolfendale
University of Durham, Durham, U. K.

I. Introduction

Following the early work carried out at great depths underground in the Kolar Gold Mines in India (Miyake et al., 1964, Menon et al., 1963a, b, Menon, 1964 and Achar et al., 1965a), we have specifically designed an experiment for the detection of the interactions of cosmic ray neutrinos; (Sreekantan, 1965 and Wolfendale, 1965). The first phase of this experiment was meant to be exploratory in character; the main objective has been to see if the fluxes of cosmic ray neutrinos, (either of terrestrial or of extra-terrestrial origin), and the high energy behaviour of neutrino interaction cross sections are consistent with the expectations based on various estimates. Preliminary results obtained in the first 3,000 m^2 days sterad of operation were reported in Physics Letters (Achar et al., 1965b, c). Here we wish to present results obtained up to the time of the Conference.

II. Experimental Details

The experiment has been in operation since March 1965 in the Kolar Gold Mines in South India at a depth of 7,600 ft. (equivalent to 7,500 m.w.e.** in the case of standard rock with $Z^2/A = 5.5$).

*Paper read at the Ninth International Conference on Cosmic Rays.

**At Kolar, $Z^2/A = 6.5$, and rock density $= 3.02$ gm/cm^3, and accordingly the depth of 7,600 ft. corresponds to 7,000 m. w. e. for Kolar rock.

We have used two identical telescopes, each consisting of 2 vertical walls of plastic scintillators, 2 m long and 3 m high, separated by 80 cm, as shown in Fig. 1. Each "scintillator element," one square metre in area, is viewed by 2 adjacent 5" diameter photomultipliers; four-fold coincidences are recorded between a pair of photomultipliers on one wall and any pair on the other wall. Between the scintillator walls there are three arrays of neon flash tubes (NFT); in each array there are 4 columns of flash tubes. There are 2 walls of lead absorber, each 2.5 cm in thickness, in between the flash tube arrays. When 4-fold coincidences occur, the photomultiplier pulses are recorded on oscilloscopes, and after a delay of about 30 microseconds a high voltage pulse is applied to the electrodes of the NFT arrays.

The effective aperture offered by one of our telescopes to atmospheric muons is ~ 0.29 m^2 sterad, if an angular distribution of the type $I_\theta = I_0 \cos^8\theta$ is assumed; and is ~ 20 m^2 sterad to an isotropically distributed radiation; and neutrino induced muons are expected to correspond approximately to this. Figure 2 shows the differential angular distributions for our telescope, with respect to projected zenith angle, for atmospheric muons with $I(\theta) = I_0 \cos^8\theta$, and for isotropic neutrino-induced muons with an amplitude (i.e., the total number of muons) equal to that of atmospheric muons.

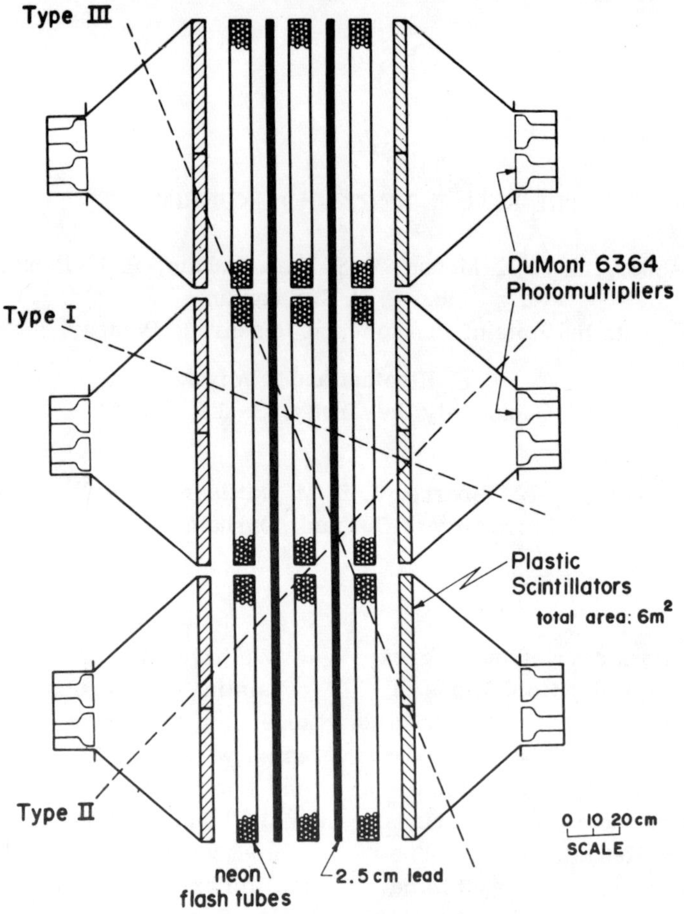

FIG. 1. Front view of one of the K. G. F. neutrino telescopes.

III. Results

A total of 13 events has been recorded so far in an effective operating period of 234 telescope days, equivalent to an exposure of $\sim$4,700 m^2 day steradians for isotropic radiation. Details relevant to these events are given in Table I (see also Figs. 1 and 3).

It can be seen from Table I that there are 2 events at very large projected zenith angles, Event 3 and Event 4. Event Nos. 4 and 13 have special characteristics and will be discussed later. There are 3 events, at 8.5°, 25° and 33°, which are all out of geometry. In these cases it is presumed that the equipment was triggered because of the existence of an accompanying soft component (in two of the cases the electrons responsible can be seen on the NFT records). Of the remaining 7 events, six have angles between 29.5° and 48° and in one case it is only possible to state that the angle is greater than 37°.

It is possible to be fairly certain that the events with large projected zenith angles (greater than about 65°) such as events Nos. 3 and 4, are due to neutrino induced muons, (see Fig. 2); on the other hand, one is not sure if an event with a zenith angle smaller than $\sim$60° is to be attributed to a neutrino interaction or to an atmospheric muon; (see Fig. 2). We have made attempts to estimate in several ways the fraction of events which can be attributed to neutrino interactions.

The relative percentages of events of Types I, II and III (as defined in Fig. 3) for events induced by neutrinos are very different from those in the case of events due to atmospheric muons; the relevant values are shown in Table II. This difference in the relative percentages arises from the different angular distributions of the two categories of muons. Let us assume that the total number of neutrino induced muons is x times that of atmospheric muons. By dividing the muons of the two categories into various types by the

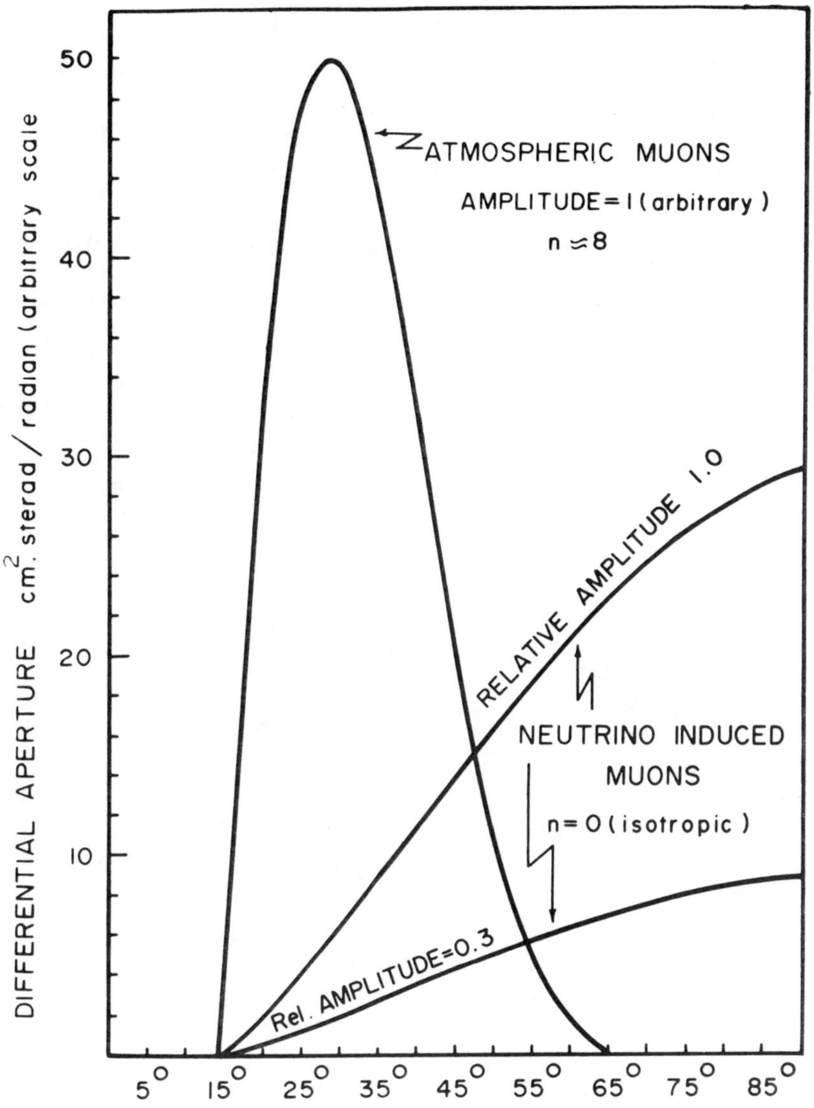

FIG. 2. The differential angular distributions for the neutrino telescopes.

respective ratios given in Table II, one can derive the expected ratios, in terms of x, for (Type I/all), (Type II/all) and (Type III/all). By comparing these ratios with the experimentally observed ratios, one can solve for x. From such an analysis, we conclude that ~5 of all the observed events are neutrino induced and the rest are due to atmospheric muons (see Fig. 4).

When a sufficiently large number of events have been observed, one can directly compare the observed projected zenith angle distributions with a mixture in known proportions of the two distributions shown in Fig. 2, and thereby obtain the best estimate of the relative amplitudes. An attempt in this direction, shown in Fig. 5, indicates that about 5-6 of the events are due to ν-induced muons. The rate of neutrino induced muons which follows from this is $\approx 1.2.10^{-12}$ particles/cm^2.sec.st. (for an isotropic distribution of neutrinos).

It should be emphasized at this stage that the curve shown in Fig. 2 is based on an assumed angular distribution $I_\theta = I_0 \cos^8\theta$ for atmospheric muons; we have used this as a conservative estimate. The indications are that the distribution is probably steeper but in view of uncertainties with regard to the scattering behaviour at very high energies there is some uncertainty in the estimate of the fraction of ν-induced muons amongst the total number observed.

Event No. 4

In this event, a sketch of which is shown in

Table I

Events obtained in the K.G.F. neutrino experiment up to the end of August, 1965.

Event No.	Date and time (Hrs. Mts.) I.S.T.	Coincidence	Event type	Projected zenith angle	Remarks
1	5.4.65 20-04	S_4 N_4 Tel. 2	I	$> 37°$	Neon flash tubes not present
2	27.4.65 18-26	S_1 N_1 Tel. 1	I	$48° \pm 1°$	Only two extreme trays of neon flash tubes present
3	25.5.65 20-03	S_6 N_6 Tel. 2	I	$75° \pm 10°$	Only the central tray of neon flash tubes present
4	3.7.65 12-30	S_1 N_1 Tel. 2	I	$96.°2 \pm 0.8°$ $99.°2 \pm 0.3°$	Double track event
5	13.7.65 16-13	$S_{3,5,6}N_4$ Tel. 2	II	$45 \pm 1°$	
6	18.7.65 02-52	$S_{1,2}N_{3,6}$ Tel. 1	Out of geom.	$8.5° \pm 1°$	
7	24.7.65 11-47	S_3 N_6Tel. 1	II	$37.°5 \pm 1°$	
8	27.7.65 03-24	S_3 N_6 Tel. 1	II	$29.°5 \pm 1°$	
9	29.7.65 19-07	S_3 N_1 Tel. 2	II	$32.°5 \pm 2.5°$	
10	1.8.65 21-00	- Tel. 1	Out of geom.	$25° \pm 1°$	Oscilloscope data not available
11	2.8.65 03-38	S_4 N_6 Tel. 1	II	$47° \pm 1°$	
12	11.8.65 17-37	S_5 N_4 Tel. 1	Out of geom.	$33° \pm 1°$	
13	12.8.65 11-38	$S_{1,2,6}N_{2,4,6}$Tel. 1 $S_{1,2,3,4,5,6}N_4$Tel. 2	?	?	Neon flash tube data; big shower in Tel. 2; nothing in Tel. 1.

Fig. 5, there are two tracks, one at a projected zenith angle of 99° and the other at 96°. By a detailed reconstruction of the event from the photographs of flash tube arrays and oscilloscopes, it can be shown that both the particles responsible for the tracks penetrated ≈ 9 radiation lengths without multiplication or large angle scattering. The tracks are therefore due to particles heavier than electrons. The possibility that this event is due to the atmospheric muon component remaining at the depth of this experiment, or due to pions in equilibrium with the atmospheric muon component, has been considered and it is concluded that it cannot be so. This event represents a clear case of the non-elastic collision of a natural neutrino (see Achar et al., 1965c for details).

The meeting point of the two tracks, within errors, could be close to the surface of the rock wall, or up to a thickness of one metre inside the rock; in the latter case, the actual distance traversed by the particles through rock would be up to ~ 1.7 metres. The uncertainty in the traversal through rock of each of the tracks, namely 0 to 1.7 m, does not allow us to say if both the tracks are due to muons, or if one of the tracks is due to a muon and the other due to a pion. The point of interest here is that if the penetration in the rock of both the tracks is several nuclear interaction mean free paths, then one would have been able to assert with high probability that both of them are due to muons; if this were so it would be probable that the event corresponds to the production of an intermediate boson.

Event No. 13

This event was primarily observed in Telescope No. 2. All 'scintillator elements' on the

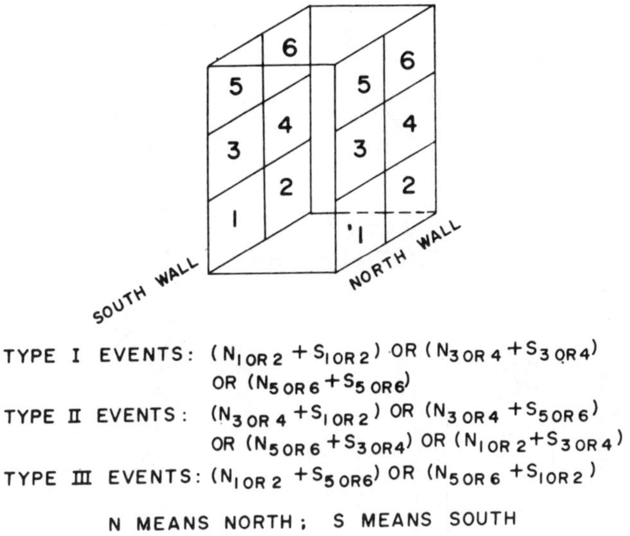

FIG. 3. Nomenclature for the various types of events.

Table II.

Category of muons	Assumed angular distribution	Type of events I	II	III
Neutrino induced	isotropic	0.67	0.31	0.02
Atmospheric	$\cos^8\theta$	0.08	0.65	0.27

southern side (S1, 2, 3, 4, 5, 6) and one scintillator element (N4) on the north side showed large pulses; other scintillator elements on the north also showed pulses, but these were small. The photograph of the NFT arrays is unfortunately incomplete; because of a defect in the film winding system a part of the picture showing details of the top section of the telescope overlapped with the previous photograph, and some of the relevant details are thus lost. A reconstruction of the event based on the photograph of the bottom two thirds of the telescope shows that practically all the flash tubes on the southern side had discharged. In the central array a few well separated large clusters of tubes which had discharged can be seen. On the north side, only a few tubes had been discharged; a few clusters can also be seen here. A remarkable feature is that three scintillator elements on the north wall (N2, 4, 6) and an equal number on the south wall (S1, 2, 6) of Telescope 1 also showed pulses in coincidence; the pulses on the north wall were rather small but those on the south wall large. The flash tube arrays of Telescope 1 were

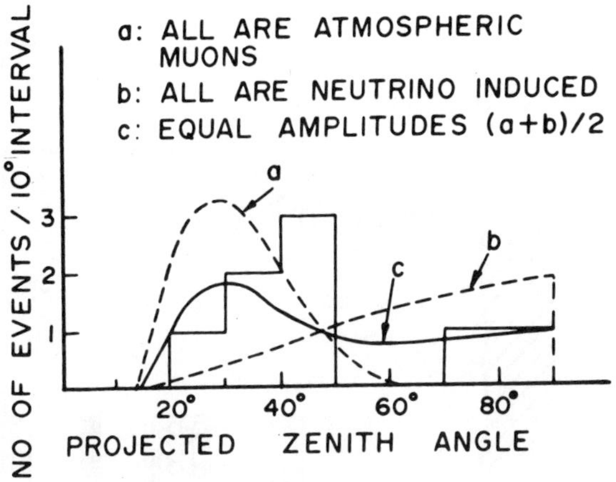

FIG. 4. The frequency distribution of measured projected zenith angles.

visible but showed very few flashes. This is the only event in which both telescopes have been triggered simultaneously.

This event is undoubtedly the most complex one which we have yet seen. It can be immediately stated, on the basis of the sizes and shapes of the scintillator pulses, and the manner in which the NFT arrays have discharged, that this event was not due to instrumental effects and that we are dealing with a genuine event involving a large number of particles, probably generated through electromagnetic processes. The most likely explanation, on the basis of a geometrical reconstruction of the event, is that it was due to an electromagnetic shower arising in the southern rock wall, and travelling at a small angle to the plane of the rock so that in addition to the primary effects in Telescope 2, it could also be seen in Telescope 1. It is difficult to assign an exact value to the zenith angle since there is no track of a clearly penetrating particle which can be unambiguously defined. It is possible to ascribe plausible directions for many of the particles in the cascade as seen in the NFT array and from this it appears likely that the zenith angle was quite large; if the zenith angle is sufficiently large the event would correspond to a neutrino interaction. From the size of the cascade, it is estimated that the total energy involved is more than several hundred GeV. We draw attention to this event to indicate the types of phenomena which are observable with arrays of the type we have used. The analysis of such events, though no doubt complicated, will be of great interest from the view-point of high energy muon and neutrino physics.

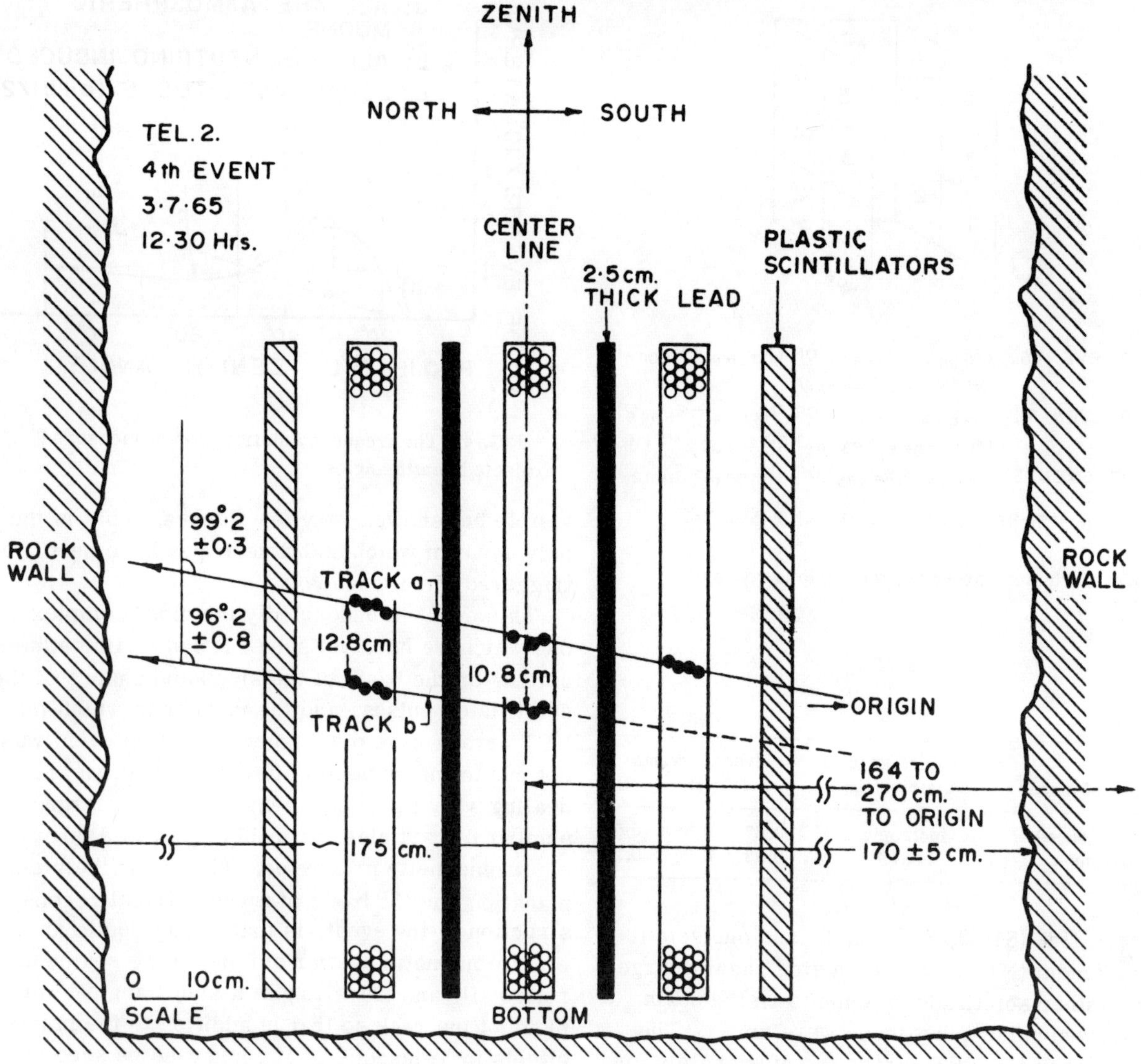

FIG. 5. Details of the double track event (event No. 4) showing the bottom one third of telescope No. 2.

IV. Discussion

The estimated number of neutrino induced muons can be compared with the theoretically expected number calculated on the basis of the atmospheric neutrino fluxes and neutrino interaction cross sections which have been measured up to energies of several GeV at CERN and Brookhaven and concerning which assumptions can be made for higher energies.

The fluxes of neutrinos produced in the terrestrial atmosphere have been calculated for various energies by Greisen (1960), Zatsepin and Kuzmin (1962), Cowsik et al. (1964, 1965) and Osborne et al. (1965). The general approach in all of these calculations has been to deduce the production spectrum of the parents of the muons from the observed energy spectrum of muons at sea level, and then to deduce the neutrino spectra from the production spectra of the parents of the muons. The neutrino intensities estimated by Zatsepin and Kuzmin (1962) were on the basis of pion and muon decays alone. In the papers of Cowsik et al. and Osborne et al. account has been taken of the production of K-mesons in high energy nuclear collisions, the intensities calculated by these two groups agree to better than 20%. The contribution from extra-terrestrial neutrinos is considered later.

We have considered the following processes for evaluating the expected interaction cross sections.

(i) Elastic collisions:

$$\nu_\mu + N \rightarrow \mu + N \quad (1)$$

(ii) Inelastic collisions:

$$\nu_\mu + N \rightarrow \mu + N + \Pi\text{'s etc.} \quad (2)$$

and

(iii) Production of intermediate boson, W:

$$\nu_\mu + P \rightarrow P + \mu + W \quad (3)$$
(incoherent production)

$$\nu_\mu + Z \rightarrow Z' + \mu + W \quad (4)$$
(coherent production)

$$\bar{\nu}_e + e^- \rightarrow W^- \quad (5)$$
(Glashow resonance)

In each of the reactions (3), (4) and (5), it is assumed that W decays within a very short time by a leptonic mode (into μ or e) or by a non-leptonic mode, i.e.,

$$W \rightarrow \mu + \nu_\mu \quad (6)$$

$$W \rightarrow e + \nu_e \quad (7)$$

$$W \rightarrow \Pi,\ K,\ Y \text{ etc.} \quad (8)$$

The cross section for process (1) has been taken to be $0.75.10^{-38}$ cm^2 per n-p pair, independent of neutrino energy for $E_\nu \geq 1$ GeV (Block et al 1964); it is assumed further that $E_\mu \approx E_\nu$. The CERN bubble chamber observations (Block et al. 1964) indicate that $\sigma_{inel} = 0.45.10^{-38}$ E cm^2/nucleon over the range from one to several GeV. Since there is no data on the behaviour of σ_{inel} at higher energies we have considered two alternatives:

(a) $\sigma_{inel} = 0.45.10^{-38}\ E_\nu$ in the range $1 < E < 10$ GeV

$= 0.45.10^{-37}$ at $E > 10$ GeV. i.e., the cross-section is proportional to E_ν upto 10 GeV, and then saturates.

(b) $\sigma_{inel} = 0.45.10^{-38} E_\nu$ at all energies.

It is further assumed that $E_\mu = E_\nu/2$ up to $E_\nu = 10$ GeV and $E_\mu = E_\nu$ above 10 GeV in case (a); and $E_\mu = E_\nu/2$ at all energies in case (b).

The cross sections for the processes (3), (4) and (5) depend on the mass of the intermediate boson, M_W. It is known from the CERN and Brookhaven experiments (Bernardini 1964, Burns et al. 1965) that $M_W > 1.8$ GeV. The W production cross-section for both incoherent and coherent interactions used here is that given by Wu et al. (as quoted by Burns et al. 1965) up to $E_\nu = 20$ GeV, together with the asymptotic expression of Von Gehlen (1963) which is valid down to about 100 GeV. One may interpolate easily between these two energies. At energies near threshold, reaction (3) dominates over (4); at higher energies the situation is reversed. Process (5) contributes very little compared to (3) and (4), mainly because at the resonance energies, ($E_\nu \approx 1{,}000\ M_W^2$ GeV if M_W is in GeV), the $\bar{\nu}_e$ fluxes are quite small. The flux of muons underground from this interaction has been calculated by Zagrebin and Zheleznykh (preprint, Lebedev Institute, Moscow 1964 and we apply their results to the K. G. F. apparatus. By folding the cross sections for processes (3) and (4) into the energy spectrum of neutrinos, it can be seen that the maximum contribution to the counting rate comes from the interactions of neutrinos with energies around 100 GeV.

There is considerable uncertainty concerning the fractional energies carried by the prompt muon, $E_{\mu p}$, in reactions (3) and (4) and by the delayed muon, $E_{\mu d}$ resulting from the decay of the W-meson in reaction (6). It may be pointed out that the counting rates are directly proportional to the effective target thicknesses, and these in turn are proportional to the energy the muon receives. It follows that the counting rates are proportional to the factor, f, given by $f = (E_{\mu p} + b\, E_{\mu d})/E_\nu$ where b is the branching ratio of W decaying by the muon mode. For production of intermediate bosons by comparatively low energy neutrinos the momentum transfer to the nucleon is large and the energy spectrum of prompt muons is strongly affected by the form factor of the nucleon. The spectrum is strongly peaked at

$$\frac{E_\nu M_\mu}{(M_W + M_\mu)}$$

corresponding to the minimum momentum transfer

$$\frac{M_W^2}{2 E_\nu}\ .$$

Table III

Expected number of neutrino induced muons in an exposure of 5,000 m^2 day steradians

Process	Expected number	
(a) Elastic	0.28	
Inelastic collisions		
(b) $\sigma_{inel} \propto E_\nu$ up to $E_\nu = 10$ GeV and saturates	0.74	
(c) $\sigma_{inel} \propto E_\nu$ up tc $E_\nu \gg 10$ GeV	1.26	
Intermediate boson, W	A	B
(d) $M_W = 1.8$ GeV	1.24	4.4
(e) $M_W = 2.5$ GeV	0.56	2.0
Glashow Resonance		
(f) $M_W = 1.8$ GeV	0.37	
(g) $M_W = 2.5$ GeV	0.08	
Total lower limit (a) + (b) σ_{inel} saturates at $E_\nu = 10$ GeV; W does not exist or $M_W \gg 2.5$ GeV	1.02	
Total upper limit (a) + (c) + (d) + (f) σ_{inel} does not saturate. $M_W = 1.8$ GeV	3.15	6.31

This is the situation which applies for $M_W = 1.8$ GeV and $E \sim 10$ GeV. For very high energy neutrinos the momentum transfer becomes smaller and this effect is less important. For extremely high energies the asymptotic formula given by Lee et al. gives

$$\bar{E}_\mu \sim \frac{E_\nu}{2} ;$$

Von Gehlen (1963) indicates that the interaction cross sections correspond to the asymptotic formula only at energies $\gtrsim 10^4$ GeV.

The median energy effective in our experiment is ~ 100 GeV; and the question we face is: what form of energy sharing between the prompt muon and W should be valid at these energies. There is no specific theoretical analysis of this problem at present. However, considering that $M_W > 1.8$ GeV and that the interaction cross section corresponds to the asymptotic formula only at $\gtrsim 10^4$ GeV, the energy sharing should be closer to the lower limit i.e.,

$$E_\mu \sim \frac{E_\nu}{20} \quad \text{for } M_W = 2 \text{ GeV}.$$

We assume that the muon decay mode of the W has a branching ratio of 40%. According to Uberall (1964) the W is strongly polarised and the muon comes out predominantly backwards in the C. M. S. If one accepts this then the delayed muon retains on the average, $\frac{1}{4}$ of the boson energy. These considerations would lead to, for $M_W = 1.8$ GeV, $f \approx 0.15$ as a lower limit. The upper limt would be $f \approx 0.55$, on the basis of the asymptotic formula; and the values calculated by Cowsik et al. 1965 are based on this.

Contributions to the expected number of neutrino induced muons by the various processes considered above for an exposure of 4,700 m^2 day sterad have all been listed in Table III. Contributions arising from W-μ production are given for the energy sharing between the prompt μ and W corresponding to the lower limiting case; these are marked (A). The true value will be greater than this but it is likely to be close to it. The maximum possible contribution corresponding to the asymptotic formula of Lee et al. is shown purely for comparison; these are marked (B). Also shown in the table are the absolute lower limits to the predicted number (obtained by assuming that W does not exist and σ_{inel} saturates around 10 GeV), and the absolute upper limit obtained by assuming that σ_{inel} does not saturate, $M_W \approx 1.8$ GeV and energy sharing is as given by the asymptotic formula. It is seen that the estimated number (≈ 5) of ν-induced muons lies in between these two limits.

On the basis of Table III we draw the following conclusions:

(i) If W does not exist, the theoretically predicted number of events is low compared to the experimentally estimated number.

(ii) This would imply that either (a) the fluxes derived by various authors are too low by a substantial extent; or (b) that σ_{inel} increases with energy much faster than $\sigma_{inel} \propto E_\nu$. Since the evidence from the accelerator experiments is that σ_{inel} does not rise much faster than $\sigma_{inel} \propto E_\nu$ up to $E_\nu = 10$ GeV it would then be necessary for σ_{inel} to rise faster than E_ν for $E_\nu > 10$ GeV.

(iii) A possible explanation of (iib) above would be that intermediate bosons exist with mass not much greater than 1.8 GeV.

It is difficult to see how the flux calculations of neutrinos of terrestrial origin can be in great error, i.e. by a factor of two or more, unless hitherto unknown processes come into play. As far as extra-terrestrial sources are concerned, the calculations (Greisen, 1960) show that they are extremely small. This view is further confirmed by the observations on primary γ-ray fluxes, which are genetically related to neutrino fluxes in the sense that both are derived from pion decays. There is of course the possibility that unknown processes exist wherein only high energy neutrinos are produced but not high energy γ-rays; or that γ-rays may have been attenuated by unexpectedly large factors before arriving at the earth. Light can be thrown on these points by studying the neutrino arrival directions in celestial coordinates; as yet, the observed number of events is too small to say if there is any dominant region in the sky from which the neutrinos could have come. Our plans over the immediate future include a study of these possibilities. Cowsik et al. (1965) have also discussed some of these aspects.

Reines et al. (1965), from their deep mine neutrino experiment, have reported events attributable to the interactions of natural neutrinos in the surrounding rock. Their observed rate, namely, $0.6.10^{-12}$ particles/cm^2. sec. sterad compares reasonably with the rate we have reported here, namely, $1.2.10^{-12}$, within the statistical weights of the two observations.

V. Vertical Intensity of Atmospheric Muons (at a depth of 7500 m. w. e. of standard rock).

As discussed earlier in §III, 8 out of 13 observed events are due to atmospheric muons. Only 5 out of these 8 are within the geometry. If the angular distribution of atmospheric muons at this depth is approximately taken to be $I_\theta = I_0 \cos^n\theta$ with n = 8 then the rate attributed to atmospheric muons divided by the aperture offered by the telescope for a radiation with n = 8 yields a value for vertical intensity

$$I_V \approx 0.9.10^{-10} \text{ particles/cm}^2 \text{ sec. sterad.}$$

at the depth of our experiment, namely, 7500 m. w. e. of standard rock with

$$\frac{Z^2}{A} = 5.5.$$

This value is in reasonable agreement with the extrapolated value from the data hitherto available only up to a depth of 6,930 m. w. e. of standard rock (Menon and Ramana Murthy 1965).

VI. Conclusions

Events which can be definitely shown to be due to the interactions of natural neutrinos in the rock surrounding the apparatus have been reported. Two events involving charged particles at large zenith angles have been seen and one can be fairly sure that these represent ν-interactions. It is estimated that a further 2-4 ν-induced muons are contained amongst the events with projected zenith angles ranging from 29.5° to 48°. Based on an estimated 5 γ-events in 4700 m^2 days sterad. the muon flux resulting from neutrino collisions deep underground is derived as $\approx 1.2.10^{-12}$ particles/cm^2 sec. sterad. This rate, when interpreted in terms of the fluxes of neutrinos produced in the terrestrial atmosphere and the high energy behaviour of neutrino interaction cross sections suggests one (or more) of the following possibilities:

(a) There are sources of high energy neutrinos (terrestrial or extra-terrestrial) which yield fluxes comparable to, (or greater than) the atmospheric fluxes which have been calculated from pion, muon and kaon decays.

(b) σ_{inel} rises much faster than $\sigma_{inel} \propto E$ at energies beyond several GeV.

(c) a possible reason for (b) would be that W-mesons exist with masses not very much greater than 1.8 GeV.

These results should be treated as tentative since this paper is only meant to be a status report on this experiment for this Conference. The statistical weight of the observations needs to be enhanced; towards this we are expanding the size of the array by a factor of 2. The great interest of the present observations is that we seem to be encountering new physical phenomena; an exact and definite interpretation is not possible at this stage; one would have to carefully evaluate various phenomena which could come into play at such depths. What is clear is that a continuation and enhancement of the experiment is of considerable value.

REFERENCES

C. V. Achar, V. S. Narasimham, P. V. Ramana Murthy, D. R. Creed, J. B. M. Pattison, and A. W. Wolfendale, 1965, (a) Proc. Phys. Soc. (in the press).

C. V. Achar, M. G. K. Menon, P. V. Ramana Murthy, B. V. Sreekantan, K. Hinotani, S. Miyake, D. R. Creed, J. L. Osborne, J. B. M. Pattison, and A. W. Wolfendale, 1965 (b) Phys. Lett. 18, 196-199. (c) Phys. Lett. (in the press).

G. Bernardini, J. Bienlein, G. Von Dardel, H. Faissner, F. Ferrero, J. Gaillard, H. Gerber, B. Hann, B. Kaftanov, F. Krienen, C. Manfredotti, M. Rienharz, R. A. Salmeron, 1964, Phys. Lett., 13, 86-91.

M. Block, H. Burmeister, D. Cundy, B. Eiben, C. Franzinetti, J. Keren, R. Møllerud, G. Myatt, A. Orkin-Lecourtois, M. Paty, D. Perkins, C. Ramm, K. Schultze, H. Sletten, K. Soop, R. Stump, M. Venus, H. Yoshiki, 1964, Phys. Lett., 12, 281-285.

R. Burns, K. Goulianos, E. Hyman, L. Lederman, W. Lee, N. Mistry, J. Rettberg, M. Schwartz, J. Sunderland, G. Danby, 1965, Phys. Rev. Lett., 15, 42-45.

R. Cowsik, Pal Yash, T. N. Rengarajan, S. N. Tandon, Proc. Int. Conf. on Cosmic Rays, Jaipur, 1963, Vol. 6, 211-214.

R. Cowsik, Pal Yash, S. N. Tandon, 1965 (to be published).

G. Von Gehlen, 1963, Nuovo Cim., 30, 859-877.

T. D. Lee, P. Markstein, C. N. Yang, 1961, Phys. Rev. Lett., 7, 429-433.

K. Greisen, 1960, Proc. Int. Conf. for Instrumentation for High Energy Physics, Berkeley (Inter Science Publishers) p. 209.

M. G. K. Menon, P. V. Ramana Murthy, B. V. Sreekantan, S. Miyake, 1963, (a) Nuovo Cim., 30, 1208-1219. (b) Phys. Lett., 5, 272-274.

M. G. K. Menon, Proc. Int. Conf. on Cosmic Rays, Jaipur, 1963, vol. 6, 152-181.

M. G. K. Menon, and P. V. Ramana Murthy, 1965, Progress in Cosmic Ray and Elementary Particle Physics, vol. 9, (in the press).

S. Miyake, V. S. Narasimham, P. V. Ramana Murthy, 1964, Nuovo Cim., 32, 1505-1540.

J. L. Osborne, S. S. Said, A. W. Wolfendale, 1965, Proc. Phys. Soc., 86, 93-99.

F. Reines, M. Crouch, T. Jenkins, W. Kropp, H. Gurr, G. Smith, J. Sellschop, B. Meyer, 1965, Phys. Rev. Lett., 15, 429-433.

B. V. Sreekantan, 1965, Proc. Informal Conf. on Experimental Neutrino Physics, CERN, (in the press).

H. Uberall, 1964, Phys. Rev., 133, B444-453.

A. W. Wolfendale, 1965, Proc. Informal Conf. on Experimental Neutrino Physics, CERN, (in the press).

G. T. Zatsepin, and V. A. Kuzmin, 1962, Soviet Physics-JETP 14, 1294-1299.

OBSERVATIONAL NEUTRINO ASTRONOMY: A ν-REVIEW*

John N. Bahcall
California Institute of Technology, Pasadena, California

I. Introduction

I would like to describe for you the theoretical predictions for some neutrino observations that can be carried out with current experimental techniques. I will emphasize the information about astronomical systems that can be obtained from these neutrino observations as well as the assumptions that underlie the theoretical expectations. The experiments that we shall discuss are all new experiments in one sense; their feasibility has been established in the last year. Some new experiments have even been suggested in the last month. Perhaps some of you can think of other, even more informative observations that can be performed.

The neutrino observations we shall discuss are difficult and interesting for the same reason: the enormous penetrating power of neutrinos. Neutrinos of all kinds are believed to interact with other particles primarily through weak interactions; thus a 10-MeV neutrino can, for example, pass through the entire Earth with only about one chance in 10^{10} of being absorbed or scattered. This enormous penetrating power makes neutrino astronomy difficult since large quantities of material (typically tons) are required to detect neutrinos; it also makes neutrino astronomy exciting since neutrinos can reach us from otherwise inaccessible regions of the universe.

We shall begin by discussing neutrino observations of the solar interior and then discuss experiments involving neutrinos from cosmic-ray secondaries and from strong radio sources. Finally, we shall examine briefly the role of neutrino observations in cosmology.

***Supported in part by the Office of Naval Research [Nonr-220(47)] and the National Aeronautics and Space Administration [NGR-05-002-028]. Much of this review is based on a non-technical account of neutrino astronomy that appeared in Science, 147, 115 (1964).**

II. Why Observe Solar Neutrinos?

The principal source of the energy radiated by main-sequence stars like the Sun is believed (Bethe and Critchfield 1938; Bethe 1939; Salpeter 1952; Burbidge, Burbidge, Fowler and Hoyle 1957; Cameron 1958) to be the burning of hydrogen nuclei (protons) in the deep interiors of the stars, where the temperatures and densities are greatest. The nuclear reactions thought to be primarily responsible for the generation of energy in the Sun are similar to the fusion reactions which occur in the explosion of a hydrogen bomb. The basic ideas of the theory of nuclear-energy generation in main-sequence stars are believed to be correct by astronomers and physicists, although no direct test of this theory has yet been possible. Since electromagnetic radiation has a mean free path of less than 1 cm under the conditions thought to exist in the Sun's interior, no light can reach us from the interior regions of the Sun. A direct test of the theory of stellar energy generation has been impossible hitherto because the collisions among nuclei that result in nuclear burning occur primarily in the deep interiors of the stars.

All conventional astronomical observations are made on electromagnetic radiation emitted from the surfaces of stars. Have we overlooked anything crucial in our theoretical extrapolations from the observed conditions on the surfaces of stars to the vastly different conditions in the interiors of stars? It would be interesting to find out.

How can one test directly the theory of nuclear-energy generation in stars? To make such a test, one needs to "see" into the deep interior of a star where the nuclear reactions are believed to occur; an information carrier with a mean free path of the order of 10^{+11} cm ($\sim$ radius of the Sun) is therefore required.

Of the known particles only neutrinos, with their enormous penetrating power, can enable us to "see" into the interior of a star. The observation of solar neutrinos would constitute the most direct test possible of the hypothesis that hydrogen-burning fusion reactions provide the principal source of the energy radiated by main-sequence stars like the Sun.

Even if one assumes that the theory of stellar energy generation is basically correct, there are other important reasons for trying to measure the intensity and energy spectrum of solar neutrinos. The long-range program should be a neutrino-spectroscopic study of the solar interior, as a means of determining quantitatively the conditions in the interior of the Sun in much the same way that astronomers have already determined the conditions on the surface of the Sun by photon-spectroscopy. One can, for example, use neutrino absorbers with different thresholds to help determine the solar neutrino spectrum. Observations of this kind, which give information regarding conditions in the interior of the Sun, are stringent tests of mathematical models describing the solar interior. Moreover, a quantitative knowledge of the Sun's neutrino spectrum is necessary before one can make full use of neutrino observations in testing various cosmological theories.

Recent theoretical (Bahcall, Fowler, Iben, and Sears 1963; Sears 1964; Bahcall 1964a, b, d) and experimental (Davis 1964; Reines and Kropp 1964; Reines and Woods 1965) results have changed qualitatively our ideas concerning the possibility of observing solar neutrinos. We shall consider first the theoretical developments.

III. Neutrino Production in the Sun

The hydrogen-burning fusion reactions in the Sun are currently believed (Fowler 1960; Parker, Bahcall, and Fowler 1964) to be initiated by the sequence $H^1(p,e^+\nu)H^2(p,\gamma)He^3$ and terminated by the sequences (i) $He^3(He^3, 2p)He^4$; (ii) $He^3(\alpha,\gamma)Be^7(e^-,\nu)Li^7(p,\alpha)He^4$; and (iii) $He^3(\alpha,\gamma)Be^7(p,\gamma)B^8(e^+\nu)Be^{8*}(\alpha)He^4$
The net result of any of these sequences can be represented symbolically by the simple formula:

$$4\,H^1 \rightarrow He^4 + 2e^+ + 2\nu_e, \qquad (1)$$

although the energies of the neutrinos emitted in each sequence are different. The carbon-nitrogen-oxygen cycle is believed to contribute only a few per cent of the energy generation in the Sun (Sears 1964) and a negligible amount to the observable solar neutrino flux (Bahcall 1964b, d).

The rates for the hydrogen-burning fusion reactions listed above can be calculated as a function of stellar temperature and density with the aid of results from many laboratory experiments in low-energy nuclear physics and some elementary facts about nuclear reactions. The rate of the $Be^7(p,\gamma)B^8$ reaction which initiates sequence (iii) is, as we shall see later, crucial for the solar neutrino observations that are currently under way. Unfortunately, the small cross-section for this reaction at low proton energies has been measured only once (Kavanagh 1960) and its value is rather uncertain.[1] This vital laboratory experiment should certainly be repeated, as should the related $Li^7(n,\gamma)Li^8$ and $Li^7(n,n')Li^7$ experiments.

R. L. Sears and his collaborators (Sears 1964; Bahcall et al. 1963; Bahcall 1964b) have calculated the neutrino fluxes (numbers/cm^2/sec) at the Earth's surface from all the nuclear reactions that are expected to be important in the Sun. To predict the neutrino fluxes, Sears constructed mathematical models of the Sun with the aid of high-speed computing machines. More specifically, Sears numerically solved the classical differential equations of stellar structure which state that at each point in the Sun the pressure due to gravitational attraction is balanced by the gas pressure; the observed luminosity of the Sun is attributed to the nuclear-fusion reactions discussed previously. These solar models involve equations of state of matter at high temperatures ($\sim 10^7$ °K) and high densities (~ 100 gm/cm^3), as well as theoretical relations that describe the transport of energy under stellar conditions. Figure 1 is a schematic representation of the predicted solar neutrino spectrum.

Sears has also calculated the uncertainties in the theoretical solar neutrino fluxes arising from

[1] Tombrello (1964) has recently reanalyzed the laboratory data for the reaction $Be^7(p,\gamma)B^8$. Tombrello suggests that the rate of the $Be^7(p,\gamma)B^8$ reaction at solar energies may be as much as 50 per cent lower than previously estimated. This would imply that, e.g., (3 ± 2) should be replaced by (1.5 ± 1) in relation (3).

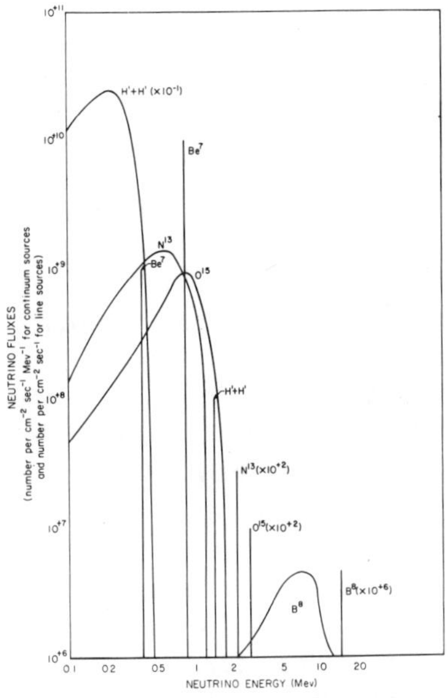

FIG. 1. Predicted neutrino spectrum from the Sun. Fluxes given here are evaluated at the Earth's surface. The neutrino lines are produced by the capture of free electrons; the small thermal widths (~ 1 keV) of these lines have been neglected in the figure.

uncertainties in our knowledge of solar age, composition, and opacity, as well as nuclear parameters. On the basis of these calculations, one can show that our present knowledge of solar and nuclear parameters enables us to predict the most important neutrino fluxes within an uncertainty of about 50 per cent, provided the theories of stellar models and nuclear-energy generation in stars are correct.

IV. Neutrino Detection with Cl^{37}

R. Davis, Jr., of Brookhaven National Laboratory has proposed (Davis 1964) a method of detecting solar neutrinos that makes use of the inverse electron-capture reaction,

$$\nu_e(\text{solar}) + Cl^{37} \rightarrow e^- + Ar^{37}. \qquad (2)$$

Pontecorvo (1964) first suggested, some years ago that this reaction might be a useful method of detecting neutrinos (Pontecorvo's epochal suggestions in nearly all new developments of neutrino astronomy and neutrino physics over the past twenty years include such diverse problems as neutrino energy loss from stars, high-energy neutrino experiments, and the role of neutrinos in cosmology). A detailed investigation of the feasibility of using Cl^{37} as a terrestrial neutrino detector was carried out by Alvarez (1949). Notice that reaction (2) is the inverse of the reaction by which radioactive argon decays in the laboratory.

On the basis of experience gained in a preliminary experiment involving two 500-gal tanks of perchlorethylene, C_2Cl_4 (an ordinary cleaning fluid), Davis and D. S. Harmer have undertaken an experiment in which a 100,000-gal tank of perchlorethylene is used as a detector. Their detection system, based on reaction (2), requires an amount of cleaning fluid that would fill an Olympic-sized swimming pool. The most important features of their detection system are: (i) tiny amounts (~ 200 atoms per month) of neutrino-produced Ar^{37} can be removed from the large volume of liquid detector (with 90% efficiency) by the simple procedure of sweeping the argon out of the liquid with helium gas; and (ii) the characteristic decay of Ar^{37} can be observed in a counter with essentially zero background.

In order to relate the observed number of Ar^{37} atoms produced per day in the proposed experiment to the solar neutrino fluxes, one must know the probability that a neutrino of a given energy incident on a Cl^{37} atom will cause reaction (2). The average neutrino cross sections have recently been calculated (Bahcall 1964a, b) for all neutrino sources believed to be important in solar-energy generation. The calculations were performed by making a mathematical model of Cl^{37} and Ar^{37} nuclei; this model includes excited states of Ar^{37} which are crucial for the solar-neutrino experiment but which have not yet been observed in the laboratory. Most importantly, the analogue state of the ground state of Cl^{37} is predicted to be at about 5.15-MeV excitation energy in Ar^{37}. The existence of this state enhances the expected neutrino capture rate by about a factor of 10 relative to expectations based only on ground-state transitions. The nuclear model employed in these calculations has also been used to make related predictions which can be tested experimentally and hence can be used to determine the accuracy of the model; a number of these experiments are currently being performed by nuclear physics groups in the United States and Canada. The most direct of these experimental tests of the nuclear model of mass 37 involves the decay of the radioactive isotope $_{20}Ca^{37}$. The same matrix elements which occur in the predicted (Bahcall 1964b; Bahcall and Barnes 1964) decay rate of Ca^{37} to $_{19}K^{37}$ also occur in the neutrino capture by $_{17}Ca^{37}$ leading to $_{18}Ar^{37}$, since these processes are essentially mirror reactions. Two experimental groups (Reeder, Poskanzer, and Esterlund 1965; Hardy and Verrall 1965), one at Brookhaven National Laboratory and one at McGill University, Montreal, have recently announced the discovery of Ca^{37} with about the predicted lifetime and mass. These important experiments confirm in a direct way the enhancement of reaction (2) because of excited-state transitions.[2]

Combining the results from the nuclear model calculations with the neutrino fluxes predicted by Sears, one finds that the neutrinos from B^8 decay [sequence (iii)], which have a maximum energy of about 14 MeV, should produce 90 per cent of the observable reactions in the proposed experiment of Davis. The B^8 neutrinos produce most of the reactions, although they constitute less than 0.1 per cent of the total solar neutrino flux (see Fig. 1), because they are the most energetic neutrinos expected to come from fusion reactions in the Sun. Calculations using the best current estimates of all recognized uncertainties yield the following predicted number of neutrino captures:

$$(3 \pm 2) \times 10^{-35} \text{ per terrestrial } Cl^{37} \text{ atom per sec.} \tag{3}$$

This corresponds to about 6 predicted solar-neutrino captures per day in the 100,000-gal experiment, with an estimated background (Davis 1964) of less than 0.5 captures per day.

Fowler (1958) in his original discussion of the solar-neutrino flux from B^8 decays, showed that the B^8 neutrino flux is a sensitive function of the central temperature of the Sun. This temperature sensitivity exists because the Coulomb barrier is large compared to solar thermal energies for the reaction $Be^7(p,\gamma)B^8$ of sequence (iii). An experimental upper limit on the central temperature of the Sun can therefore be obtained by combining the result of the preliminary experiment (Davis 1964), which provides an upper limit for the number of neutrino captures per second per Cl^{37} atom, with the predicted rate (see eq. [3]) and the known temperature dependence of the $Be^7(p,\gamma)B^8$ reaction. In calculating this upper limit on the central temperature, one can treat the solar luminosity and all other solar variables except the temperature as approximately constant, because the temperature dependence of the $Be^7(p,\gamma)B^8$ reaction rate is roughly $\exp - [150\, T_6^{-1/3})$, where T_6 is the temperature in millions of degrees Kelvin. One can show that the preliminary experiment implies that the central temperature of the Sun must be less than 20×10^6 °K and that a measurement of the B^8 neutrino flux accurate to ±50 per cent would determine the central temperature of the Sun to ±10 per cent.

V. Neutrino-Electron Scattering

Reines and Kropp (1964) have recently proposed an experiment in which the solar neutrino flux from B^8 decay is to be studied by neutrino-electron scattering, i.e.,

$$\nu_e(\text{solar}) + e \rightarrow \nu'_e + e'. \tag{4}$$

Under the experimental conditions suggested by Reines and Kropp, one can easily show (Bahcall 1964c) that the observed recoil electrons will be

[2] The measured Ca^{37} lifetime is slightly larger than the predicted value. The following non-unique model, suggested by the experimental results of Kavanagh and Goosman (1964) for K^{37} decay, yields a predicted lifetime (180 msec) in surprisingly close agreement with the measured lifetime (170 ± 5 msec): (i) the $J = (\frac{3}{2})^+$, $T = \frac{1}{2}$ level in K^{37} is at 2.74-MeV excitation (not 1.57 MeV as originally assumed); (ii) the Gamow-Teller matrix element from the ground state of Ca^{37} to the 2.74-MeV level in K^{37} is equal to the Gamow-Teller matrix element for the ground-state electron capture by Ar^{37}; (iii) all other matrix elements and energies are the same as originally assumed (Bahcall 1964b). An experimental study of the positron decay modes of Ca^{37} could eliminate the remaining uncertainties (~ 25 per cent) in the predicted Cl^{37} neutrino-capture cross-section. The experimental quantities of most interest are (i) the intensity of the ground-state positron branch (or any other branch) relative to the $T = \frac{3}{2}$ proton-unstable branch; (ii) the β-γ spectrum (which would reveal which and how much the low-lying $T = \frac{1}{2}$ states are populated by Ca^{37} decay); and (iii) an upper limit (or measurement) for the possible branching to states, other than the analogue state, that emit delayed protons.

confined to a cone with opening angle of approximately 10° with respect to the incident neutrino direction. Thus the observation of neutrino scattering by electrons can, in principle, enable one to determine the direction (presumably toward the Sun) of a low-energy (~10 MeV) extraterrestrial neutrino source.

The conserved vector current theory of weak interactions (Feyman and Gell-Mann 1958) can be used to predict the probability of reaction (4); the cross-section is, in this theory, determined primarily by the rates of the familiar O^{14} and neutron beta-decays. Reines and Kropp have already performed a preliminary experiment, which consisted of looking for counts, due to electrons with an energy greater than 8 MeV, in a 200-liter liquid scintillation detector. Their scintillation detector was surrounded by a large Cerenkov anticoincidence shield and was located 2,000 feet underground (to decrease background from cosmic-ray secondaries) in a salt mine. Their upper limit on the B^8 solar neutrino flux times the average cross-section for reaction (4) is only a factor of 15 greater than is predicted by the symbolic product: (solar model theory) × (conserved vector current theory).

This experiment should be pursued with the utmost vigor since the results are important for the basic theory of weak interactions as well as for the theory of solar models.

VI. Some New (ν)-Experiments

Two other targets have been recently suggested (Bahcall 1964d; Reines and Woods 1965) as possible detectors of the neutrino flux from B^8 decay in the sun, namely, Li^7 and B^{11}. The idea here is to count the high-energy electrons produced in a reaction such as:

$$\nu_e + Li^7 \rightarrow e^- + Be^7. \quad (5)$$

As Marx and Menyhárd pointed out some time ago (Marx and Menyhárd 1960), the direction of the outgoing electrons can be used to establish the direction of the incident neutrinos. Reines and Woods, who first proposed the use of B^{11} as a target, have recently concluded that one can detect, with current technology, solar neutrinos from B^8 decay with either Li^7 or B^{11} and that the direction of the neutrino source can also be established.

All of the experiments we have discussed so far, involving Cl^{37}, Li^7, B^{11}, and neutrino-electron scattering, are designed to detect the rare but relatively high-energy neutrinos from B^8 decay. Can we think of experiments that will enable us to study the other parts of the solar neutrino spectrum, especially the low-energy neutrinos from the basic proton-proton reaction and from the frequently occurring $Be^7(e^-,\nu)Li^7$ reaction? Some time ago Sunyar and Goldhaber (1960) suggested the reaction $Rb(\nu,e^-)Sr^{87}m$, which has a threshold of 115 keV. The lifetime of the relevant isomeric state of Sr^{87} is about 3 hours. So far no one has been able to devise a practical experimental technique which uses Rb^{87} as a target.

One of the best experiments, from the point of view of feasibility, that I have been able to think of (Bahcall 1964d) involves the reaction $H^3(\nu, e^-)He^3$, which has a threshold of -18 keV. Tritium is about equally sensitive to proton-proton neutrinos (0.43 MeV maximum energy) and Be^7 neutrinos (90 per cent having the energy 0.86 MeV). Unfortunately, 10 kg of tritium are required in order to obtain a counting rate of 300 events per year induced by solar neutrinos. Moreover, tritium is radioactive (half-life of 12 years) so that one must find a way to count the rare several hundred keV electrons from neutrino-induced transitions in the presence of a very large background of 18-keV electrons from normal tritium decay. A possible method for distinguishing between tritium-decay electrons and neutrino-produced electrons might be to cover the tritium target with a thin absorber sufficient to stop the 18-keV decay electrons without preventing the higher-energy neutrino-produced electrons from reaching a surrounding scintillator. This experiment will be greatly improved if some clever experimentalist thinks of a practical way of of taking advantage of the fact that about half of the neutrino-produced electrons have a fixed energy, 879 keV (from Be^7 decay in the Sun).

Two other targets that might be used to detect proton-proton and Be^7 neutrinos are Mn^{55} and Ga^{71}; the latter isotope was first mentioned as a possible neutrino-detector by Chiu (1964).

The relevant theoretical predictions (Bahcall 1964d) for H^3, Li^7, B^{11}, Mn^{55}, Ga^{71}, and Rb^{87} are shown in Tables 1 and 2. The calculated cross-sections for Mn^{55} and Ga^{71} do not include excited-state contributions. These corrections

Table I.

Cross-sections and asymmetry parameters.*

Target	$\sigma_{p\text{-}p}(\times 10^{45}cm^2)$	$\sigma_{Be^7}(\times 10^{45}cm^2)$	$\sigma_{B^8}(\times 10^{45}cm^2)$	$\sigma_{N^{13}}(\times 10^{45}cm^2)$	$\sigma_{O^{15}}(\times 10^{45}cm^2)$	α
H^3	5×10^1	2×10^2	7×10^3	1.5×10^2	2×10^2	- 0.1
Li^7	0	0	4.5×10^3	5	2×10^1	- .1
B^{11}	0	0	2×10^3	0	0	+ .35 ± 0.05
Mn^{55}	4×10^{-2}	2×10^{-1}	1×10^1	2×10^{-1}	3.5×10^{-1}	- .3
Ga^{71}	3	1.5×10^1	7×10^2	1×10^1	2×10^1	- .3
Rb^{87}	7	2.5×10^1	1×10^3	2×10^1	3.5×10^1	- .3

*The subscripts such as p-p or Be^7 indicate the neutrino source in the sun. The cross-sections for Rb^{87} refer to the low-lying metastable state of Sr^{87} (Sunyar and Goldhaber 1960); excited-state corrections are not included in the Mn^{55} and Ga^{71} cross-sections.

Table II.

Predicted number of neutrino-induced reactions per target particle per second*

Target	$(\varphi\sigma)_{p\text{-}p} \times 10^{35}$	$(\varphi\sigma)_{Be^7} \times 10^{35}$	$(\varphi\sigma)_{B^8} \times 10^{35}$	$(\varphi\sigma)_{N^{13}+O^{15}} \times 10^{35}$
H^3	$2.5 \pm 0.3 \times 10^2$	$2 \pm 1 \times 10^2$	$1.5 \pm 0.8 \times 10^1$	$(4 \pm 2) \times 10^1$
Li^7	0	0	$1 \pm 0.6 \times 10^1$	3 ± 1.5
B^{11}	0	0	4.5×2.5	0
Mn^{55}	$2 \pm 0.2 \times 10^{-1}$	$2.5 \pm 1.5 \times 10^{-1}$	2.5×10^{-2}	5.5×10^{-2}
Ga^{71}	$1.5 \pm 0.2 \times 10^1$	$2 \pm 1 \times 10^{-1}$	1.5	3
Rb^{87}	$4 \pm 0.4 \times 10^1$	$3 \pm 1.5 \times 10^1$	2.5 ± 1.3	6 ± 3

*The fluxes, φ, used here are adapted from the work of Sears (1964).

are, however, small for the low-energy neutrinos to which Mn^{55} and Ga^{71} are most sensitive.

VII. Neutrinos from Cosmic-Ray Secondaries

Neutrinos are produced in the Earth's atmosphere by the decay of cosmic-ray secondaries, primarily pions and kaons. The spectrum and intensity of these neutrinos can be calculated (Greisen 1960; Zatsepin and Kuz'min 1962; Lee, Robinson, Schwartz, and Cool 1963) from the well-known spectrum of cosmic-ray muons at sea level. Thus the neutrinos from cosmic-ray secondaries can be used to study properties of weak interactions at ultra-high energies, corresponding to primary cosmic-ray energies of the order of 10^{+18} or 10^{+19} eV. Two such experiments are currently under way. The University of Utah group, under the direction of J. W. Keuffel and H. E. Bergeson, is installing an experimental facility in a mine 2,000 feet deep near Park City, Utah. This facility is expected to have excellent directional resolution in detecting the reaction products of high-energy neutrinos (Keuffel and Bergeson 1964). Another group (Reines 1964), headed by F. Reines, of the Case Institute of Technology, and J. P. F. Sellschop, of the University of Witwatersrand, is testing a neutrino-detection system in a mine 10,500 feet deep near Johannesburg, South Africa. The detection systems being employed in the Utah and South African experiments are different and will provide complementary information.

Deep-mine studies of the kind mentioned above can provide unexpected information about high-energy physics. For example, Frautschi and I have recently shown (Frautschi and Bahcall 1965), using the results (Mennon, Murthy, Sreekantan, and Miyake 1963; Miyake, Narasimhan, and Murthy 1962) of an experiment performed by the Tata Institute Group in a deep mine in the Kolar Gold Fields, that there are no "resonances" in the ν_μ-nucleon systems for neutrinos with laboratory energies less than 2×10^3 BeV and no "resonances" in the ν_e-nucleon systems for laboratory neutrino energies less than 30 BeV.

VIII. Neutrinos from Strong Radio Sources

We have heard a great deal about how mysterious and interesting are the strong radio sources. Frautschi and I, stimulated by some unpublished observations of C. L. Cowan, have investigated (Bahcall and Frautschi 1964a, b) the possibility of enlarging our observational knowledge of strong radio sources by detecting neutrinos from these objects. Unfortunately, we were able to show on the basis of simple arguments involving the amount of electromagnetic energy being detected at the Earth from these objects that the neutrino fluxes which might be expected from radio sources are too small to be detected in practical experiments (10^5 tons of material required for one neutrino-induced event per day) unless some resonance exists in neutrino interactions. The only practical laboratory targets for neutrinos are electrons and nucleons. Having analyzed several deep-mine cosmic-ray studies and neutrino experiments performed with high-energy laboratory accelerators, Frautschi and I are convinced that the only resonance which can be expected to exist with the available laboratory targets is one proposed several years ago by Glashow (1960), namely, a resonance in the $\bar{\nu}_e$-e^--system.

This predicted resonance in $\bar{\nu}_e$-e^--scattering would be due to the production of the famous W^--boson (or uxl) which is hypothesized to mediate the weak interactions in much the same way as photons mediate the electromagnetic interactions. The large cross-section for a neutrino interaction associated with this resonance is expected to obtain at a neutrino energy greater than 10^3 BeV, corresponding to a mass of the intermediate boson of greater than 1 BeV.

To predict counting rates produced by neutrinos from strong radio sources, Frautschi and I made three hypotheses. The first hypothesis is that the highest energy electrons ($\sim$ several $\times 10^3$ BeV) inferred to exist in strong radio sources are produced by the Burbidge mechanism (Burbidge 1958):

$$p + p \rightarrow \text{nucleons} + \text{mesons} \qquad (6a)$$

$$\rightarrow \text{nucleons} + \text{electrons, gamma rays, and neutrinos.} \qquad (6b)$$

The second hypothesis is that magnetic fields in the strong radio sources are in the range 10^{-5} - 10^{-3} Gauss; the precise values of the fields were obtained from the theory of synchrotron radiation by the usual argument involving the minimization of the total energy computed to exist in particles and fields. Finally, we assumed that the mass of the W^--meson is of the order of 1 BeV.

Recent experiments performed at CERN have shown (Bernadini et al. 1964) that the value we assumed for the mass of the W^--mass is too small by at least a factor of 1.8. However, one can show that our predicted counting rates depend rather insensitively on the assumed mass of this intermediate boson; the predicted counting rates decrease, in fact, as $M_{W^-}^{-1.2}$ if the neutrino spectrum has the cosmic-ray form we derived. Thus if M_{W^-} is as high as 9 BeV, our original predicted counting rates from strong extragalactic radio sources should be decreased by a factor of 10. Unfortunately, for the special case of the Crab Nebula, the cutoff frequency in the optical spectrum (O'Dell 1963), which produces a cutoff in the predicted neutrino spectrum, occurs at an energy below that presently estimated to be necessary to produce the W^--resonance. Thus our predicted counting rates for the Crab Nebula should be decreased by at least a factor of 100. No such cutoff is known, however, for the quasars or for strong radio sources.

Using the three hypotheses described previously, we predicted neutrino fluxes as a function of neutrino energy for various radio sources (such as M82) that have been observed to emit, or might be emitting, optical synchrotron radiation. The reason for confining the predictions to these radio sources is that the high neutrino energies necessary to excite the W^--resonance correspond to high-energy electrons which radiate predominantly in the optical range. Since reactions (6) produce gamma rays as well as neutrinos, we were also able to predict, with the same hypotheses, gamma-ray fluxes from strong radio sources.

The neutrino fluxes from quasars and other strong radio sources can also be estimated on the basis of a different hypothesis, namely, that the energy of some strong ratio sources originates in supernovae explosions (Shklovsky 1960; Burbidge 1960; Cameron 1962; Field 1964, 1965). This hypothesis leads to estimates of the neutrino fluxes and spectra that are very similar to those Frautschi and I calculated. A recent proposal by

Field (1964, 1965) can be used to suggest the most likely sources of strong-neutrino emission on the basis of the supernovae hypothesis, namely, those sources (such as 3C 48 or 3C 273) which exhibit large variations in their electromagnetic luminosity. These variations are attributed by Field to statistical fluctuations in the number of supernova explosions occurring per year and hence can be used as a possible indication of large supernova activity.

The most promising method we proposed for testing the neutrino-predictions, and hence the hypotheses upon which the predictions are based, was to use the Earth's crust as a neutrino target. In this scheme, high-energy neutrinos from radio sources are converted by interactions with terrestrial electrons into high-energy muons which are ultimately detected. The basic resonant reaction is then

$$\bar{\nu}_e + e^- \rightarrow (W^-) \rightarrow \bar{\nu}_\mu + \mu^-. \qquad (7)$$

The angle which the muons from reaction (7) make with respect to the direction of the incident antineutrinos can be shown (Bahcall and Frautschi 1964a, b) by conservation of energy and momentum, to be $\lesssim (M_{W^-}c^2/q_{lab})$, where M_{W^-} is the mass of the W^--meson and q_{lab} is the laboratory energy at resonance. Since this angle is small ($\lesssim 10^{-3}$ radians), one can operate a neutrino telescope with the high-energy muons from reaction (7) which has excellent angular resolution ($\lesssim 10^{-6}$ steradians).

The counting rates (~ 1 to 10 high-energy muons per square meter per year at a depth of 1 km) predicted from a number of radio sources on the basis of the above considerations are detectable using equipment originally designed to observe neutrinos produced by cosmic-ray secondaries.

Since basic uncertainties exist in our knowledge of strong radio sources, attempts to observe neutrinos from these objects can furnish valuable information. Even an unsuccessful attempt to detect neutrinos would provide valuable observational evidence (provided M_{W^-} is $\lesssim$ 10 BeV) that process (6) is not currently occurring in, for example, the quasars. At least two experimental groups will be in a position to test our predictions in 1965 (Keuffel and Bergeson 1964; Reines 1964).

IX. Neutrinos and Cosmology

Neutrinos play a major role in a number of cosmological speculations (Pontecorvo and Smorodinskii 1962; Weinberg 1962a, b; Pontecorvo 1963; Marx 1963). The essential fact with which all these speculations begin is that neutrinos interact so weakly with matter that a large but as yet undiscovered background of cosmic neutrinos could exist without contradicting any known facts. The mean free path of neutrinos with energies of the order of a few MeV to scattering or absorption by matter at cosmological densities is about 10^{48} cm, i.e., the mean free path is about 10^{20} times the "radius of the universe." Because neutrinos interact so rarely with matter, the present upper limits on the possible cosmological energy density in the form of neutrinos and antineutrinos exceed by large amounts the estimates, based upon astronomical observations, of the average energy density of matter in the form of electrons and nucleons. This fact has led a number of people to speculate that the difference between the observed energy density of matter and the cosmological energy density is present in the form of neutrinos.

Weinberg (1962a, b) has pointed out that neutrinos are subect to an Olbers paradox that cannot be avoided by invoking the cosmological redshift. This paradox exists because very low-energy neutrinos (unlike very low-energy photons) can induce reactions, such as $\nu + n \rightarrow p + e^-$. The observed lifetime of the neutron then allows one to set an (admittedly very high) upper limit on the cosmic neutrino flux. The important point, emphasized by Weinberg, is that the neutrino flux is not infinite; thus one can deduce from the finite lifetime of the neutron (or any other β^--emitter) that the neutrino luminosity does not remain constant when we look at very great distances. All acceptable cosmologies must be consistent with this simple fact.

Pontecorvo and Smorodinskii (1962) have suggested that a large cosmic-neutrino background actually exists and that this background consists of neutrinos and antineutrinos in equal numbers. Such a background of neutrinos and antineutrinos would make our universe symmetric with respect to matter and antimatter except for "small fluctuations" like the matter observed in our galaxy. Note in this connection that the reaction $Cl^{37}(\nu, e^-)Ar^{37}$ can occur only with neutrinos, but neutrino-electron scattering can occur with both neutrinos and antineutrinos. Thus if a detectably large background of neutrinos (or antineutrinos) exists, one can determine the ratio of matter (neutrinos) to antimatter (antineutrinos) in the cosmic signal. This kind of observation, which distinguishes matter from antimatter at astronomical distances, cannot be made with electromagnetic radiation because the light from antiatoms is identical with the light from ordinary atoms.

REFERENCES

L. W. Alvarez, University of California Radiation Lab. Rep. 328 (unpublished).

J. N. Bahcall, 1964a, Phys. Rev. Letters, 12, 300.

____ 1964b, Phys. Rev., 135, B137.

____ 1964c, ibid., 136, B1164.

____ 1964d, Phys. Letters, 13, 332.

J. N. Bahcall, W. A. Fowler, I. Iben, Jr., and R. L. Sears, 1963, Ap. J., 137, 344.

J. N. Bahcall, and C. A. Barnes, 1964, Phys. Letters, 12, 48.

J. N. Bahcall, and S. C. Frautschi, 1964a, Phys. Rev., 135, B788.

____ 1964b, ibid., 136, B1547.

Bernadini, Bienlein, von Dardel, Faissner, Ferrero, Gaillard, Gerber, Hahn, Kaftanov, Krienen, Manfredotti, Reinharz, and Salmon 1964, Physics Letters,

H. A. Bethe, 1939, Phys. Rev., 55, 434.

H. A. Bethe, and C. L. Critchfield, 1938, Phys. Rev., 54, 248.

G. R. Burbidge, 1958, Ap. J., 127, 48.

____ 1960, Nature, 190, 1053.

E. M. Burbidge, G. R. Burbidge, W. A. Fowler, and F. Hoyle, 1957, Rev. Mod. Phys., 29, 547.

A. G. W. Cameron, 1958, Ann. Rev. Nucl. Sci., 8, 299.

____ 1962, Nature, 194, 963.

H-Y Chiu, 1964, Proc. of the Internat. Conf. on Cosmic Rays, Jaipur, India (in press).

R. Davis, Jr., 1964, Phys. Rev. Letters, 12, 302.

R. P. Feynman, and M. Gell-Mann, 1958, Phys. Rev., 109, 193.

G. B. Field, 1964, Nature, 202, 786.

____ 1965, Ap. J., (in press).

W. A. Fowler, 1958, Ap. J., 127, 551.

____ 1960, Mém. Soc. R. Sci. Liège, Ser. 5, 3, 207.

S. C. Frautschi, and J. N. Bahcall, 1965.

S. L. Glashow, 1960, Phys. Rev., 118, 316.

K. Greisen, 1960, Ann. Rev. Nucl. Sci., 10, 63.

J. C. Hardy, and R. I. Verrall, 1964, Phys. Rev. Letters, 13, 764.

R. W. Kavanagh, 1960, Nucl. Phys., 15, 411.

R. W. Kavanagh, and D. Goosman, 1964, Phys. Letters, 12, 229.

J. W. Keuffel, and H. E. Bergeson, 1964, private communication.

T. D. Lee, H. Robinson, M. Schwartz, and R. Cool, 1963, Phys. Rev., 132, 1297.

G. Marx, 1963, Nuovo Cimento, 30, 1555.

G. Marx, and Menyhård, 1960, Science, 131, 299.

M. G. K. Mennon, P. V. Ramana Murthy, B. V. Sreekantan, and S. Miyake, 1963, Phys. Letters, 5, 272.

S. Miyake, V. S. Narasimhan, and P. V. Ramana Murthy, 1962, Proc. Phys. Soc. (Japan) 3, 318.

C. R. O'Dell, 1963, Ap. J., 136, 809.

P. D. Parker, J. N. Bahcall, and W. A. Fowler, 1964, Ap. J., 139, 602.

P. Pochoda, and H. Reeves, 1964, Planet. and Space Sci., 12, 119.

B. Pontecorvo, 1964, Nat. Res. Council of Canada Rep. No. P.D.205 (unpublished), reissued by the U.S. Atomic Energy Commission as document 200-18787.

____ 1963, Soviet Phys. Uspekhi, 79, 1.

B. Pontecorvo, and Ya. Smorodinskii, 1962, Soviet Phys.—JETP, 14, 173.

O. L. Reeder, A. M. Poskanzer, and R. A. Esterlund, 1964, Phys. Rev. Letters, 13, 767.

F. Reines, 1964, private communication.

F. Reines, and W. R. Kropp, 1964, Phys. Rev. Letters, 12, 457.

F. Reines, and R. M. Woods, Jr., 1965 Phys. Rev. Letters, 14, 20.

E. E. Salpeter, 1952, Phys. Rev., 88, 547.

R. L. Sears, 1964, Ap. J., 140, 477.

I. S. Shklovsky, 1963, Soviet Astr. 6, 465.

A. W. Sunyar, and M. Goldhaber, 1960, Phys. Rev., 120, 871.

T. A. Tombrello, 1965 (to be published).

S. Weinberg, 1962a, Nuovo Cimento, 25, 15.

____ 1962b, Phys. Rev., 128, 1457.

G. T. Zatsepin, and V. A. Kuz'min, 1962, Soviet Phys.—JETP, 14, 1294.

COMMENT

H. Y. Chiu
Institute for Space Studies (NASA)

I think that neutrino astronomy really belongs to a civilization more advanced than our own. One might think along the following approach to neutrino astronomy: In order to decrease the background from the Sun one may think of going to the planet Pluto to perform Davis' experiment there and scale up his instrument by another factor of 1,000. One may also improve the sensitivity by letting all 1,000 tons of Cl^{37} be in a completely ionized state, the lifetime of Ar^{37} produced can then be increased a great deal. One can then talk about an integration time of the order of a few hundred years. Both these are necessary improvements, and as far as I can see they can only be carried out by a civilization much superior to our own, perhaps the Type II civilization recently speculated by Dr. Kardashev of the USSR. According to Kardashev, the degree of civilization can be measured by the energy used; a Type I civilization uses energy of the order of magnitude of 10^{20} ergs/sec, a Type II 10^{34} ergs/sec, a Type III 10^{44} ergs/sec. It is necessary for the Type II civilization to utilize a considerable portion of energy from the stars, and it is necessary for the Type III civilization to utilize the energy output of a galaxy.

Reprinted from Physics Today
Vol 19. No. 7. 111-112, 1966

Celestial neutrals

Neutral cosmic rays from a source on the celestial sphere have been reported in a recent issue of Physics Letters by Clyde Cowan and his collaborators at Catholic University. Neutral particles entering their detector produce muons that subsequently decayed into electrons. They found that the number of these neutral events, plotted as a function of sidereal time, peaked when the Milky Way passed over the detector. (As Cowan remarks, "No beings live on sidereal time, except maybe the angels.") Cowan believes that the detector is either seeing neutrinos acting in a surprising way or else a new variety of neutral particle.

Cowan's search for cosmic neutrinos began shortly after he and Frederick Reines made the first observations of manmade antineutrinos coming from the Savannah River reactor. That landmark experiment required many tons of detectors and long months of counting. Subsequent neutrino experiments have, in general, been on a similar scale. Groups looking for neutrinos from within the earth's atmosphere built gargantuan arrays of counters and placed them deep underground. Reines, last August, reported the first observation of cosmic-ray neutrinos. His detectors, located deep in a mine, recorded muons produced in surrounding rock by cosmic-ray neutrinos.

But Cowan's philosophy in searching for cosmic neutrinos coming from outside the earth's atmosphere was to start with a small detector and examine all background sources, then to increase detector size and repeat his experiment. In 1960 his group built its first detector in the subbasement of the physics building. A few months later the group found the first of many neutral events that produced a muon followed by a decay electron.

Since the number of events found was much greater than the expected cosmic-ray flux, the detector was moved to a cave near Washington, where it would be somewhat shielded. However, the anticipated attenuation did not occur.

Then a fortunate thing happened. The camera being used to record events needed replacement, and the new one happened to have a clock attached to it. Now that the time of an event was known the group soon noticed that the muon signal was not at all random; many more events were recorded when the Milky Way was directly over the cave.

Going back to their old laboratory, the group built a new detector. The target was a block of plastic scintillator measuring $4 \times 2 \times 1$ ft, surrounded on four sides by a plastic-scintillator anti-coincidence shield. On each of the other two ends were six photomultiplier tubes.

To detect an entering neutral particle a muon was required to decay in the detector; this limited muon energy to at most 200 MeV. An event satisfying the anticoincidence requirement produced a pulse from the muon stopping in the target. Then the decay electron produced a second pulse in delayed coincidence. Muon and electron pulses were photographed on an oscilloscope, along with Eastern Standard time.

The signal received in a given time interval, plotted as a function of Eastern Standard time, showed no significant peaks. But when the abscissa was changed to sidereal time, a marked peak always occurred at the same time in the sidereal day.

After a year of data taking, the Catholic U. physicists reported their results from this and a second, completely independent experiment (in a postdeadline paper at the APS Washington meeting and in two Physics Letters papers, 1 June 1966, by V. Acosta, G. L. Buckwalter, W. M. Carey, C. L. Cowan, D. J. Curtin, W. D. F. Ryan). Plotting the signal obtained in 1 1/2-hour time bins as a function of sidereal time, the group found (from the year-long experiment) a peak 5.7 standard deviations higher than the rest of the sidereal day. The peak always occurred when the same region of the celestial sphere passed over the detector. Since the detector was in a subbasement, it was effectively shielded by building and earth except for a "window" tilted at an angle of about 40 deg with respect to the zenith.

Using independent electronics and a liquid detector, a second experiment was done in an adjoining room; the detector looked at the same part of the sky as those in the first experiment. After two months running, the signal obtained in two-hour time bins showed a peak 6.1 standard deviations above the mean, but the peak was shifted about 15 deg in zenith angle and was somewhat broader than in the first experiment.

"What kind of neutral particle could be coming from the stars?" Cowan asks. The primary he is observing must be neutral, he says, or it would probably be smeared by magnetic fields. Neutrinos and gammas are the only known stable chargeless particles, and other kinds of neutrals would decay before reaching earth. Cowan does not think he is picking up gammas, since they would cause electron showers, and little if any correlation between star passage and air showers has been found by cosmic-ray physicists.

Cowan's group is doing several new experiments to spot other possible particles being produced and to identify the source of neutral flux. Using a liquid detector they will try to see whether the entering neutral makes a muon right away or first makes a pion that then decays to a muon. Then by adding a gadolinium salt to the liquid they will look for neutrons that might be associated with muon production. They are also looking in greater detail at the muon energy spectra.

A spark-chamber "telescope," 1.3 meters square, recently started running on the ground floor of a campus nuns' dormitory. Besides seeing everything the scintillation counter does, the "telescope" shows a spark track of the muon and any other charged particles. Two-dimensional photographs are being taken and Cowan's group is looking for correlations between the direction of muon tracks and points on the celestial sphere.

Three other experiments are under way: (1)

Neutrino Detector at Catholic University. Clyde Cowan is at left. The detector is triggered when a muon born in the stack goes downward (implying a down-going neutrino) through at least three gaps, and its electron product travels upward through at least three gaps. Track here is cosmic-ray muon.

A long water-filled slit with a lab underneath the water, a tube poked up through the water, and a scintillation detector below will act as a transit telescope. (2) Another transit telescope will start running this month, deep in a former Atlas missile site near Plattsburgh, N. Y., to obtain data at a different latitude. (3) A long thin liquid scintillation counter will be built to search for correlations between sidereal time and muons that lie in the detector plane.

REMARKS ON STATISTICS OF PARTICLES AT HIGH TEMPERATURES

G. Wataghin*
Institute of Physics, University of Turin

I report briefly the results of three papers: (1) by G. Aschieri, M. T. Reineri, and myself on Urca processes; (2) by A. Wataghin on equilibrium problems, including neutrinos at high temperatures and densities; (3) by myself on statistics at high temperatures and on the limits of applicability of the independent particles formalism. (These papers have been submitted to Nuovo Cimento for publication.)

I begin with a remark concerning the equilibrium problem at high temperatures and some aspects of the Boltzmann's method which become manifest when creation of all kinds of particles and antiparticles sets in. As was shown by one of us (A. W.; see below), also weakly interacting particles can take part in the equilibrium in conditions compatible with the limitations imposed by Schwarzschild singularity.

Let us recall some well-known formulae (G. Wataghin, 1934, 1944, 1949). Introducing the notations: n_{e^-s}, n_{e^+s}, $n_{ph,s}$, and n_{p^+s} for "occupation numbers" (numbers per cm^{-3}) of negative and positive electrons, photons, protons, etc., in the state s (where index s includes the indication of 3-momentum p_s, the spin component, isotopic spin, strangeness, etc.), one is looking for the maximum of

$$f(n_{e^-s}, n_{e^+s} \dots n_{p^+s}, n_{\nu_{es}}, \dots):$$

$$f(n_{e^-s} \dots) = \log W - \alpha' n - \beta E + \gamma' B - \\ - \delta'_e L_e - \delta'_\mu L_\mu = \max., \quad (1)$$

satisfying the subsidiary conditions (conservation laws):

$$\sum_s [(n_{e^-s} - n_{e^+s}) - (n_{p^+s} - n_{p^-s}) + (n_{\Pi^-s} - n_{\Pi^+s}) + \dots] \\ = - n = - \frac{q}{c} \text{ (charge conservation)},$$

$$\sum_s [E_{ph,s} + E_{e^-s} + E_{e^+s} + E_{p^+s} + \dots + E_{interact.} \\ + E_{gravit.} = E \text{ (energy conservation)}, \quad (2)$$

$$\sum_s [n_{Bs} - n_{\bar{B}s}] = B \text{ (baryon-number conservation)},$$

$$\sum_s (n_{e^+s} - n_{e^-s}) + \sum (n_{\bar{\nu}_e s} - n_{\nu_e s}) = \\ = L_e \text{ (1-lepton number conservation)}$$

$$\sum_s (n_{\mu^+s} - n_{\mu^-s}) + \sum (n_{\bar{\nu}_\mu s} - \nu_{\mu s}) = \\ = L_\mu \text{ (μ-lepton number conserv.)}$$

where, in the independent-particles model, one assumes that the interaction energy of particles is negligible (see the discussion in G. Wataghin 1934) and where the gravitational energy can be taken into account in the Newtonian approximation, introducing (with Ehrenfest and Tolman):

$$T^* = T g_{44}^{1/2} \sim T(1 + 2\varphi/c^2)^{1/2}$$

or $T^* = T(1 + \varphi/c^2)^{-1}$, φ being the Newtonian potential ($\varphi < 0$).

Let us write the solutions:

*C. V. Achar, M. G. K. Menon, V. S. Narasimham, P. V. Ramana Murthy and B. V. Sreekantan; Tata Institute of Fundamental Research, Bombay, India. K. Hinotani and S. Miyake; Osaka City University, Osaka, Japan, and D. R. Creed, J. L. Osborne, J. B. M. Pattison and A. W. Wolfendale, University of Durham, Durham, U. K.

$$n_{e^{\mp}} = A(T^*) \int_{x_e}^{\infty} \frac{(x^2 - x_e^2)^{1/2} x\, dx}{e^{\mp\alpha + x} + 1},$$

$$n_{p^{\pm}} = A(T^*) \int_{x_{p^+}}^{\infty} \frac{(x^2 - x_\beta^2)^{1/2} x\, dx}{e^{\mp\gamma + x} + 1},$$

$$n_{n,\bar{n}} = A(T^*) \int_{x_n}^{\infty} \frac{(x^2 - x_n^2)^{1/2} x\, dx}{e^{\mp\gamma_n^+ x} + 1},$$

$$n_{\nu_e', \bar{\nu}_e} = A(T^*) \int_0^{\infty} \frac{x^2\, dx}{e^{\mp\delta_e + x} + 1},$$

$$n_{\nu_\mu', \bar{\nu}_\mu} = A(T^*) \int_0^{\infty} \frac{x^2 dx}{e^{\mp\delta_\mu^+ + x} + 1}, \qquad (3)$$

$$n_{\pi^\pm} = A(T^*) \tfrac{1}{2} \int_{x_{\pi^\pm}}^{\infty} \frac{(x^2 - x_{\pi^\pm}^2)^{1/2} x\, dx}{e^{\mp\sigma + x} - 1},$$

$$n_{\pi^0} = A(T^*) \tfrac{1}{2} \int_{x_{\pi^0}}^{\infty} \frac{(x^2 - x_{\pi^0}^2)^{1/2} x\, dx}{e^{x} - 1},$$

$$n_{ph} = A(T^*) \int_0^{\infty} \frac{x^2\, dx}{e^{x} - 1},$$

$$A(T^*) = \frac{1}{\pi^2} \left(\frac{kT^*}{\hbar c}\right)^3, \qquad x = \frac{E}{kT^*} = \beta\left(1 + \frac{\varphi}{c^2}\right) E,$$

$$x_e = \frac{M_e c^2}{kT^*}, \qquad x_{\pi^\pm} = \frac{M_{\pi^\pm} c^2}{kT^*}, \text{ etc.,}$$

where the "statistical parameters" α, γ, δ_e, σ, . . . are simple linear function of the Lagrange multipliers α', γ', δ_e', δ'_μ, . . . :

$$\alpha = \alpha' - \delta_e', \quad \gamma_p = \gamma' - \alpha',$$

$$\sigma = \alpha', \quad \gamma_n = \gamma'^{+} \qquad (4)$$

$$\delta_\mu = \delta_\mu'^{-}, \quad \delta_e = \delta_e'.$$

It is worthwhile to point out that, after having selected the constants N, E, B, n, etc., the values of the statistical parameters α, γ_p $\delta_{e'}$, σ, . . . , are determined by the subsidiary conditions (2), which express fundamental laws of conservation and are obviously independent of the temperature. The physical significance of these statistical parameters is different from the significance and value of the usual "degeneracy parameters." Since the values of the constants n, B, L_e, L, N, etc., are not arbitrary but are fixed by the nature of the physical problem one is concerned with (e.g., n = 0, $B = n_{p^+} + n_n$ is the nucleon number per cm^{-3}) at low temperatures, $L_e = \Sigma_s n_{p^+s}$, and $L_\mu = 0$ at temperatures below the threshold of intense Urca processes and $\pm\mu$-pair production, the statistical parameters α, γ, δ_e', σ are implicit functions of each other and of the temperature T (or of $T^* = T\,[1 + \varphi/c^2]^{-1}$). An example of such dependence was given in an earlier paper (G. Wataghin 1944, p. 150, eq. [7']) where only protons, pairs of electrons, and photons are contemplated. From the conservation of charge and the assumption q = 0 follows:

$$\sum_s (n_{e^-s} - n_{e^+s}) - \sum_s n_{p^+s} = 0,$$

and one obtains below the threshold of Urca processes and $\pi^\pm$ and $\mu^\pm$ production:

$$\frac{1}{\pi^2} \sinh \alpha x_e^{-3} \int_{x_e}^{\infty} \frac{(x^2 - x_e^2)^{1/2} x\, dx}{\cosh x + \cosh \alpha} = n_{p^+} \Lambda_e^3 \qquad (5)$$

where $\Lambda_e = \hbar/m_e c$. In a similar way the statistical parameters of bosons (pions and K, η, φ, . . . , mesons* are related to the parameters of electrons, protons, neutrinos, etc. One obtains a set of equations for these parameters which can be solved only by approximation. A detailed study of these implicit functions will be published in a forthcoming paper. It is noteworthy that, if the constants N, B, and L_e are selected at low temperature and only E or T is varied, the whole problem can be solved for any T.

Let us now report the results of the first of the above-mentioned papers. The $\pm$ e-capture processes with neutrinos escaping from the stars were first studied in the pioneer work of Gamow and Schoenberg (1941). They will be re-examined in the paper of Aschieri, Reineri, and myself, taking into account the electron-pair production (G. Wataghin 1934) at high temperatures $2m_p c^2 > kT^* > 2\, m_e c^2$, where m_e is the electron mass and m_p is the proton mass. Gamow

*These parameters must satisfy an obvious limitation: the denominators in equation (3) should not become negative: e.g., for pions $[-\sigma + (m_\pi \pm c^2/kT^*)] > 0$.

and Schoenberg considered only the capture of atomic electrons, obtaining results valid in all cases in which the electron-pair production is negligible.

The distinction of three classes of interactions--strong, electromagnetic, and weak—leads us to calculate the energy losses in processes of generation of ν_e, $\bar{\nu}_e$ (electron-neutrinos and antineutrinos) making the assumption that, in a stellar body of not too high temperature, density, and extension, the weak interactions are not in a state of equilibrium, and the neutrinos escape freely from the region of the star under consideration. Starting from the formalism of pair production suggested in an earlier paper (G. Wataghin 1934) and from the general line of deduction of Gamow and Schoenberg, one finds easily the following formulae for energy loss in $\mathrm{erg/cm^3}$ in electron-capture processes:

$$W^- = \frac{2 \log_e 2}{tf} m_e c^2 x_e^{-6} n_{ZA} \int_{x_0\left(1+\frac{Q}{m_e C^2}\right)}^{\infty} \left[x - x_0\left(\frac{Q}{m_e c^2} + 1\right)\right] \frac{({}^3x^2 - x_e^2)^{1/2}\, x\, dx}{e^{-\alpha+x} + 1} \qquad (6)$$

and in positron-capture processes

$$W^+ = \frac{2 \log_e 2}{tf} m_e c^2 x_e^{-6} n_{ZA} \int_{x_e}^{\infty} x^2 + x_e \frac{Q}{m_e c^2} - 1 \; \frac{({}^3x^2 - x_e^2)^{1/2}\, x\, dx}{e^{\alpha+x} + 1} \qquad (7)$$

where $x_e = m_e c^2/kT$; tf has the usual meaning of the comparative half-life of Fermi theory; Q is the maximum kinetic energy of electron in Fermi theory; α is the statistical parameter of electrons and positrons[2] and n_{ZA} is the number per $\mathrm{cm^{-3}}$ of nuclei of charge Z and mass number A.

These formulae differ from the original ones of Gamow and Schoenberg essentially because in equations (6) and (7) the statistical laws for antiparticle creation are taken into account. Recently a similar remark was made by Pinaev (1963), who introduced assumptions different from those of the present work: he assumed that positrons and electrons were present in equal number (this hypothesis is not in accord with the charge conservation and the assumption we made that the total charge of the stellar body is vanishing). Thus he put $\alpha = 0$, whereas in our calculations α is a function of the nuclear density (see G. Wataghin 1944, eq. [7']) and is >1 in some cases quoted by Pinaev. Pinaev also introduces the assumption of equilibrium concentration of nuclei, a condition which is generally not verified in the case of neutrinos and antineutrinos escaping from the stellar body.

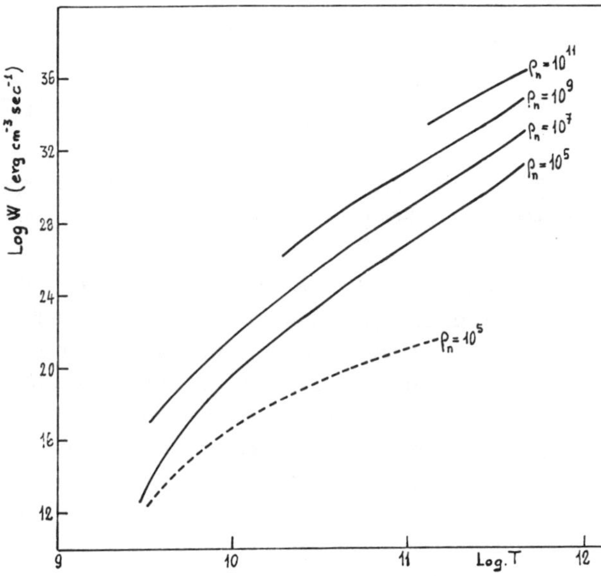

FIG. 1. Urca processes. Neutrino-energy losses by electron capture in the reaction $Fe^{56} + e^- \rightarrow Mn^{56} + \nu$. The dotted line represents (for the case $\rho_n = 10^5$) the energy losses neglecting electron pairs (Gamow-Schoenberg theory).

The results of some calculations made with formulae (6) and (7) are collected in two graphs (Figs. 1 and 2) and Tables 1 and 2. The values of energy losses are referred to the examples of nuclei of Fe_{56} and Mn_{56} (Q = 3, 7 MeV; tf ~ 3.200). The real situation one expects to encounter in the star's interior is complicated by the presence of several kinds of nuclei, many of them being in excited states. One can deduce from the tables that the e^+ capture processes of high intensity can significantly change the initial composition of isotopes at high temperatures. The Coulomb barrier for positrons does not greatly influence these results.

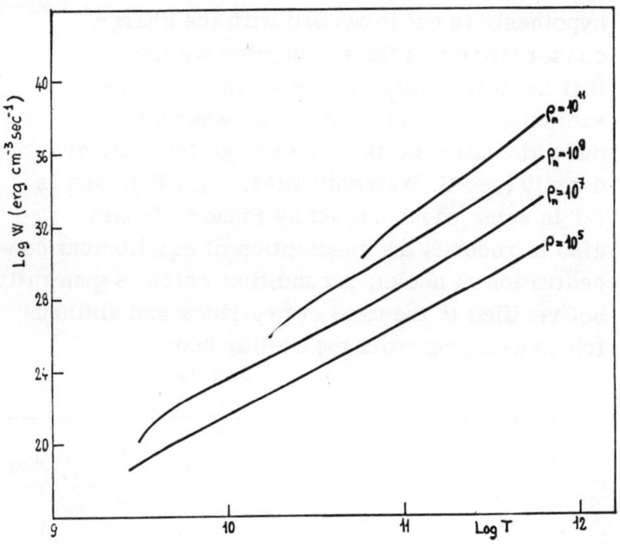

FIG. 2. Urca processes. Antineutrino-energy losses by positron capture in the reaction $Mn^{56} + e^{+} \rightarrow Fe^{56} + \nu$.

In the second paper A. Wataghin considers the equilibrium problem of weakly interacting particles and shows that such equilibrium can exist in small regions at high temperatures and densities. He calculates the energy density of e- and μ-neutrinos and antineutrinos and the corresponding fluxes, using only general statistical-thermodynamical relations and making no assumptions on the relative rates of various leptonic processes.

Following the general solution of othe problem (see eqs. [1]-[3]) and integrating some of equations (3), he finds the expression for the densities $W_{e-\nu}$, $W_{e-\bar{\nu}}$ and the analogous formulae of $W_{\mu-\nu}$, $W_{\mu-\bar{\nu}}$:

$$W_{\nu}(\delta_{e,\mu}) = 1.44 \times 10^{24}\ x_e^{-4}\ Z(\delta_{e,\mu})\ (\text{ergs cm}^{-3}) \quad (7)$$

$$W_{\bar{\nu}}(\delta_{e,\mu}) = 1.44 \times 10^{24}\ x_e^{-4}\ Z(-\delta_{e,\mu}),$$

where

$$Z(\delta) = \int_0^{\infty} \frac{x^3\, dx}{e^{-\delta + x} + 1}$$

and where $\delta_{\mu, e}$ is determined by the subsidiary conditions (2). Below the threshold of Urca processes, $\delta_e = 0$, but in more general conditions $\delta_e \neq 0$. The μ-neutrino production sets in at higher temperatures and the formulae for μ-neutrino densities and fluxes are identical to equation (7), with the difference that the value of δ_{μ} is vixed by other subsidiary conditions. In Tables 3 and 4 are given the energy densities and in Tables 5 and 6 the fluxes F (δ) are calculated by the usual formula: $F_{\nu}(\delta) = W_{\nu}(\delta)\,(c/4)$ ergs cm^{-2} sec^{-1}.

The reported values of $F_{\nu}(\delta)$, $F_{\bar{\nu}}(\delta)$ for the two kinds of neutrinos (e - ν and μ - ν) can be helpful in the calculations of fluxes to be expected if sources of neutrino emission coming from an equilibrium region of a celestial body are to be detected.

A. Wataghin cites two examples. Let us consider a central region of a stellar body with an average temperature $T \sim 5.93 \times 10^{11}$ °R ($x_e = m_e c^2/kT = 10^{-2}$), and let us assume that the resulting number per cm^{-3} of electron pairs $n_+ = n_- \sim 3 \times 10^{36}$ is much larger than the number density of nucleons ($n_{p+} + n_n$) or nuclei. Accepting the known formulae for electron-positron annihilation process and for scattering of e-neutrinos one finds the average cross-section $\sigma_{e-\nu}$ for e-neutrinos, $\sigma_{e-\nu} \sim 10^{-40}$ cm^2, and the mean free path of the neutrinos, $l_{e-\nu} \sim 10^4$ cm. Obviously the equilibrium can exist in a region of linear dimensions $10\, l_{e-\nu} = 10^5$cm, and the resulting mass is compatible with the Schwarzschild limitations. Another example is valid for both e-neutrinos and μ-neutrinos. The region contains nucleons, and the number density of nucleons is $\sim 10^{36}$ per cm^{-3} (the nucleon density is $\sim 10^{12}$ gm cm^{-3}). Experimental values of μ-neutrino and e-neutrino cross-sections are known, and for μ-neutrinos having average energy ~ 0.15 BeV (at the temperature $T \sim 6 \times 10^{11}$ °K) the cross-section is $\sigma_{\mu-\nu} \sim 10^{-40}$ cm^2, thus the mean free path is also $l_{\mu-\nu} \sim 10^4$ cm.

Finally in the third paper we discuss the existence of a fundamental limitation of the independent-particles model arising from the assumption of a non-local interaction between elementary particles. If the non-locality is a basic property of all interactions related to the value of a fundamental length l and a fundamental four-dimensional domain of interaction ($\sim l^4$), and if, e.g., one accepts the formulation of non-locality suggested by the author in a recent paper, the upper limit of the number of particles per cm^{-3} is reached in conditions in which the average spatial distance between particles is

Table I. Neutrino energy losses (by electron capture)

calculated for the reaction $Fe^{56} + e^- \rightarrow Mn^{56} + \nu$ (E_ν in ergs cm^{-3} sec^{-1}; $\rho = \rho_{Fe}$ in g^m cm^{-3})

T	2.965×10^9	3.8×10^9	5.93×10^9	1.186×10^{10}	5.93×10^{10}	1.186×10^{11}	5.93×10^{11}	5.93×10^{12}
α \ X_0	2	1.5	1	0.5	10^{-1}	5×10^{-2}	10^{-2}	10^{-3}
2	3.02×10^{14}	7.51×10^{16}	2.92×10^{20}	4.50×10^{24}	9.07×10^{31}	5.97×10^{34}	1.41×10^{41}	
	$\rho = 4.15 \times 10^6$	$\rho = 1.01 \times 10^7$	$\rho = 4.55 \times 10^7$	$\rho = 3.95 \times 10^8$	$\rho = 5.07 \times 10^{10}$	$\rho = 4.06 \times 10^{11}$	$\rho = 5.07 \times 10^{13}$	
1	3.99×10^{13}	1.02×10^{16}	4.11×10^{19}	6.46×10^{23}	1.35×10^{31}	8.97×10^{33}	2.13×10^{40}	
	$\rho = 1.49 \times 10^6$	$\rho = 3.72 \times 10^6$	$\rho = 1.74 \times 10^7$	$\rho = 1.53 \times 10^8$	$\rho = 1.98 \times 10^{10}$	$\rho = 1.59 \times 10^{11}$	$\rho = 1.99 \times 10^{13}$	
0, 5	1.09×10^{13}	2.82×10^{15}	1.14×10^{19}	1.82×10^{23}	3.85×10^{30}	2.53×10^{33}	6.06×10^{39}	
	$\rho = 6.75 \times 10^5$	$\rho = 1.70 \times 10^6$	$\rho = 8.01 \times 10^6$	$\rho = 7.13 \times 10^7$	$\rho = 9.23 \times 10^9$	$\rho = 7.39 \times 10^{10}$	$\rho = 9.24 \times 10^{12}$	
10^{-1}	1.42×10^{12}	3.67×10^{14}	1.49×10^{18}	2.39×10^{22}	5.06×10^{29}	3.33×10^{32}	7.97×10^{38}	
	$\rho = 1.30 \times 10^5$	$\rho = 3.31 \times 10^5$	$\rho = 1.55 \times 10^6$	$\rho = 1.39 \times 10^7$	$\rho = 1.80 \times 10^9$	$\rho = 1.44 \times 10^{10}$	$\rho = 1.80 \times 10^{12}$	
10^{-2}	1.29×10^{11}	3.34×10^{13}	1.36×10^{17}	2.17×10^{21}	4.60×10^{28}	3.03×10^{31}	7.65×10^{37}	
	$\rho = 1.80 \times 10^4$	$\rho = 3.28 \times 10^4$	$\rho = 1.55 \times 10^5$	$\rho = 1.38 \times 10^6$	$\rho = 1.79 \times 10^8$	$\rho = 1.43 \times 10^9$	$\rho = 1.89 \times 10^{11}$	
$<10^{-2}$	$9.84 \times 10^6 \times \rho$	$1.01 \times 10^9 \times \rho$	$8.69 \times 10^{11} \times \rho$	$1.55 \times 10^{15} \times \rho$	$2.54 \times 10^{20} \times \rho$	$2.09 \times 10^{22} \times \rho$	$4.01 \times 10^{26} \times \rho$	$4.19 \times 10^{32} \times \rho$
	$\rho < 30 \times 10^4$	$\rho < 3.28 \times 10^4$	$\rho < 1.55 \times 10^5$	$\rho < 1.38 \times 10^6$	$\rho < 1.79 \times 10^8$	$\rho < 1.43 \times 10^9$	$\rho < 1.89 \times 10^{11}$	$\rho < 1.78 \times 10^{14}$

Table II. Antineutrino energy losses in erg per cm^{-3} sec^{-1}

(by positron capture) calculated for the reaction $Mn^{56} + e^+ \to Fe^{56} + \bar{\nu}$

α \ X_0	2	1.569	1	0.5	10^{-1}	5×10^{-2}	10^{-2}	10^{-3}
2	1.49×10^{20}	1.08×10^{21}	3.56×10^{22}	6.94×10^{24}	4.21×10^{30}	1.80×10^{33}	3.05×10^{39}	
	$\rho = 4.15 \times 10^6$	$\rho = 1.01 \times 10^7$	$\rho = 4.55 \times 10^7$	$\rho = 3.95 \times 10^8$	$\rho = 5.07 \times 10^{10}$	$\rho = 4.06 \times 10^{11}$	$\rho = 5.07 \times 10^{13}$	
1	1.45×10^{20}	1.07×10^{21}	3.57×10^{22}	7.25×10^{24}	4.45×10^{30}	1.91×10^{33}	3.23×10^{39}	
	$\rho = 1.49 \times 10^6$	$\rho = 3.72 \times 10^6$	$\rho = 1.74 \times 10^7$	$\rho = 1.53 \times 10^8$	$\rho = 1.98 \times 10^{10}$	$\rho = 1.59 \times 10^{11}$	$\rho = 1.99 \times 10^{13}$	
0, 5	1.08×10^{20}	8.03×10^{20}	2.69×10^{22}	5.51×10^{24}	3.40×10^{30}	1.46×10^{33}	2.47×10^{39}	
	$\rho = 6.75 \times 10^5$	$\rho = 1.70 \times 10^6$	$\rho = 8.01 \times 10^6$	$\rho = 7.13 \times 10^7$	$\rho = 9.23 \times 10^9$	$\rho = 7.39 \times 10^{10}$	$\rho = 9.24 \times 10^{12}$	
10^{-1}	3.08×10^{19}	2.31×10^{20}	7.66×10^{21}	1.58×10^{24}	9.84×10^{29}	4.23×10^{32}	7.15×10^{38}	
	$\rho = 1.30 \times 10^5$	$\rho = 3.31 \times 10^5$	$\rho = 1.55 \times 10^6$	$\rho = 1.39 \times 10^7$	$\rho = 1.80 \times 10^9$	$\rho = 1.44 \times 10^{10}$	$\rho = 1.80 \times 10^{12}$	
10^{-2}	3.37×10^{18}	2.50×10^{19}	8.34×10^{20}	1.71×10^{23}	1.07×10^{29}	4.59×10^{31}	8.17×10^{37}	
	$\rho = 1.30 \times 10^4$	$\rho = 3.28 \times 10^4$	$\rho = 1.55 \times 10^5$	$\rho = 1.38 \times 10^6$	$\rho = 1.79 \times 10^8$	$\rho = 1.43 \times 10^9$	$\rho = 1.89 \times 10^{11}$	
$< 10^{-2}$	$2.60 \times 10^{14} \rho$	$7.68 \times 10^{14} \rho$	$5.43 \times 10^{15} \rho$	$1.25 \times 10^{17} \rho$	$6.02 \times 10^{20} \rho$	$3.23 \times 10^{22} \rho$	$4.38 \times 10^{26} \rho$	$4.28 \times 10^{32} \rho$
	$\rho < 1.30 \times 10^4$	$\rho < 3.28 \times 10^4$	$\rho < 1.55 \times 10^5$	$\rho < 1.38 \times 10^6$	$\rho < 1.79 \times 10^8$	$\rho < 1.43 \times 10^9$	$\rho < 1.89 \times 10^{11}$	$\rho < 1.78 \times 10^{14}$

Table III. Neutrino - Energy - Density

$W_{\delta X_0} = Z(\delta) \times 1.441 \times 10^{24} \times X_0^{-4}$ (erg. cm^{-3})

T	10^9	2.965×10^9	5.93×10^9	10^{10}	1.186×10^{10}	5.93×10^{10}	1.186×10^{11}	5.93×10^{11}	5.93×10^{12}
δ \ X_0	5.93	2	1	0.593	1/2	1/10	1/20	1/100	1/1000
- 10	2.95×10^{17}	2.28×10^{19}	3.65×10^{20}	2.95×10^{21}	5.85×10^{21}	3.65×10^{24}	5.85×10^{25}	3.65×10^{28}	3.65×10^{32}
- 4	1.19×10^{20}	9.20×10^{21}	1.47×10^{23}	1.19×10^{24}	2.36×10^{24}	1.47×10^{27}	2.36×10^{29}	1.47×10^{31}	1.47×10^{35}
- 1	2.34×10^{21}	1.81×10^{23}	2.90×10^{24}	2.34×10^{25}	4.64×10^{25}	2.90×10^{28}	4.64×10^{29}	2.90×10^{32}	2.90×10^{36}
$-\frac{1}{2}$	3.81×10^{21}	2.95×10^{23}	4.72×10^{24}	3.81×10^{25}	7.55×10^{25}	4.72×10^{28}	7.55×10^{29}	4.72×10^{32}	4.72×10^{36}
$-\frac{1}{10}$	5.60×10^{21}	4.33×10^{23}	6.93×10^{24}	5.60×10^{25}	1.11×10^{26}	6.93×10^{28}	1.11×10^{30}	6.93×10^{32}	6.93×10^{36}
$-\frac{1}{100}$	6.11×10^{21}	4.72×10^{23}	7.56×10^{24}	6.11×10^{25}	1.21×10^{26}	7.56×10^{28}	1.21×10^{30}	7.56×10^{32}	7.56×10^{36}

Table IV. Anti - neutrino - density

$W_\delta X_0$ (erg. cm^{-3})

T	10^9	2.965×10^9	5.93×10^9	10^{10}	1.186×10^{10}	5.93×10^{10}	10^{11}	1.186×10^{11}	5.93×10^{11}	5.93×10^{12}
δ \ X_0	5.93	2	1	0.593	1/2	1/10	0.0593	1/20	1/100	1/1000
10	3.47×10^{24}	2.68×10^{26}	4.29×10^{27}	3.47×10^{28}	6.86×10^{28}	4.29×10^{31}	3.47×10^{32}	6.86×10^{32}	4.29×10^{35}	4.29×10^{39}
4	1.78×10^{23}	1.37×10^{25}	2.20×10^{26}	1.78×10^{27}	3.52×10^{27}	2.20×10^{30}	1.78×10^{31}	3.52×10^{31}	2.20×10^{34}	2.20×10^{38}
1	1.66×10^{22}	1.28×10^{24}	2.05×10^{25}	1.66×10^{26}	3.28×10^{26}	2.05×10^{29}	1.66×10^{30}	3.28×10^{30}	2.05×10^{33}	2.05×10^{37}
$\frac{1}{2}$	1.05×10^{22}	8.10×10^{23}	1.30×10^{25}	1.05×10^{26}	2.07×10^{26}	1.30×10^{29}	1.05×10^{30}	2.07×10^{30}	1.30×10^{33}	1.30×10^{37}
$\frac{1}{10}$	7.20×10^{21}	5.56×10^{23}	8.90×10^{24}	7.20×10^{25}	1.42×10^{26}	8.90×10^{28}	7.20×10^{29}	1.42×10^{30}	8.90×10^{32}	8.90×10^{36}
$\frac{1}{20}$	6.87×10^{21}	5.30×10^{23}	8.49×10^{24}	6.87×10^{25}	1.36×10^{26}	8.49×10^{28}	6.87×10^{29}	1.36×10^{30}	8.49×10^{32}	8.49×10^{36}
0	6.62×10^{21}	5.12×10^{23}	8.19×10^{24}	6.62×10^{25}	1.31×10^{26}	8.19×10^{28}	6.67×10^{29}	1.31×10^{30}	8.19×10^{32}	8.19×10^{36}

Table V. Neutrino - Energy - flux

$F_\delta\ X_0 = Z(\delta) \times 1.441.10^{24} \times 0.07495 \times 10^{10} \times X_0^{-4}$ (erg. cm^{-2} sec^{-1})

T	10^9	2.965×10^9	5.93×10^9	10^{10}	1.186×10^{10}	5.93×10^{10}	1.186×10^{11}	5.93×10^{11}	5.93×10^{12}
δ \ X_0	5.93	2	1	0.593	1/2	1/10	1/20	1/100	1/1000
- 10	2.21×10^{27}	1.71×10^{29}	2.74×10^{30}	2.21×10^{31}	4.38×10^{31}	2.74×10^{34}	4.38×10^{35}	2.74×10^{38}	2.74×10^{42}
- 4	8.92×10^{29}	6.90×10^{31}	1.10×10^{33}	8.92×10^{33}	1.76×10^{34}	1.10×10^{37}	1.76×10^{39}	1.10×10^{41}	1.10×10^{45}
- 1	1.76×10^{31}	1.36×10^{33}	2.17×10^{34}	1.76×10^{35}	3.47×10^{35}	2.17×10^{38}	3.47×10^{39}	2.17×10^{42}	2.17×10^{46}
$-\frac{1}{2}$	2.86×10^{31}	2.21×10^{33}	3.53×10^{34}	2.86×10^{35}	5.66×10^{35}	3.53×10^{38}	5.66×10^{39}	3.53×10^{42}	3.53×10^{46}
$-\frac{1}{10}$	4.20×10^{31}	3.25×10^{33}	5.19×10^{34}	4.20×10^{35}	8.31×10^{35}	5.19×10^{38}	8.31×10^{39}	5.19×10^{42}	5.19×10^{46}
$-\frac{1}{100}$	4.96×10^{31}	3.84×10^{33}	6.14×10^{34}	4.96×10^{35}	9.82×10^{35}	6.14×10^{38}	9.82×10^{39}	6.14×10^{42}	6.14×10^{46}

Table VI. Anti - neutrino - density

$W_\delta\ X_0$ (erg. cm^{-3})

T	10^9	2.965×10^9	5.93×10^9	10^{10}	1.186×10^{10}	5.93×10^{10}	10^{11}	1.186×10^{11}	5.93×10^{11}	5.93×10^{12}
δ \ X_0	5.93	2	1	0.593	1/2	1/10	0.0593	1/20	1/100	1/1000
10	3.47×10^{24}	2.68×10^{26}	4.29×10^{27}	3.47×10^{28}	6.86×10^{28}	4.29×10^{31}	3.47×10^{32}	6.86×10^{32}	4.29×10^{35}	4.29×10^{39}
4	1.78×10^{23}	1.37×10^{25}	2.20×10^{26}	1.78×10^{27}	3.52×10^{27}	2.20×10^{30}	1.78×10^{31}	3.52×10^{31}	2.20×10^{34}	2.20×10^{38}
1	1.66×10^{22}	1.28×10^{24}	2.05×10^{25}	1.66×10^{26}	3.28×10^{26}	2.05×10^{29}	1.66×10^{30}	3.28×10^{30}	2.05×10^{33}	2.05×10^{37}
$\frac{1}{2}$	1.05×10^{22}	8.10×10^{23}	1.30×10^{25}	1.05×10^{26}	2.07×10^{26}	1.30×10^{29}	1.05×10^{30}	2.07×10^{30}	1.30×10^{33}	1.30×10^{37}
$\frac{1}{10}$	7.20×10^{21}	5.56×10^{23}	8.90×10^{24}	7.20×10^{25}	1.42×10^{26}	8.90×10^{28}	7.20×10^{29}	1.42×10^{30}	8.90×10^{32}	8.90×10^{36}
$\frac{1}{20}$	6.87×10^{21}	5.30×10^{23}	8.49×10^{24}	6.87×10^{25}	1.36×10^{26}	8.49×10^{28}	6.87×10^{29}	1.36×10^{30}	8.49×10^{32}	8.49×10^{36}
0	6.62×10^{21}	5.12×10^{23}	8.19×10^{24}	6.62×10^{25}	1.31×10^{26}	8.19×10^{28}	6.67×10^{29}	1.31×10^{30}	8.19×10^{32}	8.19×10^{36}

$< l$, (where $l \sim \hbar/2m_p c \sim 10^{-14}$ cm is the universal length), and the average mean free path of strongly interacting particles is of the same order of magnitude. The corresponding average 3-momenta result $> 2m_p c^2$ and the approximate upper limit of the temperature is $\sim 2.10^{13}$ °K. It seems noteworthy that in such conditions the Maxwell-Boltzmann asymptotic exponential decrease of probabilities with energy, exp (-E/kT), is no more valid and is substituted by a law derived from the form factors which gives rise to a decrease with energy of the type $E^2/(E^2 + a^2)^n$ (where $n \sim 2$; $a \sim 2m_p c^2$). At temperatures near the above limit the number densities (3) of all particles approach the value $\sim 2\,A(T^*)$.

REFERENCES

G. Gamow, and M. Schoenberg, 1941, Phys. Rev., 59, 539.

V. S. Pinaev, 1963, Soviet Phys.—J.E.T.P., 45, 548.

G. Wataghin, 1934, Phil. Mag., 17, 910.

____. 1944, Phys. Rev., 66, 149.

____. 1949, Nuovo Cimento, ser. 9, 6, 241.

OBSERVATIONS IN COSMOLOGY

J. KRISTIAN* AND R. K. SACHS
Relativity Center, The University of Texas, Austin, Texas
Received May 7, 1965

ABSTRACT

Observations of cosmological effects in anisotropic, inhomogeneous cosmological models are discussed in detail, with numerical estimates. The first and last sections of the paper form a self-contained unit for readers who are unfamiliar with Riemannian geometry. The other sections contain mathematical derivations.

Three assumptions are made: (i) that the universe is described by a Riemannian space time with slowly varying metric tensor; (ii) that light travels along null geodesics and obeys the usual area-intensity law; and (iii) that the gravitational field is related to the matter by Einstein's field equations for dust. The third assumption is not needed for many of the results; the second assumption is proved in the geometric optics limit, assuming a general relativistic model.

The importance of trying to observe angular variations in the various cosmological effects is emphasized. It is shown that otherwise unobservable anisotropies or inhomogeneities can easily give the observed order of magnitude and either sign for the acceleration parameter measured in any one direction via redshifts. A model-independent law for the apparent area of a distant object is given. Detailed equations for number counts and for apparent proper motions are given. It is pointed out that observations to date do not exclude the possible presence of anisotropies and inhomogeneities whose dynamical effects are comparable to the dynamical effects of the expansion.

A new effect, the "distortion effect," is discussed. In any anisotropic model, all distant objects in a particular direction on the celestial sphere may appear distorted, with a definite preferential direction for their longest dimension. The effect gives a direct measurement of space-time curvature and may be observable. But a positive result of the measurement would not favor general relativity over other theories in which assumptions (i) and (ii) above hold.

I. INTRODUCTION

In cosmology, the universe is usually assumed spatially isotropic or at least homogeneous, except for local irregularities whose distance scale is too small to have cosmological significance (Heckmann and Schücking 1959, 1962; McVittie 1956, 1959, 1962; for inhomogeneous models see Bondi 1947; Hoyle and Narlikar 1961). The observational basis of this assumption is dubious, however, in view of the scarcity of observations near the galactic plane and recent evidence pointing to possible superclustering of galaxies. Therefore, it seems reasonable to ask what observational consequences anisotropies or large-scale inhomogeneities would have. In this paper we attempt to answer this question systematically. We are mainly interested in inhomogeneities whose scale is something like 10^9 light-years or more, with smaller scale variations smoothed out as before, though we shall also briefly discuss the possibility that fluctuations whose scale is several orders of magnitude smaller might have measurable effects which would give information on the curvature of space-time. A detailed presentation of our results, with all technical details suppressed and with a Newtonian notation, is given in § III.2.

Inhomogeneities and anisotropies have not been much discussed in the literature because finding a global cosmological model which does not have a high degree of symmetry is very difficult. But a global model, with boundary conditions and everything else included, is far more than we need to discuss most of the observations that might be made in the near future (see, e.g., Sandage 1961). If we concentrate on those features of any model that have the most direct observational significance, then a very general treatment can be given using standard mathematical techniques.

We make three fundamental assumptions in this paper. In order of importance, these

* Present address: Washburn Observatory, University of Wisconsin, Madison, Wisconsin.

Reprinted from Astrophysical Journal,
143, 379-399.

are the following. (i) The geometry of space time is Riemannian, and all quantities of interest can be expanded in a power series around here-and-now as origin. This assumption amounts to saying that the distance we can see into the universe is small compared both with some reasonably defined radius of curvature and with the scale of the inhomogeneities which remain when local fluctuations are smoothed out. (ii) Light beams travel along null geodesics and obey the intensity versus area law discussed in § II.1. We shall prove this second assumption in the geometric optics limit for general-relativistic models; in other models, the assumption has usually been made *ad hoc.* (iii) The metric is governed by the Einstein field equations for dust. This third assumption is not used in deriving the general expressions of § II. But to get numerical estimates of certain effects we need some dynamical model, and we have chosen the relativistic dust model as an illustration because it is simple and currently popular. Except for these three assumptions our treatment is general; it is rigorous except for the approximations mentioned in them.

The reader interested in the conceptual framework or mathematical structure of general relativity will find nothing in this paper that is not at least implicit in the literature. We are merely concerned with applying the theory to a specific problem.

In § II, we use assumptions (i) and (ii), but not (iii), to derive expressions for five observational effects in general Riemannian space-times. These effects are given as power series, the coefficients of which describe the distribution of matter and the structure of space-time in our vicinity. In § III, we specialize these results to general-relativistic dust models, writing the equations for observable effects in three-plus-one dimensional notation, and we make some crude numerical estimates. The reader who has no knowledge of general relativity should go directly to § III.2, which is intended to be comprehensible to anyone who has read § I, knows Cartesian (i.e., Euclidean) tensors, and is willing to take the results on faith.

Three of the results discussed in § III.2 deserve special mention. (*a*) In all cosmological models (general relativistic or not) there is a universal relation

$$dA = r^2 d\Omega \,. \tag{1}$$

(R. Penrose [private communication] has since proved by a method independent of power series that this relation is exact.) Here dA is the intrinsic cross-sectional area of a distant object; r is a measured quantity, the "corrected luminosity distance," defined by equation (19); and $d\Omega$ is the measured solid angle subtended by the distant object. (*b*) Measurements of the redshift or other cosmological effects in only one direction should be interpreted with extreme caution. Only knowledge of how an effect varies over the celestial sphere gives really decisive information on the structure of the universe. (*c*) There is one effect, the "distortion effect," which deserves investigation. It turns out that, in general, we see an intrinsically spherical distant object as an ellipse on our plates. The effect, if it exists and can be measured, gives a simple, direct measurement of space-time curvature. Absorption corrections are not needed to detect the effect, and theories of galactic evolution are not needed to interpret it. It is similar to the gravitational lens effect, but is caused by the cosmological curvature rather than that of a single large body.

In § III the "incident electric-type gravitational field" $E_{\alpha\beta}$ plays an important role. We now give a simple Newtonian analogue to this quantity. Let $\phi(x,t)$ be the Newtonian gravitational potential. We set $\phi(0,0) = 0$ and expand $\phi(x,0)$ in a power series about the origin. The gradient $(\phi_{,\alpha})_0$ is zero in a freely falling frame.[1] Thus the first non-zero coefficients in the power series are the six quantities

$$(\phi_{,\alpha\beta})_0 \,. \tag{2}$$

[1] Here and throughout, Greek letters range and sum from 1 to 3, Latin letters from 1 to 4, commas denote ordinary derivatives, semicolons denote covariant derivatives, and a subscript "0" on a quantity means that the quantity is to be evaluated at $x = 0$, $t = 0$ after all indicated differentiations have been performed.

Now the trace, $\nabla^2\phi$, of this Cartesian tensor is directly determined by the local matter density ρ_0 via Poisson's equation. But the trace-free part,

$$E_{\alpha\beta} = (\phi_{,\alpha\beta})_0 - \tfrac{1}{3}\delta_{\alpha\beta}\nabla^2\phi_0 , \tag{3}$$

where $\delta_{\alpha\beta}$ is the Kronecker delta, is independent of ρ_0 and really arises from the over-all lopsidedness of the matter distribution or of the boundary conditions. For example, $E_{\alpha\beta} = 0$ if the origin is chosen at the center of a spherically symmetric mass distribution, but $E_{\alpha\beta} \neq 0$ if the origin is chosen at any off-center point. We call $E_{\alpha\beta}$ the incident electric-type gravitational field and $\nabla^2\phi_0$ the "locally determined" gravitational field.

The same kind of splitting can be carried out for the higher-order coefficients in the expansion of $\phi(x,0)$. We introduce a standard notation: round and square brackets around indices denote complete symmetrization and antisymmetrization, respectively; for example,

$$A_{(\alpha\beta\gamma)} = \tfrac{1}{6}(A_{\alpha\beta\gamma} + A_{\gamma\alpha\beta} + A_{\beta\gamma\alpha} + A_{\beta\alpha\gamma} + A_{\gamma\beta\alpha} + A_{\alpha\gamma\beta}) , \tag{4}$$

$$A_{[\alpha\beta\gamma]} = \tfrac{1}{6}(A_{\alpha\beta\gamma} + A_{\gamma\alpha\beta} + A_{\beta\gamma\alpha} - A_{\beta\alpha\gamma} - A_{\gamma\beta\alpha} - A_{\alpha\gamma\beta}) . \tag{5}$$

For the next coefficient of $\phi(x,0)$ we find that $\nabla^2\phi_{0,\alpha}$ is locally determined by $\rho_{0,\alpha}$. However, the quantity

$$(\phi_{,\alpha\beta\gamma})_0 - \tfrac{3}{5}\delta_{(\alpha\beta}\nabla^2\phi_{0,\gamma)} \tag{6}$$

again depends on the over-all situation.

In this paper we consider the relativistic analogues of the quantities in equations (3) and (6) as adjustable parameters. This attitude is quite parochial; after all $E_{\alpha\beta}$ is determined ultimately by the over-all matter distribution or the boundary conditions. Moreover, the number of adjustable parameters becomes uncomfortably large in our treatment. Finally, this type of purely local approach is not very appropriate in analyzing data on the distant past of our own Galaxy for cosmological significance (Hoyle 1962), or for discussing evolutionary or time-scale problems. Our approach also has some advantages. We work with measurable quantities rather than with boundary conditions that must at present be determined from philosophical or aesthetic arguments, and, of course, ignoring global properties is what makes the calculations manageable. Finally, the results are complete and quite general, and if any of our many adjustable parameters, which can all be measured in principle, can actually be observed, their values will be a very strong test for any proposed global cosmological model.

II. ASTRONOMICAL OBSERVATIONS IN A RIEMANNIAN SPACE-TIME

With the possible exception of very energetic cosmic rays, all of our information about the extragalactic universe comes to us as electromagnetic radiation; at present in the optical and radio regions of the spectrum, with the promise of a rapid future extension to other frequencies by observations from above the atmosphere. We must therefore begin by discussing the behavior and measurement of light in a Riemannian space-time. It might be that neutrino observations could give additional information relevant to our purposes here, but it is not obvious how. Also, we might aspire to observe the static Schwarzschild gravitational field of objects that we cannot see because they are beyond our visual horizon; such a field will in general contribute to what we have called the incident field. But it would certainly take too long to measure the incident field canonically, by measuring relative accelerations of neighboring objects, so that we must, again, look for indirect effects of the field on light beams. We will not consider here the other main body of cosmologically significant data, namely, data on the past world line of our own Galaxy (Hoyle 1962).

1. *Geometrical Optics*

We suppose that the electromagnetic tensor for the light emitted by a distant galaxy obeys Maxwell's equations for the vacuum:

$$F^{ab}{}_{;b} = 0\,, \qquad F_{[ab;c]} = 0\,. \tag{7}$$

The contribution of the light beams to the gravitational field is neglected. Unfortunately, exact solutions of equation (7) are not known for our general metrics. We will use a geometrical optics approximation, where the wavelength of the light is small compared with any reasonably defined radius of curvature of the universe. The following calculation is based on an idea of Trautman (1962). We assume that F_{ab} can be written

$$F_{ab} = A_{ab}e^{iS}\,, \tag{8}$$

where the derivatives of A_{ab} are small, say of order ϵ, and we expand A_{ab} in powers of ϵ:

$$A_{ab} = B_{ab} + \epsilon C_{ab} +, \ldots, \qquad B_{ab} \neq 0\,. \tag{9}$$

Substituting into Maxwell's equations and collecting terms of order unity and of order ϵ, we find

$$S_{,a}B^{ab} = 0\,, \qquad S_{,[a}B_{bc]} = 0\,, \tag{10}$$

$$B^{ab}{}_{;b} = -iS_{,b}C^{ab}\,, \qquad B_{[ab;c]} = -iS_{,[c}C_{ab]}\,. \tag{11}$$

By writing out each component of equations (10) and (11), or by using two-component spinors, or by using a quasi-orthonormal tetrad (Jordan, Ehlers, and Sachs 1961), the reader can verify that these equations imply

$$S_{,a}S^{,a} = 0\,, \qquad T_{ab} = \mu S_{,a}S_{,b} + 0(\epsilon)\,, \tag{12}$$

and

$$(\mu S^{,a})_{;a} = 0\,, \tag{13}$$

where T_{ab} is the stress energy tensor of the field, and $\mu(x^a)$ is some scalar.

The first of equations (12) implies that the vector field $k_a = S_{,a}$ is tangent to a set of hypersurface-orthogonal, lightlike geodesics:

$$k^a k_a = 0\,, \qquad k_{[a;b]} = 0 => k^a{}_{;b}k^b = 0\,. \tag{14}$$

The geodesics to which k^a is tangent will be called "rays." The second of equations (12) shows the algebraic structure of T_{ab} to lowest order. We will henceforth neglect all but this first term in T_{ab}. Equation (13) has a simple physical interpretation. Consider two observers who are at any two points on the same ray and who see the same (arbitrary) shift of spectral lines of the source. Suppose that each measures, at his space-time point, the infinitesimal area dA and intensity I of the same small bundle of rays. The rate of change of dA along the ray is, in general, $k^a{}_{;a}dA$ (Jordan *et al.* 1961). The intensity as measured at any point on the ray by an observer with world velocity u^a is $I = T_{ab}u^a u^b = \mu(k_a u^a)^2$. As we will show in the next paragraph, $k_a u^a$ for the two observers is the same, since they see no relative spectral shift. The rate of change of I along the ray for all such observers is therefore just proportional to $\mu_{;a}k^a$, the rate of change of μ, and this in turn, by equation (13), is equal to $-\mu k^a{}_{;a}$. It follows that the total flux of energy through dA is the same for the two observers:

$$I_1 dA_1 = I_2 dA_2\,. \tag{15}$$

This result does not hold if the observers see different spectral shifts. In the geometrical optics approximation, the rate of change of phase as measured by a single observer with world velocity u^a is $S_{,a}u^a = k_a u^a$. Therefore, if any two observers, at the same or different points on a single ray, measure frequencies f_1 and f_2, then

$$f_1/f_2 = (k_a u^a)_1/(k_a u^a)_2 \,. \tag{16}$$

But, as we mentioned in the preceding paragraph, the intensity measured by any observer is $\mu(k_a u^a)^2$. Therefore, two observers in relative motion who measure the same field at the *same* space-time point will measure different intensities, the ratio of intensities being the square of the ratio of the measured frequencies:

$$I_1/I_2 = (f_1/f_2)^2 \,. \tag{17}$$

Physically, this effect can be considered to have two causes. First, the energies of a single photon as measured by the two observers are in the ratio f_1/f_2; second, the unit intervals of time for the observers are in the same ratio, because of the time dilatation between the observers.

Now, instead of two instantaneously coincident observers measuring the intensities of a single source, consider the case of a single observer simultaneously measuring the intensities of two instantaneously coincident sources, of the same intrinsic luminosity (i.e., the luminosity as measured at the source in the rest frame of the source), the sources being in relative motion. The ratio of the measured intensities in this case is not the square, but the fourth power of the ratio of the measured frequencies (Jordan *et al.* 1961; Sachs 1961):

$$I_1/I_2 = (f_1/f_2)^4 \,. \tag{18}$$ [2]

Now suppose that an observer measures the bolometric intensity I_0 for light from a source of known intrinsic luminosity L_G, and that he measures the frequency f_0 for a spectral line of known intrinsic frequency f_G. We want to define a luminosity distance from source to observer, corrected for the spectral shift if $f_0 \neq f_G$. But from the preceding discussion, it is clear that there are two perfectly reasonable ways to make the correction, and that these give different results. We could consider the corrected luminosity to be either (i) that which would be seen by a different observer, located at the same space-time point but moving so that he sees no spectral shift, or (ii) that which would be seen by the same observer looking at a different source, intrinsically identical and located at the same space-time point, but with a different velocity, so chosen that the observer sees no spectral shift. We will arbitrarily make the second choice, and define the *corrected luminosity distance* r by

$$r^2 = \frac{I_G}{I_0}\left(\frac{f_0}{f_G}\right)^4 . \tag{19}$$

Here I_G is the intrinsic intensity at unit distance: $I_G = L_G/4\pi$. This choice for r has the advantage that two galaxies in collision will be assigned the same distance. For the Friedman cosmological models, the luminosity distance r reduces to the "distance by apparent size" defined by McVittie (1956). This name is perhaps misleading in the present context since, as we will show shortly (cf. eq. [37]), the fact that the observed area increases as r^2 is for us a theorem rather than a definition.

[2] This can also be understood physically. First, there is an intensity difference due to the relative motion of each source and the observer. As in the case of two observers and one source, this difference comes from a difference of measured photon energies and from time dilatations, and it contributes a factor $(f_1/f_2)^2$ to the intensity ratio. There is also a change of intensity due to a Lorentz contraction between the sources. It is most easily seen for an observer close enough to the sources for the relevant portion of space time to look flat. The intrinsic intensity of source 1 at the location of the observer is $L/4\pi r_1^2$, where L is the intrinsic luminosity of each source and r_1 is the distance from the source to the observer as measured in the rest frame of 1. Likewise, the intrinsic intensity of source 2 at the observer is $L/4\pi r_2^2$. But $r_2/r_1 = f_1/f_2$, so the ratio of the intrinsic intensities is $(f_1/f_2)^2$. The combination of these effects gives q. (18).

2. *The Theory of Ray Bundles*

Given a distant point source of light, we will want to calculate its intensity at our Galaxy; given a distant extended source, we will want to find its apparent size and shape at our Galaxy. Both calculations involve the theory of lightlike geodesics, whose tangents k^a obey equation (14). Since the theory has been worked out in the literature, we will just state some relevant results (Jordan *et al.* 1961; Sachs 1961).

Consider a point source with world velocity u^a. There is a one-parametric set of three-dimensional light cones emanating from the world line of the particle. On each cone lies a two-parametric set of rays, with tangent k^a, which pass through the world line of the source. Along any ray we can define a so-called preferred affine parameter, s, by the equation $dx^a/ds = k^a$. For a particular light cone, s is then determined up to an additive constant along the ray. If we consider the ray by itself, then $s' = As$, $k'^a = A^{-1}k^a$, with A constant, is also an allowed transformation. Any physically measurable quantity must be invariant under these additional transformations, a fact which is useful in checking many of the following formulae for algebraic mistakes.

We now consider the particular light cone emanating from point $x_1{}^a$ on the world line of the source. On this light cone we select an infinitesimal bundle of rays which is measured from the source to have solid angle $d\Omega$ and a circular normal cross-section. The physical picture is that someone at the source flashes a very narrow-beam flashlight on and off. Let k^a be the tangent to the central ray L of the bundle and s the affine parameter on L, with $s = 0$ at $x_1{}^a$. Now suppose that an observer who happens to cross L at some point $x_2{}^a$ places a flat two-dimensional screen normal to the ray direction in his own rest frame. It turns out that he sees an infinitesimal, instantaneous, elliptical flash of light on the screen. The key question is: What are the size, shape, and orientation of this ellipse? It can be shown that these are all independent of the world velocity of the observer (Sachs 1961). In order to describe the ellipse, we consider the connection vectors q^a between the central ray L and the rays which define the outside of the bundle. The change of each q^a along L is described by the equation of geodesic deviation:

$$D^2q^a/ds^2 = -R^a{}_{ibj}\,k^ik^jq^b\,. \tag{20}$$

Here $D(\ldots)/ds = (\ldots)_{;i}k^i$ is the absolute derivative along L: we will denote it by a dot; thus, $Dq^a/ds = \dot{q}^a$. The ellipse at $x_2{}^a$ must be found by integrating equation (20) for q^a from $x_1{}^a$ to $x_2{}^a$ along L. We need initial conditions: q^a at $x_1{}^a$ is zero because the cone comes to a point there. Without loss of generality, we can demand that $\dot{q}^a$ be orthogonal to u^a and to k^a at $x_1{}^a$ (i.e., that $\dot{q}^a$ be normal to L in the rest frame of the source). Also, we require that the ray bundle viewed from the source has a circular cross-section and solid angle $d\Omega$. The restrictions on q^a near $x_1{}^a$ are thus

$$(q^a)_1 = 0\,; \qquad (\dot{q}^a u_a)_1 = 0\,; \qquad (\dot{q}^a k_a)_1 = 0\,; \qquad \pi(\dot{q}^a\dot{q}_a)_1 = -(u_ak^a)_1{}^2 d\Omega\,. \tag{21}$$

If we now vary the initial conditions over the values allowed by equation (21), we get from equation (20) a corresponding (one-parametric) set of vectors $(q^a)_2$ which define an ellipse at $x_2{}^a$. The lengths α and β of the major and minor semiaxes of the ellipse are then a conditional minimum and maximum, respectively:

$$-\alpha^2 = \min\,(q^aq_a)_2\,; \qquad -\beta^2 = \max\,(q^aq_a)_2\,. \tag{22}$$

The area of the ellipse can be found from equation (22), and the directions at $x_1{}^a$ corresponding to the principal axes will of course have already been found in calculating α and β. Note that, since $(q^a)_2$ is a linear function of $(\dot{q}^a)_1$, there is some two-point function $X_b{}^{a'}(x_1, x_2{}')$ for which $(q^{a'})_2 = X_b{}^{a'}\,\dot{q}_1{}^b$. The maximization is a trick to avoid the need to find $X_b{}^{a'}$ explicitly in the following calculations.

To the accuracy needed in this paper we can explicitly do the calculation sketched above. Let $s = 0$ at x_1^a and $s = s_2$ at x_2^a. Expanding $q_a q^a$ in powers of s, we have

$$(q_a q^a)_2 = (q_a q^a)_1 + \left[\frac{D}{ds}(q_a q^a)\right]_1 s_2 + \tfrac{1}{2}[D^2(q_a q^a)/ds^2]_1 s_2^2 + \ldots \quad (23)$$

Using equation (20) and the initial conditions of equation (21), we obtain from equation (23)

$$(q_a q^a)_2 = s_2^2(\dot{q}^a \dot{q}^b)_1[(g_{ab})_1 - \tfrac{1}{3}(R_{aibj}k^i k^j)_1 s_2^2 - \tfrac{1}{6}(R_{aibj;n}k^i k^j k^n)_1 s_2^3 + \ldots]. \quad (24)$$

Assumption (i) of the Introduction means that this series is decreasing. Since the correction term due to the curvature of space-time,

$$t_{ab} = -\tfrac{1}{3}(R_{aibj}k^i k^j)_1 s_2^2 - \tfrac{1}{6}(R_{aibj;n}k^i k^j k^n)_1 s_2^3 + \ldots, \quad (25)$$

is assumed small, we can obtain the area and shape of the ellipse at x_2^a by considering only terms to this order in s_2. Then, as with infinitesimal linear transformations, the contribution to the area comes only from the trace, $(g^{ab})_1 t_{ab}$, and the ratio of major to minor axes comes only from the trace-free part. In fact, equations (21), (22), and (24) give for the area dA_2 at x_2^a

$$dA_2 = d\Omega s_2^2(u_a k^a)_1^2[1 - \tfrac{1}{6}(R_{ij}k^i k^j)_1 s_2^2 - \tfrac{1}{12}(R_{ij;n}k^i k^j k^n)_1 s_2^3 + \ldots]. \quad (26)$$

Similarly, if i^a is a unit vector in the direction of $(\dot{q}^a)_1$, so that $i^a u_a = i^a k_a = 0$ at x_1^a, the ratio e of major to minor axes comes out, to relevant order in s_2,

$$e = 1 - \min(i^a i^b p_{ab}) + \max(i^a i^b p_{ab}) = 1 + 2\max(i^a i^b p_{ab}), \quad (27)$$

where p_{ab} contains only the conformal tensor (see eq. [64]):

$$p_{ab} = -\tfrac{1}{6}s_2^2(C_{aibj}k^i k^j)_1 - \tfrac{1}{12}(C_{aibj;n}k^i k^j k^n)_1 s_2^3. \quad (28)$$

We shall use equations (26)–(28) to calculate luminosity distances, area distances, and the distortion effect.

We shall calculate certain observational effects which directly relate the world velocity field of the sources and the curvature to measurable quantities. In this section we shall leave the results in a form that is independent of any dynamical model or choice of coordinate system. In § III we shall specialize to general-relativistic universes and write the equations in the form that they take for an observer interpreting his data in a Newtonian, Cartesian frame of reference. In the rest of this section we consider a distant galaxy or cluster of galaxies located at a point x_G^a on the past light cone of the point $x^a = 0$, which is the present space-time position of our own Galaxy. L is the lightlike geodesic from the center of the galaxy or cluster to the center of an astronomer's telescope here and now; k^a is the tangent to L and $k_4 > 0$; s is an affine parameter along L, increasing from 0 at x_G^a to s_0 at the origin $x_0^a = 0$. A "perpendicular area" at x_G^a or at x_0^a is a two-dimensional cross-section perpendicular both to k^a and to u^a at the point in question. We define the following vector at x_0^a:

$$e^a = -(k^a/u_b k^b)_0. \quad (29)$$

The projection of e^a into the three-space orthogonal to u_0^a is a spatial unit vector in the observer's rest frame pointing in the observed direction of the source; it is given by

$$e_\perp^a = e^a - u_0^a(e_b u_0^b) = e^a + u_0^a. \quad (30)$$

3. *Redshift versus Luminosity Distance*

Suppose the astronomer measures frequency f_0 for a line with known intrinsic frequency f_G emitted by the galaxy at $x_G{}^a$. Then, from equation (16),

$$f_G/f_0 = (u_a k^a)_G/(u_a k^a)_0 . \tag{31}$$

Expanding in powers of s gives

$$f_G/f_0 = 1/(u_a k^a)_0[u_a k^a - u_{a;b}k^a k^b s + \tfrac{1}{2}u_{a;bc}k^a k^b k^c s^2 + \ldots]_0 ; \tag{32}$$

s_0 is not an empirically measurable quantity and must be replaced by the luminosity distance in a suitable way. We can use equation (26) with $s_2 = s_0$, $x_2{}^a = 0$, and $dA_2 = dA_0$, the perpendicular area of the telescope. Let the intrinsic total luminosity of the source be $4\pi I_G$ and the measured intensity be I_0. From equations (15)–(17) we get

$$I_0 dA_0(u_a k^a)_G{}^2 = d\Omega I_G(u_a k^a)_0{}^2 . \tag{33}$$

From equations (19) and (26) we therefore have

$$\begin{aligned} r^2 &= (u_a k^a)_0{}^2 s_0{}^2[1 - \tfrac{1}{6}(R_{ij}k^i k^j)_G s_0{}^2 - \tfrac{1}{12}(R_{ij;n}k^i k^j k^n)_G\, s_0{}^3 + \ldots] \\ &= (u_a k^a)_0{}^2 s_0{}^2[1 - \tfrac{1}{6}(R_{ij}k^i k^j)_0 s_0{}^2 + \tfrac{1}{12}(R_{ij;n}k^i k^j k^n)_0 s_0{}^3 + \ldots] . \end{aligned} \tag{34}$$

The reversion of this equation is

$$s_0 = (u_a k^a)_0{}^{-1} r[1 + \tfrac{1}{12}(R_{ij})_0 e^i e^j r^2 + \tfrac{1}{24}(R_{ij;k})_0 e^i e^j e^k r^3 + \ldots] . \tag{35}$$

With s_0 thus expressed in terms of r, equation (32) becomes

$$f_G/f_0 = 1 - (u_{a;b})_0 e^a e^b r - \tfrac{1}{2}(u_{a;bc})_0 e^a e^b e^c r^2 + \ldots . \tag{36}$$

This equation contains only parameters describing the velocity field of sources and empirically measurable quantities, so that it is the desired four-dimensional formulation of the redshift versus luminosity-distance law. In the three-plus-one-dimensional notation of the next section it loses its simple form. Equation (36), but not the corresponding equation (70) of the next section, is valid in all universes in which light travels along geodesics in a Riemannian space time and obeys equation (15).

4. *Area Distance versus Luminosity Distance*

We now ask what solid angle an extended distant body subtends when viewed by the astronomer. Let its intrinsic perpendicular area be dA_G. We again use equation (26), but now our own Galaxy is located at the point called $x_1{}^a$ and the distant object is located at the point called $x_2{}^a$, so the sign of k^a must be changed. A short calculation gives

$$dA_G = r^2 d\Omega . \tag{37}$$

To the accuracy we have been considering, the correction terms, which are of order r^4 and r^5, are identically zero, so that in the approximation framework, equation (37) is valid at least to terms of order r^6. R. Penrose (private communication, to be published in Hlavaty Festschrift volume) has subsequently proved, without use of power series, that equation (37) is *exact*.

5. *Distortion Effect versus Luminosity Distance*

It follows from § II.2 that a distant object whose intrinsic perpendicular cross-section is a circle will appear as an ellipse to us. Because p_{ab} is small, the magnitude and direction of this effect can be read off directly from equation (27), using the same conventions as in

§ II.4. We again designate a unit spatial vector in the astronomer's laboratory and tangent to the astronomer's unit sphere by i^a. Then, from equations (28) and (35), the ratio, e, of major to minor axis of the observed ellipse is

$$e = 1 + \tfrac{1}{3} \max\, (i^a i^b e^i e^j C_{aibj})_0\, r^2 + \ldots , \tag{38}$$

where the maximum is taken by varying the end point of i^a over the spatial unit circle orthogonal to the source direction $e_{\perp}{}^a$ in the astronomer's rest frame; the direction of the major axis is of course that i^a for which the maximum occurs.

6. *Number Counts versus Luminosity Distance*

Suppose we count the number dN of galaxies in a solid angle $d\Omega$ (measured by the astronomer) which have luminosity distances between r and $r + dr$. To dr corresponds some parameter interval ds and also some infinitesimal length dx orthogonal to the intrinsic perpendicular cross-section dA_G at $x_G{}^a$. Then for dN we find $dN = (\rho' dA dx)_G$, where ρ' is the intrinsic number of galaxies per unit volume. On the other hand, we already know $dA_G = d\Omega r^2$; moreover, a short calculation gives $dx = ds(u_a k^a)_G$. Using these equations and the differential of equation (35), we get

$$dN = (f_G/f_0) r^2 d\Omega dr[\rho_0' + (\rho'_{,a})_0 e^a r + \tfrac{1}{2} e^a e^b r^2 (\rho'_{;ab} + \tfrac{1}{2} R_{ab}\rho')_0 + \ldots] . \tag{39}$$

We could also expand the redshift correction (f_G/f_0) in powers of r, but it seems more useful to consider it as an empirical quantity as far as equation (39) is concerned.

7. *Apparent Proper Motions versus Luminosity Distance*

If an astronomer at $x^a = 0$ measures the apparent direction, with respect to a local set of non-rotating axes, of a distant object at $x_G{}^a$, it may happen that this direction changes in the course of time. Such an apparent proper motion may arise, loosely speaking, either from a "real" motion of the source or from a curvature of space time between the source and the observer. To calculate this effect, we must examine the lightlike geodesic that starts at the point $x_1{}^a = x_G{}^a + \epsilon u_G{}^a$, where ϵ is infinitesimal, and intersects the observer's world line at the point $x_2{}^a = u^a \Delta t$, where Δt is an infinitesimal increase in the observer's proper time. Note that $\Delta t/\epsilon = f_G/f_0$, and also that, by equation (16), $f_G/f_0 = (u_a k^a)_G/(u_a k^a)_0$.

Suppose that the tangent to this geodesic is $m^a = dx^a/ds$, where $s = 0$ at $x_1{}^a$ and $s = s_2$ at $x_2{}^a$. In analogy with e^a, we define the vector

$$f^a = -(m^a/u_b m^b)_2 . \tag{40}$$

This is a vector at $x_2{}^a$ whose projection, into the three-space orthogonal to $u_2{}^a$, is a unit spatial vector in the observer's rest frame pointing in the observed direction of the source at time $t_0 + \Delta t$.

We suppose henceforth that the observer moves along a geodesic, as is the case, for example, in a relativistic dust model. Consider an orthonormal tetrad $\lambda_{(a)}{}^i$ on the observer's world line, where $\lambda_{(4)}{}^i = u^i$ and the triad $\lambda_{(a)}{}^i$ $(a = 1, 2, 3)$ is parallel-transferred along the observer's world line, and therefore forms a set of non-rotating Cartesian axes for the observer. (Note that the numbers and letters in parentheses are labels for the vectors, not tensor indices.) Such a vector might be the axis direction of our own Galaxy, treated as a gyroscope, corrected for torques exerted by local irregularities. We shall return to the question of establishing a non-rotating frame in § III, and will show there that the somewhat unrealistic appeal to a local gyroscope can be avoided in practice. (See the discussion following eq. [86].) The direction cosines of the source at times t_0 and $t_0 + \Delta t$ are

$$e_{(a)} = -\, e^i(\lambda_{(a)i})_0 \quad \text{and} \quad f_{(a)} = -\, f^i(\lambda_{(a)i})_2 , \tag{41}$$

respectively. In general, $e_{(a)} \neq f_{(a)}$, and their difference measures the source's *apparent proper motion*, defined as

$$\frac{d\,e_{(a)}}{dt} = \lim \frac{f_{(a)} - e_{(a)}}{\Delta t}. \tag{42}$$

Expanding $(\lambda^i{}_{(a)})_2$ about $x^a = 0$, and noting that $f^i = e^i + 0(\Delta t)$, this becomes

$$-\frac{d\,e_{(a)}}{dt} = \lim \frac{(f^i - e^i)(\lambda_{(a)i})_0}{\Delta t} + (\Gamma_{jk}{}^i e^j u^k \lambda_{(a)i})_0. \tag{43}$$

It then remains only to expand f^i about $x^a = 0$, which we will do by expanding m^a and u_a separately, and then combining them according to equation (40).

Since the observer moves on a geodesic,

$$(u_b)_2 = (u_b)_0 + (\Gamma_{bc}{}^d u_d u^c)_0 \Delta t + O[(\Delta t)^2]. \tag{44}$$

To expand $m_2{}^a$, we first write it in terms of $\Delta x^a = x_2{}^a - x_1{}^a$. Thus, since m^a is the geodesic joining $x_1{}^a$ and $x_2{}^a$,

$$x_1{}^a = x_2{}^a - s_2 m_2{}^a - \tfrac{1}{2} s_2{}^2 (\Gamma_{bc}{}^a m^b m^c)_2 + \ldots. \tag{45}$$

The reversion of this series is

$$s_2 m_2{}^a = \Delta x^a - \tfrac{1}{2}(\Gamma_{bc}{}^a)_2 \Delta x^b \Delta x^c + O[(\Delta x)^3]. \tag{46}$$

Note further that

$$x_2{}^a = u_0{}^a \Delta t + O[(\Delta t)^2], \tag{47}$$

and that

$$x_1{}^a = x_G{}^a + \epsilon u_G{}^a + O(\epsilon^2) = x_G{}^a + (f_0/f_G) u_G{}^a \Delta t + O[(\Delta t)^2]. \tag{48}$$

The rest of the calculation is straightforward, though somewhat tedious. Equation (36) gives f_G/f_0 as a power series in r, and if we also expand $x_G{}^a$ and $u_G{}^a$ in powers of r, then equations (46)–(48) give $m_2{}^a$ as a power series in r and Δt. Substituting this and equation (44) into equation (40) gives f^a as a series in r and Δt. Finally, equation (43) then gives the apparent proper motion of the source as a series in r. The result of this calculation is

$$\frac{d\,e_{(a)}}{dt} = -h_{(a\beta)}[(u^i{}_{;b})_0 e^b + r(\tfrac{1}{2} u^i{}_{;jk} - \tfrac{1}{2} R^i{}_{jkb} u^b + u^i{}_{;b} e^b u_{j;k})_0 e^j e^k + O(r^2)](\lambda_{(\beta)i})_0. \tag{49}$$

Here $h_{(a\beta)} = \delta_{a\beta} - e_{(a)} e_{(\beta)}$ is a projection operator into the plane orthogonal to $\mathbf{e}$: the right-hand side of equation (49) is summed over β.

Equations (36)–(39) and (49) are the fundamental observational laws of this paper. In the next section we shall specialize them to general-relativistic universes and write them in a three-plus-one dimensional notation.

III. OBSERVATIONS IN DUST-FILLED GENERAL-RELATIVISTIC UNIVERSES

In § II we have derived general expressions in power series for five kinds of astronomical observations in a Riemannian space time. The coefficients of the series are the density and velocity fields of matter, the curvature of space time, and their derivatives, all evaluated here and now. These quantities are, of course, determined by the over-all structure of the universe, but they are here treated as parameters whose values must be found, if possible, from observations. For any particular gravitational theory, the coefficients will not be independent. As an example we shall consider Einstein's gravitational theory for a dust-filled universe, without making any assumptions about the symmetry of the cosmos. We suppose therefore that the Einstein tensor, $G^{ab} = R^{ab} - \frac{1}{2}g^{ab}R = R^{a\,ib}{}_{;i} - \frac{1}{2}R^{ij}{}_{;ij}g^{ab}$, has the algebraic form

$$G^{ab} = \rho u^a u^b + \Lambda g^{ab} ; \qquad u^a u_a = 1 . \tag{50}$$

Here ρ is the proper density of energy in natural units, $u^a(x^i)$ is the average world velocity of the galaxies near the point x^i, and Λ is the cosmological constant. A continuous distribution of ρ and u^a is assumed. The cosmological constant is retained merely for the sake of generality; it plays little role in the following.

We shall also use a system of Riemannian normal coordinates (Birkhoff 1923), referred to our present space-time position, with the time axis along our local time axis. That is, if $(u^a)_0$ is the average velocity of galaxies in our immediate vicinity now, then

$$(u^a)_0 = \delta_4{}^a . \tag{51}$$

Riemannian normal coordinates have two defining properties: (i)

$$(g_{ab})_0 = \eta_{ab} , \tag{52}$$

where $\eta_{ab} = \eta^{ab}$ is the diagonal Lorentz metric with components $(-1, -1, -1, +1)$; (ii) every geodesic through the origin has, at each of its points, the equation $d^2x^a/ds^2 = 0$. Detailed discussions of Riemannian coordinates will be found in the literature (Thomas 1934). Here we mention only two properties of interest in this paper. First, under very reasonable smoothness assumptions, Riemannian normal coordinates exist, but usually only in a finite region around the origin; our fundamental assumption that all quantities can be expanded in power series therefore must include the assumption that we are well within this region. Second, under the condition given by equation (51), the Riemannian normal coordinates are unique up to the Euclidean transformations

$$\bar{x}^\beta = O_\alpha{}^\beta x^\alpha , \qquad \bar{t} = t , \qquad (\alpha, \beta = 1, 2, 3) , \tag{53}$$

where $O_\alpha{}^\beta$ is a *constant* orthogonal matrix.

These special coordinates have been chosen primarily for computational convenience. They also have the advantage that the theoretical observational effects which we will write in terms of them can be compared directly with observational results as usually reported. The coordinates used by an observational astronomer are t, the time of observation of an event; $e_{(\alpha)}$ (eq. [41]), the direction cosines of the event referred to a non-rotating orthonormal triad $\lambda_{(\alpha)}{}^i$ in the observer's rest frame; and r, the luminosity distance of the source. But the tangents to the parametric lines of the space coordinates x^α of our Riemannian normal coordinates form an orthonormal triad in the observer's rest frame, and this triad can always be made to coincide with the observer's reference frame $\lambda_{(\alpha)}{}^i$ by a transformation defined by equation (53). If this is done, it is easy to show that $e_{(\alpha)} = e^\alpha$; i.e., that the contravariant component e^α is numerically the same as the corresponding direction cosine as measured by an astronomer. Therefore, when we come to write equations such as (71), e^α can be interpreted directly as a measured angle.

The calculations to follow could easily be done under assumptions other than equation (50); i.e., for other gravitational theories than Einstein's. The results, including the distortion effect discussed in § II.5, would be similar for any cosmology which assumes that light travels along geodesics in a Riemannian space-time. However, as we shall see. there are many adjustable parameters in our final results even with this assumption,

1. *The Independent Parameters*

We shall now systematically determine which of the parameters ρ, u^a, R_{abcd}, and their derivatives at $x^a = 0$ can be chosen independently. The reader who starts to get lost in the details of the calculations may find it helpful to refer to Table 1 (see below).

From the twice-contracted Bianchi identities, $G^{ab}{}_{;b} = 0$, we have

$$(\rho u^a)_{;a} = 0 \,, \qquad u^a{}_{;b} u^b = 0 \,. \tag{54}$$

The first of these is the conservation equation; the second says that matter moves along geodesics. Also, $u^a u_a = 1$ implies

$$u^a{}_{;b} u_a = 0 \,. \tag{55}$$

The first free parameter is ρ_0. Consider next in Riemannian normal coordinates $(u_{a;b})_0 = (u_{a,b})_0$. From equations (51), (52), (54), and (55) we obtain $(u_{4;a})_0 = 0$, $(u_{a;4})_0 = 0$. Consequently, only the $3 \times 3 = 9$ constants $u_{\alpha\beta}$ defined by

$$u_{\alpha\beta} = (-u_{\alpha;\beta})_0 \tag{56}$$

are at our disposal. These have a direct analogue in Newtonian theory. If v_α is the velocity of matter in Newtonian theory, then $(v_{\alpha,\beta})_0$ is fully analogous to $(-u_{\alpha;\beta})_0$. Also, if the astronomer interprets $X^\alpha = e_{(\alpha)} r$ as a Cartesian vector and defines the corresponding "velocity" $V^\alpha = dX^\alpha/dt$, then $\partial V^\alpha / \partial X^\beta = u_{\alpha\beta}$. Therefore we split $u_{\alpha\beta}$, in the familiar way, into a rotation $\omega_{\alpha\beta}$, a shear $\sigma_{\alpha\beta}$, and an expansion θ:

$$u_{\alpha\beta} = \omega_{\alpha\beta} + \sigma_{\alpha\beta} + \tfrac{1}{3}\delta_{\alpha\beta}\theta \,. \tag{57}$$

Here

$$\omega_{\alpha\beta} = u_{[\alpha\beta]} \,, \qquad \theta = u_{\alpha\alpha} \,, \qquad \sigma_{\alpha\beta} = u_{(\alpha\beta)} - \tfrac{1}{3}\delta_{\alpha\beta}\theta \,. \tag{58}$$

Note that such quantities as $\delta_{\alpha\beta}$ and $u_{\alpha\alpha}$ are well defined because of the fact that the only allowed coordinate transformations are the rigid Euclidean rotations of equation (53).

[The following is intended purely as an aside to those readers who are acquainted with representation theory. Note that what we have really done is to split $(u_{a;b})_0$ into six quantities $[(u_{4;4})_0, (u_{4;\alpha})_0, (u_{\alpha;4})_0, \omega_{\alpha\beta}, \sigma_{\alpha\beta}$, and $\theta]$, each of which forms an irreducible representation of the rotation group, equation (53). The first three vanish and the last three are at our disposal. Very similar considerations are the basis for all the remaining calculations in this section. In some cases we will find it convenient to carry the splitting only part way by combining several irreducible representations in one symbol. But any doubt concerning the arbitrariness of a given coefficient can always be settled systematically by using two-component spinors, or by using Young's diagrams and subtracting out suitable traces (Murnaghan 1938; Penrose 1960).]

Proceeding to $(\rho_{,a})_0$ we find that the three quantities $(\rho_{,\alpha})_0$ are arbitrary while $(\rho_{,4})_0$ is given by equation (54): $(\rho_{,4})_0 = -\rho_0\theta$. Next consider $(u_{a;bc})_0$. There are $3 \times (3 \times 2/2) = 9$ free quantities,

$$u_{\mu\beta\delta} = -(u_{\mu;(\beta\delta)})_0 \,. \tag{59}$$

The other components of $(u_{a;bc})_0$ are determined from our previous relations:

$$\begin{gathered} 2(u_{\mu;[\beta\delta]})_0 = (R_{4\mu\beta\delta})_0 \,, \qquad (u_{4;\beta\delta})_0 = u_{\tau\beta}u_{\tau\delta} \,, \qquad (u_{\beta;4\delta})_0 = u_{\beta\tau}u_{\tau\delta} \,, \\ (u_{\beta;\delta4})_0 = u_{\beta\tau}u_{\tau\delta} + (R_{4\beta\delta4})_0 \,, \qquad (u_{4;44})_0 = (u_{\mu;44})_0 = (u_{4;\mu4})_0 = (u_{4;4\mu})_0 = 0 \,. \end{gathered} \tag{60}$$

The Riemann tensor terms in equation (60) will be analyzed presently. For the moment we consider them as given.

The constants $u_{\mu\beta\delta}$ of equation (59) "essentially" represent differential rotations, shears, and expansions; they correspond "essentially" to $v_{\mu,\beta\delta}$ of Newtonian theory. But there is an important ambiguity in this correspondence. We could just as well draw the analogy

$$v_{\mu,\beta\delta} \leftrightarrow (-u_{\mu;(\beta\delta)})_0 + a(R_{4(\beta\delta)\mu})_0 \,, \tag{61}$$

where a is any pure number (of order unity), since $R_{4(\beta\delta)\mu}$ has the same dimensions and symmetries as $v_{\mu,\beta\delta}$ and vanishes in the correspondence limit $R_{abcd} \to 0$. In the authors' opinion it would be futile to call any particular one of the quantities in equation (61) the unique analogue of $v_{\mu,\beta\delta}$. For example, $u_{\mu,\beta\delta}$ (which is well defined because our coordinate transformations are so restricted) differs from $u_{\mu;(\beta\delta)}$ as in equation (61), but in all other respects is on the same footing. This means that any attempt to divide observational results unambiguously into curvature effects and kinematic or dynamic effects will fail for sufficiently complicated effects. Once we accept such ambiguities as inevitable they do no real harm.

Next, among the second derivatives of ρ, the six quantities

$$\rho_{\alpha\beta} = (\rho_{;\alpha\beta})_0 \tag{62}$$

are arbitrary. The remaining second derivatives are given by

$$\begin{aligned}(\rho_{;4\beta})_0 = (\rho_{;\beta 4})_0 &= -(\rho_{,\mu})_0\, u_{\mu\beta} - (\rho_{,\beta})_0\,\theta - \rho_0 u_{\mu\mu\beta}\,, \\ (\rho_{;44})_0 &= \rho_0[\theta^2 + u_{\beta\delta}u_{\delta\beta} + \tfrac{1}{2}\rho_0 - \Lambda]\,.\end{aligned} \tag{63}$$

Note that if $\omega_{\mu\beta} = 0$ and $\Lambda = 0$, then $(\rho_{;44})_0 > 0$. We have enough terms now in the expansions of ρ and u_a and can proceed to the curvature tensor.

In the lowest order, $(R_{abcd})_0$ splits into the Ricci tensor $(R_{ab})_0$ and the conformal (or Weyl) tensor $(C_{abcd})_0$; C_{abcd} is defined as

$$C^{ab}_{\cdot\cdot cd} = R^{ab}_{\cdot\cdot cd} - 2\,\delta^{[a}_{[c}R^{b]}_{d]} + \tfrac{1}{3}\delta^{[a}_{c}\delta^{b]}_{d}R\,. \tag{64}$$

The ten components of the Ricci tensor are determined by the field equations (50):

$$(R_{ab})_0 = (G_{ab})_0 - \tfrac{1}{2}\eta_{ab}G_0 = \rho_0\delta_a{}^4\delta_b{}^4 - \tfrac{1}{2}\eta_{ab}[2\Lambda + \rho_0]\,. \tag{65}$$

$(C_{abcd})_0$ is the incident field; its ten components are at our disposal. As is well known (Pirani 1962*a*, *b*), $(C_{abcd})_0$ can be further split into two quantities $E_{\mu\beta}$ and $H_{\mu\beta}$:

$$E_{\mu\beta} = (C_{4\mu 4\beta})_0\,; \qquad H_{\mu\beta} = \tfrac{1}{2}(C_{4\mu\rho\sigma})_0\epsilon_{\rho\sigma\beta}\,, \tag{66}$$

where $\epsilon_{\rho\sigma\beta}$ is the three-dimensional Levi-Civita object. $E_{\mu\beta}$ is manifestly trace-free and symmetric and, as shown in the literature (Pirani 1962*a*, *b*), $H_{\mu\beta}$ has the same properties;

$$E_{\mu\beta} = E_{\beta\mu}\,, \qquad E_{\mu\mu} = 0\,; \qquad H_{\mu\beta} = H_{\beta\mu}\,, \qquad H_{\mu\mu} = 0\,. \tag{67}$$

$E_{\mu\beta}$ is usually called the incident "electric-type" field. In the usual slow-motion, weak-field approximation, $E_{\mu\beta}$ goes into the Newtonian quantity $(\phi_{,\mu\beta})_0 - \frac{1}{3}\delta_{\mu\beta}\nabla^2\phi_0$, which we discussed in § I. $H_{\mu\beta}$ has no Newtonian analogue; it is usually called the incident "magnetic-type" field. One reason for this name is that, in linearized gravitational theory, contributions to $H_{\mu\beta}$ come primarily from rotations of the sources of the field (Sachs and Bergmann 1958).

There are, intuitively speaking, several possible sources of the incident field, such as asymmetries in the large-scale matter or velocity distribution of galaxies, or in the boundary conditions for the universe. Also, gravitational waves with periods of the order of 10^9 years would contribute to the incident field; waves whose periods are a mere 10^6 years or so would be considered in our treatment as local irregularities, and would be averaged out. Since the methods of this paper cannot be used to analyze global questions, the incident field is, from our point of view, a purely empirical quantity. In § III.2, we shall crudely estimate the maximum size of $(C_{abcd})_0$ as about 10^{-16} (light-year)$^{-2}$ on the basis of existing data.

We might point out that the list of independent parameters in Table 1 (see below) is complete to the order considered. In particular, it can be shown that the first derivatives of the metric tensor vanish at $x^a = 0$, and all of the second derivatives are calculable from the Riemann tensor.

2. *Results and Discussion*

At the cost of some repetition, this section is intended to be understandable to a reader who has no detailed knowledge of relativity theory or has not read §§ II and III.1. An understanding of Cartesian (Euclidean) tensors is assumed, but the reader who feels uneasy with tensors can get a qualitative idea of the results by thinking of quantities with more than one index as coefficients of certain spherical harmonics, as indicated in equation (72). All tensors in this section are to be interpreted as Cartesian tensors in the Euclidean frame of the astronomer's observatory. All indices run from 1 to 3; for consistency with previous sections, certain indices will be placed as superscripts rather than subscripts, but this placement has no relevance as far as this section alone is concerned.

In our units, $c = 1$, the Newtonian gravitational constant is $1/8\pi$, and the Einstein gravitational constant is consequently 1. All quantities then have the dimensions of time to some power. Thus a length of 10^9 years means a length of 10^9 light-years.

The astronomer points his telescope along the unit vector $\boldsymbol{e}$, with Cartesian components e^a, toward some distant source, such as a galaxy or cluster of galaxies. The "intrinsic" frequency of a line at the source (i.e., the frequency measured by a hypothetical observer at the source, moving with the source at the time the source emits the light) is f_G. The total intrinsic luminosity of the source is L_G. For simplicity, the source is assumed intrinsically spherical with intrinsic cross-sectional area dA_G. It is assumed that f_G, L_G, and dA_G are somehow known. We define $I_G = L_G/4\pi$.

The astronomer measures various quantities. We shall suppose that all of them have been corrected in the usual way for the motion of the Earth relative to the center of mass of the local group of galaxies; no correction of this type is needed to analyze the distortion effect because two relatively moving observers in the same place will measure the same shape (though not the same size) for the distant object.

The astronomer measures a frequency f_0. He measures an intensity and then corrects it for absorption to get the *bolometric* intensity I_0. If the intensity is measured only over some limited frequency range, then the fact that a redshift has occurred must be taken into account in extrapolating to bolometric intensity; but no additional redshift corrections are to be incorporated into I_0. He measures by Euclidean methods the solid angle $d\Omega_0$ subtended by the distant object. He sees the intrinsically spherical distant object as an ellipse and measures the orientation and shape of the ellipse on his plates; if the distant object is irregular, he measures the orientation of its longest dimension and the ratio of longest to shortest dimension for use in a statistical treatment. He counts the number dN of galaxies or groups in a given solid angle and a given intensity range. He measures the rate of change $d\boldsymbol{e}/dt$ of the source's direction relative to a non-rotating, mutually orthogonal set of directions (such as the axial direction of our own Galaxy after correction for torques exerted by members of our own galactic group; cf. eq. [86] and the discussion following it). He defines the "corrected luminosity distance" r as

$$r = (I_G/I_0)^{1/2}(f_0/f_G)^2 . \qquad (68)$$

As discussed in § II.1, it is convenient to incorporate into r the particular redshift correction given by equation (68).

The equations in this section take their simplest form if we plot all effects as functions of r. But since the redshift is by far the most accurately observed quantity, we shall also write some of the equations in more conventional form, with $z = (f_G/f_0) - 1$ as independent variable.

We assume that the gravitational field obeys Einstein's gravitational equations for dust. All small-scale variations in the density or velocity of the dust are averaged out, but inhomogeneities whose distance scales are of the order of 10^9 years or larger are allowed. We work in a frame for which the three-dimensional velocity of the dust here and now is zero. The observable effects are determined, to the order of accuracy considered, by certain parameters; these parameters are listed in Table 1. Except for the algebraic restrictions given in the table, the parameters are free and independent in a general relativistic universe without pressure, and we consider them as quantities whose values must be found from observations. The number of free parameters can of course be reduced if additional a priori assumptions are made about the character of the metric. Such assumptions greatly simplify the mathematics, but current observations do not unambiguously suggest any particular simple model, including the Friedman model (cf.

TABLE 1

THE ADJUSTABLE PARAMETERS

Symbol	Name	Newtonian Analogue	Relativistic Definition
1. ρ_0	Density	$\rho_0=\rho]_{x=t=0}$	Eq. (50)
2. Λ	Cosmological constant	None	Eq. (50)
3. ρ_τ	Density gradient	$(\rho_{,\tau})_0=\partial\rho/\partial x^\tau]_{x=t=0}$	$(\rho_{,\tau})_0$
4. $\rho_{\tau\eta}=\rho_{\eta\tau}$	Differential density gradient	$(\rho_{,\eta\tau})_0$	Eq. (62)
5. $u_{\eta\tau}=\omega_{\eta\tau}+\sigma_{\eta\tau}+\frac{1}{3}\delta_{\eta\tau}\theta$	Velocity gradient	$(v_{\eta,\tau})_0$*	Eq. (56)
a) $\omega_{\eta\tau}=-\omega_{\tau\eta}$	Rotation	$(v_{[\eta,\tau]})_0$	Eq. (58)
b) $\sigma_{\eta\tau}=\sigma_{\tau\eta}$	Shear	$(v_{(\eta,\tau)})_0-\frac{1}{3}\delta_{\eta\tau}\nabla\cdot v_0$	Eq. (58)
c) θ	Expansion	$\nabla\cdot v_0$	Eq. (58)
6. $u_{\eta\tau\chi}=u_{\eta\chi\tau}$	Differential velocity gradient	$(v_{\eta,\chi\tau})_0$	Eq. (59)
7. $E_{\eta\chi}=E_{\chi\eta}$	Incident electric-type gravitational field	$(\phi_{,\eta\chi})_0-\frac{1}{3}\delta_{\eta\chi}\nabla^2\phi_0$	Eq. (66)
8. $H_{\eta\chi}=H_{\chi\eta}$	Incident magnetic-type gravitational field	None	Eq. (66)
Restrictions: $\sigma_{\eta\eta}=E_{\eta\eta}=H_{\eta\eta}=0$			

* v is the velocity of the dust in Newtonian approximation and $v_0 = 0$.

eq. [88]). In the Friedman case, the universe is assumed to be homogeneous and isotropic, and the following equations are found to hold:

$$\rho_\mu = \sigma_{\mu\nu} = \omega_{\mu\nu} = u_{\mu\nu\gamma} = E_{\mu\nu} = H_{\mu\nu} = 0\,, \qquad \rho_{\mu\nu} = \tfrac{1}{3}\rho_0\theta^2\delta_{\alpha\beta}\,, \tag{69a}$$

so there are only three free parameters. If we make the transformation from Riemannian normal coordinates to the usual Robertson-Walker coordinates for the Friedman universe, we can establish the following identifications, which will be useful in examining our later results in the Friedman limit:

$$\theta = 3h_1\,, \qquad \rho_0 = 2(h_1^2 - h_2 + k)\,, \qquad \Lambda = h_1^2 + 2h_2 + k\,. \tag{69b}$$

The quantities h_1, h_2, and k are defined for the Robertson-Walker metric in McVittie (1956): h_1 and h_2 are the Hubble parameter and the acceleration parameter, respectively.

We shall now discuss the dependence of the redshift on r, on the direction of observation e, and on the parameters in Table 1. The r-dependence of z has the form (eq. [36])

$$z = Wr + \tfrac{1}{2}Xr^2 + \ldots\,. \tag{70}$$

The angular dependence of W is given by (cf. Ehlers 1961)

$$W = \tfrac{1}{3}\theta + \sigma_{\mu\gamma}e^\mu e^\gamma\,. \tag{71}$$

The θ-term in equation (71) is spherically symmetric (θ is the expansion of the universe, not an angle); the five linearly independent components of $\sigma_{\mu\gamma}$ correspond essentially to the five linearly independent coefficients σ_m of a sum

$$\sum_{m=-2}^{+2} \sigma_m Y_{2m}(\Theta,\phi). \tag{72}$$

Here Θ and ϕ are angles on the celestial sphere and Y_{nm} are surface harmonics. Similarly, X contains surface harmonics of order 0, 1, 2, and 3:

$$X = V + V_\mu e^\mu + V_{\mu\gamma} e^\mu e^\gamma + V_{\mu\beta\gamma} e^\mu e^\gamma e^\beta . \tag{73}$$

The tensors in this equation are defined in terms of the parameters by

$$V = \tfrac{1}{6}\rho_0 + \tfrac{1}{3}\theta^2 + \sigma_{\mu\gamma}\sigma_{\mu\gamma} - \tfrac{1}{3}(\Lambda + \omega_{\mu\gamma}\omega_{\mu\gamma}), \qquad V_\mu = \tfrac{1}{5}(2u_{\gamma\gamma\mu} + u_{\mu\gamma\gamma}),$$

$$-V_{\mu\gamma} = E_{\mu\gamma} + \omega_{\mu\lambda}\omega_{\gamma\lambda} + 2\omega_{\mu\lambda}\sigma_{\gamma\lambda} - 3\sigma_{\mu\lambda}\sigma_{\gamma\lambda} - 2\sigma_{\mu\lambda}\theta - \delta_{\mu\gamma}(\tfrac{1}{3}\omega_{\lambda\beta}\omega_{\lambda\beta} - \sigma_{\lambda\beta}\sigma_{\lambda\beta}), \tag{74}$$

$$V_{\mu\gamma\lambda} = u_{\mu\gamma\lambda} - \tfrac{1}{5}(u_{\mu\beta\beta} + 2u_{\beta\beta\mu})\delta_{\gamma\lambda} .$$

Equations (70), (71), (73), and (74) give the desired expression for z.

To compare with available observations, it is convenient to consider the Friedman models, for which equation (70) reduces to

$$z = h_1 r + \tfrac{1}{2}(3h_1^2 - h_2)r^2 + \dots . \tag{75}$$

This is equation (9.203) in McVittie (1956). In comparing these and other equations in the Friedman limit, note that

$$r = \frac{D}{2}\left[1 - 2h_1\frac{D}{c} + (4h_1^2 + h_2)\frac{D^2}{c^2} + \dots\right], \tag{76}$$

where D is defined in McVittie's equation (8.517). Although h_1 and h_2 are spherically symmetric while W and X in equation (70) have the angular dependence indicated in equations (71) and (73), we can use the empirical values of h_1 and h_2 to get some very rough estimates or upper bounds for the various parameters that appear in equations (71) and (74).

As far as the linear term W is concerned, we first remark that W has the structure of equation (71) for most current cosmological theories, not just for general relativistic dust models. If any spherical harmonic of order one, such as an asymmetry between north and south galactic poles, were observed in the linear portion of the redshift-curve, this would indicate a non-zero acceleration $(u_{a;4})_0 \neq 0$. In general relativistic models, such accelerations arise only from pressure gradients. Of course, a north-south asymmetry which is independent of r would simply indicate motion relative to the center of mass of the local group of galaxies. Since the observed linear redshift is, at least very roughly, spherically symmetric, we take the current observational value (Sandage 1962), $h_1 \approx 10^{-10}$ year$^{-1} \approx 100$ km/sec/Mpc, to indicate $\theta \approx 3 \times 10^{-10}$ year^{-1}. The authors have not been able to find observational estimates of the second spherical harmonic component in the linear redshift curve. However, a brief examination of the published cluster data (Humason, Mayall, and Sandage 1956) makes it appear likely that

$$\sigma_{\mu\gamma} \leq 5 \times 10^{-11} \text{ year}^{-1}, \tag{77}$$

since a value of $\sigma_{\mu\gamma}$ larger than this would by itself produce a larger dispersion in the redshift-magnitude relation than is observed. In the absence of a more systematic estimate we shall take equation (77) as the upper bound for $\sigma_{\mu\gamma}$; this number is to be considered with even more caution than our previous estimates.

Turning now to the quadratic term in equations (70) and (75), there seems little hope of guessing which spherical harmonics in X have actually been measured. To get numerical estimates we arbitrarily assume that no large accidental numerical cancellations have occurred in X for the particular directions in which h_2 has been measured. For the numerical value of h_2 we choose (Sandage 1961)

$$0 \geq h_2 \geq -2h_1^2 \approx -2 \times 10^{-20}\ \text{year}^{-2}\ . \tag{78}$$

This limit is of course much less accurate than the estimate of h_1. We then get the following upper bounds:

$$\Lambda,\ E_{\mu\gamma},\ u_{(\mu\gamma\lambda)} \lesssim 2 \times 10^{-20}\ \text{year}^{-2}\ , \qquad \omega_{\mu\gamma} \lesssim 10^{-10}\ \text{year}^{-1}\ ,$$
$$\rho_0 \lesssim 10^{-19}\ \text{year}^{-2} \approx 10^{-28}\ \text{gm/cm}^3\ . \tag{79}$$

It is important to the following discussion that $H_{\mu\gamma}$ does not appear in the redshift formula, so that it is not bounded by the redshift measurements. The same is true of the part of $u_{\mu\gamma\lambda}$ that is antisymmetric in the first two indices.

On occasion, much has been made of the fact that the acceleration parameter appears to be negative and of order unity. If we are willing to consider inhomogeneities or anisotropies, then the present measurements are far from decisive. It may simply be that $E_{\mu\gamma}$ or $u_{(\mu\gamma\lambda)}$ makes a contribution of the required magnitude for the particular directions in which the acceleration parameter has been measured; since measurements in the galactic plane are difficult and few, the angular variation of the second spherical harmonic term contributed by $E_{\mu\gamma}$ would not be easy to detect. Moreover the terms in $\sigma_{\mu\gamma}$ contribute with the desired sign to X and, if $\sigma_{\mu\gamma}$ were close to the limit (eq. [77]) set by the cluster data, could contribute an appreciable part of the observed acceleration parameter. In any case, it seems premature to use the observed value of h_2 to distinguish between models until better information on angular variations is available. A first step would be to improve the estimate of equation (77).

For the area of a distant object we get the very simple form

$$dA_G = r^2 d\Omega_0\ . \tag{80}$$

This relation holds in all theories, not just in general relativistic ones, and it is exact. Thus, any observed major deviation from equation (80) would be a catastrophe from the theoretician's viewpoint. Excluding this possibility, equation (80) can be combined with equations (68) and (70) in a variety of useful ways if dA_G is more accurately known than I_G, I_0, or f_G/f_0. We then eliminate the least accurate quantity from these three equations to get a relation between observable quantities and the parameters of Table 1. In fact, in practice it is often necessary to work with surface brightness rather than I_G and then equation (80) is needed. Moreover, it may sometimes be possible to measure $d\Omega_0$ but not f_0. In the case of the Friedman models, for which the distortion vanishes, the apparent shape of the galaxy remains unchanged, and equation (80) then says that its apparent angular diameter is proportional to its proper linear diameter and inversely proportional to r. This is why McVittie calls r (ξ in his notation) the distance by apparent size. If we use equation (75) to eliminate r in favor of z, we obtain the usual angular-diameter–redshift relation (McVittie 1956, eq. [9.405]).

An interesting effect is the distortion. Let $\boldsymbol{i}$ be a unit vector orthogonal to $\boldsymbol{e}$, so that it is tangent to the observer's unit sphere. With $\boldsymbol{e}$ fixed we let the Cartesian components i_μ of $\boldsymbol{i}$ vary over their one-parametric set of allowed values. We maximize the scalar

$$i^\mu i^\gamma Z_{\mu\gamma} = i^\mu i^\gamma (E_{\mu\gamma} - 2e^\lambda \epsilon_{\lambda\mu\beta} H_{\beta\gamma} - e^\lambda e^\alpha \epsilon_{\lambda\mu\beta} \epsilon_{\alpha\gamma\zeta} E_{\beta\zeta}) + O(r)\ , \tag{81}$$

where $\epsilon_{\mu\gamma\lambda}$ is the Levi-Civita symbol. The corresponding $\boldsymbol{i}$ is the major axis of the elliptical image produced by an intrinsically spherical source. The ratio e of major to minor axis is, from equation (38),

$$e = 1 + \tfrac{1}{3} \max\,(i^{\mu}i^{\gamma}Z_{\mu\gamma})r^2 + O(r^3)\,. \tag{82}$$

(Note that e is *not* the magnitude of $\boldsymbol{e}$.) If the source is not intrinsically spherical, its image will be distorted in the same proportion and in the same direction $\boldsymbol{i}$. Thus, with the direction on the celestial sphere fixed, all distant objects will show an apparent statistical tendency to line up, and the effect is proportional to r^2. Suppose that at $r \approx 10^9$ years (that is, $z \approx 0.1$), we could still detect a distortion factor $e \approx 2$. Then, according to the estimate of equation (79), $E_{\mu\gamma}$ is too small to be detectable. But a detectable contribution from $H_{\mu\gamma}$ would occur if

$$H_{\mu\gamma} \gtrsim 10^{-18}\ \text{year}^{-2}\,. \tag{83}$$

The authors have tried unsuccessfully to find theoretical or observational evidence that such large values of $H_{\mu\gamma}$ cannot occur. They feel that a direct measurement of the distortion effect would be the best way to measure or exclude them. In terms of the redshift, equation (83) reads

$$e - 1 = \max\,(i^{\mu}i^{\gamma}Z_{\mu\gamma})z^2/(\tfrac{1}{3}\theta + e^{\lambda}e^{\beta}\sigma_{\lambda\beta})^2\,. \tag{84}$$

Note that the distortion vanishes for the Friedman models.

There are several favorable aspects of the distortion effect from the observational point of view. For $\boldsymbol{e}$ and r fixed, the distortion is the same for objects of any size and shape, so that all kinds of galaxies and even galactic groups could be included in the same statistics. It does not matter whether we see a whole galaxy or only its inner part. Even if absorption estimates are in error, this would affect only our numerical estimate of $H_{\mu\gamma}$, not the presence or absence of the effect. Finally it turns out that, although distortions can also be produced by the "lens effect" (light from the distant galaxy passing near another galaxy), these individual distortions are small and random, so that they do not mask the systematic cosmological effect. (We have used the estimate that light from a galaxy at 10^9 years will not on the average come closer than 10^4 years from the center of a second galaxy with a mass of 10^{10} solar masses.)

The distortion effect is also quite interesting from the theoretical point of view. In our other equations the curvature quantities $E_{\mu\gamma}$ and $H_{\mu\gamma}$ get mixed up with kinematical quantities; but only the curvature contributes to the distortion. However, the equations for the distortion effect do not depend on the assumption that general relativity is valid. Therefore, a positive result of such a measurement would merely indicate that space-time is curved (or that distant objects are systematically intrinsically lined up for some reason).

One additional possibility should be mentioned. It is conceivable that we happen to be in a region where $H_{\mu\gamma}$ is something like 10^4 times as great as the estimate given in equation (83) but that within a distance of 10^7 years the value of $H_{\mu\gamma}$ drops to zero or changes sign. Such a local fluctuation could be detected by analyzing the shapes of nearby objects, as in the work of Wyatt, Brown, and others (see Brown 1964 and the references given there).

The next relation of interest is the number of galaxies in solid angle $d\Omega_0$ between corrected luminosity distances r and $r + dr$. If the intrinsic number density of sources is $K\rho$, where K is a constant, then the number dN of sources is, from equation (39),

$$\begin{aligned} dN = K(f_G/f_0)r^2 dr d\Omega_0 \{\rho_0 + r(\rho_0\theta + \rho_{\mu}e^{\mu}) + \tfrac{1}{2}r^2[\rho_0 - \Lambda + \tfrac{4}{3}\theta^2 + \sigma_{\mu\nu}\sigma_{\mu\nu} \\ - \omega_{\mu\nu}\omega_{\mu\nu})\rho_0 + 2e^{\mu}(\rho_{\nu}u_{\nu\mu} + \rho_{\mu}\theta + \rho_0 u_{\nu\nu\mu}) + e^{\mu}e^{\nu}\rho_{\mu\nu}] + \ldots\}\,. \end{aligned} \tag{85}$$

If evolutionary effects are important, then the transition from equation (39) to equation (85) must be suitably modified. The observational evidence on number counts does not seem sufficiently clear at the moment for this relation to yield useful numerical estimates or upper bounds (Heckmann and Schücking 1959, 1962; McVittie 1956, 1959, 1962; Zwicky 1959). A very rough guess is $\rho_\mu \lesssim 10^{-8}$ year^{-1} and $\rho_{\mu\nu} \lesssim 10^{-16}$ year^{-2}. In analyzing the data on the linear correction to the Newtonian result, attention should be paid to the possibility of a first spherical harmonic term; the second correction contains spherical harmonics of orders 0, 1, and 2. In the actual observations it is often the indefinite integral of equation (85) over dr which is found. The integration can be performed directly after substituting for f_G/f_0 from equation (70). If desired, r can then be replaced by z as independent variable by using equation (70) again. Finally, if the desired independent variable is I_0, one substitutes from equations (68) and (70) (after performing the integration) to eliminate r and z. The resulting equations contain spherical harmonics of higher order than those which appear in equation (85). For the Friedman models, equations (69a), (69b), and (76) can be used to reduce (85) to the corresponding equation (9.304) of McVittie (1956).

Finally, we consider apparent proper motions. These are of interest mainly because they could provide an independent estimate or upper bound for the incident magnetic-type field $H_{\mu\nu}$. If $\boldsymbol{e}$ is kept fixed on a distant object its rate of change relative to a local gyroscope direction is, from equation (49),

$$de^\mu/dt = h^{\mu\nu}\{e^\beta(\sigma_{\nu\beta} + \omega_{\nu\beta}) + r[e^\beta(\sigma_{\gamma\beta} + \omega_{\gamma\beta})u_{\nu\gamma} - e^\beta E_{\nu\beta} + \tfrac{1}{2}e^\beta e^\gamma(u_{\nu\beta\gamma} - \epsilon_{\nu\beta\lambda}H_{\gamma\lambda}) - e^\beta e^\gamma e^\lambda(\sigma_{\nu\gamma} + \omega_{\nu\gamma})\sigma_{\beta\lambda}] + \ldots\} . \tag{86}$$

Here $h^{\mu\nu} = \delta^{\mu\nu} - e^\mu e^\nu$. Suppose that we can measure an apparent proper motion of 1″ per century. Then, according to our previous limits on $\sigma_{\beta\nu}$ and $\omega_{\beta\nu}$, the r-independent term in equation (86) is too small to be detectable. This means that, in practice, we can avoid the somewhat unrealistic appeal to a local gyroscope, since we can choose nearby galaxies as our standard of no proper motion to a high degree of accuracy. If a large number of nearby galaxies is used, then their peculiar motions will be averaged out. Thus, apparent proper motions of distant galaxies or groups relative to nearby ones would mean that the distant galaxies, not the nearby ones, have an apparent proper motion relative to a local gyroscope. Of course it may well be that the term in equation (86) that is linear in r is also much too small to detect. Within the framework of this paper the point must be settled by observations, not by appeal to a philosophical argument that distant galaxies automatically show no proper motions.

Even if proper-motion studies could be made on objects at a distance of 10^9 years, then the upper limit we could get for quantities not estimated from the redshift is not very good:

$$H_{\nu\lambda},\ u_{[\lambda\nu]\beta} \approx 10^{-16}\ \text{year}^{-2} . \tag{87}$$

In particular, the upper limit placed by proper-motion measurements on $H_{\nu\mu}$ could probably never approach the accuracy of the distortion measurements (eq. [83]). However, the proper-motion measurements seem to be the only way to measure the part of $u_{\lambda\gamma\nu}$ that is antisymmetric in the first two indices. Note that the two terms estimated in equation (87) cannot be distinguished from each other by proper-motion observations; only the particular combination of them that enters into equation (86) can be measured by looking at de/dt. The apparent proper motion vanishes in the Friedman models.

It is often claimed that the universe in the large must be isotropic or homogeneous. Certainly this view has immense aesthetic and philosophical appeal, but is it strongly supported by current observations? Unfortunately, it is not. We can get a rough idea of whether the anisotropy and inhomogeneity parameters have observational bounds that are "small" in the dynamical sense by using the inverse Hubble constant as a time scale. We use units for which $\frac{1}{3}\theta \equiv 1$. Examination of equation (70) or similar dynamical equations arising from higher coefficients in the expansion of equation (50) now yields a rough but simple criterion: quantities that are large compared to unity are dynamically important; quantities small compared to unity are not dynamically important.

Our upper bounds in these units are

$$\sigma_{\mu\gamma} \lesssim 0.5\,, \quad \omega_{\mu\gamma} \lesssim 1\,, \quad \Lambda,\, E_{\mu\gamma},\, u_{(\mu\gamma\lambda)} \lesssim 2\,, \quad \rho_0 \lesssim 10\,, \quad \rho_\mu \lesssim 10^2\,,$$
$$\rho_{\mu\gamma},\, H_{\mu\gamma},\, U_{[\mu\gamma]\lambda} \lesssim 10^4\,. \tag{88}$$

Tentative as these numbers are, they show that, as yet, observations neither confirm nor deny the "cosmological principle" that the universe is isotropic and homogeneous, or even homogeneous, and that measurements at the present time cannot prove, but can only disprove, that particular models represent the actual structure of the universe.

In conclusion, it would be desirable to improve our considerations in several ways. There may be observable effects which we have not thought of. We can hope for more accurate numerical estimates from a careful study of existing data. And global theoretical models that are inhomogeneous should be looked for. As far as future measurements are concerned, it seems likely that an improvement of present accuracy by a factor between 10 and 100 might bring almost all the parameters in Table 1 within the range of observation. For $H_{\mu\nu}$, even an inspection of existing plates should yield a value or an improved upper bound. It may prove possible to improve our limit on $\sigma_{\mu\nu}$ by examining the angular dependence of the Hubble constant.

The idea of trying to analyze inhomogeneous models arose while the authors were attending an excellent series of lectures by Prof. G. de Vaucouleurs of the University of Texas. We are also grateful to Prof. E. Schücking of the Relativity Center for stimulating discussions on the theoretical aspects of this paper. A critical reading of the manuscript by Dr. R. H. Boyer was extremely helpful.

This research was supported in part by the Aerospace Research Laboratories, Office of Aerospace Research, and by the Office of Scientific Research (Grant 454-63), U.S. Air Force.

Note added in proof.—Some of the ideas and arguments in § II have been presented independently by David M. Zipoy ("Light Fluctuations Due to an Intergalactic Flux of Gravitational Waves" [preprint: University of Maryland Department of Physics and Astronomy, 1965]).

REFERENCES

Birkhoff, G. D. 1923, *Relativity and Modern Physics* (Cambridge, Mass.: Harvard Univrsity Press), p. 124.
Bondi, H. 1947, *M.N.*, **107**, 410.
Brown, F. G. 1964, *M.N.*, **127**, 517.
Ehlers, J. 1961, *Akad. Wiss. Mainz*, **11**, 804.
Heckmann, O., and Schücking, E. 1959, *Encyclopedia of Physics*, **53** (Berlin: Springer-Verlag), 489.
———. 1962, in *Gravitation: An Introduction to Current Research*, ed. L. Witten (New York: John Wiley & Sons).
Hoyle, F. 1962, in *Evidence for Gravitational Theories*, ed. C. Møller (New York: Academic Press), p. 141.
Hoyle, F., and Narlikar, J. V. 1961, *Observatory*, **81**, 89.
Humason, M. L., Mayall, N. and Sandage, A. R. 1956, *A.J.*, **61**, 97.
Jordan, P., Ehlers, J., and Sachs, R. K. 1961, *Akad. Wiss. Mainz*, Vol. **1**.
McVittie, G. C. 1956, *General Relativity and Cosmology* (London: Chapman and Hall).
———. 1959, *Encyclopedia of Physics*, **53** (Berlin: Springer-Verlag), 445.
——— (ed.). 1962, *Problems of Extra-galactic Research* (New York: Macmillan Co.), Part III.

Murnaghan, F. D. 1938, *The Theory of Group Representations* (Baltimore: Johns Hopkins Press).
Penrose, R. 1960, *Ann. Phys.*, **10**, 171.
Pirani, F. A. E. 1962*a*, in *Recent Developments in General Relativity* (New York: Pergamon Press).
———. 1962*b*, article in *Gravitation: An Introduction to Current Research*, ed. L. Witten (New York: John Wiley & Sons).
Sachs, R. K. 1961, *Proc. R. Soc. London*, **A264**, 309.
Sachs, R. K., and Bergmann, P. G. 1958, *Phys. Rev.*, **112**, 674.
Sandage, A. R. 1961, *Ap. J.*, **133**, 355.
———. 1962, *Problems of Extra-galactic Research*, ed. G. C. McVittie (New York: Macmillan Co.), p. 359.
Thomas, T. Y. 1934, *The Differential Invariants of Generalized Spaces* (Cambridge: Cambridge University Press).
Trautman, A. 1962, *Recent Developments in General Relativity* (New York: Pergamon Press), p. 459.
Zwicky, F. 1959, *Encyclopedia of Physics*, **53** (Berlin: Springer-Verlag), 390.

EVIDENCE FOR THE 2π DECAY OF THE K_2^0 MESON*†

J. H. Christenson, J. W. Cronin,‡ V. L. Fitch,‡ and R. Turlay§
Princeton University, Princeton, New Jersey
(Received 10 July 1964)

This Letter reports the results of experimental studies designed to search for the 2π decay of the K_2^0 meson. Several previous experiments have served[1,2] to set an upper limit of 1/300 for the fraction of K_2^0's which decay into two charged pions. The present experiment, using spark chamber techniques, proposed to extend this limit.

In this measurement, K_2^0 mesons were produced at the Brookhaven AGS in an internal Be target bombarded by 30-BeV protons. A neutral beam was defined at 30 degrees relative to the circulating protons by a $1\frac{1}{2}$-in.$\times 1\frac{1}{2}$-in.$\times$48-in. collimator at an average distance of 14.5 ft. from the internal target. This collimator was followed by a sweeping magnet of 512 kG-in. at ~20 ft. and a 6-in.$\times$6-in.$\times$48-in. collimator at 55 ft. A $1\frac{1}{2}$-in. thickness of Pb was placed in front of the first collimator to attenuate the gamma rays in the beam.

The experimental layout is shown in relation to the beam in Fig. 1. The detector for the decay products consisted of two spectrometers each composed of two spark chambers for track delineation separated by a magnetic field of 178 kG-in. The axis of each spectrometer was in the horizontal plane and each subtended an average solid angle of 0.7×10^{-2} steradians. The spark chambers were triggered on a coincidence between water Cherenkov and scintillation counters positioned immediately behind the spectrometers. When coherent K_1^0 regeneration in solid materials was being studied, an anticoincidence counter was placed immediately behind the regenerator. To minimize interactions K_2^0 decays were observed from a volume of He gas at nearly STP.

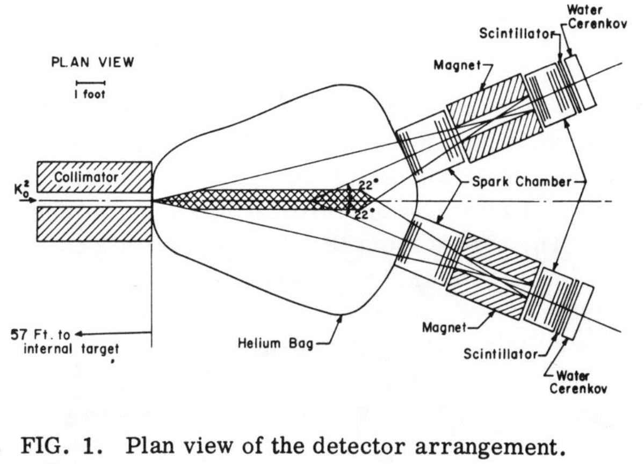

FIG. 1. Plan view of the detector arrangement.

The analysis program computed the vector momentum of each charged particle observed in the decay and the invariant mass, m^*, assuming each charged particle had the mass of the charged pion. In this detector the K_{e3} decay leads to a distribution in m^* ranging from 280 MeV to ~536 MeV; the $K_{\mu 3}$, from 280 to ~516; and the $K_{\pi 3}$, from 280 to 363 MeV. We emphasize that m^* equal to the K^0 mass is not a preferred result when the three-body decays are analyzed in this way. In addition, the vector sum of the two momenta and the angle, θ, between it and the direction of the K_2^0 beam were determined. This angle should be zero for two-body decay and is, in general, different from zero for three-body decays.

An important calibration of the apparatus and data reduction system was afforded by observing the decays of K_1^0 mesons produced by coherent regeneration in 43 gm/cm^2 of tungsten. Since the K_1^0 mesons produced by coherent regeneration have the same momentum and direction as the K_2^0 beam, the K_1^0 decay simulates the direct decay of the K_2^0 into two pions. The regenerator was successively placed at intervals of 11 in. along the region of the beam sensed by the detector to approximate the spatial distribution of the K_2^0's. The K_1^0 vector momenta peaked about the forward direction with a standard deviation of 3.4 ± 0.3 milliradians. The mass distribution of these events was fitted to a Gaussian with an average mass 498.1 ± 0.4 MeV and standard deviation of 3.6 ± 0.2 MeV. The mean momentum of the K_1^0 decays was found to be 1100 MeV/c. At this momentum the beam region sensed by the detector was 300 K_1^0 decay lengths from the target.

For the K_2^0 decays in He gas, the experimental distribution in m^* is shown in Fig. 2(a). It is compared in the figure with the results of a Monte Carlo calculation which takes into account the nature of the interaction and the form factors involved in the decay, coupled with the detection efficiency of the apparatus. The computed curve shown in Fig. 2(a) is for a vector interaction, form-factor ratio $f^-/f^+ = 0.5$, and relative abundance 0.47, 0.37, and 0.16 for the K_{e3}, $K_{\mu 3}$, and $K_{\pi 3}$, respectively.[3] The scalar interaction has been computed as well as the vector interaction

Reprinted from Physical Rev. Letters, 13, 138-140.

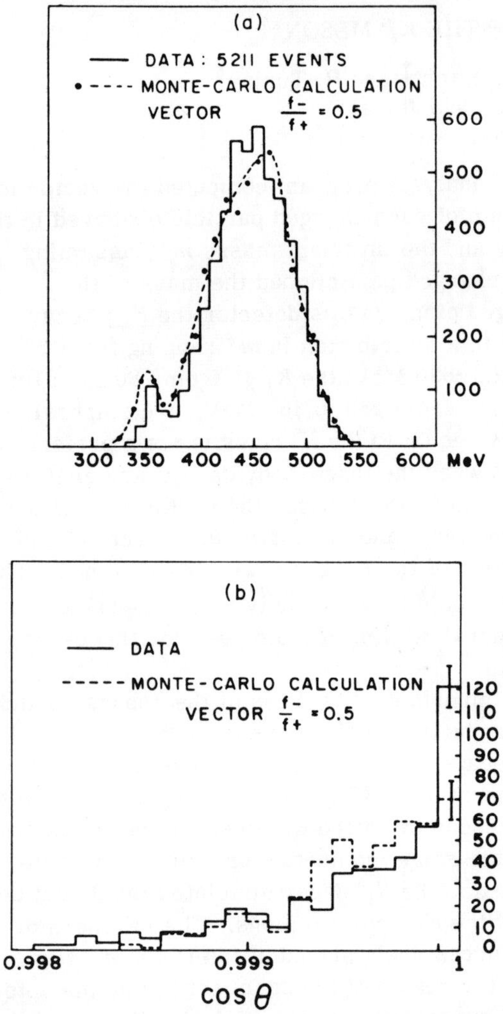

FIG. 2. (a) Experimental distribution in m^* compared with Monte Carlo calculation. The calculated distribution is normalized to the total number of observed events. (b) Angular distribution of those events in the range $490 < m^* < 510$ MeV. The calculated curve is normalized to the number of events in the complete sample.

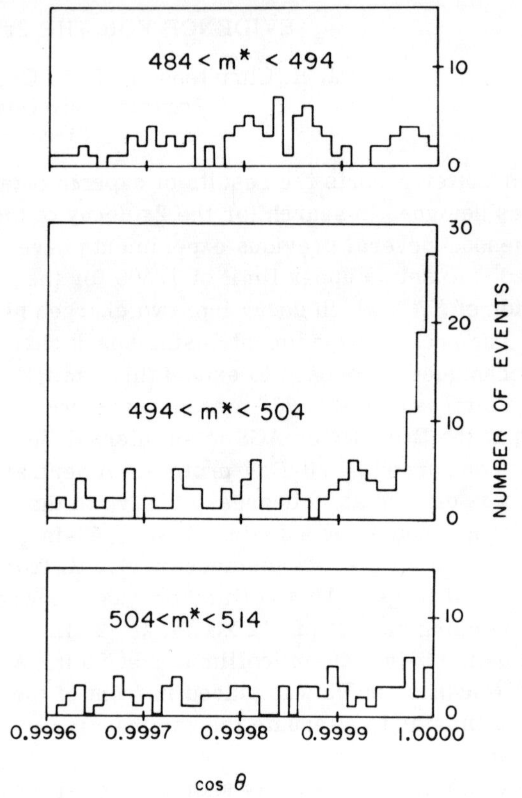

FIG. 3. Angular distribution in three mass ranges for events with $\cos\theta > 0.9995$.

with a form-factor ratio $f^-/f^+ = -6.6$. The data are not sensitive to the choice of form factors but do discriminate against the scalar interaction.

Figure 2(b) shows the distribution in $\cos\theta$ for those events which fall in the mass range from 490 to 510 MeV together with the corresponding result from the Monte Carlo calculation. Those events within a restricted angular range ($\cos\theta > 0.9995$) were remeasured on a somewhat more precise measuring machine and recomputed using an independent computer program. The results of these two analyses are the same within the respective resolutions. Figure 3 shows the results from the more accurate measuring machine. The angular distribution from three mass ranges are shown; one above, one below, and one encompassing the mass of the neutral K meson.

The average of the distribution of masses of those events in Fig. 3 with $\cos\theta > 0.999\,99$ is found to be 499.1 ± 0.8 MeV. A corresponding calculation has been made for the tungsten data resulting in a mean mass of 498.1 ± 0.4. The difference is 1.0 ± 0.9 MeV. Alternately we may take the mass of the K^0 to be known and compute the mass of the secondaries for two-body decay. Again restricting our attention to those events with $\cos\theta > 0.999\,99$ and assuming one of the secondaries to be a pion, the mass of the other particle is determined to be 137.4 ± 1.8. Fitted to a Gaussian shape the forward peak in Fig. 3 has a standard deviation of 4.0 ± 0.7 milliradians to be compared with 3.4 ± 0.3 milliradians for the tungsten. The events from the He gas appear identical with those from the coherent regeneration in tungsten in both mass and angular spread.

The relative efficiency for detection of the three-body K_2^0 decays compared to that for decay to two pions is 0.23. We obtain 45 ± 9 events in the forward peak after subtraction of background out of a total corrected sample of 22 700 $K_2{}^0$ decays.

Data taken with a hydrogen target in the beam also show evidence of a forward peak in the $\cos\theta$ distribution. After subtraction of background, 45 ± 10 events are observed in the forward peak at the K^0 mass. We estimate that ~10 events can be expected from coherent regeneration. The number of events remaining (35) is entirely consistent with the decay data when the relative target volumes and integrated beam intensities are taken into account. This number is substantially smaller (by more than a factor of 15) than one would expect on the basis of the data of Adair et al.[4]

We have examined many possibilities which might lead to a pronounced forward peak in the angular distribution at the K^0 mass. These include the following:

(i) $K_1{}^0$ coherent regeneration. In the He gas it is computed to be too small by a factor of $\sim 10^6$ to account for the effect observed, assuming reasonable scattering amplitudes. Anomalously large scattering amplitudes would presumably lead to exaggerated effects in liquid H_2 which are not observed. The walls of the He bag are outside the sensitive volume of the detector. The spatial distribution of the forward events is the same as that for the regular $K_2{}^0$ decays which eliminates the possibility of regeneration having occurred in the collimator.

(ii) $K_{\mu 3}$ or K_{e3} decay. A spectrum can be constructed to reproduce the observed data. It requires the preferential emission of the neutrino within a narrow band of energy, ±4 MeV, centered at 17 ± 2 MeV ($K_{\mu 3}$) or 39 ± 2 MeV (K_{e3}). This must be coupled with an appropriate angular correlation to produce the forward peak. There appears to be no reasonable mechanism which can produce such a spectrum.

(iii) Decay into $\pi^+\pi^-\gamma$. To produce the highly singular behavior shown in Fig. 3 it would be necessary for the γ ray to have an average energy of less than 1 MeV with the available energy extending to 209 MeV. We know of no physical process which would accomplish this.

We would conclude therefore that $K_2{}^0$ decays to two pions with a branching ratio $R = (K_2 \to \pi^+ + \pi^-)/(K_2{}^0 \to \text{all charged modes}) = (2.0 \pm 0.4) \times 10^{-3}$ where the error is the standard deviation. As emphasized above, any alternate explanation of the effect requires highly nonphysical behavior of the three-body decays of the $K_2{}^0$. The presence of a two-pion decay mode implies that the $K_2{}^0$ meson is not a pure eigenstate of CP. Expressed as $K_2{}^0 = 2^{-1/2}[(K_0 - \bar{K}_0) + \epsilon(K_0 + \bar{K}_0)]$ then $|\epsilon|^2 \cong R_T \tau_1 \tau_2$ where τ_1 and τ_2 are the $K_1{}^0$ and $K_2{}^0$ mean lives and R_T is the branching ratio including decay to two π^0. Using $R_T = \frac{3}{2}R$ and the branching ratio quoted above, $|\epsilon| \cong 2.3 \times 10^{-3}$.

We are grateful for the full cooperation of the staff of the Brookhaven National Laboratory. We wish to thank Alan Clark for one of the computer analysis programs. R. Turlay wishes to thank the Elementary Particles Laboratory at Princeton University for its hospitality.

---

*Work supported by the U. S. Office of Naval Research.

†This work made use of computer facilities supported in part by National Science Foundation grant.

‡A. P. Sloan Foundation Fellow.

§On leave from Laboratoire de Physique Corpusculaire à Haute Energie, Centre d'Etudes Nucléaires, Saclay, France.

[1]M. Bardon, K. Lande, L. M. Lederman, and W. Chinowsky, Ann. Phys. (N.Y.) 5, 156 (1958).

[2]D. Neagu, E. O. Okonov, N. I. Petrov, A. M. Rosanova, and V. A. Rusakov, Phys. Rev. Letters 6, 552 (1961).

[3]D. Luers, I. S. Mittra, W. J. Willis, and S. S. Yamamoto, Phys. Rev. 133, B1276 (1964).

[4]R. Adair, W. Chinowsky, R. Crittenden, L. Leipuner, B. Musgrave, and F. Shively, Phys. Rev. 132, 2285 (1963).

TWO-PION DECAY OF THE K_2^0 MESON*

W. Galbraith, G. Manning, and A. E. Taylor
Atomic Energy Research Establishment, Harwell, England

and

B. D. Jones and J. Malos
H. H. Wills Physics Laboratory, University of Bristol, Bristol, England

and

A. Astbury, N. H. Lipman, and T. G. Walker
Rutherford High Energy Laboratory, Didcot, Berkshire, England
(Received 15 February 1965)

The existence of the decay $K_2^0 \to \pi^+ + \pi^-$ was first observed by Christenson et al.[1] They determined the branching ratio, R, of this mode compared to all charged decay modes of the K_2^0 as $(2.0 \pm 0.4) \times 10^{-3}$ at a K_2^0 momentum of 1.1 GeV/c. This Letter reports the preliminary results of an experiment at Nimrod in which the two-pion decay mode is clearly distinguished from the background of three-body decays of the K_2^0, confirming the earlier observation.[1] The branching ratio R is found to be $(2.08 \pm 0.35) \times 10^{-3}$ averaged over the K_2^0 momentum spectrum which extends from 1.5 to 5.0 GeV/c. A comparison of this result with that of Christenson et al.[1] rules out a variation of the branching ratio with the square of the K_2^0 total energy. Such a variation has been suggested on the basis of an interaction of the K_2^0 meson with a long-range vector field.[2,3]

A neutral beam taken at zero degrees from an internal target (6 in. of copper) emerges from the shield wall. A lead converter 2 in. thick and a sweeping magnet remove γ rays and their charged products from the beam. The resulting neutral beam is 17 cm wide and 8 cm high (base widths) at entry to a vacuum chamber 40 m from the machine target. The minimum inner dimensions of the vacuum chamber are 45 cm wide and 20 cm high, its length is 5 m, and the entrance and exit windows are of Mylar (0.01 in. thick).

The charged products of decay of K_2^0's within the vacuum chamber are momentum analyzed by two spectrometer systems each comprising four sonic spark chambers and a bending magnet. All the spark chambers are triggered by a six-fold coincidence of scintillation counters. The sonic chamber data, consisting of scalar readings from which the coordinates of the sparks can be determined, are recorded on magnetic tape and later analyzed by an Orion computer. The spark chambers determine particle positions to an accuracy of ±0.3 mm.[4]

Each event is analyzed to find the decay point (x_0, y_0, z_0), the mass M_0, the momentum P_0, and the direction of motion θ_0 (with respect to the beam direction) of the decaying particle. The assumption is made that the decay is to two bodies only, both being π mesons. Only decays occurring within a fiducial volume 16 cm wide, 10 cm high, and 460 cm in length, were accepted, thus ensuring that all decays took place well within the vacuum chamber. The apparatus has an estimated mass resolution $\delta M_0 = \pm 3$

Reprinted from Physical Rev. Letters, 14, 383-386.

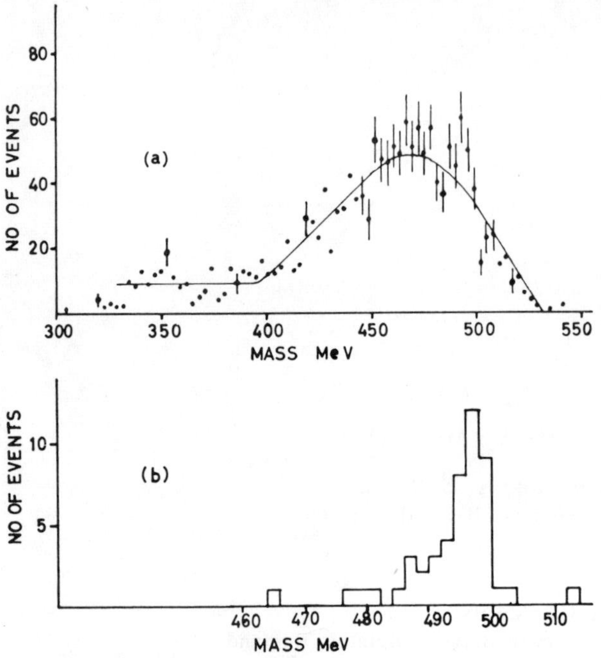

FIG. 1. Distribution of the apparent mass M_0 of the decaying particle, assuming that the two decay products are π mesons. (a) With no restriction on the angle θ_0. (b) With θ_0 restricted to values 0-2 mrad; note that the horizontal scale has been expanded in comparison with Fig. 1(a).

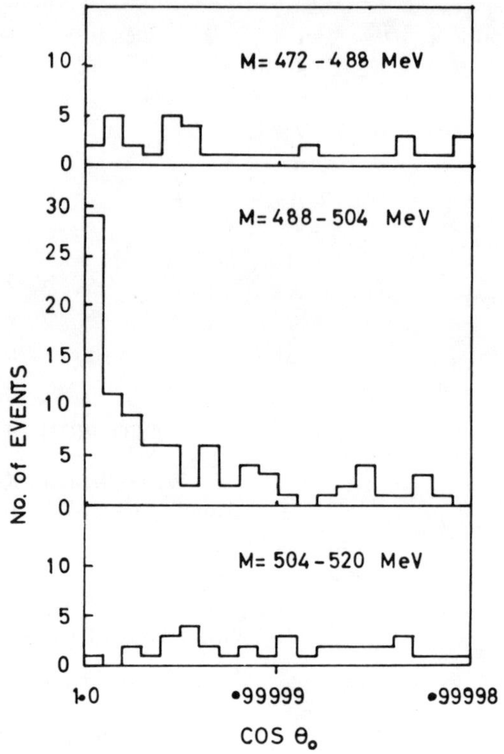

FIG. 2. Distribution in $\cos\theta_0$ for particles in the mass intervals as shown.

MeV and an angular resolution $\delta\theta_0 = \pm 2$ mrad (standard deviations) for decays to $\pi^+ + \pi^-$. The data presented in Figs. 1 and 2 were obtained with 80 000 Nimrod pulses (2 days effective running time) at an intensity of 5×10^{11} protons/pulse incident upon the internal target. Figure 1(a) shows the mass distribution, placing no restrictions on the angle θ_0. The broad peak from $M = 365$ MeV to $M = 530$ MeV results from the leptonic decays $K_{\mu 3}$ and K_{e3} which, having been analyzed as two body decays, give a spectrum of masses. The peak from 320 to 363 MeV is from the decay $K_2{}^0 \to \pi^+ + \pi^- + \pi^0$. The solid curve is the result of a Monte-Carlo calculation on the $K_{\pi 3} + K_{e3} + K_{\mu 3}$ modes, which is discussed below. The experimental data fit well to the predicted mass plot for three-body decays. Figure 1(b) shows the mass distribution for events with $\theta_0 \leq 2$ mrad, a restriction which implies that the two detected particles are coplanar with the incoming $K_2{}^0$ and have no net transverse momentum. This discriminates against most of the three-body decays, but two-body decays should be accepted. A narrow peak has been observed having a full width at half-height of 5 MeV, and centered at a mass of 497 MeV. This is equal to the mass of the K meson within the accuracy of calibration of the magnets of $\frac{1}{2}\%$. This result is strong evidence for the decay $K_2{}^0 \to 2\pi$. The peak at 497 MeV is also evident in the general mass plot [Fig. 1(a)]. The data are displayed in a different manner in Fig. 2. Here the number of detected events is plotted against $\cos\theta_0$ for three different mass ranges. A sharp peak is seen in the mass range containing the $K_2{}^0$ mass, in contrast to the flat distributions seen in the adjacent mass ranges above and below. The Monte-Carlo calculation for the three-body decays shows no such peaking in the $\cos\theta_0$ plot. The first interval used in the plot of $\cos\theta_0$ corresponds to the angular range 0-1.4 mrad.

After correction for background, there are 54 events in the forward peak. The decay points of these events occur throughout the fiducial volume and lie within the known region of $K_2{}^0$ flux. The distribution of points along the beam direction is consistent with the lifetime of the $K_2{}^0$ meson and completely inconsistent with that of the $K_1{}^0$. The possibility of these events arising from $K_1{}^0$ regeneration in the walls of the vacuum tank is ruled out, because such events would have values of $\theta_0 > 10$ mrad. Moreover the fiducial volume requirements were relaxed to include the regions close to the walls of the vacuum chamber and no decay points were found close to the walls. We conclude therefore that these events are examples of the decay $K_2^0 \to \pi^+ + \pi^-$.

The experiment has been simulated by a Monte-Carlo calculation for the charged decay modes $K\pi 2$, $K\pi 3$, $Ke3$, and $K\mu 3$, assuming a momentum spectrum of the K_2^0's calculated from yield measurements of K^+ and K^- measured at Nimrod.[5] The published branching ratios[6] for $K_{\pi 3}$, K_{e3}, and $K_{\mu 3}$ of 12%, 38%, and 27% have been used and a V-A interaction for the leptonic modes (with a ratio of the form factors $f^-/f^+ = 0.5$) has been assumed. The 3π decay spectrum was taken to be proportional to phase space. The geometry of the apparatus and all the selection criteria that were applied to the experimental data were used.

Histograms were obtained for the predicted spectra of detected events from each mode as a function of M_0, θ_0, P_0, and P_K (the K-meson momentum). In particular a comparison was made between the P_0 spectrum of the leptonic decays obtained in the experiment, and that calculated. This comparison is a sensitive test of the shape of the K_2^0 momentum spectrum fed into the Monte-Carlo calculation. Some modification of the K_2^0 spectrum was necessary to obtain exact agreement with the P_0 spectrum determined experimentally, and this modified K_2^0 momentum spectrum has been used in further calculations. Geometrical efficiencies for detecting each decay mode $K_{\pi 2}$, $K_{\mu 3}$, and K_{e3}, averaged over the K_2^0 momentum spectrum and the fiducial decay volume, were obtained from the Monte-Carlo calculation:

$$\epsilon_{\pi 2} = 1.4\times 10^{-3}, \quad \epsilon_{\mu 3} = 1.0\times 10^{-4},$$

$$\epsilon_{e3} = 0.9\times 10^{-4}.$$

The number of events observed in the present experiment were

$$N_{\pi 2} = 54 \pm 9; \quad N_{l3} = 1440 \pm 50.$$

From these figures the branching ratio

$$R = (2.08 \pm 0.35)\times 10^{-3}$$

is obtained. It should be remarked that the value of R changes by only 5% if the original K_2^0 spectrum is used, although the individual efficiencies change by a factor of 1.5. This is due to the fact that the efficiencies $\epsilon_{\pi 2}$, $\epsilon_{\mu 3}$, and ϵ_{e3} vary with momentum in a very similar way.

If the branching ratio R is proportional to the square of the total energy of the K_2^0 meson[2,3] and we use the result of Christenson et al.,[1] $(2.0 \pm 0.4)\times 10^{-3}$ at an effective momentum of 1.1 GeV/c, we would expect a value of R of $(13.4 \pm 3.0)\times 10^{-3}$. The error is predominantly determined by the scaling up of the error in R from Christenson et al.'s experiment and to a lesser extent by the uncertainty in our K_2^0 momentum spectrum. We have taken no account of the spread in energy of the K_2^0 mesons in the former experiment.

We, therefore, confirm that the K_2^0 meson decays into two charged pions in vacuum; and, further, we conclude that the decay rate is not proportional to the square of the total energy of the K_2^0 meson. Thus the source of apparent CP violation in this decay mode lies elsewhere than in the long-range vector field.[2,3]

As an alternative to CP violation, it has been suggested[7,8] that the decay $K_2^0 \to 2\pi$ can be explained by assuming the existence of a new particle K_3^0 with the same mass as the K_2^0 and decaying into two pions, but with a different lifetime. Assuming that the K_2^0 and K_3^0 are produced at the machine target and that the K_3^0 momentum spectrum at production is similar in shape to that of the K_2^0, we deduce from the present data that the lifetime is longer than 10^{-8} sec.

We are grateful to Miss M. Taylor, Mr. P. Dagley, Mr. P. Day, Mr. A. Hussri, Mr. A. G. Parham, Mr. B. T. Payne, Mr. T. J. Ratcliffe, and Mr. A. C. Sherwood, and members of the Rutherford Laboratory Computer and Electronics Groups for assistance in this experiment.

---

*Accepted without review under policy announced in Editorial of 20 July 1964 [Phys. Rev. Letters 13, 79 (1964).

[1]J. H. Christenson, J. W. Cronin, V. L. Fitch, and R. Turlay, Phys. Rev. Letters 13, 138 (1964).

[2]J. S. Bell and J. K. Perring, Phys. Rev. Letters 13, 348 (1964).

[3]J. Bernstein, N. Cabibbo, and T. D. Lee, Phys. Letters 12, 146 (1964).

[4]B. D. Jones, J. Malos, W. Galbraith, and G. Manning, Nucl. Instr. Methods 29, 115 (1964); see also

CERN Report No. 64-30, 191, 1964 (unpublished).

[5]A. J. Egginton et al., to be published.

[6]See, for example, D. Luers, I. S. Mittra, W. J. Willis, and S. S. Yamamoto, Phys. Rev. 133, B1276 (1964). The value we obtain for the branching ratio R is insensitive to the values taken for $K_{\pi 3}$, $K_{\mu 3}$, and K_{e3}.

[7]H. J. Lipkin and A. Abashian, Phys. Letters 14, 151 (1965).

[8]J. L. Uretsky, Phys. Letters 14, 154 (1965).

THE EQUIVALENCE OF INERTIAL AND GRAVITATIONAL MASS*

R. H. Dicke
Palmer Physical Laboratory, Princeton, N. J.

I propose to discuss an experiment performed with my collaborators P. Roll and R. Krotkov, and its relevance to the recently observed anomalous 2π decay of the K_2 meson (Christenson, Cronin, Fitch, and Turlay 1964.) This experiment covered a period of several years and had substantial assistance from many in our gravitational research group. Particular note should be taken of the very substantial contributions of B. Block, R. Moore, and D. Curott. The experiment was previously reported in the Annals of Physics (Roll, Krotkov, and Dicke 1964).

Experiments on the equivalence of inertial and gravitational mass have been performed many times in the past, and what may have been the first of all fundamental physical investigations was on this problem. One could readily imagine that one of our paleolithic ancestors, somewhat brighter than normal, may have noted and wondered that a fat lizard knocked from a high limb of a tree by a stone fell to earth at the same time as the stone. This old experiment (the details and accuracy of which are somewhat cloudy) was followed by numerous others. Galileo, Newton, Bessel, Eötvös, Zeeman, Potter, Southerns, and Renner all contributed. The high point in this series of classic measurements was Eötvös' determination to an accuracy of a few parts in 10^9 that a variety of substances, including the exotic schlangenholtz, fell to earth with the same acceleration (Eötvös, Pekar, and Feteke 1922). By contrast the experiment of the Princeton group showed that the acdeleration toward the Sun of gold and aluminum are equal with an accuracy of 1×10^{-11}. Some idea of the required sensitivity can be obtained by noting that this requires the detecting of a relative acceleration as small as 6×10^{-12} cm/sec^2. Starting from rest a body would reach the enormous velocity of 1.2×10^{-4} cm/sec after being accelerated a whole year at this rate.

Before discussing the experiment I shall briefly consider its significance for the $K_2 \rightarrow 2\pi$ decay. This decay scheme strongly suggests that the K_2 meson is in a state which is a superposition of the eigenstates of C.P. and hence by implication that this state is not time-reflection-invariant. A possible way of obtaining this result without violating the basic time reflection invariance of physical laws is to introduce some new long-range boson field, a field generated by distant parts of the universe, and then introduce an interaction of the K meson with this field. The basically unsymmetrical character of a matter-filled expanding universe might then be reflected in this interaction.

Without speaking directly to the problem represented by the anomalous decays of the K_2 mesons, it should be remarked that the Princeton experiment on the equivalence of inertial and gravitational mass sets stringent limits to the strengths of interactions with possible zero mass boson fields (additional to the vector electromagnetism and the tensor gravitation).

In considering other long-range fields the first possibility which might be examined would be a second-gauge invariant vector field, a field analogous to electromagnetism. Such a field having baryon number density as its source, was suggested by Lee and Yang (1955). Aside from the objection that the existence of such a field is incompatible with the symmetry requirements imposed by a uniform and isotropic universe composed of ordinary matter (Dicke 1962b), the Princeton experiments show that the strength of such an interaction must be less than 10^{-7} of the gravitational interaction. An electrostatic-type force on a body is proportional to its "charge," independent of internal motion in the body. Thus, its acceleration is inversely proportional to

*This work was supported by the National Science Foundation and the Office of Naval Research of the United States Navy.

its mass, being dependent upon the binding energy of the body (Lee and Yang 1955). By lifting the gauge invariance of the theory, the argument based on isotropy and uniformity is invalidated. Also a vector field coupled only to hypercharge could not be gauge invariant. (Hypercharge is not strongly conserved.) Nonetheless the argument concerning the strength of the interaction is still valid.

Another possibility worth considering is the existence of a second zero mass tensor field. Such a field generally causes anomalous gravitational accelerations, and the Princeton experiment sets stringent limits to their strengths also. Before discussing the interaction of a small body with two externally provided symmetric tensor fields, it is helpful to review the situation assuming only one such tensor field. General relativity is the classic example of such a theory, and it is well known that within the framework of general relativity all bodies fall with the same acceleration. The easiest way to show this formally is to choose a coordinate system locally Minkowskian at one point and then to note that if the body is sufficiently small for the effect of higher derivatives of the metric tensor to be negligible, the body thinks that it is in an inertial coordinate frame in flat space and it does not accelerate. Furthermore this non-accelerability is structure-independent (so long as the effects of higher derivatives can be neglected for all the bodies considered).

When two tensor fields are interacting with a body, there is little virtue in calling one of them a metric tensor. Furthermore either one, but only one, can be reduced to a unit diagonal form with vanishing first derivatives. The remaining tensor interacts with a particle to generate two types of force, an "inertial force"

$$\frac{d}{d\lambda} \frac{h_{ij}\, u^j}{\sqrt{(-h_{jk}\, u^j u^k)}}$$

and a "gravitational force"

$$-\frac{1}{2} \frac{h_{jk,\,i}\, u^j\, u^k}{\sqrt{(-h_{jk} u^j\, u^k)}} .$$

These forces are dependent upon the velocity and acceleration of the particle, hence upon the details of the structure of the bound system. If two such tensor fields were present, different atoms would be expected to fall with different accelerations (Peebles 1962; Peebles and Dicke 1962; Dicke 1964). Such a second tensor field permits a host of possible special cases and the acceleration anomaly depends upon details.

The scalar field is a third possibility of a new long-range boson field. This has been discussed in numerous publications (see Dicke 1964 for a discussion and a partial list of references). It has been shown that a scalar field coupling to a particle leads to the particle mass being a function of the scalar. It has also been shown that a scalar interaction between two bodies is attractive and that it must be weak, of roughly the same strength as gravitation (Dicke 1962c). It has been shown that a universal scalar interaction with all particles, the masses all having the same fundamental dependence on the scalar, cannot cause an anomaly in the gravitational acceleration (Dicke 1962a).

The argument can be put in several different forms. Perhaps the clearest is the following. Assume that a bound system of particles interacts with externally determined scalar and tensor fields φ and g_{ij}. Assume that the variations of φ and g_{ij} over the bound system are sufficiently smooth that second and higher derivatives with respect to coordinates are negligible. Assume also that the self-scalar and self-tensor interactions are negligible. Assume that equations of motion are derived from some variational principle

$$0 = \delta \int L(g_{ij},\ m(\varphi),\ \ldots)\, d^4 x,$$

where L is also a function of other field variables, coupling constants, and possibly of particle coordinates, containing integrals over the paths of the classical particles. It is assumed that m depends upon φ only through the mass dependence

$$m(\varphi) = m_0\, f(\varphi);$$

$f(\varphi)$ is the same for all particles. It is evident that for any choice of coordinate system, the equations of motion of the particles must be independent of the units of measure employed. Introducing new units of measure for the same coordinate system by the "conformal transformation" (Dicke 1962b)

$$M \rightarrow M f^{-1}, \qquad L \rightarrow K f, \qquad T \rightarrow T f,$$

leads to the transformations

$$x^i \rightarrow x^i, \quad m(\varphi) \rightarrow m_0, \quad g_{ij}\, f^2, \hbar \rightarrow \hbar, \quad c \rightarrow c.$$

The scalar field no longer appears explicitly, having been absorbed into g_{ij}. The variational equation now is identical in form with that of the corresponding equation in general relativity. The general argument given above can be used to show that all bodies fall with the same acceleration (see also Dicke 1965).

There are many possible types of coupling for a scalar field. One particularly amusing one is a coupling to the electromagnetic field through its Langrangian density in the variational equation (Dicke 1964). This results in space acquiring dielectric properties. If the Maxwell Lagrangian density is multiplied by the scalar field quantity φ, the dielectric constant ϵ and reciprocal of the permeability μ^{-1} are proportional to φ. This results in the fine-structure constant varying as φ^{-1}. As the expansion of the universe would cause φ to vary with time, this would imply a time dependence of the fine structure "constant."

The Princeton experiment sets stringent limits to this type of contribution to the weight of an object. The electrostatic energy stored in the extra-nuclear field varies as Z^2, and this force would be relatively larger in the heavier elements. It can be said that this contribution to the weight of a heavy element is no greater than 10^{-11} of the ordinary gravitational force on the body, or more exactly that this type of scalar interaction with electrostatic energy is no greater than 10^{-8} of the gravitational interaction with the same energy.

It is interesting to note in passing that there are several types of direct observational evidence that the fine-structure constant has been substantially constant over 1-2 billion years. One approach is spectroscopic. It is based on an examination of the fine-structure splitting of a spectroscopic line from a distant galaxy (Minkowski). Another approach is based on pleochroic halos, radiation damage produced in ancient mica by α-particle tracks. It is possible to argue that the range of α-particle tracks in ancient times were substantially like those today. A third approach is based upon a comparison of α- and β-decay ages of ancient rocks.

A zero-mass scalar field coupling universally to all matter would result in the gravitational coupling constant $Gm^2/\hbar c$ being a function of the scalar, hence varying with position and time. Basing units of measure on m, $\hbar$, and c (equivalent to ordinary "weights and measures"), G will be a function of coordinates. (An examination of the cosmological solutions shows that gravitation would become weaker with time.) If the gravitational self-mass of a test body were large enough, its trajectory would depend upon the gradient of such a scalar. More generally, the reason for the composition independence of the gravitational acceleration (even in the presence of such a scalar field) is the small contribution to the weight having its origin in gravitation self energy. This remark can be generalized to the following:

If the effect of a scalar field is one of changing one or more weak coupling constants (controlling a negligible amount of the self energy of the system) it causes a negligible anomaly in the gravitational acceleration.

To summarize: A zero-mass scalar field coupling universally to all matter, or affecting the weak coupling constant, is without any noticeable effect on the gravitational acceleration. By contrast, the interaction of matter with a second vector or tensor field would be expected to lead to anomalies in the gravitational acceleration, unless the interaction were extremely weak, even compared with gravitation.

To turn now to the experiment, the design of the apparatus might be usefully discussed by contrasting it with the best of the classical apparatuses, that of Eötvös et al. (1922).

Eötvös used two 30-gm weights at the ends of a 40-cm horizontal rod, the whole suspended in an air-filled chamber by a thin supporting wire. The rotation of the rod was observed using the mirror and scale technique. Unfortunately, the twofold symmetry axis was broken by lowering one of the weights 80 cm on a second wire fastened to the end of the beam. This had the effect of greatly increasing the sensitivity to the effects of gravitational gradients and served no useful purpose. The method used is most conveniently described as the balancing of a component of the gravitational force acting on the weights against the "centrifugal force" due to the Earth's rotation.

Eötvös' experiment lacked a control observation (it was impossible to turn off the Earth's rotation). Because of this intrinsic difficulty we were convinced that we should use the acceleration toward the Sun instead of toward the Earth. This small acceleration (0.6 cm/sec^2) is reversed in sign every 12 hours, providing a control. To minimize what we facetiously called the "Eötvös effect," the effect of the gravitation due to Baron Eötvös' own body acting on his balance while he sat at the telescope observing, a torque many orders of magnitude greater than the tiny twist that we were hoping to measure, we introduced a threefold symmetry axis. The balance comprised three weights (two of aluminum and one of gold), arranged at the corners of an equilateral triangle (hence with zero gravitational quadrupole moment about the rotation axis) (see Fig. 1). We made

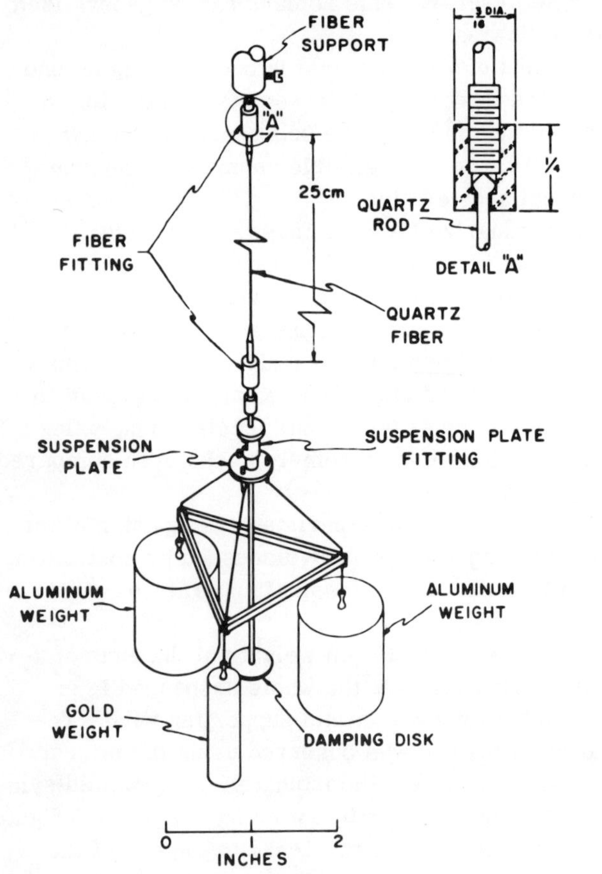

FIG. 1. Torsion balance

the dimensions of the triangle small, only 5 cm on the side, and most importantly we replaced Baron Eötvös by an impersonal optical-electronic instrument instructed to sit very quietly (Fig. 2). We would not have been concerned about the gravitational effects of Baron Eötvös' replacement but rather with the effects of such periodic phenomena as dew (with a 24-hour period), rainfall, and the 24-hour periodicity of cars in a parking lot 200 feet away.

The single most severe experimental problem was that caused by interactions between the pendulum and gas in the vacuum chamber. Some idea of the magnitude of this difficulty is seen by noting that if the chamber had been gas-filled, as was the Baron's, the velocity of flow of gas past one of the weights would have needed to have been less than 10^{-7} cm/sec (i.e., 3 cm/year). Even the introduction of a vacuum of 10^{-8} mm/Hg does not eliminate troubles, for a wind still blows, driven by temperature differences. The 24-hour period in temperature difference across the vacuum chamber must be less than 3×10^{-6} °C to make these winds of negligible importance. This was accomplished by surrounding the chamber by a series of concentric radiation shields of higher thermal conductivity, and most importantly by putting the apparatus in a specially constructed instrument well, 12 feet under ground with the top closed by an insulated thermal plug 3 feet thick. To avoid convective overturn in winter the plug was provided in its interior with an electric blanket. Also electric blankets were hung on the walls of the pit for the sole purpose of permitting temperature cycling experiments.

The severe temperature stability requirements suggested an unconventional vacuum system, a thoroughly baked stainless steel system with a tiny sealed-on "vac-ion" pump (see Fig. 3).

The requirements for freedom from ferromagnetic contaminants were very severe, and these requirements, plus those imposed by the necessity for a bakable system, suggested pure gold and aluminum for the weights. While we had no desire to employ schlangenholtz, we would have liked to have used lithium hydride for one of the weights. Dr. Block made a valiant effort to overcome the various problems but with substantially less than 100 per cent success.

The necessity of a good vacuum was absolute, but this introduced other problems. The damping of the pendulum was lost along with the air. Fortunately, the damping of the swinging mode of the pendulum was still quite good, probably because of the accompanying flexing of the suspension wires of the individual weights. However, the torsion mode of the pendulum was essentially undamped (largely because of the high Q of the quartz-fibre suspension) and rotational disturbances caused by trucks or assorted unpredictable gremlins would

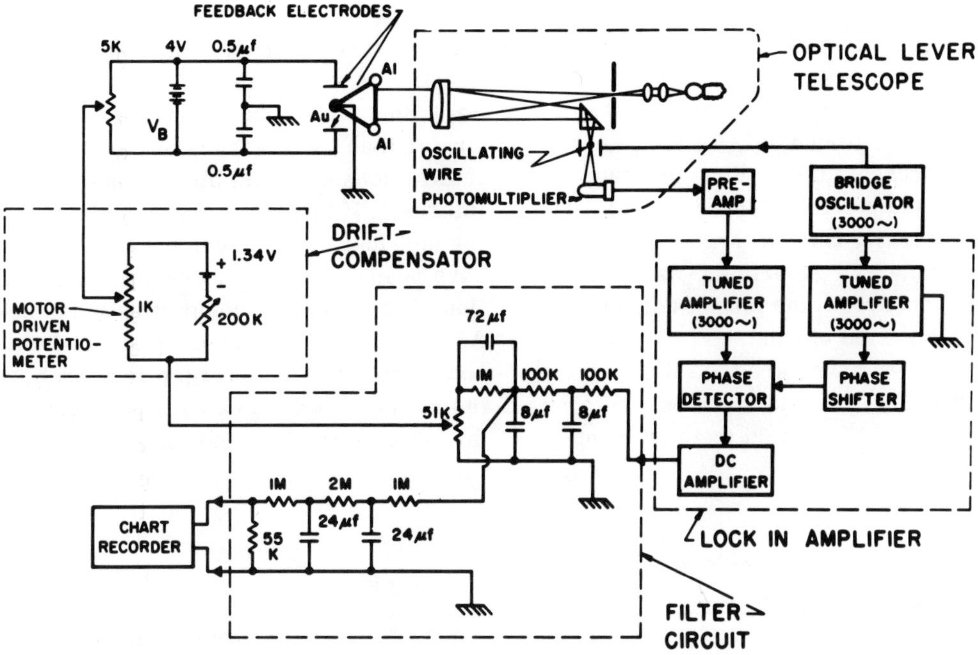

FIG. 2. Rotation detector and electrical feedback system

have persisted an inordinately long time. This difficulty was avoided by sensing the rotation, feeding back electrostatically on the pendulum rotation, and introducing the proper phase shift to damp the rotation. The feedback signal also increased the torsional oscillation frequency of the pendulum.

One incidental and somewhat amusing result of the feedback was an associated refrigeration of the torsional mode, reducing the amplitude of its Brownian motion by a factor of 10^3 implying a Brownian motion temperature of 3×10^{-4} °K. A more important result is the fact that non-Gaussian noise, in the form of short, strong, isolated disturbances could be eliminated from the data prior to filtering on the computer.

The "pit"—incidentally the experiment was sometimes called "the pit and the pendulum experiment"—was hung like a sausage shop with temperature sensors on cables. Two sensors strategically located relative to the vacuum chamber, on the supporting platform, were continuously monitored. Their readings were cross-correlated with the signal from the torsion balance to obtain measures of the temperature induced signal. Also the correlation was separately investigated by artificially cycling the temperature of the pit. Of the two sensors (T_2 and T_3), T_5 appeared to be the more reliable, giving the more consistent set of regression coefficients for this effect.

FIG. 3. Vacuum chamber

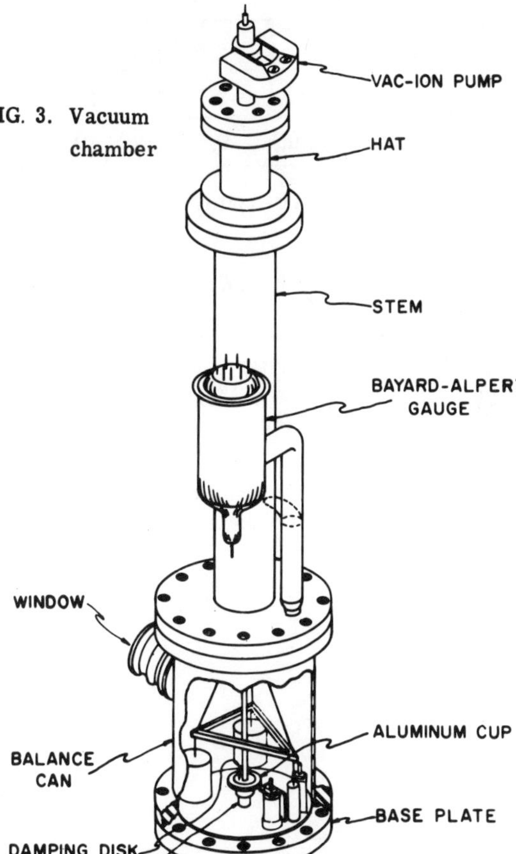

Fortunately as a result of all the precautions taken, the temperature-induced signal was sufficiently small that the data were improved only slightly by removing this contribution from the output signal from the torsion balance.

Table 1 summarizes the results of a series of data runs covering the period July 7, 1962—April 8, 1963. This represents thirty-nine separate data runs of various durations (38-86 hours). The apparatus was generally rotated through 180° after each data run. The whole experiment was remotely controlled and observed, the pit not being opened during this period of time.

Two questions often asked are the following: With a null good to 10^{-11}, would there be any reason to improve it? How could the next giant step forward be taken? My answer to the first question is a definite, "Yes." The conclusions regarding weak couplings to other fields are no firmer than the accuracy of this null. Nor are interactions of negligible strength of negligible interest! If this were a satisfactory way of judging physical importance, beta decay would never have been studied. Concerning the second question, space science could provide the next step. An experiment performed in an artificial satellite could employ the far larger acceleration toward the Earth (1,500 times as great) and still preserve the advantages of periodicity in the driving force.

Table I*

$$\text{Mean value of } \eta \text{ (Au, Al)} = \frac{(M_g/M_i)_{\text{Au}} - (M_g/M_i)_{\text{Al}}}{\langle M_g/M_i \rangle}$$

Telescope Orientation	No Correction for Temperature Variation	T_2 temperature Regression Subtracted	T_5 temperature Regression Subtracted
North	$2.2 \pm 1.4 \times 10^{-11}$	$1.4 \pm 1.3 \times 10^{-11}$	$1.3 \pm 1.25 \times 10^{-11}$
South	$-2.5 \pm 1.9 \times 10^{-11}$	$-0.3 \pm 2.7 \times 10^{-11}$	$1.3 \pm 1.73 \times 10^{-11}$
North + south	$-0.1 \pm 1.2 \times 10^{-11}$	$0.6 \pm 1.5 \times 10^{-11}$	$1.3 \pm 1.0 \times 10^{-11}$

*M_g/M_i = (gravitational mass)/(inertial mass).

REFERENCES

J. H. Christenson, J. W. Cronin, V. L. Fitch, and R. Turlay, 1964, Phys. Rev. Letters, 13, 138.

R. H. Dicke, 1962, Phys. Rev., 125, 2163.

_____. 1962b, ibid., 126, 1580.

——. 1962c, ibid., p. 1857.

_____. 1964, The Theoretical Significance of Experimental Relativity (New York: Gordon and Breach).

____. 1965, Ann. Phys., 31, 235.

R. V. Eötvös, D. Pekar, and E. Feteke, 1922, Ann. Phys., 68, 11.

T. D. Lee, and C. N. Yang, 1955, Phys. Rev. 98, 1501.

P. J. Peebles, 1962, Ann. Phys., 20, 240.

P. J. Peebles, and R. H. Dicke, 1962, Phys. Rev. 127, 629.

P. G. Roll, R. Krotkov, and R. H. Dicke, 1964, Ann. Phys. 26, 442.

CP INVARIANCE AND POSSIBLE NEW LONG RANGE INTERACTIONS

G. Feinberg
Department of Physics, Columbia University
New York, N. Y.

The experiment of Christenson et al.[1] appears to show that CP is not conserved in the decay of neutral K mesons. Nevertheless, there has been some resistance among theoretical physicists to accepting this conclusion, and correspondingly, a number of exotic alternative explanations of the experiment have been proposed[2]. I think that there are two main reasons for this resistance.

1. The magnitude of the CP violation, as measured say by the ratio of the amplitudes for $K_2 \rightarrow 2\pi$ and $K_1 \rightarrow 2\pi$ is very small, of order 10^{-3}. As a result it has been difficult to construct a theory in which CP is violated in any natural way.

2. After the non-conservation of parity was discovered, many physicists convinced themselves that CP was the "natural" space reflection operation. Such physicists are unwilling to give up CP invariance without an intensive search for alternatives.

If one wishes to preserve CP invariance of the weak interactions, it is not unreasonable to try to use the asymmetry of the environment between particles and antiparticles to account for the $K_2 \rightarrow 2\pi$ decay. Some possible effects of the local environment, such as the collimator and the helium, were considered[1] by the experimenters and others. These effects go under the general name of regeneration by interactions with matter. The conclusions are that the results cannot be accounted for by such regeneration using the known interactions of K mesons with matter[3].

The asymmetry of the environment under CP exists not only for the local environment, but also for the universe at large insofar as we have observed it. That is, the Earth, Milky Way, and presumably also the other galaxies are composed of matter rather than antimatter. If this fact is to be relevant to the decay of K_2^0 in the laboratory, some mechanism must be found by which distant bodies can act on the K_2 mesons. Such a mechanism we will refer to as a long range force. Some examples of long range forces have been known previously, i.e., electromagnetic and gravitational forces. However, these do not have the right properties to influence the K_2 decay. Specifically, the K_2 is uncharged, so that the electromagnetic forces acting on it must be very small. The gravitational effects of distant bodies are not small, but they do not distinguish between particle and antiparticle, and hence do not allow the decay $K_2 \rightarrow 2\pi$ to occur. It therefore seems that in order for there to be an influence of distant bodies on the K_2 decay, it is necessary to invoke a hitherto unknown long range interaction. From the previous discussion it can be seen that the new interaction must have the following properties:

1. The interaction with neutral K mesons must not vanish, as for electromagnetism.

2. The interaction must be different for K and $\overline{K}$, unlike gravitation.

If we make the reasonable requirement that the interaction should not vanish for a K meson at rest we are led to the conclusion that the interaction is proportional to the number of K^0 minus the number of $\overline{K^0}$. This quantity is one component of a 4 vector, the K meson current.

$$J_\mu = i\,(\overline{K}\,\partial_\mu K - \partial_\mu \overline{K}\,K) \qquad (1)$$

If we wish to preserve Lorentz invariance of the interaction, it is necessary to describe the new force by a 4 vector. This could either be a 4 vector field, or the derivative of a scalar field. As we shall see, both of these possibilities have been considered.

Let us suppose that such an interaction in fact exists and ask how it can account for the $K_2 \rightarrow 2\pi$ decay. We go into the rest system of the K meson. Then the effect of the interaction is just to produce a small mass difference between the K and

$\overline{K}^0$. The size of the mass difference is just proportional to the value of the fourth component of the vector describing the new field, which we call V. In this case $M_K - M_{\overline{K}} = 2V$. Given such a mass difference between K^0 and $\overline{K}^0$, it is straightforward to calculate which linear combination of K and $\overline{K}$ decay with simple exponentials in time. These are given by:

$$K_1 \propto K_0 + \overline{K_0}\,(x - \sqrt{1 - x^2}) \tag{2}$$

$$K_2 \propto K_0 + \overline{K_0}\,(x + \sqrt{1 - x^2}) \tag{3}$$

where

$$x = \frac{V}{M + \frac{i\Gamma}{2}} \tag{4}$$

and

$$2(M + \frac{i\Gamma}{2})$$

is the complex mass difference of K_1 and K_2 in the absence of V. Note the Γ when $x \ll 1$!

$$K_1 \propto (K_0 - \overline{K}_0) \equiv K_4 \tag{5}$$

$$K_2 \propto (K_0 + \overline{K}_0) + \frac{x}{2}(K_0 - \overline{K}_0) \tag{6}$$

So in this case the branching ratio

$$\frac{K_2 \to 2\pi}{K_1 \to 2\pi} \propto \frac{[x]^2}{4} = \frac{1}{4}\left(\frac{V}{M + \frac{i\Gamma}{2}}\right)^2 \tag{7}$$

On the other hand, for $x \gg 1$

$$K_1 \sim K_0$$
$$K_2 \sim \overline{K}_0 \tag{8}$$

and the branching ratio

$$\frac{K_2 \to 2\pi}{K_2 \to 2\pi} \sim 1$$

Already when $x \sim 1/20$, the 2π decay becomes the dominant decay mode of K_2.

We note that V was defined as the value of the fourth component of the new field in the rest system of the K^0. But if the energy of the K varies, the quantity V will vary also, since the field has a fixed value in the laboratory system. Since V is the fourth component of a vector, it follows that its value in the K rest system is proportional to the energy of the K in the laboratory, provided that the other components of the field 4 vector are zero or small in this system. The latter assumption is reasonable, since there is no preferred direction in the universe.

In the experiment of Christenson et al., the average energy of the K in the laboratory was about 1 Bev, and the value of V corresponding to the decay rate is about 10^{-8} eV. If the above hypothesis is correct, we would expect that as the energy of the K_2^0 in the laboratory is varied, the decay rate into 2π will vary as E^2, for $E \leq 10^3$ Bev. For energies of about 35 Bev, the 2π decay will be the dominant decay mode. Such an energy dependence which may be thought of as a violation of Lorentz invariance induced by the environments is a very spectacular effect, and should be easy to detect if true.

Theories of the Long Range Force

One would like to have a more detailed model of the long range interaction. The first such model proposed by Bernstein et al.[4] and by Bell et al.[5] was that of a vector field $X\mu$, coupled to the hypercharge, and of zero mass, or very small mass. Such a field would be produced by each proton, and would indeed give a $K - \overline{K}$ mass difference of the required type. One can estimate the strength of the interaction by assuming that the Compton wavelength of the field is large compared to the Hubble radius, or infinite, then the value of V is given by (in the laboratory)

$$V \frac{f^2 N_\rho}{R}$$

where f is the coupling constant of the field to hypercharge, where N_ρ is the number of protons in the universe, and R is the effective distance up to which protons contribute to V. This may be compared to the Machian equation

$$M_R = \frac{GN_\rho M_k M_\rho}{R} \tag{9}$$

$$f^2 = GM_\rho^2 \frac{V}{M_\rho} \sim 2 \times 10^{-39} \times 10^{-17}. \tag{10}$$

So in this case f^2 is much weaker than the dimensionless gravitational constant GM_k^2. If the Compton wavelength of the field is finite, then only bodies within such a wavelength contribute to V. But even if the Compton Wavelength is of the order of the diameter of the Earth, in which case only the earth would give a significant contribution to V, the coupling constant f^2 would still be ten order of magnitude weaker than gravitation.

It is interesting to note that if the Compton Wavelength of the field is taken as zero a sort of Olber's paradox arises. This is because the correct expression for V is ϑ

$$V = f^2 \int \frac{\rho_{ret}(\vec{x})}{|\vec{x}|} d^3 x \tag{11}$$

where ρ_{ret} is the hypercharge density evaluated at retarded time ($t_{ret} = t - |\vec{x}|/c$). In an infinite, static isotropic universe, this integral is quadratically divergent. Furthermore, it is not obvious that the redshift which keeps the energy flux from diverging is sufficient to keep V from diverging as well. It may therefore be that the requirement that V is finite would be a stringent condition to be satisfied by cosmological models.

The vector field theory has a serious difficulty associated with the fact that the hypercharge, or any other current containing the K^0 meson, is not conserved in weak interactions. One arrives at a contradiction in the conventional quantum theory of vector fields interacting with non-conserved currents, if one tries to make the mass of the vector field zero. Even if one makes the mass small but finite, the theory is still singular enough that various unpleasant consequences ensue[6]. Therefore, the vector field theory probably must be abandoned unless one can modify the conventional field theory. This may not be impossible.

In the meantime, an alternative theory has been proposed by Lee[7]. In this theory, a massless scalar field ϕ is introduced, whose derivative is coupled to the K^0 current only through the hypercharge. The interaction is therefore

$$g\, Y_\mu\, \partial_\mu \phi . \tag{12}$$

There is a well known theorem that such an interaction would have no effect if hypercharge were conserved, but as remarked earlier, this is not the case in weak interactions.

Lee notes that the Hamiltonian for the ϕ field does not contain ϕ at all but only its derivatives. It follows that the quantity $\pi_0 = \int \pi(x)\, d^3 x$ where $\pi(x)$ is the field conically conjugate to $\phi(x)$, is time independent. Suppose that in the world, the value of π_0 is nonzero at some time. Then it will remain nonzero. Now essentially one has $\pi_0 = \partial/\partial t \int \phi(x)\, d^3 x$. Hence the interaction Hamiltonian $g\, Y_\mu \partial_\mu \phi$ has a term proportional to

$$\frac{g\pi_0}{\Omega} (N_K{}^0 - N_K{}^0) \tag{13}$$

where Ω is the volume of space. This term is just of the type we have considered with $V = g\pi_0/\Omega$. Hence it will give a mass splitting of K and $\overline{K}$ and allow the decay $K_2 \rightarrow 2\pi$. Note that in this model, as in the vector model, the fundamental interaction is invariant under CP. The feature of the model which allows for CP violation is the non-vanishing value of π_0. If the world were CP symmetric, π_0 would necessarily have to vanish, since the quantities ϕ and π are odd under CP. Therefore, one must attribute the non-vanishing value of π_0 to the asymmetry of the world between particle and antiparticle. But in this model, it is not clear how the non-vanishing value of π_0 comes about, as this is not determined by the field equations. Note that the value of V only determines $g\pi_0/\Omega$, or the average value of $\partial\phi/\partial t$ over space multiplied by the coupling strength. An independent estimate of the coupling constant can be obtained by the requirement that the decay $K^+ \rightarrow \pi^+ + \phi$ occur at a rate less than 10^{-2} of the decay $K^+ \rightarrow \pi^+ + \pi^0$. This gives a limit for $g^2/4\pi \sim 10^{-3}$. With this value, one gets a lower limit for π_0/Ω. This lower limit corresponds to an energy density of $> 10^8$ Bev/cm^3. This is probably much too large a value to allow, and therefore one must regard this energy density as unobservable. This is not too unreasonable, since one characteristically neglects the zero point energies of all quantized fields.

It has been pointed out by Lee that independent of any particular model, the $K^0 - \overline{K^0}$ system is a particularly good analyzer for very small interactions, provided that these act differently on K^0 and $\overline{K^0}$. One can think of this system as an interferometer in which the phase difference measured is the relative phase of the K_1 and K_2 wavefunctions. If the mass difference parameter V is induced by an external field of spin J, then a simple extension of the previous argument shows that V will depend on energy at least by the factor E^J, or that the decay rate for $K_2 \rightarrow 2\pi$ will go as E^{2J}. It is therefore of great interest to perform a measurement of such energy dependence in any case. Such measurements are in progress, and hopefully, we shall soon know whether any such effects exist. If they do, one might also be able to determine the velocity of the earth relative to the "ether," defined as the coordinate system in which $\partial_\mu \phi = (0, \pi_0)$.

REFERENCES

[1] J. H. Christenson, et al. Phys. Rev. Letters 13, 138 (1964).

[2] For a review of these see G. Feinberg, Proc. of 3rd Eastern Theoretical Physical Conference, 1964.

[3] If one were willing to give the K° a small electric charge $\sim 10^{-5}$e it might be possible to explain the experiment of (1). Such an assignment would of course contradict the conservation of electric charge. This not only is unaesthetic, but also leads to difficulties similar to those described below.

[4] J. Bernstein, T. D. Lee, and N. Cabibbo, Phys. Letters 12, 146 (1964).

[5] J. Bell and J. K. Perring, Phys. Rev. Letters 13, 348 (1964).

[6] S. Weinberg, Phys. Rev. Letters 13, 495 (1964).

[7] T. D. Lee, preprint.

PART 6

GRAVITATIONAL COLLAPSE

GENERAL RELATIVISTIC HYDRODYNAMICS NEAR EQUILIBRIUM*

J. M. Bardeen
California Institute of Technology, Pasadena, California

We will discuss the behavior of spherically symmetric large masses near unstable or neutral equilibrium using a numerical method which assumes a δR for every mass element and finds how fast the velocity and kinetic energy increase as the radius changes. The method is aimed at bridging the gap between the perturbation analysis of equilibrium models in general relativity discussed by Chandrasekhar (1964a, b) and a full-scale difference equation approximation to the hydrodynamic equations.

The metric we work with is

$$DS^2 = Y^{-2}dt^2 - X^{-2}dr^2 - R^2 d\Omega^2. \tag{1}$$

Let $\dot{R} = (\partial R/\partial t)_r$, $R' = (\partial R/\partial r)_t$, $U = \dot{R}Y$, $Z = R'X$, $F = -X(Y'/Y)$, and $Q = -Y(\dot{X}/X)$. We choose r so that the baryon number inside r is independent of time. A specific volume V may then be defined so that

$$\frac{\dot{V}}{V} = -\frac{\dot{X}}{X} + 2\frac{\dot{R}}{R}, \tag{2}$$

and 1/V is the baryon number density within some factor which is independent of time. If the average molecular weight is independent of time (no nuclear reactions), it is convenient to take this constant to be the molecular weight times the atomic mass unit so the total energy density E is 1/V plus the thermal energy. Natural units, G = c = 1, will be used throughout. Let P be the pressure, which we assume to be isotropic in the baryon number rest frame. When 1/V is a rest-mass density we will define a rest mass or invariant mass M_0, a function of r only, by

$$M_0 = \int_0^r 4\pi R^2 dr/VX. \tag{3}$$

We will consider only adiabatic models here; so, E and P determine the energy-momentum tensor.

The Einstein equations (Landau and Lifschitz 1951) may be used to show (Podurets 1964; Misner and Sharp 1964)

$$Z = (1 + U^2 - 2M/R)^{1/2}, \tag{4}$$

where M is the total energy inside r,

$$M = \int_0^r 4\pi R^2 R' E dr = \int (EV) Z dM_0. \tag{5}$$

The analogue to the classical force equation is

$$Y\dot{U} = ZF - M/R^2 - 4\pi PR. \tag{6}$$

Also,

$$\dot{M} = -4\pi PR^2\dot{R}, \quad F = -\frac{XP'}{P + E}, \quad Q = \frac{XU'}{Z}. \tag{7}$$

There are two formulae for V. From equation (3)

$$V = \frac{1}{Z}\frac{\partial}{\partial M_0}\frac{4\pi}{3}R^3, \tag{8}$$

and from equation (2) and the last of equations (9)

$$Y\frac{\dot{V}}{V} = (R^2 U)'/R^2 R'. \tag{9}$$

First-order adiabatic perturbations of equation (6) away from equilibrium with time dependence $\delta R(r, t) = \delta R(r)e^{i\omega t}$ lead to the Sturm-Liouville equation

$$\frac{d}{d\Omega}\left(\frac{\Gamma P}{Y^3 Z}\frac{dg}{d\Omega}\right) - \left[\frac{4}{9}\frac{RY'}{Y}\left(1 - \frac{1}{4}\frac{RY'}{Y}\right) + \frac{8\pi}{9}\frac{PR^2}{Z^2}\right]\frac{(P + E)g}{Y^3 Z\Omega^2} + \frac{\omega^2}{9}\frac{(P + E)R^2 g}{YZ^3\Omega^2} = 0, \tag{10}$$

in which r = R, $\Omega = R^3/3$, $g = R^2 Y\delta R(r)$, and $\Gamma = -V/P(\partial R/\partial V)_S$. The boundary conditions on g are g/Ω finite at R = 0 and g finite at the surface (P = 0). The eigenfunctions of equation (10) are

***Supported in part by the Office of Naval Research [Nonr-220(47)] and the National Aeronautics and Space Administration [NGR-05-002-028].**

the normal modes of oscillation and the sign of the lowest eigenvalue determines the stability. If $\omega_0^2 = 0$ is the lowest eigenvalue the second-order variation of equation (6) will determine the stability.

If the fundamental mode is unstable ($\omega_0^2 < 0$), and the higher modes are stable, a model initially near equilibrium will evolve so most of the contribution to δR comes from the fundamental mode after a time the order of $1/\omega_0$. Even if $\omega_0 = 0$, the fundamental mode is the only one which can contribute to the overall growth of δR under the influence of the second-order perturbation. Thus the shape of δR and the velocity as a function of r will tend to that of the fundamental mode as long as $\delta R/R \ll 1$.

The first-order perturbation of equation (6) is related to virtual changes second order in δR in the total energy M of the mass and the second-order perturbation of equation (6) to third-order virtual changes in M. The compensating total kinetic energy will grow initially as $(\delta R)^2$ if $\omega_0^2 < 0$ and as $(\delta R)^3$ if $\omega_0 = 0$.

We will assume that $R^n \delta R$ maintains its shape, where n is a parameter, conveniently taken to be an integer, which can be varied to give the best extension of the calculation beyond the range where the perturbation and normal mode analysis is valid. At a moderately large number of points equally spaced in the initial radius R_0 the values of R, V, Z, M, $\dot{R}$ plus

$$\psi = R^n \delta R, \tag{11}$$

and

$$\varphi = 4\pi R_0^{2-n} \frac{\partial}{\partial M_0} (R^n \delta R) \tag{12}$$

are stored. Here ψ and φ are normalized so $\sigma = \delta R/R$ is some given small fraction of unity near the center of the mass. To go a step in time from $R = R_1$ to $R = R_2$ we calculate at each storage point

$$R_2 = R_1 + \delta R, \tag{13}$$

and

$$M_2 = M_1 - 4\pi P_1 R_1^2 \delta R (1 + \sigma - \frac{1}{2} \Gamma \frac{\delta V}{V}); \tag{14}$$

$\delta V/V$ is estimated from equation (9). The percentage change in M is of order PR^3/M compared to percentage changes in R and V, and for large masses little accuracy is lost by neglecting third- and higher-order terms in δM.

For Z_2 we have approximately

$$Z_2 = (1 + Y_1^2 \dot{R}_1^2 - 2M_2/R_2)^{1/2}$$
$$[1 + \frac{1}{2}(Y_2^2 \dot{R}_2^2 - Y_1^2 \dot{R}_1^2)]. \tag{15}$$

Assuming constant acceleration,

$$\dot{R}_2 = 2\frac{\delta R}{\delta t} - \dot{R}_1; \tag{16}$$

δr is a still adjustable parameter which will determine the magnitude of $\dot{R}_2$; the shape is specified by $\dot{R}_1$ and δR. Since Y is close to 1 in large masses, Y_2 may be set equal to Y_1 without much loss of accuracy. Or δY may be estimated for isentropic models from

$$Y = (P + E)V. \tag{17}$$

In general Y must be found by integration.

The correction term to the square root in equation (15) is carried along to first order in calculating

$$V_2 = (1+\sigma)^2 \left[V_1 Z_1 (1 - n\sigma) + \varphi \left(\frac{R_1}{R_0}\right)^{2-n} \right] \Big/ Z_2, \tag{18}$$

and $(EV)_2$ from V_2 by the adiabatic equation of state. The use of φ in equation (18) means that V_2 is calculated to the accuracy of the V of the initial model and the assumptions about δR, avoiding any finite difference approximation to $\partial R/\partial M_0$.

The parameter δt is determined by requiring

$$\int_0^{\text{surface}} [(EV)_2 Z_2 - (EV)_1 Z_1]\, dM_0$$
$$= \delta W_a + \delta W_b/\delta t + \delta W_c/(\delta t)^2 = 0. \tag{19}$$

The three integrals δW_a, δW_b, and δW_c are evaluated by Simpson's rule and sufficient accuracy is obtained with a few hundred radial points. The critical integral is δW_a, and this is similar to the integral for the binding energy, which can be checked in the initial model. Here δW_a is the change in the total energy holding $\dot{R}$ constant or, equivalently, is the negative of the increase in kinetic energy in going from time 1 to time 2. The total kinetic energy T is minus the running sum of the δW_a.

During the first few steps away from equilibrium the acceleration is not even approximately constant and δt is not a very good estimate of the time for the displacement to take place, which technically is infinite for the first step.

It is hard to devise a satisfactory test of the accuracy of the method when $PV \ll 1$. A calculation of the acceleration using a finite difference approximation for the pressure gradient is too sensitive to small errors in V and R, since typically the acceleration is only a small fraction of the free fall acceleration. Another check is the consistency of M with a calculation of $M_0 - M$ by Simpson's rule at an interior mass point. However, usually the kinetic energy is only a small fraction of $M_0 - M$ and a much smaller fraction of M. In models with $3/4\ \Gamma - 1 = 0.1$, 0.3, and 0.5 the accuracy seems to be best when n is 2 or 3. The larger the value of n, the faster any deviation of $\delta R/R$ from its value at $R = 0$ is accentuated in collapse and the faster $\delta R/R$ approaches uniformity in expansion. The quantity $\delta R/R$ is less in the outer part of the mass than in the center, and difference equation calculations show that $|U/R|$ decreases rapidly exterior to the location of the maximum value of $2M/R$ when this maximum value approaches one. One might expect the accuracy to be greatest for $n = 2$ because then ψ is closest to the eigenfunction g of the fundamental mode.

Our numerical calculations have been carried out on the IBM 7094 computer at the California Institute of Technology. The usual model has been an isentropic model in which Γ is a constant independent of radius and time. This is a good approximation to large masses, for which the ratio of gas to gas-plus-radiation pressure, β, is much less than 1 and almost a constant at constant entropy. Γ is a function only of β,

$$\frac{3}{4}\Gamma \approx 1 + \beta/8. \tag{20}$$

In our models

$$EV = 1 + \frac{PV}{\Gamma - 1}, \tag{21}$$

which differs from the gas-plus-radiation pressure result only in order β^2. With $\beta \ll 1$ the total mass is a function only of β, $M_0 \propto \beta^{-2}$. At $M_0 = 10^6\ M_\odot$ the value of Γ is given by $3/4\ \Gamma = 0.001$, assuming pure hydrogen composition with molecular weight 0.5.

The constant Γ models have a free scale factor such that, if α is the ratio of the masses of two models with the same Γ and the same value of PV at the center, corresponding radii are in the ratio α, corresponding specific volumes in the ratio α^2, and corresponding pressures and energy densities in the ratio α^{-2}. This allows more flexibility than when gas and radiation pressure are included explicitly.

Figure 1 is a plot of the fractional binding energy and the frequency of oscillation of the fundamental mode against central PV for a series of models with $3/4\ \Gamma - 1 = 0.001$, assuming a mass of $10^6\ M_\odot$ to obtain ω in $\sec^{-1}$. The fundamental mode becomes unstable at a maximum of the binding energy, a result true for any series of isentropic models with the same rest mass M_0.

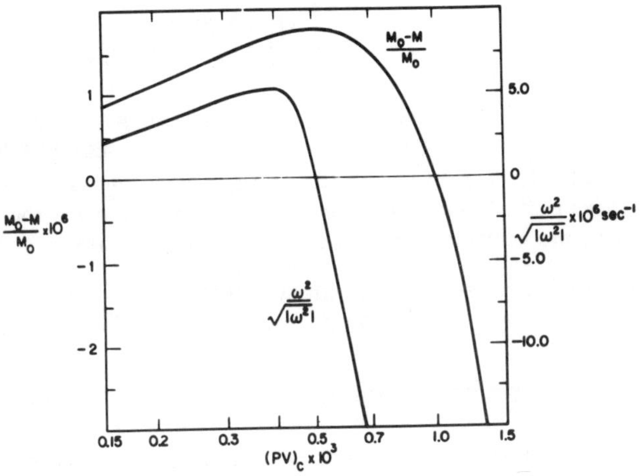

FIG. 1. Fractional binding energy and frequency of fundamental mode of radial oscillation are plotted against the central value of PV for a sequence of models with $M_0 = 10^6 M_\odot$ and $3/4\ \Gamma - 1 = 1.0 \times 10^{-3}$, corresponding to $\beta = 6 \times 10^{-3}$.

The numerical method described above was used to obtain the curves in Fig. 2, starting from models of Fig. 1 with $(PV)_C = 0.507 \times 10^{-3}$, 0.7×10^{-3}, and 0.85×10^{-3} and going in ($\delta R < 0$), out($\delta R > 0$), and in and out, respectively. The negative of the kinetic energy per unit mass, T/M_0, is plotted against the radius of a given mass point, at M_0 about 0.997 of the total, adjusted so the ratios between the initial radii for the three models are those of the total radii.

The first model is one for which $\omega_0 = 0$, and as expected T is approximately proportional to $(\delta R/R)^3$ for small $\delta R/R$. T soon starts increasing somewhat faster than this, once $\delta R/R$ is larger than a few per cent. A rough rule giving the initial

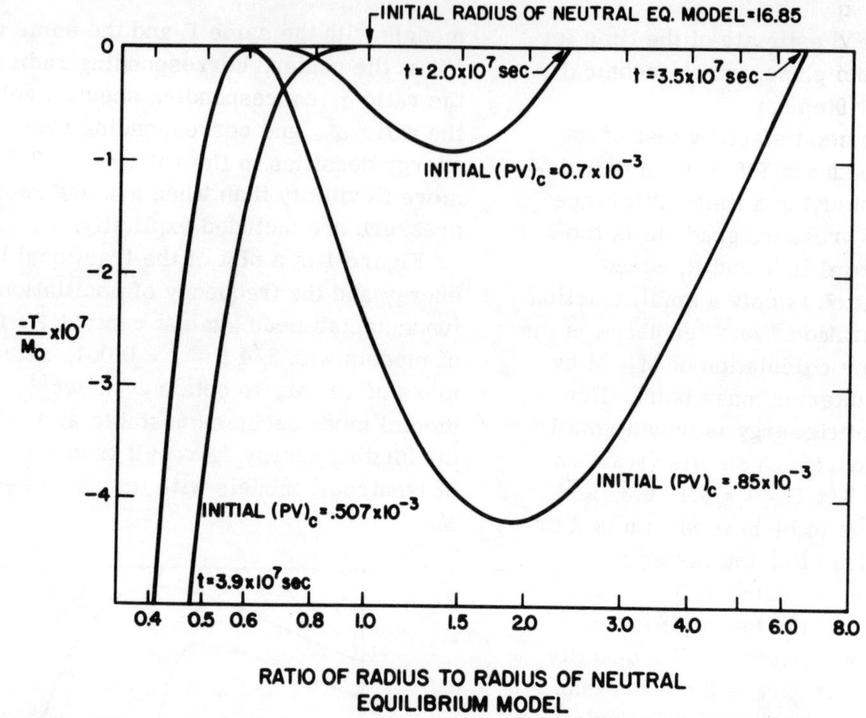

FIG. 2. Negative of kinetic energy per unit mass as a function of logarithmic change in radius is plotted for three unstable models of Fig. 1. The times given are rough estimates of the collapse or expansion time from near the unstable equilibrium. The initial radii are given relative to the radius of the neutral equilibrium model.

rate of increase of T for models with $3/4\,\Gamma - 1 \ll 1$ in neutral equilibrium is

$$\frac{T}{M_0 - M} / (\delta R/R)^3 \approx 0.3. \qquad (22)$$

Our calculations show explicitly that large masses in neutral equilibrium are in fact unstable to collapse, not expansion.

The initial equilibrium models for the other two cases are unstable in both directions. The peak in "potential energy" at the unstable equilibrium is a rather unsymmetrical one; the rate of increase of T rapidly becomes greater than that of $(\delta R/R)^2$ in collapse and less in expansion. The value of $T/(\delta R/R)^2$ near equilibrium can be estimated from the variational integral (Chandrasekhar 1964b) derivable from equation (10). The value of T in the first step away from equilibrium is, if $\delta R/R$ is small enough, equal to

$$-\int_{\text{all } M_0} \delta_2 (EVZ)\, dM_0, \qquad (23)$$

where δ_2 (EVA) is the second-order variation in δR. This is calculable, with the result,

$$\delta_2 (EVZ) = \frac{d}{dM_0} (\delta_2 M) + YZ$$

(integrand of variational integral before integration by parts).

The quantity $\delta_2 M$ is the actual second-order change in M during the displacement and is zero at the limits of integration. The factor of YZ occurs because the virtual energy for the virtual change in EVZ to test the stability comes from infinity and is blue-shifted before it makes its contribution to $\delta_2 (EVZ)$. The stability depends on the balance of the virtual energy at infinity. Except for the factor YZ, then, the integral (23) is the same as one of the integrals in the variational principle for ω^2 and the results check reasonably well. For $(PV)_C = 0.85 \times 10^{-3}$, $(T/M_0)/(\delta R/R)^2 = 2.2 \times 10^{-6}$ at $\delta R/R = -0.04$ and 1.6×10^{-6} at $\delta R/R = +0.10$. The value expected from the variational integral is 2.0×10^{-6}.

One would expect that the expansion from an unstable equilibrium would eventually stop if $M_0 - M < 0$ and would go to infinity if $M_0 - M > 0$. The calculations we have made show that the stopping point of the expansion is certainly becoming very large as $M_0 - M$ goes to zero.

The maximum velocity reached in the expansion of the $(PV)_c = 0.85 \times 10^{-3}$ model is 2.8×10^{-3} at the surface, well below the speed of sound in the bulk of the material of the object. For these models the average kinetic energy per unit mass is about 5.4 per cent of the square of the surface velocity.

The times shown in Fig. 2 are rough numerical estimates of the time required to go from near the unstable equilibrium to the end of the line on the graph.

The results of numerical integration of time and radially centered difference equations based on using equation (9) for V are presented in Fig. 3 for a model in which gas and radiation pressure

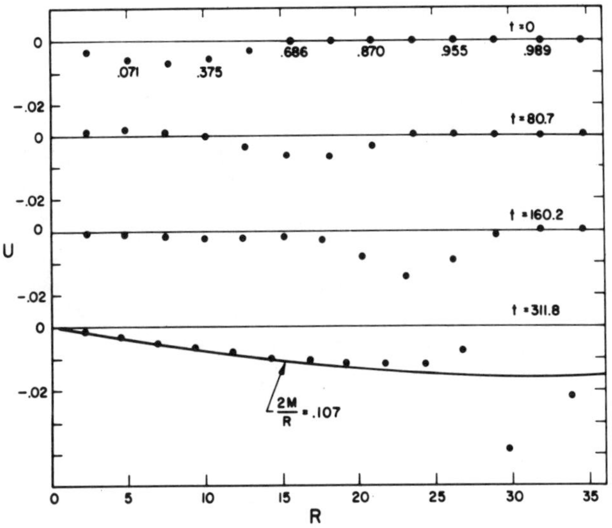

FIG. 3. The radial positions and velocities of several mass points are given at four different times in the collapse of an unstable mass with $M_0 = 1.09 \times 10^8 M_\odot$, molecular weight $\mu = 1.33$, $\beta = 2.67 \times 10^{-4}$, initial $T_c = 10^8\,°K$, and time constant for the unstable fundamental mode $1/|\omega| = 172$. Velocities, radii, and times are in natural units based on a unit mass of $10^8\ M_\odot$. The continuous curve is the shape of the fundamental mode.

are explicitly included, with molecular weight 1.33, rest mass 1.0875×10^8 solar masses, and initial central temperature $10^8\,°K$. The fundamental mode of the initial model is unstable, with a time constant, if unit mass is 10^8 solar masses and unit time is 500 sec, equal to 172. All higher modes of radial oscillation are stable. Thus even though the core of the object is given an initial inward velocity it cannot start collapsing until a rarefaction wave proceeding outward at the speed of sound reaches the surface, allowing the whole mass to collapse together with a velocity distribution roughly that of the eigenfunction of the fundamental mode, as is seen at the last time displayed.

The integration was carried out further, and when the largest value of 2M/R was about 0.2 at $t \approx 500$ the core started to collapse faster than the envelope until at $t \approx 560$ 2M/R became greater than 1 at a mass fraction of 0.15, with the velocity very sharply peaked near this radial position. The core never separated from the envelope in the sense that there was a region where the density was decreasing.

The difference equation result tends to support the use of our fixed δR method near equilibrium. One of the main applications of the latter method might be to use it as a starting program for a difference equation calculation, thus avoiding the expenditure of a lot of computer time when velocities are much less than the speed of sound. Fowler (1964) has considered the stability and initial collapse of large masses in the post-Newtonian approximation, obtaining results consistent with ours.

REFERENCES

S. Chandrasekhar, 1964a, Phys. Rev. Letters, 12, 114.

______. 1964b, Ap. J., 140, 417.

W. A. Fowler, 1964, Proceedings of the Second Texas Symposium on Relativistic Astrophysics,

L. Landau, and E. Lifschitz, 1951, The Classical Theory of Fields (Cambridge, Mass.: Addison-Wesley Press), p. 311.

C. W. Misner and D. H. Sharp, 1964, Phys. Rev., 136, B571.

M. A. Podurets, 1964, Soviet Astr., 8, 19.

THE EQUATIONS OF RELATIVISTIC SPHERICAL HYDRODYNAMICS

Charles W. Misner
Department of Physics and Astronomy
University of Maryland, College Park, Maryland
and
David H. Sharp
Palmer Physical Laboratory
Princeton University, Princeton, New Jersey

Hydrodynamics with spherical symmetry is not immensely more complex with relativistic gravitational fields than with Newtonian fields. In contrast, all dynamic problems with less than spherical symmetry can give rise to gravitational radiation, so the relativistic problem would contain entirely new degrees of freedom which are not present in the corresponding Newtonian problem. The reduction of the Einstein equations to a Newtonian-like form in the spherical case has been carried out independently by the authors (1964) and by J. M. Bardeen (see his paper in this volume). Instead of merely restating these equations in the previously published form (Misner and Sharp 1964) we shall present them in a rearranged form which May and White have developed and found suitable as a basis for numerical integration (see their paper in this volume).

We begin by stating the basic physical laws which we will subsequently specialize to the case of spherical symmetry. As a description of matter we take a fluid described by the stress-energy tensor

$$T^{\mu\nu} = \rho(1+\epsilon)u^{\mu}u^{\nu} + p(g^{\mu\nu} + u^{\mu}u^{\nu}) \tag{1}$$

$$= \begin{vmatrix} \rho(1+\epsilon), & 0 & 0 & 0 \\ 0, & p & & \\ 0, & & p & \\ 0, & & & p \end{vmatrix} .$$

Here u^{μ} is the 4-velocity of the fluid, and the second form of $T^{\mu\nu}$ above refers to a local Lorentz frame where $u^{\mu} = \delta^{\mu}_{0}$, thus showing that p is the pressure and $\rho(1+\epsilon)$ the energy density in the rest frame. The "matter density" ρ is just the baryon number density times the atomic mass of hydrogen, and ϵ is then the internal energy per mole of baryons. If other particles (electrons, photons, neutrinos, etc.) do not have an isotropic momentum distribution in the $u^{\mu} = \delta_0^{\mu}$ rest frame at the baryons, then equation (1) must be modified, and we would add to it a term

$$\Delta T^{\mu\nu} = \begin{vmatrix} \rho\Delta\epsilon, & q & 0 & 0 \\ q, & \Delta p & & \\ 0, & & \Delta p & \\ 0, & & & \Delta p \end{vmatrix} \tag{2}$$

arising from these anisotropies, as will be discussed in a second paper by us in this volume.

We can now list the laws of physics from which we shall begin:

1. Einstein's equations:

$$R_{\mu\nu} - \frac{1}{2} g_{\mu\nu}R = 8\pi(T_{\mu\nu} + \Delta T_{\mu\nu}). \tag{3}$$

2. Local energy and momentum conservation:

$$(T^{\mu\nu} + \Delta T^{\mu\nu})_{;\nu} = 0. \tag{4}$$

3. Conservation of matter (of baryon number)

$$(\rho u^{\mu})_{;\mu} = 0. \tag{5}$$

4. Equations of state:

$$p = p(\rho, \epsilon). \tag{6}$$

5. Equations of energy transport (here the adiabatic condition):

$$\Delta T^{\mu\nu} = 0. \tag{7}$$

In specializing these laws to the spherically symmetric case there are two critical steps: the choice of an appropriate comoving coordinate system, and the explicit integration of the Einstein constraint equations (those involving T^0_μ).

Our choice of coordinates is defined by taking the metric to be in diagonal form:

$$ds^2 = -e^{2\Phi}dt^2 + e^{\lambda}dr^2 + R^2(r, t)\,d\Omega^2, \tag{8}$$

with

$$d\Omega^2 = d\theta^2 + \sin^2\theta d\varphi^2, \tag{9}$$

and by requiring that these be comoving coordinates so that

$$u^\mu = (e^{-\Phi};\ 0,\ 0,\ 0). \tag{10}$$

The comoving proper-time derivative is then

$$D_t \equiv u^\mu \frac{\partial}{\partial x^\mu} = e^{-\Phi}\frac{\partial}{\partial t}. \tag{11}$$

The Einstein constraint equations (eq. [3] with $\nu = 0$) give a linear equation for $e^{-\lambda} = (g_{rr})^{-1}$ which can be solved (Misner and Sharp 1964) in the form

$$e^{\lambda} = R'^2/\gamma^2, \tag{12}$$

with

$$\gamma \equiv (1 + U^2 - 2mR^{-1})^{1/2}, \tag{13}$$

$$U \equiv D_t R, \tag{14}$$

$$m(r, t) = 4\tau \int_0^r \rho(1 + \epsilon)R^2 R' dr. \tag{15}$$

We will see that this quantity m enters various equations in a way which allows us to think of it as the total mass-energy enclosed in sphere of fixed r.

The equation of continuity also gives an expression for e^{λ}, for it states that the total number of baryons A(r) inside a sphere defined by the comoving coordinate r is independent of t, and gives for A(r) the expression

$$A(r) = \int_0^r \rho u^0 (-g)^{1/2} dr d\theta d\varphi$$
$$= 4\pi \int_0^r \rho e^{\lambda/2} R^2 dr, \tag{16}$$

so that

$$e^{\lambda/2} = A'/4\pi\rho R^2. \tag{17}$$

Note that since r is an arbitrary label attached to a fluid element, the choice r = A is always possible and gives A' = 1. May and White introduce the quantity

$$\Gamma \equiv 4\pi\rho R^2 (\partial R/\partial A) \tag{18}$$

where $(\partial/\partial A) = (1/A')(\partial/\partial r)$ to rewrite Eq. (17) in the form

$$\gamma = |\Gamma|, \tag{19}$$

where from Eq. (12) one has written $e^{\lambda/2} = |R'|/\gamma$. Negative values of R' which make the distinction between γ and Γ significant appear in the computations which May and White report, and also in the analytic examples such as the spherical Friedmann cosmology where $p = 0 = \epsilon$ and $\rho' = 0$.

By differentiating Eq. (17) and using the R^0_r Einstein equation to evaluate $D_t\lambda$ one obtains a generalization of the Newtonian equation of continuity, namely,

$$D_t\left(\frac{1}{\rho}\right) = \frac{1}{\Gamma}\frac{\partial}{\partial A}(4\pi R^2 U) \tag{20}$$

The statement of local energy conservation is

$$-u_\mu T^{\mu\nu}{}_{;\nu} = 0, \tag{21}$$

which reduces to

$$D_t\epsilon = -pD_t\left(\frac{1}{\rho}\right), \tag{22}$$

and thus gives, from the thermodynamic relation

$$d_\epsilon = Tds - pd\left(\frac{1}{\rho}\right), \tag{23}$$

the adiabatic flow condition

$$D_t s = 0. \tag{24}$$

Because of our use of comoving coordinates, the equation

$$T^\mu_r\,;_\mu = 0, \tag{25}$$

which is the Euler equation of hydrodynamics,

appears as a simple "hydrostatic" balance of forces

$$\frac{\partial \Phi}{\partial r} = -\frac{1}{\rho w}\frac{\partial p}{\partial r} \tag{26}$$

with

$$w = 1 + \epsilon + (p/\rho). \tag{27}$$

In subsequent equations, then, we must recognize grad Φ as representing the non-gravitational forces.

The true dynamics of the matter is given by the behavior of the metric component $R(r, t)$ which describes how the circumference of a given sphere of matter varies with time. From the $R_{\theta\theta}$ or $R_{\varphi\varphi}$ Einstein equation one finds

$$D_t U = \Gamma D_r \Phi - \frac{m + 4\pi R^3 p}{R^2}, \tag{28}$$

where

$$D_r = e^{-\lambda/2}\frac{\partial}{\partial r} = 4\pi\rho R^2 \frac{\partial}{\partial A} \tag{29}$$

is the derivative with respect to proper radial distance. The first term on the right in Eq. (28) represents the mechanical forces through Eq. (26), while the second term generalizes the Newtonian $-m/R^2$ gravitational force.

From the above equations, or in the course of deriving them, one finds the interesting relations

$$D_t m = -4\pi R^2 pU \tag{30}$$

and

$$D_t \Gamma = U D_r \Phi. \tag{31}$$

The first can be interpreted as stating that the work done by pressure forces acting on the moving surface of a ball of fluid gives the increase of mass-energy m associated with that ball. To interpret the second equation, we note that Eq. (15) can be written

$$\frac{\partial m}{\partial A} = (1 + \epsilon)\Gamma. \tag{32}$$

Thus a mole of nucleons, whose total mass measured locally in a comoving frame would be $(1 + \epsilon)$, contributes $\Gamma(1 + \epsilon)$ to the total energy m. The factor Γ includes then the kinetic energy and the gravitational potential energy per unit mass, as is also suggested by its Newtonian limit, obtained from Eqs. (19) and (13),

$$\Gamma \sim 1 + \frac{1}{2}\frac{v^2}{c^2} - \frac{Gm}{c^2 R}. \tag{33}$$

Equation (31) then states that the rate of change of this energy per unit mass Γ is given by the work done by non-gravitational forces per unit mass (cf. eq. [26]).

To summarize, we collect a complete system of equations for this problem using $r = A$ as our comoving coordinate. The dynamic equations are

$$\frac{\partial R}{\partial t} = e^{\Phi} U, \tag{34-p}$$

$$\frac{\partial U}{\partial t} = -e^{\Phi}\left(\frac{\Gamma}{w}4\pi R^2 \frac{\partial p}{\partial A} + \frac{m}{R^2} + 4\pi p R\right) \tag{34-U}$$

$$\frac{\partial}{\partial t}\left(\frac{1}{\rho}\right) = \frac{e^{\Phi}}{\Gamma}\frac{\partial}{\partial A}(4\pi R^2 U), \tag{34-ρ}$$

$$\frac{\partial \epsilon}{\partial t} = -p\frac{\partial}{\partial t}\left(\frac{1}{\rho}\right), \tag{34-ϵ}$$

corresponding to a choice of R, U, ρ, and ϵ as basic variables. In order to compute the right-hand sides of the above equations, given R, U, ρ, and ϵ at some time t, one makes use of the following subsidiary equations

$$\Gamma = 4\pi\rho R^2 (\partial R/\partial A), \tag{35-Γ}$$

$$m(A, t) = \int_0^A (1 + \epsilon)\Gamma dA, \tag{35-m}$$

$$w = 1 + \epsilon + (p/\rho), \tag{35-w}$$

$$(e^{\Phi})' = e^{\Phi}(-p'/\rho w), \tag{35-Φ}$$

$$p = p(\rho, \epsilon). \tag{35-p}$$

The quantities $R(A, t)$, $U(A, t)$, $\rho(A, t)$ and $\epsilon(A, t)$ may not be chosen arbitrarily at $t = 0$, but are subject to one initial constraint which will be automatically preserved by the dynamic equations. It is $\gamma = |\Gamma|$ or

$$\left(4\pi\rho R^2 \frac{\partial R}{\partial A}\right)^2 = 1 + U^2 - \frac{2m}{R}. \tag{36}$$

If $2m/r \ll 1$ this constraint simply gives ρ in terms of R and U, but it becomes non-trivial when general relativity effects are important. Thus computations will most conveniently start from Newtonian initial conditions.

REFERENCE

C. W. Misner, and D. H. Sharp, 1964, Phys. Rev., 136 B571.

ENERGY TRANSPORT BY RADIATIVE DIFFUSION IN RELATIVISTIC SPHERICALLY SYMMETRIC HYDRODYNAMICS

Charles W. Misner
Department of Physics and Astronomy, University of Maryland
College Park, Maryland
and
David H. Sharp
Palmer Physical Laboratory, Princeton University
Princeton, New Jersey

One of the most interesting aspects of the gravitational collapse of stars involves the question of the dynamics of radiative energy transport from part of the star to another, or out of the star. A complete discussion of this problem, when gravitational forces are important, would involve the study of the general relativistic Boltzmann equation, which has been formulated by Tauber and Weinberg (1961) and by Chernikov (1963), combined with Einstein's equations. Such a study has been initiated by J. M. Bardeen (private communication), Tauber and Weinberg (private communication), and by R. Lindquist (private communication). For purposes of practical computations and of gaining some insight into the physical content of the equations, however, it seems useful to discuss some relatively simple approximations to radiative transport theory first. Consequently, in this paper we shall briefly discuss the theory of radiative heat diffusion including the effects of relativistic gravitation theory. A discussion of radiative diffusion based on transport theory within the framework of special relativity has been given by Thomas (1930).*

The equations given here can be used to study the effects of the changing gravitational field of the collapsing matter on the radiation, as well as the effect of the radiation on the energy balance. Aside from these basic questions, the diffusion approximation appears to be applicable to at least one of the practical problems which prompted this investigation. This concerns the extension to the relativistic regime of models of supernova explosions, such as the one developed by Colgate and White (1965).

In the case of supernovae, Newtonian calculations indicate that free gravitational collapse, initiated by the Fe-He transition in heavy stars ($M > 5M_{\odot}$) or by inverse beta-decay at a somewhat later stage in lighter stars, is first halted in a central core whose temperature is $\sim$60 MeV and whose radius is $r_{core} \sim 8 \times 10^5$ cm. The slowing down of the collapse is accompanied by copious neutrino production in a region of strong gravitational fields.

In the core, the neutrino mean free path is of the order of 450 cm, whereas a scale height in density is typically $r_{core}/10 \sim 8 \times 10^4$ cm. Moreover, the neutrino collision time $\tau_\nu \sim$(mean free path) /(mean velocity) $\sim 1.3 \times 10^{-8}$ sec is much smaller than the time scale characteristic of the hydrodynamic changes which is τ_{hydro} $\sim$ (core radius) /(sound velocity in the core) $\sim$ 3×10^{-3} sec. Thus in the core, one finds the conditions for thermal equilibrium satisfied, to a good approximation, and it will be appropriate to treat the neutrino energy transport in a diffusion approximation using a black-body spectrum for the neutrinos and the concepts of opacity and emission from a surface.

In this paper we shall first outline a relativistic theory of radiative diffusion in which the form of the diffusion law is determined, apart from the expression for the coefficient of radiative transport, by the thermodynamic requirement that the entropy production rate be non-negative. Our discussion is modeled after that of Eckart (1940) and Landau

*We thank J. W. Weinberg for calling our attention to this important paper.

and Lifshitz (1959), but shows the diffusion mechanism more explicitly so that the approximations involved in using the diffusion model can be pointed out clearly. We do not consider any transport equations, but simply take from special-relativistic theory (Thomas 1930) the expression for the (scalar) radiative transport coefficient in terms of the temperature and opacity.

Next, we shall write out the Einstein equations in the case of spherical symmetry assuming local thermodynamic equilibrium and the diffusion approximation and give the form, for spherical symmetry, of the equation for the radiation flux and the equation expressing local (thermodynamic) energy balance.

Let us consider the form of the stress-energy tensor. Photons or neutrinos moving in the direction of the vector k^μ give a contribution to the stress-energy tensor which is proportional to $k^\mu k^\nu$;

$$T^{\mu\nu}_{\text{rad}} \propto k^\mu k^\nu , \tag{1}$$

where k^μ is a 4-vector of length zero,

$$k^\mu k_\mu = 0. \tag{2}$$

For a wave vector distribution of radiation f(k), the radiation contribution to the stress energy tensor then has the general form

$$T^{\mu\nu}_{\text{rad}} = \int f(k, x) \; k^\mu k^\nu \, d^3 k. \tag{3}$$

We can decompose f(k) into a part f_0 which is isotropic in the rest frame of the material particles plus a term f_1 which depends on the direction of the radiation and which will give a net flux of energy. Thus we may write

$$f = f_0 (\mid k \mid) \; + f_1 (k) . \tag{4}$$

The energy flux vector (Poynting vector) is given by

$$q^i = \int f k^0 k^i \, d^3 k = \int f_1 \, k^0 k^i \, d^3 k. \tag{5}$$

The total stress-energy tensor we will write as $T^{\mu\nu} + \Delta T^{\mu\nu}$, where $\Delta T^{\mu\nu}$ contains contributions from only the anisotropic, f_1, part of the radiation.

We have, first of all,

$$T^{\mu\nu} = \rho \, (1 + \epsilon \,) \, u^\mu u^\nu + \tau^{\mu\nu} \tag{6}$$

$$= \begin{pmatrix} \rho(1+\epsilon) & 0 \\ 0 & 0 \end{pmatrix} + \begin{pmatrix} 0 & 0 \\ 0 & p\delta_p \end{pmatrix}$$

In equation (6), u^μ is the 4-velocity field of the fluid, which appears in the continuity equation

$$(\rho u^\mu)_{;\mu} = 0, \tag{7}$$

and the scalar function ρ is the baryon number density times the atomic mass of a hydrogen atom. Also, ϵ is the internal energy per baryon including an appropriate contribution from the energy density of the isotropic radiation. Similarly, p includes the fluid pressure and the pressure of the isotropic radiation. In the first line of equation (6) we have written $T^{\mu\nu}$ in an arbitrary system of coordinates, in the second line we have written the same equation in a local Lorentz frame in which $u^i = 0$.

Next, $\Delta T^{\mu\nu}$ is given by

$$\Delta T^{\mu\nu} = \rho \Delta \epsilon \, u^\mu u^\nu + q^\mu u^\nu + q^\nu u^\mu + \Delta T^{\mu\nu} \tag{8}$$

$$= \begin{pmatrix} \rho\Delta\epsilon & 0 \\ 0 & 0 \end{pmatrix} + \begin{pmatrix} 0 & q \\ q & 0 \end{pmatrix} + \begin{pmatrix} 0 & \\ & \Delta p \end{pmatrix}. \tag{9}$$

The symbols have meanings similar to those used above, but in giving the explicit forms in the second line we do not mean to imply that the radial and transverse components of the anisotropic radiation need be the same.

It is easy to see from the explicit matrix forms (6) and (9) that u^μ, q^μ and $\tau^{\mu\nu}$ satisfy

$$u^\mu u_\mu = -1, \; q^\mu q_\mu = q^2 \geq 0, \; q^\mu u_\mu = 0,$$

$$\tau^{\mu\nu} u_\nu = 0 = \Delta \tau^{\mu\nu} u_\nu \; . \tag{10}$$

The above expressions for the stress-energy tensor can be substituted into the law of local energy conservation:

$$-u_\mu \, (T^{\mu\nu} + \Delta T^{\mu\nu})_{;\nu} = 0, \tag{11}$$

to find

$$[\rho(1+\epsilon+\Delta\epsilon)\, u^{\mu}]_{;\mu} + u_{\mu;\nu}(\tau^{\mu\nu}+\Delta\tau^{\mu\nu})$$
$$= -q^{\mu}_{\ ;\mu} - q^{\mu} u_{\mu;\nu} u^{\nu}. \tag{12}$$

To interpret this equation we consider the thermodynamic statement of energy conservation. Considering energy on a "per particle" basis this reads

$$d\epsilon + p d\left(\frac{1}{\rho}\right) = T ds. \tag{13}$$

However, to compare with equation (12) we need the corresponding statement based on a unit volume. This reads

$$[\rho(1+\epsilon+\Delta\epsilon)\, u^{\mu}]_{;\mu} + u_{\mu;\nu}(\tau^{\mu\nu}+\Delta\tau^{\mu\nu})$$
$$= T(\sigma u^{\mu})_{;\mu}. \tag{14}$$

The first term of equation (14) represents the rate at which internal energy is produced in a unit volume of fluid. The second term can be understood as the rate at which work is done by the fluid in a unit volume in a frame of reference where one point of the fluid element is at rest. In the right-hand side of this equation, σ is the entropy density.

Comparing equations (12) and (14) we find

$$T(\sigma u^{\mu})_{;\mu} = -q^{\mu}_{\ ;\mu} - q^{\mu} u_{\mu;\nu} u^{\nu}, \tag{15}$$

which may be rewritten as

$$\left(\sigma u^{\mu} + \frac{q^{\mu}}{T}\right)_{;\mu} = -\frac{q^{\mu}}{T^2}(T_{,\mu} + T u_{\mu;\nu} u^{\nu}). \tag{16}$$

On the left-hand side of the equation stands the divergence of the total entropy current 4-vector. This total current describes not only the convective flow of entropy σu^i but also the transfer of entropy from one region to another by reversible heat flow. According to equation (16), this current is not conserved and the right-hand side represents the rate of entropy production per unit volume by irreversible processes.

Requiring that this rate of entropy production be non-negative, that q^{μ} be linear in the gradients of thermodynamic quantities and of the velocity, and that $q^{\mu}u_{\mu} = 0$, one is led uniquely to the form (Tauber and Weinberg, private communication):

$$q^{\mu} = -\chi(g^{\mu\nu} + u^{\mu}u^{\nu})(T_{,\nu} + T u_{\nu;\alpha} u^{\alpha}). \tag{17}$$

Here χ is an as-yet undetermined positve scalar function, the coefficient of radiative transport. We take its value as obtained from special-relativistic theory in the diffusion approximation with local thermodynamic equilibrium (Thomas 1930);

$$\chi = \frac{4ac}{3}\frac{T^3}{\kappa\rho}, \tag{18}$$

assuming that a general relativistic theory of radiative transfer would give the same result. Note that in equation (18) a is the radiation density constant and κ the opacity.

The linearity assumption used in deriving equation (17) and the assumption of local thermodynamic equilibrium imply that the radiation distribution is nearly isotropic ($f_1 \ll f_0$), so that

$$\Delta\epsilon,\ \Delta\rho \lll P_{rad} < p. \tag{19}$$

Now let us suppose that the metric has the spherically symmetric form

$$ds^2 = -e^{2\Phi} dt^2 + (R'^2/\gamma^2)\, dr^2 + R^2 d\Omega^2, \tag{20}$$

as described in another paper by us in this volume (see also Misner and Sharp 1964). In this case, the specific form of equation (17) becomes

$$q_r = -\chi e^{-\Phi} \frac{\partial}{\partial r}(Te^{\Phi}). \tag{21}$$

If we define a luminosity L by

$$L = 4\pi R^2 q = 4\pi R^2 |R'/\gamma| q_r, \tag{22}$$

and use equation (18) for χ we may write

$$L = \frac{a}{3}\frac{e^{-4\Phi}}{\kappa}(4\pi R^2)^2 \frac{\partial}{\partial A}(T^4 e^{4\Phi}). \tag{23}$$

In this equation we have also introduced as the independent variable A, which is the number of baryons enclosed in the coordinate sphere r.

The above form of the equation of radiative diffusion may be incorporated into a full system of equations which can be summarized as follows:

Dynamic equations:

$$D_t R = U, \tag{24a}$$

$$D_t U = -\frac{\Gamma}{w}(4\pi R^2)\frac{\partial p}{\partial A} - \frac{m + 4\pi R^3 p}{R^2}, \tag{24b}$$

$$D_t\left(\frac{1}{\rho}\right) = \frac{1}{\Gamma}\left[\frac{\partial}{\partial A}(4\pi R^2 U) - \frac{L}{\rho R}\right], \tag{24c}$$

$$D_t\epsilon = -p\, D_t\left(\frac{1}{\rho}\right) - e^{-2\Phi}\frac{\partial}{\partial A}(L\, e^{2\Phi}). \tag{24d}$$

Subsidiary equations:

$$\Gamma = 4\pi\rho R^2 R'\ , \quad w = 1 + \epsilon + (p/\rho)\,; \tag{25}$$

$$\frac{\partial m}{\partial A} = (1+\epsilon)\ \Gamma; \tag{26a}$$

$$\frac{\partial \Phi}{\partial A} = -\frac{1}{\rho w}\ \frac{\partial p}{\partial A}\,; \tag{26b}$$

$$L = -\frac{a}{3\kappa}(4\pi R^2)^2\ e^{-4\Phi}\frac{\partial}{\partial A}(T^4\ e^{4\Phi}), \tag{26c}$$

Equations of state for p, T, and κ.
Initial constraint:

$$\Gamma^2 = 1 + U^2 - 2mR^{-1}. \tag{27}$$

In the above set of equations, we have introduced the comoving proper-time derivative

$$D_t = u^{\mu}\frac{\partial}{\partial x^{\mu}} = e^{-\Phi}\left(\frac{\partial}{\partial t}\right)_r . \tag{28}$$

Of the dynamic equations, only (24c) and (24d) differ from the corresponding equations in the adiabatic case.

One will note in the energy balance equation, (24d), that the heat input to a shell of matter is proportional to the difference in the values of $Le^{2\Phi}$ at the inner and outer surfaces of the shell. Thus, if no heat is absorbed or given off, the quantity $Le^{2\Phi}$ is constant across the shell, and L will decrease corresponding to the increase in $e^{2\Phi}$. One factor of e^{Φ} represents the energy redshift, while the second represents the time dilation of the radiation pulse.

Consistent with our previous approximation $f_1 \ll f_0$ we have neglected in equations (24b), (24d), and (26b) the forces Δp associated with the anisotropic component of the radiation.

In the absence of nuclear-energy generation the above set of equations, together with appropriate boundary conditions, completely define the problem. To include nuclear-energy generation one must supplement these equations with equations governing the rate of change of the chemical composition. Equation (24d) is still correct, provided one understands the average internal energy per nucleon, ϵ, to include its nuclear binding energy. (Thus ϵ can become negative when a sufficient amount of hydrogen has converted into helium.)

REFERENCES

N. A. Chernikov, 1963, Acta Phys. Polonica, 23, 629.

S. A. Colgate, and R. H. White, 1965, "The Hydronamic Behavior of Supernovae Explosions" (UCRL-7777) (in press).

C. Eckart, 1940, Phys. Rev., 58, 919

L. D. Landau, and E. M. Lifshitz, 1959, Fluid Mechanics (London: Pergamon Press), sec. 127.

C. W. Misner, and D. H. Sharp, 1964, Phys. Rev., 136, B571.

G. E. Tauber, and J. W. Weinberg, 1961, Phys. Rev., 122, 1342.

L. H. Thomas, 1930, Quart. J. Math. (Oxford), 1, 239.

THE ACCELERATION OF MATTER TO RELATIVISTIC ENERGIES IN SUPERNOVA EXPLOSIONS*

Stirling A. Colgate
New Mexico Institute of Mining and Technology
and
Richard H. White
Lawrence Radiation Laboratory

The mass ejected with relativistic energy from a supernova is usually considered in terms of cosmic-ray production (Colgate and Johnson 1960; Colgate and White 1963); however, the recent interest in quasars (Robinson, Schild, and Schucking 1965) brings attention to the efficiency for producing relativistic electrons that meet the energy requirements of radio-frequency emission. In this paper we describe our view of the mechanism of supernova explosions with emphasis upon the efficiency for the ejection of matter at relativistic energy (Colgate and White 1964).

The results of the hydrodynamic calculations (Figs. 1-6) can be summarized as follows: If the mass of the evolved Fe core of the star, M_{core}, is $\gtrsim 5\ M_\odot$, the core may be unstable to the Fe-He transition and initiate a dynamical implosion starting at a density of 10^7 to 10^8 gm/cm^3. A somewhat smaller core will evolve quasi-estatically (stably) to a density of 10^{11} gm/cm^3 and then become unstable. Regardless of the prior evolutionary history, once $M_{core} \geq M_{cr} \simeq 1.16\ M_\odot$ and $\rho \geq 10^{11}$ gm/cm^3 there occurs a dynamic implosion that proceeds independently of the evolution prior to this state. The instability occurs because thermal decomposition to partial free protons and subsequent neutrino emission by inverse beta-decay to neutron-rich matter removes heat (and degeneracy pressure) faster than quasi-static contraction can supply it. The resulting implosion continues in approximate free fall until the neutron degeneracy pressure in the core becomes high enough to stop the radial implosion. This occurs only when the equation of state of the core matter becomes "stiff" enough to counterbalance the gravitational force. The requisite restriction in the degrees of freedom of matter occurs only in the limit of "unbound" nucleons where the density is therefore at least an order of magnitude greater than nuclear and the composition almost entirely neutrons ($\rho \simeq 5 \times 10^{14}$ gm/cm^3). A very small fraction ($\approx$ 5 per cent) of the neutron core forms adiabatically and cold. The outer layers of matter fall onto this core and accumulate as a shock wave. The heat generated behind this shock will necessarily be emitted in the form of neutrinos but, because of the high shock temperature and high local density, the neutrino mean free path is small and a diffusion wave of neutrinos deposits the energy throughout the rest of the star. Since this total energy is of the order of the gravitational potential of the neutron core, it is independent of the initiating instability and it is more than adequate to eject a much larger mass (the stellar envelope) from its lesser gravitational potential.

The deposition of neutrino energy gives rise to a radially outgoing shock which traverses the envelope of the star giving each radial region a different velocity and internal energy. In general, the velocity and internal energy increase toward the surface, becoming relativistic for a small fraction of the envelope ($\approx 10^{-4}$ M), and thus leading to cosmic rays (Colgate and Johnson 1960).

The explosion phase from neutrino deposition (Figs. 1-6) develops into a radially outgoing shock because the material closer to the core shock receives more neutrino deposited energy and has a higher temperature and higher sound speed. The shock wave speeds up (becomes stronger) in the density gradient of the mantle (Colgate and Johnson 1960; Ono, Sakashita, and Ohyama 1961) as can be seen by the positive curvature in Figs. 2 and 5.

*This work was partially supported by the U.S. Atomic Energy Commission.

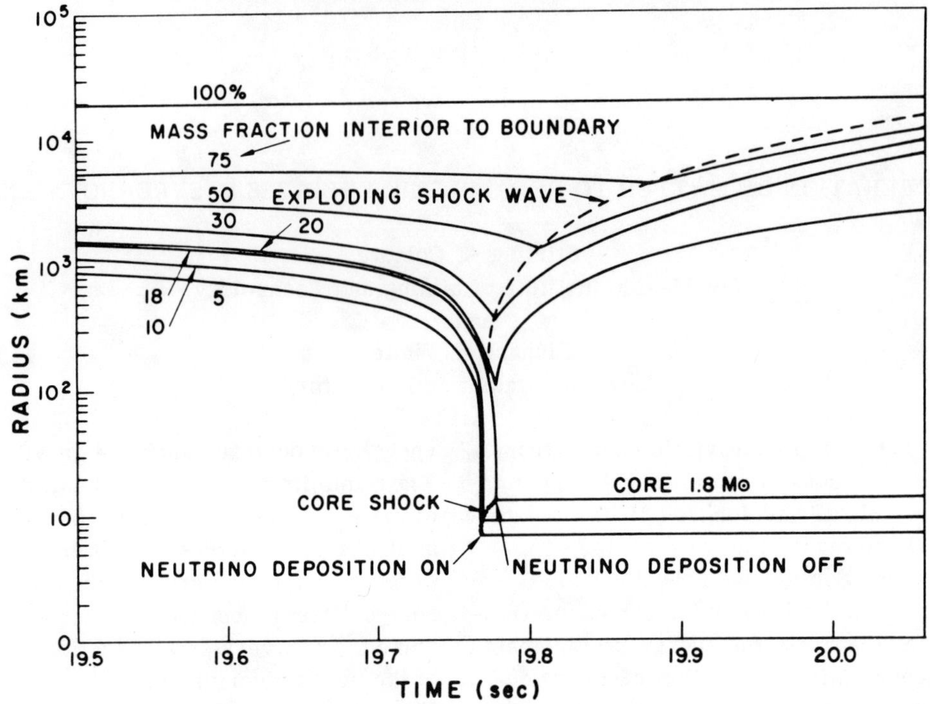

FIG. 1. Radius versus time for 10 $M_{\odot}$ supernova with neutrino deposition. During the initial collapse the neutrino energy is assumed lost from the star, but at the time of formation of a core shock wave a fraction of the neutrino energy is deposited in the envelope. The deposition ceases when the explosion terminates the imploding shock wave on the core.

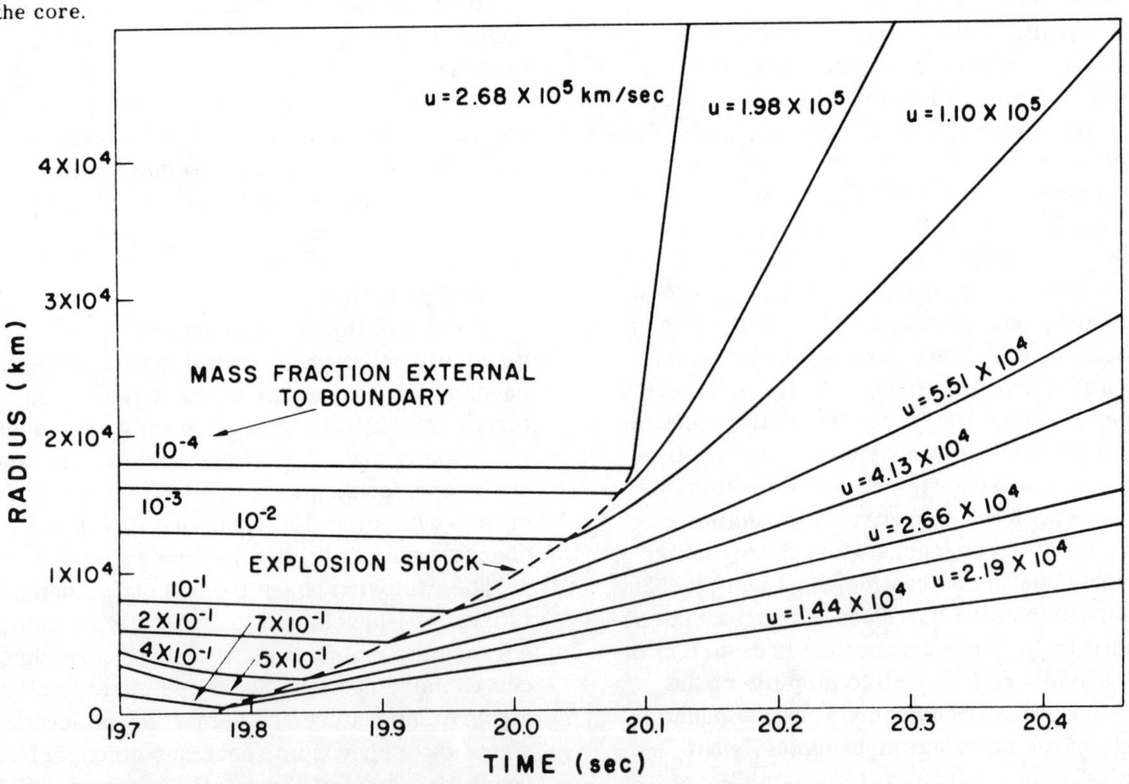

FIG. 2. 10 $M_{\odot}$ supernova radius versus time. The linear plot shows the increasing velocity of the outward explosion shock wave reaching the relativistic limit for 10^{-4} mass fraction.

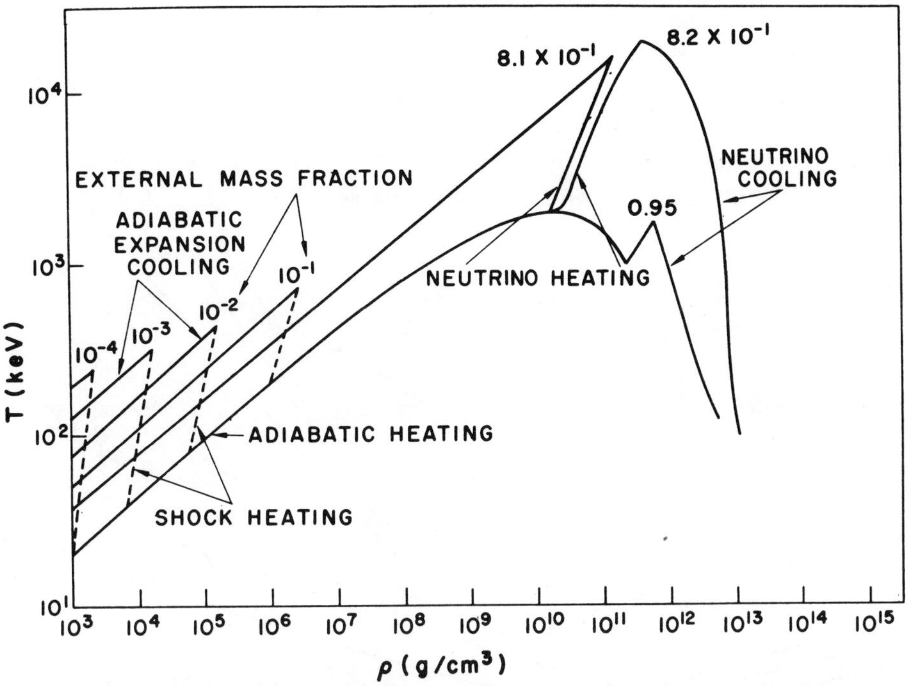

FIG. 3. 10 $M_\odot$ supernova temperature versus density with neutrino deposition. The lower line of slope 4/3 corresponds to the initial adiabatic compression during implosion. The innermost zones cool by neutrino emission; the intermediate zones heat by neutrino deposition; and the outermost zones are shock-heated. The expanding matter then cools adiabatically.

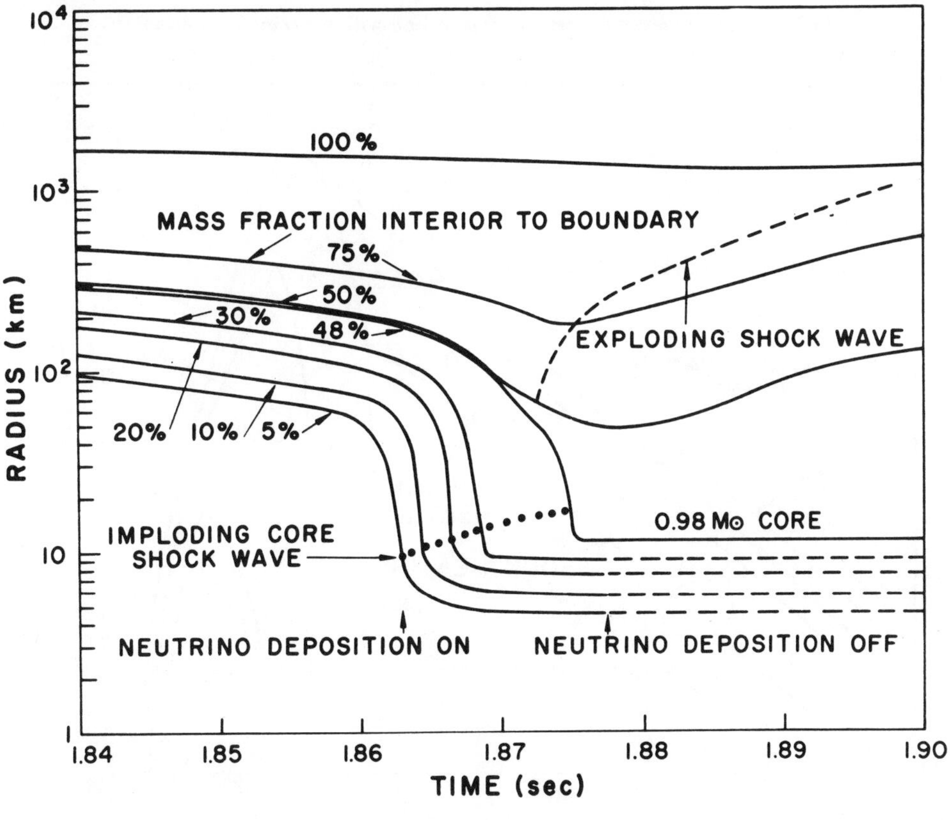

FIG. 4. 2$M_\odot$ supernova radius versus time with neutrino deposition. The instability occurs due to neutrino emission and nucleon binding in the equation of state with $\rho > 2 \times 10^{11}$ gm/cm^3.

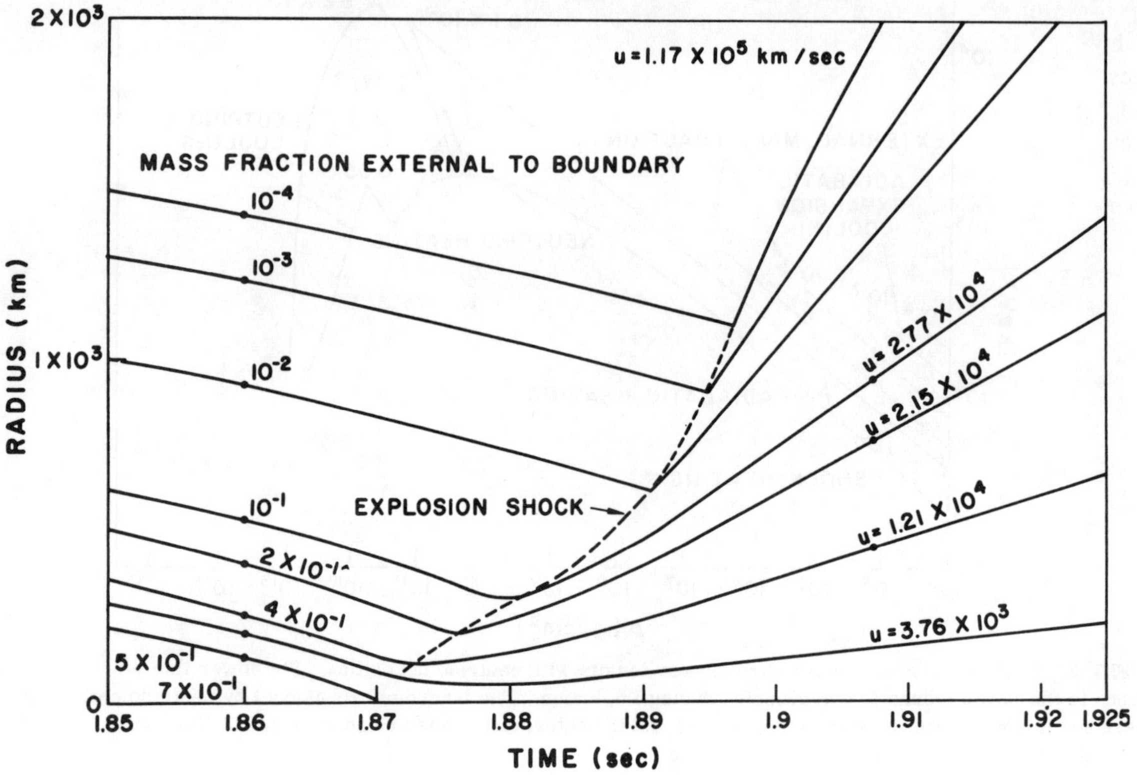

FIG. 5. $2M_\odot$ supernova radius versus time with neutrino deposition.

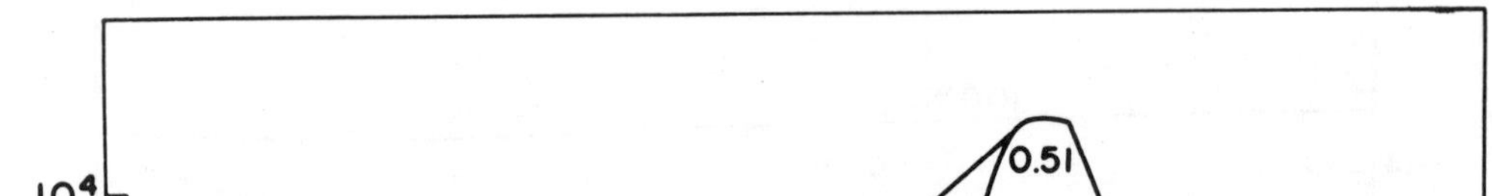

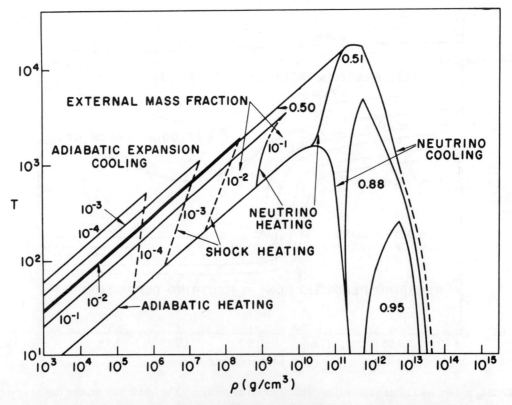

FIG. 6. $2M_\odot$ supernova temperature versus density with neutrino deposition.

The radial matter velocity following expansion is shown in Fig. 7 as a function of external mass fraction F. The slope $u \approx (F)^{-1/6}$ is in agreement with the similarity solution of Ono et al. (1961) if the plane parallel mass element dr is replaced by the spherical element $4\pi r^2 \rho dr$ which, in an exponential atmosphere, is proportional to F.

In Fig. 2 the velocity of 2.6×10^{10} cm/sec for the 10^{-4} mass fraction corresponds to the special relativistic energy $2M_0c^2$, i.e., the rest-mass energy equals the kinetic energy and so matter ejected external to this radius we identify with cosmic rays. The hydrodynamic computing code will not yet perform special relativistic

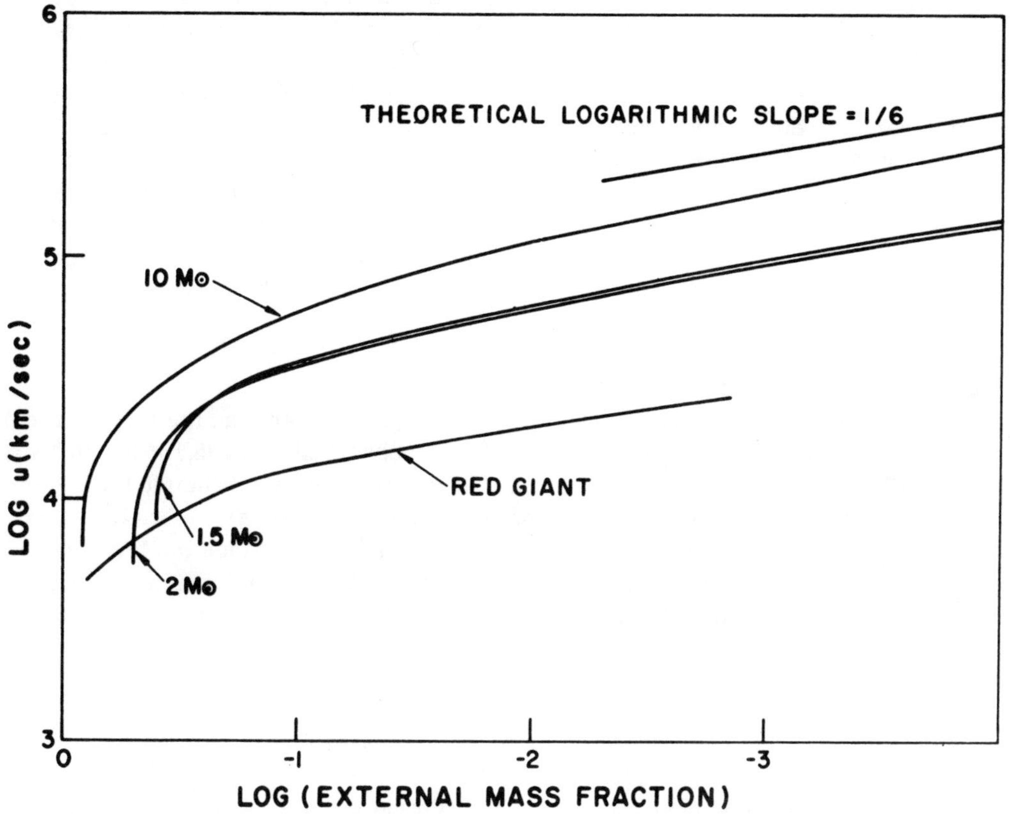

FIG. 7. Log expansion versus log mass fraction for supernova envelopes of 10, 2, and 1.5 $M_\odot$ polytrope 3 initial stars and a red giant. The logarithmic slope corresponds to the theoretical value of 1.6.

hydrodynamics, but similarity solutions of a quite general nature derived by Johnson (Colgate and Johnson 1960) lead to an energy spectrum of ejected matter that agrees, within rather narrow limits with the observed cosmic-ray spectrum. The total cosmic-ray energy injected into the galaxy becomes $MF_{cr}c^2 = 2 \times 10^{51}$ ergs.

The above example of the cosmic-ray acceleration in the 10 $M_\odot$ star can be derived using a more general argument.

As pointed out above, the fluid velocity behind the shock, $u_f \sim F^{-1/6}$, where F is the external mass fraction. This proportionality is valid for the ejected matter so that $(u_f)^2 = u_0^2 F^{-1/3}$, where $u_0 = u_f$ at the mass ejection radius. Therefore, the total kinetic energy of matter whose specific kinetic energy immediately behind the shock is greater than $(u_f)^2/2$ becomes

$$M_0 F(u_f)^2/2 = M_0 \frac{(u_f)^2}{u_f} \left(\frac{u_0}{u_f}\right)^6 .$$

In the expansion following the passage of the shock wave, the fluid expands from the moving frame and since the original internal energy equals the kinetic, the final post expansion velocity is doubled and the final kinetic energy is 4 times the shocked fluid kinetic energy. Therefore

$$W = 2M_0(u_f)^2 \left(\frac{u_0}{u_f}\right)^6 .$$

The quantity u_0 is determined by the shock strength at the ejection point in the star. Since for ejection

$$u_0{}^2/2 \cong \frac{M_0(1 - F_j)G}{r_j},$$

where F_j and r_j are the ejected mass fraction and radius, respectively, then the efficiency achieved in ejected matter whose energy is greater than $(u)^2/2$ becomes

$$\epsilon = \frac{4}{F_j} \frac{u_0}{u}^4.$$

In the above 10 $M_\odot$ example, the ejection mass cut occurs where the gravitational potential

$$\frac{M_0(1 - F_j)}{r_j} G \simeq 3 \times 10^{-2} c^2$$

so that the efficiency for cosmic-ray acceleration where $u^2 \simeq c^2$ becomes $\epsilon = 3\times10^{-3}$ in agreement with the hydrodynamic calculations.

In these calculations the ejection mass "cut" occurs at a radius $r = 10^7$ cm corresponding to a gravitational potential $V = 0.03\ c^2$ whereas the core shock wave at $4 = 10^6$ cm corresponds to a potential $V = c^2/10$. The very crude treatment of neutrino deposition rather than diffusion leaves the possibility open that a more realistic treatment that includes diffusion with the implied greater confinement of heat may initiate the ejection cut at a lower gravitational potential with a stronger shock and consequent greater efficiency. The ingenious model of a very massive star $\sim 10^6\ M_\odot$ by Fowler and Houle (1964) suffers from the difficulty of a relatively weak gravitational potential at "turn-around" and shock formation--corresponding to the temperatures at which carbon burns: $V \simeq 10^{-5} c^2$ and so consequently $\epsilon \simeq 4 \times 10^{-10}$, and the energy required for radio emission becomes excessively large. Furthermore, if we recognize that the defined efficiency includes the total mass of the fluid, then the efficiency for acceleration of electrons must be reduced by the mass ratio 1/1836.

One can envisage a mechanism for converting almost all the fluid kinetic energy into electron energy and thereby neglect the above mass ratio if one considers the collision of the accelerated matter with the magnetic fields plus matter of interstellar space. Present theories of collisionless magnetic shock structure (Rosenbluth 1957; Kellogg 1964) depend almost entirely upon the charge separation between the heavier ions and light electrons. The resulting electric field always maintains a value such that the drift velocity of the electrons corresponds to a kinetic energy equal to the change in kinetic energy of the ions. The limiting energy for the validity of this structure must occur where the relativistic mass of the electrons equals that of ions, namely, 10^9 eV per particle or where the fluid kinetic energy equals its rest mass. The resulting BeV electrons we associate with the radio emission from quasars.

Using the multiple supernova model of quasars of Colgate and Cameron (1963), where the observed light fluctuations of several weeks' minimum period are ascribed to frequent supernova (weeks to months) occurring in a medium of "high" (10^6-10^7 p/cm^3) interstellar density, then the rate of producing BeV electrons for radio emission from a 10 $M_\odot$ supernova per week becomes

$$R = \frac{M_0 F\ c^2}{\tau} = 3 \times 10^{45}\ \text{ergs/sec}.$$

This is slightly larger than the largest radio luminosity observed, 3C 295, of 2×10^{45} ergs/sec (Mathews, Morgan, and Schmidt 1964). If electrons of energy less than 10^9 eV can radiate their energy in less than 10^6 years, then the lower energy ejecta can contribute to radio emission and still higher radio luminosities are predicted.

REFERENCES

S. A. Colgate and A. G. W. Cameron, 1963, Nature, 200, 870.

S. A. Colgate and H. J. Johnson, 1960, Phys. Rev. Letters, 5, 235.

S. A. Colgate and R. H. White, 1963 "Cosmic Rays from Large Supernova," Internt. Cong. on Cosmic Rays, Jaipur.

______. 1964, "The Hydrodynamic Behavior of Supernovae Explosions" (UCRL-7777) (in press).

W. A. Fowler, 1964 Rev. Mod. Phys. 36, 545, 1964.

W. A. Fowler, and F. Hoyle, "Gravitational Collapse," Univ. of Chicago Press, 1964.

Paul J. Kellogg, 1964, Phys. Fluids, 7, 1555, and to be published.

T. A. Mathews, W. W. Morgan, and M. Schmidt 1964, Ap. J., 140, 35.

I. Robinson, A. Schild and E. L. Schucking, (eds.) 1965, Quasi-stellar Sources and Gravitational Collapse (Chicago University of Chicago Press).

M. Rosenbluth, 1957, Magnetohydrodynamics, ed., R. Landshoff (Stanford, Calif.: Stanford University Press), Chap. 5.

Ono, Sakashita, and Ohyama 1961.

GRAVITATIONAL FORCES ACCOMPANYING BURSTS OF RADIATION*

Robert A. Schwartz
Institute of Space Studies, Columbia University

Richard W. Lindquist
The University of Texas

and

Charles W. Misner
University of Maryland

As everybody knows, the electromagnetic field of a moving charge distribution is different from that of a stationary one; in particular, there is a long-range $1/r$ part of the field which is absent in the static case. Well, if there is any justice in the world, the same sort of things ought to be true for gravitational fields, and there ought to be some sort of $1/r$ gravitational fields in the universe. One type has been much discussed previously, the gravitational waves, and I will not deal with them. I would like to discuss another type of "induction" field: that which accompanies any spherically symmetric pulse of radiation.

The original idea behind this calculation was to determine the effect of a strong burst of radiation on the metric ouside a collapsing supernova core. Stirling Colgate has discussed the large neutrino flux that is expected from such an object, and it is natural to ask what the effect of such radiation is on the geometry of space-time outside the star.

First, let me briefly discuss how to obtain the metric for this situation. We treat the case of spherical symmetry, where the metric has the form

$$ds^2 = - e^{\Phi}dt^2 + e^2 \xi \, dr^2 + r^2 \, d\Omega^2 .$$

Any spherically symmetric geometry can be put into this form.

Now, inspired by the treatment of Bondi (1960; see also Bondi, van der Burg, and Metzner 1962). and Sachs (1962) of gravitational radiation, we introduce a retarded time coordinate, which is the general relativistic analogue of

$$t_{ret} = t - (r/c)$$

in electromagnetism,

$$du = (1/f) \; (e^{\Phi/2} \, dt - e^{2\,\xi/2} dr).$$

Here, f is an integrating factor, which is needed to make du an exact differential. The metric then looks like

$$ds^2 = - c^{\psi} du^2 - 2e^{\xi} \, dudr + tr^2 \, d\Omega^2 .$$

We also notice that $u^{j\nu}$ is a null vector.

Now, the stress-energy tensor of pure radiation is of the form $T_{\mu\nu} = qk_{\mu}k_{\nu}$, where k_{ν} is a null vector, and you may easily convince yourself that $T_{\mu\nu} = q \, u_{j\mu} \, u_{j\nu}$ represents outward, spherically symmetric radiation. Thus, T_{uu} is the only non zero component for this situation.

With this preparation, we may now look at Einstein' s equation

$$R_{\mu\nu} - \frac{1}{2} g_{\mu\nu} R = -8\pi T_{\mu\nu} \; (G = c = 1).$$

When you do the arithmetic, the metric turns out to be

$$ds^2 = - [1 - (2 \, m(u) /r)]du^2 - 2dr\,du + r^2 \, d\Omega^2 ,$$

***Based on a paper published in Physical Review, 137, B1364 (1965).**

where m is an arbitrary non-increasing function of u. This metric seems to have been written down first by Vaidya (1951, 1952) about ten years ago.

For the case dm/du = 0, this metric describes the Schwarzschild exterior solution (Eddington 1924; Finkelstein 1958), and the constant m is just the mass.

In the case when m is changing, dm/du is the rate of change of the total energy of the system, just as would be expected. Let q be the energy flux measured by an observer moving with 4-velocity v^μ. Then

$$q = v^\mu v^\nu T_{\mu\nu}.$$

For radially moving observers, moving with velocity $v^r = (dr/d\tau) \equiv U$, $v^u = (du/d\tau)$, one finds

$$q = -\frac{dm/du}{4\pi r^2} \quad \frac{1}{(\gamma + U)^2},$$

where

$$\gamma \equiv \sqrt{[1 + U^2 - 2(m/r)]}.$$

If we then define a luminosity $L = 4\pi r^2 q$, the luminosity seen by a stationary observer at infinity is

$$L_\infty = \lim_{r \to \infty} (4\pi r^2 q) = -(dm/du).$$

This, of course, is precisely what one would guess by regarding m as the total energy of the body.

Our form of the metric also exhibits the Schwarzschild surface very nicely. At r = 2m(u), there is no singularity in the geometry and none in the coordinate system. In Kruskal's form of the Schwarzschild metric, one sees that there are really two distinct Schwarzschild surfaces r = 2m; one coincides with $t = +\infty$, the other with $t = -\infty$. In our metric, the region $r > 2m$, for m = constant, corresponds to the fourth quadrant in Fig. 1. The surface r = 2m is the positive w-axis. Its character can be seen most easily in Fig. 2 where we have sketched light cones bounded by the radial null vectors

$$k = k^\mu (\partial/\partial x^\mu) = \partial/\partial r$$

and

$$l = l^\mu (\partial/\partial x^\mu) = -(1 - 2(m/r)) \frac{\partial}{\partial r} + 2 \frac{\partial}{\partial u}$$

in the u-r plane.

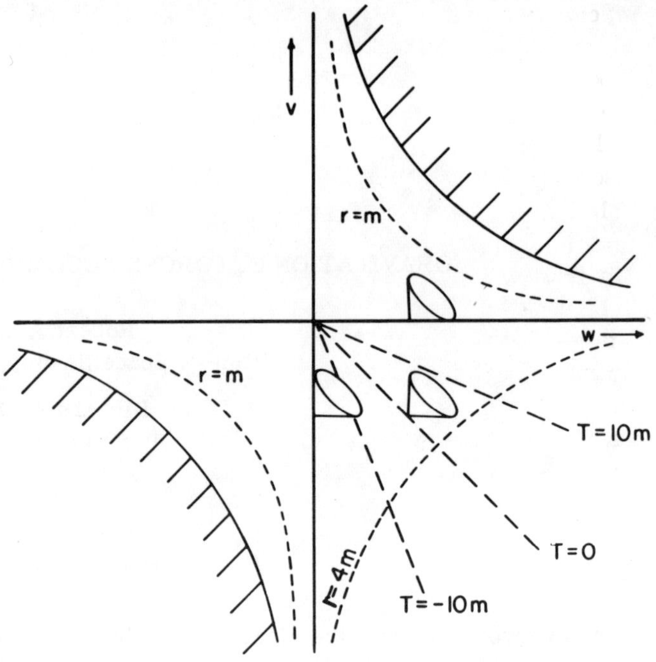

FIG. 1.

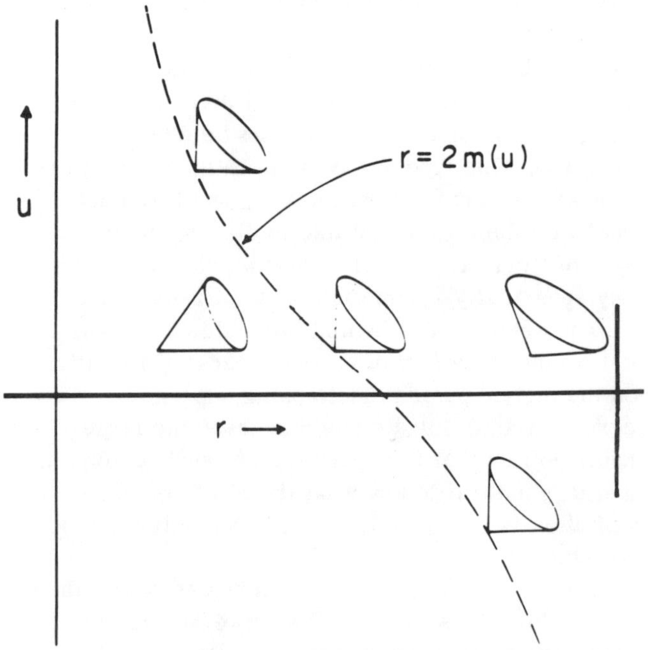

FIG. 2.

From this sketch one sees that, for dm/du $\neq 0$, the surface r = 2m(u) lies outside the light cone; i.e., it is a spacelike hypersurface. This is also obvious from the form of the induced metric on the hypersurface

$$(ds^2)_{r=2m(u)} = 2(-dm/du)\, du^2 + r^2\, d\Omega^2,$$

which has signature (+, +, +) whenever $dm/du < 0$. We may properly consider it to be a spacelike hypersurface lying in the past of the region $r > 2m(u)$, since no light ray starting in this region will intersect the $r = 2m(u)$ hypersurface. It is also evident from the figure that no material particle following a timelike path can reach the $r = 2m(u)$ hypersurface starting from the outside. For this reason we consider the region $r \leq 2m(u)$ to be unphysical: the sources of the strong gravitational fields there cannot be objects which once existed in an $r > 2m(u)$ region and were then assembled into something in the region $r < 2m(u)$. As these regions cannot, in principle, be produced experimentally, we ignore them. Every physical situation must consequently contain a boundary hypersurface $r = f(u) > 2m(u)$, with the interior metric in the region $r \leq f(u)$ differing from our metric because of the presence of matter or other fields.

The hypersurface $r = 2m(\infty)$ at $u = \infty$ in our metric is analogous to the Schwarzschild hypersurface $r = 2m$ at $T = +\infty$ in Kruskal's metric. It is reasonable to suppose that such surfaces are sometimes formed in the gravitational collapse of the cores of stars at supernova stage. In any case they can, in principle, be produced.

Now, let us see how particles behave in the field of such a radiating object. A test particle obeys the geodesic equations of motion

$$\frac{du^{\mu}}{d\tau} + \Gamma^{\mu}_{\nu\eta} u^{\nu} u^{\eta} = 0.$$

Spherical symmetry lets us omit θ from the problem, and gives us a constant of the motion

$$l = r^2 \dot{\phi} \qquad (\dot{\phi} = d\phi/d\tau).$$

The equation for the radial motion is found to be

$$\frac{d^2 r}{d\tau^2} = -\frac{L}{r} - \frac{m}{r^2} + \frac{l^2}{r^3} - \frac{3ml^2}{r^4}$$

The term m/r^2 is just the ordinary Newtonian force, l^2/r^3 is the centrifugal force, while $3ml^2/r^4$ is the relativistic term which causes the perihelion precession. (This is only part of the precession; you also need the distinction between proper time and coordinate time.)

The new term is L/r, the non-Newtonian field associated with the radiated power L emitted from the central source. If you like, you can call it a gravitational induction field associated with the changing Gm/r^2 field.

We see that the induction field L/r is always attractive, and one can easily verify that it acts to increase the energy at the test particle. For a normal star, it is completely negligible; it is 10^{-11} of the term which causes the 43'' precession in the orbit of Mercury.

Although the induction field L/r may be comparable to the Newtonian m/r^2 field at sufficiently large distances from a strongly radiating source, it will not have important effects there because of its limited duration in time. The maximum time a constant luminosity L can be maintained by a mass m is $\Delta t = mc^2/L$, so the maximum velocity change a test particle can have due to the $GL/c^3 r$ acceleration during this interval is $\Delta U = \Delta(v/c) = a_L \Delta t/c = Gm/c^2 r$. But if the test particle were originally in a Newtonian orbit with $v^2 \sim Gm/r$, this gives

$$\Delta(v/c) \sim v^2/c^2 \ll v/c.$$

REFERENCES

H. Bondi, 1960, Nature, 186, 535.

H. Bondi, M. G. J. van der Burg, and A. W. K. Metzner, 1962, Proc. R. Soc., A269, 21.

A. S. Eddington, 1924, Nature, 113, 192.

D. Finkelstein, 1958, Phys. Rev., 110, 965.

R. Sachs, 1962, Proc. R. Soc., A270, 103.

P. C. Vaidya, 1951, Proc. Indian Acad. Sci., A33, 264.

———. 1952, Current Sci., 21, 96.

RELATIVISTIC GAS SPHERES AT CONSTANT ENTROPY*

J. N. Snyder
Digital Computer Laboratory, University of Illinois
Urbana, Illinois
and
A. H. Taub
Computer Center and Department of Mathematics
University of California
Berkeley, California

I. Introduction

It is the purpose of this paper to discuss numerical solutions of the Einstein field equations for the metric tensor associated with a spherically symmetric static distribution of a relativistic gas at constant specific entropy and occupying a limited region of space time. In the treatment given below, account is taken of both the special theory of relativity and of the general theory. The former theory enters in that the equations used for determining the relations between the pressure, the rest density, and the temperature for a gas at constant specific entropy are not those given by pre-relativity thermodynamics but are those that obtain in the special theory of relativity.

The formulae used are those described by Synge (1957), namely,

$$\rho = \rho_0 \frac{k_2}{x} \exp \frac{k_3}{k_2} \quad , \tag{1.1}$$

$$x = \frac{mc^2}{kT} = \frac{\rho c^2}{p} \quad , \tag{1.2}$$

that is,

$$\rho = \rho_0 c^2 \frac{k_2}{x^2} \exp x \frac{k_3}{k_2} \quad , \tag{1.3}$$

$$w = \rho(c^2 + \epsilon) = \rho c^2 \frac{k_3}{k_2} - \frac{1}{x} = p \quad x \frac{k_3}{k_2} - 1 \quad , \tag{1.4}$$

$$\rho(c^2 + i) = xp \frac{k_3}{k_2} \quad , \tag{1.5}$$

*This work was supported in part by the National Science Foundation

where k is Boltzmann's constant, m is the mass of the gas molecule, c is the special relativity velocity of light in vacuum, ρ_0 is a constant related to the entropy with the dimension of a density, T is the absolute temperature, ρ is the rest density, p is the pressure, ϵ is the rest-specific internal energy whose functional dependence on two thermodynamic variables, say,

$$\epsilon = \epsilon(p, \rho),$$

gives the caloric equation of state of the gas and may change if the gas is changed, and i is the specific enthalpy, i.e.,

$$i = \epsilon + \frac{p}{\rho} \tag{1.6}$$

The functions $k_n (n = 2, 3)$ are Bessel functions defined by the equation

$$k_n(x) = \int_0^\infty \exp(-x \cosh \theta) \cosh n\theta \, d\theta. \tag{1.7}$$

For x large, that is, for low temperature,

$$k_n(x) = \sqrt{\left(\frac{\pi}{2x}\right)^{\frac{1}{2}}} e^{-x} \left[1 + \frac{4n^2 - 1^2}{1!\, 8x} + \frac{(4n^2 - 2^2)(4n^2 - 3^2)}{2!\, (8x)^2} + \dots \right]$$

and hence

$$\frac{k_3}{k_2} = 1 + \frac{5}{2x} + \dots .$$

Equations (1.1) and (1.3) then become

$$\rho = \lambda T^{3/2} + \dots , \quad p = KT^{5/2} + \dots .$$

If only the first terms in these expressions are used, we have

$$p = K'\rho^{5/3}.$$

That is, in the approximation of large x, equations (1.1) and (1.3) represent the classical equation of state for a monoatomic gas at constant entropy with a ratio of specific heats $\gamma = 5/3$. We further note that in this approximation

$$\rho(c^2 + \epsilon) = xp + \ldots = \rho c^2 + \ldots .$$

That is, ϵ/c^2 is negligible compared to ρ. Thus x large corresponds to the non-relativistic case.

For x small, that is for high temperature, and $n > 0$,

$$k_n(x) = \frac{1}{2}(n-1)!\left(\frac{2}{x}\right)^n + \ldots .$$

Thus

$$k_2(x) = \frac{2}{x^2} + \ldots, \quad k_3(x) = \frac{8}{x^3} + \ldots ;$$

and

$$\frac{k_3}{k_2} = \frac{4}{x} + \ldots .$$

Hence

$$\rho = \lambda T^3 + \ldots, \tag{1.8}$$

$$p = KT^4 + \ldots, \tag{1.9}$$

$$w = 3p + \ldots . \tag{1.10}$$

as follows from equations (1.1), (1.3), and (1.4). From equations (1.8) and (1.9) we see that, if the additional terms are neglected, that we may write

$$p = K'\rho^{4/3}.$$

That is, this case corresponds to a classical gas with the ratio of specific heats $\gamma = 4/3$ at constant specific entropy. Note that it follows from equation (1.10) that ρ is negligible in comparison with ϵ/c^2 in this case. We shall refer to the case of x small as the extreme relativistic case.

It is a consequence of the equations to be discussed later and equations (1.1)-(1.5) that the temperature of the spherical distribution of the relativistic gas at constant entropy is a monotonically decreasing function of the distance from the center. Hence if the central temperature is low, then the gas behaves classically throughout insofar as the relations between pressure, density, and temperature are concerned. That is, the gas may be said to have a constant $\gamma = 5/3$ throughout the material. If, however, the central temperature is large, then for the region of the material for which it remains large, the gas behaves as though it had a $\gamma = 4/3$ and it goes through a transition region after which it behaves as though $\gamma = 5/3$. Since the pressure and temperature of the gas must vanish at the boundary of the occupied region, the gas must always have a classical region near the boundary and may have a transition region and an extreme relativistic region if the central temperature is great enough. Thus for low central temperature, we shall be dealing with the general relativistic behavior of an Emden polytrope of index $n = 3/2$. For high central temperatures, the equation of state given by equations (1.1)-(1.5) describes a body of material such that in the vicinity of the center of the body the material behaves as if $\gamma = 4/3$ and near the boundary it behaves as if $\gamma = 5/3$.

Tooper (1964) has discussed another type of generalization of the Emden polytropes to general relativity. He has integrated the static spherically symmetric Einstein field equations supplemented by the equation

$$p = Kw^{1+1/n}, \tag{1.11}$$

where K is a constant and p and w are defined as above. He has treated a range of values of n. Equation (1.11) holds for a classical polytrope if w is taken to be the mass density. It is clear that for low central temperature and for $n = 3/2$, the results given below and those obtained by Tooper should agree. We shall see that this is so and compare the results described below with those obtained by Tooper for high central temperatures.

The systems treated here and those treated by Tooper are of interest because they provide a method of studying those effects of the Einstein theory of gravitation which do not occur in the Newtonian theory. Although systems in which the presence of radiation and energy sources are neglected are not realistic models of astronomical bodies, one expects that the differences between the predictions of the two theories of gravitation displayed in the study of such systems are significant and are of importance in astrophysics.

II. Conservation Theorems and Energy Content

A perfect fluid is described by the stress-energy tensor

$$T^{\mu}_{\nu} = \rho(1 + \frac{i}{c^2})\, U^{\mu} U_{\nu} - \frac{p}{c^2}\, \delta^{\mu}_{\nu} \tag{2.1}$$

where U^{μ} is the 4-velocity vector field describing the motion of the fluid and satisfies

$$g_{\mu\nu} U^{\mu} U^{\nu} = 1, \tag{2.2}$$

and the other quantities entering into equation (2.1) have the definitions given in Section I. The units of T^{μ}_{ν} are those of mass per unit volume, i.e , of a mass density.

The quantities p, ρ, U^{μ} are not all prescribed in our problem but must be determined and must be such that the conservation equations

$$T^{\mu}_{\nu;\mu} = 0 \tag{2.3}$$

are satisfied, where the semicolon denotes the covariant derivative with respect to the metric tensor of space-time. The metric tensor must also satisfy the Einstein field equations which will be stated later.

It follows from equation (2.1) and (2.2) that

$$T^{\mu}_{\nu}\, U^{\nu} = wU^{\mu}, \tag{2.4}$$

where w is defined by equation (1.4).
In view of equations (2.1) and (2.2), the four equations (2.3) may be written as

$$(w + \frac{p}{c^2})\, U^{\mu} U^{\nu}_{;\mu} = \frac{p_{,\mu}}{c^2}(g^{\mu\nu} - U^{\mu}U^{\nu}) \tag{2.5}$$

and

$$(w + \frac{p}{c^2})\, U^{\mu}_{;\mu} + w_{,\mu}\, U^{\mu} = 0. \tag{2.6}$$

The latter equation may be also be written as

$$(1 + \frac{i}{c^2})\, (\rho U^{\mu})_{;\mu} + \frac{\rho}{c^2}\, TS_{,\mu} U^{\mu} = 0, \tag{2.7}$$

where T is the temperature and S is the specific entropy defined by the equation

$$T\, dS = d\epsilon + p d(\frac{1}{\rho}). \tag{2.8}$$

If one postulates an "equation of state," that is, a relation expressing p as a function of ρ, we may then define a quantity α by the equation

$$\log \alpha = \int \frac{dw}{w + p/c^2} \tag{2.9}$$

It then follows from equation (2.6) that

$$(\alpha\, U^{\mu})_{;\mu} = 0, \tag{2.10}$$

that is, the quantity α is conserved during the motion. Note that if and only if the motion is isentropic, i.e.,

$$d\epsilon + p d(\frac{1}{\rho}) = 0,$$

it follows from equation (2.9) that

$$\alpha = \rho. \tag{2.11}$$

The conservation law described by equation (2.10) results from equations (2.3) and the equation of state. Equivalent forms of equations (2.3) lead to other conservation theorems. Thus if λ^{ν} is an arbitrary vector field, it is a consequence of equation (2.3) that

$$(\lambda^{\nu} T^{\mu}_{\nu})_{;\mu} = T^{\mu\nu} \lambda_{\mu;\nu}$$

and hence

$$\int \lambda^{\nu}\, T^{\mu}_{\nu}\, N_{\mu}\, dv = \int T^{\mu\nu}\, \lambda_{\mu;\nu} \sqrt{(-g)}\, d^4x, \tag{2.12}$$

where the integral on the right-hand side is carried out over a four-dimensional volume and the integral on the left is carried out over a three-dimensional closed hypersurface bounding this volume. If the parametric equations of the hypersurface are given by

$$x^{\mu} = x^{\mu}(\xi^1, \xi^2, \xi^3),$$

then

$$N_{\mu} dv = \sqrt{(-g)}\, \epsilon_{\sigma\tau\lambda\mu} \frac{\partial x^{\sigma}}{\partial \xi^1} \frac{\partial x^{\tau}}{\partial \xi^2} \frac{\partial x^{\lambda}}{\partial \xi^3} d\xi^1 d\xi^2 d\xi^3. \tag{2.13}$$

If we set $\lambda^{\mu} = U^{\mu}$ in equation (2.12), we find from equation (2.4) that

$$\int wU^{\mu} N_{\mu}\, dv = -\frac{1}{c^2} \int p\, U^{\mu}_{;\mu} \sqrt{(-g)}\, d^4 x. \tag{2.14}$$

This equation is of course just another form of equation (2.6) which may be written as

$$(w\ U^\mu)_{;\mu} + \frac{p}{c^2} U^\mu{}_{;\mu} = 0.$$

If the 4-velocity vector of the fluid is known as a function of x^μ, we may determine the world lines of the elements of the fluid by integrating the equations

$$\frac{dx^\mu}{ds} = U^\mu(x)$$

and obtaining the solutions as

$$x^\mu = x^\mu(\xi^i;\ s)\ (i = 1,\ 2,\ 3), \tag{2.15}$$

where the ξ^i are a parametrization of an initial hypersurface, the hypersurface s = 0. Thus

$$x^\mu = x^\mu(\xi^i;\ 0)$$

are the points of some hypersurface specified initially.

Equations (2.15) may be interpreted as defining a coordinate transformation between the Lagrange coordinates ξ^i and the proper time s to the general coordinate system x^μ. The volume element dv in the hypersurface s = constant is given by equations (2.15) where the $\partial x^\mu/\partial\xi^i$ entering therein are computed from equation (2.15). Hence

$$U^\mu N_\mu\ dv = \sqrt{(g)}\epsilon_{\sigma\tau\lambda\mu} \frac{\partial x^\sigma}{\partial\xi^1}\frac{\partial x^\tau}{\partial\xi^2}\frac{\partial x^\lambda}{\partial\xi^3}\frac{\partial x^\mu}{\partial s} d\xi^1 d\xi^2 d\xi^3$$

$$= \sqrt{(g^*)}\ d\xi^1\ d\xi^2\ d\xi^3$$

where g* is the value of the determinant of the metric tensor in the Lagrange coordinate system.

If equation (1.16) is applied to the region of space time between the hypersurface $s = s_1$ and $s = s_1 + ds$, it may be written as

$$\frac{s}{ds}\int w\sqrt{(g^*)}\ d\xi^1 d\xi^2 d\xi^3 = -\frac{1}{c^2}\int p\frac{\partial\sqrt{(g^*)}}{\partial s}d\xi^1 d\xi^2 d\xi^3$$

since $U^{*\mu} = \delta^\mu_4$. If the integral on the left is considered as the energy content of the fluid elements in the hypersurface s = constant, then this equation may be interpreted as stating that the rate of change of the energy in the hypersurface is equal to the integral of the pressure times the change in volume.

We shall call

$$w = \int w\ U^\mu N_\mu\ dv, \tag{2.16}$$

where N_μ is the normal to a hypersurface the proper energy content of the hypersurface in space-time. In the general case equation (2.14) holds for this quantity. Thus w will not be independent of which hypersurface is chosen from a set of hypersurfaces intersecting the world lines of the elements of the fluid.

On the other hand, if we define

$$N = \int \rho U^\mu N_\mu\ dv \tag{2.17}$$

and if the motion is such that mass is conserved, that is, if

$$(\rho U^\mu)_{;\mu} = 0, \tag{2.18}$$

then N is independent of which hypersurface is chosen in such a set. Note that (2.18) may be imposed in addition to equation (2.3). Then (2.7) reduces to

$$\rho\ T\ S_{,\mu}\ U^\mu = 0, \tag{2.19}$$

which is the form that the equations of conservation of energy take in classical hydrodynamics. If we impose equation (2.18) and prescribe the caloric equation of state of the gas, then we deduce that the specific entropy is a constant along the world line of the fluid. Thus imposing equations (2.18) is a weaker assumption than imposing an equation of state for the fluid.

The quantity N may be interpreted as the rest-mass content of the hypersurface with normal N_μ. It will differ from W when ϵ is non-vanishing over the hypersurface. In fact

$$U = W - Nc^2 = \int \rho\ \epsilon\ U^\nu N_\nu\ dv \tag{2.20}$$

is the total internal energy content of the hypersurface.

It should be noted that whenever λ^ν is a Killing vector, that is, whenever

$$\lambda_{\mu;\nu} + \lambda_{\nu;\mu} = 0,$$

and hence determines a one-parameter group of isometries in space-time, then equation (2.12) reduces to

$$\int \lambda^\nu\ T^\mu_\nu\ N_\mu\ dv = 0$$

when taken over a closed hypersurface in space-time.

Thus the integral taken over a hypersurface intersecting the world lines of the fluid elements is independent of the hypersurface.

In a static problem

$$\lambda^{\mu} = \delta^{\mu}_{4}$$

is a Killing vector and hence

$$E = \int T^{\mu}_{4} N_{\mu} \, dv \tag{2.21}$$

is independent of the hypersurface. We may call E the energy content of the hypersurface.

III. The Mass

It is well known that in the static spherically symmetric case one may introduce a coordinate system in which the line element is given by

$$ds^2 = e^{\nu} dt^2 - \frac{1}{c^2}(e^{\lambda} dr^2 + r^2(d\theta^2 + \sin^2\theta \, d\varphi^2)), \tag{3.1}$$

where ν and λ are functions of r and

$$U^{\mu} = e^{-\nu/2} \delta^{\mu}_{4}. \tag{3.2}$$

Note that equation (2.2) is satisfied as a consequence of equations (3.1) and (3.2). The coordinate system in which equation (3.1) holds will be said to be a Schwarzschild coordinate system. In this coordinate system the gas will be assumed to be confined to the region

$$0 < r < r_1, \tag{3.3}$$

where r_1 is a constant and the hypersurface $r = r_1$ will be said to be the boundary of the gas.

For $r \geq r_1$, we assume that $T^{\mu}_{\nu} = 0$ and have the well-known Schwarzschild metric

$$e^{\nu} = e^{-\lambda} = 1 - \frac{2Gm_1}{rc^2}, \tag{3.4}$$

where the constant G is Newton's constant of gravitation and the constant m_1 will be said to be the gravitational mass of the gas.

In the interior region, that is, where (3.3) holds, we shall write

$$e^{-\lambda} = 1 - \frac{2Gm(r)}{rc^2} \tag{3.5}$$

and replace the variable $\lambda(r)$ by the variable $m(r)$. The Einstein field equations will provide a differential equation for $m(r)$. The variable $m(r)$ will be called the gravitational mass contained in the sphere with coordinate radius r. The boundary condition for this differential equation is

$$m(r_1) = m_1. \tag{3.6}$$

It follows from equations (3.1) and (3.2) that equations (2.5) reduce to the single equation

$$(wc^2 + p)\nu_r = -2p_r, \tag{3.7}$$

where the subscript denotes differentiation with respect to r. This equation is the general-relativistic analogue of the equation of hydrostatic equilibrium. It is a consequence of equations (3.7) and (2.9) that, when an equation of state exists, then

$$e^{\nu/2} = \frac{A\alpha}{w^2 + p/c^2} = \frac{A\alpha}{\rho(1 + i/c^2)}, \tag{3.8}$$

where A is a constant and α is the conserved quantity defined by equation (2.9). In the isentropic case we have

$$e^{\nu/2} = \frac{A}{1 + i/c^2} \tag{3.9}$$

as follows from equation (2.11).

The boundary conditions we shall impose on the Einstein field equations together with the requirement $i(r_1) = 0$ will enable us to write equation (3.9) as

$$e^{\nu/2} = \frac{[1 - (2Gm_1/r_1c^2)]^{1/2}}{1 + i/c^2}. \tag{3.10}$$

In the remainder of this section we shall give explicit formulae for the various quantities discussed in the preceding section in the Schwarzschild coordinate system for the hypersurface t = constant. For this hypersurface.

$$N_{\mu} dv = e^{(\nu+\gamma)/2} r^2 \sin\theta \, d\theta \, d\varphi \, dr \, \delta^{4}_{\mu},$$

and hence equation (2.16) becomes

$$W_1 = 4\pi \int_0^{r_1} e^{\lambda/2} w r^2 \, dr$$

or

$$\frac{W_1}{c^2} = 4\pi \int_0^{r_1} \frac{\rho(1 + \epsilon/c^2)}{\{1 - [2Gm(r)/rc^2]\}^{1/2}} r^2 \, dr. \tag{3.11}$$

Equation (2.17) becomes

$$N_1 = 4\pi \int_0^{r_1} \frac{\rho r^2 \, dr.}{\{1 - [2Gm(r)/rc^2]\}^{1/2}}$$

and (2.20) becomes

$$U_1 = W_1 - N_1 c^2 = \int_0^{r_1} \frac{\rho \epsilon r^2 \, dr}{\{1 - [2Gm(r)/rc^2]\}^{1/2}} \quad (3.12)$$

Equation (2.21) becomes

$$\frac{E_1}{c^2} = 4\pi \int_0^{r_1} \frac{\rho(1 + \epsilon/c^2)e^{\nu/2}}{\{1 - [2Gm(r)/rc^2]\}^{1/2}} r^2 d\mathrm{r}, \quad (3.13)$$

and in the isentropic case this reduces to

$$\frac{E_1}{c^2} = 4\pi\left(1 - \frac{2Gm_1}{r_1 c^2}\right)^{1/2} \int_0^{r_1} \frac{\rho(1 + \epsilon/c^2)}{(1 + i/c^2)} \frac{r^2 \, dr}{\{1 - [2Gm(r)/rc^2]\}^{\frac{1}{2}}}. \quad (3.14)$$

It will be shown below that

$$m_1 = 4\pi \int_0^{r_1} \rho(1 + \epsilon/c^2) r^2 \, dr. \quad (3.15)$$

It then follows that

$$-\Omega = W_1 - m_1 c^2 = 4\pi \int_0^{r_1} \rho c^2 (1 + \epsilon/c^2) \left[\left(1 - \frac{2Gm}{rc^2}\right)^{-1/2} - 1\right] r^2 \, dr \quad (3.16)$$

or

$$-\Omega = c^2 \int_0^{r_1} \left[\left(1 - \frac{2Gm}{rc^2}\right)^{-1/2} - 1\right] dm. \quad (3.17)$$

If the integrand is expanded in powers of $1/c^2$ and the limit of the ensuing expansion is taken as $1/c^2 \to 0$, we obtain

$$-\Omega = G \int_0^{r_1} \frac{m dm}{r},$$

which is the classical expression for the gravitational potential energy of the gas. Hence we will use equations (3.10) and (3.17) to define the relativistic gravitational potential energy in the static spherically symmetric case. Note that it is an invariant; namely it is the difference between the proper energy content and the gravitational mass as measured by an observer far removed from the matter.

It is a consequence of equations (3.12) and (3.16) that

$$N_1 c^2 - m_1 c^2 = -\Omega - U. \quad (3.18)$$

This equation states that the difference between the proper rest mass of the gas and the gravitational mass is equal to the gravitational potential energy minus the proper internal energy. Classically $\Omega + U$ is called the total energy of the matter. If we call the left-hand side of the equation (3.18) the binding energy of the matter, this equation states that it is the negative of the total energy. Equation (3.18) is of course a consequence of the definitions of Ω and U, but does have a classical counterpart.

Equation (2.21) becomes

$$\frac{E_1}{c^2} = 4\pi \int_0^{r_1} \frac{\rho(1 + \epsilon/c^2)e^{\nu/2}}{\{1 - [2Gm(r)/rc^2]\}^{1/2}} r^2 \, dr \quad (3.19)$$

where $e^{\nu/2}$ is given by equation (3.8) in case a general equation of state obtains and by (3.10) in the isentropic case. If we write

$$\frac{\nu}{2} = \frac{V}{c^2},$$

Then it is a consequence of equation (3.7) that in the limit as $1/c^2 \to 0$, the quantity V approaches the Newtonian potential, that is, V_r is the gravitational force. Equation (3.19) may then be written as

$$E_1 = c^2 \int_0^{r_1} e^{V/c^2} \left(1 - \frac{2Gm}{rc^2}\right)^{-1/2} dm$$
$$= c^2 \int_0^{r_1} \left(1 + \frac{V}{c^2} + \frac{Gm}{rc^2} + \ldots\right) dm$$
$$= c^2 m_1 + \Omega + \ldots, \quad (3.20)$$

since in the classical limit

$$\Omega = -G \int_0^{r_1} \frac{m dm}{r} = \frac{1}{2} \int V dm.$$

Equation (3.20) is valid in the classical limit. Equation (3.9) and (3.10) enable us to relate E_1 to $N_1 c^2$ as follows:

In the isentropic case we have as a consequence of equation (3.10)

$$\frac{E_1}{c^2} = \left(1 - \frac{2Gm_1}{r_1 c^2}\right)^{1/2} \int_0^{r_1} \frac{\rho(1 + \epsilon/c^2) r^2 \, dr}{(1 + i/c^2)\sqrt{(1 - 2Gm/rc^2)}}. \quad (3.21)$$

Then

$$\frac{E_1}{c^2} \leq \left(1 - \frac{2Gm_1}{r_1 c^2}\right)^{1/2} N_1 \leq N_1. \tag{3.22}$$

IV. The Field Equations

The Einstein field equations are

$$R^{\mu}_{\nu} - \frac{1}{2} R\delta^{\mu}_{\nu} = - 8\pi G T^{\mu}_{\nu}, \tag{4.1}$$

where R^{μ}_{ν} and R are the Ricci tensor and the curvature scalar, respectively, computed from the metric tensor $g_{\mu\nu}$, and T^{μ}_{ν} is the stress-energy tensor creating the gravitational field represented by the latter tensor. In the case being considered here T^{μ}_{ν} is given by equations (2.1), the metric is assumed to be of the form given by equation (3.1), and equations (3.2) and (1.1)-(1.5) are assumed to hold.

When the function $\lambda(r)$ is replaced by the function m(r) by means of equations (3.5), equations (4.1) reduce to

$$m_r = 4\pi\rho(1 + \epsilon/c^2)r^2 \tag{4.2}$$

$$\nu_r = \frac{2G[m + 4\pi(pr^3/c^2)]}{c^2 r^2 [1 - (2Gm/rc^2)]} \tag{4.3}$$

and equation (3.7), namely,

$$\rho c^2 (1 + i/c^2)\nu_r = -2p_r. \tag{4.4}$$

We may write equation (4.3) as

$$p_r = - \frac{G\rho(1 + i/c^2)[m + 4\pi(pr^3/c^2)]}{r^2 [1 - (2Gm/rc^2)]} \tag{4.5}$$

Equations (4.2), (4.3), and (4.4) hold for all values of r. For

$$0 \leq r \leq r_1,$$

equations (1.1)-(1.5) serve to relate ρ, ϵ, and p. For $r \geq r_1$, we assume that $p = \rho = \epsilon = 0$, that is, $T^{\mu}_{\nu} = 0$. The solution of the equations in the exterior of the boundary is given by equation (3.4). On the boundary that is at $r = r_1$ we must have

$$p(r_1) = 0 \tag{4.6}$$

and λ and ν continuous functions of r. Further ν_r must also be continuous but λ_r need not be.

Equation (4.6) provides one boundary condition for the pair of first-order differential equations (4.2) and (4.5). The second boundary condition results from the requirement that λ be non-singular in the interior region.

If a general equation of state obtains for which $T^4_4 > 0$ and $T^4_4 = \rho(1 + \epsilon/c^2)$ is analytic function of $T^1_1 = -p/c^2$ which is finite for finite values of T^1_1, and if T^1_1 is finite at r = 0, the origin, it may be shown that the metric will be non-singular in the interior if and only if

$$m(0) = 0. \tag{4.7}$$

It is also a consequence of these assumptions that

$$\lim_{r \to 0} p_r = 0. \tag{4.8}$$

This requirement of non-singularity of the metric tensor and the requirement that p, ρ, and ϵ be positive in the interior imply that the function m(r) increase monotonically from zero at the origin to a value $m_1 = m(r_1)$ at $r = r_1$. Further p_r vanishes at the origin. It will vanish elsewhere in the interior or on the boundary if and only if $\rho(1 + 1/c^2)$ vanishes. This situation obtains on the boundary, $r = r_1$, in case equations (1.1)-(1.5) hold for, in the case, $\rho(1 + i/c^2)$ vanishes when p does.

In view of the above, the problem of solving the field equations in the static spherically symmetric case for a relativistic gas may be stated as follows: The functions ρ, ϵ, and i are determined as functions of p by means of equations (1.1)-(1.5). Equations (4.2) and (4.5) are then a closed system of first-order differential equations which may be solved numerically by choosing numbers m_1 and r_1 such that

$$\frac{2GM_1}{r_1 c^2} < 1.$$

It is further required that

$$p(r_1) = 0, \quad m(r_1) = m_1.$$

Only those solutions are acceptable for which

$$0 \leq 2Gm(r) < c^2 r$$

in the interior. In this procedure the mass m_1 and the coordinate radius r_1 are prescribed and the equations (4.2) and (4.5) are integrated inward from the boundary.

This procedure was not used in obtaining the results reported below for it is numerically unstable. Instead the integration was done integrating outward from the center. A constant

$$u_c = \frac{1}{x_c} = \frac{p_c}{\rho_c c^2}, \tag{4.9}$$

where p_c and ρ_c are the values of the pressure and density at the origin, is specified as is the value of ρ_0 in equations (1.1)-(1.5). Thus u_c and ρ_0 together with the conditions $m(0) = 0$, $p_r(0) = 0$, determine the initial conditions for equations (4.2) and (4.5). These equations are numerically integrated up to the value of r, say, r_1 such that

$$p(r_1) = 0$$

Then

$$m_1 = m(r_1)$$

is determined. The solutions are studied as functions of the two parameters u_c, which is simply related to the central temperature, and ρ_0. The dependence on ρ_0 is relatively simple as will be seen when the equations are written in dimensionless form in the next section.

When the outward integration is performed, it is convenient to compute other quantities of interest such as W(r), N(r), E(r), and the physical radius $\bar{r}(r)$, where

$$W_r = \frac{4\pi\rho(1+\epsilon/c^2)r^2c^2}{\sqrt{(1-2Gm/rc^2)}}, \tag{4.10}$$

$$N_r = \frac{4\pi\rho\, r^2}{\sqrt{(1-2Gm/rc^2)}} \tag{4.11}$$

$$E_r = \frac{4\pi c^2\rho(1+\epsilon/c^2)\sqrt{(1-2Gm_1/r_1c^2)}}{\sqrt{(1+i/c^2)}\sqrt{(1-2Gm/rc^2)}}r^2, \tag{4.12}$$

and

$$\bar{r}_r = \frac{1}{\sqrt{(1-2Gm/rc^2)}}. \tag{4.13}$$

The values of these quantities at $r = r_1$ then be obtained as functions of u_c and ρ_0.

V. Dimensionless Variables

We may write the differential equations (4.2) and (4.5) in dimensionless form by defining new dimensionless variables M, P, and R as

$$r = \beta R, \quad m = \frac{\beta c^2}{G}M, \quad p = p_c P, \tag{5.1}$$

where β is a constant with the dimensions of a length which will be specified later and $p_c = p(0)$ is the central pressure. The above differential equations then become

$$M_R = 4\pi\rho(1+\epsilon/c^2)\frac{G\beta^2}{c^2}R^2 \tag{5.2}$$

and

$$P_R = \frac{-\rho(1+i/c^2)c^2}{p_c}\,\frac{[M+4\pi(p_c\beta^2G/c^4)PR^3]}{R^2[1-(2M/R)]} \tag{5.3}$$

We set

$$\beta^2 = \frac{u_c^4c^4}{4\pi Gp_c} = \frac{c^4}{4\pi Gx_c^4p_c}, \tag{5.4}$$

where the constant u_c is defined by equation (4.9) and hence

$$p_c = \rho_0 c^2\frac{k_2(x_c)}{x_c^2}\exp\left[x_c\frac{k_3(x_c)}{k_2(x_c)}\right].$$

Note that in the extreme relativistic case (u_c large)

$$\frac{p_c}{u_c^4} = x_c^4p_c = \rho_0c^2\, x_c^2k_2(x_c)\exp x_c\frac{k_3(x_c)}{k_2(x_c)}$$

$$\approx 2e^4\rho_0c^2,$$

and β remains finite.

In view of equations (1.4) and (1.5) and the definition

$$Y + 1 = \frac{k_3}{k_2}, \tag{5.5}$$

we may write equations (5.2) and (5.3) as

$$M_R = u_c^4\,P[x(y+1)-1]R^2 \tag{5.6}$$

and

$$P_R = -x(1+y)P\frac{(M+u_c^4\,PR^3)}{R^2[1-(2M/R)]}, \tag{5.7}$$

respectively. It is evident from the last of these equations that $P_R = 0$ when $P = 0$. Hence it is difficult to determine by numerical procedures the value of $R_1 = r_1/\beta$ for which

$$P(R_1) = 0.$$

We may however replace the variable P by the variable y. It follows from the definition of the functions involved that

$$y_R = \frac{P_R}{xP} = -(1 + y)\,\frac{(M + u_c^4\, PR^3)}{R^2\,[1 - (2M/R)]}. \qquad (5.8)$$

One therefore integrates equations (5.6) and (5.8) from R = 0, with

$$M(0) = 0, \quad y(0) = y_c,$$

to $R = R_1$ where

$$y(R_1) = 0.$$

The initial value y_c is determined from equation (5.5) in terms of u_c and P is a function of y and y_c, given by equation (1.3).

It is actually more convenient to deal with another set of variables, namely,

$$\mathcal{M} = M/R, \quad Y = \log(1 + y), \quad \mathcal{P} = u_c^4 PR^2, \qquad (5.9)$$

and

$$s = \log R. \qquad (5.10)$$

It may readily be verified that in terms of these variables

$$\mathcal{M}_s = (xe^y - 1) - \mathcal{M}, \quad Y_s = -\frac{\mathcal{M} + \mathcal{P}}{1 - 2\mathcal{M}},$$

with the initial values

$$\mathcal{M}(-\infty) = 0, \quad Y(-\infty) = \log(1 + y_c).$$

We define

$$N = \frac{GN}{rc^2}, \quad W = \frac{GW}{rc^4}, \quad E = \frac{GE}{rc^4}, \quad R = \frac{\bar{r}}{r}; \qquad (5.12)$$

then equations (4.10)-(4.13) become

$$W_s = \frac{P(xe^y - 1)}{\sqrt{(1 - 2M)}} - W, \quad N_s = \frac{x\,p}{\sqrt{(1 - 2M)}} - N,$$

$$E_s = \sqrt{(1 - 2M_1)}\,\frac{(x - e^{-y})p}{\sqrt{(1 - 2M)}} - E,$$

$$R_s = \frac{1}{\sqrt{(1 - 2M)}} - R. \qquad (5.13)$$

Equations (5.11) and (5.13) were integrated outward from the center ($s = -\infty$) to the value $s_1 = \log R_1$. The initial values for the quantities W, N, and E were taken to be zero and that for R was

$$R(-\infty) = 1.$$

VI. The State of the Interior

The results of integrating equations (5.11) for $u_c = \frac{1}{4}$ and $u_c = 4$ are given in Figs. 1, 2, and 3 respectively where P, y, and M are plotted as functions of R. In addition the quantity,

$$\theta = \frac{1}{xu_c} = \frac{T}{T_c},$$

where T is the absolute temperature at R and T_c is the absolute temperature at the center (R = 0), is also plotted. Figure 1 is typical of the results obtained for small u_c, the classical case, and Figs. 2 and 3 are typical of those for large u_c, the relativistic case.

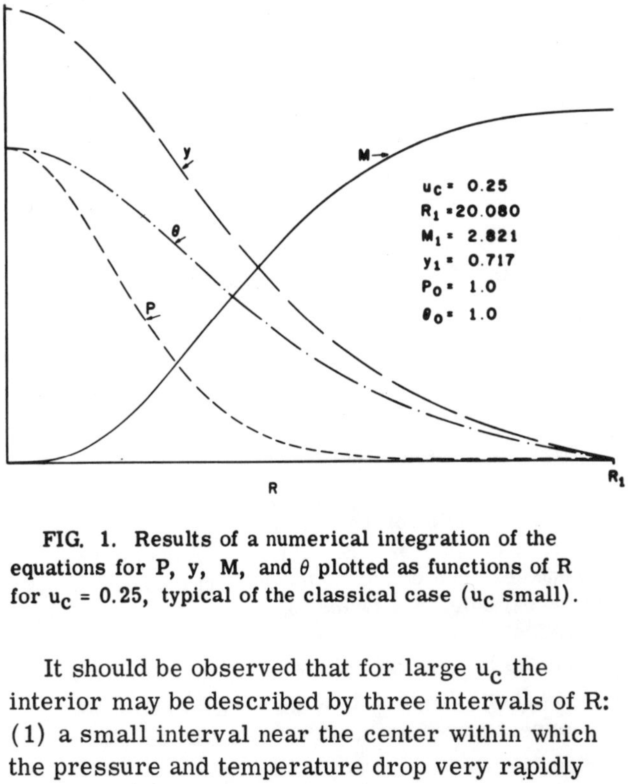

FIG. 1. Results of a numerical integration of the equations for P, y, M, and θ plotted as functions of R for u_c = 0.25, typical of the classical case (u_c small).

It should be observed that for large u_c the interior may be described by three intervals of R: (1) a small interval near the center within which the pressure and temperature drop very rapidly in value, (2) a relatively large interval in which these quantities are essentially constant, and (3) a smaller interval in which they drop to zero. Thus for large u_c the relativistic gaseous sphere behaves as a uniform sphere except for a very hot core and a non-uniform envelope within which the temperature and pressure decrease monotonically to zero with increasing coordinate radius.

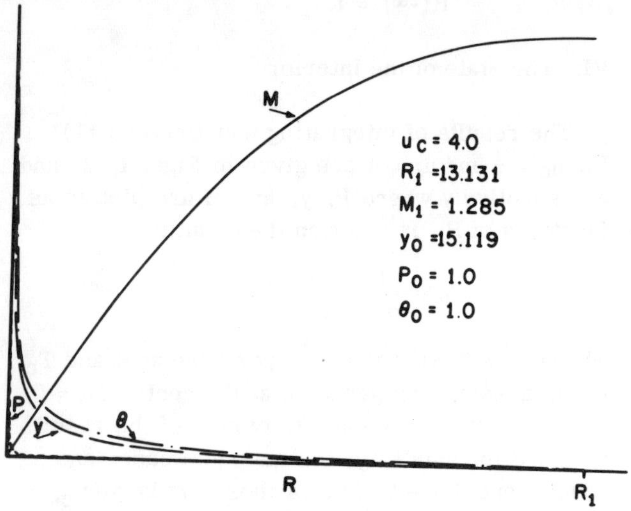

FIG. 2. Results of a numerical integration of the equations for P, y, M, and θ plotted as functions of R for $u_c = 4$, typical of the relativistic case (u_c large).

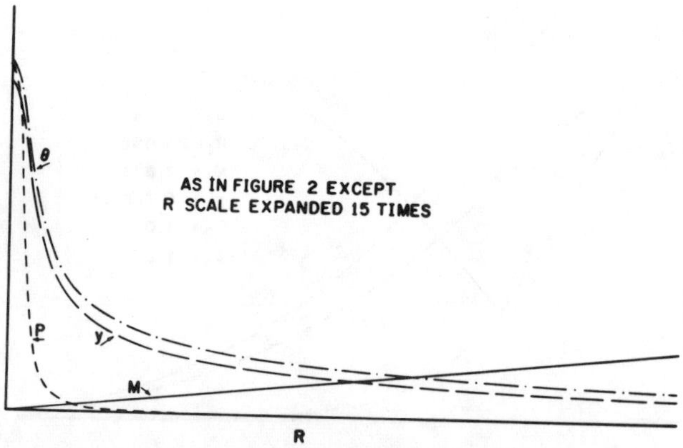

FIG. 3. Behavior of the quantities P, y, M, and θ of Fig. 2 for small values of R. The scale of the ordinate is the same as that of Fig. 2, the scale of the abscissa has been expanded by a factor 15.

The existence and nature of the hot core may be seen from the fact that when u_c is large we may expect a region in which u is large. For such a region we have

$$\rho\left(1 + \frac{i}{c^2}\right) = \frac{4p}{c^2}, \qquad \rho\left(1 + \frac{\epsilon}{c^2}\right) = \frac{3p}{c^2},$$

and equations (5.6) and (5.7) may be written as

$$M_s = 3P - M$$
$$P_s = \frac{2P(1 - 4M - 2P)}{1 - 2M}. \tag{5.14}$$

These equations have the solution

$$P = \frac{1}{14}, \quad M = \frac{3}{14}, \tag{5.15}$$

that is

$$m = \frac{3}{14}\frac{c^2}{G} r, \quad p = \frac{c^4}{14 \times 4\pi\, Gr^2}, \tag{5.16}$$

which is singular at $r = 0$ corresponding to an infinite pressure and temperature at the origin. Equations (5.14) give the internal distribution of mass and pressure for the case $u_c = \infty$. For large u_c equations (5.14) are a good approximation of equations (5.11) and admit solutions which have as initial values $M = P = 0$ and which tend to the solution (5.15) as s increases. Hence for large u_c and for r small but bounded away from zero we may expect m and p to behave as functions of r given by equations (5.16). It is evident from Fig. 3 that in the first interval PR^{-2} is essentially constant, that is, P is essentially constant.

VII. The Mass-Radius Relation

The integration of equations (5.11) and (5.13) results in a sequence of values of the dependent variables for a sequence of values of u_c; that is the dependent variables are determined as functions of u_c. Before discussing these functions we shall show how to relate these variables to such quantities as the gravitational mass m_1 and the physical radius $\bar{r}_1$ of a gaseous sphere. In particular we shall see how we may determine u_c and ρ_0 for a gaseous sphere with given m_1 and $\bar{r}_1$ from a knowledge of $M_1(u_c)$ and $R_1(u_c)$.

It follows from the definitions of the quantities involved that

$$\frac{M_1}{R_1} = \frac{Gm_1}{\bar{r}_1 c^2}. \tag{7.1}$$

Hence, given m_1 and $\bar{r}_1$ and M_1, R_1 as a function of u_c, we may determine u_c for a given configuration. Next consider the function of u_c defined by the equation

$$V(u_c) = \frac{u_c^4 R_1^{-4}}{4\pi M_1^2} = \frac{u_c^4 R_1^4}{4\pi M_1^2} R_1^2. \tag{7.2}$$

This function is also known as a result of the integrations. It follows from the definitions of the quantities involved that

$$p_C = \frac{Gm_1^2}{\bar{r}_1^4} V(u_C). \qquad (7.3)$$

Hence, given m_1 and $\bar{r}_1$ we may determine u_C and knowing $V(u_C)$ we may determine p_C. Equation (1.3) then enables us to determine ρ_0.

The gravitational mass m_1 and the physical radius r_1 of a gaseous sphere at constant entropy may be determined as functions of u_C and ρ_0 from the variables M_1 and R_1 as follows

$$m_1 = \frac{\beta c^2}{G} M_1 R_1, \quad \bar{r}_1 = \beta R_1 R_1. \qquad (7.4)$$

It should be remembered that β, which is given by equations (5.4) and (1.3), is a function of ρ_0 and u_C. The dependence of both m_1 and $\bar{r}_1$ on the variable ρ_0 is quite simple, namely, $m_1 \rho_0^{1/2}$ and $\bar{r}_1 \rho_0^{1/2}$ are independent of ρ_0 and functions of u_C alone. The dimensionless quantity $M_1/f = (G/16c^2)\, m_1 (4\pi G\rho_0/c^2)^{1/2}$ is plotted as a function of u_C in Figs. 4 and 5. The quantity $\bar{R}_1/f = (\bar{r}_1/32)\,(4\pi\rho_0/c^2)^{1/2}$ is similarly plotted in Figs. 6 and 7. From these curves we may construct Figs. 8 and 9 which plot M_1/f as a function of $\bar{R}_1/f$ and hence relate the gravitational mass to the physical radius.

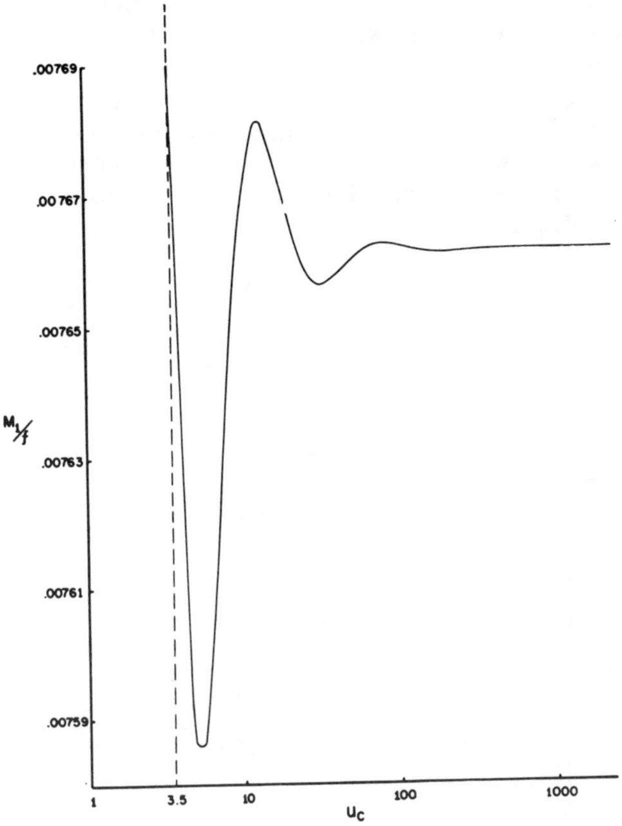

FIG. 5. M_1/f plotted as a function for $u_c = u_c \geq 3.5$ (as in Fig. 4) but with the scale of the ordinate expanded 100-fold.

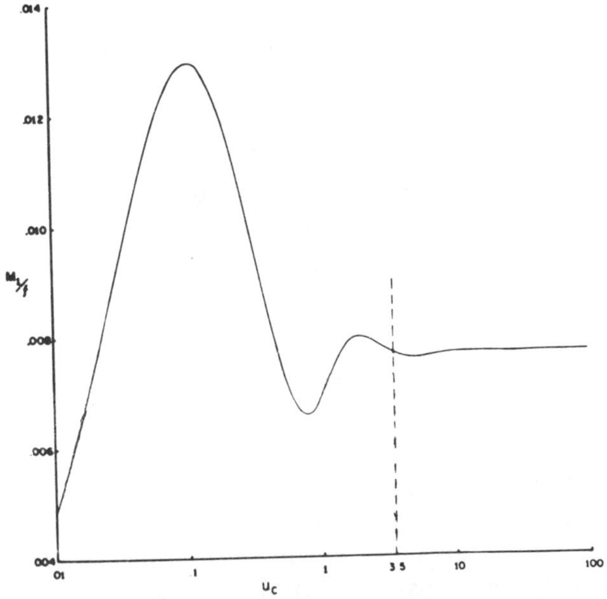

FIG. 4. The total equilibrium mass plotted as the dimensionless scaled quantity M_1/f as a function of u_C. As u_C approaches infinity, M_1/f approaches 0.0076615.

The values of M_1/f and of $\bar{R}_1/f$ for $u_C = \infty$ approach 0.0076615 and 0.048041, respectively.

Figures 4 and 5 are similar to those obtained by Oppenheimer and Volkov (1939), by Tooper (1964), and by Misner and Zapolsky (1964), who integrated the Einstein field equations for static spherically symmetric distributions of matter satisfying different equations of state. The similarity involves the existence of a maximum value of m_1, and the fact that m_1 is a multivalued function of u_C for a range of values of m_1 when ρ_0 is constant. Since each point on each of these figures corresponds to a different value of u_C, the multivaluedness of m_1 implies that for a given value of m_1 a number of equilibrium configurations of the gaseous sphere at constant entropy exist. Each of these configurations has a different central temperature. For m_1 greater than the maximum, no equilibrium configuration exists.

The stability of these equilibrium configuration may be determined by applying the method described by Chandrasekhar (1964). The calculations involved have not yet been performed, but the results reported here are so similar in nature to those referred to above that it seems quite likely that the stability results obtained by Misner and Zapolsky (1964) and by Chandrasekhar and Tooper (1964) hold in this case too.

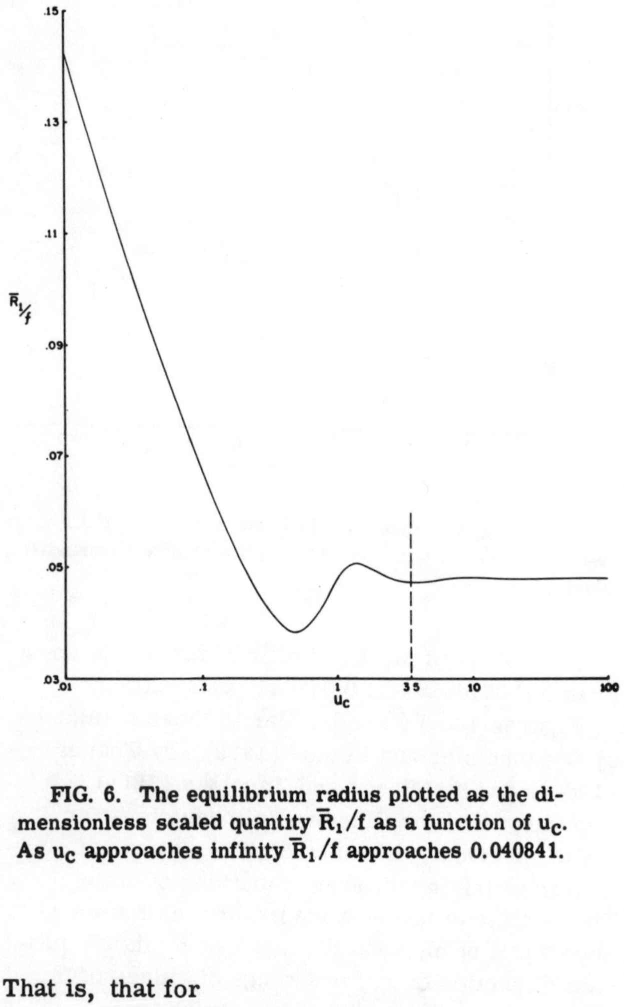

FIG. 6. The equilibrium radius plotted as the dimensionless scaled quantity $\overline{R}_1/f$ as a function of u_c. As u_c approaches infinity $\overline{R}_1/f$ approaches 0.040841.

That is, that for

$$0 < u_c < u_{c\,max} \cong 0.10$$

the configurations are stable, and for

$$u_c > u_{c\,max} \cong 0.10$$

the configurations are unstable.

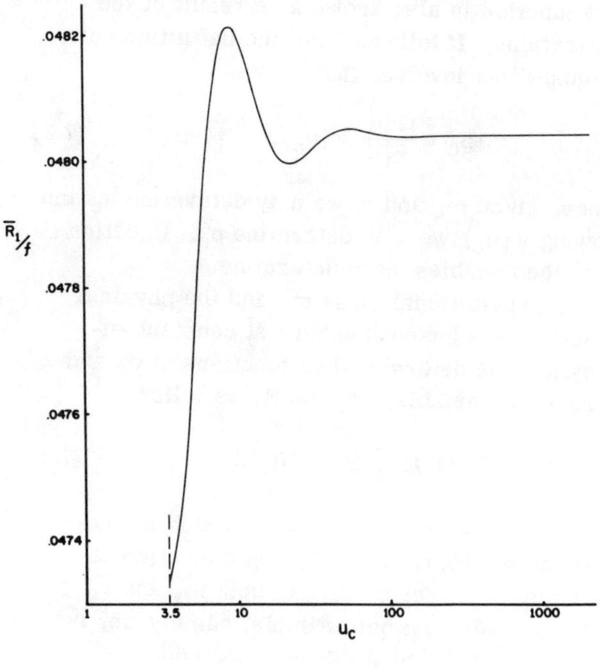

FIG. 7. $\overline{R}_1/f$ plotted as a function of u_c for $u_c \geq 3.5$ (as in Fig. 6) but with the scale of the ordinate expanded 100-fold.

VIII. Configurations with Constant Rest Mass

Figures 4-9 describe the sequence of equilibrium configurations each of which has the same constant entropy determined by the constant ρ_0. The sequence of such configurations for which the total rest mass, N_1, is the same is described by Fig. 10. It follows from equations (5.1), (5.9), and (5.12) that

$$\frac{M_1}{N_1} - 1 = \frac{m_1}{N_1} - 1 = \frac{m_1 - N_1}{N_1} = \frac{\Omega + U}{N_1 c^2}$$

Hence for constant N_1, $M_1/N_1 - 1$ in Fig. 10 represents the total energy of the configuration as well as the negative of the binding energy as a function of u_c. It is evident that for $u_c < 0.28$, $N_1 > m_1$, that is, the gravitational mass, is less than the rest mass and that the binding energy is positive. Even for $u_c < 0.28$ there may be two, one, or no values of u_c for a given m_1/N_1. Hence considerations of the binding energy, that is, the total energy, cannot distinguish between

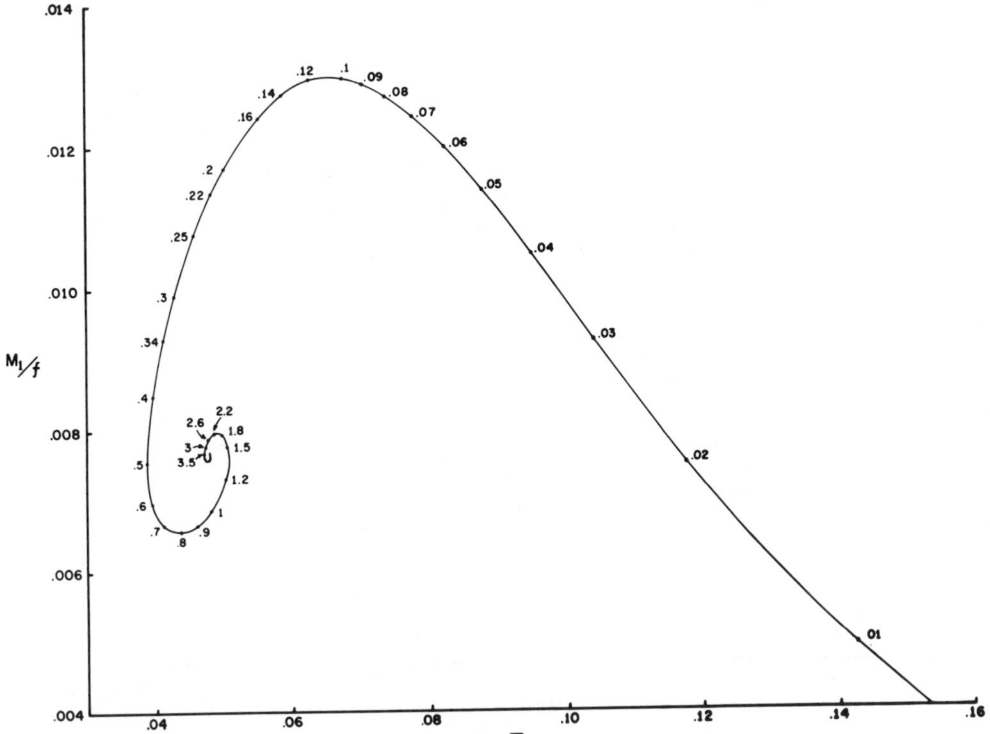

FIG. 8. M_1/f plotted as a function of $\bar{R}_1/f$. The numbers labeling points on the curve are the pertinent values of u_c.

stable and unstable states of equilibrium.

The minimum in the binding energy for constant rest mass (Fig. 10) occurs at $u_c \cong 0.10$, the same value of u_c for which the gravitational mass for constant entropy (constant ρ_0) has a maximum (Fig. 5). As has been remarked earlier, this value of u_c is probably the greatest value of this variable for which the equilibrium solutions are stable. For this value of u_c the maximum of the m_1 versus $\bar{r}_1$ curve also occurs (Fig. 8). A comparison of Figs. 4, 5, and 10 shows that the extrema of $M_1/N_1 - 1$ as a function of u_c coincide with the extrema of M_1 as a function of u_c except that maxima and minima are interchanged. As u_c approaches infinity, the quantity $M_1/N_1 - 1$ approaches and oscillates about the value 0.066887.

The calculations reported above were performed on the IBM 7094 of Brookhaven National Laboratory and the one at the Digital Computer Laboratory, University of Illinois. The authors wish to thank the Brookhaven National Laboratory, its Applied Mathematics Division, and, particularly, Dr. Y. Shimamoto and the late Dr. John W. Calkin for their generous hospitality.

We wish also to thank Mrs. J. Blankfield, who

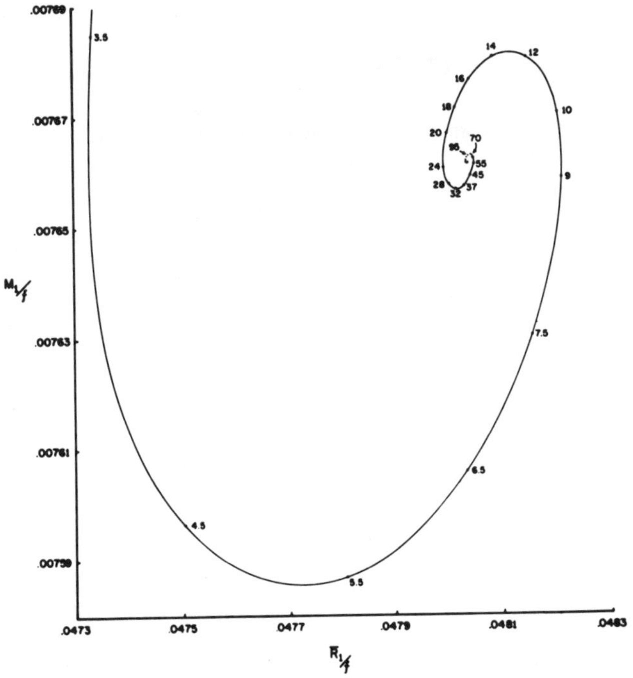

FIG. 9. M_1/f plotted as a function of $\bar{R}_1/f$ for values of $u_c \geq 3.5$ (the central portion of the spiral in Fig. 8). The scales of both ordinate and abscissa have been expanded 100-fold.

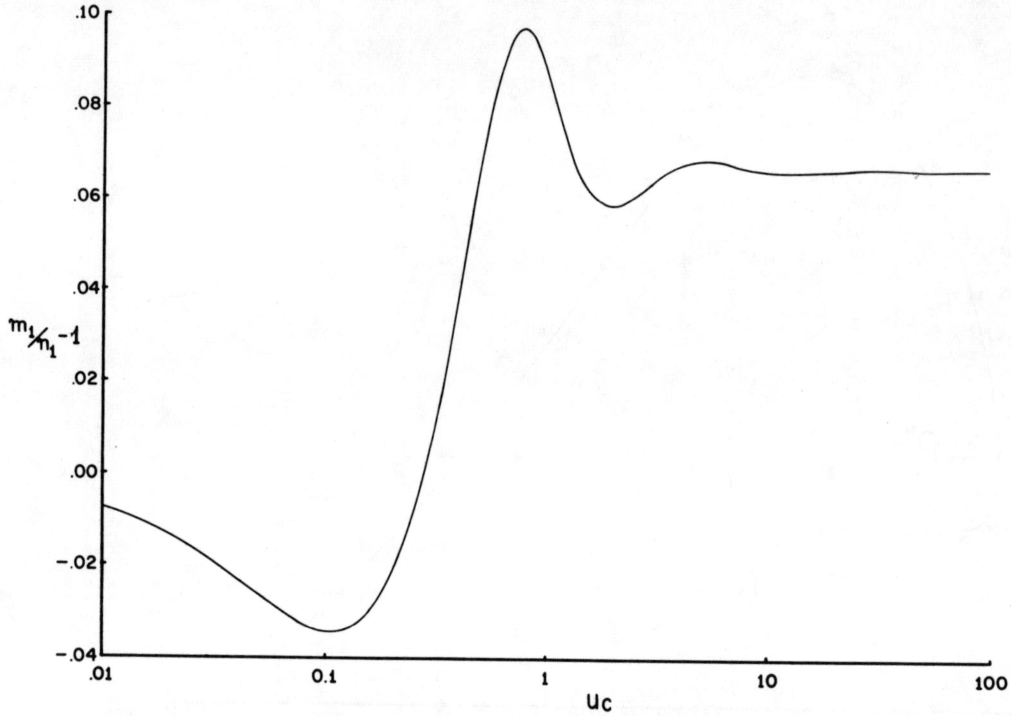

FIG. 10. The quantity M_1/N_1 - 1 describing the total energy of the configuration for constant rest mass N_1 plotted as a function of u_C. As u_C approaches infinity, this quantity approaches 0.068887.

supervised some of the early calculations at the Digital Computer Laboratory, University of Illinois.

REFERENCES

S. Chandrasekhar, 1964, Phys. Rev. Letters, 12, 114 and 437.

S. Chandrasekhar, and R. F. Tooper 1964, Ap. J., 140, 417.

C. W. Misner, and H. S. Zapolsky, 1964, Phys. Rev. Letters, 12, 635.

J. R. Oppenheimer, and G. M. Volkoff, 1939, Phys. Rev., 53, 374.

J. L. Synge, 1957, The Relativistic Gas (Amsterdam: North-Holland Publishing Co.), p. 36.

R. F. Tooper, 1964, Ap. J., 140, 434.

HYDRODYNAMIC CALCULATIONS OF GENERAL RELATIVISTIC COLLAPSE

M. M. May and R. H. White
Lawrence Radiation Laboratory, Livermore, California

I. INTRODUCTION

THE gravitational collapse of spherically symmetric masses under conditions where the general theory of relativity is expected to apply has been calculated by solving the field equations in finite-difference approximation on digital computers. This paper presents results of this calculation for materials with simple equations of state and for simple initial and boundary conditions. The purpose is to provide a description of the collapse in the presence of nonzero pressures, and to verify current estimates of maximum stable mass. The calculation can be extended to take into account pair production, heat transfer, and other mechanisms of interest for the description of astrophysical processes.

* Work performed under the auspices of the U. S. Atomic Energy Commission.

II. EQUATIONS

We consider an ideal fluid and neglect all heat transfer except that due to the motion of the fluid itself. The specific entropy of the fluid at a given mass point is then constant except when the mass point goes through a shock. We neglect pair production and annihilation, and the interaction of the fluid with external fields, so that rest mass is conserved. Assuming spherical symmetry leads to the metric[1]

$$ds^2 = a^2(\mu,t)c^2dt^2 - b^2(\mu,t)d\mu^2 - R^2(\mu,t)d\Omega^2, \qquad (1)$$

[1] L. D. Landau and E. M. Lifshitz, *The Classical Theory of Fields* (Addison-Wesley Publishing Company, Reading, Massachusetts, 1962), 2nd ed., pp. 331–332.

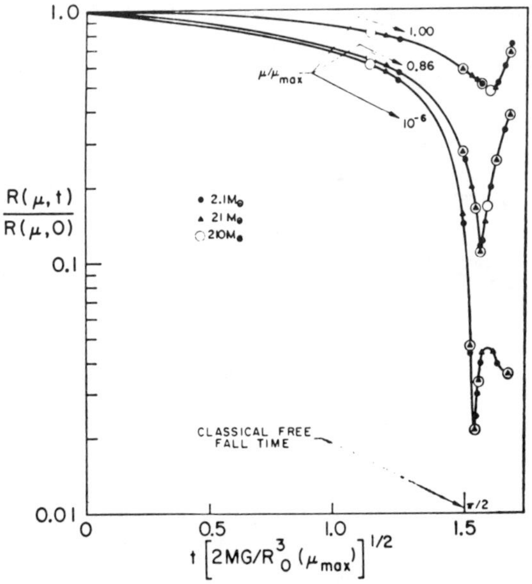

FIG. 1. $R(\mu,t)/R(\mu,0)$ versus $t[2MG/R^3(\mu\max,0)]^{1/2}$ for various mass fractions during the collapse and bounce of 2.1, 21, and 210 $M\odot$, $\gamma=5/3$ spheres. The initial conditions in each case were $R_s/R=6.2\times10^{-3}$, $\epsilon_0/c^2=3.84\times10^{-5}$ corresponding to a K (Eq. 22) of 2.

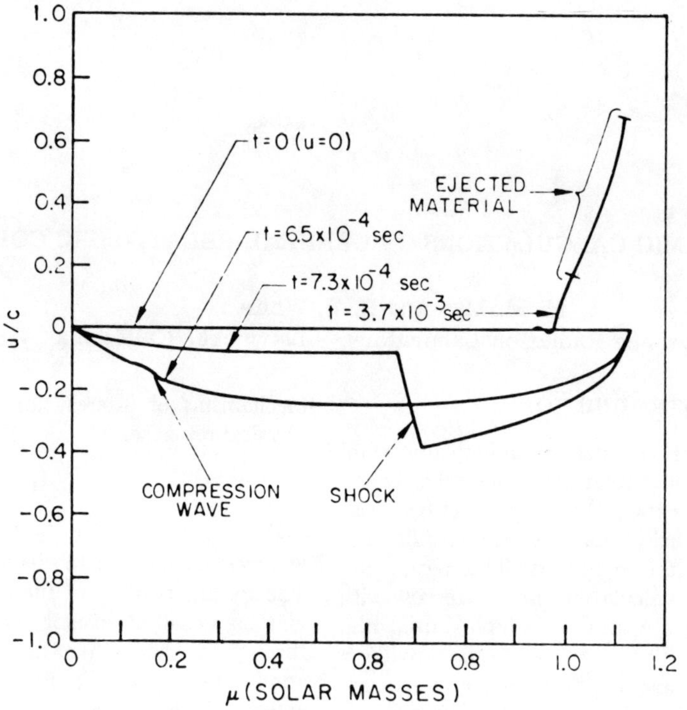

FIG. 2. u/c versus μ at several times during the collapse and bounce of a 1.1 $M\odot$ sphere, equation of state (iii), with initially uniform density ($\rho_0 = 10^{+13}$) and specific internal energy ($\epsilon_0/c^2 = 4.04\times10^{-3}$) corresponding to $K=1.06$ [Eq. (22)].

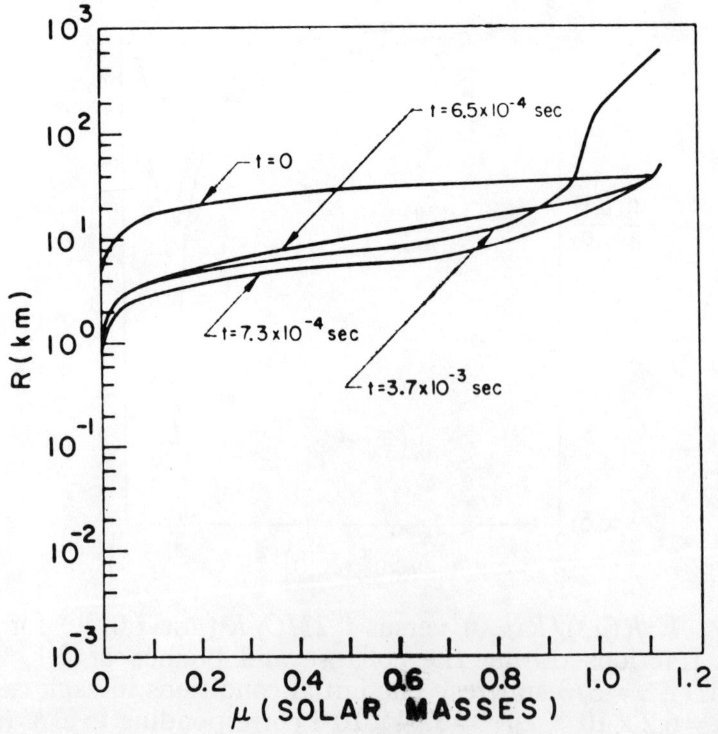

FIG. 3. R versus μ at several times during the collapse and bounce of the 1.1 $M\odot$ sphere (see caption, Fig. 2, for initial conditions).

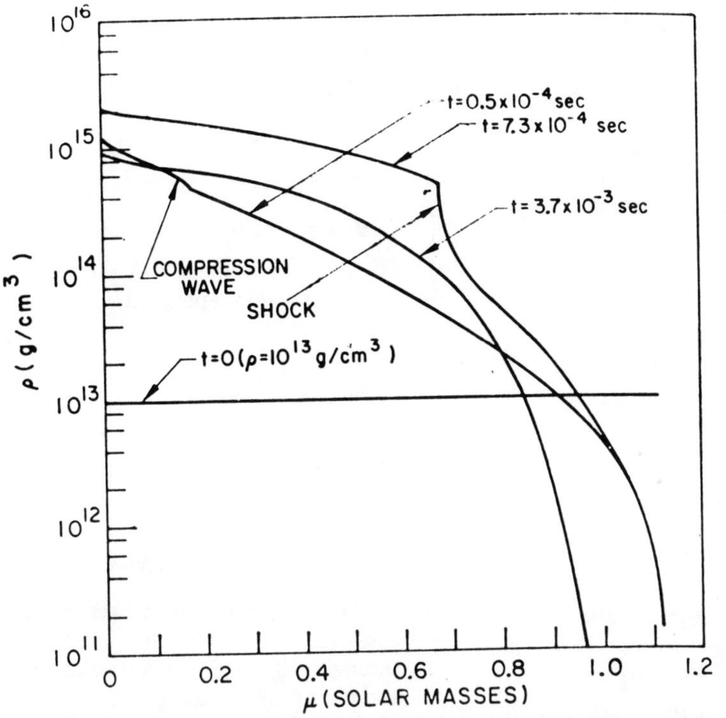

FIG. 4. ρ versus μ at several times during the collapse and bounce of the 1.1 $M_\odot$ sphere (see caption, Fig. 2, for initial conditions).

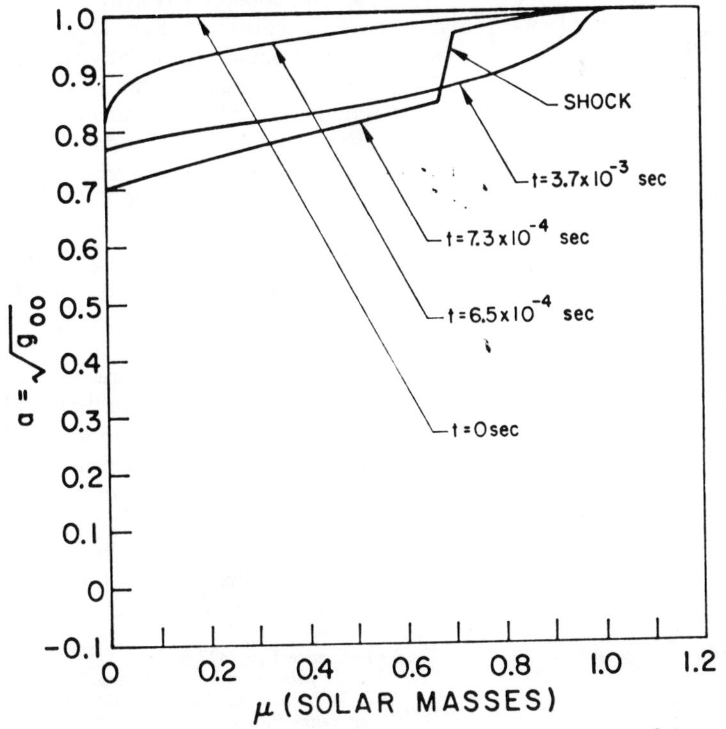

FIG. 5. $a = g_{00}^{1/2}$ at several times during the collapse and bounce of the 1.1 $M_\odot$ sphere (see caption, Fig. 2, for initial conditions).

where $d\Omega^2 = d\theta^2 + \sin^2\theta d\phi^2$, μ is a radial coordinate, and $2\pi R(\mu,t)$ is the circumference of a circle going through points of a given μ at time t. Choosing a coordinate system which moves with the fluid leads to the energy momentum tensor:

$$T_1^1 = T_2^2 = T_3^3 = P, \quad T_0^0 = -\rho(c^2+\epsilon), \tag{2}$$

where P is the pressure, ϵ the internal energy per gram, and ρ the rest mass density of the separated particles making up the fluid, i.e., the rest mass excluding ϵ/c^2. If μ is defined as the rest mass between the point labeled and the center, the assumption of rest-mass conservation becomes

$$4\pi\rho R^2 b = 1, \tag{3}$$

and the Einstein field equations become:

$$(T_0^0) \quad 4\pi G\rho\left(1+\frac{\epsilon}{c^2}\right)R^2R' = \frac{c^2}{2}\left(R+\frac{R\dot{R}^2}{a^2c^2}-\frac{RR'^2}{b^2}\right)' \equiv m'G, \tag{4}$$

$$(T_1^1) \quad \frac{4\pi G}{c^2}PR^2\dot{R} = -\frac{c^2}{2}\left(R+\frac{R\dot{R}^2}{a^2c^2}-\frac{RR'^2}{b^2}\right)^{\cdot} = -\dot{m}G, \tag{5}$$

$$(T_2^2, T_3^3) \quad 4\pi G\rho w R^3 = c^2\left(R+\frac{R\dot{R}^2}{a^2c^2}-\frac{RR'^2}{b^2}\right) + \frac{R^3c}{ab}\left[\left(\frac{a'c}{b}\right)' - \left(\frac{\dot{b}}{ac}\right)^{\cdot}\right], \tag{6}$$

$$(T_0^1, T_1^0) \quad 0 = \frac{a'\dot{R}}{a} + \frac{\dot{b}R'}{b} - \dot{R}', \tag{7}$$

where $\cdot$ means $\partial/\partial t$, $'$ means $\partial/\partial\mu$, and w, the specific enthalpy, is

$$w = (1+\epsilon/c^2+P/\rho c^2). \tag{8}$$

Shown in parentheses to the left of Eqs. (4) through (7) are the energy momentum tensor components involved. We have introduced into (4) and (5) the total mass up to point μ

$$m(\mu,t) = 4\pi\int_0^\mu \rho(1+\epsilon/c^2)R^2R'd\mu. \tag{9}$$

Defining also

$$u = \dot{R}/a, \tag{10}$$

$$\Gamma = 4\pi\rho R^2R', \tag{11}$$

Eq. (4) can be integrated from 0 to μ, giving

$$\Gamma^2 = 1+(u/c)^2-(2mG/Rc^2). \tag{12}$$

u is the 1-component of the 4-velocity in a Schwarzchild coordinate system[2] defined by the metric

$$ds^2 = e^\nu c^2 d\tau^2 - e^\lambda dR^2 - R^2 d\Omega^2. \tag{13}$$

If G were zero, Γ would become the $\gamma = (1-v^2/c^2)^{-1/2}$ of special relativity corresponding to the Lorentz transformation connecting, at a given time, the coordinate systems defined by (1) and (13).

The divergence equations $T_i^k{}_{;k} = 0$ are

$$(T_0^k{}_{;k}) \quad [\rho(1+\epsilon/c^2)]^{\cdot} + [(\dot{b}/b)+(2\dot{R}/R)]\rho w = 0, \tag{14}$$

$$(T_1^k{}_{;k}) \quad P' + (a'/a)\rho wc^2 = 0. \tag{15}$$

Of the six equations: (4) through (7), (14), and (15), only four are independent. Using (3) in (14) gives

$$\dot{\epsilon} + P(1/\rho)^{\cdot} = 0. \tag{16}$$

Using (12), (7) for $\dot{R}'$, and (15) for a', (5) becomes the equation of motion:

$$\dot{u} = -a(4\pi P'R^2(\Gamma/w) + (mG/R^2) + (4\pi PGR/c^2)). \tag{17}$$

Using (3), (7) becomes the equation of mass conservation

$$(\rho R^2)^{\cdot}/\rho R^2 = -a(u'/R'). \tag{18}$$

These equations have been obtained independently and discussed by Misner and Sharp.[3] Equations (15) through (18), together with the definitions of u, Γ, m, and w, and the equation of state, determine the solution of the problem of spherically symmetric fluid motion, with rest mass conservation and no heat transfer. Taking derivatives with respect to local clock time, $(1/a)\partial/\partial t$, to be equivalent to time derivatives in the nonrelativistic limit, Eqs. (16) through (18) are seen to differ from the nonrelativistic equations of energy conservation, motion, and continuity, respectively (in Lagrangian coordinates), only through the introduction, into the equation of motion (17), of the factor Γ/w multiplying the pressure gradient and of the added gravitational term $4\pi PGR/c^2$.

These equations were solved on digital computers (mainly the CDC-3600), using a finite-difference method similar to the one generally used for solving hydrodynamic problems in a Lagrange coordinate system.[4] Mathematical details of the method will be made available elsewhere. Shocks were treated by means of an artificial viscosity, similar to that of Richtmyer and von Neumann,[5] that is, by means of a scalar stress which is zero where no pressure or density discontinuity tends to form, and spreads the discontinuity over three or four zones where it does tend to form. For the description of complete collapse, it makes no difference whether an artificial viscosity is introduced or not; such an artificial viscosity is never operative.

[2] Reference 1 pp. 327 ff.
[3] C. W. Misner and D. H. Sharp, Phys. Rev. **136**, B571 (1964).
[4] R. D. Richtmyer, *Difference Methods for Initial Value Problems* (Interscience Publishers, New York, 1957) Chap. X.
[5] J. von Neumann and R. D. Richtmyer, J. Appl. Phys. **21**, 232 (1950).

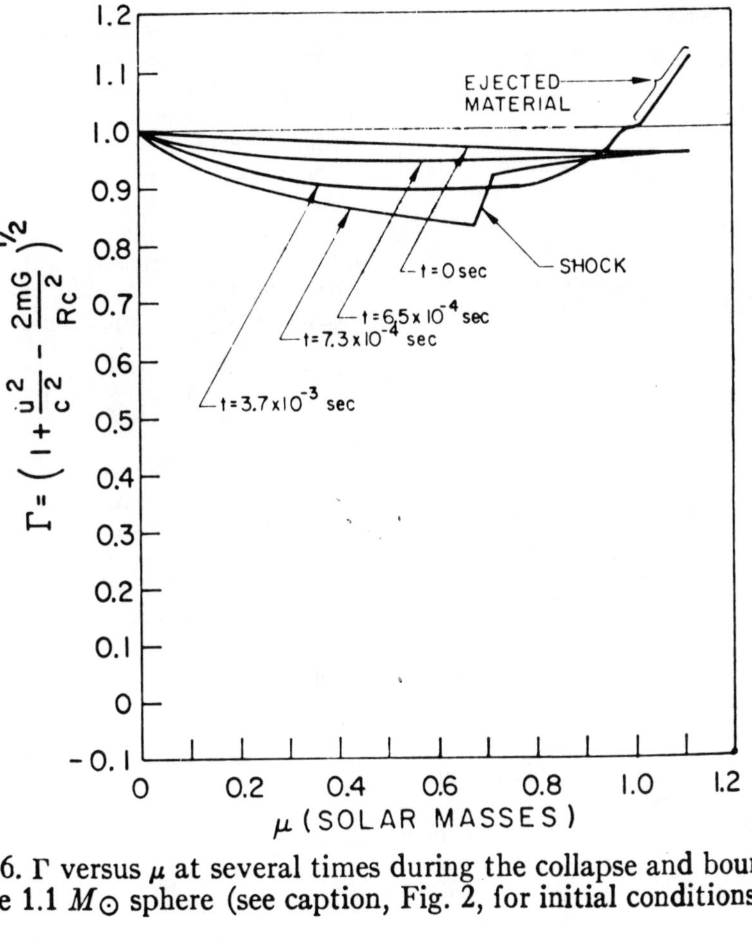

FIG. 6. Γ versus μ at several times during the collapse and bounce of the 1.1 $M_\odot$ sphere (see caption, Fig. 2, for initial conditions).

Initial conditions for these equations were varied. The boundary conditions used were

$$P=0,\quad a=1 \text{ at } \mu=\mu_{max},$$
$$u=0,\quad R=0 \text{ at } \mu=0. \qquad (19)$$

The condition $a=1$ at $\mu=\mu_{max}$ makes coordinate time equal to the clock time of an observer moving with the outer boundary. This boundary, in the problems run so far, did not come into a region where general relativistic effects were important, even when the inner part of the material collapsed completely.

The conditions at $\mu=0$ lead to $\Gamma(\mu=0)=1$, if ρ and ϵ are not to be infinite there.

III. RESULTS

A. Collapse versus "Bounce." Mass Loss

Consider material at rest and in local thermodynamic equilibrium. Let the total rest mass of the separated particles making up the material be μ_{max} and its total energy be Mc^2. If the assembly is not in hydrostatic equilibrium and $M<\mu_{max}$, gravitational collapse begins. If there is no external interaction, and no radiation or other mass loss, both M and μ_{max} remain constant, as does therefore their difference, the binding energy

$$B=(\mu_{max}-M)c^2=\int_0^{\mu_{max}}[c^2-\Gamma(c^2+\epsilon)]d\mu. \qquad (20)$$

Even if M is below the maximum stable mass for the material in question (e.g., in the case of cold neutrons, below 0.72 $M_\odot$, the Oppenheimer-Volkoff limit[6]), the assembly is not expected to settle into the equilibrium configuration appropriate to its mass and initial entropy. It will have too little binding energy. Except for radiation and mass loss, it seems likely to oscillate about an equilibrium point, until viscous damping transforms enough of the excess kinetic energy into heat for the material to come into hydrostatic equilibrium at a higher entropy than the initial one.

If M is above the maximum stable equilibrium mass, there may still be a range of masses for which collapse will not occur. Since the binding energy of an assembly (with a polytropic equation of state) initially in a nonequilibrium state is less than that of a polytrope of the same mass, its material will not settle into as compact a configuration as the polytrope and the gravitational pull due to general relativity will not be as pronounced. Such a range was found to exist for the simple cases studied. It is not very wide, but may be of importance in stellar evolution.

A rough criterion for predicting from the initial conditions whether or not such an assembly will collapse can be obtained by estimating whether an outer radius can be found for which the force due to the pressure gradient just balances the gravitational attraction. If we indicate quantities at the outer boundary at this turnaround time by an asterisk, we are asking whether the equation of hydrostatic equilibrium

$$\left(\frac{1}{\rho w}\frac{dP}{dR}\right)^*\left(1-\frac{2MG}{R^*c^2}\right)=-\frac{MG}{R^{*2}}-\frac{4\pi G}{c^2}P^*R^* \qquad (21)$$

can be satisfied for any R^*. Equation (21) follows immediately from (17) if time derivatives are set to zero and the independent space variable is changed from μ to R. Barring negative pressures, it has no solution if $\Gamma=0$, i.e., if $2MG/Rc^2=1$, anywhere.

Under our boundary condition, $P^*=0$. We estimate $(1/\rho w)^*(dP/dR)^*$ by assuming first, that the internal energy of the assembly increases approximately adiabatically according to a γ law (the shocks which do occur change the adiabat very little), and second, by writing the expression as a constant, n, times the ratio of the average specific internal energy at turnaround, $\epsilon_0(R_0/R^*)^{3(\gamma-1)}$, to R^*. n will depend on the form of $P(R)$ at turnaround time and on γ, and should be of order unity. We define α as the ratio of R^* to the Schwarzchild radius R_s. With these assumptions, (21) becomes an algebraic equation for α:

$$\frac{\alpha^{3\gamma-3}}{\alpha-1}=n\left[\frac{2\epsilon_0}{c^2}\left(\frac{R_0}{R_s}\right)^{3\gamma-3}\right]\equiv nK. \qquad (22)$$

K is specified by the initial conditions. This method fails for $\gamma=\frac{4}{3}$, for which, in any case, no stable relativistic polytrope exists.

Equation (22) is clearly only indicative of the situation to which the fall will lead. In particular, n is not constant. Furthermore the radius at which the balance between pressure gradient and force of gravity is most precarious, in critical cases, is not that of the outer mass point, which has been kept large by the zero boundary pressure. It occurs in the outer half of the material. Nevertheless, the minimum value of nK for which (22) has a real solution should give the order of magnitude of the minimum radius inside of which most of the mass can come to hydrostatic equilibrium. Physically, if $\alpha\sim 1$, the quantity Γ which multiplies the pressure gradient term in (17) will fall significantly below unity at the time at which the mass should turn around; the term u^2/c^2 in Γ [see Eq. (12)] will have been reduced by the cumulative effect of the outward pressure gradient below its free-fall value of $2MG/Rc^2$. Since $R_s\sim R$, the difference between Γ and 1 will not be negligible. As a result, the pressure gradient becomes ineffective, and the assembly continues to collapse. Since the pull of gravity increases as the pressure gradient decreases, the effect is expected to be catastrophic. On the other hand, if α is significantly greater than 1, these effects will be small and collapse will be avoided.

[6] J. R. Oppenheimer and G. M. Volkoff, Phys. Rev. 55, 374 (1939).

The behavior of assemblies of 2.1 $M_\odot$, 21 $M_\odot$, and 210 $M_\odot$ was calculated, using the equation of state $P=2\rho\epsilon/3$, corresponding to an adiabatic index $\gamma=\frac{5}{3}$. The assemblies were started from rest and from uniform rest mass and internal energy densities. The choice of uniform initial conditions was made, in part, because assemblies which start from complete Newtonian equilibrium were found to be more difficult to follow on the computer, the regions of extreme curvature occurring close to the center thereby necessitating fine zoning and long computing times; and, in part, because gravitationally unstable configurations of astrophysical interest, while not known, are now held likely to occur at the center of highly evolved stars,[7,8] where the assumptions of uniform density and temperature may be no worse than any other guess.

All the assemblies collapsed for $K\leq 1$ and bounced for $K\geq 2$, where bounce is defined as a state in which all the material is either at rest or moving outward (Fig. 1). For uniform initial conditions, if R_0 is the initial outer radius, the values of ϵ_0/c^2 and of $2MG/R_0c^2$ determine the path of a given mass fraction μ/μ_{max} in the plane of R/R_0 and $t(2MG/R_0^3)^{1/2}$.

Calculations were then made on the collapse of cold neutron spheres, varying both the mass and the equation of state. The neutrons were started from densities for which the Fermi energy is well in the nonrelativistic range, and were assumed to follow the equation of state of a nonrelativistic degenerate Fermi gas ($\gamma=\frac{5}{3}$) to begin with. The value of K for which collapse occurred was then bracketed for three equations of state:

$$P=\tfrac{2}{3}\rho\epsilon(\gamma=5/3) \text{ throughout the density range} \qquad \text{(i)}$$

$$\begin{aligned} P&=\tfrac{2}{3}\rho\epsilon, & \rho&<\rho^* \\ &=(1/13)(\rho-\rho^*)\epsilon+\tfrac{2}{3}\rho^*\epsilon, & \rho&>\rho^* \qquad \text{(ii)} \\ &\text{with } \rho^*=1.7\times 10^{15} \text{ g/cm}^3. \end{aligned}$$

(ii) represents a "soft" high density neutron equation of state, $\gamma=14/13$. It is approximately the one used by Misner and Zapolsky[9] to fit calculations of Ambartsumyan and Saakian on processes converting a high kinetic neutron into a baryon at rest. As would be expected for $\gamma<\frac{4}{3}$ (Ref. 6), an assembly in which ρ exceeds ρ^* does not come into equilibrium. The innermost region collapses to ever increasing densities.

$$\begin{aligned} P&=\tfrac{2}{3}\rho\epsilon \quad \rho<\rho^* \\ &=(\rho-\rho^*)\epsilon+\tfrac{2}{3}\rho^*\epsilon,\ \rho>\rho^*, \qquad \text{(iii)} \\ &\text{with } \rho^*=5\times 10^{14} \text{ g/cm}^3. \end{aligned}$$

This choice represents a "hard core" high-density-neutron equation of state, $\gamma=2$. When $\gamma=2$, at very relativistic thermal energies, the speed of sound approaches the speed of light. However, at lower thermal energies, stiffer equations of state have been envisaged, notably one by Salpeter[10] discussed briefly in Sec. IV.

The results of these calculations are summarized in Table I. For equations of state (i) and (iii), $K\sim 1$ indicates the demarcation line between collapse and bounce. If masses of the same entropy and equation of state are compared, $\epsilon_0 R_0^{3(\gamma-1)}$ is constant, and K depends on the mass alone: for cold, nonrelativistic neutrons in particular:

$$K=\left(\frac{2\times 10^{33}}{M}\right)^{4/3}, \quad =1 \text{ when } M\cong 1M_\odot.$$

Misner and Zapolsky[9], using $\gamma=5/3$, obtained a maximum stable polytropic mass of about 0.8 $M_\odot$.

Figures 2 through 7 show profiles of the 4-velocity, the circumference variable R, the rest mass density ρ, the quantities a and Γ, and the three terms in the equation of motion, Eq. (17), versus the radial variable μ at various times during the implosion and subsequent bounce of a 1.1 $M_\odot$ assembly with equation of

TABLE I. Results of calculations on cold neutron spheres.

Equation of state	M ($M_\odot$)	μ_{max} ($M_\odot$)	B ($M_\odot c^2$)	R_0 (km)	R_s/R_0	K	Collapse?	Final (g/cm³)	Maximum (g/cm³)
(i)	0.6890	0.6999	0.0109	32.0	0.0638	2.0	No	6×10^{14}	2×10^{15}
(i)	0.8924	0.9099	0.0175	34.88	0.0758	1.4	No	3×10^{15}	10^{16}
(i)	1.1024	1.1285	0.0261	37.426	0.0873	1.06	Yes	...	...
(i)	1.6321	1.6845	0.0524	42.656	0.1134	0.62	Yes	...	...
(ii)	0.8924	0.9099	0.0175	34.88	0.0758	1.4	Yes	...	...
(iii)	1.1024	1.1285	0.0261	37.426	0.0873	1.06	No	0.9×10^{15}	3×10^{15}
(iii)	1.3039	1.3388	0.0349	39.58	0.0976	0.84	Yes	...	...
				$\rho_0=10^{13}$ g/cm³ $\epsilon_0=3.64\times 10^{18}$ erg/g					

[7] F. Hoyle and W. A. Fowler, Astrophys. J. **132**, 565 (1960).
[8] H. Y. Chiu, Phys. Rev. **123**, 1040 (1961).
[9] C. W. Misner and H. S. Zapolsky, Phys. Rev. Letters **12**, 635 (1964). Also, H. S. Zapolsky (private communication).
[10] E. E. Salpeter, Ann. Phys. (N. Y.) **11**, 393 (1960).

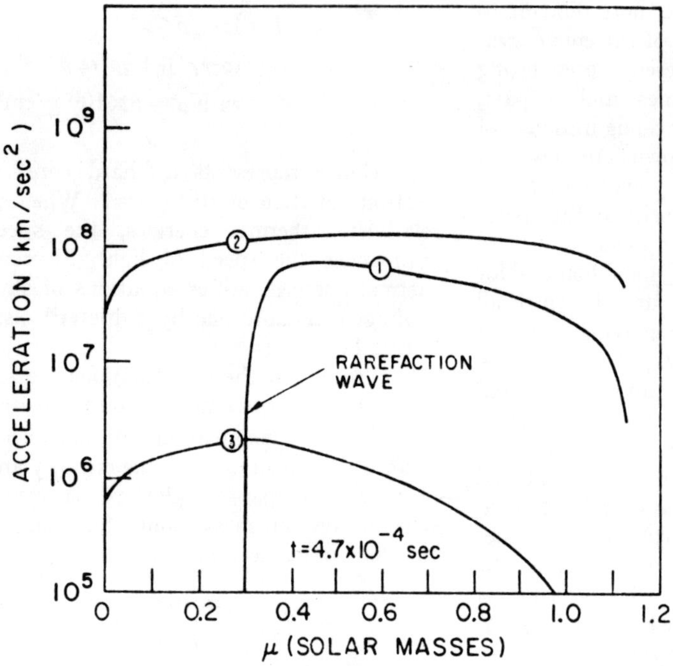

(a)

FIG. 7. The terms of Eq. (17) ($1=-4\pi aP'R^2\Gamma/w$, $2=amG/R^2$, $3=4\pi aPGR/c^2$) at various times during the collapse of the 1.1 $M_\odot$ sphere. (Initial conditions are given in the caption of Fig. 2). Figure 7(a) ($t=4.7\times10^{-4}$ sec) shows the initial rarefaction wave proceeding from the outside toward the center. In 7(b) ($t=6.5\times10^{-4}$ sec) the rarefaction has reached the center and a reflected compression wave can be seen. In 7(c) ($t=7.3\times10^{-4}$ sec) the rarefaction has steepened into a shock; the oscillations are numerical in origin. In 7(d) ($t=3.7\times10^{-3}$ sec) the shock has reached the outer surface and blown off $\sim$0.1 $M_\odot$. The acceleration terms are in equilibrium.

state (iii). Figures 8 through 13, which show the variations of similar quantities in a case of collapse, are discussed in part B of this section.

In several of the assemblies which bounced, some material was moving at more than escape velocity at the time of the last configuration calculated (Fig. 2). The upper limit to the amount of material which can escape from an assembly of initial binding energy B_0 and initial mass μ_0 is given from the conservation of energy and of rest mass by

$$B_P(\mu)=B_0(\mu_0)\,, \tag{23}$$

where $B_P(\mu)$ is the binding energy of a relativistic polytrope of rest mass μ, the highest binding energy possible for an assembly of that rest mass. For $\gamma=5/3$, this polytropic binding energy is about $0.04c^2M_\odot$ at the maximum stable polytropic mass of about $0.8M_\odot$, and goes to $\sim$0 at $0.3M_\odot$.[9]

While not enough relativistic polytropic configurations are available at this time to make a thorough survey, the problems run appear to allow escape of an amount of the order of the maximum mass allowed by Eq. (23). For instance, the $0.7M_\odot$ problem, which has an initial binding energy of $0.01c^2M_\odot$ allows escape of at least $0.052M_\odot$, together with about $0.003c^2M_\odot$ of excess kinetic energy over that needed for escape. The remaining $0.65M_\odot$ has a binding energy of

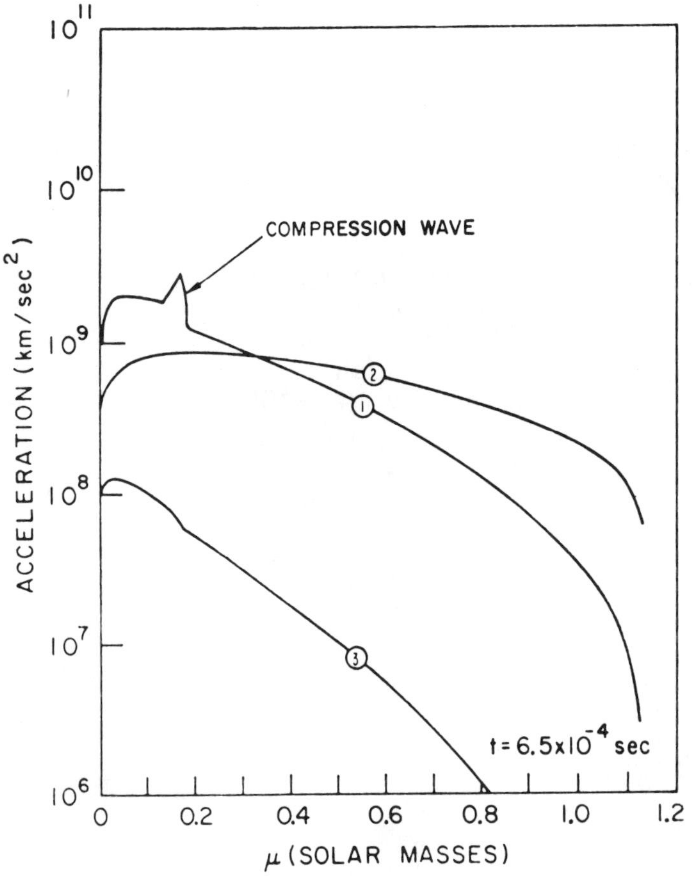

COMPRESSION WAVE
ACCELERATION (km/sec²)
μ (SOLAR MASSES)
t = 6.5x10⁻⁴ sec

(b)

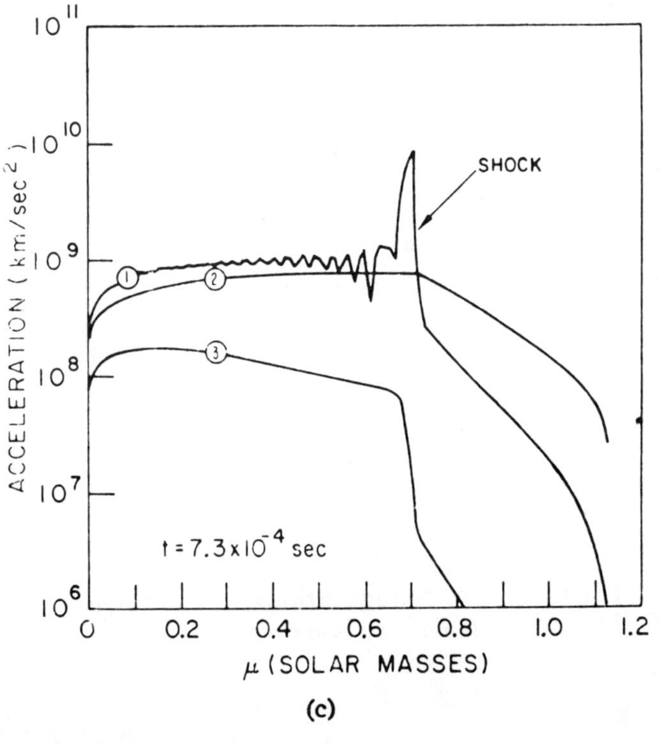

SHOCK
ACCELERATION (km/sec²)
μ (SOLAR MASSES)
t = 7.3x10⁻⁴ sec

(c)

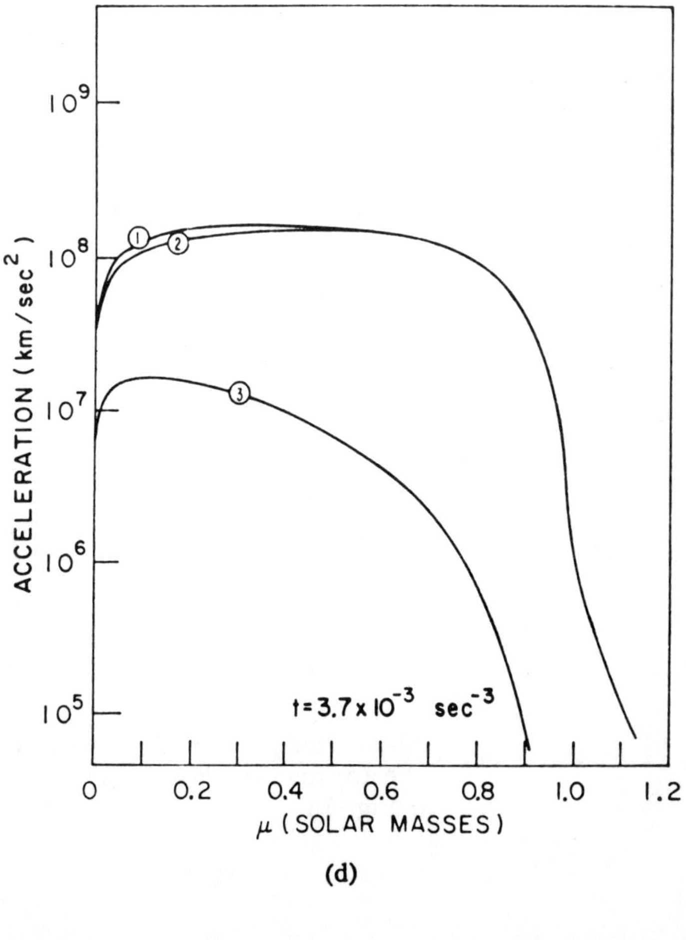

ACCELERATION (km/sec²)
μ (SOLAR MASSES)
t = 3.7x10⁻³ sec⁻³

(d)

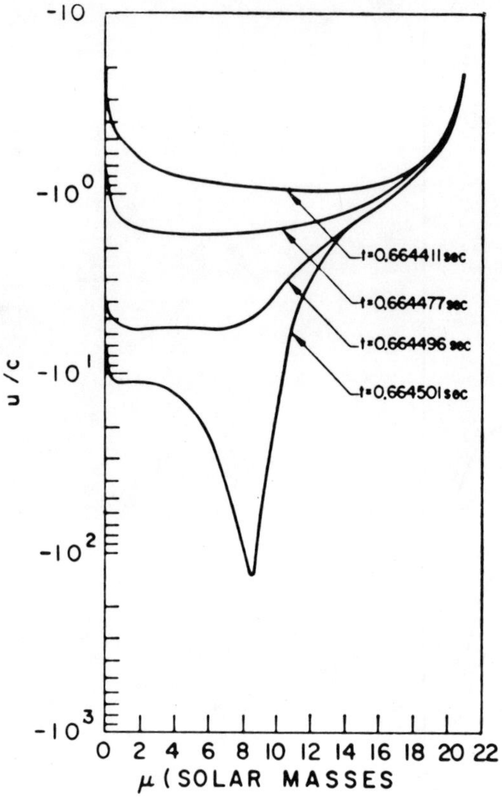

FIG. 8. u/c versus μ at various times during the collapse from rest of a $\gamma=5/3$, 21 $M_\odot$ sphere having an initially uniform density ($\rho_0=10^7$ g/cm^3) and specific internal energy ($\epsilon_0/c^2 = 9.61\times10^{-7}$). These initial conditions give $2MG/R_0c^2=0.0062$ and K [Eq. (22)]=0.05.

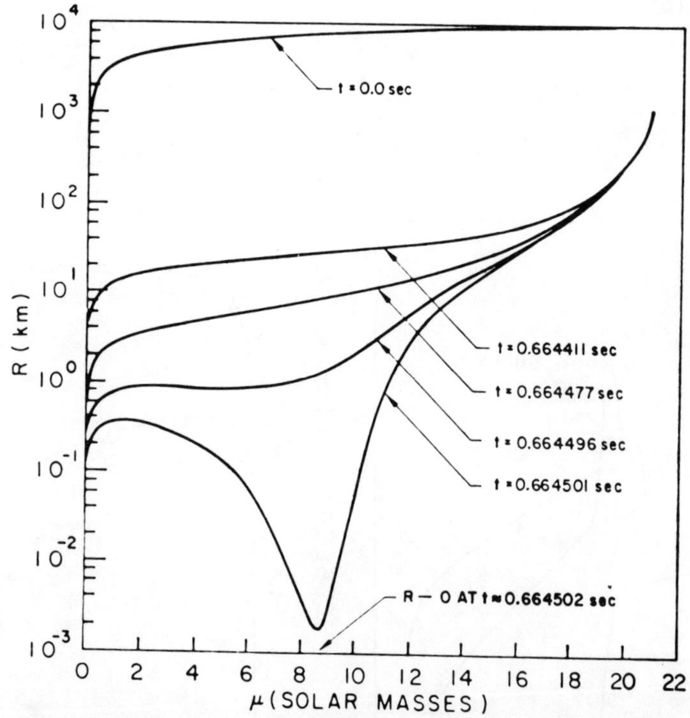

FIG. 9. R versus μ at times during the collapse of the 21 $M_\odot$ sphere (see caption, Fig. 8 for details of the initial configuration) The approach of R to zero at $\mu=8.5\ M_\odot$ and $t\approx0.6645$ sec is shown in Fig. 15.

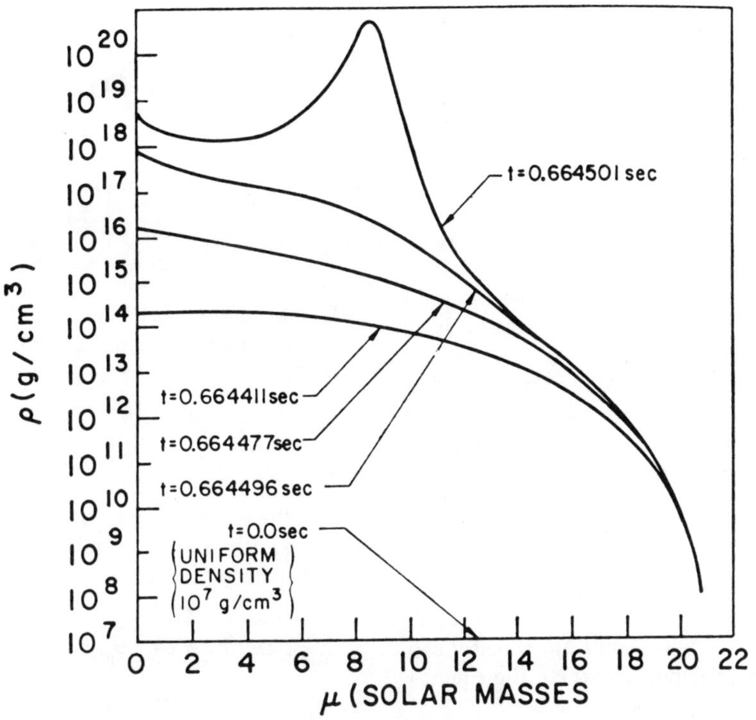

FIG. 10. ρ versus μ at times during the collapse of the 21 $M_\odot$ sphere (see caption, Fig. 8 for details of the initial configuration).

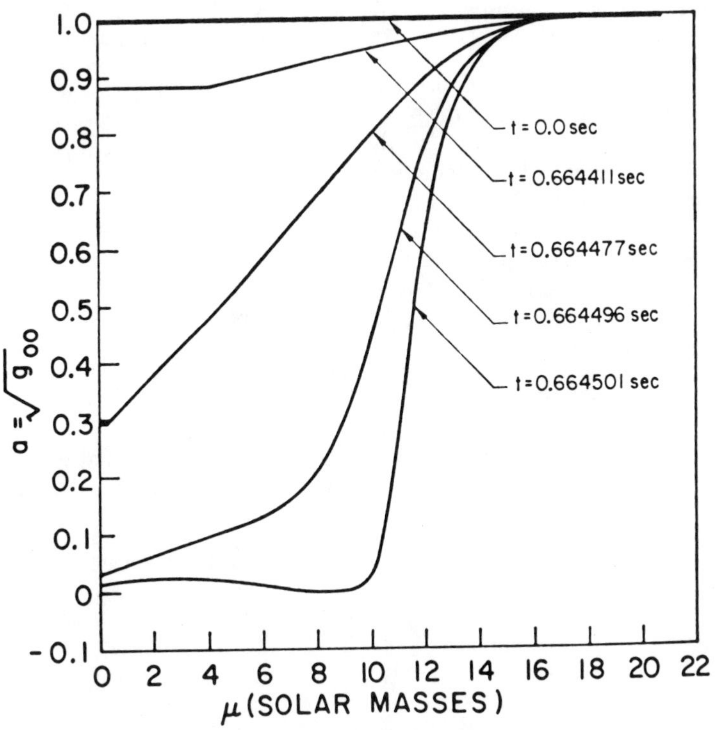

FIG. 11. $a = \sqrt{g_{00}}$ versus μ at times during the collapse of the 21 $M_\odot$ sphere (see caption, Fig. 8 for details of the initial configuration).

$0.013c^2M_\odot$, as compared with $0.018c^2M_\odot$ for a polytrope of that mass.

B. Description of Continued Collapse

Continued collapse is characterized by material falling inside "its own" Schwarzchild radius,

$$R_s(\mu)=[2m(\mu)G]/c^2>R(\mu). \qquad (24)$$

This event occurs first, in our coordinate time, at a point inside the assembly. The Schwarzchild radius there does not have the same clear meaning which it has when calculated for the mass of the whole assembly. The rate at which light signals are propagated inside the assembly is discussed in Sec. IV.

While the continuation of collapse of any part of the assembly past its Schwarzchild radius may take an infinite time as measured by clocks at infinity, it is of some theoretical interest to see how the hydrodynamic quantities and the geometry evolve during this collapse. As was suggested by Oppenheimer and Snyder,[11] this behavior is qualitatively similar to that which occurs when the pressure is zero.

The decrease in the pressure gradient term of Eq. (17) that occurs when Γ falls significantly below unity leads quite rapidly to a condition where the equations governing the motion of the material are

$$\dot{u}\cong -amG/R^2, \quad u^2\cong 2mG/R, \qquad (25)$$

replacing (17) and (12), respectively Eq. (25) has the solution

$$u^2R=\frac{\dot{R}^2R}{a^2}=2mG, \quad \text{a constant in time.} \qquad (26)$$

Each shell, $d\mu$, of material for which R_s has become much larger than R therefore falls freely since the pressure gradient term through which neighboring shells interact has become vanishingly small. The interaction by means of Eq. (15), for $T_1{}^k{}_{;k}$ is also small. In any case, since P' no longer affects the motion, (15) can only affect the way in which proper times compare in neighboring regions. In this connection, we recall that $\dot{R}$ is the time rate of change of the circumference variable. The sum of local measurements of radial distances,

$$r_p=\int_0^\mu \frac{d\mu}{4\pi\rho R^2}=\int_0^R \frac{dR}{\Gamma} \qquad (27)$$

is becoming larger rather than smaller.

Differentiating (26) with respect to μ and using (18), we obtain

$$\dot{\rho}/\rho=-\tfrac{3}{2}(\dot{R}/R)-\tfrac{1}{2}(\dot{R}m'/R'm). \qquad (28)$$

m'/m can be evaluated by using the $T_0{}^0$ and $T_1{}^1$ equations, (4) and (5), which give

$$\dot{R}m'/R'=-\dot{m}\rho(c^2+\epsilon)/P.$$

In our approximation, $\dot{m}/m\rightarrow 0$ while $\rho(1+\epsilon/c^2)/P$ remains finite. Hence, (28) gives

$$\rho R^{3/2}=\text{constant in time.} \qquad (29)$$

So long as there is no interaction between neighboring materials, no shock forms, and the entropy at each mass point remains constant. Hence

$$\epsilon R^{3(\gamma-1)/2}=\text{constant in time},$$
$$aR^{-3(\gamma-1)/2}=\text{constant in time}, \qquad (30)$$

where the second equation above also requires that the specific enthalpy w becomes very large compared with 1. Substituting for a from (30) into (26) and integrating gives:

$$R^{3-3/2\gamma}=R_0^{2-3/2\gamma}[R_0-(3-\tfrac{3}{2}\gamma)|\dot{R}_0|], \quad \gamma<2$$
$$R=R_0\exp(-|\dot{R}_0|t/R_0), \quad \gamma=2 \qquad (31)$$

where R_0, $\dot{R}_0$ are the values of R and $\dot{R}$ measured at the given mass point sometime after free fall begins, t being measured from that time.

For $\gamma<2$, the collapse therefore takes place in a finite interval of coordinate time, which is different for each shell of material. The fact that densities and pressures go to ∞ as R goes to 0 does not alter these conclusions, as the pressure gradient term, $P'R^2\Gamma/w$, goes to zero there as $RR'-\text{const } R'^2$.

All of these extrapolations are verified by the machine calculations. Figures 8 through 13 show, for the following case

$$M=21M_\odot, \quad R_0=10^4 km, \quad K=0.05, \quad p=\tfrac{2}{3}\rho\epsilon,$$

the same quantities as were shown in Figs. 2 through 7 for an assembly which bounced. We note that eventually Γ and R' become negative, and the surface area of concentric spheres decreases as one moves outward in μ space.

This curvature can also be obtained if it is assumed that ρ and ϵ are spatially uniform. The integration of the field equations then leads to an implicit equation

[11] J. R. Oppenheimer and H. Snyder, Phys. Rev. 56, 455 (1939).

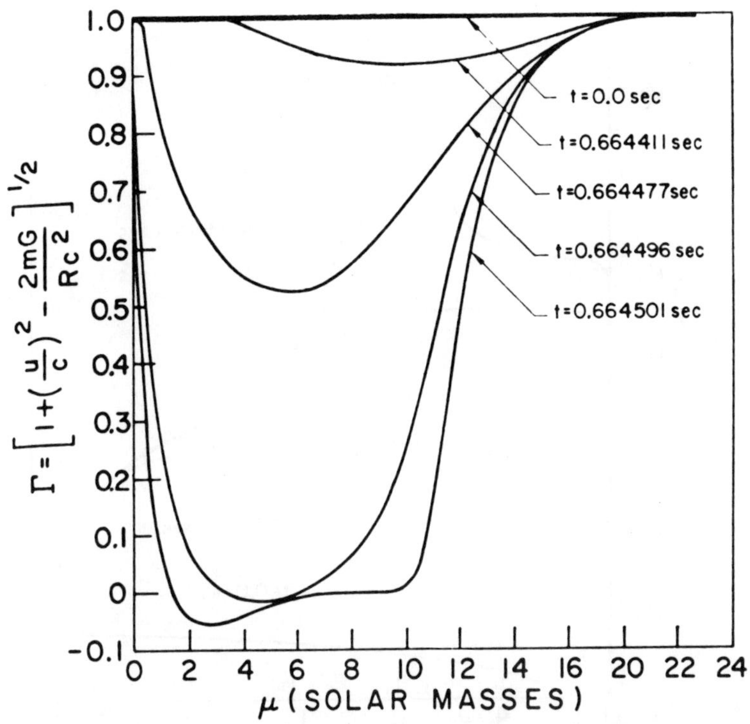

FIG. 12. Γ versus μ at times during the collapse of the 21 $M_\odot$ sphere (see caption, Fig. 8 for details of the initial configuration).

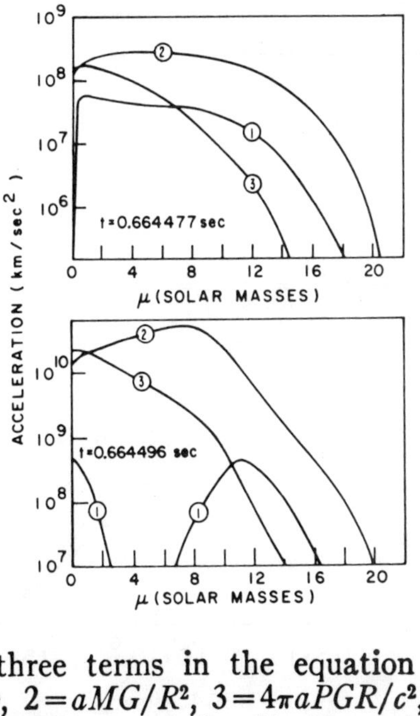

FIG. 13. The three terms in the equation of motion (17), $1 = -4\pi aP'R^2\Gamma/w$, $2 = aMG/R^2$, $3 = 4\pi aPGR/c^2$, versus μ at two times during the collapse of the 21 $M_\odot$ sphere (see caption, Fig. 8). The progressive failure of the pressure gradient term is clearly seen by comparing the upper and lower figures.

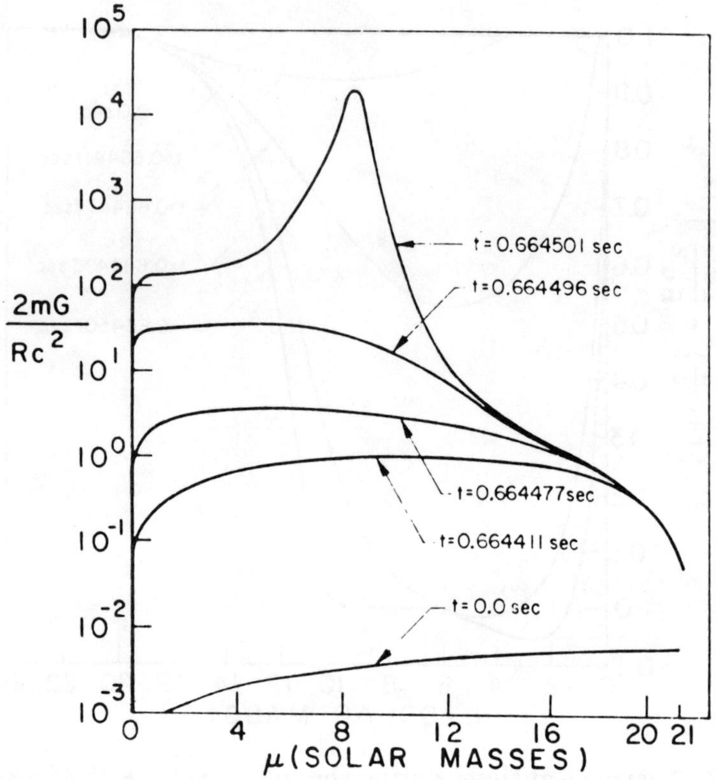

FIG. 14. $2MG/Rc^2 = R_s(\mu)/R(\mu)$ versus μ at various times during the collapse of the 21 $M_\odot$ sphere (see caption, Fig. 8 for details of initial conditions).

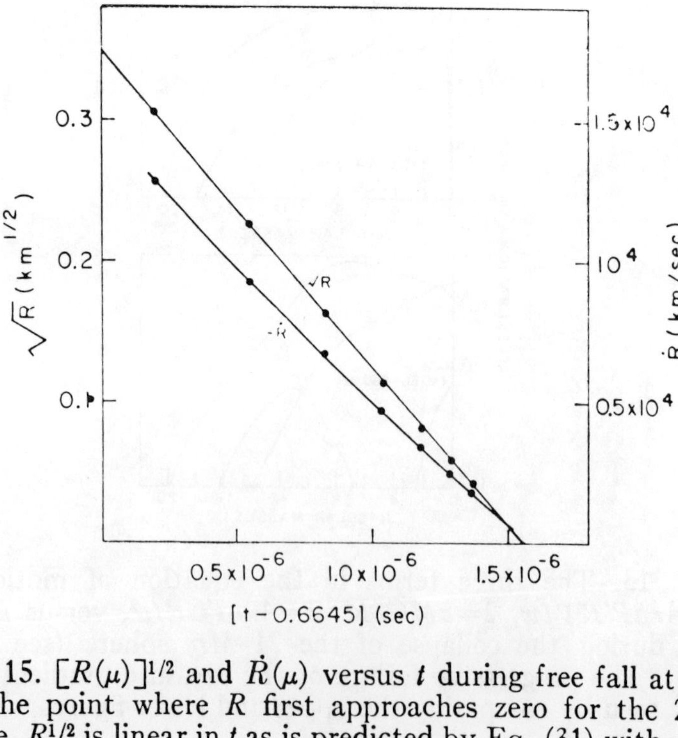

FIG. 15. $[R(\mu)]^{1/2}$ and $\dot{R}(\mu)$ versus t during free fall at $\mu = 8.5$ $M_\odot$, the point where R first approaches zero for the 21 $M_\odot$ collapse. $R^{1/2}$ is linear in t as is predicted by Eq. (31) with $\gamma = 5/3$.

for R as a function of μ:

$$\mu=2\pi\rho R_c[-R(R_c^2-R^2)^{1/2}+R_c^2\sin^{-1}(R/R_c)],$$

$$R_c=\left[\frac{3c^2}{8\pi G\rho(1+\epsilon/c^2)}\right]^{1/2}. \quad (32a)$$

For assemblies of rest mass exceeding some critical value

$$\mu_c=\pi^2\rho R_c^3, \quad (32b)$$

R decreases with increasing μ outside $\mu=\mu_c$ while, at $\mu=\mu_c$, $R=R_s$. The Schwarzchild radius of the assembly decreases with increasing μ even more rapidly than R, so that for $\mu>\mu_c$ we have $R>R_s$ and the addition of rest mass causes the Schwarzchild radius to retreat back inside the assembly. If negative pressures are allowed, this configuration can be static, and is then one of the Schwarzchild interior solutions.

Figure 14 shows the ratio of the Schwarzchild radius $R_s(\mu)$, defined by (24), to R, as a function of μ at various times during the collapse. We note, by comparing Figs. 12 and 14, that the region of negative Γ is inside the region where $2mG/Rc^2>1$. Figure 15 shows R and $\dot{R}$ versus coordinate time for the point at which R first approaches zero.

IV. DISCUSSION

The results described, although preliminary, lead to the conclusion that the upper mass limit for stability of a bound system against relativistic gravitational collapse is of the same order of magnitude whether calculated for falling material with a finite amount of kinetic energy or for a static configuration. While the excess kinetic energy of the falling material does offer the possibility of throwing off mass and leaving behind a core light enough to be stable, in order for this possibility to be realized, the pressure gradient must overcome the pull of gravity at the point of separation between what is to become the core and what is to be thrown off. If the pressure does not become high enough to do the job until the point of separation is at about the Schwarzchild radius for the mass enclosed, then the mechanism for turning material around fails. Static configurations near the limit of stability are only a few Schwarzchild radii in extent, so that our criterion for failure of the pressure gradient will lead to about the same maximum masses as the equilibrium calculations.

For the case of complete collapse described above, $m(\mu)$ approaches a finite limit at the value of μ where first $R\rightarrow 0$. This limit is, however, greater than any value that m has had at that point earlier. This is required by Eq. (5), $\dot{R}$ being negative everywhere or, in the limit, 0. Therefore, in spite of the fact that there are negative contributions to the mass enclosed between the two values of μ for which $R=0$ (see Figs. 9 and 16), it does not seem possible, at least under the circumstances of our problem, for this mass to disappear and leave flat space behind.

The path of light signals through the collapsing assembly was calculated by integrating numerically:

$$(d\mu/dt)_{ds=0}=\pm 4\pi\rho R^2 ac \quad (33)$$

during the collapse, for signals started at regular intervals from the center and from the outside of the assembly. The results are shown in Fig. 17. Light signals do not leave the region where $2mG/Rc^2>1$, although they enter it in finite coordinate time. The region where $\Gamma<0$ is entirely enclosed by the region where $2mG/Rc^2>1$.

If $\Gamma<0$ in some element of rest mass, the total energy there

$$c^2dm=\Gamma(c^2+\epsilon)d\mu \quad (34)$$

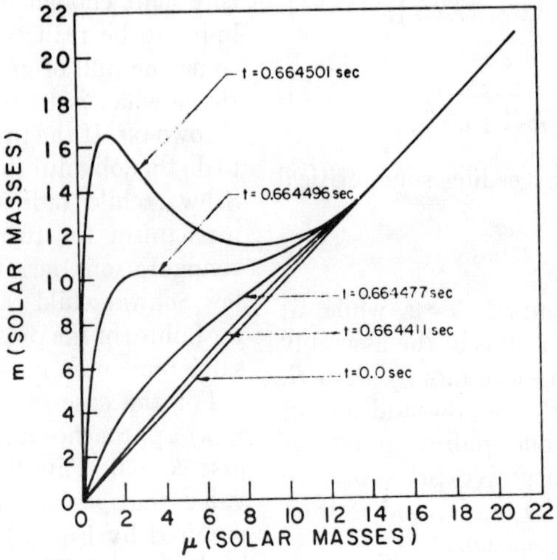

FIG. 16. $M(\mu)$ 'versus μ at various times during the collapse of the 21 $M_\odot$ sphere (see caption, Fig. 8 for details of the initial conditions).

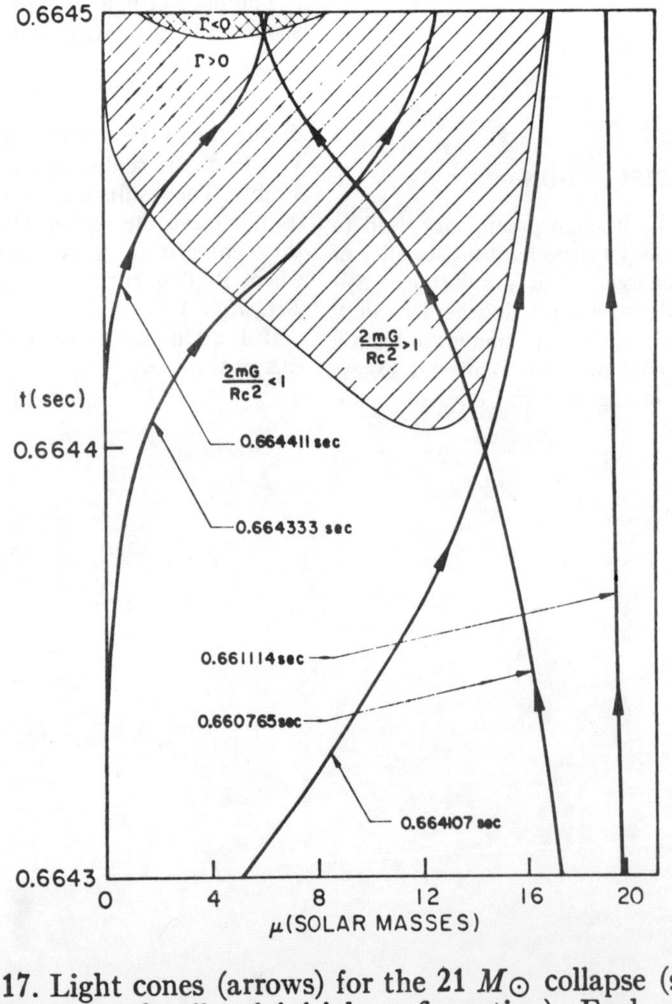

FIG. 17. Light cones (arrows) for the 21 $M_\odot$ collapse (see caption, Fig. 8 for details of initial configuration). Each signal is labeled with its time of origin.

is negative. The partial annihilation of $d\mu$, if permitted, would not change the sign of the total energy. Neither would an increase of ϵ. Since $\Gamma=0$ at the boundary of regions of negative Γ, pressure gradients could not push the material out of such regions. The annihilation of material at the center of the assembly, where $\Gamma>0$, and the subsequent transport of radiation, either into or possibly through the region of negative Γ, may provide a means for reducing m [see Eqs. (9), (11), (12)] but the effect of such reduction on the gravitational binding of a region of negative total energy is unclear.

Whether radiation can proceed from a region where $\Gamma<0$ to a region where $2mG/Rc^2<1$ and hence escape the assembly is also unclear. Configurations described by Eq. (32) (for $\mu>\mu_c$) do contain a region next to the outer boundary where both $\Gamma<0$ and $2mG/Rc^2<1$ obtain but such a region might not be physically realizable as it requires Γ to change sign at its boundary. We have not obtained such regions in our calculations. They would seem to be ruled out by Penrose's proof[12] that the null geodesics issuing from a trapped 2-surface converge toward the future, together with Hernandez and Misner's proof[13] that the Schwarzchild surface inside the matter is such a trapped surface.

The relevance of these calculations to astrophysical questions is now being studied. The effects of asphericity, of heat transfer by neutrinos or radiation, of pair creation and annihilation, and of interaction with outlying material must in general be treated. One of the simple results obtained so far may nevertheless be applicable. Supernovae are now presumed to be occasioned by the inward fall of a fraction of the highly evolved core of certain stars.[7,8] According to at least one model of this evolution, due to Colgate and White,[14] the amount of material which falls far enough so that its gravitational field must be corrected for general relativistic effects is of the order of magnitude of one to two solar masses. This occurs because it is mass of this order of magnitude which first fails to support itself in a Newtonian gravitational field upon cooling. We have followed the fall of a 1 $M_\odot$ core of this type, using the initial conditions obtained by Colgate and White from the classical evolution of a star of total mass $2M_\odot$, and also using a fit to Salpeter's equation of state.[10] The core still bounced. Whether enough of the energy released by neutrinos during the collapse and bounce would be absorbed in the (very high density) surrounding medium to prevent its falling on the core and thereby collapse it, remains to be calculated.

ACKNOWLEDGMENTS

We would like to thank S. A. Colgate, J. Fletcher, R. Lindquist, C. W. Misner, E. Teller, and, most particularly, J. A. Wheeler, for their encouragement and for very helpful discussions.

Earl Tech gave us considerable assistance with the computations.

---

[12] R. Penrose, Phys. Rev. Letters **14**, 57 (1965).
[13] Hernandez and C. W. Misner, Astrophys. J. (to be published).
[14] S. A. Colgate and R. H. White, University of California Lawrence Radiation Laboratory Report UCRL-7777 (1964) (to be published).

THE RESISTANCE OF MAGNETIC FLUX TO GRAVITATIONAL COLLAPSE*

Kip S. Thorne†
Palmer Physical Laboratory, Princeton University
Princeton, New Jersey

I. Introduction and Summary

One of the most significant characteristics of Einstein's general theory of relativity is the extent to which physical singularities pervade solutions to the field equations: Among all matter-filled cosmological models constructed to date within the framework of Einstein's theory (sans cosmological constant), not one is free of singularities both in the remote past and in the remote future.[1] On a smaller scale, general relativity tells us[2] that any non-rotating star, which has reached the end point of thermonuclear evolution and contains more than $A_{crit} \approx 10^{57}$ baryons, must gravitationally collapse to a singularity in a proper time of the order of seconds. Even spherical configurations of cold matter containing much less than 10^{57} baryons cannot escape collapse if they are subjected to sufficient external pressure.

At this point in the development of the theory of gravitational collapse, it is important to determine precisely how inevitable the evolution of singularities is: To what extent can rotation of a massive object or a cosmological model impede or prevent its collapse to a singularity?

The purpose of this paper is to propose a partial answer to the last of these questions: All evidence now available suggests that magnetic and electric field lines resist gravitational collapse; no matter how tightly they are compressed, the gravitational attraction between field lines can never overcome their Maxwell-Faraday repulsion. Let us put this point more precisely:

Principle of Flux Resistance to Gravitational Collapse: In a configuration of electromagnetic fields gravitationally collapsing to a singularity, the total electric and magnetic flux across each 2-surface in the collapsing region must vanish as the singularity is reached—a non-zero flux will stop the collapse. In more mathematical terminology: Let S_2 be an arbitrary 2-surface passing through the region in which collapse is occurring, just before the singularity is reached; and let $\bar{S}_2$ be that portion of S_2 which is in the collapsing region. Then, the principle of flux resistance to gravitational collapse states that

$$\int_{\bar{S}_2} f^{ij}\, dS_{ij} = 0 = \int_{\bar{S}_2} {}^*f^{ij}\, dS_{ij}, \qquad (1)$$

where ${}^*f^{ij}$ is the dual of the electromagnetic field tensor f_{ij}. This principle is illustrated in Fig. 1.

Comments on the principle: (1) At our present stage of knowledge, the principle of flux resistance to gravitational collapse can only be a conjecture; there is as yet no hard-and-fast proof of its validity within the framework of Einstein's theory. However, considerable evidence for it can be evoked, as we shall see in Section II. (2) As an example, if this principle is valid, then a toroidal bundle of magnetic field lines (geon)[3] of minor radius a and major radius b cannot gravitationally collapse to its guiding line (a → 0, b remain finite), but it might collapse to its center (a → 0 and b → 0 simultaneously). In Section II we will see that the dynamical behavior of a toroidal magnetic geon is in accord with this prediction. (3) As a second

*Expanded version of a paper presented at the Second Texas Symposium on Relativistic Astrophysics, Austin, Texas, December 15-19, 1964. This work was supported in part by the U. S. Air Force Office of Scientific Research.

†N. S. F. Predoctoral Fellow.

[1] For a discussion of this point see, e.g., the contribution of Shepley (1965) to this volume.

[2] For a review of the evidence see Harrison, Thorne, Wakano, and Wheeler (1965).

[3] The concept of a geon was first introduced by J. A. Wheeler and is discussed extensively in Wheeler (1963).

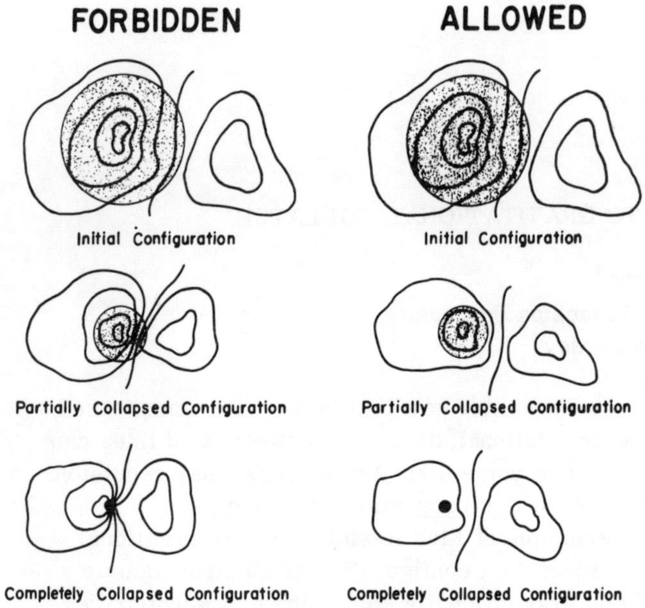

FIG. 1. A schematic illustration of the principle of flux resistance to gravitational collapse. Two possible histories are shown for the gravitational collapse of a system composed of electromagnetic fields. One history is allowed by the principle of flux resistance to gravitational collapse; the other is forbidden. In the figure each history depicts the electromagnetic field on a succession of spacelike hypersurfaces approaching the singularity. The shaded areas represent the region in which the electromagnetic field is gravitationally collapsing, and the solid lines represent magnetic field lines. The electric field is not depicted. In the forbidden mode of collapse, magnetic flux threads the collapsing region at the moment when the singularity is reached, but in the allowed mode, every magnetic field line is either swallowed in its entirety by the singularity or left entirely free—there is no net flux through the singularity.

example, the principle of flux resistance to gravitational collapse does not forbid the collapse of a cloud of electromagnetic radiation or of a radiation-filled universe. This is fortunate, since Tolman (1934) has constructed a radiation-filled cosmological model which both explodes from a singularity and collapses to a singularity. (4) This principle is meant to apply only when electromagnetic fields alone are present. We exclude from attention systems in which particles or neutrinos or other fields contribute to the stress-energy. However, if the principle is valid, then electric and magnetic flux probably also show a partial (or even complete) resistance to collapse in the presence of particulate matter;[4] and magnetic fields might, consequently, play an important role in the collapse of astrophysical objects.[5] (5) The principle asserts that both magnetic and electric flux resist collapse because, in the absence of charges and currents, the electric field and the magnetic field are dynamically equivalent. (6) This principle treats only classical electromagnetic fields interacting with classical gravitational fields in accordance with the laws of general relativity. The quite new phenomena which enter when the electromagnetic field strength reaches the critical value $F_{crit} = mc^2/[e(\hbar/mc)] = 4.4 \times 10^{13}$ gauss $= 1.3 \times 10^{18}$ volt/m, where vacuum polarization effects become important, are not taken into account.

The remainder of this paper is a presentation of evidence supporting the principle of flux resistance to gravitational collapse. We briefly summarize that evidence before presenting it in detail:

A prototype for spherically symmetric gravitational collapse is the collapse of the Einstein-Rosen bridge of the Schwarzschild solution. This collapse is characterized by the pinching-off of the throat of the bridge as its time development is followed. As our first evidence for the principle of flux resistance to gravitational collapse, we review the discovery by Graves and Brill (1960) that, if the Einstein-Rosen bridge is threaded by electric or magnetic flux, then the resistance of that flux to collapse causes the throat to pulsate rather than pinch off.

A second piece of evidence is the existence of many different model electromagnetic universes which do not undergo gravitational collapse, and no known ones that do collapse—except the Tolman universe (cf. comment (3) above) and Lindquist's toroidal magnetic universe (cf. Sec. IIC), whose dynamics are compatible with our conjecture. (Contrast this with matter-filled cosmological

[4] Ginzburg (1964), Ginzburg and Ozernoy (1964), Novikov (1964), and Kardashev (1964) have recently discussed the role of magnetic fields in the gravitational collapse of massive stars. Ginzburg and Ozernoy find that, as a massive magnetic star collapses through its Schwarzschild radius, it pulls its magnetic field lines into its surface and carries them all, in their entirety, into the singularity. This result is what would be expected if magnetic flux resists gravitational collapse in the presence of matter.

[5] Colgate (1965) argues that magnetic field cannot play an important role in the gravitational collapse which initiates supernova explosions.

models, of which there are none known that do not either collapse to a singularity or explode from one.) Of all known model electromagnetic universes, one due to Melvin (1964) is of particular interest. Melvin's universe, which consists of a cylindrically symmetric magnetic field pointing along the axis of symmetry, has been proved to be stable against large as well as small radial perturbations (Melvin 1965; Thorne 1965b); it cannot be induced to evolve a singularity as the result of any finite perturbation.

These first two pieces of evidence for the principle of flux resistance to gravitational collapse are not as satisfying as would be the analysis of more physical electromagnetic systems, around which spacetime is asymptotically flat. Fortunately, the third piece of evidence has what the others lack: It is an analysis of the dynamics of a toroidal bundle of magnetic field lines (geon)[3] residing in asymptotically flat spacetime. This analysis reveals that, in keeping with the principle of flux resistance, a toroidal magnetic geon with magnetic field initially uniform inside the torus cannot collapse to its guiding line (minor radius go to zero, major radius stay finite).[6]

II. Evidence for the Principle of Flux Resistance to Gravitational Collapse

A. Dynamics of the Einstein-Rosen Bridge

The Schwarzschild solution to the vacuum field equations of general relativity has been known for nearly fifty years, but only in the last five years has it been really understood. This is because Schwarzschild's line element

$$ds^2 = (1 - 2m/r)dt^2 - (1 - 2m/r)^{-1}dr^2 - r^2(d\theta^2 + \sin^2\theta\, d\varphi^2) \tag{2}$$

has a coordinate singularity at $r = 2m$ and is incomplete—from any point (r, t) there are spacelike and timelike geodesics to $(r = 2m, t = +\infty)$ or to $(r = 2m, t = -\infty)$ with finite proper length.[7] Our modern understanding of the Schwarzschild solution stems from the work of Kruskal (1960).[8] Kruskal completed the Schwarzschild solution in a manner which exhibits the true nature of the region $r = 2m$. We will not be concerned here with the relationship between the Schwarzschild solution and the Kruskal completion of it, nor with the nature of the region $r = 2m$; rather, we shall confine ourselves to a review of the dynamics of Kruskal's solution.

Kruskal's completion of the Schwarzschild solution is expressed in terms of a new time coordinate, v, and a new radial coordinate, u, in the form

$$ds^2 = f^2(dv^2 - du^2) - r^2(d\theta^2 + \sin\theta\, d\varphi^2). \tag{3}$$

Here f and r are given in terms of u and v by

$$[(r/2m) - 1]\, e^{r/2m} = u^2 - v^2,$$

$$f^2 = (32m^3/r)e^{-r/2m}. \tag{4}$$

The r appearing here is Schwarzschild's r coordinate; and Schwarzschild's t coordinate is related to u and v by

$$\tanh(t/2m) = 2uv/(u^2 + v^2). \tag{5}$$

Fuller and Wheeler (1962) have made clear the dynamical behavior of Kruskal's solution. The spacelike hypersurface $v = 0$ ($t = 0$, $r \geq 2m$ in Schwarzschild coordinates) is a bridge or "wormhole" between two asymptotically flat spaces (see Fig. 2). (It is often called the "Einstein-Rosen bridge," since Einstein and Rosen (1935) discussed it extensively.) One can follow the dynamical evolution of this wormhole in Kruskal's solution by looking at the geometry of a succession of spacelike hypersurfaces, each lying to the future (+v direction) of the preceding one. Such an analysis reveals that the throat of the wormhole undergoes gravitational collapse; it pinches off, disconnecting the asymptotically flat spaces originally linked by the wormhole (see Fig. 3).

Now, suppose that a magnetic field were made to thread the wormhole. The collapse of the wormhole would provide a means for squeezing the magnetic field into a smaller and smaller region. But if the principle of flux resistance to gravitational collapse is correct, the magnetic field should protest against this squeeze; it should actually halt the collapse before the throat pinches off.

[6] For independent evidence that a toroidal magnetic geon may not collapse to its guiding line, see Thorne (1964). For evidence that collapse to the center of the torus should occur for sufficiently massive geons, see Thorne (1964) and Wheeler (1964).

[7] For a discussion of geodesics in the Schwarzschild solution, see Fuller and Wheeler (1962).

[8] See also Fronsdal (1959).

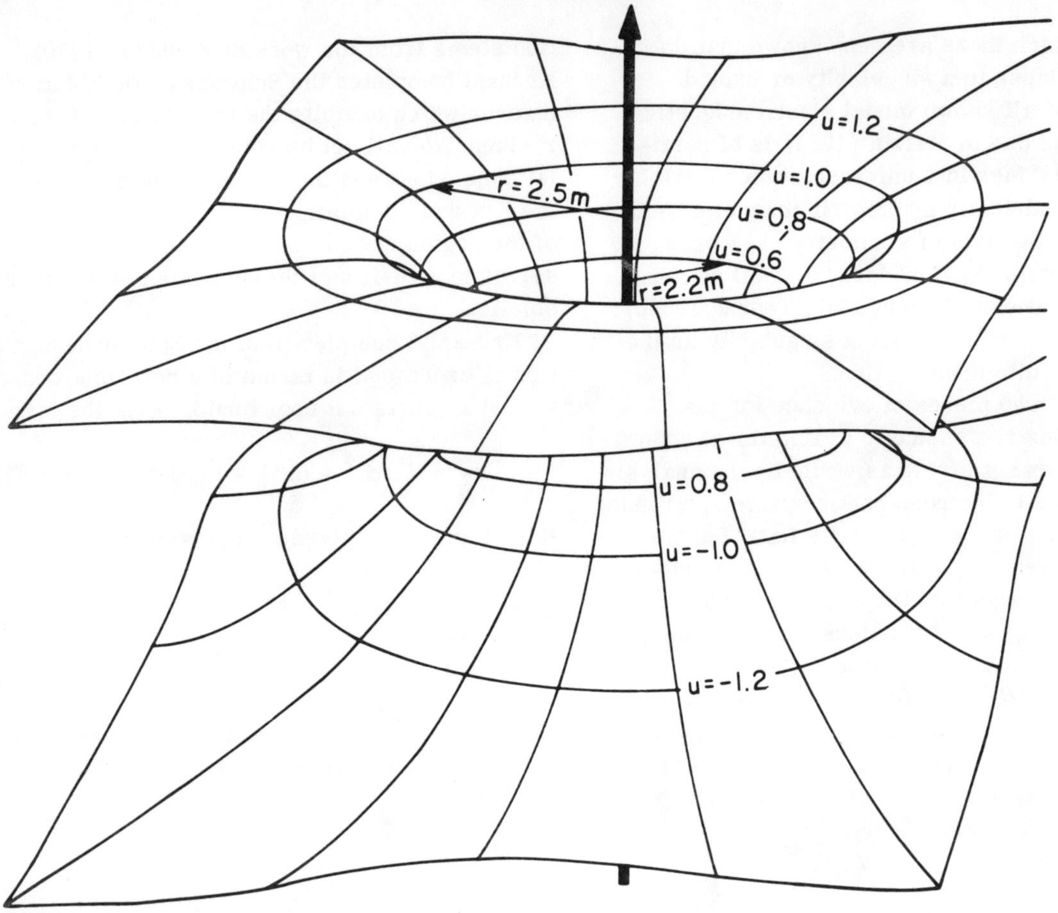

FIG. 2. The 2-surface (v = 0, ϕ = constant) of Kruskal's completion of the Schwarzschild solution, as it appears when embedded in 3-dimensional Euclidean space. The hypersurface v = 0 is the 3-dimensional analogue of this surface, embedded in a 4-dimensional Euclidean space. The bridge or wormhole connecting the upper and lower asymptotically flat surfaces is often called the "Einstein-Rosen bridge," since Einstein and Rosen (1935) discussed it extensively; however, Weyl (1917) described it much earlier. (This figure was kindly provided by J. A. Wheeler.)

Graves and Brill (1960) have given the solution to the Einstein-Maxwell field equations for a wormhole threaded by electric or magnetic flux. Their solution is a completion of the Reissner-Nordström solution for a "charged, point mass" in the same way as Kruskal's solution is a completion of the Schwarzschild solution for an "uncharged point mass." Graves and Brill find that even a very minute amount of magnetic flux threading the wormhole will cushion its collapse. Rather than pinching off, the throat oscillates in and out between its initial radius r_{max} and a minimum radius

$$r_{min} = \frac{G}{16\pi^2 c^4} \times (\text{Total flux threading the throat})^2 \times \frac{1}{r_{max}}$$

(see Fig. 4). Hence, in this particular example, the (ex post facto) predictions of the principle of flux resistance to gravitational collapse are borne out.

B. Non-Collapsing Model Electromagnetic Universes

The easiest test of the principle of flux resistance to gravitational collapse which can be performed is to search the literature for counterexamples to the principle. The author's search has revealed none.

Nearly all electromagnetic systems with strong gravitational fields which have been studied are non-asymptotically flat at spatial infinity. We call such systems "model electromagnetic universes." There has been much interest in model electromagnetic universes recently (Bertotti 1959; Misra and Radhakrishna 1961; Brill 1964; Melvin 1964, 1965; Thorne 1965a, b; Harrison 1965. Of the model universes recently exhibited, several deserve special mention in connection with the principle of flux resistance to gravitational collapse:

Brill (1964) has given a family of model

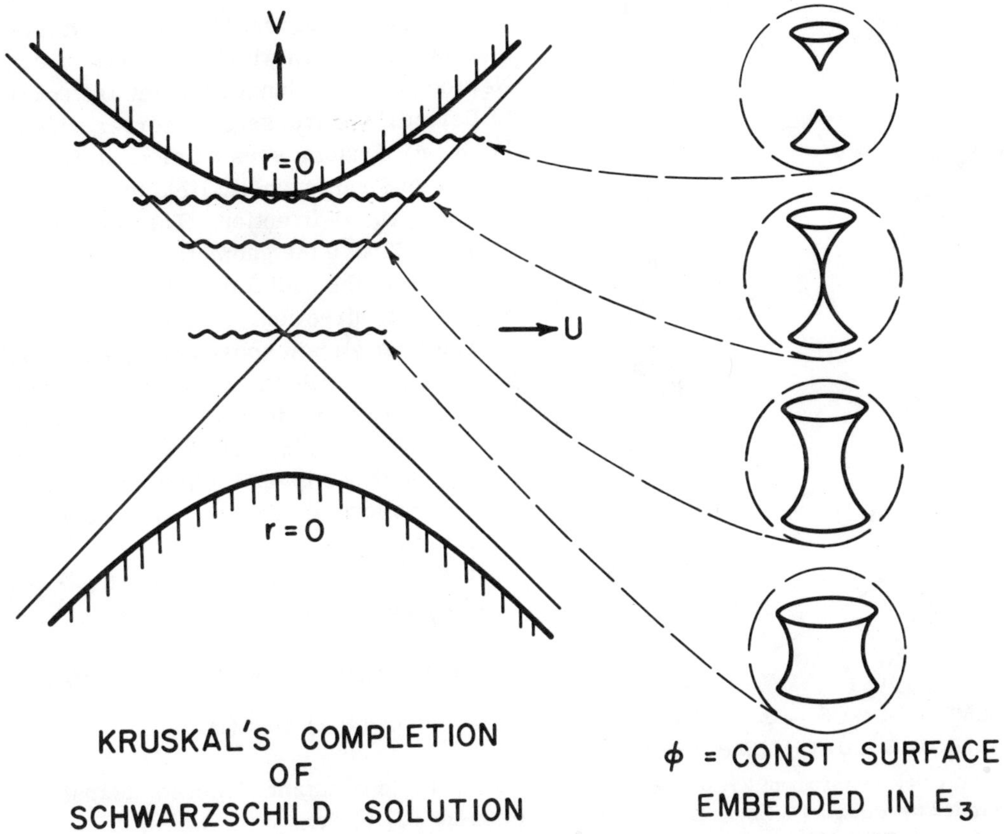

FIG. 3. The dynamics of the throat of the Einstein-Rosen bridge. On the left is a diagram of the Kruskal u-v coordinate plane, showing a succession of spacelike hypersurfaces, each one to the future of the preceding one. On the right is a picture of each of these hypersurfaces embedded in a 3-dimensional Euclidean space. (The angle of rotation, ϕ, is suppressed.) These successive "snapshots" of the throat of the wormhole reveal that it pinches off; it gravitationally collapses to a singularity.

electromagnetic universes which are generalizations of the Taub-NUT vacuum solution to Einstein's equations. Like the Taub-NUT solution, Brill's universes do not possess any physical singularities; however, the dust-filled generalizations of the Taub-NUT solution, which have been given by Behr (1961) and by Shepley (1965), all evolve singularities.

Melvin (1964) has described a static, cylindrically symmetric magnetic universe in which the magnetic field points along the axis of symmetry. This model universe not only does not gravitationally collapse, but no large or small radial perturbation can cause it to collapse (Melvin 1965; Thorne 1965b).

Thorne (1965a) has shown that no cylindrical electromagnetic universe, which is non-singular in a certain canonical coordinate system on some initial spacelike hypersurface, can undergo gravitational collapse.

C. Toroidal Magnetic Geons

In our discussion of evidence supporting the principle of flux resistance to gravitational collapse, we turn now to toroidal magnetic geons residing in asymptotically flat spacetime. We shall consider the family of all geons whose initial configuration has the following properties: (1) It is a momentarily static configuration ("configuration of time-symmetry"). (2) It contains only a magnetic field; the electric field vanishes everywhere. (3) The magnetic field is contained inside a thin-ring torus (torus with minor radius much smaller than major radius), where it is uniform and parallel to the guiding line of the torus. (4) There is no gravitational radiation present anywhere (see Fig. 5).

The members of this family will range from geons with such dilute magnetic field that their dynamics can be treated with great accuracy in the

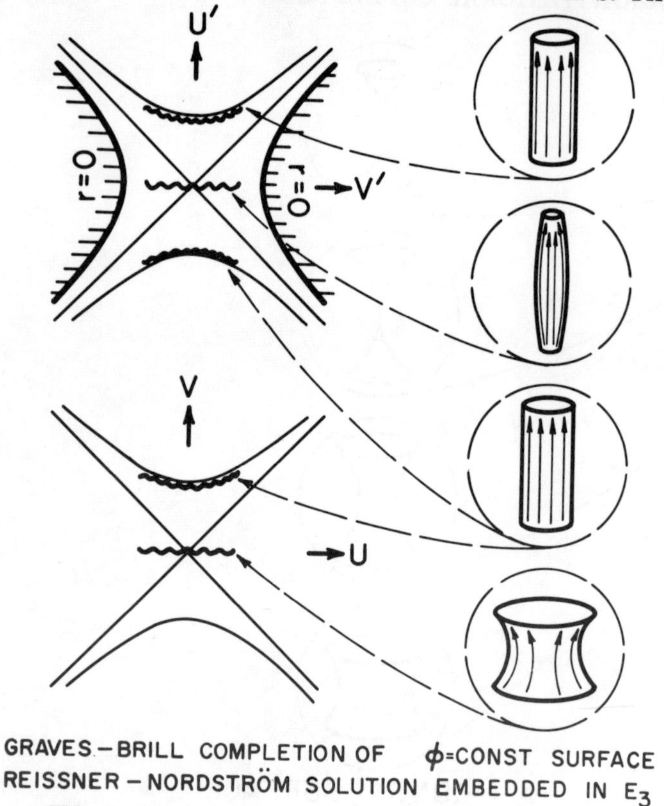

FIG. 4. The dynamics of the throat of Kruskal's wormhole when threaded by a magnetic field. On the left is a diagram of the Graves-Brill (1960) coordinate system which is characterized by an infinite sequence of pairs of coordinate patches identical to the pair shown. A succession of spacelike hypersurfaces, each one to the future of the preceding one, is shown in the coordinate diagram. On the right is a picture of each of these hypersurfaces embedded in a 3-dimensional Euclidean space. (The angle of rotation, ϕ, is suppressed; and the magnetic field lines are represented by arrows.) These successive "snapshots" of the throat of the wormhole reveal that, instead of pinching off, the throat pulsates. What is shown here is one period of the pulsation—from maximum radius to minimum radius, and then back out.

special relativity approximation, to geons with such intense magnetic fields that they wrap space up into closure around the ring of the torus. We shall show, in accordance with the resistance of magnetic flux to gravitational collapse, that none of the geons in this family will undergo collapse to the guiding line of the torus as it evolves in time.

The initial configuration of each of these geons can be constructed from Bertotti's (1959) static cylindrical magnetic universe. Bertotti's universe is described by the line element

$$ds^2 = (1 + B_0^2 z^2)\,dT^2 - \frac{dz^2}{1 + B_0^2 z^2} - \frac{1}{B_0^2}(d\eta^2 + \sin^2\eta\, d\varphi^2). \tag{8}$$

(Here, and throughout this paper, we use "geometrized units" in which the speed of light and Newton's gravitational constant are equal to 1.) In Bertotti's universe an observer with world line (z, η, φ) constant sees no electric field, but he sees a uniform magnetic field of strength B_0 pointing along the z-direction. The static surfaces of constant T have the geometry $E_1 \times S_2$; they are closed up in the radial direction (η- direction) but not in the z-direction.

The first step in constructing a toroidal magnetic geon from Bertotti's cylindrical universe is to bend it around into a toroidal universe (give it the geometry $S_1 \times S_2$ rather than $E_1 \times S_2$). Lindquist (1960) has given the prescription for doing this; we follow this discussion closely in the next paragraph.

Introduce new time and longitudinal coordinates, t and μ, defined by

$$B_0 z = \cos(B_0 t)\sinh(B_0 b\mu), \quad \tan B_0 T = \cot(B_0 t)\cosh(B_0 b\mu), \tag{9}$$

where b is a constant. Thereby transform the line element (8) to read

$$ds^2 = dt^2 - (\cos^2 B_0 t)\, b^2 d\mu^2 - (1/B_0^2)(d\eta^2 + \sin^2\eta\, d\varphi^2). \tag{10}$$

If μ is now interpreted as an angular coordinate of period 2π, then equation (10) is the line element for a toroidal universe with major circumference $2\pi b \cos(B_0 t)$ and minor circumference $1/B_0$. An observer with world line (μ, η, φ) = constant in this universe sees a uniform magnetic field of strength B_0 pointing in the μ-direction. (The only non-vanishing components of the electromagnetic field tensor are $f_{\eta\varphi} = -f_{\varphi\eta} = 1/B_0 \sin\eta$). The hypersurface t = 0 of the toroidal universe is a hypersurface of time-symmetry. As the universe evolves away from this momentarily static configuration, it gravitationally collapses along the μ-direction (its major circumference, $2\pi b \cos(B_0 t)$, decreases to zero).[9]

[9] None of the invariants of the Riemann tensor are infinite at $t = \pm\pi/2$. Hence, the singularities there can be removed by an appropriate choice of coordinates—e.g., by introducing Bertotti's coordinates (8). However, the singularities are removable only at the expense of destroying the periodicity of the coordinate μ; if we insist that (t, μ, η, φ) and $(t, \mu + 2\pi, \eta, \varphi)$ correspond to one and the same point, then we cannot remove the singularities at $t \pm \pi/2$.

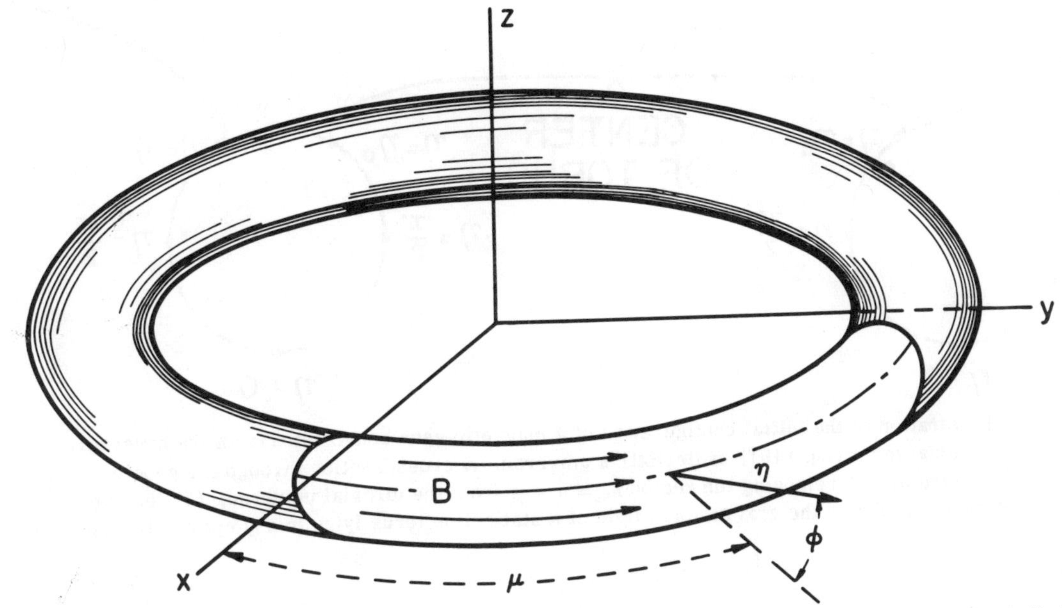

FIG. 5. The initial configuration of the magnetic geon whose subsequent dynamical evolution is studied. A uniform magnetic field threads the ring of the torus. There is no electric field or gravitational radiation in the initial configuration; but as the magnetic-field distribution changes with time it generates them. No currents or charges are present.

Note that this mode of collapse is perfectly compatible with the resistance of flux to gravitational collapse; the length of each closed field line decreases to zero, but the distance between adjacent field lines remains constant.

The next step in constructing a toroidal magnetic geon from Bertotti's universe is to take Lindquist's toroidal form of the universe (10) at the moment of time-symmetry t = 0, remove the region $\eta_0 \le \eta \le \pi$, and join what is left onto the gravitational field of a static line ring (see Fig. 6). The mathematical details of this procedure will be given elsewhere (Thorne 1965c).

The toroidal geon, which is thereby constructed, is a solution to the initial-value equations of general relativity for a hypersurface of time-symmetry.[10] The geon consists of a torus, which contains a uniform magnetic field of strength B_0. The surface of the torus is at $\eta = \eta_0$; its proper major circumference is $2\pi b$; its proper minor circumference is $2\pi B_0^{-1} \sin \eta_0$; and its proper minor radius is $\eta_0 B_0^{-1}$. If $\eta_0 \ll \pi$, then the geon is so dilute that its subsequent dynamics can be treated in the special relativity approximation; but if $\eta_0 \approx \pi$, then the geon is so massive that it wraps space up around itself almost into closure. For $\eta_0 = \pi$, space is completely closed up around the geon, and we have Lindquist's toroidal magnetic universe. This initial configuration of a toroidal geon contains no gravitational radiation, in the following sense: The dynamical evolution at any point inside or outside the geon is static until information that stresses were not initially balanced at the geon's surface (internal pressure $= B_0^2/8\pi$, external pressure = 0) has propagated to the point in question. Spacetime is static outside the geon because the initial external gravitational field is that of a static ring mass. It is locally static inside the geon because the dynamical evolution is initially that of the toroidal universe (10), which can be put into a static form by introducing Bertotti's coordinates locally. (Bertotti's coordinates cannot be introduced globally if we insist that μ be an angular coordinate of period 2π.)

Let each of these geons be followed as it evolves away from its initial configuration. Will gravitational collapse to the guiding line occur, in violation of the principle of flux resistance to gravitational collapse? No! The way in which the toroidal universe evolution (10), which involves no motion in the radial (η) direction, will be modified is this: Explosion away from the guiding line, rather than collapse to the guiding line, will be induced by the lack of balance of the magnetic pressure at the geon's surface.[11]

[10] For a discussion of the time-symmetric initial value equations, see Brill(1959).

[11] For geons with surfaces at $\eta_0 < \pi/2$, one can alternatively prove the impossibility of collapse to the guiding line by means of C-energy arguments, similar to those used by Thorne (1965a) to rule out the collapse of certain cylindrical electromagnetic universes.

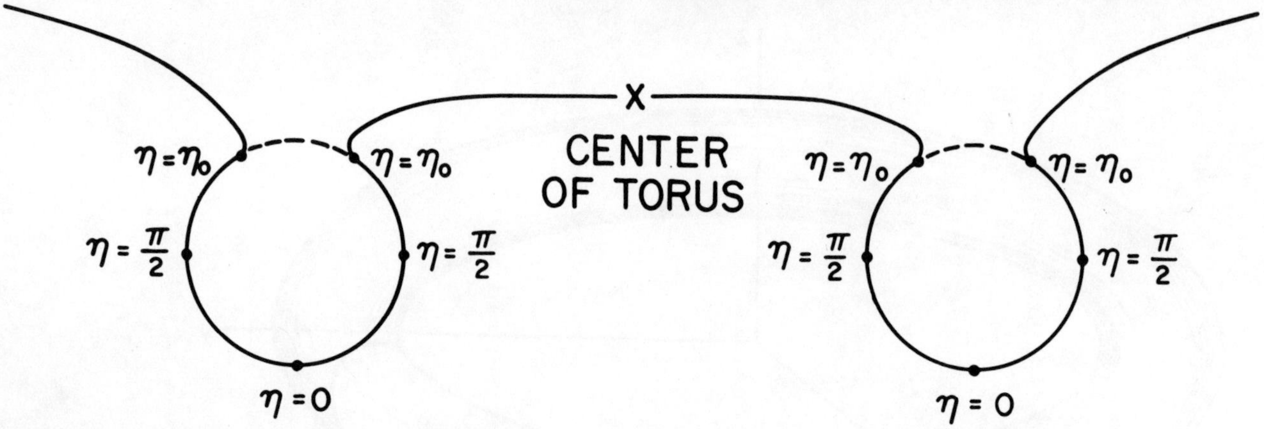

FIG. 6. A schematic illustration of the initial configuration of a magnetic geon constructed from the hypersurface t = 0 of Lindquist's toroidal form (eq. [10]) of Bertotti's universe. A cross-section through the geon is shown. The geon is constructed by (1) removing the region $\eta_0 \leq \eta \leq \pi$ from the toroidal universe; (2) joining the remaining configuration ($\eta < \eta_0$) onto the gravitational field of a static-line torus lying in asymptotically flat space.

III. Conclusions

We have suggested in this paper that, in vacuo, electric and magnetic flux resist gravitational collapse; and we have given a number of examples which support this viewpoint. A concerted effort should be made to prove or disprove this conjecture, not only because it would give us deeper insight into the nature of gravitational collapse in general relativity theory, but also for the following reason: If this conjecture is correct, and if flux also partially or completely resists collapse when matter is present, then magnetic fields could play an important role in astrophysical processes based on gravitational collapse.

REFERENCES

B. Bertotti, 1959, Phys. Rev., 116, 1531.

C. Behr, 1961, Mathematische Diplom-Arbeit, University of Hamburg (unpublished).

D. R. Brill, 1959, Ann. Phys., 7, 466.

_____. 1964, Phys. Rev., 133, B845.

S. Colgate, 1965, in Proceedings of the Second International Symposium on Relativistic Astrophysics.

A. Einstein, and N. Rosen, Phys. Rev., 48, 73.

R. W. Fuller, and J. A. Wheeler, 1962, Phys. Rev., 128, 919.

C. Fronsdal, Phys. Rev., 116, 778.

V. L. Ginzburg, 1964, Doklady Akad. Nauk SSSR, 156, 43 (translation: Sov. Phys.--Doklady, 9, 329 [1964]).

V. L. Ginzburg, and L. M. Ozernoi, (1964), Zh. Eksp. i. Teoret. Fiz., 47, 1030 (translation: Sov. Phys.--J.E.T.P. [to be published]).

J. C. Graves, and D. R. Brill, 1960, Phys. Rev., 120, 1507.

B. K. Harrison, 1965, paper submitted for publication.

B. K. Harrison, K. S. Thorne, M. Wakano, and J. A. Wheeler, 1965, Gravitation Theory and Gravitational Collapse (Chicago University of Chicago Press).

N. Kardashev, (1964), Astr. Zh., 41, xxx (translation: Soviet Astr.--A. J. [in press]).

M. D. Kruskal, 1960, Phys. Rev., 119, 1743.

R. Lindquist, (1960), "The Two Body Problem in Geometrodynamics," unpublished Ph. D. thesis, Princeton University.

M. A. Melvin, 1964, Phys. Letters, 8, 65.

_____. 1965, paper submitted for publication.

M. Misra, and L. Rhadakrishna, 1962, Proc. Nat. Inst. Sci. India, 28A, 632.

I. Novikov 1964, Astr. Tsirk., No. 290.

L. Shepley, 1965, chapter in Proceedings of the Second International Symposium on Relativistic Astrophysics.

K. S. Thorne, in Quasi-stellar Sources and Gravitational Collapse, ed. I. Robinson, A. Schild, and E. Schucking (Chicago: University of Chicago Press).

_____. 1965a, Phys. Rev., to be published.

_____. 1965b, submitted to Phys. Rev.

_____. 1965c, in preparation.

R. C. Tolman, 1934, Relativity, Thermodynamics, and Cosmology (Oxford: Clarendon Press).

H. Weyl, 1917, Ann. d. Phys., 54, 117.

J. A. Wheeler, 1963, Geometrodynamics (New York: Academic Press).

_____. 1964, in Relativity, Groups, and Topology, ed. C. and B. DeWitt (New York: Gordon and Breach).

CAN A DUST-FILLED COSMOLOGY BOUNCE?

L. C. Shepley*
Palmer Physical Laboratory, Princeton University
Princeton, New Jersey

In the early days of cosmology theory, researchers were disturbed by the fact that relativistic models were not static. As knowledge of the universe grew and non-static models were found to be physically realistic, another, more sophisticated problem took shape. Relativity seemed to predict that the universe had existed for only a finite time. This global question, whether a cosmological model in general relativity must have a singularity or whether there are theoretical models which are non-singular, remains with us, unanswered.

This paper will discuss this problem and will answer part of it. A very general class (but not the most general!) of cosmological models will be examined. These are interesting in that rotation is allowed, but are special in that they have a three-dimensional group of symmetries, the group SO(3, R). All such universes are found to be singular: That is, in each model there is a point, which is reached by a geodesic of finite length, at which the metric becomes irreparably singular. The mathematics involved in the calculation will be given in some detail, since we will thereby be able to discuss this important question of the existence of singularities fairly concisely.

I. A Picture of the Universe

We must first decide how the universe is to be portrayed. General relativity, with the field equations

$$R_{ij} - \frac{1}{2}g_{ij} R = T_{ij}, \tag{1}$$

is certainly the accepted theory (without a cosmological constant).

Let us first ask for the most natural and easily visualized models of the universe, to see if they are singular. The simplest realistic cosmology is the Friedmann universe (Tolman 1962). The galaxies are treated as dust, without interaction other than gravity:

$$T_{ij} = \rho u_i u_j \tag{2}$$

(ρ is the proper matter density, and u_i the components of the matter velocity). The Friedmann universe, however, inevitably collapses.

Let us now ask if there are metrics quaranteed to be non-singular. The Gödel universe (Gödel 1949), invariant under a four-dimensional group of symmetries, does represent a non-singular manifold. Matter is present, so it might be thought that this solution is a cosmological model. But pressure, to an unreasonably large degree, $p = \rho$, is required. We should not, however, discard the model. Perhaps an equation of state in which the pressure p becomes a large as the mass density ρ as ρ increases may stave off the end of a collapsing universe. Such equations of state are certainly allowed by causality considerations (Zel'dovich 1962).

We must therefore turn to more complicated models, since simple realistic universes are singular and since naturally non-singular solutions are unphysical.

The zeroth approximation to the cosmos is a vacuum. The Taub-NUT-Misner (T-NUT-M) universe (Taub 1951; Newman, Tamborino, and Unti 1963; Misner 1963; Misner and Taub[in press]) is a good candidate for a cosmological model: Space-like sections are closed and expand and contract. Moreover, no singularity, at least no geometric singularity, ever occurs. However, matter placed in this universe inevitable destroys it! This approximation, then, is not adequate.

*National Science Foundation Predoctoral Fellow

But the Gödel universe, which has rotating matter, and the empty Taub-NUT-Misner universe, which is non-singular, point out a path to take. We will treat a universe that has the same spatial homogeneity as T-NUT-M space but in which dust exists and may rotate. We will even add pressure, in the form of an equation of state. But this path, so promising at first, leads to a sad end: All such universes become singular.

II. Grandiose Symmetries

Let us now describe the cosmological models we are concerned with. The notation and methods of differential geometry, especially the calculus of differential forms, will be used (Helgason 1962; Misner 1963; Misner and Taub [in press]).

The metrics are to be invariant under a three-dimensional Lie group (in particular, the group SO(3, R)). That is, each group element is a transformation carrying a space-time point into another point at which the metric is the same (an isometry). Using all the group elements, any one space-time point traces out a three-dimensional set on which the metric is the same everywhere. This set will be called an "invariant" or "homogeneous" hypersurface.

Since space-time is four-dimensional, it is filled by a one-parameter family of these hypersurfaces. Let us call these invariant subspaces $H(\tau)$, τ being some labeling parameter. As τ varies, the hypersurfaces will be said to "evolve." Some of these homogeneous subspaces will be spacelike, but not necessarily all.

On any one invariant hypersurface, the metric is the same everywhere. We are able to exhibit this fact by using a set of differential forms ω^μ (μ = 1, 2, 3 labels the three differential forms) to express the three-dimensional metric. The ω^μ obey

$$d\omega^\mu = \frac{1}{2} C^\mu_{\sigma\tau}\, \omega^\sigma \wedge \omega^\tau. \tag{3}$$

The $C^\mu_{\sigma\tau}$ are the structure constants of the group of isometries, and $\wedge$ is the wedge (antisymmetric) product of differential geometry.

The full four-metric is obtained by adding a coordinate t to label hypersurfaces. The curl of t, dt, is then the fourth differential form used, ω^0 = dt. The metric is given by

$$ds^2 = g_{ij}\, \omega^i \omega^j, \tag{4}$$

where $g_{ij} = g_{ij}(t)$ only (Taub 1951).

But we are interested in one special group. This is the group of Bianchi type nine, denoted SO(3, R) (the group of special [unit determinant] orthogonal 3 × 3 matrices of real numbers). This group is of paramount importance for the following three reasons:

First, its structure constants are

$$C^\mu_{\sigma\tau} = \epsilon_{\mu\sigma\tau} \tag{5}$$

or

$$C^1_{23} = \epsilon_{123} = 1 \text{ et cyc.}$$

Put another way, the differential forms ω^μ obey

$$d\omega^1 = \omega^2 \wedge \omega^3 \text{ et cyc.} \tag{6}$$

If we were to take as a new basis to express the metric the forms

$$\hat{\omega}^\mu = \alpha^\mu_\sigma\, \omega^\sigma, \tag{7}$$

where α^μ_σ are constants, then

$$d\hat{\omega}^\mu = \frac{1}{2}\epsilon_{\mu\sigma\tau}\, \hat{\omega}^\sigma \wedge \hat{\omega}^\tau \tag{8}$$

if (α^μ_σ) is an orthogonal 3 × 3 matrix of unit determinant. In other words, the group properties of the invariant subspaces are invariant under rotations. Various directions are equivalent in this way, although not necessarily equivalent metrically. So that these three equivalent directions may correspond to equivalent directions in the real world, we will require that at least one of the invariant hypersurfaces be spacelike.

Second, the invariant hypersurfaces are three-spheres, S^3, topologically. Thus every direction is the same topologically and homogeneous sections are closed. Finite spacelike sections to correspond to the real cosmos are certainly desirable. It is not necessary, however, to resort to this explanation to explain facts such as the dark midnight sky, and observational data have not conclusively indicated that the real universe is closed. Thus this second point merely hints that our choice of group is good; it is not a deciding criterion.

Third, the resulting space-time will be general enough to allow rotation. Rotation is expressed in the following way: The covariant velocity vector u of cosmic matter can be expressed as a differential form:

$$u = u_i\, \omega^i, \tag{9}$$

where $u_i = u_i(t)$ only.
The curl of u is

$$du = \frac{du_i}{dt}\, dt \wedge \omega^i + u_i\, d\omega^i. \tag{11}$$

This is rotation: If $du \neq 0$, then cosmic matter rotates with respect to an inertial coordinate system; if $du = 0$, no rotation exists (Gödel 1950).

It has been shown that universes without rotation inevitably become singular (Raychaudhuri 1955). The fact that our models do allow arbitrarily great rotation leads us to hope that this class includes some non-singular universes (see Raychaudhuri 1955), a wish that will be disappointed.

The group T_3, the translation group of flat Euclidean three-space (Bianchi type one), gives the same equivalence to spacelike directions as SO(3, R) does: The structure constants of T_3 are all zero. The invariant hypersurfaces are all open (but may be closed by identifying points in a lattice and imposing periodic boundary conditions). However, no rotation is allowed. It has been shown that all cosmological models with T_3 as the group of symmetries have a singular point (Heckmann and Schucking 1962). Thus the group SO(3, R) remains at the focus of our attention.

For good luck, we will display a concrete representation of the ω^μ of equation (6) in a coordinate system (but only for good luck! A specific representation in terms of coordinates is never needed). The three-sphere S^3 is the set of all real numbers x, y, z, w such that

$$x^2 + y^2 + z^2 + w^2 = 1. \tag{12}$$

In the hemisphere N, where $w > 0$, the numbers x, y, z adequately describe all points; that is x, y, z is a coordinate system on N. We use ω^1, ω^2, ω^3 as a basis of differential forms on N, where

$$\omega^1 = (2w + \frac{2x^2}{w})\,dx + (\frac{2xy}{w} - 2z)\,dy + (\frac{2xz}{w} + 2y)\,dz,$$

$$\omega^2 = (\frac{2xy}{w} + 2z)\,dx + (2w + \frac{2y^2}{w})\,dy + (\frac{2yz}{w} - 2X)\,dz, \tag{13}$$

$$\omega^3 = (\frac{2xz}{w} - 2y)\,dx + (\frac{2yz}{w} + 2x)\,dy + (2w + \frac{2Z^2}{w})\,dz,$$

w being $\sqrt{(1 - x^2 - y^2 - z^2)}$. To see that equation (6) holds, that $d\omega^1 = \omega^2 \wedge \omega^3$ et cyc., we need merely the formula

$$d\omega^\mu = \alpha^\mu_{\sigma,\tau}\, dx^\tau \wedge dx^\sigma, \tag{14}$$

where

$$\omega^\mu = \alpha^\mu_\sigma\, dx^\sigma. \tag{15}$$

III. The Cartan Equation

Having found the form of the metric we will use in equation (4) we would like to compute the gravitational field equations. To do this, we must find the affine connections and then the Riemann curvature tensor. The Cartan structural equations (Misner 1963; Misner and Taub [in press]; Helgason 1962) allow us to make this computation solely within the framework of the differential forms that so concisely express the symmetry structure.

In a coordinate system, the metric is expressed as

$$ds^2 = h_{ij}(x^a)\, dx^i dx^j, \tag{16}$$

where the metric coefficients h_{ij} are functions of the coordinates x^a. The affine connection coefficients or Christoffel symbols Γ^i_{jk} are uniquely derived from the two equations

$$h_{ij;k} = h_{ij,k} - \Gamma^s_{ki} h_{sj} - \Gamma^s_{kj} h_{is} = 0, \tag{17}$$

$$\Gamma^i_{jk} - \Gamma^i_{kj} = 0. \tag{18}$$

We have, however, a basis of differential forms ω^i, in which

$$ds^2 = g_{ij}\, \omega^i \omega^j. \tag{19}$$

In this basis the equation corresponding to equation (17) is of the same form:

$$g_{ij,k} - \Gamma^s_{ki} g_{sj} - \Gamma^s_{kj} g_{is} = 0. \tag{20}$$

Equation (20) can be expressed as

$$g_{ij;k}\, \omega^k - \Gamma^s_{ki}\, \omega^k g_{sj} - \Gamma^s_{kj}\, \omega^k g_{ij} = 0. \tag{21}$$

Since the curls (or gradients) of the ten functions g_{ij} are

$$dg_{ij} = g_{ij,k}\,\omega^k, \tag{22}$$

equation (21) becomes

$$dg_{ij} = \omega_{ij} + \omega_{ji}, \tag{23}$$

where we have defined the connection forms ω^i_j as

$$\omega^i_j = \Gamma^i_{kj}\,\omega^k \tag{24}$$

and have set

$$\omega_{ij} = g_{is}\,\omega^s_j\,. \tag{25}$$

The equation which corresponds to equation (18) is the first Cartan equation, and reflects the facts that ω^i need not have zero curl, as dx^i does:

$$d\omega^i = -\,\omega^i_s \wedge \omega^s. \tag{26}$$

(Technically, this equation gives the unique "torsionless" connection corresponding to the metric [Helgason 1962]). Clearly the antisymmetry of Λ causes equation (26) to reduce to equation (18) when $d\omega^i = 0$.

To find the Riemann curvature tensor in a coordinate system, the first derivatives of the affine connections and a quadratic combination of these symbols are added. The second Cartan equation generalizes this process to a basis of forms:

$$\frac{1}{2} R^i_{jkl}\,\omega^k \wedge \omega^l = d\omega^i_j + \omega^i_s \wedge \omega^s_j\,. \tag{27}$$

The Cartan equations (26) and (27), and equation (23) (which are derived in detail in Helgason 1962) allow us to compute the curvature without ever resorting to explicit coordinate representations of the ω^i such as we gave in equation (13). For example, we can use a basis dt, ω^μ, where

$$\begin{aligned} d(dt) &= 0 \\ d\omega^\mu &= \frac{1}{2}\,\epsilon_{\mu\sigma\tau}\,\omega^\sigma \wedge \omega^\tau. \end{aligned} \tag{28}$$

in which the metric is expressed as

$$ds^2 = g_{ij}(t)\,\omega^i\omega^j. \tag{29}$$

Equations (28) and (29) are all we need to know to make use of equations (23), (26), and (27). In this case the dg_{ij} are given by

$$dg_{ij} = \frac{dg_{ij}}{dt}\,dt = \dot{g}_{ij}\,dt. \tag{30}$$

Alternatively we may "diagonalize" g_{ij} by using as basis forms σ^i. These forms will be given as time dependent combinations of the ω^i:

$$\sigma^i = B^i_s(t)\,\omega^s, \tag{31}$$

and their curls are expressed in terms of the derivatives of B^i_j and the curls of the ω^i:

$$d\sigma^i = \dot{B}^i_s\,dt \wedge \omega^s + B^i_s\,d\omega^s. \tag{32}$$

Equation (31) is then re-employed to express the ω^s in terms of the σ^s, and so find $d\sigma^i$ as a linear time-dependent combination of $\sigma^s \wedge \sigma^t$. By properly choosing the σ^s, equation (29) becomes

$$\begin{aligned} ds^2 &= -(\sigma^0)^2 + (\sigma^1)^2 + (\sigma^2)^2 + (\sigma^3)^2 \\ &= \eta_{ij}\,\sigma^i\sigma^j. \end{aligned} \tag{33}$$

Hence equation (23) reduces very simply to

$$\sigma_{ij} + \sigma_{ji} = 0. \tag{34}$$

Situations will arise where it will be convenient not to confine the time dependence of the metric to either the forms σ^i or the metric coefficients g_{ij}. Even in that case, the time dependence appearing both in σ^i and in g_{ij}, the Cartan equations will prove to be a big time saver over the usual coordinate-system computation process.

IV. The Synchronous System

The direction of the evolution coordinate t on which the metric depends is free to be chosen. The ω^μ, μ = 1, 2, 3, satisfying equations (6) and (4), may then be found. A very useful choice for the direction of t is to take the timelike direction perpendicular to the invariant spacelike surfaces spanned by ω^μ, and parametrize by proper time (Taub 1951). The metric is thus

$$ds^2 = -\,dt^2 + g_{\mu\nu}(t)\,\omega^\mu\omega^\nu. \tag{35}$$

This system is called "synchronous," as it indicates that clocks are synchronized throughout the spacelike surfaces (Lifshitz and Khalatnikov 1963).

It will be shown that in this basis the determinant of $g_{\mu\nu}$ goes to zero in a finite time (Lifshitz and Khalatnikov 1963). This, however, does not necessarily indicate the occurrence of a singularity: The $H(\tau)$ may simply change from spacelike to lightlike (and then to timelike), directing our attention to the fact that the synchronous system is not a universally valid basis to use. This changeover occurs in T-NUT-M space. But no such alteration of signature occurs when matter is present, and a singularity does arise when det $|g_{\mu\nu}| \to 0$.

Since the field equations will include the special cases of the Friedmann universe and the T-NUT-M (vacuum) solution, we will derive them in some detail.

Rather than using the form of the metric in equation (35), it is convenient to switch to a basis dt, σ^μ, where

$$\sigma^\mu = b^\mu_\nu(t)\,\omega^\nu. \tag{36}$$

The matrix $B = (b^\mu_\nu)$ is chosen to be the symmetric square root of $G = (g_{\mu\nu})$. Therefore

$$b^\mu_\nu = b^\nu_\mu, \tag{37}$$

and the metric is in diagonal, or Minkowski, form:

$$ds^2 = -dt^2 + (\sigma^1)^2 + (\sigma^2)^2 + (\sigma^3)^2. \tag{38}$$

Put another way, the 4×4 matrix of metric coefficients, g_{ij}, is the constant matrix

$$g_{ij} = \eta_{ij} = \begin{vmatrix} -1 & 0 & 0 & 0 \\ 0 & 1 & 0 & 0 \\ 0 & 0 & 1 & 0 \\ 0 & 0 & 0 & 1 \end{vmatrix} \tag{39}$$

The curls of the σ^i will be needed:

$$d\sigma^0 = d(dt) = 0, \quad d\sigma^\mu = \dot b^\mu_\nu\, dt \wedge \omega^\nu + b^\mu_\nu\, d\omega^\nu \tag{40}$$

Let us now use equation (36) to give ω^μ in terms of σ^μ:

$$\omega^\mu = a^\mu_\nu(t)\sigma^\nu; \quad a^\alpha_\nu b^\nu_\beta = \delta^\alpha_\beta. \tag{41}$$

Thus equation (40) becomes

$$d\sigma^\mu = k^\mu_\nu \sigma^0 \wedge \sigma^\nu + \frac{1}{2} d^\mu_{\alpha\beta}\, \sigma^\alpha \wedge \sigma^\beta, \tag{41a}$$

where

$$k^\mu_\nu = \dot b^\mu_\alpha a^\alpha_\nu \tag{42}$$

$$d^\alpha_{\beta\gamma} = b^\alpha_\rho\, \epsilon_{\rho\sigma\tau}\, a^\sigma_\beta a^\tau_\gamma. \tag{43}$$

Note that

$$d^\mu_{\sigma\tau} = -d^\mu_{\tau\sigma}. \tag{44}$$

The $\epsilon_{\rho\sigma\tau}$ appears because

$$d\omega^\mu = \frac{1}{2}\epsilon_{\mu\rho\sigma}\,\omega^\rho \wedge \omega^\sigma. \tag{45}$$

We are now ready to find the connection forms σ^i_j. First equation (23) becomes

$$\sigma_{ij} \ni \sigma_{ji} = 0. \tag{46}$$

In terms of the σ^i_j, equation (46) states

$$\sigma^0_0 = 0, \ \sigma^0_\mu = \sigma^\mu_0, \quad \sigma^\mu_\nu = -\sigma^\nu_\mu. \tag{47}$$

Consequently we only use σ^i_j with $i < j$ in the first Cartan equation (26). It will be convenient to express equation (26) in terms of the Christoffel coefficients Γ^i_{jk}:

$$0 = -\Gamma^0_{ij}\ \sigma^i \wedge \sigma^j; \ k^\mu_\nu \sigma^0 \wedge \sigma^\nu + \frac{1}{2} d^\mu_{\alpha\beta}\sigma^\alpha \wedge \sigma^\beta = -\Gamma^\mu_{ij}\sigma^i \wedge \sigma^j. \tag{48}$$

Equating the coefficients of $\sigma^i \wedge \sigma^j$, $i < j$ (which is a basis for the two-forms), yields

$$k^\mu_\nu = -\Gamma^\mu_{0\nu} + \Gamma^\mu_{\nu 0}; \ d^\mu_{\sigma\tau} = -\Gamma^\mu_{\sigma\tau} + \Gamma^\mu_{\tau\sigma}\ (\sigma < \tau);$$
$$0 = -\Gamma^0_{ij} + \Gamma^0_{ji}. \tag{49}$$

The solution of these equation is

$$\Gamma^0_{0\mu} = 0 \tag{50}$$

$$\Gamma^0_{\mu\nu} = \frac{1}{2}(k^\mu_\nu + k^\nu_\mu) \equiv l^\mu_\nu \tag{51}$$

$$\Gamma^\mu_{0\nu} = \frac{1}{2}(k^\nu_\mu - k^\mu_\nu) \equiv m^\nu_\mu \tag{52}$$

$$\Gamma^\mu_{\sigma\tau} = -\frac{1}{2}(d^\mu_{\sigma\tau} + d^\tau_{\mu\sigma} + d^\sigma_{\mu\tau}). \tag{53}$$

We have defined the symmetric and antisymmetric parts of k^μ_ν as

$$l^\mu_\nu = \frac{1}{2}(k^\mu_\nu + k^\nu_\mu) \text{ and } m^\mu_\nu = \frac{1}{2}(k^\mu_\nu - k^\nu_\mu). \quad (54)$$

In matrix language with $K = (k^\mu_\nu)$, $L = (l^\mu_\nu)$, $M = (m^\mu_\nu)$,

$$L = \frac{1}{2}(K + K^T) \text{ and } M = \frac{1}{2}(K - K^T). \quad (55)$$

Equation (42) in matrix language is

$$K = \dot{B} B^{-1}. \quad (56)$$

Consequently

$$BL - LB = BM - MB. \quad (57)$$

Equation (57) allows us to determine m^μ_ν from a knowledge of l^μ_ν and b^μ_ν! We will therefore find that L, B, given at an initial time t_0, form a set of initial data. L is the second fundamental form of the homogeneous hypersurface $H(\tau)$ imbedded in the four-dimensional space-time.

We can now determine the Ricci tensor. Rather than computing every term, we will compute R_{00} and $R_{0\mu}$ and merely exhibit the $R_{\mu\nu}$. By definition

$$R_{00} = R^s_{0s0}. \quad (58)$$

By the symmetry of R_{ijkl}, equation (58) can be written

$$R_{00} = - R^0_{101} - R^0_{202} - R^0_{303}. \quad (59)$$

To compute $R^0_{\mu 0 \mu}$, we use the second Cartan equation (27): We compute $d\sigma^0_\mu$ and $\sigma^0_s \wedge \sigma^s_\mu$, add, and take the coefficient of $\sigma^0 \wedge \sigma^\mu$. Thus, since

$$\sigma^0_\mu = l^\mu_\nu \sigma^\nu, \quad (60)$$

the coefficient of $\sigma^0 \wedge \sigma^\mu$ in $d\sigma^0_\mu$ is

$$(d\sigma^0_\mu)_{0\mu} = \dot{l}^\mu_\mu \quad l^\mu_\nu k^\nu_\mu (\text{NS on } \mu), \quad (61)$$

using equation (40). Similarly, since σ^0_μ has no dt $= \sigma^0$ term ($\Gamma^0_{0\mu} = 0$),

$$(\sigma_s \wedge \sigma^s_\mu)_{0\mu} = - (\sigma^s_\mu)_0 (\sigma^0_s)_\mu = - m^\sigma_\mu l^\mu_\sigma (\text{NS on } \mu). \quad (62)$$

Adding the contributions for μ = 1, 2, 3 gives

$$R_{00} = - \dot{l}^\mu_\mu - l^\mu_\nu l^\nu_\mu. \quad (63)$$

Similarly

$$\begin{aligned} R_{0\mu} &= R^0_{ss\mu} \\ &= (d\sigma^0_\alpha)_{\alpha\mu} + (\sigma^0_\beta \wedge \sigma^\beta_\alpha)_{\alpha\mu} \quad (64) \\ &= l^\alpha_\beta (d\sigma^\beta)_{\alpha\mu} + (\sigma^0_\beta)_\alpha (\sigma^\beta_\alpha)_\mu \\ &\quad - (\sigma^0_\beta)_\mu (\sigma^\beta_\alpha)_\alpha \\ &= l^\alpha_\beta d^\beta_{\alpha\mu} + l^\alpha_\beta \Gamma^\beta_{\mu\alpha} - l^\beta_\mu \Gamma^\beta_{\alpha\alpha}. \end{aligned}$$

Since $l^\alpha_\beta = l^\beta_\alpha$ and $\Gamma^\alpha_{\mu\beta} = - \Gamma^\beta_{\mu\alpha}$, then $l^\alpha_\beta \Gamma^\beta_{\mu\alpha} = 0$. Equation (53) shows that

$$\Gamma^\beta_{\alpha\alpha} = - d^\alpha_{\beta\alpha}.$$

But equation (43) shows that

$$d^\alpha_{\beta\alpha} = b^\alpha_\rho \epsilon_{\rho\sigma\tau} a^\sigma_\beta a^\tau_\alpha = \epsilon_{\rho\sigma\rho} = 0, \quad (65)$$

since $a^\mu_\sigma b^\sigma_\nu = \delta^\mu_\nu$. Therefore,

$$R_{0\mu} = l^\alpha_\beta d^\beta_{\alpha\mu}. \quad (66)$$

At this point let us note that all of this computation could have been carried out for a space invariant under T_3. In that case, $C^\alpha_{\beta\gamma} = 0$ and $d^\alpha_{\beta\gamma} = 0$. Equation (66) then shows that $R_{0\mu} = 0$. Since $R_{0\mu} = \rho u_0 u_\mu$,

$$u_\mu = 0 \text{ (in } T_3 \text{ case)} \quad (67)$$

$$\text{or } u = u_0 \sigma^0 = u_0(t)\, dt.$$

(Actually $u_0 = 1$, constant, here.) Consequently,

$$du = 0 \text{ (in } T_3 \text{ case)}, \quad (68)$$

so that, indeed, rotation is impossible in the T_3-invariant universes. We redirect our attention to models invariant under SO(3, R), where rotation is possible.

The other components, the spatial ones, of R_{ij} are

$$\begin{aligned} R_{\mu\nu} = \dot{l}^\mu_\nu &+ l^\mu_\nu l^\sigma_\sigma + l^\mu_\sigma m^\sigma_\nu + l^\nu_\sigma m^\sigma_\mu \\ &- \frac{1}{2} d^\sigma_{\mu\tau} d^\sigma_{\nu\tau} - \frac{1}{2} d^\sigma_{\mu\tau} d^\tau_{\nu\sigma} + \frac{1}{4} d^\mu_{\sigma\tau} d^\nu_{\sigma\tau}. \end{aligned} \quad (69)$$

Having computed the Ricci tensor, we are ready

to look at the field equations

$$R_{ij} - \frac{1}{2} g_{ij} R = \rho u_i u_j. \tag{70}$$

(Remember that $g_{ij} = \eta_{ij}$.) Contracting and using the unit property of u yields

$$R = \rho. \tag{71}$$

We will therefore use the field equations

$$R_{ij} = \rho u_i u_j + \frac{1}{2} \rho \eta_{ij} \tag{72}$$

instead of equation (70).

V. The Cosmos

We have derived the Ricci tensor for a special basis of one-forms, the synchronous system, in a space obeying the following assumptions: The metric is invariant under S0(3, R); at least one of the homogeneous hypersurfaces is spacelike. The metric may thus be written in the form

$$ds^2 = - dt^2 + (\sigma^1)^2 + (\sigma^2)^2 + (\sigma^3)^2, \tag{73}$$

where

$$\sigma^\mu = b^\mu_\nu(t)\ \omega^\nu \tag{74}$$

and

$$d\omega^\mu = \frac{1}{2} \epsilon_{\mu\rho\sigma}\ \omega^\rho \wedge \omega^\sigma. \tag{74a}$$

The Cartan equations were then used to derive the Ricci tensor.

The (00) component of R_{ij} was found to be

$$R_{00} = - \dot{l}^\sigma_\sigma - l^\sigma_\tau l^\tau_\sigma. \tag{75}$$

The (00) component of the field equations (72) is

$$R_{00} = \rho u_0^2 - \frac{1}{2} \rho. \tag{76}$$

But u is timeline and unit, and $g_{ij} = \eta_{ij}$; hence:

$$u_0^2 - u_1^2 - u_2^2 - u_3^2 = 1, \tag{77}$$

so that

$$u_0^2 \geq 1. \tag{79}$$

Therefore

$$R_{00} > 0 \tag{79}$$

when $\rho \neq 0$.

We can rewrite equation (79) as

$$-\dot{l}^\sigma_\sigma - \frac{1}{3}(l^\sigma_\sigma)^2 - \frac{1}{3}\left[(l^1_1 - l^2_2)^2 + (l^1_1 - l^3_3)^2 + (l^2_2 - l^3_2)^2\right] - \left[(l^1_2)^2 + (l^1_3)^2 + (l^2_3)^2\right] > 0. \tag{80}$$

Thus

$$\frac{d}{dt}\left(\frac{1}{l^\sigma_\sigma}\right) \geq \frac{1}{3}. \tag{81}$$

(We were allowed to divide through by l^σ_σ, since the strict inequality in equation (80) implies that $l^\sigma_\sigma \neq 0$ at some time.) Thus $|l^\sigma_\sigma| \to \infty$ in a finite interval of time t.

Since $l^\sigma_\sigma = k^\sigma_\sigma$, by the definition of K, equation (42),

$$l^\sigma_\sigma = \frac{1}{\det(B)} \frac{d(\det(B))}{dt}, \tag{82}$$

where B is the matrix (b^μ_ν). Therefore equation (81) implies that

$$\det(B) \to 0 \tag{83}$$

in a finite (proper) time. This "collapse" of the volume of spacelike homogeneous hypersurfaces may either be toward the past or toward the future.[1]

The "collapse," $\det(B) \to 0$, may merely indicate that the synchronous coordinate system has reached the end of its usefulness. However, the uniqueness of timelike directions perpendicular to spacelike invariant subsapces shows that $H(\tau)$ must therefore become lightlike: If $H(\tau)$ remained spacelike the synchronous system could still be used. The only alternative to $H(\tau)$ becoming lightlike is that a true singularity develops when $\det(B) \to 0$. And we will show that this alternative is the correct one by demonstrating that the existence of a lightlike $H(\tau)$ is contrary to the existence of matter.

When we demonstrate that $\det(B) = 0$ corresponds to a singularity in matter-filled space (but not necessarily in empty space), we evidently show that the synchronous system is useful for describing the entire range of the cosmology. Let

[1] This theorem was given by Lifshitz and Khalatnikov (1963) who remarked that it was first proved by Landau. It has also been proved by Raychaudhuri (1955) and Komar (1956).

us therefore develop some facts about the universe in this basis.

The "conservation of matter" equation

$$T^{ij}_{;j} = 0 \tag{84}$$

yields two laws. First is the geodesic property of u:

$$u^{i}_{;j}\ u^{j} = 0. \tag{85}$$

The second is the "continuity law"

$$(\rho\ u^{i})_{;i} = 0. \tag{86}$$

Equation (86) may be explicitly integrated: Written out, it becomes

$$(\rho\ u^{i})_{;i} = (\rho\ u^{i})_{,i} + \rho\ u^{s}\Gamma^{i}_{is} = 0, \tag{87}$$

But $\Gamma^{i}_{is} = 0$ unless s = 0 and

$$\Gamma^{i}_{i0} = l^{\sigma}_{\sigma} = \frac{1}{\det(B)}\ \frac{d(\det(B))}{dt}. \tag{88}$$

Hence, equation (86) implies that

$$\frac{d(\rho u^{0})}{dt} + \frac{\rho u^{0}}{\det(B)}\ \frac{d(\det(B))}{dt} = 0 \tag{89}$$

since ρ and u^0 are functions of time only. Hence

$$\rho\ u^{0}\ \det(B) = \text{const.} \tag{90}$$

Some consequences of equation (90) will be discussed later. The fact that det (B) → 0 clearly implies that ρu^0 becomes infinite in a finite proper time, but it is not yet clear that this infinite value represents a true singularity.

Before transforming to a new basis and proving that det(B) = 0 is truly a singular point, let us look at the form of the field equations. First, we will form the quantity

$$S = \frac{1}{2}(R_{00} + R_{11} + R_{22} + R_{33}). \tag{91}$$

From the field equations (72),

$$S = \rho\ u_0^2. \tag{92a}$$

Remember that

$$R_{0\mu} = \rho\ u_0\, u_\mu. \tag{92b}$$

Now suppose we are given two 3 × 3 symmetric matrices of constants, A_1 and A_2. Suppose further that A_1 is positive definite and non-singular. There is nothing to prevent us from giving the names B_0 to A_1 and L_0 to A_2. We can then form $d^{\alpha}_{0\beta\gamma}$ from B_0 (that is, A_1) and K_0 from B_0 and L_0, using equation (57), and can compute $R_{0\mu}$ and S, since these quantities depend only on B and L and not on $\dot{L}$ = dL/dt.

We then look to see if S and $S^2 - (R_{01}^2 + R_{02}^2 + R_{03}^2)$ are both positive. If so, then it is a straightforward process to find values for ρ and u_i (since $u^i u_i = -1$) from the $R_{0\mu}$ and S.

Having found ρ and u_i from the two given matrices, we use the $(\mu\nu)$ components of the field equations to yield a value for dL_0/dt (since $R_{\mu\nu}$ contains $\dot{l}^{\mu}_{\nu}$). In other words, B_0 and L_0, given at an initial time t_0, act as initial-value data.

(The form of the equations is ideally suited for numerical computation. Several interesting examples of rotating dust-filled universes have been computed, including some in which the matter density ρ oscillates for a short time. The volume of the universes, det(B), inevitably becomes zero, of course, and numerical work does not throw light on the existence of a singularity.)

We can use this explicit form for the initial data to count the number of different universes we have. There are twelve independent numbers that make up the two symmetric 3 × 3 matrices. But some of these reflect the freedom of making coordinate transformations. In particular, equation (8) shows that a rotation among the ω^μ to give new $\hat{\omega}^\mu$ results in exactly the same form of the metric. But this rotation may be used to diagonalize b^{μ}_{ν} at a given initial time t_0. We are left with nine parameters.

We can also multiply B_0 by a constant α and divide L_0 by the same constant. The result is a new universe which develops in precisely the same way from αB_0 and L_0/α as the old universe evolved from B_0 and L_0 with one change: The new ρ is every where equal to the old ρ divided by α^2. In this manner we may freely change scale without changing the physics of the universe, and may choose α so that the constant value of ρu^0 det(B) is 1.

Nor is this all we may do: The R_{00} equation showed us that a static universe is impossible, that B_0, L_0 given at time t_0 change to completely different B_1, L_1, at a later time t_1. This is because l^{σ}_{σ}, which is independent of rotations among the ω^μ, must change in value. Thus the eight parameters we are so far left with correspond to only seven different universes.

The seven parameters describing the seven different freedoms in our class of spatially homogeneous universes may not be chosen arbitrarily. B_0 must be positive definite and non-singular. We must also check that $S > 0$ and $S^2 - R_{01}^2 - R_{02}^2 - R_{03}^2 > 0$. If these inequalities are satisfied, then B_0, L_0 may be used as initial data to yield a member of our class of universes.

How can we picture one such universe? The symmetric matrix B^2 corresponds to the metric imposed on the homogeneous sections, which are three-spheres topologically. B^{-2} also may be used to represent the surface of a solid ellipsoid (imbedded in Euclidean three-space), with volume det(B). Hence, for illustrative purposes only, one of our universes may be pictured as a rotating, non-isotropic solid ellipsoid (Fig. 1). After a finite time the ellipsoid will collapse to a disk (or perhaps into the special cases of a disk, a line segment or a point).

Does this collapse in the volume correspond to the development of a true singularity? Yes, it does, as we now show.

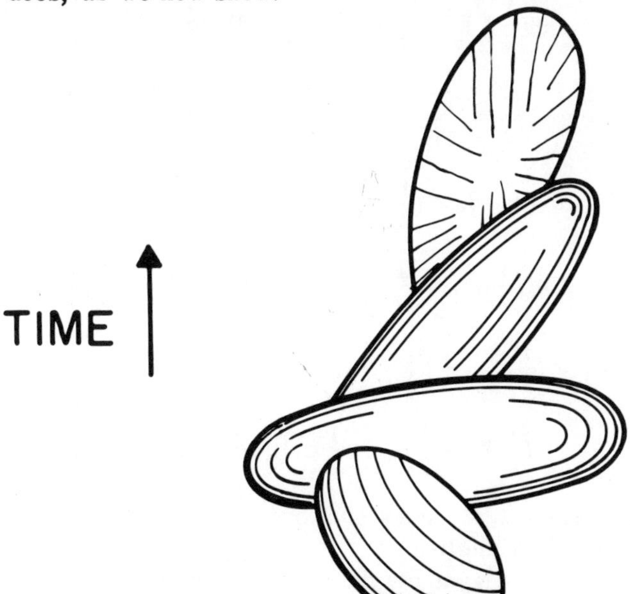

FIG. 1. A spheroidal universe may be represented by a rotating, non-isotropic solid ellipsoid. This ellipsoid is not the form of spacelike sections of the universe but simply graphically represents the metric on a hypersurface invariant under S0(3, R). The solid ellipsoid collapses to a disk after a finite amount of proper time, and the space-time manifold is singular there.

VI. The Twilight of the Gods

We have seen that the $H(\tau)$, the invariant hypersurfaces under S0(3, R), start out spacelike but evolve to a zero volume after a finite proper time. This zero volume may correspond to a lightlike geometry in $H(\tau_0)$, τ_0 being the value of the parameter τ when this collapse occurs, without a singularity appearing. Or the zero volume may correspond to a true singularity.

Are these two possibilities all that may happen? Yes: The $H(\tau)$, which begin as spacelike three-spheres, may change their metric, but may not change their topology without a singularity occurring: The existence of the matter velocity field u, for the non-colliding particles, allows us to find a homeomorphism between any non-singular invariant $H(\tau)$ and any other one. Since the topology of non-singular $H(\tau)$'s is invariant, only the two alternatives above can exist when the synchronous system breaks down.

In T-NUT-M space the first alternative occurs: The lightlike geometry of the $H(\tau_0)$ in the "Misner boundary" between Taub space and NUT space causes no trouble (Misner 1963). At this boundary the geometry of $H(\tau)$ changes from spacelike to timelike without a singularity.

The other alternative, the occurrence of a true singularity, is what comes to pass in the Friedmann universe. We now will prove that this second alternative is correct in any of our matter-filled universes.

To do this we must first find a suitable basis for any manifold, invariant under S0(3, R), with any stress-energy tensor, in which a lightlike $H(\tau)$ occurs. This has been done in detail elsewhere (Shepley 1964), so we omit the proof. The basis is $d\hat{t}$, $\hat{\sigma}^1$, $\hat{\sigma}^2$, $\hat{\sigma}^3$, where

$$\hat{\sigma}^\mu = \hat{b}^\mu_\nu\,(\hat{t})\;\omega^\nu. \tag{93}$$

Here $\hat{b}^\mu_\nu$ is not necessarily symmetric, but is a non-singular matrix and is diagonal at $\hat{t} = 0$, when $H(\hat{t} = 0)$ has a lightlike geometry.

In this basis, the metric is given by

$$ds^2 = 2\,d\hat{t}\,\hat{\sigma}^1 + g(\hat{t})\;(\hat{\sigma}^1)^2 + (\hat{\sigma}^2)^2 + (\sigma^3)^2. \tag{94}$$

The function $g(\hat{t})$ is zero at $\hat{t} = 0$:

$$g(0) = 0. \tag{95}$$

Put in different words, the matrix $\hat{g}_{ij}$ is not

constant, but has the values

$$\hat{g}_{ij} = \begin{Vmatrix} 0 & 1 & 0 & 0 \\ 1 & g & 0 & 0 \\ 0 & 0 & 1 & 0 \\ 0 & 0 & 0 & 1 \end{Vmatrix} \qquad (96)$$

When the function $g(\hat{t})$ is positive, $H(\tau)$ has a spacelike geometry; this region overlaps the area in which the synchronous basis may be used. When $g < 0$, $H(\tau)$ has a timelike direction.

The "time" direction is a lightlike direction, since $g_{00} = 0$. This basis may be called the "lightlike evolution" system.

As before

$$d\omega^{\mu} = \frac{1}{2} \epsilon_{\mu\sigma\tau}\, \omega^{\sigma} \wedge \omega^{\tau}, \qquad (97)$$

and therefore

$$d\hat{\sigma}^{\mu} = \hat{k}^{\mu}_{\nu}\, d\hat{t} \wedge \hat{\sigma}^{\nu} + \frac{1}{2} \hat{d}^{\mu}_{\alpha\beta}\, \hat{\sigma}^{\alpha} \wedge \hat{\sigma}^{\beta}, \qquad (98)$$

where k^{μ} and $d^{\mu}_{\alpha\beta}$ have the same values as before (equations [42] and [43]).

The two Cartan equations now beckon. We will show that the computed values of R_{ij} are inconsistent with the stress-energy tensor of dustlike matter at $\hat{t} = 0$, when $g = 0$. This will show that it is impossible for a matter-filled universe to allow a lightlike $H(\tau)$; thus a true singularity must exist at $\hat{t} = 0$.

Equation (23) is now somewhat complicated, since $\hat{g}_{ij}$ is not constant. We write it in the form

$$d\hat{g}_{ij} - \hat{\sigma}^{s}_{i}\, \hat{g}_{sj} - \hat{\sigma}^{s}_{j}\, \hat{g}_{is} = 0. \qquad (99)$$

Remembering that $d\hat{g}_{ij} = 0$ except for

$$d\hat{g}_{11} = \dot{g}\, d\hat{t}, \qquad (100)$$

we can "solve" equation (99) by exhibiting six $\hat{\sigma}^{i}_{j}$'s from which all the other $\hat{\sigma}^{i}_{j}$'s may be obtained. Thus,

$$\hat{\sigma}^{A}_{B} = - \hat{\sigma}^{B}_{A}, \qquad (101)$$

$$\hat{\sigma}^{A}_{0} = - \hat{\sigma}^{1}_{A}, \qquad (102)$$

$$\hat{\sigma}^{A}_{1} = - \hat{\sigma}^{0}_{A} - g\, \hat{\sigma}^{1}_{A}; \quad (A,\ B = 2,\ 3) \qquad (103)$$

$$\hat{\sigma}^{1}_{0} = 0,$$

$$\hat{\sigma}^{0}_{1} = g\, \hat{\sigma}^{0}_{0} + \frac{1}{2} \dot{g}\, d\hat{t}, \qquad (104)$$

$$\hat{\sigma}^{1}_{1} = - \hat{\sigma}^{0}_{0}.$$

Therefore the first Cartan equation (26) need involve only $\hat{\sigma}^{2}_{3}$, $\hat{\sigma}^{0}_{A}$, $\hat{\sigma}^{1}_{A}$ (A = 2, 3), and $\hat{\sigma}^{0}_{0}$.

The first Cartan equation (26) expands to

$$0 = - \hat{\Gamma}^{0}_{st}\, \hat{\sigma}^{s} \wedge \hat{\sigma}^{t}$$

$$\hat{k}^{\mu}_{\nu}\, \hat{\sigma}^{0} \wedge \hat{\sigma}^{\nu} + \frac{1}{2} \hat{d}^{\mu}_{\alpha\beta}\, \hat{\sigma}^{\alpha} \wedge \hat{\sigma}^{\beta} = - \hat{\Gamma}^{\mu}_{ij}\, \hat{\sigma}^{i} \wedge \hat{\sigma}^{j}. \qquad (105)$$

The solution will be different from the synchronous case, of course, because of the different symmetry of the $\hat{\Gamma}^{i}_{jk}$. The final results for the connection forms $\hat{\sigma}^{i}_{j}$ are

$$\hat{\sigma}^{0}_{0} = \hat{k}^{1}_{1}\, \hat{\sigma}^{0} + (g\hat{k}^{1}_{1} + \frac{1}{2} \dot{g})\hat{\sigma}^{1} + (\frac{1}{2} \hat{k}^{B}_{1} + \frac{1}{2} g\hat{k}^{1}_{B} - \frac{1}{2} g\, \hat{d}^{1}_{1})\hat{\sigma}^{B} \qquad (106)$$

$$\hat{\sigma}^{0}_{A} = (\frac{1}{2} \hat{k}^{A}_{1} + \frac{1}{2} g\hat{k}^{1}_{A} - \frac{1}{2} \hat{d}^{1}_{1A})\hat{\sigma}^{0} + (\frac{1}{2} g\, \hat{k}^{A}_{1} + \frac{1}{2} g^{2} \hat{k}^{1}_{A} - \frac{1}{2} g\, \hat{d}^{1}_{1A})\hat{\sigma}^{1} + [\frac{1}{2} g\, (\hat{k}^{A}_{B} + \hat{k}^{B}_{A}) - \frac{1}{2}(\hat{d}^{A}_{1} + \hat{d}^{B}_{1A})]\, \hat{\sigma}^{B} \qquad (107)$$

$$\hat{\sigma}^{1}_{A} = - \hat{k}^{1}_{A}\, \hat{\sigma}^{0} - (\frac{1}{2} \hat{d}^{1}_{1A} + \frac{1}{2} \hat{k}^{A}_{1} + \frac{1}{2} g\, \hat{k}^{1}_{A})\hat{\sigma}^{1} + (\frac{1}{2} \hat{d}^{1}_{AB} - \frac{1}{2} \hat{k}^{A}_{B} - \frac{1}{2} \hat{k}^{B}_{A})\hat{\sigma}^{B} \qquad (108)$$

$$\hat{\sigma}^{2}_{3} = (\frac{1}{2} \hat{k}^{3}_{2} - \frac{1}{2} \hat{k}^{2}_{3} - \frac{1}{2} \hat{d}^{1}_{23})\hat{\sigma}^{0} + (\frac{1}{2} \hat{d}^{3}_{12} - \frac{1}{2} \hat{d}^{2}_{13} - \frac{1}{2} g\, \hat{d}^{1}_{23})\hat{\sigma}^{1} - \hat{d}^{2}_{23}\hat{\sigma}^{2} - \hat{d}^{3}_{23}\hat{\sigma}^{3} \qquad (109)$$

where A = 1, 2, and a sum over the repeated B (= 1, 2) is implied.

Now we may apply the second Cartan equation (27) to find the Riemann tensor and then the Ricci tensor. However, we may save a lot of work, and will compute only one Ricci component, by making a few simple observations. As in the synchronous case, there are four field equations which are "constraints" in initial conditions and six which are "evolution" equations. In the synchronous system, the (0μ) equations and the S relation (equations [92a] and [92b]) served to define ρ, u^i in terms of a given initial B_0 and L_0. If all of these four equations were satisfactory then the $(\mu\nu)$ field equations allowed us to integrate from B_0 and L_0 at an initial t_0 to other times: No additional obstructions showed up.

In the lightlike evolution basis, equation (11) will prove to be such an equation relating initial data to initial values of ρ and u^i (at $\hat{t} = 0$). This equation, however, will prove contradictory unless $\rho \leq 0$, which is impossible. Thus, it will prove to be sufficient to calculate $\hat{R}_{11}$ when $\hat{t} = 0$, so that $g = 0$ and $\hat{b}^{\mu}_{\nu}$ is diagonal. (Recall that when $\hat{b}^{\mu}_{\nu}$ is diagonal, $d^{\mu}_{\mu\nu}$ (NS) is zero.)

We first express $\hat{R}_{11}$ in terms of the $\hat{R}^i_{jkl}$:

$$\hat{R}_{11} = \hat{R}^0_{101} + \hat{R}^2_{121} + \hat{R}^3_{131} = g\hat{R}^0_{001} + \hat{R}^0_{212} + g\,\hat{R}^1_{212} + \hat{R}^0_{313} + g\,\hat{R}^1_{313}. \tag{110}$$

The second line in equation (110) uses the fact that $\hat{R}_{ijkl} = -\hat{R}_{jikl}$. When $g = 0$,

$$\begin{aligned}\hat{R}_{11} &= \hat{R}^0_{212} + \hat{R}^0_{313} \quad (\text{at } g = 0)\\ &= (d\hat{\sigma}^0_2)_{12} + (d\hat{\sigma}^0_3)_{13} + (\hat{\sigma}^0_S \wedge \hat{\sigma}^S_2)_{12} + (\hat{\sigma}^0_S \wedge \hat{\sigma}^S_3)_{13},\end{aligned} \tag{111}$$

where the notation $(\ )_{12}$ indicates that the coefficient of $\hat{\sigma}^1 \wedge \hat{\sigma}^2$ is to be taken. Expanding equation (111) still further yields

$$\hat{R}_{11} = (d\hat{\sigma}^0_2)_{12} + (d\hat{\sigma}^0_3)_{13} + (\hat{\sigma}^0_0 \wedge \hat{\sigma}^0_2)_{12} + (\hat{\sigma}^0_3 \wedge \hat{\sigma}^3_2)_{12} + (\hat{\sigma}^0_0 \wedge \hat{\sigma}^0_3)_{13} + (\hat{\sigma}^0_2 \wedge \hat{\sigma}^2_3)_{13} \quad (\text{at } g = 0). \tag{112}$$

We have used the fact that $\hat{\sigma}^0_1 = \frac{1}{2}\,\dot{g}\, d\hat{t}$ at $g = 0$, and so cannot contribute either to a $\hat{\sigma}^1 \wedge \hat{\sigma}^2$ or $\hat{\sigma}^1 \wedge \hat{\sigma}^3$ term.

We now use equation (106)-(109) to compute R_{11}. First of all, note that the derivatives of the coefficients of $\hat{\sigma}^i$ in the expressions for $\hat{\sigma}^i_j$ do not contribute: They only affect the $\hat{\sigma}^0 \wedge \hat{\sigma}^{\mu}$ terms in $d\hat{\sigma}^i_j$, and these terms are not used in equation (112). Therefore, at $g = 0$,

$$(d\hat{\sigma}^0_A)_{1A} = -\frac{1}{2}(\hat{d}^A_{1B} + \hat{d}^B_{1A})\hat{d}^B_{1A} \ (\text{at } g = 0). \tag{113}$$

Summing on A as well as B yields

$$(d\hat{\sigma}^0_2)_{12} + (d\hat{\sigma}^0_3)_{13} = -\frac{1}{2}(\hat{d}^2_{13} + \hat{d}^3_{12})^2, \ (g = 0) \tag{114}$$

remembering that $\hat{d}^A_{1A} = 0$ (NS) since $\hat{b}^{\mu}_{\nu}$ is diagonal.

The other terms are easily computed

$$(\hat{\sigma}^0_0 \wedge \sigma^0_A)_{1A} = 0\ )(g = 0) \tag{115}$$

since $\hat{d}^A_{1A} = 0$. Similarly,

$$(\hat{\sigma}^0_3 \wedge \hat{\sigma}^3_2)_{12} + (\hat{\sigma}^0_2 \wedge \hat{\sigma}^2_3)_{13} = 0 \ (g = 0). \tag{116}$$

The result is that

$$\hat{R}_{11} = -\frac{1}{2}(\hat{d}^2_{13} + \hat{d}^3_{12})^2 \quad (\text{at } g = 0) \tag{117}$$

At $g = 0$, $\hat{b}^{\mu}_{\nu}$ is diagonal, with components β_1, β_2, β_3 (all of which are positive). By using equation (43) for $\hat{d}^{\mu}_{\alpha\beta}$, we find

$$\hat{R}_{11} = -\frac{1}{2}[(\beta_2^2 - \beta_3^2)/(\beta_1\beta_2\beta_3)]^2 \ (g = 0). \tag{118}$$

Therefore

$$\hat{R}_{11} \leq 0 \ (g = 0). \tag{119}$$

But the field equations yield

$$\hat{R}_{11} = \rho\,\hat{u}_1^2 \geq 0 \ (g = 0). \tag{120}$$

And we will further prove that ρ and $\hat{u}_1$ are $\neq 0$.

$$R_{11} = \rho\,\hat{u}_1^2 > 0 \ (g = 0), \tag{121}$$

in contradiction to equation (119). This inconsistency shows the impossibility of having a lightlike $H(\tau)$ in a dust-filled universe.

Do ρ and $\hat{u}_1$ have to be non-zero at $\hat{t} = 0$ if we did not have a singularity? Yes. The "conservation law" $T^{ij}_{;j} = 0$ implies the "continuity law," $(\rho\hat{u}^i)_{;i} = 0$. And again, as in the synchronous basis, this equation is

$$d(\rho\hat{u}^0)/d\hat{t} + \rho\hat{\Gamma}^i_{is}\,\hat{u}^S = 0. \tag{122}$$

Since $\hat{d}^{\sigma}_{\sigma\tau} = 0$ (sum on σ), as in the synchronous case, equation (122) reduces to

$$d(\rho\hat{u}^0)/d\hat{t} + \rho\,\hat{u}^0\,\hat{k}^{\sigma}_{\sigma} = 0. \tag{123}$$

Here

$$\hat{k}^{\sigma}_{\sigma} = \frac{1}{\det(B)}\,\frac{d(\det(\hat{B}))}{d\hat{t}} + f(\hat{t}) \tag{124}$$

where $f(\hat{t}) = 0$ at $\hat{t} = 0$. But since f is continuous, equation (123) may be integrated to give

$$\rho \hat{u}^0 \det(\hat{B}) = \text{const} \cdot F(\hat{t}) \qquad (125)$$

where $F(\hat{t})$ is positive and non-zero at $\hat{t} = 0$.

We know, however, that $\rho \neq 0$ when $g > 0$. Hence equation (125) shows that ρ and $\hat{u}^0$ are non-zero at $\hat{t} = 0$. Since $\hat{u}_1 = \hat{u}^0$ at $\hat{t} = 0$, we find that it is impossible for $\rho \hat{u}_1^2$ to be less then or equal to zero when $H(\tau)$ is lightlike.

That is to say, if our space-time manifold did not contain a singularity when $H(\tau)$ is lightlike, at $\hat{t} = g = 0$, then

$$\rho \hat{u}_1^2 > 0 \quad (g = 0), \qquad (126)$$

and equations (119) and (121) could not be reconciled. Thus the only alternative is to admit that these dust-filled spaces always become singular when the volume of space collapses:

VII. Gödel and Götterdämmerung

We have just seen that all universes, with or without rotation, which are invariant under $SO(3, R)$ and which start out with closed spacelike hypersurfaces, eventually come to an unavoidable end. To throw some light on this unfortunate behavior we now shall look at other models.

But, first, let us investigate a special subclass in the family of universes that we have studied. This subclass will be the non-rotating universes. And we are free to look at these universes with the synchronous basis, since by the time that this basis is no longer valid a singularity has appeared.

The effects of non-rotation are studied by adapting equation (11) to the synchronous basis:

$$du = \dot{u}_\mu \, dt \wedge \sigma^\mu + u_\mu \, d\sigma^\mu. \qquad (127)$$

The sum in equation (127) is over the spatial components of u only, since u_i is a function of t only, and $d(\sigma^0) = 0$. Equation (127) implies that $du = 0$ if and only if $u_\mu = 0$ ($\mu = 1, 2, 3$):

$$du = 0 \rightleftharpoons u_\mu = 0 \; (\mu = 1, 2, 3). \qquad (128)$$

Because u is unit and $g_{ij} = \eta_{ij}$, equation (128) states:

$$du = 0 \rightleftharpoons |u_0| = 1, \text{ constant}. \qquad (129)$$

Another consequence of $du = 0$ is that B may be diagonalized at all times. This is seen in the following way: Equation (92b) is

$$R_{\dot{0}\mu} = l^\alpha_\beta \, d^\beta_{\alpha\mu} = \rho \, u_0 u_\mu = 0 \quad 9(\text{if } du = 0). \qquad (130)$$

At an initial time t_0, B_0 may, of course, be presumed to be diagonal. Equation (130) tells us that L_0 is also diagonal (or can be made so). With both initial data matrices B_0 and L_0 diagonal, $B(t)$ is seen to be diagonal at all times. Of course, the converse remark is true: If $B(t)$ is always diagonal then $R_{0\mu} = 0$ and thus $u_\mu = 0$. Consequently:

$$du = 0 \rightleftharpoons B(t) \text{ diagonal at all times.} \qquad (131)$$

It may also be noted that, if rotation vanishes at any instant of time, t_0, B_0 and L_0 are both diagonalizable then. Hence

$$du = 0 \text{ at one time } t_0 \rightleftharpoons du = 0 \text{ always.} \qquad (132)$$

The physical effects of non-rotation are brought out by equation (90). With $u_0 = 1$, constant, the integrated continuity law becomes

$$\rho \det(B) = \text{const.} \; (\text{if } du = 0). \qquad (133)$$

Since $\det(B) \to 0$ in a finite time,

$$du = 0 \Rightarrow \rho = R \to \infty \text{ in finite time.} \qquad (134)$$

The fact that $\rho = R$, the scalar curvature, shows that this collapse without rotation produces an easily interpretable consequence.

Collapse with mass density rising to infinity is what occurs in the Friedmann universe (Tolman 1962): This universe has a metric which is not only invariant under $SO(3, R)$ but is also isotropic. The metric may be put in the form

$$ds^2 = -dt^2 + \beta^2(t) \left[(\omega^2)^2 + (\omega^3)^2 \right] \qquad (135)$$

from which we see that the model is representable by an expanding and collapsing solid three-dimensional ball (Fig. 2). This ball does not collapse to a disk; its symmetry requires that it collapse to a point.

In more general models, $du \neq 0$, and u_0 must be > 1. In fact the departure of u_0 from one may be thought of as measuring the amount of rotation in the universe. And this is rotation of the cosmic dust (which interacts only through gravity and thus

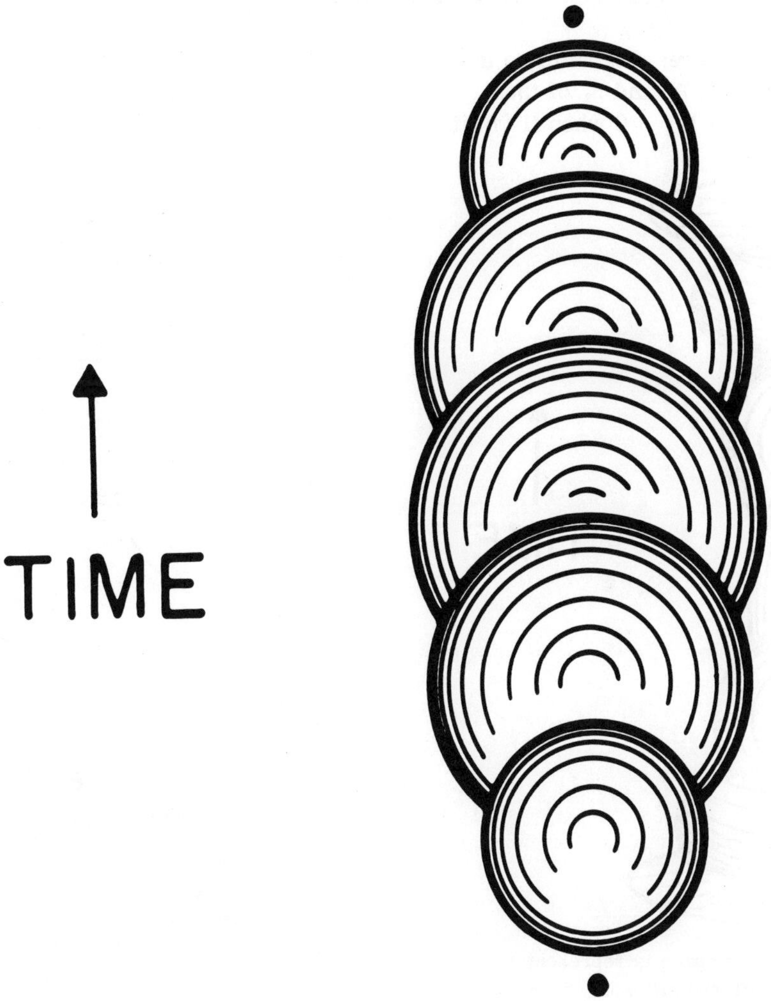

FIG. 2. The Friedmann universe is both spatially homogeneous (invariant under SO(3, R)) and isotropic. Consequently it may be represented as an expanding and contracting solid ball. This ball collapses to a point after a finite amount of time. Since the Friedmann universe is non-rotating, the matter density ρ becomes infinite at the time of collapse.

is "freely falling") with respect to inertial coordinate systems! (Gödel 1950). Equation (90) shows that $\rho u_0 \to \infty$ in a finite time. This may be thought of as a rise in rotational energy of the cosmos. This energy clearly is not completely a coordinate-dependent quantity, since its rise to infinity heralds the approach of a true singularity.

But let us leave matter-filled universes, with their solemn reminder of death, to look at Taub-NUT-Misner space. As we mentioned before, this space lets $H(\tau)$ change the signature of its geometry without a singularity in the metric. We may ask why this universe is non-singular and why putting in matter causes so much trouble.

T-NUT-M space is invariant under SO(3, R), and so may be described by equations (35) and (36). However, since there is no matter there cannot be rotation $R_{0\mu} = 0$. Hence B(t) is always diagonal. Moreover, two of the elements of this diagonal B are equal:

$$ds^2 = -dt^2 + \gamma_{11}^2(\omega^1)^2 + \gamma_{22}^2[(\omega^2)^2 + (\omega^3)^2] \quad \text{(T-NUT-M)}. \tag{136}$$

In the synchronous system, therefore, T-NUT-M space is an ellipsoid of revolution, non-rotating (Fig. 3). It collapses to a disk in a finite interval of time, when $\gamma_{11} \to 0$, with γ_{22} remaining positive.

Let us therefore look at T-NUT-M space in the lightlike system:

$$ds^2 = 2\,dt\,\omega^1 + g(t)(\omega^1)^2 + \gamma^2[(\omega^2)^2 + (\omega^3)^2] \quad \text{(T-NUT-M)}. \tag{137}$$

As we saw, equation (119) and (121) were contradictory unless $R_{11} = 0$ at $g = 0$ and also $\rho = 0$.

This is certainly the case here: There is no matter, and the two eigenvalues β_2 and β_3 of the matrix $\hat{B}$ are equal, so that $R_{11} = 0$ when $H(\tau)$ is lightlike.

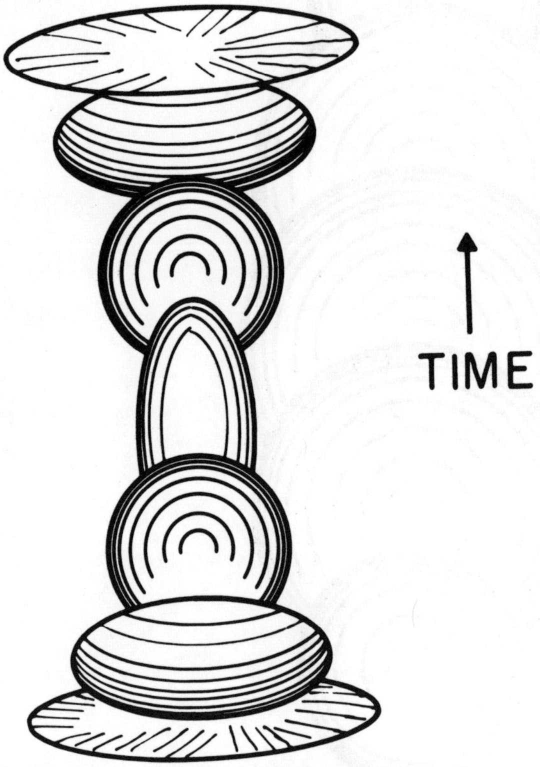

FIG. 3. Taub-NUT-Misner space is representable as a disk, which becomes a flattened ellipsoid of revolution, thickens to a cigar-shaped ellipsoid as its volume reaches a maximum, and collapses again to a disk. This is for the synchronous system, which is no longer a valid descriptive device after the collapse to zero volume has taken place. T-NUT-M space does not have a geometric singularity at the time of collapse.

Figure 4 illustrates the way the light cones "roll" on their sides as the homogeneous hypersurfaces of T-NUT-M space develop. Taub space is the region in which invariant subspaces are spacelike: The "one" direction, which runs around the three-sphere section, is spacelike. NUT space encompasses the regions in which this "one" direction lies within the interiors of the light cones.

At the Misner boundary (where $H(\tau)$ is lightlike) the light cones have rolled over so that one edge lies along the "one" direction. This rollover allows the following startling possibility: A timelike geodesic may run around the three-sphere so fast, approaching a lightlike direction so quickly, that it has finite total length, and yet goes nowhere. It wraps around so quickly that it has as limiting points all places on the lightlike circle it approaches (Misner 1963; Misner and Taub 1965). A particle on this geodesic would have nowhere to go!

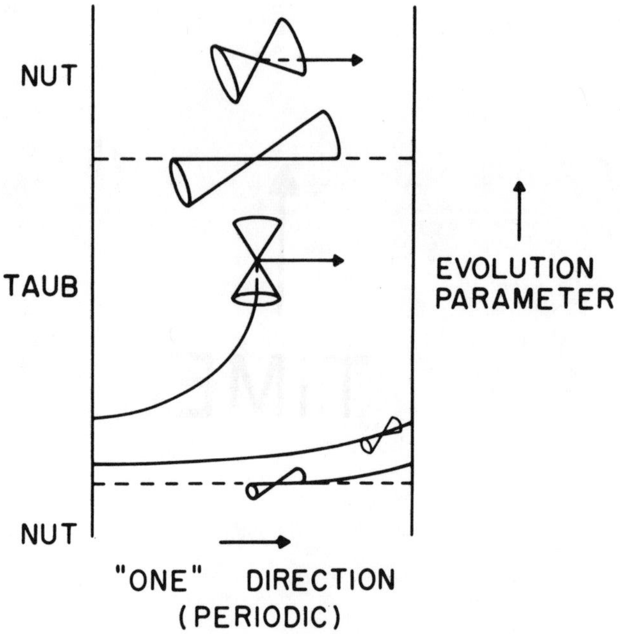

FIG. 4. The Taub-NUT-Misner universe, in a basis which shows the evolution parameter τ in a non-singular fashion, exhibits the "rollover" of light cones. At the Misner boundary between the Taub space and NUT space regions, one direction around the three-sphere invariant hypersurfaces is lightlike. (In the figure, every point on one vertical line is identified with the opposite point on the other vertical line. This is to show the periodicity in the "one" direction, which is a path on a 3-sphere.) Although T-NUT-M space has no geometric singularity (the metric is regular everywhere), it is incomplete: A timelike geodesic is shown which wraps around the "one"-direction circle, approaching the Misner boundary in such a way that the total length of the geodesic is finite. Because every point on the Misner boundary is a limiting point of this geodesic, the geodesic is inextendible.

There are also timelike geodesics which enter NUT space, travel around the "one" direction circle and reenter Taub space. A particle might thereby influence its own past.

The incompleteness of T-NUT-M space, in which not every geodesic may be extended to infinite values of its canonical path variable, and also the closed timelike lines, make T-NUT-M space unsuitable as a cosmological model. Nor, as we have seen, are the special conditions which allow Taub space to be extended into NUT space adaptable for use in matter-filled universes.

But perhaps we may adopt some features of the Gödel model (Gödel 1949) to stop the collapse of our universes. The Gödel cosmos is invariant under a four-dimensional, simply transitive Lie group. Because the metric is non-singular at one point, it is non-singular at all other points, since the group invariance states that the metric is the same everywhere.

To express the metric of the Gödel universe, we may use the same trick we employed to represent the metric of the $H(\tau)$. Equation (3) illustrated that in the invariant hypersurfaces we could choose one-forms whose curls reflected the structure constants of the invariance group. Equation (4) showed that, in terms of these forms, the metric in the $H(\tau)$ was independent of position.

In the Gödel universe, the invariant space is the entire space-time manifold. Hence, we may choose four one-forms σ^i, i = 0, 1, 2, 3, such that

$$d\sigma^i = \frac{1}{2} C^i_{jk}\, \sigma^j \wedge \sigma^k \text{ (Gödel)} \tag{138}$$

and such that

$$ds^2 = g_{ij}\, \sigma^i \sigma^j \tag{139}$$

with g_{ij} constant. Clearly we may assume that $g_{ij} = \eta_{ij}$, so that

$$ds^2 = -(\sigma^0)^2 + (\sigma^1)^2 + (\sigma^2)^2 + (\sigma^3)^2 \quad \text{(Gödel)}. \tag{140}$$

We may also presume that

$$u_0 = 1, \; u_\mu = 0. \tag{141}$$

The Cartan equations (23), (26), and (27) are then used to find the Ricci tensor. The Ricci tensor is expressed in terms of the structure constants C^i_{jk} of equation (138), and the field equations are solved by finding values for the C^i_{jk}.

Gödel discovered that the field equations

$$R_{ij} - \frac{1}{2} g_{ij} R = \rho\, u_i u_j \tag{142}$$

were not solvable in this manner. However, the equations

$$R_{ij} - \frac{1}{2} g_{ij} R = \hat{\rho}\, u_i u_j + \lambda\, g_{ij} \tag{143}$$

were found to be solvable, and a solution was found with cosmological constant $\lambda < 0$. Equation (143) may be written in the form

$$R_{ij} - \frac{1}{2} g_{ij} R = (\rho + p)\, u_i u_j + p g_{ij}, \tag{144}$$

in which the right-hand side is the stress-energy of a fluid with pressure $p = \lambda$.

Since Gödel's solution had $\lambda = \frac{1}{2}\hat{\rho}$ in equation (143), in equation (144)

$$p = \rho \text{ (Gödel)}.$$

(Remember that p and ρ, being group invariant functions, are constants, independent of position on the manifold.) Thus Gödel's universe has a positive, albeit an unreasonably large, pressure.

Perhaps if we added a positive pressure to the dust within our universes we might prevent a collapse from occurring. We allow ourselves the freedom of using an equation of state $p = p(\rho)$ such that, if necessary,

$$p(\rho)/\rho \to 1 \quad \text{as} \quad \rho \to \infty. \tag{146}$$

It has been shown that such equations of state are not contrary to the principle of causality (Zel'dovich 1962). Consequently, while it may be unreasonable to accept a model in which $p = \rho$ at all matter densities, it may be helpful to allow p to approach ρ at high densities.

But this does not help: And pressure is no aid for the same reason that the Friedmann universe filled with radiation pressure collapses faster than the Friedmann universe with dust (Tolman 1962). In our models, invariant under SO(3, R), in which pressure is added, the (00) Ricci component obeys the equation

$$R_{00} = -l^\sigma_\sigma - l^\sigma_\tau l^\tau_\sigma = \rho(u_0^2 - \tfrac{1}{2}) + p(u_0^2 + \tfrac{1}{2}). \tag{147}$$

Thus positive pressure only increases the rate of collapse. We can also see that the larger u_0 is, that, is, the greater the rotation, the faster the collapse! In other words, the energy that rotation and pressure add to the universe more than offsets any resistance these quantities may offer against a singularity.

But the Gödel universe has positive pressure and positive rotation, and these quantities prevent a singularity! What is different about the Gödel model that allows this behavior? The Gödel cosmos has a timelike differential form σ^0 which has non-zero curl. But more important, it is impossible to choose a time coordinate (dt, with zero curl) to express the group invariance of the space.

This fact is reflected in the behavior of timelike lines. More explicitly, the Gödel universe is of the topological form $\hat{H} \times R$. Here $\hat{H}$ is a three-space, invariant under the three-dimensional group L_3.[2] $\hat{H}$ has a timelike direction. Recall that NUT space is of the form $H \times R$, where R is again a spacelike parameter direction and H again is timelike. However, in NUT space H is invariant under the group of Bianchi type nine (SO(3, R)). Thus timelike lines in H are closed. Some timelike lines in $\hat{H}$ are closed, but not all (Gödel 1949) .[3] Consequently, as in the NUT portion of T-NUT-M space, a point Q in the future of a point P in Gödel' s model may communicate to a point R in the future of Q which is in the past of P.

In our models, this communication into the past is impossible, and a singularity cannot be avoided. It is possible that in order to prevent a singularity by using rotation or pressure, causality may have to be dropped by allowing time to curl around as in the Gödel universe. Would this be too high a price to pay for a classical, non-singular model of the universe?

VIII. Summing Up

It is unknown whether or not the general-relativity field equations require the existence of a singularity in a cosmological model. The only non-singular models known are unphysical; they either cannot have matter or have too much pressure (Table 1) . Physically reasonable models, the Friedmann universe and the spheroidal universes studied in detail in this paper, become singular in a finite proper time.

TABLE 1. A Comparison of Various Cosmological Models[a]

MODEL	MATTER	ROTATION	SINGULARITY
FRIEDMANN	DUST , $p = 0$	NO	YES
GÖDEL	YES , $p = \rho$	YES	NO
T-NUT-M	EMPTY	NO	
SPHEROIDAL	YES , $p = p(\rho)$	YES	YES
GENERAL	YES	YES	?

[a]The T-NUT-M universe is geometrically non-singular but is incomplete (there are geodesics which cannot be extended to infinite values of their affine parameter) . Both the Gödel and T-NUT-M models are acausal: the past may be influenced by the future.

Two candidates for a non-singular cosmological model, the Gödel universe and T-NUT-M space, are acausal. Thus we may speculate on the possibility that, if relativity allows a non-singular cosmological model at all, it will be one lacking the principle of causality.

As a final remark, let us mention the connection between cosmological and stellar models. As we should not take the issue of the final state in collapsing stars lightly, for our universe may suffer a similar fate (Wheeler 1964) , so the techniques and some conclusions of cosmological theory may aid in the understanding of stellar collapse. In particular, the rotating models studied in detail here may help in understanding the effects of rotation in collapsing stars.

The author is greatly indebted to Professor Charles Misner, of the University of Maryland, and to Professor John Wheeler, of Princeton University; to the former for having suggested this problem, to the latter for having suggested the title, and to both for encouragement and many helpful discussions.

[2] L_3 is the three-dimensional group of Bianchi type eight (Taub 1951), which is isomorphic to the three-dimensional Lorentz group (see Bargmann 1947). The basis directions of $\hat{H}$ do not have the topological and group similarity that directions in a space invariant under T_3 or SO(3, R) do. In particular in the covering space of L_3, there are both infinite and closed directions.

[3] The timelike geodesics in Gödel' s universe have all been calculated by Chandrasekhar and Wright (1961). There are closed timelike lines in Gödel' s model, but not closed timelike geodesics.

V. Bargmann, 1947, Ann. Math., 48, 568.

S. Chandrasekhar, and J. P. Wright, 1961, Proc. Nat. Acad. Sci., 47, 341.

K. Gödel, 1949, Rev. Mod. Phys., 21, 447.

_____. 1950, Proc. 1950 Int. Cong. Math., 1, 175.

O. Heckmann, and E. Schucking, 1962, in Gravitation: An Introduction to Current Research, ed. L. Witten (New York: John Wiley & Sons).

S. Helgason, 1962, Differential Geometry and Symmetric Spaces (New York: Academic Press).

A. Komar, 1956, Phys. Rev., 104, 544.

E. M. Lifshitz, and I. M. Khalatnikov, 1963, Adv. in Phys., 12, 185 (trans. J. L. Beeby).

C. W. Misner, 1963, J. Math. Phys. 4, 924.

C. W. Misner, and A. H. Taub, 1965 (in press).

E. Newman, L. Tamborino, and T. Unti, 1963, J. Math. Phys., 4, 915.

A. Raychaudhuri, 1955, Phys. Rev. 98, 1123.

L. C. Shepley, 1964, Proc. Nat. Acad. Sci., 52, 000.

A. H. Taub, 1951, Ann. Math., 53, 472.

R. C. Tolman, 1962, Relativity Thermodynamics and Cosmology (Oxford: Clarendon Press).

J. A. Wheeler, 1964, in Relativity, Groups and Topology, ed. C. DeWitt and B. DeWitt (New York: Gordon & Breach).

Ya. B. Zel'dovich, 1962, Soviet Phys.—J.E.T.P., 14, 1143 (in Russian 41, 1609 [1961]).

NUMERICAL RESULTS ON THE EQUILIBRIUM OF HIGHLY COLLAPSED BODIES

L. Gratton
Laboratorio di Astrofisica dell' Università di Roma
4th Sezione del Centro di Astrofisica del CNR

1. Introduction

The scope of this paper is to give a short account of the investigations which are going on at the Astrophysical Laboratory of the University of Roma and "Laboratorio Gas Ionizzati" of the Italian Commission for Atomic Energy (CNEN) on the subject of the equilibrium structure of a spherical body whose mass and density are such that the body is near the Schwarzschild limit of General Relativity. Four papers on this subject are in print and a fifth paper by Dr. Giannone and the present writer will be published very soon (Gratton 1964 a, b and c, 1965, Giannone and Gratton 1965).

Although the main interest during this meeting seems to have been stressed upon the dynamical aspect of gravitational collapse, I feel that the relatively simpler static problem well deserves consideration and, perhaps, might shed more light upon the physics involved. This is quite independent of the possible identification of the structures considered here with existing physical bodies; nevertheless, it is obvious that if we wish to get some results which may have some physical interest we must state our problem in realistic terms. This, I believe, is especially true in what concernes the equation of state of the matter or, better, the relation between (mass) density and pressure, which one must assume in order to solve the problem.

In section 2 and 3 this relation is discussed for two specially interesting cases: a cold neutron gas and a mixture of gas and radiation in adiabatic equilibrium. In section 4 and 5, some details are given on the models which arise from these two cases; in section 6 I put forward some speculations upon the physical implications of these models. More details will be found in the papers quoted above.

2. The Equation of State of a Cold Neutron Gas

The equation of state of neutron matter at zero temperature was considered first by Landau more than thirty years ago (Landau, 1932); more recent contributions are due to Cameron (1959) and Salpeter (1960), who considered the effect of nuclear forces, and to Ambartsumian and Saakyan (1960, 1961) and also to Saakyan and Vartanian (1963) who discussed the formation of hyperons at very high density, according to a suggestion made first, I believe, by Cameron.

The whole problem was treated anew by Szamosi and myself (Gratton and Szamosi, 1964) with the following results.

Ordinary nuclear forces play some role at densities comparable with that of nuclear matter, but their effect is by no means large; it leads to a slight decrease of the pressure relative to that of a free Fermi gas. This effect is greatest at $n = 10^{37}$ cm^{-3} (n = number of neutrons per unit volume).

At this density, the energy per particle is still much smaller than the rest-energy; but at densities of $n = 10^{39}$ cm^{-3} and larger, Special Relativity effects become important. At the same time the repulsive character of nuclear forces (hard core) as shown by scattering experiments with high energy protons, must also be taken into account.

The corresponding many-body problem becomes then exceedingly difficult, especially because of the change of the frame of reference when one is changing from the elementary two-body interaction to the complete many-body case. For this reason a quantum-mechanical treatment similar to that of Huang and Yang (1957, see also Yang and Lee 1957, de Dominicis and Martin 1957) was found impracticable. This difficulty was partly overcome by us by adopting a semi-classical treatment, whose justification lies in the fact that at these high densities the average energy per particle in a Fermi gas is so high that the corresponding De Broglie wave-length becomes smaller than the diameter of the hard core.

If the hard core interaction is assumed to be such as would correspond to perfectly rigid spheres the equation of state becomes that of a Fermi gas with excluded volumes and is easily found to be represented by the following set of equations.

Define

$$x = p_F/mc \tag{1}$$
$$h(x) = 3\int_0^x \frac{\xi^2\, d\xi}{\sqrt{1+\xi^2}} = \frac{3}{2}\left(x\sqrt{1+x^2} - \sinh^{-1}x\right),$$

where p_F is the Fermi momentum of the particles and m is the neutron mass; then the number of particles is related to x by the equation

$$n = \frac{8\pi}{3}\left(\frac{mc}{h}\right)^3 \frac{x^3}{1+\alpha h(x)}, \tag{2}$$

and

$$\alpha = \frac{8\pi}{3}\left(\frac{mc}{h}\right)^3 v_0,$$

v_0 being the (rest) volume of the hard core. Equation (2) replaces the well known equation for an ordinary Fermi gas to which it reduces when the hard core constant α is equal to zero.

The average energy per particle and the pressure are found to be

$$E = mc^2\,\frac{g(x)}{x^3},$$
$$P = \frac{\pi}{3}\, mc^2\left(\frac{mc}{h}\right)^3 \frac{f(x)}{1+\alpha\, h(x)}, \tag{3}$$

where

$$g(x) = \frac{3}{8}\left[(x+2x^3)\sqrt{1+x^2} - \sinh^{-1}x\right],$$
$$f(x) = (2x^3 - 3x)\sqrt{1+x^2} + 3\sinh^{-1}x. \tag{4}$$

For $\alpha = 0$ or for x small (that is for low densities), these equations are identical with those of an ordinary Fermi gas at zero temperature, but for $\alpha \neq 0$ and large x, it is easily seen that

$$n \simeq \frac{2}{3v_0}\,x,$$
$$E \simeq \frac{3}{4}\,mc^2\,x = \frac{9}{8}\,mc^2\,v_0 n, \tag{5}$$
$$P \simeq \frac{1}{6}\,\frac{mc^2}{v_0}\,x^2 = \frac{1}{3}\,nE.$$

Therefore when $x \to \infty$ the number density goes to infinity notwithstanding the hard core. This is due to the fact that when the number density increases, the particles move faster and faster; hence their volumes (in a frame of reference at rest with the gas) become smaller and smaller because of the Lorentz contraction and eventually the average volume vanishes at infinite n.

The mass density is, obviously, $\zeta = nE/c^2$; at very high densities we have, then, as it ought to be

$$P = \frac{1}{3}\,\zeta\, c^2. \tag{6}$$

I may note in passing that in my opinion the equation for very high density $P = \zeta c^2$ obtained by Zel'dovich (1961) by means of a repulsive potential of the type e^{-u}/r is probably incorrect because he does not take into account the change of the energy of the elementary two-body interaction when one changes from a system of reference in which the center of gravity of a pair of particles is at rest to the system at rest with the whole gas.

At somewhat smaller densities the equation of state of a cold neutron gas may be expressed in a parametric form by

$$\zeta = \frac{8\pi}{3}\, m\left(\frac{mc}{h}\right)^3 \frac{g(x)}{1+\alpha\, h(x)},$$
$$P = \frac{\pi}{3}\, mc^2\left(\frac{mc}{h}\right)^3 \frac{f(x)}{1+\alpha\, h(x)}. \tag{7}$$

Let us finally note for later use that for any reasonable value of α, equations (7) can be represented in a very good approximation by the much simpler interpolation formula

$$c^2\zeta = 3P + Q_n P^{3/5}, \tag{8}$$

where

$$Q_n = 20^{3/5}\left(\frac{\pi}{3}\right)^{2/5}\left(\frac{m}{h^2}\right)^{3/5} mc^2 = 1.304\times 10^{15}$$

(c.g.s. units)

The two terms at the right in equation (8) become comparable at densities of the order of 10^{16} g cm^{-3}.

3. Adiabatic of a Mixture of Gas and Radiation

The P, ζ-relation corresponding to an adiabatic of a mixture of a monoatomic gas and black body radiation is, of course, well known (Chandrasekhar

1939). However usually one neglects the contribution of the black body radiation to the mass density, which is in normal cases exceedingly small.

The correct equation in the general case is easily shown to be

$$\frac{d \log P}{d \log \zeta} = \frac{\Gamma_1}{1 + \frac{P}{c^2 \zeta}} . \tag{9}$$

Here

$$\Gamma_1 = \beta + \frac{2(4 - 3\beta)^2}{3(8 - 7\beta)} , \tag{10}$$

and β, the ratio of the partial gas pressure to the total pressure, is given by

$$\frac{c^2\zeta}{3P} = 1 - \frac{1}{2}\beta + \frac{\mu H c^2}{3k} \left(\frac{a}{3}\right)^{1/4} \frac{\beta}{P^{1/4}(1 - \beta)^{1/4}} . \tag{11}$$

Furthermore μ is the mean molecular weight of the gas and H the mass of unit atomic weight; the other constants have their usual meanings.

These equations hold if the thermal energy of the particles is much smaller than their rest energy, or

$$\frac{kT}{\mu H c^2} = \frac{k}{\mu H c^2} \left(\frac{3}{a}\right)^{1/4} (1 - \beta)^{1/4} P^{1/4} \ll 1. \tag{12}$$

To discuss the adiabatic relation, let us introduce the numerical variables x and y defined by

$$P = \left(\frac{\mu H c^2}{3k}\right)^4 \frac{a}{3} x^4, \qquad \frac{P}{c^2\zeta} = \frac{1}{3} y. \tag{13}$$

From (9), (11) and (13) we obtain, then,

$$\frac{d \log y}{d \log x} = 4\left(1 - \frac{3 + y}{3\Gamma_1}\right), \qquad x\left(\frac{1}{y} - 1\right) = \frac{\beta}{(1 - \beta)^{1/4}} - \frac{x}{2}\beta. \tag{14}$$

The solutions of these equations represent the family of the adiabatic curves in the x, y-plane; only values of x much less than 1 come into consideration. Furthermore $0 \le y \le 1$ (because of special relativity, and $0 \le \beta \le 1$.

Now $y = 1$, $\beta = 0$ is a solution of (14) as it can be seen at once; indeed it can be shown that this is a singular solution, towards which all other solutions converge when $x \to \infty$. Physically this means that at very large pressure all the adiabatics converge towards the relation $P = c^2\zeta/3$, the extreme relativistic equation of state.

When $x \to 0$, the solutions are found to be, $\beta = 1$, $y \sim x^{8/5}$, or $P = K\zeta^{5/3}$ (K = constant), the adiabatic law for a monoatomic gas.

We are however especially interested in the solutions corresponding to $\beta \ll 1$, since we want later to apply the results to very massive bodies in adiabatic equilibrium and it is well known that in this case the gas pressure is negligible compared with the radiation pressure.

Hence, we put $\Gamma_1 = 4/3$ and from the first of equations (14), we obtain by integration

$$y = \frac{1}{1 + \beta_0/x} . \tag{15}$$

β_0 being an integration constant. β is given by

$$\beta = \frac{\beta_0}{1 - x/2} . \tag{16}$$

Notice that the present set of equations holds only if $x \ll 1$, so that we must not worry about the possibility of β becoming negative if $x > 2$; furthermore $\beta \simeq \beta_0$.

From (15) by reverting to the physical variables ζ and P, we obtain

$$c^2 \zeta = 3 P + Q_{ad} P^{3/4}, \tag{17}$$

where

$$Q_{ad} = \frac{Hc^2}{k} \left(\frac{a}{3}\right)^{1/4} \beta_0 . \tag{18}$$

We see, thus, that also in this case we obtain an equation of the general form

$$\zeta c^2 = 3P + QP^{\frac{n}{n+1}} \tag{19}$$

as in the case discussed previously. Notice however that, apart from the polytropic index n, which is 3/2 for a cold neutron gas and 3 for the adiabatics of gas and radiation, there is an essential difference between the two cases, since in the first $Q = Q_n$, a universal constant, whereas in the second Q_{ad} is a parameter which can be used to classify any particular adiabatic.

4. The R-polytropic Model

I call an R-polytropic sphere, a self-gravitating body possessing spherical symmetry, in which ξ and P are related throughout by equation (19). If $Q = Q_n$, $n = 3/2$, the corresponding R-polytrope is an approximate model for a neutron star.

On the other side the case $n = 3$, seems to be a reasonable model for a massive body in adiabatic equilibrium, in the sense that if we take a mass element and let it move throughout the body in such a way that no energy is added to or subtracted from the element, we shall find it always in thermal equilibrium with its surrounding. This is the sort of equilibrium which is usually called adiabatic or indifferent. The condition that β be very small, means that the R-polytropic model in this case can be applied only to very large masses.

It is unnecessary to give here details of the numerical integrations. I have found useful to represent the solutions by means of the well-known "homologic" variables U and V, although they are no longer homology invariants, if the equilibrium equations of General Relativity are used. U and V are defined by

$$U = \frac{d \log m}{d \log r}, \qquad V = -\frac{d \log P}{d \log r}. \tag{20}$$

Figure 1 and 2 give a few solutions for $n = 3/2$ and $n = 3$ respectively in the U, V-plane.

Notice that the point $U = 3$, $V = 0$ corresponds to the center and the point at infinity of the V-axis to surface of the R-polytrope. All the curves represented correspond to a finite density at the center; hence the whole family of curves for a given n correspond to the single Emden curve of the newtonian case. The latter is the limit which one obtains when Q increases and at the same time P decreases, in such a way that the second term at the right in equation (19) remains finite while the first term tends to zero.

The opposite limit, corresponding to $Q \to 0$, is of particular interest. It corresponds to a complete relativistic gas or to pure radiation, and is, of course, the same for all n.

The corresponding equation in the U, V-variables is

$$\frac{d \log V}{d \log U} = -\frac{6 - 10\,U - 3V - UV}{2(3 + U)(3 - U - V)} \tag{21}$$

which is very similar to the classical equation of an isothermal sphere.

The solution of (21) corresponding to a finite central density is also shown in Figs. 1 and 2.

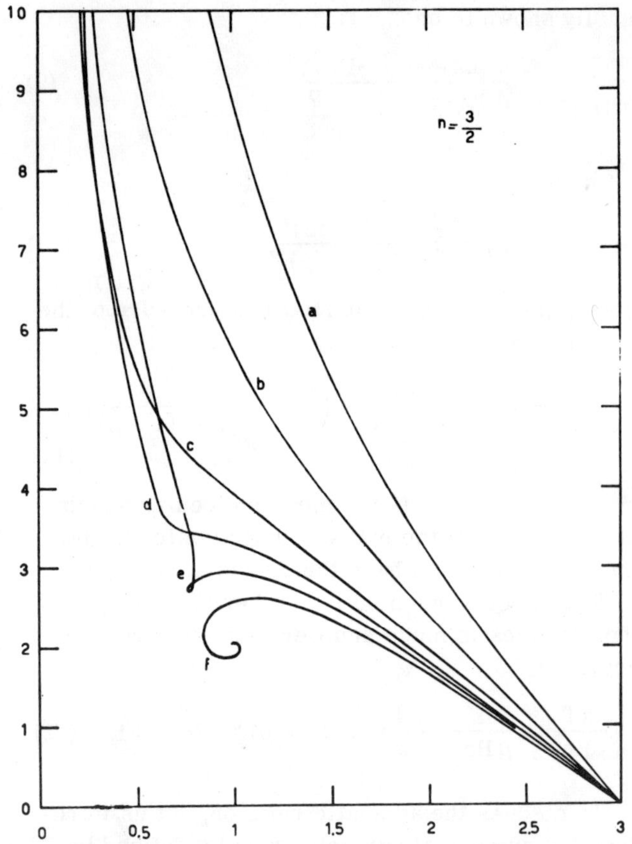

FIG. 1. Solutions of the R-polytropic model for $n = 3/2$ in the U, V-plane. The curves are labelled as follows: a, $\theta_c = 0.03$; b, $\theta_c = 0.3$; c, $\theta_c = 1.0$; d, $\theta_c = 2.0$; e, $\theta_c = 4.0$; f = extreme relativistic case.

It is seen, and it can be easily shown directly, that the point $U = 1$, $V = 2$ is a singular point of equation (21). This singular point is a vortex; the integral curve spirals around it, without being able to reach it for any finite value of the radius.

Physically this means that a spherical distribution of extreme relativistic matter (or pure radiation) having a finite density at the center, has no boundary, but must extend to infinity; its mass is also infinite.

However the newtonian potential is finite at each point. In General Relativity a somewhat similar role as that of the newtonian potential is played by the so-called Schwarzschild's radius

$$r_s = \frac{2G}{c^2} m(r), \tag{22}$$

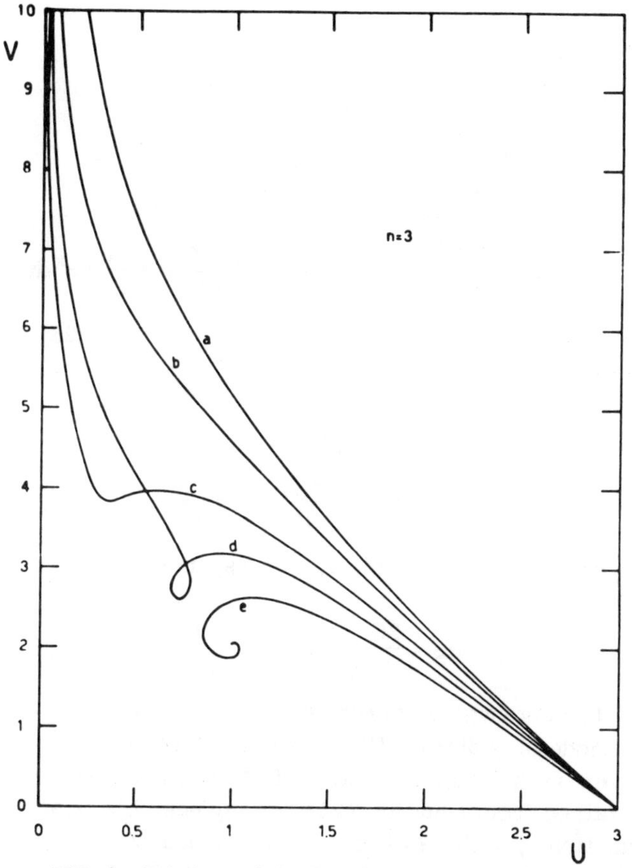

FIG. 2. Solutions of the R-polytropic model for n = 3 in the U, V-plane. The curves are labelled as follows: a, $\theta_c = 0.01$; b, $\theta_c = 0.1$; c, $\theta_c = 0.4$; d, $\theta_c = 1.0$; the spiral curve is the solution for the extreme relativistic case.

where $m(r)$ is the mass inside a certain radius r and G the gravity constant and it is well known that if at any point $r = r_s$, some very peculiar effects occur, which are currently described by saying that at that point a Schwarzschild singularity exists.

Now it is found that inside an extreme relativistic sphere r_s/r starts from zero at the center, grows to a maximum, whose value is found to be 0.493, and then oscillates indefinitely with decreasing amplitude around the limiting value $3/7 = 0.429$ which corresponds to the singular point U = 1, V = 2.

Considering the limiting character of the extreme relativistic case, I take this as strongly suggestive of the fact that a Schwarzschild singularity cannot occur inside a body in equilibrium, if the compressibility of the matter is properly taken into account. Very likely, the ratio r_s/r will never be found to be greater than 0.493 or, roughly, 1/2, the maximum value obtained for the extreme relativistic matter.

The existence of a (considerably larger) upper limit,

$$\frac{r_s}{r} < \frac{8}{9}$$

was proved rigorously by Buchdahl (1956, see also Chandrasekhar 1964).

Some numerical results concerning the R-polytropic models are contained in Tables 1 and 2.

Table I. Constants of the R-polytropes for n = 3/2

θ_c	$\theta_c^{3/2}(1 + 3\theta_c)$	ξ_1	ϕ_1	μ_1	m_1	$\zeta_c/\bar{\zeta}$
0.01	1.030×10^{-3}	25.00	0.03435	0.8587	0.8718	6.247
0.03	5.664×10^{-3}	17.92	0.08930	1.600	1.667	6.790
.1	4.111×10^{-2}	11.42	0.1983	2.265	2.504	9.012
.2	1.431×10^{-1}	8.438	0.2586	2.182	2.551	13.13
.3	3.122×10^{-1}	7.146	0.2768	1.978	2.380	19.20
.5	0.8839	5.993	0.2743	1.644	2.042	38.58
.7	1.816	5.616	0.2555	1.435	1.814	74.73
1.0	4.000	5.617	0.2245	1.261	1.611	1.874×10^2
1.5	10.10	6.195	0.1879	1.164	1.487	6.876×10^2
2.0	19.80	6.845	0.1724	1.180	1.403	1.794×10^3

Table II. Constants of the R-polytrope for n = 3

θ_c	$\theta_c^3(1 + 3\theta_c)$	ξ_1	ϕ_1	μ_1	m_1	$\zeta_c/\bar{\zeta}$	α_1
.001	1.003×10^{-9}	19495	.002315	45.13	45.26	53.9	45.1
.002	8.048×10^{-9}	9744	.004577	44.60	44.80	55.6	44.6
.003	2.724×10^{-8}	6493	.00679	44.09	44.35	56.4	44.1
.01	1.030×10^{-6}	1946	.02095	40.76	41.58	62.0	40.72
.03	2.943×10^{-5}	653.5	.05085	33.23	34.79	82.4	33.06
.1	1.300×10^{-3}	227.5	.08544	19.44	21.74	2.63×10^2	18.75
.2	1.280×10^{-2}	194.3	.06279	12.20	14.25	2.56×10^3	11.21
.3	5.130×10^{-2}	377.2	.02675	10.09	11.90	9.10×10^4	9.06
.4	1.408×10^{-1}	1052	.01137	11.96	13.71	4.57×10^6	10.94
.43	1.8208×10^{-1}	1148	.01134	13.02	14.77	5.02×10^6	12.01
.5	3.125×10^{-1}	1060	.01409	14.94	16.74	8.31×10^6	13.95
.7	1.0633	690.7	.02367	16.35	18.30	8.23×10^6	15.37
1.0	4.000	517.7	.02996	15.51	17.55	1.19×10^7	14.52

In both tables the columns 1 to 7 contain the following quantities:

- col. 1, θ_c a central parameter classifying the solution;
- col. 2, $\theta_c^n(1 + 3\theta_c)$ the density at the center;
- col. 3, ξ_1, the radius;
- col. 4, φ_1, the ratio r_s/r at the boundary;
- col. 5, μ_1, the total observable mass;
- col. 6, m_1, the total proper mass;
- col. 7, $\zeta_c/\bar{\zeta}$, the ratio of the central to the mean density.

All the quantities are given in numerical units; they may be reduced to physical units if one of the quantities is given. Suppose, for instance, that the radius R of the structure is given; then if

$$A = R/\xi_1 \quad (R \text{ in cm})$$

the mass is

$$M = \frac{c^2}{2G} A\mu_1 = 3.384\frac{\mu_1}{\xi_1} R \quad (\text{solar masses})$$

and the central density

$$\zeta_c = \frac{c^2}{8\pi G} \xi_1^2 \theta_c^n (1 + 3\theta_c) \frac{1}{R^2}.$$

The proper mass M_{pr} is defined as the integral of the density times the element of proper volume; it is given by

$$M_{pr} = 4\pi \int_0^R \frac{\zeta r^2 \, dr}{\sqrt{1 - r_s/r}},$$

while the observable mass is, of course,

$$M = 4\pi \int_0^R \zeta r^2 \, dr.$$

The quantity m_1 bears to the proper mass the same relation μ_1 does to the observable mass, thus

$$M_{pr} = \frac{c^2}{2G} A \, m_1 = \frac{m_1}{\mu_1} M.$$

Finally, in the case n = 3, another quantity is important, the proper mass corresponding to matter alone. This is given by

$$M_0 = 4\pi \int_0^R \frac{\zeta_{mat} \, r^2 dr}{\sqrt{1 - r_s/r}} = 4\pi Q \int_0^R \frac{P^{3/4} \, r^2 \, dr}{\sqrt{1 - r_s/r}}.$$

Putting

$$M_0 = \frac{\alpha_1}{\mu_1} M,$$

we obtain the values of α_1, contained in the 8th column of Table 2.

A very important point concerns the stability of these configurations. A general criterion for stability against radial oscillations was derived in the papers of the series mentioned above (Gratton 1964 a and c); this is essentially the same as that given by Chandrasekhar (1964). In its variational form it may be written as follows. Consider the equation

$$\sigma^2 \int_0^R \exp\left(\frac{3}{2}\lambda - \frac{1}{2}\nu\right) \frac{\eta^2}{r^2 (P + c^2\zeta)} \, dr$$
$$= \int_0^R \exp\left(\frac{\lambda}{2} + \frac{\nu}{2}\right) \frac{1}{r^2 (P + c^2\zeta)} \frac{dP}{d\zeta}\left(\frac{dy}{dr}\right)^2 dr$$
$$- \frac{4TG}{c^2} \int_0^R \exp\left(\frac{5}{2}\lambda + \frac{1}{2}\nu\right) \left(\frac{8\pi G}{c^4} P + \frac{1}{r^2}\right) \eta^2 \, dr \qquad (23)$$

where

$$\lambda = -\log\left(1 - \frac{r_s}{r}\right),$$
$$\nu = -2\log\left(1 - \frac{R_s}{R}\right) \int_0^R \frac{dP}{P + c^2\zeta}, \qquad (24)$$

R_s being the value of the r_s corresponding to r = R, or

$$R_s = \frac{2G}{c^2} M.$$

The eigenvalue corresponding to the fundamental mode of radial oscillations of the structure is given by the minimum value of σ^2 which can be obtained from equation (23) for any possible trial function η, which is finite and continuous and satisfies the boundary conditions $\eta = 0$ at r = 0 and r = R. The stability condition is, of course, that this minimum be positive.

By means of this criterion it was found that only the first 3 solutions of Table 1 for n = 3/2 are probably stable. All the other solutions for n = 3/2 and for n = 3 are unstable. We shall comment later upon these results.

5. An Incomplete Polytropic Model

The equilibrium of a sphere of gas following the simple polytropic law

$$P = K\zeta^{(n+1)/n} \qquad (25)$$

was studied in General Relativity by Tooper (1964, this paper arrived when the present work was practically finished and was used to check some results).

Obviously, if $\zeta > (c^2/3K)^n$, equation (25) gives $P > c^2\zeta/3$, a physical impossibility. Now from the results of sections 2 and 3 it seems plausible to assume that in all cases at very high densities $P = c^2\zeta/3$, that is the equation of state of an extreme relativistic gas holds.

A more reasonable model is, therefore, that which is obtained by assuming an equation of state of the type

$$P = \begin{cases} K\zeta^{(n+1)/n}, & \zeta < \zeta_{\text{lim}}, \\ \dfrac{c^2}{3}\zeta, & \zeta > \zeta_{\text{lim}}, \end{cases} \tag{26}$$

where

$$\zeta_{\text{lim}} = \left(\frac{c^2 k}{3}\right)^n \tag{27}$$

In such a model we have an incomplete polytropic region extending from the outer boundary to the point at which $\zeta = \zeta_{\text{lim}}$, and an extreme relativistic core from this point to the center. If the center is attained before the limit density is reached, we have the complete polytropic model of Tooper.

Physically this means that, if we consider only the case n = 3, we assume in the envelope an adiabatic equilibrium neglecting the contribution of the gas to the pressure and that of the radiation to the mass density; in the core the radiation alone contributes to the pressure and to the mass density, the gas being "blown off" by radiation pressure.

The conditions of "fit" at the interface between the core and the envelope are very easily derived; they express the continuity of r and P and, moreover, the fact that at the interface $\zeta = \zeta_{\text{lim}}$.

In the U, V-plane the solutions are obvious; we start from the center (U = 3, V = 0) with the extreme relativistic solution and at a certain (arbitrary) point change to the solution of the equation corresponding to the polytropic relation (25) passing through that point. Some of the solutions thus obtained for n = 3 are shown in Fig. 3, together with some "complete" polytropic solutions in which the polytropic region extends right to the center (Tooper's model).

An interesting point arises in connection with the special character of the singular point of the extreme relativistic curve. As the "fitting point" is moved along this curve approaching the singular point, it is obvious that the solutions begin to "oscillate" around a limiting solution corresponding to that particular curve for which the "polytropic branch" starts from the singular point U = 1, V = 2. This limiting solution is physically impossible since it would correspond to infinite density and pressure at the center.

Its importance lies in the fact that it defines a limiting configuration towards which the solutions

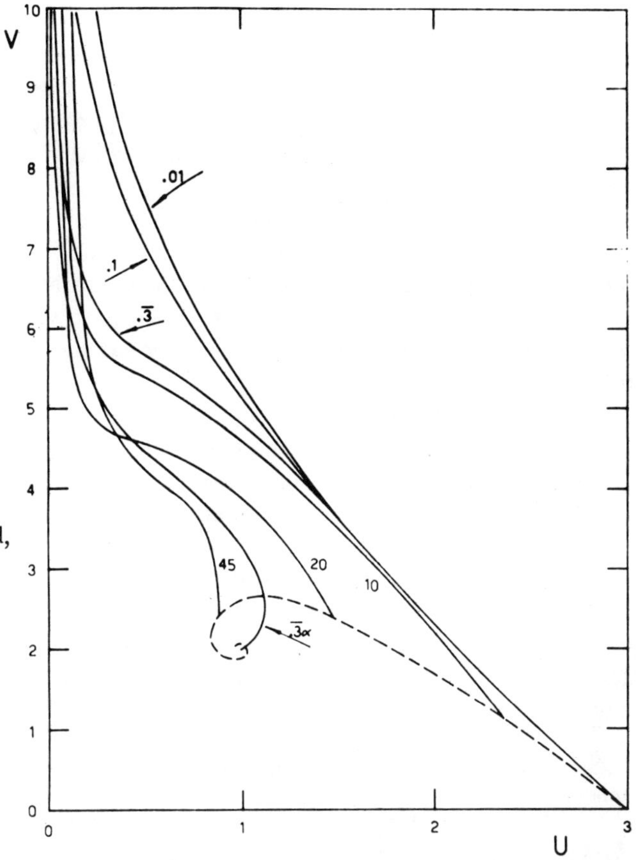

FIG. 3. Solutions of the incomplete polytropic model in the U, V-plane. The curves are labelled according to a conventional number; 0.01, 0.1, 0.3 are the corresponding θ_c for the degenerate models; 10, 20, 45 correspond respectively to ξ_i = 1.0, 2.0, 4.5 for the incomplete model; α is the limiting solution. The dashed curve is the extreme relativistic solution.

of the incomplete polytropic model tend when the central condensation represented by the numerical parameters $\zeta_c/\bar{\zeta}$ grows beyond all limits.

Table 3a and b contain the numerical parameters of interest for the polytropic model of index 3. The

Table. IIIa. Constants of the complete polytropic model (degenerate case)

θ_c	ξ_1	ϕ_1	μ_1	α_1	$\zeta_c/\bar{\zeta}$
0.001	19486	2.325×10^{-3}	45.3	45.3	54.5
.002	9733	4.619	45.0	45.0	54.7
.003	6482	6.883	44.6	44.6	55.0
.01	1932	2.190×10^{-2}	42.3	42.3	56.8
.03	635.6	5.764	36.7	36.4	63.0
.1	193.1	1.264×10^{-1}	24.4	23.1	98.3
.2	112.4	1.435	16.13	13.40	235
.3	102.1	1.193	12.18	8.48	787
.3̄	106.0	1.073	11.37	7.44	1293

following quantities are tabulated.

Col. 1: a parameter classifying the solution. This is different in Table 3a and Table 3b. In Table 3a the parameter θ_c is proportional to $P^{1/4}$; notice that when $\theta_c = 1/3 = 0.\overline{3}$ the extreme relativistic core just begins to form. In Table 3b the parameter was chosen equal to the ratio between the radius of the core and the radius of the external boundary; notice again that this ratio increases at first to a maximum and then begins to oscillate with decreasing amplitude around the limiting value 0.0150.

Col. 2. The external radius ξ_1;

Col. 3. The ratio r_s/r at the boundary φ_1;

Col. 4. The total observable mass μ_1;

Col. 5. The total proper mass corresponding to matter alone α_1;

Col. 6. The central condensation $\zeta_c/\overline{\zeta}$.

The stab ility of this model was not investigated, but there can be little doubt from Chandrasekhar's general results, that all the structures considered must be unstable.

Table IIIb. Constants of the incomplete polytropic model.

ξ_i/ξ_1	ξ_1	ϕ_1	μ_1	α_1	$\zeta_c/\overline{\zeta}$
0.0220	108.7	10.20×10^{-2}	11.09	7.08	1.68×10^3
.0311	123.4	8.557	10.56	6.43	3.97
.0256	167.6	6.178	10.35	6.26	1.85×10^4
.0173	243.3	4.457	10.84	6.91	1.00×10^5
.0123	305.4	4.069	12.43	8.80	4.74
.0147	216.9	5.849	12.69	9.12	2.84×10^6
.0154	223.0	5.479	12.22	8.56	4.54×10^7
.0148	231.3	5.348	12.37	8.74	2.93×10^8
.0149	226.9	5.470	12.41	8.80	3.65×10^9
.0150	226.6	5.460	12.37	8.74	2.88×10^{10}
.0150	227.2	5.450	12.38	8.76	∞

6. Physical Interpretations

We wish, now to investigate the physical implications, if any, of the previous models.

The model corresponding to equation (8) is a reasonable approximation for a neutron star. In this connection it was discussed also by Misner and Zapolsky (1964) and by myself (Gratton 1964a); it is, therefore, unnecessary to repeat here the results obtained. They do not give anything essentially new relative to the well known results obtained by Oppenheimer and Volkoff (1939) several years ago.

The R-polytropic model of index 3 and the incomplete polytropic model are essentially unstable against radial oscillations. This unpleasant property might lead one to discard them altogether in connection with physical problems.

There is however, some possibility that this instability may prove to be not quite as fatal as it might seem at first.

For one thing the period of the instability might be long enough to permit the existence of an object built according to these models for a significantly long time. We did not actually compute the eigenvalues, but contented ourselves with the result that their squares are negative. Considering that in the newtonian case n = 3 corresponds to a kind of "indifferent" equilibrium in the sense that the radius is not a function of the mass, one may guess that the instability of an R-polytrope of index 3 is only a very slight one.

Secondly, one may argue that a very small amount of rotational momentum, although too small to alter significantly the equilibrium, might nevertheless have a stabilizing effect upon the model.

Finally there is the possibility that the external layers, which in adiabatic equilibrium would correspond to a polytrope of index 3/2 and which have not been taken into account might also have a stabilizing effect upon the whole configurations.

Until these effects have not been proved to be ineffective we feel, therefore, justified in speculating upon the possible implications of the models, notwithstanding their instability.

To this purpose we observe that the difference between the energy corresponding to the proper mass of the matter alone and that corresponding to the observable mass,

$$\Delta E = (M_0 - M)c^2 = \frac{\alpha_1 - \mu_1}{\alpha_1} M_0 c^2. \qquad (28)$$

is equal to the energy which can be obtained from a mass M_0 initially cold and dispersed to infinity, if it is brought to the state considered, leaving unchanged the total number of heavy particles, or conversely, to the energy which must be given to a body built according to this model if its mass is dispersed to infinity and cooled down to zero temperature. In other words ΔE is the binding energy of the structure, in the usual sense.

Now, from Tables 2 and 3, and from Figs. 4 and 5 it is seen that $-\Delta E$ is always positive and may reach very large values. This shows that the states corresponding to an R-polytrope and to an

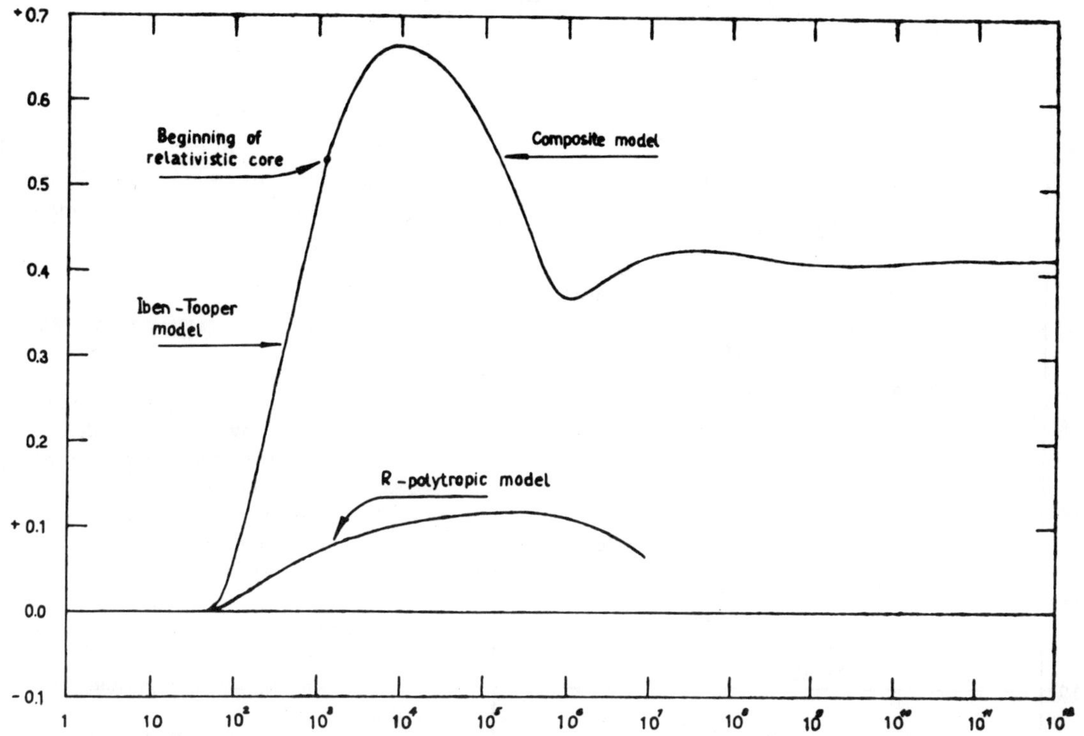

FIG. 4. Excess energy versus central condensation. Abscissae are central condensation, $\zeta_c/\bar{\zeta}$, ordinates give $\mu_1/\alpha_1 - 1$.

incomplete polytrope of index 3 are unbound states; a body built on these models, left alone, would expand spontaneously and cool down. It cannot be in any case a permanent structure, at least if its central condensation $\zeta_c/\bar{\zeta}$ is less than, say 10^4 in the case of the composite model and 10^6 (more or less) in the case of the R-polytrope. If the central condensation is larger than these limits, although its binding energy is positive, the body cannot expand through a series of equilibrium configurations.

In any case it is impossible to obtain a structure similar to that described by these models by letting a certain amount of matter contract from an infinitely dispersed state, through a series of equilibrium states built according to them, since the gravitational energy which is liberated by contraction is not enough to heat the body and to provide the energy which must be radiated in space at all times. If structures like these exist at all they must have originated in a different way.

Still, if we leave aside the problem of the origin, and consider a body of a central condensation, say, $\zeta_c/\bar{\rho} = 80$, $\theta_c = 0.03$, for which the difference between the R-polytropic and the composite model is not important, assuming a mass of 10^9 solar masses, the following physical parameters are obtained (the actual values correspond to the R-polytropic model)

$R = 6 \times 10^{15}$ cm
$P_c = 5 \times 10^{15}$ erg cm^{-3}
$T_c = 4 \times 10^7$ °K
$\zeta_c = 2 \times 10^{-5}$ gr cm^{-3}
$\beta_0 = 1.2 \times 10^{-4}$

At this temperature and density no nuclear reactions can contribute to the energy generation, so that all the energy radiated must be due to the cooling of the matter. The actual rate of emission can be readily computed by assuming that the opacity is due to Thomson scattering. A general argument due to Hoyle and Fowler (1963) shows then that roughly

$$L = 5 \times 10^4 M \tag{29}$$

if both L and M are expressed in solar units. For $M = 10^9$ solar masses one obtains

$$L = 5 \times 10^{13}\ L_\odot = 2 \times 10^{47}\ \text{ergs/sec},$$

which together with the value obtained for the radius gives a surface temperature $T_c = 55\,000$ °K.

The bolometric magnitude corresponding to the luminosity is around -30 and the visual magnitude

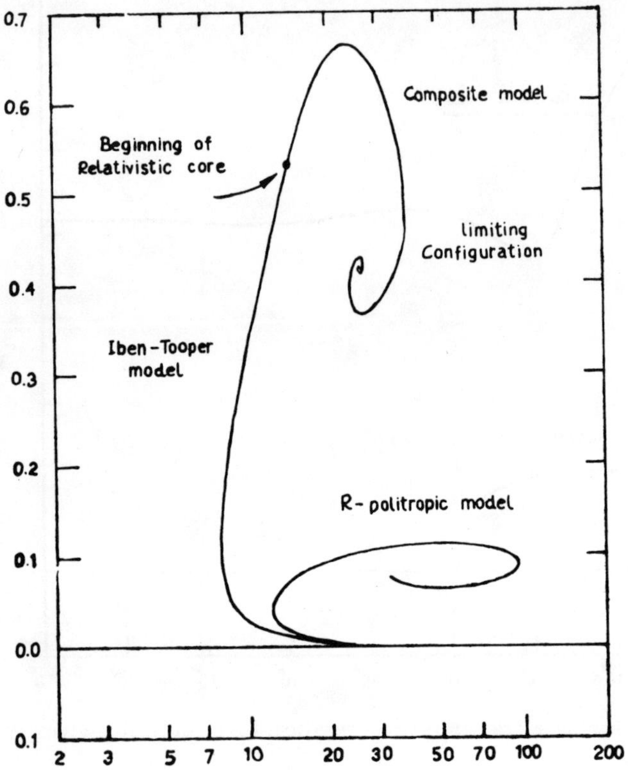

FIG. 5. Excess energy versus radius (at constant M_0). Abscissae are ξ_1/α_1, ordinates $\mu_1/\alpha_1 - 1$. Note that ξ_1/α_1 gives the radius in units of the Schwarzschild radius $2GM_0/c^2$.

-25, with a bolometric correction estimated from the surface temperature.

The values of the radius, temperature and luminosity obtained for $M = 10^9$ are remarkably similar to those suggested for the central bodies of the quasi-stellar radio sources 3C 48 and 3C 273 (Greenstein and Schmidt 1964), but of course cannot be taken too seriously. I consider this result as a mere suggestion that the real structure of Quasars might not be too different from that of an R-polytrope of index 3.

The time necessary for this body to radiate away its whole excess energy is of the order of $\Delta E/L \simeq 10^7$ years. After a time of this order it will have become a cold extended spherical cloud of gas of the size of a galaxy.

REFERENCES

V. A. Ambartsumian, and G. S. Saakyan, 1960—A. J. USSR 37, 193; Sov. Astr. 4, 187.

V. A. Ambartsumian, and G. S. Saakyan, 1961—A. J. USSR 38, 785; Sov. Astr. 5, 601.

H. A. Buchdahl, 1959—Phys. Rev. 116, 1027.

A. G. W. Cameron, 1959—Ap. J. 130, 884.

S. Chandrasekhar, 1939—Stellar Structure, pag. 55.

S. Chandrasekhar, 1964—Ap. J. 140, 417.

C. De Dominicis and P. C. Martin, 1957—Phys. Rev. 105, 1417.

P. Giannone, and L. Gratton, 1965—Mem. Soc. Astr. It. (in press).

L. Gratton, 1964a—Padova Symposium on Cosmology (in press).

L. Gratton, 1964b—Rend. Acc. Lincei—Nota I (in press).

L. Gratton, 1964c—Rend. Acc. Lincei—Nota II (in press).

L. Gratton, 1965—Rend. Acc. Lincei—Nota III (in press).

L. Gratton, and G. Szamosi, 1964—Nuovo Cimento 33, 1056.

J. Greenstein, and M. Schmidt, 1964—Ap. J. 140, 1.

K. Huang, and C. N. Yang, 1957—Phys. Rev. 105, 757.

I. Iben, 1963—Ap. J. 138, 1090.

L. D. Landau, 1932—Phys. Z. Sowjetunion—1, 285.

C. W. Misner, and H. S. Zapolsky, 1964—Phys. Rev. Lett. 12, 635.

J. R. Oppenheimer and G. M. Volkoff, 1939—Phys. Rev. 55, 374.

E. E. Salpeter, 1960—Ann. Phys. 11, 393.

R. F. Tooper, 1964—Ap. J. 140, 434.

C. N. Yang, and T. D. Lee, 1957—Phys. Rev. 105, 1119.

J. B. Zel'dovich, 1961—J. Expt. Theor. Phys. 41, 1609-Sov. Phys. 14, 1143.

APPENDIX

ABSTRACT

The following excerpt from the thesis "An X-Ray Survey of the Sky from Balloon Altitudes" supplements the article of Boldt et al. (pp. 265-268) with a description of several recent X-ray observations of the vicinity of the north galactic pole and a discussion of possible implications of the results. An increase in the counting rate during the first experiment was interpreted as the X-ray flux from an extended source in the direction of the Coma cluster of galaxies, which is a few degrees away from the direction of the north galactic pole. Subsequent observations failed to detect a similar feature and set upper limits to the photon flux from celestial objects near the direction of the north galactic pole, including the Coma cluster of galaxies. It is concluded that, if the signal first observed were due to a cosmic source of X-radiation, then the source intensity must have varied in time during the initial observation. The hypothesis that a supernova explosion at the distance of the Coma cluster of galaxies caused the signal detected in the first experiment is shown to lead to an energy output in the X-ray region which might not be excessively large for such a phenomenon.

APPENDIX

RESULTS OF THE OBSERVATIONS*

(in connection with E. Boldt et al., pp. 265-268)

The Vicinity of the North Galactic Pole

4. Summary of experimental information. The counting-rate pattern observed during a balloon experiment on 6 December 1965 (Boldt et al., 1966) was interpreted as the response to an extended source of X-radiation near the direction of the north galactic pole, e.g. from the Coma cluster of galaxies. Two similar subsequent experiments, on 2 September 1966 and 18 November 1966 (Table 1), placed upper limits to the X-ray flux from objects near the direction of the north galactic pole. (Details of these experiments are described in sections 1-3 of this chapter).

During a rocket experiment on 25 November 1964, Friedman and Byram (1967a) found that no signal with an intensity as great as 7.4% of that from the Crab Nebula was observed in the 1-10 A wavelength interval from point objects near the north galactic pole, e.g. at the position of the Coma cluster of galaxies. If the spectra of the objects are assumed to be power-laws with the same index as that of the Crab Nebula,

$$J(E) \propto E^{-2}$$

then the extrapolated upper limit to the photon flux at the top of the atmosphere is

$$J(E = 25\,\text{keV}) \lesssim 9.10^{-4} \text{ photons/cm}^2\text{-sec-keV}.$$

In an observation on 13 February 1967, Lewin et al. (1968) placed an upper limit to the flux from the same region of the celestial sphere. From the reported information we deduce that

$$J(E = 25\,\text{keV}) \underset{\alpha}{<} 15.10^{-4} \text{ photons/cm}^2\text{-sec-keV}.$$

The available experimental information on the photon flux from the vicinity of the north galactic pole, for example from the direction of the Coma cluster of galaxies, is summarized in Table 1.

5. Discussion. The observation of the Coma cluster as an extended source of X-rays could be relevant to the interpretation of the measured diffuse cosmic X-ray background $dj/d\Omega$. Several models have been proposed for this background; one of them assumes that the observed flux is due to the sum of many unresolved galaxies. If $\langle J_g \rangle$ is the mean emission from a galaxy, then the observed flux due to all unresolved galaxies up to a radius $R \approx R_{Hubble}$ can be estimated from

$$\frac{dj}{d\Omega} \approx \frac{n_g \langle J_g \rangle \cdot R}{4\pi}$$

where n_g is the mean density of galaxies ($n \approx 10^{-75}\ \text{cm}^{-3}$). When $\langle J_g \rangle$ is set equal to the observed emission of our galaxy, then the resulting value of $dj/d\Omega$ is about two orders of magnitude lower than the measured diffuse X-ray flux (Oda, 1965; and Gould and Burbidge, 1967). However, the assumption that our galazy is typical with respect to its X-ray emission is somewhat arbitrary. If we assume that X-radiation from the Coma cluster of galaxies was observed in the first experiment, then we obtain a better estimate for $\langle J_g \rangle$ since the Coma cluster contains about 1500 galaxies at a mean distance of 90 Mpc (Abell, 1965). Felten et al. (1966) have, however, pointed out that the resulting diffuse X-ray flux would be too high by a factor of 30 to account for the observed background spectrum.

The fact that subsequent observations failed to detect comparable X-radiation from the Coma cluster of galaxies has somewhat removed this discrepancy, but has not solved the problem. The upper limits of Friedman and Byram (1967a) and Lewin et al. (1968) reduce the discrepancy factor from 30 to approximately 3. Considering the uncertainty in the number of galaxies in the Coma cluster, the various parameters may actually combine to yield the observed diffuse X-ray flux. In this case the "unresolved galaxies"-model for the cosmic X-ray background is not yet ruled out by the limits on the X-ray emission from galaxies of the Coma cluster. The conclusion would then be that the mean emission per galaxy is approximately two orders of magnitude higher than that of our own galaxy. In this connection it is interesting to note that, according to recent measurements (Bradt et al., 1967), the X-ray luminosity of the radio galaxy M 87 is two to three orders of magnitude higher than that of our galaxy (Friedman et al., 1967b).

In view of the fact that other experiments failed to observe a significant photon flux from the vicinity of the north galactic pole we may tentatively attribute the observation of 6 December

*Excerpt from the doctoral thesis of Guenter Riegler, University of Maryland, titled "An X-Ray Survey of the Sky from Balloon Altitudes," to be published by NASA-Goddard Space Flight Center in December 1968.

1965 to a transient phenomenon associated with a discrete object.

No information from optical observations during the time of our first experiment is available. Any unusual activity in the vicinity of the north galactic pole would probably have remained undetected since this region of the celestial sphere was in a very unfavorable position for reliable optical observations.

Four flare stars were observed in 1964 and 1965 in a region in Coma Berenices, approximately 6° away from the direction of the north galactic pole (Haro, 1966). The number of flares that were observed per hour in that region is 0.05. We are not aware of flare observations during the time when the first experiment was conducted.

Pencil-beam radio observations of the Coma cluster of galaxies (Bozyan, 1968) have recently resolved a small-diameter source, Coma A, which is centered at $\alpha = 12.51.8$, $\partial = 27° 53.8'$. The energy spectrum of Coma A has an index of $\Gamma = 0.6$. An extrapolation of the power-law radio spectrum can account for the optical luminosity (Matthews and Sandage, 1963; Schmidt, 1965; Wyndham, 1966) and yields a photon flux of $2.5 \cdot 10^{-4}$ photons/cm^2-sec-keV at E = 25 keV. Table 1 shows that this flux is approximately two orders of magnitude below the previously reported value (Boldt et al., 1966) and about one order of magnitude below the upper limits found in other experiments.

It is of interest to discuss the hypothesis that the observed feature in the first experiment was due to a supernova explosion in the Coma cluster of galaxies. The duration of X-ray emission during a supernova outburst has not yet been established. Colgate's hydrodynamic model (Colgate, 1968) suggests a time-span for X-ray emission of the order of 10^{-3} sec. The fluorescence theory of optical emission from a supernova (Sartori and Morrison, 1967) assumes that the explosion takes place within a few seconds. If the above hypothesis were correct, the observation on 6 December 1965 would indicate a period of at least one hour. The photographic light curve for a type I supernova suggests a time-scale of about 20 days for the main phase. The absence of a significant flux of soft X-ray emission within about 30 days after a supernova explosion (Bradt et al., 1968) indicates that an upper limit of about 30 days can be set on the duration of X-ray emission. The nova-like behavior of the low-energy X-ray emission from the variable source Cen XR-2 (Harries et al., 1967; Cooke et al., 1967; Chodil et al., 1967) indicates an e-folding time of about 30 days.

Table 1

PHOTON FLUX FROM THE COMA CLUSTER OF GALAXIES

Balloon-borne measurements at 25 keV, extrapolated to the top of the atmosphere.

Observation date	Photon flux photons/cm^2-sec-keV	Reference
25 Nov. 1964	$< 9.4 \cdot 10^{-4}$*	Friedman and Byram (1967a)
6 Dec. 1965	$(140 \pm 42) \cdot 10^{-4}$	Experiment #1. (Boldt et al., 1966)
2 Sept. 1966	$< 72 \cdot 10^{-4}$	Experiment #2.
18 Nov. 1966	$< 50 \cdot 10^{-4}$	Experiment #3.
13 Feb. 1967	$< 15 \cdot 10^{-4}$	Lewin et al. (1968)

*Extrapolation from lower-energy rocket measurements assuming a power-law of index 2.

With the assumption that the Coma cluster contains 1500 galaxies, about 4 supernova explosions per year are expected when a rate of 1 event per 400 years per galaxy is adopted. Considering the uncertainty of the observations, this estimate is supported by the fact that 12 supernovae were observed in the Coma cluster during a recent 10 year period (Zwicky, 1967). The probability of observing the main phase of a supernova explosion during a one-hour observation is therefore 20% if the main phase lasts 20 days, and is 0.04% if it is assumed to last 1 hour.

We can also estimate the energy output in order to evaluate the plausibility of the supernova hypothesis. We assume a power-law photon spectrum of index 2 and compute the energy flux in the X-ray interval from 1 to 100 keV. If the distance to the source is 90 Mpc, then we arrive at a total X-ray energy output of $q = 8 \cdot 10^{51}$ ergs emitted over a period of 20 days. This is not unreasonably high. In fact, the fluorescence model (Sartori and Morrison, 1967) requires an output of 10^{52} ergs in the ultraviolet band. If the emission is assumed to last only one hour, however, then we find an output of $q = 2 \cdot 10^{49}$ ergs. The X-ray energy output increases by an order of magnitude if thin-source bremsstrahlung at $T = 5 \cdot 10^7$ °K is assumed. Colgate's model (Colgate, 1968) predicts an X-ray pulse with a total energy output of only $q \approx 5 \cdot 10^{47}$ ergs within a fraction of a second and can therefore not be invoked to explain our observation.

REFERENCES

Abell, G. O., Ann. Rev. Astron. Astrophys. 3, 1 (1965).

Boldt, E., McDonald, F. B., Riegler, G., and Serlemitsos, P., Phys. Rev. Letters 17, 447 (1966).

Bozyan, E. P., Ap. J. 152, L155 (1968).

Bradt, H., Mayer, W., Naranan, S., Rappaport, S., and Spada, G., Ap. J. 150, L199 (1967).

Bradt, H., Naranan, S., Rappaport, S., Zwicky, F., Ogelman, H., and Boldt, E., Nature 218, 856 (1968).

Chodil, G., Mark, H., Rodrigues, R., Seward, F., Swift, C. D., Hiltner, W. A., Wallerstein, G., and Mannery, E. J., Phys. Rev. Letters 19, 681 (1967).

Colgate, S. A., Can. J. Phys. 46, 476 (1968).

Cooke, B. A., Pounds, K. A., Stewardson, E. A., and Adams D. J., Ap. J. 150, L189 (1967).

Felton, J. E., Gould, R. J., Stein, W. A., and Woolf, N. J., Ap. J. 146, 955 (1966).

Friedman, H., and Byram, E. T., Ap. J. 147, 399 (1967a).

Friedman, H., Byram, E. T., and Chubb, T. A., Science 156, 374 (1967b).

Gould, R. J., and Burbidge, G. R., Handbuch der Physik, S. Flügge, Ed, Vol 46/2, p. 265, Springer Verlag, Berlin (1967).

Haro, G., Nebulae and Interstellar Matter, B. M. Middlehurst and L. H. Aller, Ed., p. 141, University of Chicago Press, Chicago (1968).

Harries, J., McCracken, K. G., Francey, R. J., and Fenton, A. G., Nature 215, 38.

Lewin, W. H. G., Clark, G. W., and Smith, W. B., Can. J. Phys. 46, 409 (1968).

Matthews, T. A., and Sandage, A. R., Ap. J. 138, 30 (1963).

Oda, M., Proc. Int. Conf. Cosmic Rays, London, p. 68 (1965).

Sartori, L., and Morrison, P., Conference on Supernovae held at NASA/Goddard Institute for Space Studies, New York (1967).

Schmidt, M., Ap. J. 141, 1 (1965).

Wyndham, J. D., Ap. J. 144, 459 (1966).

Zwicky, F., Conference on Supernovae held at NASA/Goddard Institute for Space Studies, New York (1967).